Mathematics: A Universal Lang

Sir Ronald Aylmer Fisher

John Maynard Keynes

Richard Courant

John von Neumann

Ol'ga Arsen'evna Oleinik

L.S. Pontryagin

United Kingdom

Switzerland

France

Germany
Poland
Austria

Soviet Union (U.S.S.R.)

China

India

Chen Ning Yang

Jean Piaget

Benoit Mandelbrot

Mark Kac

Srinivasa Ramanujan

FORMULAS FROM GEOMETRY

Triangle

$$\text{Area} = \frac{1}{2} bh$$

$$\text{Area} = \frac{1}{2} ab \sin \theta$$

Parallelogram

$$\text{Area} = bh$$

Trapezoid

$$\text{Area} = \frac{a+b}{2} h$$

Circle

$$\text{Circumference} = 2\pi r$$

$$\text{Area} = \pi r^2$$

Sector of Circle

$$\text{Arc length } s = r\theta$$

$$\text{Area} = \frac{1}{2} r^2 \theta$$

Polar Rectangle

$$\text{Area} = \frac{R+r}{2} (R-r)\theta$$

Right Circular Cylinder

$$\text{Lateral area} = 2\pi rh$$

$$\text{Volume} = \pi r^2 h$$

Ball

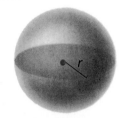

$$\text{Area} = 4\pi r^2$$

$$\text{Volume} = \frac{4}{3} \pi r^3$$

Right Circular Cone

$$\text{Lateral area} = \pi rs$$

$$\text{Volume} = \frac{1}{3} \pi r^2 h$$

Frustum of Right Circular Cone

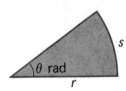

$$\text{Lateral area} = \pi s(r + R)$$

$$\text{Volume} = \frac{1}{3} \pi (r^2 + rR + R^2)h$$

General Cone

$$\text{Volume} = \frac{1}{3} (\text{area } B)h$$

Wedge

$$\text{Area } A = (\text{area } B)\sec \theta$$

4/2007

Calculus
with
Applications

Calculus with Applications

Dale Varberg
Hamline University

Walter Fleming
Hamline University

This Instructor's Edition is exactly like the student text, except that it contains answers to all problems and exercises. The student text contains answers to odd-numbered problems only. When ordering the text for your students please use ISBN 0-13-110826-3.

Prentice Hall,
Englewood Cliffs, NJ 07632

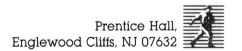

ISBN 0-13-110834-4

Editorial and production supervision: *Ruth Cottrell*
Interior design: *Lorraine Mullaney*
Cover design: *Bruce Kenselaar*
Cover art: *Salem Krieger*
Pre-press buyer: *Paula Massenaro*
Manufacturing buyer: *Lori Bulwin*
Photo editor: *Lorinda Morris-Nantz*
Photo research: *Barbara Scott*

Photo credits: **Chapter 1**, Laima Druskis; **p. 1**, Historical Pictures Service, Chicago; **46**, Ed Lettau/Photo Researchers; **47**, Historical Pictures Service, Chicago; **84**, Tcherevkoff Studio/The Image Bank; **85**, Historical Pictures Service, Chicago; **120**, FORTUNE, © 1989 The Time Inc. Magazine Company. All rights reserved; **121**, Massachusetts Institute of Technology; **160**, Fundamentals of Anatomy and Physiology (Frederic Martini), Englewood Cliffs, N.J.: Prentice Hall, 1989, p. 40; **161**, University of Cambridge; **188**, NASA; **189**, S. Preobrazhensky/Sovfoto; **216**, Mel Digiacomo/The Image Bank; **217**, Historical Pictures Service, Chicago; **244**, from "Mathematical Games," Martin Gardner. Copyright © 1986 by SCIENTIFIC AMERICAN, Inc. All rights reserved; **245**, UPI/Bettmann Newsphotos; **274**, from "Climate Modeling," Stephen H. Schneider. Copyright © 1987 by SCIENTIFIC AMERICAN, Inc. All rights reserved; **275**, University of California, Berkeley; **310**, "The Allocation of Resources by Linear Programming," by Robert G. Bland. Copyright © 1981 by SCIENTIFIC AMERICAN, Inc. All rights reserved; **311**, Stanford University; **346**, Michaelangelo/"Creation of Adam"/Art Resource; **347**, Dr. Chen Ning Yang; **368**, image by H. Jurgens, H.-O. Peitgen, D. Saupe, from *The Beauty of Fractals* (H.-O. Peitgen, P. Richter, Heidelberg: Springer-Verlag, 1986; **369**, IBM Research; **406**, Sovfoto/Eastfoto; **438**, AT&T Archives; **439**, U.S. Navy; **470**, from *The Mathematical Intelligencer* 7(3): cover, New York: Springer-Verlag, 1985; **471**, Cambridge University; **500**, Jim Pickerell; **501**, Rockefeller Archive Center.

Printed in the United States of America
10 9 8 7 6 5 4 3 2 1

ISBN 0-13-110834-4

Prentice-Hall International (UK) Limited, *London*
Prentice-Hall of Australia Pty. Limited, *Sydney*
Prentice-Hall Canada Inc., *Toronto*
Prentice-Hall Hispanoamericana, S.A., *Mexico*
Prentice-Hall of India Private Limited, *New Delhi*
Prentice-Hall of Japan, Inc., *Tokyo*
Simon & Schuster Asia Pte. Ltd., *Singapore*
Editora Prentice-Hall do Brasil, Ltda., *Rio de Janeiro*

Contents of Teaching Outlines

(The Contents for the text, as it appears in the students' edition, is on pages v–x.)

Introduction

The version of the book that you are examining has special features that we hope will make it especially attractive to teachers. On the following pages there are teaching outlines keyed by section number to the text. These outlines lay out in terse form the concepts we believe should be emphasized and suggest examples to use in illustrating them.

Of course, most experienced calculus teachers neither need nor want advice on how to do their jobs. Fine. Such teachers may ignore our teaching outlines or (if they choose to read them) send us their suggestions on how to improve them. However, we thought that a young teaching assistant or even a harried older professor might shorten lesson preparation time by considering how two well-worn college teachers view the task.

In addition to the teaching outlines, this instructor's edition contains the answers to all problems, rather than just the answers to odd-numbered problems that appear in the student edition.

Finally, a sample of the available overhead projector transparencies has been inserted in this volume.

Teaching Outlines

1 Slopes and Rates of Change

1.1 FUNCTIONS (p. 2)

1. Begin with the definition of a function. Give simple examples such as (i) the volume V of a cube of side x, (ii) the area A and perimeter P of a right triangle with legs x and $2x$.

2. Introduce functional notation and illustrate with examples such as

$$f(x) = x^2 - 6$$

and

$$g(x) = x + \sqrt{2x - 4}$$

Find $f(3)$, $f(\pi)$, $g(2)$, and $g(10)$. Work all parts of Problem 10.

3. Use f and g above to illustrate the notions of domain and range. As another example, use

$$F(x) = \frac{1}{x^2 - 9}$$

4. Stress the concept of composite function using, for example

$$f(x) = 2x + 3 \qquad g(x) = \frac{4}{x - 2}$$

Find $f(g(1))$, $g(f(1))$, $f(g(x))$, and $g(f(x))$. Work Problem 24.

5. For a good applied example, use Problem 36.

6. Work Problem 38 noting the stages in the calculation: 7, $\sqrt{7}$, $\sqrt{7} + 3$, $(\sqrt{7} + 3)^2$, $(\sqrt{7} + 3)^2 - 10$

7. If time permits, show how a (composite) function can be decomposed into its component parts. For example, $\sin \sqrt{x}$ can be regarded as $f(g(x))$ where $f(x) = \sin x$ and $g(x) = \sqrt{x}$. How about $\sin \sqrt{1/x}$?

1.2 GRAPHS OF FUNCTIONS (p. 7)

1. Draw the graph of the function

$$y = f(x) = 2x^2 - 3$$

following the sequence of three steps suggested in the text.

2. Define the terms polynomial and degree of a polynomial. Illustrate with examples such as

$$\pi x - 10 \qquad \text{(degree 1)}$$
$$x^3 - 3x + 3 \qquad \text{(degree 3)}$$

Sketch the graph of

$$y = F(x) = x^3 - 3x^2 + 3$$

on $-1 \le x \le 3$

3. Stress the notion of symmetry for graphs using
   (i) $f(x) = 2x^4 - x^2 + 5$
       Since $f(-x) = f(x)$, the graph is symmetric with respect to the y-axis.
   (ii) $g(x) = 2x^3 - 5/x$
        Since $g(-x) = -g(x)$, the graph is symmetric with respect to the origin.
   (iii) $F(x) = x^3 - 3x^2 + 3$. The graph was drawn in part 2 above. It has neither kind of symmetry.

4. Work Problem 4.

5. As an example of a function with a multipart rule, use

$$G(x) = \begin{cases} x^2 - 4 & \text{if} \quad x < 2 \\ 2x - 2 & \text{if} \quad x \ge 2 \end{cases}$$

Find $G(-1)$, $G(2)$ and $G(4)$. Sketch the graph.

6. For good applied problems, try Problems 34 and 38.

1.3 LINEAR FUNCTIONS: CONSTANT SLOPE (p. 16)

1. Introduce the notion of linear functions by sketching the graphs of $f(x) = 2x - 3$, $g(x) = -x + 5$, $h(x) = 4x$, and $k(x) = 4$.

2. Define and discuss the notion of slope. In the picture below, determine the slope of the line through A and B; through A and C.

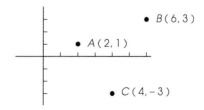

3. Discuss the point-slope form of the equation of a line. Determine these equations for lines AB and AC above.

4. Introduce the slope-intercept form $y = mx + b$. Write an equation of the line (i) with slope 3 and y-intercept 4 and (ii) passing through $(2, 1)$ and $(0, -5)$. Determine the slope and y-intercept of the line with equation $4x + 3y = 9$ by first writing this equation in the slope-intercept form.

5. Review the concept of parallel lines and note that non-vertical lines are parallel if and only if they have the same slope. Which of the following lines are parallel? $y = 3x - 2$, $x = 3y + 1$, $6x - 2y = 9$. Write an equation of the line through $(2, -1)$ which is parallel to the line $4x - 2y = 8$. Perhaps discuss the perpendicularity of lines $(m_2 = -1/m_1)$ and illustrate.

6. Find the linear function $f(x) = ax + b$ which satisfies $f(-1) = 3$ and $f(3) = -5$.

7. End the lesson with Problem 34.

1.4 APPLICATIONS OF LINEAR FUNCTIONS (p. 22)

1. A process is linear if the change in output is proportional to the change in input, that is, if

$$\Delta y = k \, \Delta x$$

Give examples from everyday life and then work Problem 4. Sketch the corresponding graph and note that

$$\Delta y = 30 \, \Delta x$$

2. As an example of straight-line depreciation, work Problem 8. Sketch the graph of $y = C(n)$. Find k in the equation

$$\Delta y = k \, \Delta n$$

3. Use Problem 10 to illustrate the cost of production. Add to the problem by supposing that each freezer can be sold for $400. How many freezers must then be sold to break even? Sketch the cost and revenue functions on the same graph and then shade the regions representing profit and loss.

4. Problems 16, 18, and 20 nicely illustrate the concept of *rate of change*.

5. Finish the lesson with Problem 24.

1.5 NONLINEAR FUNCTIONS: VARIABLE SLOPE (p. 27)

1. The graph of a nonlinear function is a curve rather than a straight line. Accordingly, the change in output Δy is much more complicated than just a constant multiple of the change in input Δx.

2. Use the function $f(x) = 3x - x^2$ as an example of a nonlinear function. Sketch its (parabolic) graph and draw the tangent lines at $(-1, -4)$, $(1, 2)$, and $(3, 0)$. Estimate the slope in each case.

3. Work Problem 14, using it to show that we can find the exact slope of a curve at a given point.

4. Show that the slope of the curve $y = x^4$ at x is $4x^3$ by paralleling the derivation in the text of the slope formula for $y = x^n$ at x. Start with

$$\frac{\Delta y}{\Delta x} = \frac{(x + \Delta x)^4 - x^4}{\Delta x}$$

5. Stress the formulas for $d(ax^n)/dx$ and $d(ax^n + bx^m)/dx$. Use them to find dy/dx for:

$$y = 2x^4 - 5x^3$$

and

$$y = 3x^5 + 4x^3 - 3x^2 + 5x$$

6. Write the equation of the tangent line to the curve

$$y = 2x^4 - 5x^3 + 4x + 7$$

at the point $(2, 7)$,

7. End the lesson with Problem 36.

1.6 VARIABLE RATES OF CHANGE (p. 34)

1. Introduce the concept of variable rate of change using Problem 10 as an example.

2. Discuss the concepts of speed and velocity informally by referring to automobiles. Note that cars have speedometers, not velocity meters.

3. Use Problem 8 to illustrate velocity. Also, calculate the speed at $t = -3, -1, 0, 1, 3$.

4. Clearly distinguish between average rate of change on an interval and actual (instantaneous) rate of change at a point. Use Problem 16 to illustrate these two concepts on a graph.

5. Introduce the concept of *marginal* from economics by working Problem 20.

6. For a good application from physics, try a falling body problem like Problem 26; for an application from the life sciences, try Problem 28.

7. Summarize the lesson by saying that for $y = f(x)$, the change in y caused by a change Δx in x is given by

$$\Delta y = f(x + \Delta x) - f(x)$$

the average rate of change is

$$\frac{\Delta y}{\Delta x} = \frac{f(x + \Delta x) - f(x)}{\Delta x}$$

and the actual (instantaneous) rate of change at x is

$$\frac{dy}{dx} = \lim_{\Delta x \to 0} \frac{f(x + \Delta x) - f(x)}{\Delta x}$$

2 The Derivative

2.1 THE LIMIT CONCEPT (p. 48)

1. Talk about the word *limit* in an informal way. Mention that the idea goes back to the Greeks who in effect defined the circumference of a circle as the limit of the perimeters of inscribed regular polygons.

2. Discuss the compound interest problem that begins the section leading to the question: What does $(1 + x)^{1/x}$ approach as x approaches zero?

3. Give the definition of $\lim_{x \to a} f(x)$. Then find $\lim_{x \to 4} f(x)$ for
   (i) $f(x) = x^2 - 4$
   (ii) $f(x) = (3\sqrt{x} + 1)/x^2$

4. Introduce right- and left-hand limits and do a problem like Problem 12.

5. Work Problems 14 and 22. Emphasize that for $\lim_{x \to a} f(x)$ to exist it is not necessary for $f(a)$ to exist.

6. State the Main Limit Theorem, perhaps writing it on the board or displaying it on an overhead. Use this theorem to justify each step in finding the following limits.

   (i) $\lim_{x \to 3} \dfrac{2x - 1}{3x - 4}$

   (ii) $\lim_{x \to 2} [(4x - 1)^2 - 3\sqrt{x^2 + 5}]$

7. Show that neither of the following limits exist.

   (i) $\lim_{x \to 4} \dfrac{|x - 4|}{x - 4}$

   (ii) $\lim_{x \to 3} \dfrac{4}{x - 3}$

8. End the lesson with Problem 44.

2.2 THE DERIVATIVE (p. 55)

1. State the definition of the derivative of y with respect to x and introduce the two common notations dy/dx and $f'(x)$ for it.

2. Find $f'(x)$ for $f(x) = x^3 + 3x$ using the definition

$$f'(x) = \lim_{h \to 0} \frac{f(x + h) - f(x)}{h}$$

Note that this limit cannot be found by inspection, since the limit of the denominator is 0. Rather a good bit of algebra must be done before the limit can be obtained.

3. Work Problem 6.

4. Discuss the Power Rule and point out that (though we don't prove it now) it actually holds for all real exponents. Use this rule to find the derivatives of $x^{5/3}$, $1/x^3$, and $\sqrt[7]{x}$.

5. Use the Constant Multiple Rule to find the derivatives of $4x^{3/4}$ and $\frac{3}{2}x^{2\sqrt{2}}$.

6. Review the laws of exponents. Use them to simplify

$$y = (4x^3)^{1/2} \frac{\sqrt{x}}{8x^{-2}}$$

Then find dy/dx.

7. For more practice with exponents, use Problem 26.

8. Problems 32 and 38 offer applications of what has been learned.

2.3 RULES FOR DIFFERENTIATION (p. 61)

1. It would be a good idea to list all the rules for differentiation on the board or on an overhead for quick reference.

2. Use these rules to find dy/dx for:
   (i) $y = 3x^{-2} - \frac{1}{3}x^{2.4} + 6\sqrt{x}$
   (ii) $y = (x^2 + 3x - 5)(x^3 - 4x + 1)$
   (iii) $y = (x - 5)/(2x + 3)$

3. Prove the Product Rule as in the text. Then illustrate it further by doing Problems 14 and 22.

4. Discuss the Quotient Rule and use it to find the derivative of $(x^2 - 1)/x$. Then find this same derivative another way and reconcile your results.

5. Find $f'(x)$ for

$$f(x) = \frac{2x^2 - 3x + 1}{x^2 + 5x - 2}$$

and simplify the answer.

6. Find the equation of the tangent line to the curve $y = 2x^3/(\sqrt{x} + 1)$ at the point $(1, 1)$

7. Finish the lesson with an applied problem, say Problem 40.

2.4 THE GENERAL POWER RULE (p. 69)

1. It may be helpful to begin by finding $d(x^4)/dx$, $d(t^4)/dt$, and $d(u^4)/du$, thereby preparing the way for the General Power Rule.

2. Write the following pairs of functions on the board or on an overhead.

$$f(x) = x^{15} \qquad F(x) = (2x + 1)^{15}$$
$$g(x) = x^{3/2} \qquad G(x) = (2x^2 - 4)^{3/2}$$
$$h(x) = x^{\pi} \qquad H(x) = \left(\frac{2x}{x + 5}\right)^{\pi}$$

Note that we know the derivative of the first member of each pair, but that we need a new rule to handle the second member in each case.

3. State carefully the General Power Rule. We think it is helpful to make use of the variable u to stand for the function that is being raised to a power. Write down the derivatives of each pair of functions above, thereby illustrating the General Power Rule.

4. Work Problems 4, 8, and 16 for more practice with this new rule.

5. Find $f'(0)$ for each of the following:
   (i) $f(x) = [(2x + 1)\sqrt{x^2 + 1}]^5$
   (ii) $f(x) = (3x - 1)^5/(x^2 + x - 1)^4$

6. Problem 36 is an appropriate tangent line problem.

7. For an application of the General Power Rule, use Problem 46.

2.5 HIGHER ORDER DERIVATIVES (p. 75)

1. Note that the derivative of the derivative (called the second derivative) makes perfectly good sense. Use $y = f(x) = x^4 - 2x$ to illustrate this notion. Introduce the two notations d^2y/dx^2 and $f''(x)$ for this new object and the analogues for higher derivatives. Work Problems 6 and 8. For $f(x) = 4/\sqrt{x - 4}$, find $f'''(8)$.

2. Point out that d^2y/dx^2 represents the rate of change in slope. Give meaning to this by using a graph like the one below.

3. As in the opening paragraphs of this section, interpret d^2s/dt^2 as acceleration. Use a graph of $s = f(t)$ to help you with this interpretation but also refer to the accelerator on an automobile to make this notion come alive for students.

4. Work Problem 16 as an example of a distance-velocity-acceleration problem. Stress that if $s = s(t)$ is the equation of motion, $s(t)$ represents position (directed distance from a fixed point), $s'(t)$ represents velocity, and $s''(t)$ represents acceleration, all at time t.

5. Complete the lesson with more applications, say, Problems 18 and 30.

3 Applications of the Derivative to Graphing

3.1 CONTINUITY AND DIFFERENTIABILITY (p. 86)

1. Mention how the word *continuous* is used in everyday life. Then use graphs like those in Figures 1–4 to show how this word is used in mathematics.

2. Give the definitions of continuity at a point and on an interval. Show that $f(x) = x^3$ is continuous everywhere, as is any polynomial function.

3. Show that
$$f(x) = (x^2 - 25)/(3x - 15)$$
is continuous at $x = 2$ but is discontinuous at $x = 5$. Graph this function. How should $f(5)$ be defined to make f continuous everywhere?

4. Work Problems 14 and 18.

5. Consider a function like
$$f(x) = \begin{cases} 4 - x^2 & \text{if } x \le 1 \\ 5x - 3 & \text{if } x > 1 \end{cases}$$

Sketch its graph and then discuss its continuity properties. Work Problem 30.

6. Discuss the concept of differentiability by means of a graph like Figure 13. At least mention (and maybe prove) that differentiability implies continuity.

7. Conclude the lesson with Problems 40 and 46.

3.2 MONOTONICITY AND CONCAVITY (p. 93)

1. Introduce interval notation by means of many examples.

2. Solve the inequality $7x - 9 < 3x + 3$ and write the solution set in interval notation. Then work Problem 10.

3. Solve the quadratic inequality
$$x^2 - x - 12 \ge 0$$
using the split point method. Write the solution set in interval notation and show it graphically on the real line. Work Problem 20.

4. Relate the notions of increasing and decreasing to the sign of the derivative. Then determine where
$$f(x) = 2x^3 - 3x^2 - 36x + 2$$
is increasing and where it is decreasing.

5. Relate the notions of concave up and concave down to the sign of the second derivative. Then find where the graph of the function f above is concave up and where it is concave down.

6. Work examples like Problems 28, 32, and 38 to solidify these ideas.

7. Do examples like Problems 40 and 42 to show what can be read from a graph.

8. Write the following conditions on the board and then draw the graph of a function that satisfies them.
   (i) f is continuous everywhere.
   (ii) $f(2) = 10$ and $f(6) = 4$.
   (iii) $f'(2) = f'(6) = 0$
   (iv) $f''(x) < 0$ on $(-\infty, 4)$ and $f''(x) > 0$ on $(4, \infty)$

9. Conclude with examples like Problems 60 and 62.

3.3 EXTREME POINTS AND INFLECTION POINTS (p. 100)

1. Use a figure like Figure 26 to introduce the notions of global and local extrema.

2. Sketch the graph of
$$f(x) = \begin{cases} 4 - x^2 & \text{if } -2 \le x \le 1 \\ x + 2 & \text{if } 1 < x \le 4 \end{cases}$$

Identify the local max and min values as well as the global max and min values of f on $[-2, 4]$.

3. State the Max-Min Existence Theorem and the Max-Min Location Theorem. Clarify the three types of critical points. Find the global max and min values of

$$f(x) = 2x^3 - 9x^2 + 12x + 4$$

on $[-1, 5]$. As a second example, work Problem 24.

4. Use Figure 28 to make the First Derivative Test reasonable. Work Problem 28 and/or 34 as an example.

5. State the Second Derivative Test and rework Problem 28 and/or 34 to illustrate it.

6. Introduce the concept of *inflection points* and show how to find them using Problem 44 and/or 48.

7. Bring all the ideas of the section together by sketching the graph of

$$f(x) = x^3 - 3x^2 - 9x + 2$$

and finding the coordinates of all extreme points and inflection points.

8. We recommend that you include Problems 57, 59, 61, 63, and 65 in your assignment for students.

3.4 LIMITS INVOLVING INFINITY AND ASYMPTOTES (p. 109)

1. Stress that the symbols ∞ and $-\infty$ do not represent numbers. Rather they are used to communicate information about the behavior of a function. Tell in words what each of the following statements means and draw sketches to illustrate each of them.

(i) $\lim\limits_{x \to \infty} f(x) = 4$

(ii) $\lim\limits_{x \to 3^+} f(x) = -\infty$

(iii) $\lim\limits_{x \to 3^-} f(x) = \infty$

(iv) $\lim\limits_{x \to 2} f(x) = \infty$

2. Find each of the following limits.

(i) $\lim\limits_{x \to \infty} \dfrac{1}{\sqrt{x}}$

(ii) $\lim\limits_{x \to \infty} \sqrt{x}$

(iii) $\lim\limits_{x \to 4^+} \dfrac{5}{x - 4}$

(iv) $\lim\limits_{x \to 3} \dfrac{12}{(x - 3)^2}$

3. Show how to find the limits at infinity when both numerator and denominator become large. Note the significance of the highest-degree terms. Illustrate with many examples such as Problems 24–38.

4. Discuss infinite limits using an example like

$$f(x) = \frac{3x^2 - 27}{x^2 - 4}$$

Note the significance of points where the denominator becomes 0.

5. Relate limits at infinity to horizontal asymptotes, and infinite limits to vertical asymptotes.

6. Discuss the five-step graphing process and illustrate it by working Problems 54 and 56.

7. See if you can get the class to help you do Problem 64.

4 Other Applications of the Derivative

4.1 OPTIMIZATION: GEOMETRIC PROBLEMS (p. 122)

1. A class discussion of Examples A, B, and C would be a good way to start the lesson. Emphasize the four-step procedure.

2. Try Problem 6 to illustrate how to finish a maximization problem after it has been expressed in mathematical symbols. Note that you must express z in terms of one variable before you can maximize or minimize.

3. For an example of a "best container" problem, consider an open tin box that is three times as long as wide and has an outer surface area of 225 square centimeters. Find the dimensions of this box if its volume is a maximum.

4. Work Problem 26 as an example of shortest route problem. Then rework the problem assuming that Ole can use a motorboat that travels at 10 miles per hour. This will illustrate the importance of checking endpoints.

5. We suggest Problems 28 and/or 30 as possible concluding examples. An appropriate diagram for Problem 30 is shown below.

Note that $x^2 + y^2 = 144$ and that the volume of the cone is $V = \frac{1}{3}\pi r^2(12 + x)$

4.2 OPTIMIZATION: BUSINESS PROBLEMS (p. 131)

1. It is important that students be able to state the meaning of $C(x)$, $R(x)$, $P(x)$, and their derivatives in words. For example, $R'(x)$ is the rate of change in total revenue with respect to the number x of units sold. Thus, $R'(x)$ is approximately the increase in revenue resulting from increasing x by one unit. What is the meaning of $C(x)/x$ and its derivative?

2. Work Problem 4.

3. Work Problem 12. Emphasize that

$$R(x) = \left(\begin{array}{c}\text{number of} \\ \text{units sold}\end{array}\right)\left(\begin{array}{c}\text{unit} \\ \text{price}\end{array}\right)$$

$$= xp(x) = 75x - 0.025x^2$$

4. Work Problem 14 which has the new wrinkle that you have to find $p(x)$. Since $[12 - p(x)]/0.2$ is the number of 20¢ reductions in price,

$$x = 3000 + \frac{200[12 - p(x)]}{0.2}$$

5. If you choose to do an inventory problem, try Problem 18. Note that

$$I(x) = 5\left(\frac{x}{2}\right) + (250 + 2x)\left(\frac{24,000}{x}\right)$$

6. Problem 24 can serve as a good banking problem. For the profit function, you should get

$$P(x) = (0.1 - x)k(x - 0.04)^2$$

7. As an alternative, try Problem 30. The revenue function is

$$R(x) = (1000 + 30x)(2.5 - 0.05x)$$

where x is the number of days before John sells.

8. Consider assigning Problems 27 and 29.

4.3 *MORE APPLICATIONS (p. 138)

1. This section will be rather difficult for most students. Go slowly. It would be better to cover two or three problems well than five or six superficially. Reading and rereading a problem is the important first step. Then:
   (i)   Decide what the independent variable x shall represent.
   (ii) Determine what quantity q is to be maximized (minimized)
   (iii) Express q in terms of x.
   (iv) Do the necessary calculus.
2. Follow the outline above in solving Problem 2.
   (i)   Let x be the number of trees per acre.
   (ii) Let q be the yield per acre.
   (iii) $q = x[60 - (x - 40)] = 100x - x^2$
   (iv) $q' = 100 - 2x$, giving a critical number of $x = 50$. Check that this really does maximize q.
3. Try Problem 8, a best speed problem. The hardest part is to express the cost C of the 200-mile run in terms of x, the speed in miles per hour. The correct expression is

$$C(x) = \frac{2000}{x} + 0.60(x + 2)$$

4. Problem 14 can be patterned after Example D. You need to maximize the annual harvest $h(x) = f(x) - x$.

5. If you have plenty of time, you might attempt Problem 16, 18, or 22.

4.4 IMPLICIT DIFFERENTIATION AND RELATED RATES (p. 144)

1. When we solve the equation

(i) $yx^2 - 3x = 2y + 2$

This section may be omitted without loss of continuity.

for y in terms of x, we get

(ii) $y = \dfrac{3x + 2}{x^2 - 2}$

Both equations describe the same function; the first does it implicitly, and the second does it explicitly. Find dy/dx at $x = 2$ and $y = 4$ by implicit differentation of (i). [It may help to replace y by $f(x)$ before you do this.] Do the same problem by explicit differentiation of (ii). Both methods should yield $dy/dx = -\frac{13}{2}$.

2. Work Problem 20. As your first step, you should get

$$4y\frac{dy}{dx} - 3\left(x^2\frac{dy}{dx} + 2xy\right) = 0$$

3. Find the equation of the tangent line to

$$(x^2 - 2xy^2)^3 = 5x + 2y + 10$$

at $(3, 1)$.

4. Problem 36 is a good related rates problem. An appropriate picture is shown below. We are given that $dx/dt = 30$ feet per second and are to find dy/dt when $y = 500$. Use implicit differentiation with respect to t on the equation

(i) $x^2 + (295)^2 = y^2$

to obtain an equation involving x, dx/dt, y, and dy/dt. Evaluate when $y = 500$. You will need to use (i) to find x at this instant.

5. Other problems you might consider working are Problems 40 and 44.

4.5 *APPROXIMATIONS AND THE DIFFERENTIAL (p. 151)

1. Think of Δy as the change along the curve $y = f(x)$ when x changes by an amount $dx = \Delta x$, and dy as the corresponding change on the tangent line (a picture will be helpful). Find Δy and dy for the function

$$y = f(x) = x^3 + 2x$$

Show that $\Delta y - dy = (3x + \Delta x)(\Delta x)^2$. Point out that if Δx is small, so are Δy and dy but that their difference $\Delta y - dy$ has a higher degree of smallness due to the factor $(\Delta x)^2$.

2. Note that $f(x + dx) \approx f(x) + dy$. Use this fact to do Problem 16 with $f(x) = \sqrt[3]{x}$, $x = 64$, and $dx = -0.2$.

3. In a similar vein, you might approximate $f(15.98)$ where

$$f(x) = 32\sqrt[4]{x} + \frac{x}{4} - 5$$

4. Work Problems 22 and 24. You should be able to show that for any values of x (side length) and dx,

$$\frac{dV}{V} = 3\frac{dx}{x} \quad \text{and} \quad \frac{dS}{S} = 2\frac{dx}{x}$$

5. If you have lots of business students, you will want to motivate and define *elasticity of demand, elastic case,* and *inelastic case.* Then work Problem 28.

6. If time permits, you may want to conclude the lesson by doing Problem 36.

5 Exponential and Logarithmic Functions

5.1 EXPONENTIAL AND LOGARITHMIC FUNCTIONS (p. 162)

1. Use $f(x) = 3^x$ to introduce the notion of an exponential function. Distinguish between $f(x) = 3^x$ and $g(x) = x^3$. Sketch their graphs using the same axes. Note that g ultimately grows much faster than f. In fact, 3^x ultimately grows faster than any power of x.

2. Review the properties of exponents with your class. Have them help you simplify expressions like the following:

(i) $3^2 3^4$ (ii) $3^2 3^{-4}$

(iii) $(3^2)^{-4}$ (iv) $(3^2 3^{-4})^{3/2}$

(iii) $(3^2 3^{1/2})^{1/5}/3^{-3/2}$

3. Stress the definition of logarithm. Point out that the exponential equivalent of $y = \log_4 x$ is $x = y^4$ and that the logarithmic equivalent of $m = 4^{4.2}$ is $\log_4 m = 4.2$. Use the definition of logarithm to evaluate:

(i) $\log_7 49$ (ii) $\log_5(\frac{1}{125})$

(iii) $\log_5(5\sqrt{5})$ (iv) $\log_9 27$

4. Write the properties of logarithms on the board. Given that $\log_8 a = 2.42$ and $\log_8 b = -2.58$, find:

(i) $\log_8(a/b)$ (ii) $\log_8(8a^2)$

(iii) $\log_8\sqrt{a}$ (iv) $\log_8(a^3 b^2)$

5. Sketch the graphs of $f(x) = 2^x$ and $g(x) = \log_2 x$ using the same axes. What is special about their graphs? About their domains and ranges? Discuss the concept of inverse functions.

6. Work Problem 50 and (if there is time) try Problem 72.

5.2 THE NATURAL EXPONENTIAL AND LOGARITHMIC FUNCTIONS (p. 168)

1. Introduce the number e with some historical remarks. Suggest that it may be the most important number in calculus. One reason is that e^x is its own derivative, a fact that will be discussed later in the lesson.

2. We use $\ln x$ (as shorthand for $\log_e x$) to denote the function inverse to e^x. Thus, $e^{\ln x} = x$ and $\ln e^x = x$. Have

the class help you simplify:

(i) $\ln e^{1.2}$ (ii) $e^{\ln 3}$

(iii) $\ln e^{3x}$ (iv) $e^{3\ln x}$

3. Most scientific calculators have $\ln$ and e^x keys. Demonstrate their use in doing problems like Problems 8–14.

4. Solve the exponential equation

$$4^{2x+1} = 120$$

by equating the natural logarithms of both sides and solving the resulting logarithmic equation. Do the same for

$$3^{2x-1} = 5^x$$

5. State (and maybe derive) the differentation formulas for $\ln x$ and e^x. Use them to find dy/dx in:

(i) $y = e^{\sqrt{x}}$ (ii) $y = \ln(x^2 + 1)$

(iii) $y = x^2 e^{1/x}$ (iv) $y = \frac{1}{x}\ln(2x + 1)$

6. Work Problem 30.

7. Consider the function

$$f(x) = x^2 e^{-2x}$$

(i) Where is it increasing? Decreasing?

(ii) Determine its maximum and minimum values.

5.3 THE CHAIN RULE (p. 174)

1. Review the concept of composite function. Have the class help you rewrite each of the following in the form $y = f(u)$ where $u = g(x)$.

(i) $y = e^{3x}$

(ii) $y = \ln(x^2 + 1)$

(iii) $y = (\ln x)^3$

2. State the Chain Rule.

3. Now find dy/dx for each of the composite functions in part 1 above.

4. Work Problems 4, 8, and 12. Note that Problem 12 can be simplified by writing

$$\ln\left(\frac{x^2}{2x - 1}\right) = 2\ln x - \ln(2x - 1)$$

5. Try Problems 20 and 22.

6. Point out that the General Power Rule (Section 2.4) is a special case of the Chain Rule. Do Problems 36 and 38.

7. Refer to Example D which requires repeated use of the Chain Rule. Then illustrate by finding dy/dx for:

(i) $y = \ln[(3x + 2)^4 + x^2]$

(ii) $y = (e^{\sqrt{x}} + 2x)^3$

(iii) $y = [\ln(\ln x)]^4$

8. You may want to discuss the proof of the Chain Rule. Begin by noting that if A can run twice as fast as B and if B can run three times as fast as C, then A can run six

$(2 \cdot 3)$ times as fast as C. Then give the partial proof outlined in the text.

9. If time permits, conclude by doing Problem 50.

5.4 *GENERAL LOGARITHMIC AND EXPONENTIAL FUNCTIONS (p. 180)

1. We have now met the natural logarithmic function $\ln x$ and the natural exponential function e^x. Other logarithmic and exponential functions can be expressed in terms of them.

2. The equations $\log_2 x = y$ and $x = 2^y$ are equivalent. Take $\ln$ of both sides of the latter to obtain $\ln x = y \ln 2$ which means

$$\log_2 x = y = \frac{\ln x}{\ln 2}$$

In a similar manner, show that

(i) $\log_b x = \dfrac{\ln x}{\ln b}$

3. Use (i) and calculator to evaluate $\log_3 14$.

4. In a similar manner, let $y = 5^x$. Take $\ln$ of both sides to obtain $\ln y = x \ln 5$ which implies that $y = e^{x \ln 5}$ Thus, $5^x = e^{x \ln 5}$. More generally,

(ii) $b^x = e^{x \ln b}$

5. From (i) and (ii) above, we learn that

(iii) $\dfrac{d}{dx} \log_b x = \dfrac{1}{x \ln b}$　　　　(iv) $\dfrac{d}{dx} b^x = b^x \ln b$

6. Do Problems 16, 18, 20, and 22.

7. Note that we now know how to find the derivatives of both x^4 and 4^x. But how do we find the derivative of x^x or x^{4x-1}? We suggest logarithmic differentiation, as illustrated in Example C. Try it on $y = x^x$ and then on Problems 30 and 32.

8. Illustrate logarithmic differentiation for complicated products as in Problem 42.

6 Applications of Exponential and Logarithmic Functions

6.1 EXPONENTIAL GROWTH AND DECAY (p. 190)

1. Begin by showing that

(i) $y = 10e^{0.2t}$

satisfies the differential equation $dy/dt = 0.2t$ and that

(ii) $y = 10e^{-0.3t}$

satisfies the differential equation $dy/dt = -0.3t$. Sketch the graphs of (i) and (ii). The first is an example of exponential growth, the second of exponential decay.

2. State the general result that $y = y_0 e^{kt}$ is the solution to the differential equation $dy/dt = ky$.

3. Work Problems 4 and 8.

4. Work Problem 10. You will need to explain how to evaluate the constant k.

5. Discuss world population growth as in Example C. Then work Problem 14. When will the world's population reach 10 billion under this model?

6. Define the concept of half-life in radioactive decay. Work Problem 20. It will require that you solve the equation $0.5 = e^{-0.1022t}$ for t.

7. Discuss carbon dating and then work Problem 22.

8. If you have time, try Problem 32. You will first need to find k; we get $k \approx -0.000121$. For part (b), you will need to solve the equation

$$627.3 = 1013.5 e^{-0.000121h}$$

for h. We get $h \approx 3965$ meters.

6.2 OTHER GROWTH MODELS (p. 195)

1. Discuss the concepts of exponential growth, bounded growth, and logistic growth and help students understand why they are modeled by the differential equations $dy/dt = ky$, $dy/dt = k(L - y)$, and $dy/dt = ky(L - y)$. Write these equations and their solutions on the chalkboard and draw their graphs. We think students do not need to memorize these solutions, at least the last two; it is enough to be able to apply them.

2. Work Problem 6 as an example of bounded growth. Note that k is obtained by solving the equation

$$9600 = 24,000(1 - e^{-10k})$$

Problem 10 would be a good second problem of this type.

3. For an illustration of logistic growth, try Problem 16. To find k, solve the equation

$$2000 = \frac{40,000}{1 + 79e^{-40,000k}}$$

We get $k \approx 0.00003563$. Then use this k to solve

$$40,000 = 10,000(1 + 79e^{-40,000kt})$$

for t. We get $t \approx 2.3$ years.

4. As an alternate or additional example of logistic growth, try Problem 20.

6.3 MATHEMATICS OF FINANCE (p. 201)

1. The simple-interest formulas

$$I = Prt \quad \text{and} \quad F = P + I$$

are easy to understand. Work a few simple examples such as:

(i) What will \$100 be worth after 8 years if invested at 7% simple interest?

(ii) How long will it take for \$1200 to accumulate to \$2253 at 6.75% simple interest?

2. Give careful derivations of the compound interest formulas

$$F = P(1 + r)^t$$

and

$$F = P(1 + i)^n = P\left(1 + \frac{r}{k}\right)^{kt}.$$

the first corresponding to annual compounding, the second to compounding k times per year. Make up examples; try Problem 12.

3. Derive the continuous interest formula $F = Pe^{rt}$; work Problem 16.

4. Introduce the concept of effective rate of interest; work Problem 22. Find the effective rate of interest corresponding to 12% compounded continuously.

5. Determine how long it takes money to double at 7.5% compounded semiannually and at 7.25% compounded continuously.

6. Try Problem 38 as an illustration of a problem with an unknown interest rate. You will need to solve the equation

$$3000 = 1000(1 + i)^{180}$$

for i (use logarithms) and then multiply by 12 to get r.

6.4 MORE ON THE MATHEMATICS OF FINANCE (p. 207)

1. Discuss accumulating money and discounting money.

2. To accumulate \$4000 for 6 years at 8% interest compounded quarterly is to calculate

$$4000(1.02)^{24} = \$6433.75$$

Work problem 2.

3. To discount \$1500 for 7 years at 8% compounded quarterly is to calculate

$$1500(1.02)^{-28} = \$861.56$$

Work problem 4.

4. Suppose that I owe \$4000 due 1 year from now and \$1500 due 14 years from now. What single payment 7 years from now would completely pay off my debt if interest is at 8% compounded quarterly? Use a time line.

The answer is

$$4000(1.02)^{24} + 1500(1.02)^{-28}$$

which is the sum of the two answers in parts 2 and 3 above, namely,

$$6433.75 + 861.56 = \$7295.31$$

5. Work problems 10 and 12.

6. To illustrate the effect of inflation, work Problem 20. The true growth rate g is

$$g = \frac{0.09 - 0.05}{1.05} \approx 0.038095$$

You need to find P such that

$$P(1 + g)^6 = 60{,}000$$

7. Use problem 24 to illustrate the double-declining-balance method of depreciation. When will the depreciated value be \$33,000?

7 The Antiderivative

7.1 ANTIDERIVATIVES (p. 218)

1. Introduce the terms *antidifferentiate*, *antiderivative*, and *general antiderivative*. illustrate.

2. Be sure to emphasize that the general antiderivative always involves an arbitrary constant C. Students should get used to writing $2x^3 + C$, not just $2x^3$, for the general antiderivative of $6x^2$.

3. Introduce the notation $\int \ldots dx$ for the general antiderivative (indefinite integral) and state the formulas for $\int x^r dx \, (r \ne -1)$ and $\int (1/x) dx$. Work Problems 10, 12, 14, and 18.

4. Note that $\int e^x dx = e^x + C$. Then use the linearity properties of the indefinite integral to work Problems 20, 24, 30, 34, 36, and 40. In Problem 36, first expand the integrand, and in Problem 40, rewrite the integrand as

$$3x^{1/2} - 5x^{-3/2}$$

5. Find the particular antiderivative $f(x)$ of $f(x) = (x - 2x^{-2})^2$ for which $f(1) = 5$. You should discover that $C = 6$.

6. Problem 50 is a good applied problem with a side condition.

7. For an example involving two arbitrary constants C_1 and C_2, try Problem 52.

7.2 THE SUBSTITUTION RULE (p. 224)

1. Mention that antidifferentiation is intrinsically more difficult than differentiation. Many antidiffentiation problems are hard; some are impossible (for example, $\int e^{x^2} dx$

cannot be expressed in terms of elementary functions). We need to develop techniques to help us.

2. Give a careful discussion of the most important antidifferentiation technique—the Substitution Rule. Use it to antidifferentiate

$$x^2(x^3 + 4)^9 \quad \text{and} \quad \frac{e^{\sqrt{x}}}{\sqrt{x}}$$

In the first let $u = x^3 + 4$; in the second, let $u = \sqrt{x}$. Check the answers by differentiation

3. Work Problems 10, 18, 22, and 24.

4. Sometimes substitution works in an unexpected way. For example, in $\int x(2x - 1)^5 dx$, let $u = 2x - 1$ so $du = 2 dx$. Note that $x = \frac{1}{2}(u + 1)$. The integral takes the form

$$\int \frac{1}{2}(u + 1)u^5 \cdot \frac{1}{2} du = \frac{1}{4} \int (u^6 + u^5) du$$

which is doable.

5. Sometimes substitution doesn't work at all, as in

$$\int \left(x^2 - \frac{2}{x}\right)^2 dx.$$ But here the integrand can be expanded
and is then easy to antidifferentiate.

6. Rational functions can sometimes be rewritten to prepare for antidifferentiation. For example, use long division to write $2x^2/(x + 1)$ as $2x - 2 + 2/(x + 1)$. Then antidifferentiate. Use this technique on Problem 48.

7.3 SOME INTEGRATION TECHNIQUES (p. 230)

1. Introduce the words *integrate* and *integral* as synonyms for antidifferentiate and antiderivative. Substitution is the chief integration technique; integration by parts (double substitution) is perhaps the second most important technique. Show how the form

(i) $\int u \, dv = uv - \int v \, du$

follows from the Product Rule for derivatives.

2. Use (i) on $\int xe^{2x} dx$ with $u = x$ and $dv = e^{2x} dx$. Note that $du = dx$, whereas $v = \int e^{2x} dx = \frac{1}{2}e^{2x}$ (we can omit the constant C here). We obtain for the right side of (i)

$$\frac{1}{2}xe^{2x} - \frac{1}{2} \int e^{2x} dx = \frac{1}{2}xe^{2x} - \frac{1}{4}e^{2x} + C$$

Now use this technique on Problem 8.

3. In applying this technique to $\ln x$ or to $x^r \ln x$, always let $u = \ln x$. Try it on Problems 4 and 6 and, appropriately modified, on Problem 12.

4. Go over Example C, a problem that uses integration by parts twice.

5. Introduce the partial fraction decomposition technique. Illustrate it by determining A and B in

$$\frac{x + 11}{(x + 2)(x - 1)} = \frac{A}{x + 2} + \frac{B}{x - 1}$$

You should get $A = -3$ and $B = 4$.

6. Use the result above to find

$$\int \frac{x + 11}{x^2 + x - 2} dx$$

7. Finish the lesson by working Problem 44 which requires integration of $x^2 \ln x$ (integration by parts)

7.4 *DIFFERENTIAL EQUATIONS (p. 236)

1. Explain what is meant by a differential equation and what is meant by its solution. Illustrate by showing that $y = x^3 - 4x + C$ is a solution to $dy/dx = 3x^2 - 4$. A differential equation of the form $dy/dx = f(x)$ has the solution $y = \int f(x)dx$. But what about an equation of the form

$$\frac{dy}{dx} = f(x)g(y)$$

Such an equation is said to be separable because it can be rewritten as

$$\frac{dy}{g(y)} = f(x)dx$$

and then integrated.

2. Use the separation of variables technique to solve $dy/dx = y^2 e^x$.

3. Solve the equation

$$\frac{dy}{dx} - 2y = 5$$

subject to the side condition $y = \frac{13}{2}$ when $x = 0$.

4. Solve the growth-decay equation $dy/dt = ky$ subject to $y = y_0$ at $t = 0$, thereby fulfilling a promise from Section 6.1. Work Problem 18.

5. Solve the bounded growth and/or the logistic growth equations

$$\frac{dy}{dt} = k(L - y) \quad \text{and} \quad \frac{dy}{dt} = ky(L - y)$$

thereby fulfilling promises from Section 6.2. Work Problems 22 and/or 24.

6. If time permits, try Problem 38. The appropriate model is

$$\frac{dp}{dt} = k(100 - p)$$

8 The Definite Integral

8.1 AREA AND THE DEFINITE INTEGRAL (p. 246)

1. Review the notion of area and its properties, recalling familiar formulas. Find the area of the figure below.

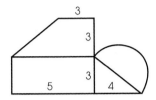

2. We define the definite integral in terms of area; that is,

$$\int_a^b f(x)\, dx = A(R_u) - A(R_d)$$

3. Sketch the graph of

$$f(x) = \begin{cases} x & -2 \le x \le 2 \\ 2 - \frac{1}{3}x & 2 < x \le 6 \end{cases}$$

Then calculate $\int_{-2}^6 f(x)\, dx$. Work Problems 14 and 20.

4. Calculate $\int_{-2}^2 \sqrt{4 - x^2}\, dx$ by thinking geometrically.

5. Introduce the notion of even and odd functions. Ask the class to decide which of the following are even, which are odd, and which are neither.

   (i) $f(x) = x^4 - 1$ (ii) $f(x) = x^5 - \sqrt[3]{x}$
   (iii) $f(x) = |x|$ (iv) $f(x) = x^4 + x$

6. Given the formula

$$\int_0^a x^n\, dx = \frac{a^{n+1}}{n + 1}$$

use symmetry to evaluate each of the following.

   (i) $\int_{-3}^3 x^4\, dx$ (ii) $\int_{-4}^4 x^3\, dx$
   (iii) $\int_{-4}^4 |x|\, dx$ (iv) $\int_{-2}^2 |x|^3\, dx$

7. Evaluate the definite integral

$$\int_0^6 |2x - 4|\, dx$$

8. If time permits, discuss the relation between velocity and total distance traveled and work Problem 52.

8.2 THE FUNDAMENTAL THEOREM OF CALCULUS (p. 252)

1. State the Fundamental Theorem of Calculus. Emphasize two things:
   (i) It establishes an intimate relationship between definite integrals and antiderivatives.
   (ii) It makes possible the evaluation of a wide range of definite integrals.

You will want to prove this theorem either now or perhaps after you have illustrated it in part 2 below.

2. Illustrate the theorem by evaluating $\int_{-1}^3 x^2\, dx$, $\int_0^4 \sqrt{x}\, dx$, and $\int_{-2}^2 (x + 2)^2\, dx$. Work Problems 10 and 20.

3. Recall the relation between the definite integral and area. Then work Problem 22.

4. Emphasize the three properties of the definite integral listed just before Example F (linearity and additivity). Then work Problems 30, 32, 36, and 38.

5. Discuss the concept of average value of a function on an interval. Ask the class to guess the average value of $f(x) = x^2$ on $[0, 3]$. Then calculate this average value.

6. If there is time, work Problem 62.

8.3 CALCULATION OF DEFINITE INTEGRALS (p. 259)

1. Review antidifferentiation by substitution. Let $u = 3x^2 + 1$ and $du = 6x\, dx$ to transform the integral below.

$$\int x(3x^2 + 1)^3\, dx = \frac{1}{6} \int u^3\, du = \frac{1}{24} u^4 + C$$
$$= \frac{1}{24} (3x^2 + 1)^4 + C$$

2. Evaluate $\int_0^1 x(3x^2 + 1)^3\, dx$ directly from the Fundamental Theorem of Calculus and part 1 above. You should get

$$\left[\frac{1}{24} (3x^2 + 1)^4 \right]_0^1 = \frac{255}{24}$$

3. Now evaluate the same definite integral by making a complete u-substitution in this integral, including a change in the limits.

$$\int_0^1 x(3x^2 + 1)^3\, dx = \frac{1}{6} \int_1^4 u^3\, du = \left[\frac{1}{24} u^4 \right]_1^4 = \frac{255}{24}$$

We suggest that students use the second method except in cases where they can do the required antidifferentiation in their heads.

4. Work Problems 8, 12, 16, and 24.

5. Illustrate integration by parts in definite integrals. Let $u = x$ and $dv = e^{2x}$ in the following integral.

$$\int_0^3 x e^{2x}\, dx = \left[\frac{1}{2} x e^{2x} \right]_0^3 - \frac{1}{2} \int_0^3 e^{2x}\, dx = \frac{3}{2} e^6 - \frac{1}{4}(e^6 - 1)$$
$$= \frac{5}{4} e^6 + \frac{1}{4}$$

Work Problem 28.

6. Use the table of integrals to evaluate $\int_3^4 [1/(x\sqrt{25 - x^2})]\, dx$. Work Problem 36.

7. End with Problem 44.

8.4 THE DEFINITE INTEGRAL AS A LIMIT (p. 264)

1. While some instructors might choose to skip this section, students do not really understand the definite integral without it. Also, a good understanding will help with the applications in the next chapter.

2. Discuss the rigorous definition of area (as in the first two pages of the section), perhaps by considering the area of the region under the curve $y = 5 - x^2$ between $x = 0$ and $x = 2$.

3. Give the limit of sum definition of the definite integral.

4. Introduce sigma notation and work Problems 2, 6, 10, 16, and 18.

5. Discuss the properties of Σ.

(i) $\displaystyle\sum_{i=1}^{n} ca_i = c \sum_{i=1}^{n} a_i$

(ii) $\displaystyle\sum_{i=1}^{n} (a_i + b_i) = \sum_{i=1}^{n} a_i + \sum_{i=1}^{n} b_i$

(iii) $\displaystyle\sum_{i=1}^{n} c = nc$

Work Problems 24 and 26.

6. Write the definite integral as the limit of a sum using sigma notation.

7. Evaluate $\int_0^2 x^2\,dx$ using the limit of sum definition (Example E). Then work Problem 36 in complete detail.

8. If there is time, try some of the miscellaneous problems, say Problems 46, 48, and 50. In Problem 50, note that

$$\ln\frac{1}{2} + \ln\frac{2}{3} + \ln\frac{3}{4} + \cdots + \ln\frac{19}{20}$$

can be written as a single logarithm.

9 Applications of the Definite Integral

9.1 AREA BETWEEN TWO CURVES (p. 276)

1. If $f(x) \geq 0$, then the area of the region under the curve $y = f(x)$ between a and b is given by the integral $\int_a^b f(x)\,dx$. For example, the area of the region under the curve $y = 1 + x^2$ between $x = 0$ and $x = 2$ is given by

$$\int_a^b f(x)\,dx = \int_0^2 (1 + x^2)\,dx$$

Think of it this way (see Figure 3). To obtain the area of the region, slice it into narrow strips, each of area approximately $f(x)\,dx$, add together all these areas, and take the limit. We abbreviate this process with the three-word motto *slice*, *approximate*, *integrate*.

2. Draw a picture of the region R bounded by the curves $y = g(x)$ and $y = f(x)$ where $g(x) \leq f(x)$ and $a \leq x \leq b$. Show a typical narrow strip. Applying our motto, we approximate the area of the strip by $[f(x) - g(x)]\,dx$ and then integrate to get

$$A(R) = \int_a^b [f(x) - g(x)]\,dx$$

We recommend remembering the process rather than the final formula.

3. Illustrate the above process by working Problems 14 and 16.

4. Problem 24 asks for the area of the region R bounded by $y = x^2 - 1$ and $y = 2x + 2$. The new twist is that we must determine the limits, that is, the x-coordinates of the intersection points (solve simultaneously, getting $x = -1$ and $x = 3$).

$$A(R) = \int_{-1}^{3} [2x + 2 - (x^2 - 1)]\,dx$$

5. Illustrate horizontal slicing by working Problem 40.

6. If time permits, finish with Problem 46.

9.2 VOLUMES OF SOLIDS OF REVOLUTION (p. 283)

1. Draw a three-dimensional picture of a solid of revolution showing a typical thin slice (a disk) as in Figure 13. Note that the formula for the volume of a disk of radius $f(x)$ and thickness dx is $V = \pi[f(x)]^2\,dx$. Then discuss Example A.

2. Work Problem 2. First draw the given picture on the board. Ask the students to visualize the disk generated by a typical rectangle; its volume is $\pi(x^2 + 2)^2\,dx$. The volume of the whole solid is $\int_0^2 \pi(x^2 + 2)^2\,dx$.

3. Ask the students to visualize the solid obtained when the region in Problem 2 is revolved about the line $y = -2$. Draw this solid as best you can, showing the slice generated by a typical rectangle (a washer). The volume of this washer is

$$\pi[(x^2 + 4)^2 - 2^2]\,dx$$

and the volume of the whole solid is

$$\int_0^2 \pi(x^4 + 8x^2 + 12)\,dx$$

4. Try Problem 14 and perhaps Problem 18.

5. Revolving about a vertical line leads to horizontal slicing. Draw the two-dimensional region in Problem 20 with a typical horizontal rectangle. When revolved about the y-axis, this rectangle generates a disk of volume

$$\pi x^2\,dy = \pi y^{2/3}\,dy$$

The corresponding solid of revolution has volume $\int_0^8 \pi y^{2/3}\,dy$.

6. Now work the harder Problem 20. A typical horizontal rectangle generates a washer of volume

$$\pi\,(x_{\text{big}}^2 - x_{\text{small}}^2)\,dy = \pi\,[(4 - y) - \tfrac{1}{4}(4 - y)^2]\,dy$$

9.3 SOCIAL SCIENCE APPLICATIONS (p. 290)

1. The Fundamental Theorem of Calculus tells us that
(i) $\quad Q(b) - Q(a) = \int_a^b Q'(x)\,dx$
Thus, the change in a quantity over the interval $[a, b]$ equals the integral of its derivative over this interval. Apply this to Problems 2 and 4.

2. If we apply (i) to the case where $Q'(x)$ is the marginal cost of producing x units of a quantity, then the total cost of producing b units is

$$Q(b) = Q(0) + \int_0^b Q'(x)\,dx$$

Here, $Q(0)$ is the cost of producing 0 units; that is, it is the fixed cost. Use this to work Problem 6.

3. Make clear the meaning of demand function, supply function, and equilibrium price using a figure like Figure 21. Work Problem 14.

4. The area of the region under the demand curve $p = D(x)$ from $x = 0$ to $x = b$ represents the total number of dollars a consumer would pay for b units if buying according to the demand curve. Thus, the area of the region under this curve and above the line $y =$ equilibrium price (called the consumer surplus) is the amount the consumer would save by buying at the constant-equilibrium price. Similarly explain the producer surplus. Work Problems 20 and 24.

5. Discuss the Lorenz curve and the Gini coefficient using Figure 23. Then work Problem 30.

9.4 CONTINUOUS MONEY STREAMS (p. 296)

1. Start with this question: What is the present value P of $5000 due 12 years from now if interest is at 8% compounded continuously? Now

$$Pe^{0.08(12)} = 5000$$

so

$$P = 5000\,e^{-0.08(12)} = \$1914.46$$

The latter is a special case of the general formula $P = Fe^{-rt}$. Work Problems 4 and/or 6.

2. Tell what is meant by a continuous money stream. Go over Example B leading to the important formula

(i) $\qquad PV = \int_0^T R(t)\,e^{-rt}\,dt$

Here, $R(t)$ is the annual rate of income of the money stream at time t, r is the interest rate, and T is the period. Use (i) to work Problems 8 and 10.

3. What happens to (i) if the money stream is a perpetual stream, that is, lasts forever? This leads to the important concept of improper integrals. Give the appropriate definitions using $\int_0^\infty (1/t^2)\,dt$ and $\int_0^\infty (1/\sqrt{t})\,dt$ as examples (the first converges, the second diverges). Work Problems 12 and 14.

4. Find the present value of a perpetual money stream which continues at the rate of P dollars per year if the interest rate is r compounded continuously. The answer (Example G) is $\int_0^\infty Pe^{-rt}\,dt$ which simplifies to the elegant formula P/r. Illustrate with $P = \$10,000$ and $r = 0.05$. Then work Problem 18.

9.5 *CONTINUOUS PROBABILITY (p. 301)

1. We suggest reviewing elementary probability theory with a simple example. Toss a fair die and let $\mathbf{X}$ denote the number of dots that appear. We call $\mathbf{X}$ a discrete random variable because it takes on discrete values. Calculate prob$\{\mathbf{X} = 5\}$ and prob$\{\mathbf{X} < 5\}$.

2. Next discuss the notion of a continuous random variable, that is, a random variable which can take on any value on an interval. The stopping point of the spinner in Figure 26 is a good example. Associated with each continuous random variable is a probability density function f that is nonnegative with total integral 1. Furthermore,

$$\text{prob}\{c < \mathbf{X} < d\} = \int_c^d f(x)\,dx$$

3. As an example of a uniform distribution, work Problem 2. For a nonuniform example, but still on a finite interval, work Problem 6. For an example on an infinite interval, let $\mathbf{X}$ have the density function $f(x) = ke^{-2x}$ on $[0, \infty]$ and $f(x) = 0$ elsewhere. Determine k and then calculate prob$\{\mathbf{X} \geq 2\}$.

4. Define the mean μ of a random variable by

$$\mu = \int_{-\infty}^{\infty} xf(x)\,dx$$

and give its physical interpretation as a balance point. Calculate the mean for the three random variables in part 3 above.

5. Work Problem 14 and, if you have time, try Problem 22.

10 Multivariate Calculus

10.1 FUNCTIONS OF TWO OR MORE VARIABLES (p. 312)

1. Begin by contrasting the function $f(x) = x^2$ with the function
$$f(x, y) = x^2 - 3y$$

The first is a function of one variable, and the second is a function of two variables. Help students with the notation by calculating $f(2, -1)$, $f(w, w^2)$, and $f(2x, 3y)$. Work Problem 10.

2. Sketch the picture of a rectangular solid of width x, length $2x$, and height y. Develop the formulas for its surface area $S(x, y)$ and volume $V(x, y)$. Calculate $S(3, 2)$ and $V(4, 5)$. Work Problem 14.

3. You might discuss domains. For example, what is the natural domain of $f(x, y) = \sqrt{4 - x^2 - y^2}$?

4. Introduce three-dimensional coordinate space and plot a few points.

5. Show how we graph functions of two variables in three-space by first finding the traces in the three coordinate planes (or other planes). Begin with the linear example

$$f(x, y) = 8 - 4x - 2y$$

Follow with examples like

$$g(x, y) = 16 - x^2 - y^2$$

$$h(x, y) = x^2 + 2y^2$$

$$F(x, y) = x^2$$

6. What are the traces of g in the planes $z = k$? The traces of h in the planes $z = k$?

7. End the lesson with Problem 38. Note that $R(p, q) = xp + yq$. Express this in terms of p and q alone.

10.2 PARTIAL DERIVATIVES (p. 318)

1. Let $z = f(x, y)$. Note that $f(x, 6)$ is a function of x alone. More generally, if we regard y as fixed, $f(x, y)$ becomes a function of x alone and it makes sense to ask for its derivative with respect to x. This (partial) derivative is denoted by $\partial z/\partial x$. Illustrate by finding $\partial z/\partial x$ for
   (i) $z = xy^3 + 3x^2$
   (ii) $z = 2x - x/y$
Similarly, $\partial z/\partial y$ denotes the derivative of z with respect to y holding x fixed. Find $\partial z/\partial y$ for (i) and (ii) above. Then work Problems 6, 12, and 18.

2. Explain the notation $f_x(x, y)$ and also $f_x(2, 1)$. Then work Problem 22.

3. As in Figures 12 and 13, draw the graph of a surface $z = f(x, y)$ and interpret $f_x(x_0, y_0)$ and $f_y(x_0, y_0)$ as the slopes of tangent lines at (x_0, y_0). Work Problem 28.

4. Discuss Example E and/or Problem 32.

5. Problem 36 gives an opportunity to discuss partial derivatives as rates of change.

6. Higher-order derivatives present no new difficulties. To illustrate, find $f_{yz}(2, -1, 1)$ for $f(x, y, z) = xy^2/z^3$. Note that

$$f_y(x, y, z) = \frac{2xy}{z^3}$$

$$f_{yz}(x, y, z) = \frac{-6xy}{z^4}$$

Now evaluate $f_{yz}(2, -1, 1)$.

7. If time permits, work Problem 56.

10.3 OPTIMIZATION (p. 326)

1. On the board or on an overhead, draw a picture something like Figure 17 and use it to discuss maximum points, minimum points, and saddle points. Distinguish between extrema (global extrema) and local extrema. You might note that a little common sense can sometimes substitute for a lot of complicated calculus. For example, the maximum value of

$$f(x, y) = 24 - x^2 - y^2$$

is clearly $f(0, 0) = 24$ and the minimum value of

$$F(x, y, z) = (x - 2)^2 + 4(y + 1)^2 + z^2 + 9$$

is $F(2, -1, 0) = 9$.

2. For harder problems, the Second Partials Test is the tool, but we don't think students should be required to memorize it. Use it to work Problem 10. You will find two critical points: $(0, 0)$ is a saddle point, and $(18, 6)$ is a maximum point.

3. Problem 18 is a good geometric problem. If x, y, and z denote length, width, and height, respectively, then $xyz = 8$. You must minimize

$$M = 4yz + 2xz + xy$$

$$= \frac{32}{x} + \frac{16}{y} + xy$$

4. Work Problem 22 as an example of an economics problem. Note that

$$P = x(p - 60) + y(q - 70)$$

Replace x and y by their given expressions and then maximize P.

5. If you have time, work Problem 24. The function to be minimized is

$$f(x, y) = 2x + 2y + \frac{24}{xy}$$

10.4 *CONSTRAINED OPTIMIZATION AND LAGRANGE'S METHOD (p. 332)

1. Distinguish between a free optimization problem and a constrained optimization problem as in the opening remarks of the section. The latter type of problem is very common,

especially in economics. It is the problem of optimizing $f(x, y)$ subject to a constraint $g(x, y) = 0$. (Example: Minimize $x^4 + y^2 + 2$ subject to $x + y - 1 = 0$.) If the constraint can be solved for y (or x), the result can be substituted in $f(x, y)$, thereby reducing the problem to optimizing a function of one variable, something we studied way back in Section 3.3. As an alternative that often works well when $g(x, y) = 0$ can't be solved for x or y, introduce Lagrange's Method. Why does it work? Unfortunately, this requires a lot of background best left to more advanced courses.

2. Illustrate Lagrange's Method using Problems 4 and/or 8.

3. Explain how Lagrange's Method works for a function of three variables.

4. Illustrate by working Problem 14. If x, y, and z denote length, width, and height, respectively, then the problem is to minimize

$$C = C(x, y, z) = 7xy + 6z(x + y)$$

subject to

$$g(x, y, z) = xyz - 252 = 0$$

5. For an economics application, try Problem 16.

10.5 *APPLICATION: THE LEAST SQUARE LINES (p. 338)

1. A social scientist claims that a variable y can be predicted from a variable x by using the linear equation $y = -1.2x + 3.1$. How did she come to this conclusion? She probably began by plotting some (x, y) data pairs that she had available or had gathered, obtaining a graph of the type shown in Figure 28. (Put such a figure on the board or on an overhead.) Then she tried to fit a straight line to these data as in Figure 29. But how does one fit a line to a set of data pairs?

3. Explain what is meant by the least squares line (see Figure 30). Then work Problem 4.

4. Now consider the general problem where n data pairs are given. Go through the derivation of the least squares line $y = mx + b$; that is, derive the formulas for m and b. It may be wise to review the properties of Σ before beginning the derivation.

5. Work Problem 8 using the formulas just derived.

6. Work Problem 12. Note that it asks you to make a prediction based on the data.

7. As an alternative or an addition to Problem 12, bring some real data to class (for example, midterm grades and final grades for an earlier math class) and use it to make a prediction.

11 Double Integrals

11.1 DOUBLE INTEGRALS OVER RECTANGLES (p. 348)

1. Review the definition of the definite integral $\int_a^b f(x)\, dx$. Then by analogy, define the double integral as:

$$\iint_D f(x, y)\, dy\, dx = V(S_u) - V(S_d)$$

where V denotes volume and S_u and S_d are defined appropriately as in Figure 2.

2. Calculate $\iint_D \frac{1}{3}(6 - 2x)\, dy\, dx$ where $D = \{(x, y): 0 \le x \le 5, 0 \le y \le 4\}$. Do this by drawing a good sketch, noting that S_u and S_d are prisms of volume 12 and $\frac{16}{3}$, respectively. Thus the double integral has a value of $\frac{20}{3}$.

3. Discuss the fact that a double integral can be written as a limit of sums, appealing to Figure 4. But here is the big question: Is there a simple analytic way of evaluating double integrals?

4. Introduce the concept of a partial integral by evaluating

$$\int_1^e \frac{3x^2}{y}\, dy.$$ Do another example, say Problem 4.

5. Use $\int_0^2 \left(\int_1^e \frac{3x^2}{y}\, dy \right) dx$ to initiate a discussion of iterated integrals (its value is 8). Work Problems 8 and 12.

6. With the help of Figures 5 and 6, show that a double integral can be written as an iterated integral. Then evaluate the following double integral:

$$\iint_R (4xy + 3y^2)\, dx\, dy$$

where $R = \{(x, y): 0 \le x \le 2, 0 \le y \le 1\}$ Write this double integral as

$$\int_0^1 \left[\int_0^2 (4xy + 3y^2)\, dx \right] dy$$

and evaluate (the value is 6).

7. Work Problems 14 and/or 18.

11.2 DOUBLE INTEGRALS OVER GENERAL REGIONS (p. 354)

1. For later use, begin by evaluating the following iterated integral in which the limits on the inner integral are variables.

$$\int_0^1 \left[\int_{x^2+1}^{5+x} 12x\, dy \right] dx$$

2. Describe x-simple and y-simple sets. Sketch the region $D = \{(x, y): x^2 + 1 \leq y \leq 5 + x, 0 \leq x \leq 1\}$ and note that it is y-simple.

3. Convince your class with the help of Figures 10 and 11 that a double integral

$$\iint_D f(x, y) \, dy \, dx$$

over a y-simple region D can be written as an iterated integral

$$\int_a^b \left[\int_{g(x)}^{h(x)} f(x, y) \, dy \right] dx$$

Illustrate by evaluating

$$\iint_D 12x \, dy \, dx$$

where D is the y-simple region of part 2, leading to the iterated integral of part 1.

4. Work Problem 8 and perhaps Problem 20.

5. Write the iterated integral

$$\int_0^1 \left[\int_{y/4}^{\sqrt{y}/2} x \, dx \right] dy$$

with the order of integration reversed. Then evaluate. You will first need to sketch the appropriate region.

6. If there is time, work Problem 28 in which the region D consists of two parts.

11.3 APPLICATIONS OF DOUBLE INTEGRALS (p. 361)

1. Sketch the first octant portion of the plane $x + 3y + 2z = 6$. Use integration to find the volume of the solid bounded by this plane and the coordinate planes. Check the answer by using the simple formula for volumes of tetrahedra.

2. Find the volume of the solid under the surface $z = f(x, y) = 25 - x^2 - y^2$ and above the region

$$D = \{(x, y): 0 \leq y \leq \tfrac{4}{3}x, 0 \leq x \leq 3\}$$

Note that $f(x, y) \geq 0$ on D.

3. Use Problem 10 to talk about finding the mass of a lamina with variable density. Note that $2x^2 \, \Delta x \, \Delta y$ approximates the mass of a small rectangular piece centered at (x, y).

4. Problem 16 can be used to discuss population problems.

5. Give the definition of the average value of a function $f(x, y)$ on a region D. Make the result plausible by pointing out the analogy with the corresponding result for a function of one variable. Work Problem 20.

6. If there is time, work Problems 22 and/or 24.

12 The Trigonometric Functions

12.1 ANGLES, ARCS, AND TRIANGLES (p. 370)

1. Discuss angles and their measurement, both degree measure and radian measure. Point out that the relation

$$\pi \text{ radians} = 180°$$

implies: To convert degrees to radians multiply by $\pi/180$; to convert radians to degrees multiply by $180/\pi$. Convert $420°$ to radians; then convert $-5\pi/6$ radians to degrees. Discuss the special angles: $30°$, $45°$, $60°$, and $90°$.

2. Derive the formulas $s = tr$ and $A = \tfrac{1}{2}tr^2$ for the arc length and area of the sector of a circle of radius r, having angle t radians. Work Problems 24 and 28.

3. Define the sine, cosine, and tangent for an acute angle t of a right triangle.

4. With the help of a calculator, find the two legs of a right triangle with hypotenuse 22 and one acute angle $16.5°$. It may be worthwhile to spend a little time explaining the use of a calculator in trigonometry.

5. Work Problem 40. Note that the belt makes contact with each pulley for half of its circumference and each of the straight-line segments is the hypotenuse of a right triangle.

12.2 THE SINE AND COSINE FUNCTIONS (p. 377)

1. There are six trig functions, but we put the emphasis on sine and cosine. To define $\sin t$ and $\cos t$ for general t, put t in standard position, pick any point $P(x, y)$ on the terminal side, and let r be its distance from the origin. Then $\sin t = y/r$ and $\cos t = x/r$. This makes sense for any t, large or small, positive or negative.

2. Work Problems 4 and 8. Then find $\sin(5\pi/4)$ and $\cos \pi$. Illustrate finding several sines and cosines using a calculator, including some values like $\cos(\sin 2.12)$ and $\cos^2(\pi/9)$.

3. Discuss the important boxed identities that appear in this section.

4. Draw the graphs of $w = \sin t$ and $w = \cos t$ and discuss their characteristics. Note how the identities relate to the graphs.

5. Spend considerable time discussing the graphs of $w = A \sin t$, $w = \sin Bt$, and $w = A \sin Bt$ for various values of A and B. Introduce the concepts of amplitude and period and relate them to the above. Note that the same ideas apply to the cosine function.

6. Graph $w = 4 \cos t + \sin 2t$ on $[0, 3\pi]$ using the method of adding ordinates. If time permits, work Problem 36.

12.3 DERIVATIVES OF THE TRIGONOMETRIC FUNCTIONS (p. 385)

1. The key results of the section are:

$$\frac{d}{dt}\sin t = \cos t \qquad \frac{d}{dt}\cos t = -\sin t$$

Make the first one plausible by first graphing $w = \sin t$ and then superimposing the graph of its slope function. Consider trying to derive this formula in the novel way proposed in the text. (Note: The standard way of deriving this formula uses several ideas which have not been discussed in the book.)

2. Taking the first formula above for granted, derive the second one using identities and the Chain Rule as shown in the text.

3. Practice using these two formulas together with the Chain Rule by finding $\dfrac{d}{dt}\cos(t^2 - 2t)$ and $\dfrac{d}{dx}(2\sin 3x)$. Then work Problems 8, 12, 20, and 26.

4. Introduce the other four trig functions and derive some or all of their derivative formulas.

5. Find the equation of the tangent line to the curve $y = 4\tan 2x$ at the point $(\pi/8, 4)$.

6. If time permits, consider working Problem 40 as a way of tying together trigonometry with ideas studied much earlier.

12.4 INTEGRATION OF THE TRIGONOMETRIC FUNCTIONS (p. 391)

1. Start by writing the differentiation and integration formulas for sine and cosine on the board, side by side. Then find:

$$\frac{d}{dx}\sin 2x \qquad \int \sin 2x\, dx$$

$$\frac{d}{dx}\cos(1 - 3x) \qquad \int \cos(1 - 3x)\, dx$$

2. Making a substitution (changing variables) is a critical technique in this section. Use it to work Problems 2, 12, and 18.

3. Emphasize the need to change the limits when making substitutions in definite integrals. Illustrate with Problems 16 and 24.

4. Recall the technique of integration by parts. Find the value of

$$\int_0^{\pi/4} x\sec^2 x\, dx$$

Then evaluate

$$\int_0^{\sqrt{\pi/4}} x\sec^2(x^2)\, dx$$

What looks harder may be easier. Always look for a simple substitution before trying integration by parts.

5. Problems 26 and 28 offer a chance to review topics that may have been forgotten.

12.5 APPLICATIONS OF THE TRIGONOMETRIC FUNCTIONS (p. 396)

1. Before plunging into applications of sines and cosines, it may be well to emphasize some basic graphical ideas. Sketch the graphs of $y = \sin x$, $y = \sin 2x$, $y = 3\sin x$, $y = \sin(x - \frac{\pi}{6})$, $y = 3\sin 2(x - \frac{\pi}{6})$, and $y = 4 + 3\sin 2x$. Make sure that the terms *amplitude* and *period* are understood.

2. Work Problems 10 and 16.

3. Illustrate the application to blood pressure by going through Example A. Be sure that students are asked to work Problem 17.

4. Problem 18 offers a chance to study an animal population that displays a cyclical growth pattern. For part (b), you will need to find the smallest positive value of t for which $\sin(\pi t/12) = -1$.

5. Problem 20 (cyclical sales) or Problem 22 (simple harmonic motion) are appropriate applications with which to end the lesson.

13 Differential Equations

13.1 FIRST-ORDER EQUATIONS (p. 408)

1. Define the terms *differential equation, general solution,* and *particular solution*.

2. Explain what it means for a differential equation to be separable by considering the equation $dy/dx = x^2 y^3$. Find the general solution to this equation and also the particular solution satisfying $y = 1$ when $x = 0$.

3. Separate variables in

$$3(x^2 + 1)\frac{dy}{dx} - \frac{x}{\sqrt{y}} = 0$$

Then find the general solution (answer: $2y^{3/2} = \frac{1}{2}\ln(x^2 + 1) + C$. Next find the particular solution for which $y = 4$ when $x = 0$.

4. Mention that an equation of the form

$$\frac{dy}{dx} + p(x)y = q(x)$$

is called a first-order linear equation. To solve such an equation, one need only remember to multiply both sides by the integrating factor.

$$\exp\left[\int p(x)\, dx\right]$$

The result is an equation that can be integrated since the left side takes the form $\dfrac{d}{dx}\left\{y\,\exp\left[\displaystyle\int p(x)\,dx\right]\right\}$. Illustrate by solving the equation

$$\frac{dy}{dx} + \frac{2y}{x} = 4x$$

subject to the condition $y = 7$ at $x = 2$.

5. Work, as time allows, Problems 22, 28, and 40.

13.2 APPLICATIONS OF FIRST-ORDER EQUATIONS (p. 414)

1. You won't have time to illustrate every topic mentioned in this section. We suggest that you emphasize Examples A, B, C, D, and F.

2. Work Problem 4 as a parallel to Example A.

3. Problem 6 is analagous to Example B. The appropriate differential equation is $dT/dt = k(22 - T)$ with side conditions $T(0) = -18$ and $T(20) = 0$.

4. Problem 8 is a good parallel to Example C. For the modification that corresponds to Example D, try Problem 10.

5. Problem 14 provides a way to discuss Example F.

6. For a challenging problem consider Problem 24. You should discover that the amount Q of salt in tank B at the end of t seconds, satisfies

$$dQ/dt = 100e^{-0.02t} - 0.01Q$$

with $Q(0) = 10,000$.

13.3 HOMOGENEOUS SECOND-ORDER LINEAR EQUATIONS (p. 423)

1. Show that $y = e^{rx}$ is a solution to the differential equation

$$\frac{d^2y}{dx^2} + a\frac{dy}{dx} + by = 0$$

provided r satisfies the auxiliary equation

$$r^2 + ar + b = 0$$

State the results about the general solution to the differential equation when the solutions to the auxiliary equation are real, treating the two cases of distinct and repeated roots.

2. Write the following differential equations on the board together with their auxiliary equations.

$$\frac{d^2y}{dx^2} - 4\frac{dy}{dx} + 3y = 0$$

$$\frac{d^2y}{dx^2} - 4\frac{dy}{dx} = 0$$

$$\frac{d^2y}{dx^2} - 4\frac{dy}{dx} + 4y = 0$$

$$\frac{d^2y}{dx^2} - 4\frac{dy}{dx} + y = 0$$

Solve these equations.

3. Consider next the differential equation

$$\frac{d^2y}{dx^2} + 4y = 0$$

together with its auxiliary equation $r^2 + 4 = 0$. Note that the roots of the latter are the imaginary numbers $\pm 2i$. Then check that $\sin 2x$ and $\cos 2x$ are solutions to the differential equation and therefore that

$$y = C_1 \sin 2x + C_2 \cos 2x$$

is its general solution.

4. Solve

$$\frac{d^2y}{dx^2} - 2\frac{dy}{dx} + 4y = 0$$

5. Work Problems 16, 32, and 36.

13.4 NONHOMOGENEOUS SECOND-ORDER LINEAR EQUATIONS (p. 429)

1. Discuss carefully the opening paragraph of the section leading to

$$y = y_h + y_p$$

as the general solution of a nonhomogeneous differential equation. Illustrate with

$$\frac{d^2y}{dx^2} - 4y = 10e^{3x}$$

For this equation,

$$y_h = C_1 e^{2x} + C_2 e^{-2x}$$

Try substituting $y_p = Ae^{3x}$ and discover that $A = 2$. Thus,

$$y = C_1 e^{2x} + C_2 e^{-2x} + 2e^{3x}$$

2. Work Problems 4 and 12. Use $y_p = Ax + B$ and $y_p = A \sin x + B \cos x$ as the respective trial solutions.

3. Work Problem 16 which involves a new wrinkle. Because e^{5x} is a solution to the homogeneous equation, use $y_p = Axe^{5x}$ rather than $y_p = Ae^{5x}$ as the trial solution.

4. Consider working Problem 22 to illustrate solving a differential equation with side conditions.

5. Discuss Example F and then work Problem 26 as a further illustration.

14 Numerical Calculus: Approximation

14.1 NUMERICAL INTEGRATION (p. 440)

1. Use $\int_0^6 \frac{1}{1+x}\, dx$ (Problem 4) to illustrate three types of approximation: right endpoint Riemann sum, Trapezoidal Rule, and Parabolic Rule. Use $n = 6$ in each case and draw pictures to show what you are doing. Finally, find the exact value using the Fundamental Theorem of Calculus.

2. Sketch the derivation of the Trapezoidal Rule and/or the Parabolic Rule.

3. Discuss the two kinds of error, namely, discretization error and roundoff error, and explain why they are in competition. Use Figure 6 to compare the Trapezoidal and Parabolic Rules and to illustrate the two kinds of errors.

4. State the results about bounds for the discretization errors given in the text. Then illustrate by working Problems 6 and 12. You should get 0.0625 and 0.00625 as the two error bounds.

5. Problem 16 is a good applied problem. It calls for approximating $\int_0^{20} v(t)\, dt$ where we only know v at 2-second intervals.

14.2 APPROXIMATING FUNCTIONS BY POLYNOMIALS (p. 446)

1. Say something about why it is desirable to approximate general functions by polynomial functions.

2. Illustrate with $f(x) = x + 2\sqrt{x}$. Find Taylor's linear approximation $P_1(x)$ and quadratic approximation $P_2(x)$, both based at 4. In the same coordinate plane, show the graphs of f, P_1, and P_2.

3. Use P_1 to estimate $f(3.9)$ and $f(4.01)$. Then use P_2 to do the same.

4. Write down the formula for the Taylor polynomial of order n based at a and emphasize the special case, called the Maclaurin polynomial, corresponding to $a = 0$.

5. Find the Maclaurin polynomial of order n for $f(x) = e^x$.

6. Work Problem 8. Then discuss the bound for the discretization error and work Problem 18. Note: For $f^{(5)}(x)$, you should get $(-6x - 30)/(x + 1)^5$.

7. If time permits, work Problems 26 and/or 28.

14.3 SOLVING EQUATIONS NUMERICALLY (p. 452)

1. Discuss Newton's method for approximating a root of the equation $f(x) = 0$, using Figures 13 and 14 to make it seem legitimate. Illustrate by working Problem 4.

2. Work Problem 6. Your graph should suggest $x_1 = 2.5$ as a seed value.

3. Next work Problem 10. Your graphs of $y = e^x$ and $y = 1 + \cos x$ should suggest $x_1 = 0.6$ as a good seed value.

4. Use the method of Example D to work Problem 14. The required equation is

$$5x^3 - 2x^2 - 2x - 2 = 0$$

and an appropriate seed value is 1.09.

5. Show how Newton's method can fail.

6. Consider working Problem 20.

14.4 THE METHOD OF SUCCESSIVE SUBSTITUTIONS (p. 457)

1. Note that the method of successive substitution requires the equation to be in the form $x = g(x)$. Try it on the equation $x = \frac{1}{3}e^x$ using 0.6 as a seed value.

2. Discuss Example B, noting that the method diverges when the equation is written in one form, but converges when written in another form. Explain what happens in various cases by discussing Figure 7. For convergence, we need $|g'(x)| < 1$ in a neighborhood of the root, and the smaller the magnitude of this derivative, the faster the convergence will be.

3. Illustrate by working Problems 6 and 8. In Problem 8, you will find that $f'(0.85) \approx 1.77$, while $g'(0.85) \approx 0.39$.

4. If time permits, work Problem 16. Write the equation as

$$x = g(x) = \sqrt{\tfrac{1}{3}\cos x}$$

Use $x_1 = 0.5$ as the seed value and note that the process will converge since $|g'(0.5)| \approx 0.15$.

14.5 NUMERICAL SOLUTION OF DIFFERENTIAL EQUATIONS (p. 463)

1. The procedures described in this section for solving the differential equation $dy/dx = g(x, y)$, with $y(a) = y_0$, are straightforward but exceptionally time-consuming. Thus, you will want to put your emphasis on explaining the two methods and perhaps working just one example both ways.

2. We suggest Problem 6 for this example. You will want to prepare a table as in Figure 5. The exact solution to the equation is $y(x) = \frac{1}{4}x^3 - \frac{2}{x}$ which give $y(2.2) = 1.7529$ and $y(2.4) = 2.6267$. We get the corresponding approximations to be $y_1 = 1.7$, $y_2 = 2.5135$, $y_1 = 1.7567$, and $y_2 = 2.6321$, giving you a check that you are off to a correct start.

3. Problems 3, 5, and 9 may be sufficient for a student assignment. If students have access to a computer and know how to program, you might consider assigning Problem 11.

15 Infinite Series

15.1 INFINITE SEQUENCES AND SERIES (p. 472)

1. Define the term *sequence* and illustrate with several examples such as the following:
(a) $\frac{1}{2}, \frac{2}{3}, \frac{3}{4}, \frac{4}{5}, \ldots$
(b) $-1, 1, -1, 1, \ldots$
(c) $1, 4, 9, 16, \ldots$
Sequences are often specified by displaying enough terms to establish a pattern. A better way is to give an explicit formula, for example,

$$a_n = n/(n + 1), \ b_n = (-1)^n, \text{ and } c_n = n^2,$$

which define the same sequences as the patterns above.
2. Give the definition of convergence for a sequence and note that $\{a_n\}$ converges to 1 but that both $\{b_n\}$ *and* $\{c_n\}$ diverge.
3. Work several parts of Problems 4 and 6.
4. Define the term *infinite series*. Then clearly distinguish between:

the sequence $a_1, a_2, a_3, \ldots$

the series $a_1 + a_2 + a_3 + \cdots$

the sequence $S_n = a_1 + a_2 + \cdots + a_n$

Illustrate these ideas with the first two sequences in part 1.
5. Represent an infinite series and a partial sum using sigma notation. Then define convergence for an infinite series. Work Problems 10 and 12.
6. Discuss the nth Term Test for divergence and the Bounded Sum Test.
7. Show that the harmonic series diverges.
8. Work Problems 16, 22, and 24.

15.2 GEOMETRIC SERIES (p. 479)

1. Define the term *geometric series* and illustrate with:
(a) $1 + \frac{1}{3} + \frac{1}{9} + \frac{1}{27} + \cdots$
(b) $2 + \frac{3}{2} + \frac{9}{8} + \frac{27}{32} + \cdots$
(c) $100 + 100(1.08) + 100(1.08)^2 + \cdots$
2. State (perhaps develop) the formula for S_n, the sum of the first n terms of a geometric series. Calculate S_5 for each of the above series. Interpret your result for (c) as in Example B.
3. Show that a geometric series converges if and only if $|r| < 1$ and that in this case the sum S of the whole series is given by $S = a/(1 - r)$. Apply this to the series in part 1 and then work Problems 16, 20, and 24.

4. Review present value and then work Problem 30.
5. As time allows, work Problems 28 and/or 32.

15.3 POSITIVE SERIES (p. 486)

1. Emphasize that the tests of this section apply to positive series.
2. Discuss the Comparison Test. Note that it is useful only when we have some known series with which to compare other series. Establish the convergence of

$$\tfrac{1}{2} + \tfrac{1}{4} + \tfrac{1}{10} + \tfrac{1}{28} + \cdots$$

by noting that the kth term is $1/(3^{k-1} + 1)$ and comparing with $\sum_{n=1}^{\infty} \dfrac{1}{3^{k-1}}$. Work Problem 6.
3. Discuss the Ratio Test and use it to test the series $\sum_{k=1}^{\infty} \dfrac{3^k}{k!}$. Then work Problems 12 and 14. Use Problem 12 to state the fact that if Σa_n and Σb_k both converge, so does $\Sigma(a_k + b_k)$.
4. Discuss the Integral Test. Show how it implies that $\sum_{k=1}^{\infty} \dfrac{1}{k^p}$ converges or diverges according as $p > 1$ or $p \le 1$. Work Problems 18 and 20.
5. Work as many of Problems 44, 46, 54, and 56 as time permits.

15.4 POWER SINES (p. 491)

1. We first generalize to series of mixed signs. Define *absolute convergence* and then state the Absolute Convergence Test. Work Problem 4.
2. Discuss the Alternating Series Test, noting its three conditions. Work Problem 6. Then define conditional convergence and work Problem 16.
3. Next we generalize to power series, that is, to series of the form $\sum_{k=0}^{\infty} a_k x^k$. An important question to ask is: For what values of x does the series converge? The answer is—always on an interval centered at 0 which may or may not include the endpoints. We can determine this interval by use of the Ratio Test (checking the endpoints separately). Work Problems 22 and 26.
4. Mention the five important Maclaurin series given in the text and show how they can be used to find power series representations for other functions. Work Problems 30, 36, 38, and 40.
5. If time permits, consider working Problem 42.

16 Calculus and Probability

16.1 DISCRETE DISTRIBUTIONS (p. 502)

1. Introduce the terminology of probability theory: *experiment, outcome, sample space, event, probability*. Illustrate with the example of tossing two fair tetrahedral dice. List all 16 outcomes: $(1, 1), (1, 2), \ldots, (4, 3), (4, 4)$. Find the probability of getting a total of 5. Note the properties of probability and then use them to find the probability of the following events.

$$E: \text{get a total of 3 or 4}$$

$$F: \text{get a total less than 8}$$

2. Define the terms *random variable* $\mathbf{X}$, *probability function* $f_i = \text{prob}\{\mathbf{X} = i\}$, and *bar graph*; then work Problems 4 and 6. Note that the heights of the bars sum to 1, that is $\Sigma f_i = 1$.

3. Introduce the terms *mean* μ and *variance* σ^2 and explain what they measure. Work Problems 8 and 10.

4. Discuss the concept of expected value and work Problem 12 (answer: $0.70).

5. Work Problem 18 or 22.

6. Prepare the way for the next lesson by distinguishing between discrete random variables and continuous random variables.

16.2 CONTINUOUS DISTRIBUTIONS (p. 508)

1. Use a spinner example to introduce students to the concept of a continuous random variable $\mathbf{X}$ and its associated probability density function f. Note that this leads to a so-called uniform distribution.

2. State the properties that a probability density function must have and note the analogies with a discrete probability function ($\int$ versus Σ).

3. Find the value of k so that the function $f(x) = k(4x - 3x^2)$ on $[0, 2]$ and $f(x) = 0$ elsewhere, is a probability density function. Graph f and then calculate $\text{prob}\{1 \leq \mathbf{X} \leq 1.2\}$.

4. Discuss the concept of mean and variance for a continuous random variable. Then find these two numbers for $\mathbf{X}$ in part 3.

5. Find the mean and variance for the uniformly distributed spinner random variable in part 1.

6. Work Problem 20 which requires calculation of $\text{prob}\{0.7 \leq \mathbf{X} \leq 1\}$.

16.3 THE EXPONENTIAL AND NORMAL DISTRIBUTIONS (p. 513)

1. Review the concept of improper integral using $\int_0^\infty e^{-3x} \, dx$ as an example. Show if that $k > 0$,

$$\int_0^\infty ke^{-kx} \, dx = 1$$

2. $\mathbf{X}$ is said to be exponentially distributed if its probability density function is $f(x) = ke^{-kx}$ on $[0, \infty)$ and $f(x) = 0$ elsewhere, with k depending on the phenomena being modeled. Write down the integrals for μ and σ and state that both have the value $1/k$.

3. Work Problem 4 for which $k = \frac{1}{5}$ and you are to calculate $\int_{15}^\infty 0.2e^{-0.2x} \, dx$.

4. Define the normal density function, perhaps using parameters p and q initially. Show a typical graph and then state the theorem which says that $\mu = p$ and $\sigma = q$. Give meaning to the standard deviation σ using a figure like Figure 15.

5. A standard normal variable $\mathbf{Z}$ has $\mu = 0$ and $\sigma = 1$. Illustrate the use of the standard normal table (Table E) by working Problems 8 and 10.

6. The transformation $\mathbf{Z} = (\mathbf{X} - \mu)/\sigma$ standardizes an arbitrary normal variable $\mathbf{X}$ and allows us to use Table E in calculating probabilities for it. Illustrate by working Problems 14 and 16.

7. If time permits, work Problems 18 and 22.

16.4 CONTINUOUS BIVARIATE DISTRIBUTIONS (p. 519)

1. We generalize to a bivariate random variable $(\mathbf{X}, \mathbf{Y})$ with a bivariate probability density function $f(x, y)$, having the property that the volume of the solid underneath $z = f(x, y)$ is 1.

2. A bivariate random variable is uniformly distributed if $f(x, y)$ is constant k on a set D and is 0 elsewhere. Illustrate by working Problem 2.

3. As an example of a nonconstant probability density function, work Problem 6. Answers: (a) $k = \frac{1}{32}$, (b) $\frac{15}{16}$.

4. Discuss Example D and work Problem 8 and perhaps Problem 10.

5. Consider the calculation of means and variances for $\mathbf{X}$ and $\mathbf{Y}$ and then introduce the concepts of covariance and correlation coefficient. Use Example F or Problem 14 as an illustration.

Calculus
with
Applications

Calculus with Applications

Dale Varberg

Hamline University

Walter Fleming

Hamline University

Prentice Hall,
Englewood Cliffs, NJ 07632

Library of Congress Cataloging-in-Publication Data

Varberg, Dale E.
   Calculus with applications / Dale E. Varberg, Walter Fleming.
      p.   cm.
   ISBN 0-13-110826-3
   1. Calculus.   I. Fleming, Walter.   II. Title.
QA303.F58362   1991
515—dc20   90–33019
                                                    CIP

Editorial and production supervision: *Ruth Cottrell*
Interior design: *Lorraine Mullaney*
Cover design: *Bruce Kenselaar*
Cover art: *Salem Krieger*
Photo editor: *Lorinda Morris-Nantz*
Photo research: *Barbara Scott*
Pre-press buyer: *Paula Massenaro*
Manufacturing buyer: *Lori Bulwin*

Photo credits: **Chapter 1**, Laima Druskis; **p. 1**, Historical Pictures Service, Chicago; **46**, Ed Lettau/Photo Researchers; **47**, Historical Pictures Service, Chicago; **84**, Tcherevkoff Studio/The Image Bank; **85**, Historical Pictures Service, Chicago; **120**, FORTUNE, © 1989 The Time Inc. Magazine Company. All rights reserved; **121**, Massachusetts Institute of Technology; **160**, Fundamentals of Anatomy and Physiology (Frederic Martini), Englewood Cliffs, N.J.: Prentice Hall, 1989, p. 40; **161**, University of Cambridge; **188**, NASA; **189**, S. Preobrazhensky/Sovfoto; **216**, Mel Digiacomo/The Image Bank; **217**, Historical Pictures Service, Chicago; **244**, from "Mathematical Games," Martin Gardner. Copyright © 1986 by SCIENTIFIC AMERICAN, Inc. All rights reserved; **245**, UPI/Bettmann Newsphotos; **274**, from "Climate Modeling," Stephen H. Schneider. Copyright © 1987 by SCIENTIFIC AMERICAN, Inc. All rights reserved; **275**, University of California, Berkeley; **310**, "The Allocation of Resources by Linear Programming," by Robert G. Bland. Copyright © 1981 by SCIENTIFIC AMERICAN, Inc. All rights reserved; **311**, Stanford University; **346**, Michaelangelo/"Creation of Adam"/Art Resource; **347**, Dr. Chen Ning Yang; **368**, image by H. Jurgens, H.-O. Peitgen, D. Saupe, from *The Beauty of Fractals* (H.-O. Peitgen, P. Richter, Heidelberg: Springer-Verlag, 1986; **369**, IBM Research; **406**, Sovfoto/Eastfoto; **438**, AT&T Archives; **439**, U.S. Navy; **470**, from *The Mathematical Intelligencer* 7(3): cover, New York: Springer-Verlag, 1985; **471**, Cambridge University; **500**, Jim Pickerell; **501**, Rockefeller Archive Center.

© 1991 by Prentice-Hall, Inc.
A Division of Simon & Schuster
Englewood Cliffs, New Jersey 07632

Printed in the United States of America
10 9 8 7 6 5 4 3 2 1

ISBN 0-13-110826-3

Prentice-Hall International (UK) Limited, *London*
Prentice-Hall of Australia Pty. Limited, *Sydney*
Prentice-Hall Canada Inc., *Toronto*
Prentice-Hall Hispanoamericana, S.A., *Mexico*
Prentice-Hall of India Private Limited, *New Delhi*
Prentice-Hall of Japan, Inc., *Tokyo*
Simon & Schuster Asia Pte. Ltd., *Singapore*
Editora Prentice-Hall do Brasil, Ltda., *Rio de Janeiro*

Contents

4   121

Other Applications of the Derivative

Biography of *Paul Anthony Samuelson*

5   161

Exponential and Logarithmic Functions

Biography of *Sir Ronald Aylmer Fisher*

6   189

Applications of Exponential and Logarithmic Functions

Biography of *Ol'ga Arsen'evna Oleinik*

*Sections marked with stars are optional and can be omitted without loss of continuity.

7   217

The Antiderivative

Biography of *Jean Piaget*

8   245

The Definite Integral

Biography of *Richard Courant*

9   275

Applications of the Definite Integral

Biography of *David Blackwell*

*Sections marked with stars are optional and can be omitted without loss of continuity.

Calculus and Probability

Appendix

*Sections marked with stars are optional and can be omitted without loss of continuity.

Preface

Audience

This is a book for those people who want to see the power and usefulness of the calculus but who feel they do not have time to work through the massive tomes that are common in engineering courses. In particular, it was written for students in the management, social, and life sciences; however, we believe it can serve as an introduction to calculus for students in all disciplines. Our goal was to make the book *lean and lively*; readers will judge whether we have succeeded.

Algebra Review

Our book has a number of novel features that we want to describe. Since we assumed that our readers would want to get to the meat of the subject quickly, we have foregone the general review of algebra one finds at the beginning of many books. Of course, algebra is foundational to calculus. Therefore, we review essential topics as they are needed at various points throughout the book. A further discussion of algebraic concepts can be found in the Appendix. But calculus is our subject, and we introduce the notions of slope and rate of change in the first chapter.

Pacing

Our general procedure is to postpone discussion of a topic until it will make sense to the student. Thus, the concept of continuity waits until it can be discussed along with differentiability and other properties of curves (Chapter 3). More significant is our decision to leave the Chain Rule until Chapter 5 where the exponential and logarithmic functions are available for examples. Of course, the special case that we call the General Power Rule occurs early (Chapter 2). We have always felt that the Chain Rule falls flat if it can be illustrated only with power functions.

Conceptual Emphasis

Like most good authors, we attempt to be clear and informative rather than encyclopedic. We give only those proofs that are easy to understand and which provide genuine insight. We try to make all of our descriptions conceptual in nature and, if a diagram will help, we display it. A notable example is our treatment of the definite integral. We think that a rigorous definition based on upper and lower sums is too abstract for our audience.

On the other hand, the common practice of defining the definite integral as $F(b) - F(a)$ gives no insight into the real meaning of the integral and obscures the significance of the Fundamental Theorem of Calculus. Therefore, we define the definite integral geometrically, taking area to be an intuitively obvious notion. Later, we show that the definite integral is the limit of Riemann sums.

Road Signs

We think of our book as providing a road map through difficult and sometimes confusing terrain. In line with this metaphor, our margins have road signs of four kinds. There are CAUTION signs warning readers of dangerous mistakes. There are RECALL signs reminding students of facts that may have been forgotten. A few signs carry the words SHARP CURVE, suggesting that the argument being made has a subtle turn that requires careful navigation. Finally, there are many signs with the message POINT OF INTEREST. When you see one of these signs, it is time to stop and enjoy a historical note or a bit of mathematical lore.

Applications

Our book is called *Calculus with Applications* for good reason. Four of our chapters are devoted in toto to applications of the calculus. In addition, most sections contain an application or two, and the problem sets introduce many more. These applications come from everywhere, but we tend to emphasize those that are important in the management, social, and life sciences. Economic and business applications occur the most frequently.

Biographies

To support our applied calculus emphasis, we begin each chapter with a picture and brief biography of an applied mathematician. These people come from a wide variety of disciplines but have several characteristics in common. All have made notable contributions to their fields and all have used mathematics in a significant way. They come from many different countries, demonstrating that mathematics (including calculus) is a universal language. They did their major work in this century, suggesting that calculus is not dead. There are sixteen of them, corresponding to the sixteen chapters in this book. The front endpapers indicate their diverse national backgrounds.

Examples and Problems

Our book is notable for the wealth of examples and problems. We chose our examples with two criteria in mind: They must illustrate the major ideas of calculus, and they should be realistic. The problem sets that conclude each section are carefully graded in difficulty thereby giving students a chance to build their confidence before hitting the hard nuts. First, there is a collection of relatively easy problems, all clearly related to the examples of the section. Then (following a horizontal bar), there is a set of miscellaneous problems. These latter problems are designed to integrate all the concepts of the section and usually culminate in 3 or 4 substantial applications. Each chapter ends with a lengthy set of review problems. The symbol [c] attached to a problem indicates that a calculator will be helpful. A very few problems are identified with the label [pc] suggesting that their solution requires a programmable calculator or a computer. Such problems can be omitted without loss of continuity.

Acknowledgements

We acknowledge with thanks the generous help of many people. Louis Guillou (St. Mary's College, Winona, MN) critiqued the whole manuscript, checked our answers to all the problems, and prepared the instructor's and student's solution manuals. Stephen Mondy (Normandale Community College, Bloomington, MN) read the entire manuscript, checked all the examples, and helped correct the galley proofs. A team of four mathematicians (Frisinger, Montgomery, Davitt, and Kasube—identified by school below) met with authors over a two-day period, offering suggestions for improvements in the manuscript. Finally, each of the following people served as a reviewer for all or part of the book.

Philip Crooke—Vanderbilt University, Nashville, TN
Richard Davitt—University of Louisville, Louisville, KY
Howard Frisinger—Colorado State University, Fort Collins, CO
Chaitan P. Gupta—Northern Illinois University, DeKalb, IL
Herbert E. Kasube—Bradley University, Peoria, IL
Walter Thomas Kyner—University of New Mexico, Albuquerque, NM
Philip R. Montgomery—University of Kansas, Lawrence, KA
Robert Moreland—Texas Technical University, Lubbock, TX
S. F. Neustadter—California State University, San Francisco, CA
James W. Newsom—Tidewater Community College, Portsmouth, VA
David Nilsson—University of Texas, Austin, TX
Clifton Thomas Wyburn—University of Houston, Houston, TX
Melvin R. Woodard—Indiana University of Pennsylvania, Indiana, PA

We express our deep appreciation to Prentice Hall, to mathematics editor Robert Sickles, designer Lorraine Mullaney, project editors John Morgan and Ruth Cottrell, and to all others who helped produce this book.

Dale Varberg
Walter Fleming

OUR
COMMITMENT
TO EXCELLENCE

Prentice Hall has taken
every possible step to ensure the precision and accuracy of this text. Experts reviewed content, checked exercises, and proofread technical material. They assisted the authors and Prentice Hall in producing an outstanding text of the highest quality.

You can help Prentice Hall maintain these high standards by perusing this text and relaying pertinent information found, including errors, to Mathematics Editor, Prentice Hall, Englewood Cliffs, NJ 07632 or to your local Prentice Hall representative. We look forward to serving your text needs now and in the future.
Thank you very much for your support.

Calculus
with
Applications

It is said that to do well in the stock market, a person must have the nerves of a gambler, the sixth sense of a clairvoyant, and the courage of a lion. But one must also have a deep understanding of economic principles. John Maynard Keynes had all the qualifications.

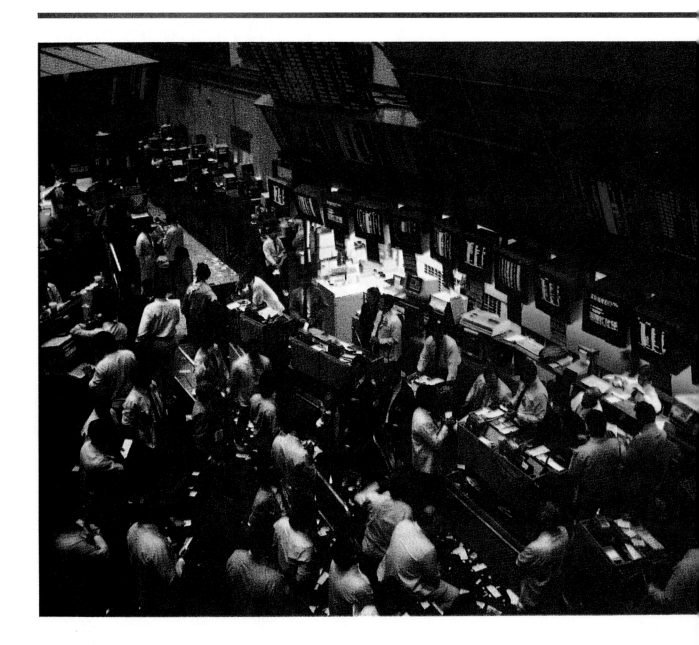

1

Slopes and Rates of Change

John Maynard Keynes
(1883–1946)

To say that John Maynard Keynes was versatile is an understatement. In addition to being the most influential economist of the twentieth century, he was a high official in the British government, the editor for 33 years of Britain's most distinguished economics journal, a teacher at Cambridge University, a collector of rare books, a patron of the arts, and a member of the literary guild known as the Bloomsbury Group. Unlike most academic types, he was able to parlay his knowledge and insight into a large fortune. He did this by analyzing financial reports and telephoning orders to his broker before getting out of bed in the morning.

Like his countryman and contemporary, Bertrand Russell, Keynes got his early training in mathematics. In fact, he mastered the subject so well that he was able to write a treatise on the foundations of probability, a book still read with profit by students of mathematics and philosophy. It is in mathematics, perhaps more than any other subject, that one learns to think about assumptions, to construct models, to analyze arguments, and to draw rigorous conclusions. Dare we suggest that it was this training that prepared Keynes to make even greater contributions when he turned his attention to the field of economics?

In the middle of the great world depression, Keynes published his masterpiece: *The General Theory of Employment, Interest and Money* (1936). In it, he argued that government could and should influence the economic welfare of the state by reducing interest rates, by equalizing incomes (through progressive taxation), and by stimulating the economy with such "pump-priming" measures as public work projects. These ideas had a profound impact on Franklin Delano Roosevelt and inspired many of the programs of the New Deal. It is safe to say that the United States has been permanently affected by Keynes, and even the conservative economist, Milton Friedman, has been quoted as saying, "We are all Keynesians now."

FUNCTIONS

Much of science involves studying the relationship between two variables. An economist believes that the number of housing starts in a year depends in a definite way on the magnitude of the prime interest rate. A business executive knows that the price charged for a certain product must be adjusted to the supply available. A physiologist observes that the response time of an automobile driver increases with alcohol intake. In each of these examples, the size of one variable determines the size of another. Mathematicians use the term *function* to describe such relationships.

DEFINITION

A **function** is a rule that associates with each number x in one set a definite number y from another set.

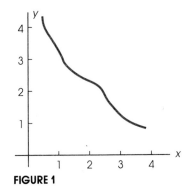

FIGURE 1

In newspapers, magazines, and many social science textbooks, functions are specified by drawing graphs (as in Figure 1). However, in mathematics, our normal starting point is a formula or equation. Consider, for example, the equation

$$y = x^2$$

which readers will recognize as the formula for the area of a square. As soon as the side length x is given, the area y is determined. If $x = 5$, then $y = 5^2 = 25$; if $x = 3.12$, then $y = (3.12)^2 = 9.7344$.

There is nothing sacred about the use of x and y as variables. In geometry, we learned that the volume V of a sphere of radius r (Figure 2) is given by

$$V = \tfrac{4}{3} \pi r^3$$

Here π is that often appearing constant with approximate value 3.1416 (we write $\pi \approx 3.1416$). Knowing a value for r, we can easily determine the corresponding value of V. If $r = 3$, then $V = \tfrac{4}{3} \pi 3^3 = 36\pi \approx 113.0973$. Since r can be given any (positive) value, it is called the **independent variable**. V depends on r; it is called the **dependent variable.**

Functional Notation

A single letter like f (or g or F) will be used to name a function. Then $f(x)$, read "f of x" or "f at x", denotes the value that f assigns to x. Thus, if f names the "area of a square" function mentioned earlier, it is clear that

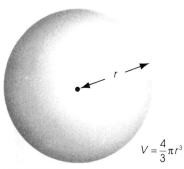

$$V = \frac{4}{3}\pi r^3$$

FIGURE 2

$$f(x) = x^2$$

and

$$f(2) = 2^2 = 4$$

$$f(4.2) = (4.2)^2 = 17.64$$

$$f(a) = a^2$$

$$f(2m) = (2m)^2 = 4m^2$$

In the latter four cases, we simply replaced x by 2, 4.2, a, and $2m$, respectively.

Somewhat more complicated, but of critical importance later, are calculations like the following.

$$f(2 + h) = (2 + h)^2 = 4 + 4h + h^2$$

$$f(2 + h) - f(2) = 4 + 4h + h^2 - 4 = 4h + h^2$$

$$\frac{f(2 + h) - f(2)}{h} = \frac{4h + h^2}{h} = \frac{h(4 + h)}{h} = 4 + h$$

Because functional notation plays such a large role in calculus, we pause for three more examples.

EXAMPLE A

If $f(x) = x^3 - 2x + 1$, find (a) $f(3)$, (b) $f(-2)$, (c) $f(\frac{1}{2})$.

Solution Perhaps it will help to think of the formula for f in the following way

$$f(\ \ ) = (\ \ )^3 - 2(\ \ ) + 1$$

Then putting the same number inside each pair of parentheses, we obtain

(a) $f(3) = (3)^3 - 2(3) + 1 = 27 - 6 + 1 = 22$
(b) $f(-2) = (-2)^3 - 2(-2) + 1 = -8 + 4 + 1 = -3$
(c) $f(\frac{1}{2}) = (\frac{1}{2})^3 - 2(\frac{1}{2}) + 1 = (\frac{1}{8}) - 1 + 1 = \frac{1}{8}$ ■

EXAMPLE B

For $g(x) = x^2 - 4x$, find and simplify (a) $g(3)$, (b) $g(3 + h)$, (c) $[g(3 + h) - g(3)]/h$, (d) $[g(x + p) - g(x)]/p$.

Solution

(a) $g(3) = (3)^2 - 4(3) = 9 - 12 = -3$
(b) $g(3 + h) = (3 + h)^2 - 4(3 + h) = 9 + 6h + h^2 - 12 - 4h$
$$= -3 + 2h + h^2$$

(c) $\dfrac{g(3 + h) - g(3)}{h} = \dfrac{-3 + 2h + h^2 - (-3)}{h}$

$$= \dfrac{h(2 + h)}{h} = 2 + h$$

(d) $\dfrac{g(x + p) - g(x)}{p} = \dfrac{(x + p)^2 - 4(x + p) - (x^2 - 4x)}{p}$

$$= \dfrac{x^2 + 2xp + p^2 - 4x - 4p - x^2 + 4x}{p}$$

$$= \dfrac{2xp + p^2 - 4p}{p}$$

$$= \dfrac{p(2x + p - 4)}{p} = 2x + p - 4 \qquad \blacksquare$$

EXAMPLE C

For $F(x) = 1/x$, find and simplify $[F(a + h) - F(a)]/h$.

Solution

> **CAUTION**
> The common error in Example C is to write
> $$F(a + h) = \dfrac{1}{a} + h$$
> which is wrong. Rather
> $$F(a + h) = \dfrac{1}{a + h}$$

$$\dfrac{F(a + h) - F(a)}{h} = \dfrac{\dfrac{1}{a + h} - \dfrac{1}{a}}{h} = \dfrac{\dfrac{a}{a(a + h)} - \dfrac{a + h}{a(a + h)}}{h}$$

$$= \dfrac{a - (a + h)}{a(a + h)} \cdot \dfrac{1}{h}$$

$$= \dfrac{-h}{a(a + h)h} = \dfrac{-1}{a(a + h)} \qquad \blacksquare$$

Domain and Range

> **RECALL**
> If $a \geq 0$, then $\sqrt{a}$ denotes the non-negative square root of a. For $G(x) = \sqrt{x - 1}$,
> $$G(5) = \sqrt{5 - 1} = 2$$
> On the other hand,
> $$G(-3) = \sqrt{-3 - 1} = \sqrt{-4}$$
> is left undefined.

The **domain** for a function is the set of allowable values for the independent variable; the **range** is the set of values attained by the dependent variable. Thus for the "area of a square" function $y = f(x) = x^2$, we take the domain to be the set of positive numbers (a square can have any positive length). In this case, the range would also be the set of positive numbers. If no domain is specified either explicitly or implicitly (that lengths should be positive is an implicit restriction), then we take the domain to be the largest set of real numbers for which the formula for the function makes sense. Thus in the case of $F(x) = 1/x$ (Example C), we take the domain to be the set of all real numbers except 0 (0 is excluded to avoid division by zero). If $G(x) = \sqrt{x - 1}$, we take the domain to be the set of all real numbers greater than or equal to 1. Numbers smaller than 1 are not allowed because we cannot take the square root of a negative number (imaginary numbers play no role in this book).

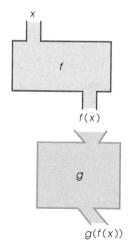

FIGURE 3

FIGURE 4

FIGURE 5

Functions Viewed as Machines

A helpful way to think of a function f is as a machine (Figure 3). The f machine accepts an input x, works on it, and produces the output $f(x)$. The set of all input numbers is the domain of the function, while the set of output numbers is its range.

Your hand-held calculator nicely illustrates this view (Figure 4). Enter an input 3 and press the x^2 key, and the calculator displays 9 as output. Scientific models have built-in keys for many functions ($\sqrt{x}$, $1/x$, $\log x$, $\sin x$, and so on).

If you do not own a calculator, we urge you to purchase one. Be sure to get a scientific model and, if you can afford it, one that is programmable.

Almost every section of this book wil have some problems that are simplified by the use of a calculator. Such problems are marked with the symbol $\boxed{C}$. A few problems are best done on a programmable calculator or a personal computer. They bear the code $\boxed{PC}$.

Composite Functions

Simple machines can be hooked together in tandem to form more complicated machines; so can functions. Let the function f take the input x and produce the output $f(x)$. Then let $f(x)$ become the input for a second function g which accordingly produces the output $g(f(x))$ as in Figure 5. The resulting coupled function $g(f)$ is called the **composite of g with f.**

EXAMPLE D

Let $f(x) = (x - 1)^2$ and $g(x) = 1/x$. Find (a) $g(f(3))$, (b) $g(f(\frac{1}{2}))$, (c) $g(f(1))$, (d) $g(f(x))$, (e) $g(g(x))$.

Solution Always begin by calculating the value of the function that first operates on the input; that is, begin by evaluating the inner function.

(a) $g(f(3)) = g(4) = \frac{1}{4}$
(b) $g(f(\frac{1}{2})) = g(\frac{1}{4}) = 4$
(c) $g(f(1)) = g(0) = \frac{1}{0}$, which is undefined
(d) $g(f(x)) = g[(x - 1)^2] = 1/(x - 1)^2$
(e) $g(g(x)) = g(1/x) = 1/(1/x) = x$

Now we are in for a mild surprise: $g(f)$ and $f(g)$ are in general not the same; composition of functions is not commutative.

EXAMPLE E

Let $f(x) = (x - 1)^2$ and $g(x) = 3x + 1$. Find $g(f(x))$ and $f(g(x))$.

Solution

$$g(f(x)) = g[(x - 1)^2] = 3(x - 1)^2 + 1 = 3x^2 - 6x + 4$$
$$f(g(x)) = f(3x + 1) = (3x + 1 - 1)^2 = 9x^2 \qquad \blacksquare$$

In using your calculator, you will often, perhaps without thinking, calculate the composite of several functions. The next example actually involves the composite of $h(x) = 1/x$ with $g(x) = x^2$ with $f(x) = \log x$.

EXAMPLE F

Use your calculator to evaluate $1/(\log 5)^2$.

Solution Press the following keys.

$$5 \;\; \boxed{\log} \;\; \boxed{x^2} \;\; \boxed{1/x}$$

Our calculator gives 2.0468354. Note that this result is really $h(g(f(5)))$. $\blacksquare$

Problem Set 1.1

In Problems 1–4, write an equation for the function that is described.

1. The area y of a rectangle that is twice as long as wide, in terms of the width x.

2. The distance s in miles traveled by a car going 50 miles per hour, in terms of t, the number of hours spent traveling.

3. The volume V of a box with a square base and height twice the base length, in terms of the base length x.

4. The total surface area S of the box described in Problem 3, in terms of the base length x, assuming the box has a lid.

5. For $f(x) = 7x - 3$, evaluate each expression.
   (a) $f(2)$, (b) $f(-2)$, (c) $f(0)$, (d) $f(\frac{8}{7})$, (e) $f(\frac{3}{14})$

6. For $g(x) = 5x + 2$, evaluate each expression.
   (a) $g(1)$, (b) $g(-1)$, (c) $g(10)$, (d) $g(\frac{-2}{5})$, (e) $g(\frac{3}{10})$

7. For $F(x) = 2x^3 - 4x + 3$, evaluate each of the following.
   (a) $F(1)$, (b) $F(-2)$, (c) $F(5)$, (d) $F(\frac{1}{2})$, (e) $F(-\frac{3}{2})$

8. For $G(x) = x^4 - 2x^2 - 6$, evaluate each of the following.
   (a) $G(1)$,   (b) $G(-2)$,   (c) $G(3)$,   (d) $G(-\frac{1}{2})$,
   (e) $G(\sqrt{3})$

9. For $f(x) = 3x - 2x^2$, find and simplify:
   (a) $f(2)$ (b) $f(2 + h)$
   (c) $[f(2 + h) - f(2)]/h$
   (d) $[f(x + d) - f(x)]/d$

10. For $g(x) = 4x - x^2$, find and simplify
    (a) $g(3)$ (b) $g(3 + h)$
    (c) $[g(3 + h) - g(3)]/h$
    (d) $[g(x + p) - g(x)]/p$

11. For $f(x) = (4/x) - 2$, find and simplify
$$[f(a + h) - f(a)]/h$$

12. For $f(x) = 2/x^2$, find and simplify
$$[f(a + h) - f(a)]/h$$

13. For $g(x) = 5 - 2x + 3x^2$, find and simplify
$$[g(x + h) - g(x)]/h$$

14. For $g(x) = x^3$, find and simplify
$$[g(c + h) - g(c)]/h$$

15. Let $f(x) = 2x - 4$ and $g(x) = 2/x^2$. Find:
    (a) $g(f(1))$ (b) $g(f(\frac{5}{2}))$ (c) $g(f(2))$
    (d) $g(f(x))$ (e) $g(g(x))$

16. Let $f(x) = x^2 - 2x - 3$ and $g(x) = 2/(x - 5)$. Find:
    (a) $g(f(0))$ (b) $g(f(2))$ (c) $g(f(-2))$
    (d) $g(f(x))$ (e) $g(g(x))$

17. Let $f(x) = x^2 + 1$ and $g(x) = \sqrt{x - 3}$. Find:
    (a) $g(f(4))$ (b) $g(f(\sqrt{2}))$
    (c) $g(f(\frac{3}{2}))$ (d) $g(f(x))$

18. Let $f(x) = \sqrt{x} + 1/\sqrt{x}$ and $g(x) = x^2$. Find:
    (a) $g(f(1))$ (b) $g(f(4))$
    (c) $g(f(2))$ (d) $g(f(x))$

19. Let $f(x) = 2x^2 + 1$ and $g(x) = 1 - x$. Find $f(g(x))$ and $g(f(x))$.

20. Let $f(x) = 3x + 1$ and $g(x) = x^2 - x$. Find $f(g(x))$ and $g(f(x))$.

21. For $f(x) = x^3 + 1$ and $g(x) = x^2$, find $g(f(x))$ and $f(g(x))$.

22. For $f(x) = (1/x) + 2$ and $g(x) = 1/(x - 2)$, find $g(f(x))$ and $f(g(x))$.

23. If $f(x) = 2x - 1$, $g(x) = x^2$, and $h(x) = 4/(x + 1)$, find $g(h(f(x)))$ and $f(h(g(x)))$.

24. If $f(x) = 1/x^2$, $g(x) = 4x^2$, and $h(x) = \sqrt{x}$, find $h(g(f(x)))$ and $f(g(h(x)))$.

C 25. Calculate (with your calculator in radian mode):
    (a) $\ln (\sin \sqrt{0.5})$ (b) $\sqrt{\cos (\ln 3.2)}$
    (c) $\ln \sqrt{\cos 1.2}$

C 26. Calculate:
    (a) $\sqrt{1/\ln \pi}$ (b) $1/\ln \sqrt{\pi}$
    (c) $1/(\ln \pi)^2$

27. If $h(g(f(x))) = \sqrt{\cos (\ln x)}$, give formulas for $f(x)$, $g(x)$, and $h(x)$.

28. If $h(g(f(x))) = 1/\ln \sqrt{x}$, give formulas for $f(x)$, $g(x)$, and $h(x)$.

---

29. Give the domain of each of the following functions.
    (a) $f(x) = (x - 1)^2 + 3$ (b) $g(x) = 5/(x - 2)^2$
    (c) $F(x) = 3/(x^2 - 9)$
    (d) $G(x) = 3/[(x^2 + 1)^2 - 4]$
    (e) $H(x) = \sqrt{1/(x - 1)}$

30. Let $f(x) = (x + 1)^2$, $g(x) = \sqrt{x}$, and $h(x) = 1/(x^2 + 1)$. Find and simplify:
    (a) $f(-3)$ (b) $g\left(\frac{25}{4}\right)$
    (c) $h\left(\frac{5}{3}\right)$ (d) $h(g(f(0)))$
    (e) $f(g(h(\sqrt{5}/2)))$ (f) $h(g(f(x)))$
    (g) $f(f(x))$

31. If $f(x) = x^4$, find and simplify
$$[f(x + h) - f(x)]/h$$

32. In a certain city, you can rent a car for $18.00 per day and $0.20 per mile. Write a formula for $f(x)$, the cost of renting a car for 10 days and driving x miles.

33. Write a formula for the daily cost $f(x)$ of manufacturing x pairs of speed skates if there are fixed costs (utilities, taxes, rent, and so on) of $700 per day and variable costs (material and labor) of $60 per pair of skates manufactured.

34. Susan sells major appliances. Her monthly salary is $1200 plus $20 for each appliance she sells. The store deducts 26% of her total salary for taxes and retirement and then $10 that she has pledged to the United Way. Write a formula for $f(x)$, her take home pay, if she sells x appliances in a month.

35. Annual union dues at a certain firm are $\frac{1}{2}\%$ of the first $18,000 of salary and $1\frac{1}{2}\%$ of the portion of salary which exceeds $18,000. Write a formula for the annual dues $f(x)$ of an employee whose annual salary is x dollars assuming this salary exceeds $18,000.

36. A tray is to be made from a square piece of sheet metal measuring 14 inches on a side by cutting equal squares out of the four corners and turning up the sides. If these squares are x inches on a side, what is the formula for the volume $V(x)$ of the tray and what is the domain for this function?

37. A function g takes an input x, doubles it, adds 3, squares the result, subtracts 4, and finally takes the square root. Write a formula for $g(x)$.

C 38. A function F takes a positive input x, finds its square root, adds 3, raises the result to the third power, and finally subtracts 10.
    (a) Evaluate $F(7)$ and $F(19.6)$.
    (b) Write a formula for $F(x)$.

C 39. A study of a certain hospital showed that the percentage $p(t)$ of its patients discharged within t days is given by $p(t) = 100\{1 - [10/(10 + t)]^3\}$.
    (a) Calculate $p(0)$, $p(3)$, $p(20)$.
    (b) Within how many days will 99% of its patients be discharged?

40. Let $f(x) = \dfrac{1}{1 - 1/(1 - x)}$. Find and simplify $\dfrac{f(1/x)}{f(x)}$.

1.2

GRAPHS OF FUNCTIONS

Two Frenchmen, René Descartes and Pierre de Fermat, are responsible for the main idea in this section. Although Descartes is better known and usually gets credit for it, Fermat introduced the concept first.

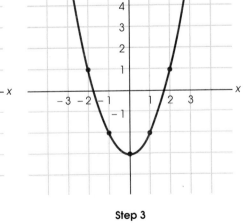

FIGURE 6

Place two real lines in a plane, one horizontal and the other vertical. Call the first the **x-axis** and the second the **y-axis**; call their intersection point the **origin.** Now any point in the plane is uniquely specified by its coordinates (x, y). The first, called the **x-coordinate** or abscissa, measures the directed distance from the vertical axis. The second, called the **y-coordinate** or ordinate, measures the directed distance from the horizontal axis. A few examples of points and their coordinates are shown in Figure 6. Note that the two axes divide the plane into four quadrants, labeled I, II, III, and IV as shown.

The Graph of a Function

The graph of the function f is the set of all points whose coordinates (x, y) satisfy the equation $y = f(x)$. But how does one draw a graph? We suggest a sequence of three steps.

1. Obtain the coordinates of a few points.
2. Plot these points in the plane.
3. Connect the points with a smooth curve in the order of increasing x-values.

$y = x^2 - 3$

x	y
-3	6
-2	1
-1	-2
0	-3
1	-2
2	1
3	6

Step 1

Make a table of values

FIGURE 7

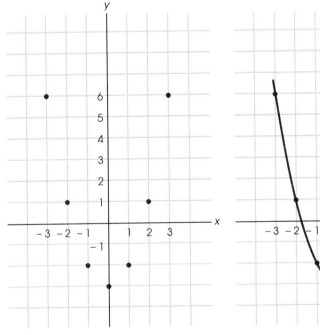

Step 2

Plot those points

Step 3

Connect those points with a smooth curve

The best way to do Step 1 is to make a table of values. Assign values to x and determine the corresponding values of $y = f(x)$; then list the pairs of values in tabular form. The whole three-step procedure is illustrated in Figure 7 for $y = f(x) = x^2 - 3$.

Of course, you must use some common sense in carrying out the three-step procedure. Be sure to plot enough points so that the special features of the graph are evident.

A Partial Catalog of Functions

Our first class of functions is the class of linear functions, that is, functions of the form $f(x) = ax + b$. These functions will be the subject of the next section; here we give one example.

EXAMPLE A

Sketch the graph of $y = f(x) = \frac{1}{2}x - 2$.

Solution A small table of values and the graph are shown in Figure 8. The graph is a straight line. ■

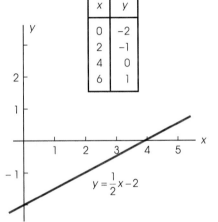

x	y
0	-2
2	-1
4	0
6	1

FIGURE 8

Next we mention the class of **quadratic functions,** that is, functions of the form $f(x) = ax^2 + bx + c$, $a \neq 0$. One such function was graphed in Figure 7; the graph was a cup-shaped curve called a **parabola.** The graph of a quadratic function is always a parabola, but it will open down if the leading coefficient is negative, as we note in our next example.

EXAMPLE B

Sketch the graph of $y = f(x) = -x^2 + 4x$.

Solution A table of values and the graph are shown in Figure 9. ■

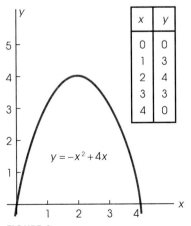

x	y
0	0
1	3
2	4
3	3
4	0

FIGURE 9

Linear and quadratic functions are special cases of what are called **polynomial functions,** functions whose formula consists of a sum of terms of the form cx^m, c being a constant and m a nonnegative integer. A typical example is

$$f(x) = 2x^4 - x^3 + 3x^2 + 19$$

In general, any function of the form

$$f(x) = a_n x^n + a_{n-1} x^{n-1} + \cdots + a_1 x + a_0, \qquad a_n \neq 0$$

where the a's are real numbers and n is a nonnegative integer, is a polynomial function of degree n.

The graph of a polynomial function is always a smooth curve with a number of hills and valleys—usually one less than the degree. For example, the graph of a typical third-degree polynomial has one hill and one valley for a total of two; the graph of a typical fourth-degree polynomial has a total of three hills and valleys.

EXAMPLE C

Sketch the graphs of $y = f(x) = x^3 - 3x$ and $y = g(x) = x^4 - 2x^2 - 1$

Solution The required graphs are shown in Figures 10 and 11.   ◾

The quotient of two polynomial functions is called a **rational function.** An example is

$$f(x) = \frac{3x^4 - 2x^3 + 11x - 9}{3x^2 - 7x + 2}$$

Graphs of rational functions tend to be complicated; we postpone discussion of them until Chapter 3.

Symmetry Properties

The graphs in Figures 10 and 11 illustrate two important kinds of symmetry. The graph of $y = f(x) = x^3 - 3x$ (Figure 10) is **symmetric with respect to the origin;** that is, for each point on the graph there is another point directly opposite on the other side of the origin and equally far away. More precisely, if (x, y) is on the graph, so is $(-x, -y)$.

The graph of $y = g(x) = x^4 - 2x^2 - 1$ (Figure 11) is **symmetric with respect to the y-axis.** For each point on the graph, there is another directly

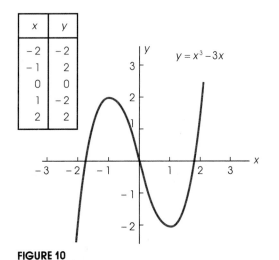

x	y
-2	-2
-1	2
0	0
1	-2
2	2

$y = x^3 - 3x$

FIGURE 10

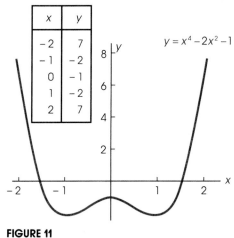

x	y
-2	7
-1	-2
0	-1
1	-2
2	7

$y = x^4 - 2x^2 - 1$

FIGURE 11

opposite on the other side of the y-axis and equally far away. Here the coordinate condition is that if (x, y) is on the graph, so is $(-x, y)$.

We could have predicted the symmetry properties of f and g before we drew their graphs. If $f(-x) = -f(x)$, the graph will be symmetric with respect to the origin; if $f(-x) = f(x)$, the graph will be symmetric with respect to the y-axis. Note that for $f(x) = x^3 - 3x$,

$$f(-x) = (-x)^3 - 3(-x) = -x^3 + 3x = -(x^3 - 3x) = -f(x)$$

while for $g(x) = x^4 - 2x^2 - 1$,

$$g(-x) = (-x)^4 - 2(-x)^2 - 1 = x^4 - 2x^2 - 1 = g(x)$$

A function f satisfying $f(-x) = -f(x)$ is called an **odd function,** probably because a polynomial with only odd powers is an odd function. If $f(-x) = f(x)$, then f is called an **even function.** A polynomial with only even powers is an even function.

EXAMPLE D

In each case, decide whether the given function is odd, even, or neither.

(a) $f(x) = 2/(x^2 + 1)$
(b) $g(x) = 3x^3 + 2x^2 + 4$
(c) $h(x) = (2x^6 + 1)/x$

Solution

(a) $f(-x) = 2/[(-x)^2 + 1] = 2/(x^2 + 1) = f(x); f$ is even.
(b) $g(-x) = 3(-x)^3 + 2(-x)^2 + 4 = -3x^3 + 2x^2 + 4; g$ is neither even nor odd.
(c) $h(-x) = [2(-x)^6 + 1]/(-x) = (2x^6 + 1)/(-x) = -(2x^6 + 1)/x = -h(x);$ therefore h is an odd function.

Note that of these three functions, only g is a polynomial. ■

Functions with Multipart Rules

A special function that will play a significant role in this book is the **absolute value function** $|. . .|$, defined by

$$|x| = \begin{cases} x & \text{if } x \geq 0 \\ -x & \text{if } x < 0 \end{cases}$$

Its graph is shown in Figure 12. Note the sharp corner at the origin.

Occasionally, we will meet other functions defined by different rules on separate parts of their domains.

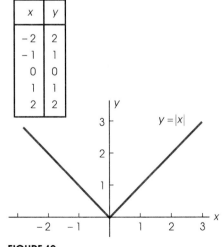

x	y
-2	2
-1	1
0	0
1	1
2	2

FIGURE 12

EXAMPLE E

Sketch the graph of

$$f(x) = \begin{cases} -x + 1 & \text{if } x < 0 \\ -x^2 + 3x & \text{if } 0 \le x \le 3 \\ x - 2 & \text{if } x > 3 \end{cases}$$

Solution The three-part graph is shown in Figure 13. Note that at 0 and 3, it is $-x^2 + 3x$ that determines the value of the function. This is the reason for the open circles and the heavy dots at these points on the graph.

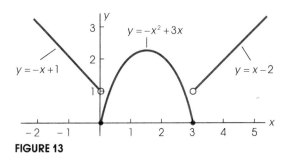

FIGURE 13   ■

EXAMPLE F

Dues for a certain professional organization are based on one's annual salary S according to the following scheme: (a) \$30 if $S \le 20{,}000$, (b) \$30 plus 0.1% of salary in excess of 20,000 if $20{,}000 < S \le 50{,}000$, (c) \$60 plus 0.2% of salary in excess of 50,000 if $S > 50{,}000$. Write a multipart formula for the dues $D(S)$ and then graph the resulting function.

Solution

$$D(S) = \begin{cases} 30 & \text{if } S \le 20{,}000 \\ 30 + 0.001(S - 20{,}000) & \text{if } 20{,}000 < S \le 50{,}000 \\ 60 + 0.002(S - 50{,}000) & \text{if } S > 50{,}000 \end{cases}$$

$$= \begin{cases} 30 & \text{if } S \le 20{,}000 \\ 10 + 0.001S & \text{if } 20{,}000 < S \le 50{,}000 \\ -40 + 0.002S & \text{if } S > 50{,}000 \end{cases}$$

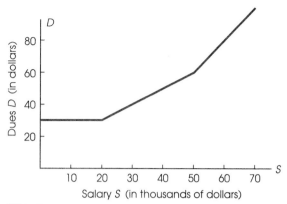

FIGURE 14

The graph, which consists of three line segments, is shown in Figure 14. ■

A Final Note

We encourage you to study the chart that appears as Figure 15 at the end of the problem set. This chart supplements what we have said about graphing and illustrates how a graph changes under various transformations.

Problem Set 1.2

1. Which of the following are polynomial functions?
   (a) $f(x) = \frac{3}{2}x^2 + 5x - 6$
   (b) $g(x) = 4x^3 - 5/x$
   (c) $h(x) = \pi x^4 - \sqrt{2}x + 3$
   (d) $F(x) = \sqrt{x^2 + 4}$
   (e) $G(t) = 2.4t^6 - 1.5t^2 + 9$
   (f) $H(s) = \frac{1}{5}s^5 - (2/\sqrt{3})s + 4$

2. Which of the following are polynomial functions?
   (a) $f(x) = 2x^3 - \frac{4}{5}x - 13$
   (b) $g(x) = 4x^2 + 2/x^2$
   (c) $h(x) = 3$
   (d) $F(x) = 2x + \sqrt{x} - 2$
   (e) $G(t) = (t^2 + 4)/t + 3t^2$
   (f) $H(s) = \sqrt{3}\,s^{10} - (15/\pi)s^2 + 10$

3. Identify each of the following functions as to whether it is even, odd, or neither even nor odd.
   (a) $f(x) = 5x^3 - 7x$
   (b) $g(x) = 5x^4 - 2x^2 + 11$
   (c) $h(x) = 3x^3 + 5x - 4$
   (d) $F(x) = x^6 - 2x^4 + 2x - 5$
   (e) $G(x) = 4/(x^2 + 5)$
   (f) $H(x) = (x^2 + 5)/(2x^3 + x)$

4. Identify each of the following functions as to whether it is even, odd, or neither.
   (a) $f(x) = x^5 + 14$
   (b) $g(x) = 3x^{11} - 2x^9 + 7x$
   (c) $h(x) = x^6 - 2x^2 + 14$
   (d) $F(x) = (x^3 - 2x)^2$
   (e) $G(x) = x^3/(x^4 - 2x^2)$
   (f) $H(x) = (x^4 + 3x^2 + 1)(x^3 - 3x)$

In Problems 5–26, graph each function showing enough of the graph to bring out its essential features. Note either of the two kinds of symmetry discussed in the text.

5. $f(x) = 2x - 5$
6. $f(x) = -3x + 2$
7. $f(x) = -\frac{3}{4}x + 2$
8. $f(x) = \frac{3}{2}x - 3$
9. $f(x) = -3$
10. $f(x) = 5$
11. $f(x) = -x^2 + 4$
12. $f(x) = -x^2 + 4x$
13. $f(x) = x^2 - 5x$
14. $f(x) = x^2 - 5$
15. $f(x) = x^2 + x - 4$
16. $f(x) = -x^2 + 3x + 2$
17. $f(x) = x^3 - 4x$
18. $f(x) = 9x - x^3$

19. $f(x) = x^4 - 4x^2 + 3$

20. $f(x) = x^4 - 5x^2 + 6$

21. $f(x) = x^3 - x^2$

22. $f(x) = x^3 - 2x^2 + 1$

23. $f(x) = x^5 - x$

24. $f(x) = 16x - x^5$

25. $f(x) = |x - 3|$

26. $f(x) = |2x + 4|$

As in Example E, Problems 27–32 define functions by means of multipart rules. Sketch the graph of each function.

27. $f(x) = \begin{cases} 2x & \text{if } x \le 1 \\ -3x & \text{if } x > 1 \end{cases}$

28. $f(x) = \begin{cases} x + 2 & \text{if } x \le 0 \\ 1 & \text{if } x > 0 \end{cases}$

29. $f(x) = \begin{cases} x^2 & \text{if } x < 1 \\ 2x & \text{if } x \ge 1 \end{cases}$

30. $f(x) = \begin{cases} 2x + 2 & \text{if } x < 0 \\ 1 - x^2 & \text{if } x \ge 0 \end{cases}$

31. $f(x) = \begin{cases} x & \text{if } x < -1 \\ x^2 - 2 & \text{if } -1 \le x \le 2 \\ -x^2 + 6 & \text{if } x > 2 \end{cases}$

32. $f(x) = \begin{cases} -x^2 & \text{if } x < 0 \\ -x^2 + 3x & \text{if } 0 \le x < 3 \\ x - 2 & \text{if } x \ge 3 \end{cases}$

33. In 1984, the United States domestic rates for first-class mail were 20 cents for the first ounce or fraction thereof, and 17 cents for each additional ounce or fraction. Let $C(x)$ be the cost of mailing x ounces. Sketch the graph of $C(x)$ for $0 < x \le 6$.

34. In a certain city, the United Way offered the following guidelines for giving to its annual fund: (a) 1% for incomes under $30,000, (b) $300 plus 2% of income exceeding $30,000 for incomes between $30,000 and $50,000, (c) $700 plus 3% of income exceeding $50,000 for incomes over $50,000. Graph $G(I)$, the suggested gift corresponding to income I.

---

35. Let f be an even function, g an odd function. Decide whether h is even or odd or neither in each of the following.
    (a) $h(x) = f(x) \cdot g(x)$
    (b) $h(x) = f(x)/g(x)$
    (c) $h(x) = [g(x)]^2$
    (d) $h(x) = [f(x)]^3$
    (e) $h(x) = [f(x)]^2 + [g(x)]^2$
    (f) $h(x) = f(x) + x^2 + 4$
    (g) $h(x) = [f(x) + g(x)]^2$
    (h) $h(x) = g(x) + g(-x)$

36. Graph each of the following.
    (a) $f(x) = (x - 1)(x + 2)$
    (b) $f(x) = 3(x - 1)^2$
    (c) $f(x) = (x + 2)(x - 1)^2$

(d) $f(x) = x(x^2 - 4)$

(e) $f(x) = \begin{cases} -2 & \text{if } x \le -2 \\ 1 + x & \text{if } -2 < x \le 0 \\ 1 - x & \text{if } 0 < x \le 2 \\ -2 & \text{if } x > 2 \end{cases}$

37. Graph $y = x^2$, $y = x^2 + 2$, and $y = (x - 2)^2$ using the same axes. Then explain how the graphs of $y = f(x) + k$ and $y = f(x - k)$ are related to the graph of $y = f(x)$, assuming k is a constant.

38. The RS Bus Company offered to take a church group to a lake resort for $150 per person if 30 or fewer sign up. If more than 30 persons make the trip, each fare will be reduced by $15. The capacity of the bus is 70 people. Write a two-part formula for $R(x)$, the total fare revenue if x persons make the trip. Then calculate $R(35)$ and $R(70)$.

39. The UV Bus Company will charge the church group in Problem 38 a fare of $160 each if 30 or fewer people sign up but will reduce the fare for everyone by $2 for each person exceeding 30 who makes the trip. Write a two-part formula for $R(x)$, the total fare revenue if x persons make the trip. Then calculate $R(35)$ and $R(70)$.

40. The accompanying table is taken from the 1983 Internal Revenue Service (IRS) tax form. It gives a multipart rule for a function representing tax T in terms of taxable income I for a single taxpayer. Sketch the graph of T for $0 \le I \le 15,000$.

Schedule X
Single Taxpayers

Use this Schedule if you checked Filing Status Box 1 on Form 1040—

If the amount on Form 1040, line 37 is: Over—	But not over—	Enter on Form 1040, line 38	of the amount over—
$0	$2,300	—0—	
2,300	3,400	11%	$2,300
3,400	4,400	$121 + 13%	3,400
4,400	8,500	251 + 15%	4,400
8,500	10,800	866 + 17%	8,500
10,800	12,900	1,257 + 19%	10,800
12,900	15,000	1,656 + 21%	12,900
15,000	18,200	2,097 + 24%	15,000
18,200	23,500	2,865 + 28%	18,200
23,500	28,800	4,349 + 32%	23,500
28,800	34,100	6,045 + 36%	28,800
34,100	41,500	7,953 + 40%	34,100
41,500	55,300	10,913 + 45%	41,500
55,300	. . . .	17,123 + 50%	55,300

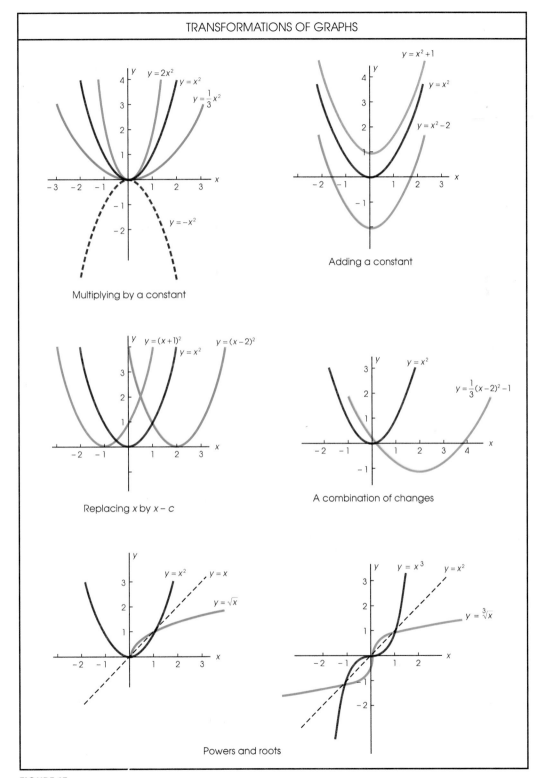

TRANSFORMATIONS OF GRAPHS

Multiplying by a constant

Adding a constant

Replacing x by $x - c$

A combination of changes

Powers and roots

FIGURE 15

C 41. Use the information in Problem 40 to find the tax paid by a single person whose taxable income was (a) $8751, (b) $12,390, (c) $34,981.

PC 42. In Utopia, there are no taxes on incomes of $50,000 or less and only 6% taxes on that part of one's income which exceeds $50,000. Write a program for computing the tax on any income in Utopia.

43. The 1986 Tax Reform Act sets federal taxes for single taxpayers (beginning in 1988) at 15% on a taxable in-

come of $17,850 or less and at 28% on the portion of a taxable income exceeding $17,850. Work Problem 41 under this plan.

PC 44. (For computer buffs only.) Write a program for computing the tax on any income using the table in Problem 40.

1.3
LINEAR FUNCTIONS: CONSTANT SLOPE

The simplest kind of function is a **linear function,** that is, a function of the form

$$f(x) = ax + b$$

where a and b are constants. Linear functions get their name from the fact that their graphs are lines. Thus, our study of linear functions begins with a complete discussion of lines.

We assume that you have a good intuitive feeling for the concept of line from looking along a taut string or the edge of a ruler. Let us agree on one thing: Two points determine a unique line. In Figure 16, there is exactly one line through points A and B.

FIGURE 16

The Slope of a Line

Consider the line in Figure 16. A move from A to B produces a vertical change of $7 - 1 = 6$ units and a corresponding horizontal change of $4 - 2 = 2$ units. The ratio $\frac{6}{2} = 3$ of these two numbers is called the slope of the line.

In general, if y changes by an amount Δy (read "delta y") while x is changing by Δx (Figure 17), we say that the slope of the line is $m = \Delta y / \Delta x$, that is,

FIGURE 17

$$m = \frac{\text{change in } y}{\text{change in } x} = \frac{\Delta y}{\Delta x} = \frac{y_2 - y_1}{x_2 - x_1}$$

A careful reader will ask an immediate question. Since a line has many points, does the value we get for the slope depend on which pair of points we use for A and B? The answer is no. The similar triangles in Figure 18 show us that

$$\frac{y_2' - y_1'}{x_2' - x_1'} = \frac{y_2 - y_1}{x_2 - x_1}$$

The points A' and B' give the same value for m as do A and B. It does not

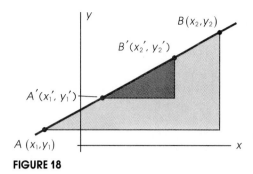

FIGURE 18

even matter whether A is to the left or right of B, since

$$\frac{y_1 - y_2}{x_1 - x_2} = \frac{y_2 - y_1}{x_2 - x_1}$$

All that matters is that we subtract the coordinates in the same order in the numerator and the denominator.

The feature just noted, that $m = \Delta y/\Delta x$ is constant, distinguishes lines from all other curves. In fact, m is a measure of the steepness of a line. A horizontal line has zero slope, and a line that rises to the right has positive slope. The larger this positive slope is, the more steeply the line rises. A line that falls to the right has negative slope. The concept of slope is meaningless for vertical lines, since it would involve division by zero. In Figure 19, we have computed the slope for a number of lines all going through the point (2, 1).

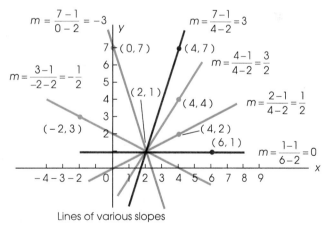

Lines of various slopes

FIGURE 19

EXAMPLE A

Find the slope of the line passing through $(-2, 6)$ and $(4, 3)$.

Solution

$$m = (3 - 6)/[4 - (-2)] = -3/6 = -1/2 \qquad \blacksquare$$

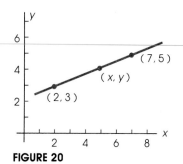

FIGURE 20

The Point-Slope Form

Our goal is to write an equation for the line in Figure 20. Note that this line

(i) passes through (2, 3).

(ii) has slope $(5 - 3)/(7 - 2) = \frac{2}{5}$.

Consider any other point (x, y) on this line. When we use this point together with (2, 3) to measure slope, we must get $\frac{2}{5}$, that is,

$$\frac{y - 3}{x - 2} = \frac{2}{5}$$

or, after multiplying by $x - 2$,

$$y - 3 = \tfrac{2}{5}(x - 2)$$

This last equation is satisfied by all points on the line, even by (2, 3). Moreover, only points on the line can satisfy this equation.

What we have just done in an example can be done in general. The line passing through the fixed point (x_0, y_0) with slope m (Figure 21) has equation

$$y - y_0 = m(x - x_0)$$

FIGURE 21

We call this equation the **point-slope form** of the equation of a line.

EXAMPLE B

Find the point-slope equation of the line through $(-2, 4)$ and $(4, 7)$. Show that you get equivalent equations using either $(-2, 4)$ or $(4, 7)$ as the fixed point (x_0, y_0).

Solution First, we calculate the slope $m = (7 - 4)/(4 + 2) = \frac{1}{2}$. Then, using $(-2, 4)$ as the fixed point gives the equation

$$y - 4 = \tfrac{1}{2}(x + 2)$$

whereas (4, 7) yields

$$y - 7 = \tfrac{1}{2}(x - 4)$$

Solve either equation for y and you will get $y = \frac{1}{2}x + 5$. ∎

The Slope-Intercept Form

Suppose we are given the slope m for a line and the y-intercept b, meaning that the line intersects the y-axis at $(0, b)$ as in Figure 22. Choosing $(0, b)$ as the fixed point of the point-slope form, we get

$$y - b = m(x - 0)$$

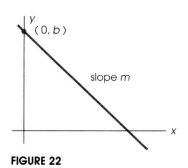

FIGURE 22

which we can write as

$$y = mx + b$$

The latter equation is called the **slope-intercept form.**

Of what use is this new form? Anytime we see an equation written this way, we immediately recognize it as an equation of a line and can read off its slope and y-intercept.

EXAMPLE C

Find an equation for the line through $(2, -3)$ that is parallel to the line with equation $3x - 4y = 7$.

Solution When $3x - 4y = 7$ is solved for y, we get successively

$$-4y = -3x + 7$$
$$y = \tfrac{3}{4}x - \tfrac{7}{4}$$

The latter is an equation of a line with slope $m = \tfrac{3}{4}$.

Now we call on a basic fact: *Two nonvertical lines are parallel if and only if they have the same slope.* It follows that the line we want has slope $\tfrac{3}{4}$. Since this line goes through $(2, -3)$, its equation in point-slope form is

$$y + 3 = \tfrac{3}{4}(x - 2) \qquad\blacksquare$$

A derivation similar to that in Example C shows that any equation of the form $ax + by = c$ with $b \neq 0$ represents a nonvertical line. We will find plenty of uses for this fact.

EXAMPLE D

Find an equation for the line through the intersection of the lines $3x - 2y = 10$ and $2x + 6y = 14$ which has slope $m = -\tfrac{2}{3}$. Write your answer in the form $ax + by = c$.

Solution Our first task is to solve the two given equations simultaneously. Begin by multiplying the first equation by 3 and adding it to the second equation.

$$
\begin{array}{r}
9x - 6y = 30 \\
\underline{2x + 6y = 14} \\
11x \qquad\;\; = 44 \\
x = 4
\end{array}
$$

Now substitute $x = 4$ in either of the original equations and solve for y. You will get $y = 1$. The intersection point is $(4, 1)$.

The required line passes through $(4, 1)$ and has slope $-\frac{2}{3}$. In point-slope form, its equation is

$$y - 1 = -\tfrac{2}{3}(x - 4)$$

Finally, multiplying both sides by 3 and simplifying gives

$$3y - 3 = -2x + 8$$

or

$$2x + 3y = 11$$

The geometric interpretation of this example appears in Figure 23. ■

FIGURE 23

(Graph showing lines $3x - 2y = 10$, $2x + 6y = 14$, Required line, intersecting at $(4, 1)$.)

Lines and Linear Functions

We claimed in our introduction that the graph of the linear function $f(x) = ax + b$ is a line. Our discussion shows that it is the line with slope a and y-intercept b. For example, the graph of the function $f(x) = 4x + 3$ is the line with slope 4 and y-intercept 3. Conversely, each nonvertical line has a slope m and a y-intercept b and so is the graph of the linear function $f(x) = mx + b$.

EXAMPLE E

Find the formula for the linear function f that satisfies $f(-1) = 4$ and $f(3) = 12$. Then calculate $f(2)$.

Solution The graph of f (a line) goes through $(-1, 4)$ and $(3, 12)$ and so has slope $m = (12 - 4)/(3 + 1) = 2$. In point-slope form, the equation of this line is

$$y - 4 = 2(x + 1)$$

or solved for y,

$$y = 2x + 6$$

Thus, $f(x) = 2x + 6$ and $f(2) = 10$. ■

FIGURE 24

Two Simple Examples of Slopes

In building a house, a carpenter may say he is putting up a roof with a 7:12 pitch (Figure 24). What he means is that the roof will rise 7 feet in a run of 12 feet; that is, the roof has slope $\frac{7}{12}$.

In the mountains, one often sees a road sign like that in Figure 25, indicating what is called the grade. A grade of 12% is simply a slope of $\frac{12}{100}$. Of course, if you are driving downhill, the slope is really $-\frac{12}{100}$.

12 %

FIGURE 25

Problem Set 1.3

In Problems 1–6, find the slope of the line passing through the indicated points.

   1. $(3, 1)$ and $(5, 9)$ 2. $(2, 3)$ and $(10, 15)$
   3. $(-4, 2)$ and $(3, 0)$ 4. $(2, -4)$ and $(0, -6)$
   [c] 5. $(2.563, -9.108)$ and $(\sqrt{5}, -\sqrt{3})$
   [c] 6. $(\pi, \sqrt{3} + \sqrt{5})$ and $(1.642, \sqrt{2})$

Find an equation for the indicated line.

   7. The line through $(2, 1)$ with slope 3.
   8. The line through $(3, 5)$ with slope $\frac{3}{2}$.
   9. The line through $(3, 1)$ and $(5, 9)$.
 10. The line through $(2, 3)$ and $(10, 15)$.
 11. The line through $(1, -2)$ and $(3, 7)$.
 12. The line through $(2, 3)$ and $(-10, 15)$.

In Problems 13–16 find an equation for the indicated line. Write your final answer in the form $ax + by = c$.

 13. The line with slope 4 and y-intercept -3.
 14. The line with y-intercept 4 and slope $-\frac{2}{3}$.
 15. The line through $(2, -1)$ that is parallel to the line $5x - y = 12$.
 16. The line through $(5, 0)$ that is parallel to the line $2x + 3y = 13$.

Here is an important fact. **Two nonvertical lines are perpendicular if and only if their slopes are negative reciprocals.** *For example, the lines $y = 2x - 7$ and $y + 1 = \frac{-1}{2}(x - 5)$ are perpendicular because 2 and $-\frac{1}{2}$ are negative reciprocals. Use this information to obtain an equation for the indicated lines in Problems 17–20. Write your final answer in the form $ax + by = c$.*

 17. The line through $(2, -1)$ that is perpendicular to the line $y = -\frac{1}{2}x + 4$.
 18. The line through $(4, 1)$ that is perpendicular to the line $y = 3x - 5$.
 19. The line through $(2, 4)$ that is perpendicular to the line $x - 4y = 21$.
 20. The line through $(7, -3)$ that is perpendicular to the line $5x + 2y = -10$.

Follow the ideas of Example D to find an equation for the line described in each of Problems 21–24. Write your final answer in the form $ax + by = c$.

 21. The line through the intersection of $x + 2y = -1$ and $3x + y = 7$ which has slope $\frac{3}{4}$.
 22. The line through the intersection of $4x + 3y = 2$ and $x + 5y = -8$ which has slope $-\frac{2}{5}$.
 23. The line through the intersection of $x + 2y = 5$ and $3x - 2y = -9$ which is parallel to the second line.
 24. The line through the intersection of $2x + 4y = 0$ and $3x + 8y = -4$ which is perpendicular to the first line.

In Problems 25–28, find a formula of the form $f(x) = mx + b$ for the linear function f satisfying the given conditions. Then find $f(5)$. See Example E.

 25. $f(1) = 3, f(3) = 9$
 26. $f(-2) = 5, f(8) = 25$
 27. $f(-2) = 4, f(3) = 2$
 28. $f(1) = -2, f(6) = 7$

 29. Find the slope and y-intercept of each line.
    (a) $3y = -2x + 6$ (b) $4x + 3y = -36$
    (c) $y + 2 = \frac{3}{4}(x - 1)$
    (d) $3y + 2x = -7x + 9y + 16$
 30. Find an equation for the line with y-intercept 2 and x-intercept 3.
 31. Write (in the form $ax + by = c$) an equation of the line through $(2, -5)$ which is:
    (a) parallel to the line $2y = -3x + 2$.
    (b) parallel to the line through $(0, 4)$ and $(3, -2)$.
    (c) parallel to the x-axis.
    (d) perpendicular to the line $3x - 5y = 6$.
 32. Show that the equation of the line with y-intercept b and x-intercept a can be put in the form $x/a + y/b = 1$. Assume a and b are both nonzero.
 33. A farm tractor costs \$20,000 and will have a value of \$2000 at the end of 10 years. Let $f(x)$ be the value of the tractor after x years ($0 \le x \le 10$) and assume that f is a linear function. Find a formula for $f(x)$.
 34. Find the value of k for which the line $4x + ky = 5$
    (a) passes through the point $(1, -1)$.
    (b) is parallel to the line $y = 2x + 4$.
    (c) has equal x- and y-intercepts.
    (d) has y-intercept -10.

35. The Celsius and Fahrenheit temperature scales are related in a linear way; that is, $F = aC + b$.
    (a) Use the fact that $F = 32$ when $C = 0$ and $F = 212$ when $C = 100$ to determine a and b.
    (b) Calculate F when $C = 32$.
    (c) Calculate C when $F = 81$.

36. Sketch the graph of the line $y = -2x + 4$ and the point $(7, 5)$. Find:
    (a) an equation of the line through $(7, 5)$ perpendicular to the given line.
    (b) the point of intersection of the given line and the line in part (a).
    (c) the distance between $(7, 5)$ and the given line.

C 37. In pedaling up a mountain road with a grade of 12%, a bicyclist increased her altitude 1000 feet. How far did she pedal? *Hint:* Draw a right triangle and use the Pythagorean Theorem $a^2 + b^2 = c^2$.

C 38. A rectangular house 24 feet by 36 feet has a gable roof with a 7:12 pitch (to have a gable roof means the roof line on the shorter side of the house looks like a wide inverted vee). Assuming the roof has no overhanging eaves, calculate the area of the roof. *Hint:* You will need the Pythagorean Theorem (see Problem 37).

1.4
APPLICATIONS OF LINEAR FUNCTIONS

The concept of linearity finds numerous applications in the world of experience. A process is **linear** if a change Δx in input produces a change Δy in output that is a constant multiple of Δx; that is, $\Delta y = k \Delta x$. Examples abound. A change in temperature produces a proportional change in the height of mercury in a glass tube (the principle of the thermometer). A change in the depth of an underwater diver produces a proportional change in the pressure on his skin. A change in the number of refrigerators produced by a manufacturer normally produces a proportional change in the revenue received. Recently, a major automobile manufacturer advertised that its cars behave in a linear way. The claim was that an increase in force on the brakes or steering wheel produces a proportional response from the car which would therefore be very predictable.

Linear processes are the easiest of all processes to describe mathematically; this is because they are represented by linear functions. A number of such processes are discussed in this section. Later (Sections 1.5 and 1.6), we will take up the more difficult concept of nonlinear processes.

Business Applications

Many readers of this book will take a job in sales; our first formal example should seem relevant to them.

EXAMPLE A (WORKING ON COMMISSION)

Alice Drinkwater has accepted a position at Compudyne selling personal computers. Her salary will be $1500 per month plus a commission of 3% of her sales. Express her total salary $T(x)$ in terms of x, her monthly sales in dollars. Then calculate her salary based on sales of $25,000 per month.

Solution

$$T(x) = 1500 + (0.03)x$$

$$T(25,000) = 1500 + (0.03)(25,000) = \$2250$$

Note that $T(x)$ does have the linear form $ax + b$. ∎

The Internal Revenue Service allows several methods for figuring depreciation of equipment, one being *straight-line depreciation*.

EXAMPLE B (STRAIGHT-LINE DEPRECIATION)

A milling company bought a machine costing \$120,000, expecting it to depreciate at a constant yearly rate and to have a junk value of \$3000 after 5 years. What was its value at the end of x years ($0 \le x \le 5$)? At the end of 3.5 years?

Solution The graph of V (Figure 26) is the line through (0,120,000) and (5,3000) and so has slope

$$m = (3000 - 120,000)/(5 - 0) = -23,400$$

Thus

$$V(x) = -23,400x + 120,000$$

and

$$V(3.5) = (-23,400)(3.5) + 120,000 = \$38,100$$

Note that the slope $m = -23,400$ means that the machine loses \$23,400 in value each year. ∎

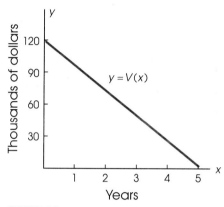

FIGURE 26

EXAMPLE C (COST OF PRODUCTION)

A manufacturer of vacuum cleaners has fixed monthly costs (warehouse rent, office utilities, insurance, and so on) of \$4500 and estimates that labor and material for each cleaner cost \$115. Find the cost $C(x)$ of producing x vacuum cleaners per month.

Solution $C(x) = 4500 + 115x$. Note that when graphed, the line would have slope $m = 115$, exactly the cost of one vacuum cleaner. ∎

EXAMPLE D (BREAK-EVEN ANALYSIS)

The manufacturer in Example C expects to receive \$145 for each vacuum cleaner it sells. Assuming the manufacturer can sell all it makes, what is its monthly revenue $R(x)$ and how many units must it sell per month to just break even?

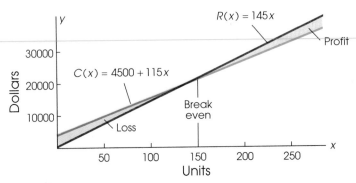

FIGURE 27

Solution $R(x) = 145x$. To exactly break even, $R(x) = C(x)$; that is,

$$145x = 4500 + 115x$$

$$30x = 4500$$

$$x = 150$$

Making and selling 150 units will allow the manufacturer to break even. Making and selling more than 150 units will yield a profit. The situation is shown graphically in Figure 27. ∎

Rates of Change

About half of calculus (the part called *differential calculus*) deals with rates of change. Given two related variables x and y, we often want to know how fast one is changing with respect to the other. To ask how fast a car is going is to ask how the distance traveled s is changing with respect to the elapsed time t. The price p of pork is dependent on the number h of hogs being raised on farms. An economist wants to know how p changes with respect to h. These rates of change are constant precisely when the underlying functions are linear.

EXAMPLE E

John Whitfield is 50 miles from Chicago and driving away from that city at 55 miles per hour. Express his distance s from Chicago in terms of the elapsed time t starting right now.

Solution

$$s = 50 + 55t$$

Be sure to note something. The speed (that is, the rate of change in s with respect to t) 55 is exactly the slope of the line $s = 50 + 55t$. This is no accident. ∎

FIGURE 28

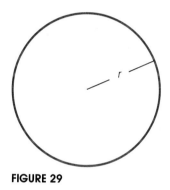

FIGURE 29

EXAMPLE F

The pressure at the surface of a lake is 15 pounds per square inch, while at a depth of 33 feet it has doubled to 30 pounds per square inch (Figure 28). If pressure p increases linearly with depth d (it does), find the rate of change in p with respect to d. Then write a formula for p in terms of d.

Solution

$$\text{rate} = \frac{\Delta p}{\Delta d} = \frac{30 - 15}{33 - 0} = \frac{15}{33} \approx 0.4545 \text{ pounds per square inch per foot}$$

$$p = 0.4545d + 15$$

Note again that the rate 0.4545 is the same as the slope of the line. It tells us that the pressure increases 0.4545 pounds per square inch with each additional foot of depth. ∎

EXAMPLE G

A circular ring (Figure 29) is heated, thus increasing the radius r and circumference C of the ring. How does C change with respect to r?

Solution Since $C = 2\pi r$ (formula from geometry) is a linear relationship, the desired rate is simply the coefficient of r, namely, $2\pi \approx 6.2832$. If r increases by 1 unit, the circumference will increase by a little more than 6 units. Note that this statement is true no matter what the size of the original ring (the ring on your finger or one of the rings of Saturn). ∎

Summary

We have said much about linear functions in the last two sections, but here is the essence of it all. A linear function is determined by an equation of the form $y = mx + b$. The graph of this equation is a straight line. The number m (a constant) has two interpretations. It measures the slope (steepness) of the line, and it tells us how fast y is changing with respect to x. That the geometrical notion of slope and the physical notion of rate of change are really the same idea is worth highlighting.

---

slope = rate of change

This fact, here observed for linear functions, is also true for nonlinear functions, the subject we take up in the next two sections.

Problem Set 1.4

C 1. Anthropologists use the linear function $H(x) =$ 2.89x + 70.64 to estimate the height (in centimeters) of a male whose humerus (bone from elbow to shoulder) is x centimeters long. Among the old bones discovered at a desert site was a male humerus 43.52 centimeters long. How tall was this man?

C 2. Biologists use the linear function $E(R) = 1.235R + 3.805$ to estimate the energy yield (in kilocalories per liter of oxygen) where R is the so-called respiratory quotient. Compute the energy yield for a person on a high-fat diet for which R is 0.74.

3. Neil Thackwood sells appliances. He makes $1200 per month plus a commission of 4% of his sales.
   (a) Express his total salary $S(x)$ in terms of x, his monthly sales in dollars.
   (b) Calculate his total salary for a month in which his sales amounted to $36,000.

4. Marie Howser sells dresses at an exclusive women's shop. Her monthly salary is $1100 plus $30 for each dress she sells.
   (a) Express her monthly salary $f(x)$ in terms of x, the number of dresses she sells.
   (b) Calculate her salary for a month in which she sold 38 dresses.
   (c) If Marie hopes to earn $2600 next month, how many dresses must she sell?

5. Airbus leases cars for $150 per month plus 12¢ per mile.
   (a) Find the formula for $F(x)$, the cost of leasing a car for a month and driving it x miles.
   (b) What is the total monthly cost for a person who leases a car and plans to drive 8000 miles each month?

6. A natural gas exploration company offered a Montana rancher $10,000 for the right to drill on his ranch plus 9¢ for each thousand cubic feet of gas extracted. Find the formula for the amount $A(x)$ the rancher will receive expressed in terms of x, thousands of cubic feet of gas.

7. A road grader purchased today for $75,000 will depreciate linearly to a scrap value of $8000 after 10 years. See Example B.
   (a) Write a formula for its value $V(n)$ after n years $(0 \leq n \leq 10)$.
   (b) Sketch the graph of $V(n)$ for $0 \leq n \leq 10$.
   (c) What is the slope of this graph and what is its significance?

8. A farmer purchases a combine today for $120,000, knowing that it will depreciate (linearly) to a value of only $10,000 at the end of 5 years.

   (a) Write a formula for the value $C(n)$ of the combine at the end of n years $(0 \leq n \leq 5)$.
   (b) Calculate the value of the combine after 3.5 years.
   (c) When will the combine be worth $80,000?

9. A manufacturer of color television sets has fixed monthly costs (utilities, insurance, and so on) of $42,000, and the cost of labor and material for each set is $250. Find the cost $C(x)$ of producing x television sets per month. See Example C.

10. A manufacturer of freezers has fixed costs of $28,000 per month, and the cost of labor and material is $300 per freezer. Find the cost $C(x)$ of manufacturing x freezers per month.

11. The manufacturer in Problem 9 sells each color television set at $375. Assuming that every set that is made is sold, what is the monthly revenue $R(x)$ if x sets are sold? How many sets must be sold per month in order to just break even? See Example D.

12. The manufacturer in Problem 10 sells the freezers at $475. If every freezer that is made is sold, what is the monthly revenue from selling x freezers? How many freezers must be sold per month for the company to just break even?

13. A sports company is planning to introduce a new line of cross-country skis. Set-up and overhead costs for the first year are estimated at $20,000, and it will cost $40 in labor and materials to make and market each pair of skis. The company expects to sell a pair of skis for $100.
   (a) Determine $C(x)$, the cost of making and marketing x pairs of skis during the first year.
   (b) Determine $R(x)$, the revenue from selling x pairs of skis.
   (c) Determine $P(x) = R(x) - C(x)$, the profit realized from the production and sale of x pairs of skis.
   (d) Calculate $P(1000)$, the profit from making and selling 1000 pairs of skis.

14. A consultant has warned the sports company in Problem 13 that it is unrealistic to expect to sell 1000 pairs of skis at $100 per pair. She suggests that the demand x for skis depends on the price p (in dollars) according to the formula $x = 2000 - 15p$. Assume the consultant is right.
   (a) How many pairs of skis can be sold at $p = 100$?
   (b) What is the profit when $p = 100$?
   (c) Express the profit P in terms of the price p.

15. Joan Appleseed is 480 miles from Los Angeles and is driving toward that city at 60 miles per hour. Express

her distance s from Los Angeles in terms of t, the number of hours from now ($0 \le t \le 8$).

16. Jonathon Witherslack is 300 miles from home and driving away from it at 54 miles per hour. Express his distance s from home t hours from right now.

17. Sarah Newburg and John Plank are traveling east from Hays, Kansas. Sarah started at noon and is driving at a constant speed of 55 miles per hour. John started 2 hours later and is driving at a constant speed of 70 miles per hour.
    (a) Express the distance each has traveled as a function of t, the number of hours after 12:00 noon.
    (b) At what time does John pass Sarah?

18. The R-rating of fiberglass insulation varies linearly with its thickness. A 5-inch batt of insulation has an R-rating of 16. Roger Figham's house has an uninsulated ceiling with an R-rating of 4.
    (a) If he puts in t inches of fiberglass insulation, what will the R-rating be?
    (b) To obtain an R-rating of 40, how thick will the insulation need to be?

C 19. A piece of steel railroad track was 64 feet long at 0°F and expanded to 64.114 feet at 50°F. Assume that the expansion of steel under heating is linear.
    (a) Express the length L of the track in terms of the temperature T.
    (b) Determine L when $T = 100°F$.
    (c) Find the rate of change in L with respect to T.

20. The percentage of tomato seeds that germinate depends in a linear way on the temperature. At 12°C, 40% germinate, while at 15°C, 70% germinate.
    (a) Express the percentage P that germinate in terms of the temperature T.
    (b) Find P at $T = 17°C$.
    (c) Determine the rate of change in P with respect to T.

21. The Fahrenheit and Celsuis temperature scales are related by the equation $F = \frac{9}{5}C + 32$.
    (a) Find the rate of change in F with respect to C.
    (b) Solve for C in terms of F and find the rate of change in C with respect to F.

22. An object moves in a straight line so that its velocity is $v = -88t + 100$ feet per second t seconds from now. Find the rate of change in velocity with respect to time (that is, find the acceleration).

23. Records show that egg production in Halman County is growing linearly. In 1965 it was 1,200,000 cases, and in 1975 it was 1,680,000 cases.
    (a) Write a formula for $f(n)$, the number of cases produced n years after 1965.
    (b) What is the rate of change in $f(n)$ with respect to n?

24. In a certain state 12% of the population was in college in 1935. By 1939 this number had risen to 17.4%. Assume that $P(t)$, the percentage of the population in college t years after 1935, increases linearly with t.
    (a) Express $P(t)$ in terms of t.
    (b) Find the rate of change in $P(t)$ with respect to t.

25. The total cost of manufacturing x snowsuits of a certain kind in one month was found to be a linear function of x. It cost $30,000 to produce 500 snowsuits and $82,500 to produce 2000 snowsuits.
    (a) Find the formula for $C(x)$, the total cost of manufacturing x snowsuits in 1 month.
    (b) What are the fixed costs for the month?
    (c) Find the rate of change in the total monthly costs with respect to the number of snowsuits made.
    (d) What are the labor and material costs per snowsuit?

26. At 6:00 A.M. two motorists are 600 miles apart driving in opposite directions (on the same freeway). One motorist is driving at a constant speed of 62 miles per hour and the other at a constant speed of 43 miles per hour.
    (a) Find the formula for $s(t)$, the distance in miles between the two motorists t hours after 6:00 A.M.
    (b) What is the rate of change in $s(t)$ with respect to t?

1.5

NONLINEAR FUNCTIONS: VARIABLE SLOPE

Most functions are nonlinear, and most physical processes, when examined closely, seem to be nonlinear too. For a function to be nonlinear means simply that its graph is a curve rather than a line. A typical example is shown in Figure 30. One look at this figure suggests a question. What do we mean by

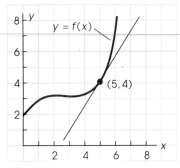

FIGURE 30

the slope of a curve? In contrast to a line, the steepness of a curve keeps changing. We can't talk about the slope of the whole curve; we will have to talk about its slope at a specific point. At such a point, it seems reasonable to define the slope of the curve to be the slope of the corresponding **tangent line.** For now, it is enough to say that the tangent line is the line that hugs the curve most closely at the point in question. We will be more precise about this concept shortly.

EXAMPLE A

Find the slope of the curve in Figure 30 at the point (5, 4).

Solution The tangent line at (5, 4) is shown in Figure 30. This line appears to go through (3, 1) and (7, 7) and so has slope $m = (7 - 1)/(7 - 3) = \frac{6}{4}$. We conclude that the curve has slope $\frac{3}{2}$ at (5, 4). ∎

The Tangent Line

But what precisely is a tangent line? Euclid's notion of a tangent as a line touching a curve at just one point is all right for circles (Figure 31) but unsatisfactory for many other curves (Figure 32). The idea of a tangent to a curve at P as the line which best approximates the curve near P is better but still too vague for our purposes.

Let P be a fixed point on a curve and let Q be a nearby movable point on the curve. Consider the line, called a **secant line,** that goes through P and Q. The tangent line at P is the limiting position (if it exists) of the secant line as Q moves toward P along the curve (Figure 33).

FIGURE 31

FIGURE 32

FIGURE 33

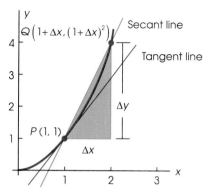

FIGURE 34

The Slope of the Curve $y = x^2$

Consider as a first concrete example of a nonlinear curve, the graph of $y = x^2$ (Figure 34). Our goal is to find the slope of this curve at the point $P(1, 1)$. The nearby point Q with x-coordinate $1 + \Delta x$ has y-coordinate $(1 + \Delta x)^2$. The secant line joining P and Q has slope

$$\frac{\Delta y}{\Delta x} = \frac{(1 + \Delta x)^2 - 1}{\Delta x} = \frac{1 + 2\Delta x + (\Delta x)^2 - 1}{\Delta x}$$

$$= \frac{\Delta x(2 + \Delta x)}{\Delta x} = 2 + \Delta x$$

Now let Δx get smaller and smaller, thereby forcing Q to approach P along the curve and also forcing the secant line to rotate toward the tangent line. Correspondingly, the quantity $\Delta y/\Delta x = 2 + \Delta x$ will get nearer and nearer to 2. This limiting value of 2 is the slope of the curve at P.

What we have just done at the point $P(1, 1)$ can be done as well at any point $P(x, x^2)$ on the curve. The neighboring point Q will have coordinates $[x + \Delta x, (x + \Delta x)^2]$, and the slope of the secant line PQ will be

$$\frac{\Delta y}{\Delta x} = \frac{(x + \Delta x)^2 - x^2}{\Delta x} = \frac{2x\,\Delta x + (\Delta x)^2}{\Delta x} = \frac{\Delta x(2x + \Delta x)}{\Delta x} = 2x + \Delta x$$

As Δx gets smaller and smaller, the quantity $\Delta y/\Delta x$ tends toward a limiting value of $2x$. In symbols, we write

$$\lim_{\Delta x \to 0} \frac{\Delta y}{\Delta x} = 2x$$

and say "the limit as Δx approaches 0 of $\Delta y/\Delta x$ is $2x$."

We have achieved something very significant, our first true calculus result. We have shown that *the slope of the curve $y = x^2$ at x has the value $2x$*. As we vary x, we vary the slope m. At $x = -1$, $m = 2(-1) = -2$; at $x = \frac{1}{2}$, $m = 2(\frac{1}{2}) = 1$; and at $x = 2$, $m = 2(2) = 4$ (Figure 35).

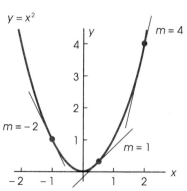

FIGURE 35

Find an equation for the tangent line to the curve $y = x^2$ at $x = -1$.

Solution The point on the curve corresponding to $x = -1$ is $(-1, 1)$, and the slope there (as we saw above) is $2(-1) = -2$. Thus, the equation of the tangent line (point-slope form) is

$$y - 1 = -2(x + 1)$$

The Slope of the Curve $y = x^n$

Our goal for $y = x^n$ is to carry out the procedure that worked so well on $y = x^2$. Almost immediately we are faced with expanding $(x + \Delta x)^n$, something we should know how to do but may well have forgotten. It is the problem of raising a binomial (a two-term expression) to a power and goes by the name—the Binomial Theorem. Rather than state this theorem formally, we offer the following outline.

$$a + b = a + b$$

$$(a + b)^2 = a^2 + 2ab + b^2$$

$$(a + b)^3 = a^3 + 3a^2b + 3ab^2 + b^3$$

$$(a + b)^4 = a^4 + 4a^3b + 6a^2b^2 + 4ab^3 + b^4$$

$$\vdots$$

$$(a + b)^n = a^n + na^{n-1}b + \frac{n(n-1)}{2}a^{n-2}b^2 + \cdots + nab^{n-1} + b^n$$

Now for $y = x^n$, the slope of the secant line from $P(x, x^n)$ to $Q[x + \Delta x, (x + \Delta x)^n]$ is

$$\frac{\Delta y}{\Delta x} = \frac{(x + \Delta x)^n - x^n}{\Delta x}$$

$$= \frac{x^n + nx^{n-1}\,\Delta x + \frac{1}{2}n(n-1)x^{n-2}\,(\Delta x)^2 + \cdots + (\Delta x)^n - x^n}{\Delta x}$$

$$= \frac{\Delta x[nx^{n-1} + \frac{1}{2}n(n-1)x^{n-2}\,\Delta x + \cdots + (\Delta x)^{n-1}]}{\Delta x}$$

We conclude that

$$\lim_{\Delta x \to 0} \frac{\Delta y}{\Delta x} = nx^{n-1}$$

that is,

> slope of $y = x^n$ at x is nx^{n-1}

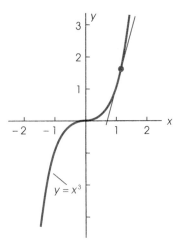

FIGURE 36

EXAMPLE C

Find the slope of the curve $y = x^3$ at the point where $x = 1.2$ (Figure 36).

Solution The general formula for slope in this case is $3x^2$. At $x = 1.2$, the value of the slope is $3(1.2)^2 = 4.32$. ∎

Slope for Arbitrary Polynomials

We need a notation for the slope of a curve $y = f(x)$ at x. Following Leibniz, one of the founders of the calculus, we use the symbol dy/dx. Thus,

$$\frac{dy}{dx} = \lim_{\Delta x \to 0} \frac{\Delta y}{\Delta x}$$

For now, think of d/dx as an operator. It operates on y and gives the slope at x, namely, dy/dx. In particular, our slope formula may be written

$$\frac{d(x^n)}{dx} = nx^{n-1}$$

The operator d/dx has two nice properties, properties that will be demonstrated in the next chapter. If a and b are constants, then

(i) $\dfrac{d(ax^n)}{dx} = a\dfrac{d(x^n)}{dx} = anx^{n-1}$

(ii) $\dfrac{d(ax^n + bx^m)}{dx} = \dfrac{d(ax^n)}{dx} + \dfrac{d(bx^m)}{dx} = anx^{n-1} + bmx^{m-1}$

These two results allow us to find the slope at x for any polynomial. For example, if

$$y = 3x^4 + 2x^3 - 3x = 3x^4 + 2x^3 + (-3)x$$

then

$$\frac{dy}{dx} = \frac{d(3x^4)}{dx} + \frac{d(2x^3)}{dx} + \frac{d(-3x)}{dx}$$

$$= 3 \cdot 4x^3 + 2 \cdot 3x^2 + (-3)1$$

$$= 12x^3 + 6x^2 - 3$$

Note one other fact. If c is a constant, then the graph of $y = c$ is a horizontal line and so $dc/dx = 0$.

EXAMPLE D

Find dy/dx if $y = 2x^{10} - 11x^3 + 3x^2 - 5x + 11$.

Solution

$$\frac{dy}{dx} = \frac{d(2x^{10})}{dx} + \frac{d(-11x^3)}{dx} + \frac{d(3x^2)}{dx} + \frac{d(-5x)}{dx} + \frac{d(11)}{dx}$$

$$= 2 \cdot 10x^9 + (-11)3x^2 + 3 \cdot 2x + (-5)1 + 0$$

$$= 20x^9 - 33x^2 + 6x - 5$$

∎

EXAMPLE E

Find an equation for the tangent line to the curve $y = 2x^4 - 3x^2 + 5x$ at the point $(1, 4)$.

Solution The slope at x is

$$\frac{dy}{dx} = 8x^3 - 6x + 5$$

and so the slope at $x = 1$ is $8 - 6 + 5 = 7$. The equation of the tangent line in point-slope form is

$$y - 4 = 7(x - 1) \qquad \blacksquare$$

EXAMPLE F

Find two points on the curve $y = x^3 - 2x$ where the slope is 1 (Figure 37).

Solution

$$\frac{dy}{dx} = 3x^2 - 2$$

We want this slope to have the value 1, that is,

$$3x^2 - 2 = 1$$
$$3x^2 = 3$$
$$x^2 = 1$$
$$x = \pm 1$$

The y-coordinate corresponding to $x = 1$ is $y = -1$ and for $x = -1$ the y-coordinate is $y = 1$. We conclude that the required two points are $(1, -1)$ and $(-1, 1)$. $\qquad \blacksquare$

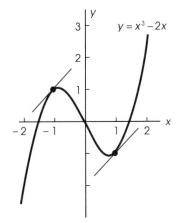

FIGURE 37

Problem Set 1.5

In Problems 1–8, you are to estimate the slope of the curve at the indicated point (see Example A). Note that you will first need to draw the tangent line in Problems 5–8.

1.

2.

3.

7.

4.

8.

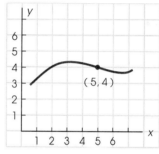

In problems 9 and 10, indicate at which of the points A, B, C, D, and E, the curve has (a) positive slope, (b) negative slope, (c) zero slope.

5.

9.

6.

10.

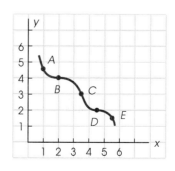

11. Sketch the graph of a function which has positive slope for $x < 2$, zero slope for $2 \le x \le 3$, and negative slope for $x > 3$.

12. Sketch the graph of a function which has negative slope for $x < -1$, positive slope for $-1 < x < 2$, and zero slope for $x \ge 2$.

13. Find an equation for the tangent line to the curve $y = x^2$ at the point where $x = 3$. See Example B.

14. Find an equation for the tangent line to the curve $y = x^2$ at the point where $x = -5$.

In Problems 15–18, find the slope of the given curve at the indicated value of x (see Example C).

15. $y = x^3$ at $x = 4$
16. $y = x^4$ at $x = -2$
17. $y = x^5$ at $x = -1$
18. $y = x^8$ at $x = 2$

As illustrated in Example D, find dy/dx in Problems 19–24.

19. $y = 5x^6 + 4x^3 - 12x$
20. $y = 3x^4 - 2x^3 + 5x - 11$
21. $y = 2x^4 - 3x^3 + 5x^2 + \pi$
22. $y = 2x^6 - 11x^2 + \pi^2$
23. $y = -5x^3 + \frac{1}{2}x^2 - \frac{3}{4}$
24. $y = \frac{3}{4}x^{12} - \frac{1}{3}x^6 + \frac{5}{7}$

In Problems 25–28, find an equation for the tangent line to the given curve at the indicated point (see Example E).

25. $y = 2x^5$ at $(-2, -64)$
26. $y = 4x - x^3$ at $(3, -15)$

27. $y = x^6 - 12x^3 + 5x - 8$ at the point where $x = 2$.
28. $y = 2x^5 + 11x^2 + 37$ at the point where $x = -2$.
29. Find two points on the curve $y = 2x^3 - 15x + 6$ where the slope is 9. See Example F.
30. Find two points on the curve $y = 2x^3 - 10x - 22$ where the slope is 44.

---

31. Consider the curve $y = 2x^3$. The points $P(x, 2x^3)$ and $Q[x + \Delta x, 2(x + \Delta x)^3]$ are both on this curve.
    (a) Find the slope $\Delta y/\Delta x$ of the secant line PQ and simplify your answer.
    (b) Find the slope dy/dx of the tangent line at P by letting Δx approach zero in your answer to part (a).

32. Follow the procedure in Problem 31 to find dy/dx for the curve $y = 3x^2 - x$.

33. Find equations of the two tangent lines to the curve $y = x^3 - 9x + 10$ that have slope 3.

34. Find an equation for the line perpendicular to the curve $y = 3x^2 - 7x + 2$ at the point $(0, 2)$. (You should be able to decide what the phrase "perpendicular to a curve" means).

35. At what values of x is the slope of the curve $y = x^3 + 2x$ positive?

36. For what values of x is the slope of the curve $y = 2x^2 + 4x - 5$ positive?

37. At what values of x is the tangent line to the curve $y = 2x^2 - 9x + 7$ parallel to the line $y = 3x + 1$?

38. Sketch the two parabolas $y = x^2$ and $y = x^2 - 2x + 1$ and then show that they intersect at right angles.

C 39. Find the slope of the curve $y = \pi x^3 - \sqrt{2}x^2 + \sqrt{3}$ at $x = 1.592$.

1.6
VARIABLE RATES OF CHANGE

For linear functions, slope and rate of change turned out to be identical concepts, as we saw in Section 1.4. The same is true for nonlinear functions, as we shall try to make clear in this section. In contrast to the linear case where slope and rate of change were constant, slope and therefore rate of change are variable for nonlinear functions.

For a first concrete example, consider the (nonlinear) relationship between the area A of a circle and its radius r, namely, $A = \pi r^2$. If we increase the radius of a small circle by a little, we increase the area by a certain amount. But if we increase the radius of a large circle by that same bit, we increase the area by a much larger amount (Figure 38). Quite clearly,

FIGURE 38

the rate of change in the area with respect to the radius is variable; this rate increases as r increases.

If r increases from r to $r + \Delta r$, then the area increases from πr^2 to $\pi (r + \Delta r)^2$. The change in area ΔA divided by the change in radius Δr is called the **average rate of change** in A with respect to r on the given interval. Thus

$$\text{Average rate of change} = \frac{\Delta A}{\Delta r} = \frac{\pi (r + \Delta r)^2 - \pi r^2}{\Delta r}$$

The **rate of change** in area with respect to radius at r is the limiting value of this average rate as Δr shrinks to 0; that is,

$$\text{rate of change at } r = \lim_{\Delta r \to 0} \frac{\Delta A}{\Delta r}$$

But this latter expression is what we called slope at r and denoted by dA/dr in the previous section. We conclude that

$$\frac{dA}{dr} = \text{slope at } r = \text{rate of change at } r$$

Since we know how to calculate dA/dr, we can calculate rates of change with very little effort as we now illustrate.

EXAMPLE A

Find for a circle the rate of change in area with respect to radius at $r = 4$ inches and at $r = 10$ inches.

Solution

$$A = \pi r^2$$

$$\frac{dA}{dr} = \pi \cdot 2r = 2\pi r$$

At $r = 4$, $dA/dr = 2\pi (4) \approx 25.13$ square inches per inch; at $r = 10$, $dA/dr = 2\pi (10) \approx 62.83$ square inches per inch.

Here is how to interpret these results. At $r = 4$, a small numerical increase in radius will produce a corresponding numerical increase in area of about 25 times that amount. However, at $r = 10$, a small numerical increase in radius produces almost 63 times that numerical increase in area. ∎

Speed and Velocity

College students don't need to be told about cars and speedometers; they have known about such things since they were children. But what does a speedometer measure?

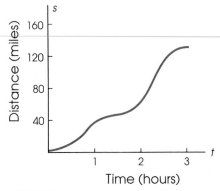

FIGURE 39

If I drove my car a distance of 135 miles in 3 hours, some would say that I drove at a speed of $\frac{135}{3} = 45$ miles per hour. That, in fact, would be my *average speed,* but this average speed has little to do with the speedometer reading. At the start of my trip, the speedometer registered 0; at times, it rose as high as 57; at the end, it fell back to 0. To put it simply, speed kept changing; it was a variable.

Speed (sometimes called instantaneous speed) is the rate of change in distance with respect to time. If I had charted s (distance I traveled) in terms of t (time elapsed), I would have obtained a graph something like Figure 39. My speed ds/dt corresponds to the slope of the curve. Note that this slope is 0 at $t = 0$ and $t = 3$ (when I started and stopped). At $t = 2$, this slope measures about 55; I was going about 55 miles per hour at that instant.

What we have said so far assumes that distance s is increasing with time t (a car has a reverse gear but normally you don't use it on a highway). If the moving object can go backward as well as forward, then ds/dt will be negative whenever distance is decreasing with time. In this situation, ds/dt is given the name **velocity,** and **speed** is defined to be $|ds/dt|$. Thus, velocity can be positive, negative, or zero, but speed is always nonnegative.

EXAMPLE B

Suppose an object moves along a line so that its directed distance s (in feet) from a fixed point is given by

$$s = -t^3 + 6t^2$$

t being measured in seconds. (a) Sketch the graph of s for $0 \le t \le 6$. (b) Find the velocity $v = ds/dt$ at $t = 1, 2,$ and 5.

Solution (a) The graph is shown in Figure 40. (b) Since $v = ds/dt = -3t^2 + 12t$,

$$v(1) = -3 + 12 = 9 \text{ feet per second}$$
$$v(2) = -12 + 24 = 12 \text{ feet per second}$$
$$v(5) = -75 + 60 = -15 \text{ feet per second}$$

Note that the object is moving in reverse at $t = 5$. ■

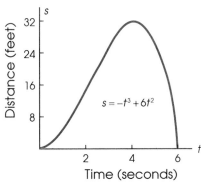

FIGURE 40

EXAMPLE C

A ball is thrown upward from an initial height of 100 feet and with an initial velocity of 64 feet per second. From physics, we learn that its height s (in feet) after t seconds is given by

$$s = 100 + 64t - 16t^2$$

(a) Find the ball's velocity at $t = 1, 2,$ and 3.
(b) When did the ball reach maximum height?
(c) Determine the maximum height.

Solution

(a) Since $v = ds/dt = 64 - 32t$,

$$v(1) = 32 \text{ feet per second}$$
$$v(2) = 0 \text{ feet per second}$$
$$v(3) = -32 \text{ feet per second}$$

(b) The ball reaches maximum height when its velocity is 0, namely, at $t = 2$.

(c) $s(2) = 100 + 64(2) - 16(2)^2 = 164$ feet   ■

Other Time Rates

Since the independent variable is time, we refer to velocity as a *time rate of change*. Time rates of change come in many forms. A chemist wants to know how fast a reaction is occurring; a biologist is interested in how fast a drug is being absorbed; a sociologist studies the rate at which a rumor is spreading. Our next example might interest a city manager.

EXAMPLE D

A small eastern city pumps water into its reservoir between 12:00 midnight and 6:00 A.M. (when electricity is cheapest and most people are asleep). During a recent day, a recording device charted the amount of water in the reservoir as shown in Figure 41. At what time were people using water most rapidly and what was the rate of usage at that time?

Solution The water level was falling most rapidly at about 8 A.M. At that time the slope of the curve is about -50, which means that water was being used at 50,000 gallons per hour.   ■

FIGURE 41

Marginal Rates in Economics

Economists are especially interested in how certain variables (revenue, cost, profit) change with respect to the number of units produced. They call these rates **marginal** rates. If in manufacturing and selling x units of its product, a company has total cost $C(x)$, total revenue $R(x)$, and total profit $P(x) = R(x) - C(x)$, then dC/dx, dR/dx, and dP/dx are called marginal cost, marginal revenue, and marginal profit, respectively.

EXAMPLE E

The ABC Company has fixed costs (office utilities, real estate taxes, and so on) of $8000 each month and must pay $48 in labor and materials for each tape recorder it builds. It expects to receive total revenue $R(x)$ from selling x recorders per month, where

$$R(x) = 75x - 0.01x^2$$

(a) Find expressions for the marginal cost, marginal revenue, and marginal profit.
(b) Evaluate the marginal profit at $x = 100$, $x = 1000$, $x = 2000$ and interpret your results.

Solution

(a) The total cost is the sum of the fixed cost $8000 and the variable cost $48x$; that is,

$$C(x) = 8000 + 48x$$

Since

$$R(x) = 75x - 0.01x^2$$

we see that

$$P(x) = R(x) - C(x) = 27x - 0.01x^2 - 8000$$

We conclude that the required rates are

$$\frac{dC}{dx} = 48$$

$$\frac{dR}{dx} = 75 - 0.02x$$

$$\frac{dP}{dx} = 27 - 0.02x$$

(b) At $x = 100$, $\dfrac{dP}{dx} = 27 - 0.02(100) = 25$

At $x = 1000$, $\dfrac{dP}{dx} = 27 - 0.02(1000) = 7$

At $x = 2000$, $\dfrac{dP}{dx} = 27 - 0.02(2000) = -13$

CAUTION
That marginal profit is only approximately equal to the increase in profit due to one unit increase in production should be apparent from the graph below.

These marginal profits have the following interpretation. Increasing production by one unit when $x = 100$ will yield (approximately) an additional profit of $25. When $x = 1000$, this additional profit will be only $7, and when $x = 2000$, there will be a reduction in profit of $13. One might ask why very high production leads to lower profit. Mathematically, it is due to the term $-0.01x^2$ in the revenue function. This term reflects the reduction in price that usually comes from flooding the market. ∎

EXAMPLE F (MARGINAL TAX RATES)

The graph in Figure 42 relates federal income tax to taxable income for a single taxpayer in 1983. Estimate the marginal tax rate for Susan Smith who had a taxable income of $44,000 and interpret the result. (See Problem 40 in Section 1.2 for the basis of this graph.)

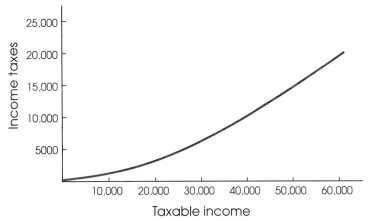

FIGURE 42

Solution The slope at 44,000 is about 0.45. Thus, Susan's marginal tax rate is 45%, which means that for an additional dollar of income she will pay an additional 45¢ in federal taxes. Note that the marginal tax rate increases with income. This is what we mean when we say that the United States had a progressive income tax in 1983. Some politicians argue for a flat income tax, meaning that the marginal tax rate should be constant. The 1986 Tax Reform Act has only two rates (15% and 28%) and thus federal income taxes are less progressive than formerly. ∎

Problem Set 1.6

1. The volume V of a sphere of radius r is given by $V = \frac{4}{3}\pi r^3$.
   (a) Find the rate of change in volume with respect to radius at $r = 2$ inches.
   (b) If the radius increased from 2 to 2.1 inches, about how much would the volume change?

2. The surface area S of a sphere of radius r is given by $S = 4\pi r^2$.
   (a) Find the rate of change in surface area with respect to radius at $r = 9$ centimeters.
   (b) If the radius increased from 9 to 9.2 centimeters, about how much would the surface area change?

3. A swimming pool is being filled in such a way that after t minutes there are $100t^3 - 25t^2 + 50$ gallons of water in the tank. At what rate (in gallons per minute) is water entering the tank at the end of 2 minutes?

4. Between midnight and noon the temperature T (in degrees Fahrenheit) rose according to the formula $T = \frac{1}{10}t^2 + 3t + 30$, where t is the time in hours after midnight. At what rate was the temperature rising at 10.00 A.M.?

In Problems 5 and 6, estimate the rate of change in y with respect to x at (a) x = 2, (b) x = 4.

5.

6.

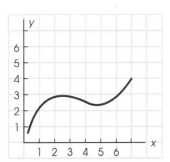

7. An object is moving along a line so that its directed distance s (in feet) from a fixed point is given by

$$s = t^3 - 3t^2 + 2$$

where t is measured in seconds.
(a) Sketch the graph of s for $0 \le t \le 4$.
(b) Find the velocity $v = ds/dt$ when $t = 0, 1, 2, 4$.
(c) When is the velocity equal to zero?

8. Suppose that a particle moves along a line in such a way that its directed distance s (in meters) from a fixed point after t seconds is given by

$$s = 12t - t^3$$

(a) Sketch the graph of s for $-3 \le t \le 3$.
(b) Find the velocity when $t = -3, -1, 0, 1, 3$.
(c) When is the velocity equal to zero?

C 9. Suppose an object moves along a line so that its directed distance s (in feet) from a fixed point is given by $s = f(t)$, t being measured in seconds. The **average velocity** of the object on the interval t to $t + \Delta t$ is

$$v_{ave} = \frac{\Delta s}{\Delta t} = \frac{\text{change in } s}{\text{change in } t} = \frac{f(t + \Delta t) - f(t)}{\Delta t}$$

Thus, for $s = t^2$, the average velocity on the interval $t = 3$ to $t = 3.5$ is

$$v_{ave} = \frac{\Delta s}{\Delta t} = \frac{(3.5)^2 - 3^2}{3.5 - 3} = \frac{3.25}{0.5} = 6.5 \text{ feet per}$$

second

Find:
(a) v_{ave} on the interval 3 to 3.1.
(b) v_{ave} on the interval 3 to 3.01.
(c) v_{ave} on the interval 3 to 3.001.
(d) $v = ds/dt$ at $t = 3$. (Do you see the point of this problem?)

C 10. Follow the directions in Problem 9 for $s = 6t - t^2$.

11. When a ball is thrown upward from ground level at 128 feet per second, its height s (in feet) after t seconds is given by $s = 128t - 16t^2$. See Example C.
(a) Find the ball's velocity when $t = 2$ and $t = 6$.
(b) At what time does the ball reach its maximum height?
(c) What is this maximum height?

12. A ball is thrown downward at 96 feet per second from a helicopter which is 3000 feet above the ground. The height s in feet of the ball after t seconds is given by $s = -16t^2 - 96t + 3000$.
(a) Find the ball's velocity when $t = 2$ and $t = 4$.
(b) At what time is the ball falling at 224 feet per second?
(c) Find the height and the velocity of the ball when $t = 10$.

13. A ball thrown upward from the ground is at height s (in feet) after t seconds, where $s = 80t - 16t^2$. With what velocity does it hit the ground? (It hits the ground when $s = 0$.)

14. A ball is thrown upward from the rim of a building 96 feet high at 80 feet per second. Its height s (in feet) after t seconds is $s = -16t^2 + 80t + 96$. With what velocity does the ball hit the ground?

15. The accompanying graph (Figure 43) gives the population $P(t)$, in thousands, of a certain city at time t, the number of years after 1950. See Example D.

(a) Calculate the average rate of increase in population from 1960 to 1970.
(b) When was the rate of increase in population greatest?
(c) Estimate this maximum rate of increase.

FIGURE 43

16. The graph in Figure 44 gives the number $f(t)$ of family farms in a certain Iowa county t years after 1970.
(a) Calculate the average rate of change in the number of farms from 1970 to 1980.
(b) At what time was the number of farms decreasing most rapidly?
(c) At what rate was the number of farms changing at that time [the time in part (b)]?

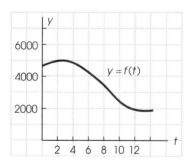

FIGURE 44

17. A small sports equipment manufacturer makes x baseball bats per day. It estimates that its total profit (in cents) from the production and sale of these bats is $P(x) = 250x - x^2$. See Example E.
(a) What is the total profit per day at a production level of 80 bats per day?
(b) Find the marginal profit when the level of production is 80 bats per day.
(c) Approximately how much additional profit would be generated by increasing production to 81 bats per day?
(d) At what production level would it become foolish to consider increasing production?

18. A manufacturer of reclining chairs finds that the cost (in dollars) of manufacturing x chairs per week is given by

$$C(x) = -0.05x^2 + 250x + 3000$$

(a) Calculate the marginal cost when the level of production is 500 chairs per week.
(b) Estimate the cost of making the 501st chair.

19. A manufacturer of toy wagons finds that the weekly cost and weekly revenue functions for manufacturing and selling x wagons are given (both in dollars) by

$$C(x) = 400 + 3.5x - \tfrac{1}{300}x^2$$

$$R(x) = 7x - \tfrac{1}{200}x^2$$

(a) Find the marginal cost, marginal revenue, and marginal profit.
(b) What is the rate of change in cost when the level of production is 500 wagons per week?
(c) How fast does the revenue change when 500 wagons per week are sold?
(d) Calculate the rate of change in profit when $x = 500$ and when $x = 1000$.

20. A shoe manufacturer finds that the weekly cost and weekly revenue functions for manufacturing and selling x pairs of shoes are given (both in dollars) by

$$C(x) = 1000 + 20x - 0.02x^2$$

$$R(x) = 50x - 0.10x^2$$

(a) Find the marginal cost, marginal revenue, and marginal profit.
(b) Estimate the cost of manufacturing the 201st pair of shoes.
(c) Find the rate of change in revenue when the level of production is 120 pairs of shoes.
(d) Find the marginal profit when 250 pairs of shoes are made (and sold) per week.
(e) At what level of production is it no longer profitable to increase production?

21. Suppose John MacDougal had a taxable income of $35,000 in 1983. According to the graph in Example F:
(a) How much income tax did he pay that year?
(b) What is the marginal tax (rate) at that income level?

22. Answer the same question as in Problem 21 for Lu Wong whose taxable income for 1983 was $48,000.

---

23. An object is moving along a line in such a way that its position (measured in centimeters) after t seconds relative to a fixed point on the line is given by

$$s = \tfrac{1}{3}t^3 - 4t + 5$$

(a) Find the velocity when $t = 1$ and $t = 4$.
(b) At what instant is the velocity 96 centimeters per second?
(c) At what instant does the object change directions?
(d) Find the average velocity of the object from $t = 3$ to $t = 6$.

24. If a ball was thrown upward from the rim of a building 160 feet high with an initial speed of 48 feet per second, then its height above the ground after t seconds is given by

$$s = -16t^2 + 48t + 160$$

(a) When did it reach maximum height?
(b) What was the maximum height?
(c) At what instant did the ball hit the ground?
(d) At what speed did the ball hit the ground?

25. Two particles moving along a line have the following positions relative to a fixed point on the line after t seconds.

$$s_1 = 4t - 3t^2 \qquad s_2 = t^2 - 2t$$

(a) When do they have the same position?
(b) When do they have the same velocity?
(c) When do they have the same speed?

26. A manufacturer of blue jeans has learned that at p dollars per pair he can sell x pairs of blue jeans per month, where $x = 4000 - 125p$. The cost of manufacturing x pairs of jeans per month (in dollars) is given by

$$C(x) = 25x + 1000$$

(a) Find the monthly revenue $R(x)$.
(b) Find the marginal cost.
(c) Find the marginal profit.
(d) Find the additional profit (or loss) that results from increasing production by one pair when the level of production is 200 pairs per month.
(e) At what level of production does it become inadvisable to increase production?

27. Suppose that the bacteria population in a certain medium t hours from now is given by

$$N(t) = 25{,}000 + 1400t - 500t^2$$

(a) What is the present bacteria count?
(b) What is the growth rate of the population 30 minutes from now?
(c) At what time does the population start to decrease?

28. The radius of a spherical cancerous tumor is growing linearly at a rate of $\frac{1}{10}$ centimeter per week.
(a) If its radius is 3 centimeters now, find an expression for its radius t weeks from now.
(b) Find a formula for the volume of the tumor t weeks from now.
(c) At what rate will the volume of the tumor be increasing 5 weeks from now?

29. Here is the meat market manager's response to an angry customer: "I didn't say the price of meat was tapering off. I said the rate of increase was tapering off." Let $P = f(t)$ denote the price of meat at time t. What had the manager said about P in the language of mathematics? What did the customer think he had said?

CHAPTER 1 REVIEW PROBLEM SET

1. Let $f(x) = x^2 - 3x + 5$ and $g(x) = \sqrt{3x - 2}$.
   (a) Find the domain of the function g.
   (b) Calculate $f(g(2))$ and $g(f(1))$.
   (c) Simplify $[f(x + h) - f(x)]/h$.
2. Let $f(x) = 4\sqrt{x - 2}$ and $g(x) = x^2 + 2$.
   (a) Determine the domain of f.
   (b) Calculate $g(f(3))$ and $f(g(2))$.
   (c) Simplify $[g(x + h) - g(x)]/h$.
3. Identify each of the following functions as to whether it is even, odd, or neither.
   (a) $f(x) = x^3 + 1$
   (b) $g(x) = \sqrt{x^2 + 1}/x$
   (c) $h(x) = |x|(2x^2 + 4)$

4. Identify each of the following functions as to whether it is even, odd, or neither.
   (a) $f(x) = \sqrt{|x| + 1}$
   (b) $g(x) = x/(x + 1)$
   (c) $h(x) = \sqrt{x^3 + x}$
5. Sketch the graph of each of the following functions.
   (a) $f(x) = -3x + 9$
   (b) $g(x) = 9x - x^3$
   (c) $h(x) = \begin{cases} 4 - x^2 & \text{if } x < 2 \\ x - 3 & \text{if } x \geq 2 \end{cases}$
6. Sketch the graph of each of the following functions.
   (a) $f(x) = 4x - 12$
   (b) $g(x) = x^3 - 4$

(c) $h(x) = \begin{cases} x^2 - 3 & \text{if } x \le 0 \\ 4 - x & \text{if } x > 0 \end{cases}$

7. Give formulas for $f(x)$, $g(x)$, and $h(x)$ if
$$h(g(f(x))) = \sqrt{(2x + 1)^2 + 3(2x + 1)}$$

8. Give formulas for $f(x)$, $g(x)$, and $h(x)$ if
$$h(g(f(x))) = 4/\sqrt{x^2 + 3x + 5}$$

9. Find the slope and y-intercept of each of the following.
(a) $3x - 2y = 12$
(b) $y - 2 = 3(x + 1)$
(c) $\dfrac{x}{3} + \dfrac{y}{4} = 1$

10. Find the slope and y-intercept of each of the following lines.
(a) $x + 2 = 4(y - 1)$
(b) $6x - 5y = 15$
(c) $2(x + y - 1) = 3(x - 2)$

11. Write in slope-intercept form the equation of the line that passes through $(-2, 1)$ and is parallel to the line $2x - 5y = 7$.

12. Write in slope-intercept form the equation of the line that passes through $(-2, 4)$ and is parallel to the line $3x - y = 8$.

13. Find the value of k for which the line $kx - 3y = 10$
(a) Passes through the point $(2, 4)$.
(b) Is parallel to the x-axis.
(c) Is perpendicular to the line $y - 2 = 2(x - 1)$.

14. Find the value of k for which the line $2x + ky = 15$
(a) is parallel to the line $x - y = 4$;
(b) passes through the point $(3/2, -4)$;
(c) is perpendicular to the line $y - 1 = (-1/2)(x + 4)$.

15. Find an equation for the tangent line to the curve
$$y = 2x^5 - 10x^3 + 3x^2 - 2$$
at the point where $x = 1$.

16. Find an equation for the tangent line to the curve $y = -2x^3 + 3x^2 - 4x + 6$ at the point where $x = 2$.

17. Frank Wheeler leased a car for $180 per month plus 15¢ per mile.
(a) Write a formula for $F(x)$, the cost in dollars for a month in which Frank drove x miles.
(b) What will the car cost Frank for a month in which he drives 6000 miles?

18. When the Environmental Protection Agency found a certain company dumping sulfuric acid into the Missouri River, it fined the company $320,000 plus $2000 per day until the company complied with the agency's pollution regulations. Write a formula $F(x)$

for the total fine in terms of x, the number of days it took for compliance.

19. Phyllis, working at a department store, makes $1800 per month plus a commission of 5% of her sales.
(a) Express her total salary $S(x)$ in terms of x, her monthly sales in dollars.
(b) What must her sales be in order for her to make $2300 per month?

20. A medical insurance company pays $1275 for a tonsillectomy and $250 per day for a semiprivate room.
(a) Write a formula $F(x)$ for the total amount the company will pay for a tonsillectomy that requires x days of hospitalization.
(b) Howard Hong, who carries this insurance, was billed $3,200 after an operation and 4 days in the hospital. How much will he have to pay out of his own funds?

21. A bus, purchased today for $90,000, will have a scrap value of $2000 after 8 years. Let $f(t)$ be the value of the bus t years from now ($0 \le t \le 8$). Write a formula for $f(t)$ assuming that it is a linear function.

22. The amount in dollars $F(x)$ that the Great Plains Gas Company pays a property owner for extracting x thousand cubic feet of natural gas from his or her land is a linear function of x. The company paid Helen $8800 for extracting 5 million cubic feet of gas from her land; it paid John $11,200 for extracting 20 million cubic feet of gas from his land.
(a) Write a formula for $F(x)$.
(b) How much does the company pay just for the *right* to extract natural gas from private property?

23. Figure 45 shows the graph of $y = f(x)$. For what values of x is
(a) $f(x) < 0$
(b) $f(x) = 1$
(c) the slope positive
(d) the slope equal to 0
(e) the slope equal to -1?

FIGURE 45

24. Sketch the graph of a function $y = f(x)$ satisfying all of the following conditions.

(a) $f(-1) = 2$, $f(3) = 4$, and $f(x) = 2$ for $x \geq 4$.

(b) The slope is 0 at $x = -1$ and at $x = 3$.

(c) The slope is negative for $x < -1$ and for $3 < x < 4$; the slope is positive for $-1 < x < 3$.

25. When a ball is thrown upward from ground level at 192 feet per second, its height s in feet after t seconds is given by $s = 192t - 16t^2$.

   (a) Determine the ball's velocity at $t = 3$ and $t = 10$.
   (b) At what time does the ball reach its maximum height?
   (c) What is the maximum height?
   (d) At what time does the ball hit the ground?
   (e) With what speed does it hit the ground?

26. When a ball was thrown upward from the top of a 512-foot building at 64 feet per second (Fig. 46), its height s (in feet) above the ground after t seconds was given by

$$s = -16t^2 + 64t + 512$$

   (a) Determine the ball's velocity at $t = 1$ and at $t = 5$.
   (b) At what time did the ball reach its greatest height?
   (c) When did the ball hit the ground?
   (d) With what speed did it hit the ground?

FIGURE 46

27. Figure 47 shows the percentage $f(t)$ of the total number of residents in a certain community that are 60 years old or older, t years after January 1, 1975.

   (a) Calculate the average rate of increase in this percentage from January 1, 1977, to January 1, 1982.
   (b) Estimate the (instantaneous) rate of change in this percentage on January 1, 1982.

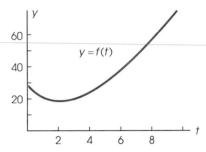

FIGURE 47

28. Mario made a 100-mile trip in his new car. Figure 48 shows s, the number of miles traveled, during the first t hours.

   (a) What was his average velocity for the whole trip?
   (b) What was his average velocity during the first 30 minutes?
   (c) When was his velocity greatest?
   (d) Estimate his maximum velocity.
   (e) Estimate his velocity at $t = 1.5$ hours.

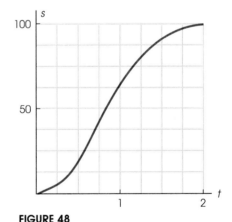

FIGURE 48

29. A manufacturer of tricycles estimates that the weekly cost and revenue functions for making and selling x tricycles are given (in dollars) by

$$C(x) = 1200 + 43x - 0.008x^2$$

$$R(x) = 55x - 0.012x^2$$

   (a) Find the marginal cost, marginal revenue, and marginal profit when 1000 tricycles are made per week.

(b) Approximately what does it cost to make the 1001st tricycle?

(c) How many tricycles should be made and sold per week for maximum profit?

30. A manufacturer of ice skates estimates that the weekly cost and revenue (in dollars) for making and selling x pairs of skates are given by

$$C(x) = 1500 + 46x - 0.01x^2$$

$$R(x) = 70x - 0.018x^2$$

(a) Find the marginal cost, marginal revenue, and marginal profit when 1200 pairs of skates are made per week.

(b) How many pairs should be made per week for maximum profit?

31. Figure 49 shows a square topped by a semicircle.

(a) Write formulas for the area $A(x)$ and perimeter $P(x)$ of the enclosed region in terms of x, the side length of the square.

(b) Find dA/dx and dP/dx when $x = 10$.

FIGURE 49

32. A rectangle twice as high as wide as topped by an equilateral triangle.

(a) Write a formula for the area $A(x)$ and the perimeter $P(x)$ of the enclosed region, given that it is x units wide.

(b) Find dA/dx and dP/dx when $x = 12$.

(c) Use your answer to part (b) to estimate how much the area would increase if x changed from 12 to 13.

FIGURE 50

That game theory is today an important branch of applied mathematics is largely due to the influence of John von Neumann. He codified the principles that underlie parlor games such as chess and checkers and showed that these principles apply to competitive situations that arise in the world of business, politics, and the military.

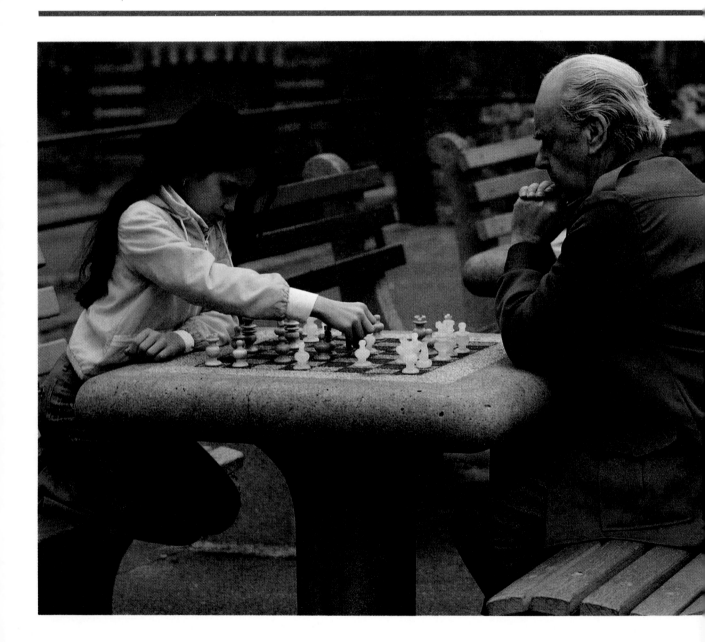

2

The Derivative

Hungary is small and land-locked but, for its size, it has produced in this century more first-rate mathematicians and physicists that any other country. Perhaps, the most brilliant of these was John von Neumann, born December 28, 1903, in Budapest. His work in pure mathematics was certainly of Nobel prize quality (unfortunately, there is no Nobel prize in mathematics), but that is not the reason for his place on our list. We honor rather his contributions to economics, physics, astronomy, biology, and computer science.

Because his genius was recognized early, he received private tutoring in mathematics and by age 17 had written his first paper with his teacher (Fekete). Before age 21, he had obtained two doctorates—one in mathematics, the other in chemistry.

After teaching at the Federal Institute of Technology in Zurich, the University of Berlin, and the University of Hamburg, he came to Princeton University in 1930. In 1933, he and Einstein were appointed two of the founding permanent members of the Institute for Advanced Study at Princeton, and there he continued working until his death.

His contributions to physics include *Mathematical Foundations of Quantum Mechanics* (1936). His pioneering work on game theory led in 1944 to the epoch-making treatise *Theory of Games and Economic Decisions* written jointly with the economist Oskar Morgenstern. Acting as a consultant to the U.S. Government, he worked at Los Alamos on the atomic bomb project, and this in turn pushed him into the development of the electronic computer. He is credited with the concept of the internally stored program. The analogy between the computer and the human brain inspired him to create a new field called automata theory.

Outside of science, his main interest was in history where he had an unbelievably detailed knowledge supported by a prodigous memory. When he was appointed to the U.S. Atomic Energy Commission in 1954, one of his colleagues is reported to have said, "Johnny will do all right, because he is probably the smartest man in the world."

John von Neumann
(1903–1957)

THE LIMIT CONCEPT

Almost all that we do in calculus rests ultimately on the concept of limit. The word *limit* was used informally in Section 1.5. It is time to make the meaning of this word more precise. Perhaps an illustration from the business world can help us think about our task.

A few years ago, a Minneapolis bank announced with much fanfare that it would no longer compound interest on savings as most banks do—quarterly or monthly or daily. It would henceforth compound interest *continuously,* that is, instantaneously. What did the bank mean by this statement?

Suppose a bank's interest rate is 100% per year (this is unrealistically high, but it makes the calculations easier without spoiling the idea). Then in 1 year, $1.00 grows to $1 + 1 = \$2.00$ if interest is compounded annually, $1.00 grows to $(1 + \frac{1}{12})^{12} = (1.08333)^{12} = \$2.61$ if interest is compounded monthly, and $1.00 grows to $(1 + \frac{1}{365})^{365} = (1.00274)^{365} = \$2.71$ if interest is compounded daily. In general, if interest is compounded at the ends of periods of length x, $1.00 will grow to

$$A(x) = (1 + x)^{1/x}$$

dollars at the end of 1 year.

In Figure 1, we have summarized a number of calculations made on a hand-held calculator. The question to be answered is, What is happening in column 3? That is, what number L are the numbers in Column 3 approaching? Be sure to note that we can't get the answer by substituting $x = 0$ in the formula $A(x) = (1 + x)^{1/x}$. Why?

In Figure 2, we have displayed the same information in graphical form. The problem is to predict where the graph will hit the vertical axis based on its behavior to the right of this axis.

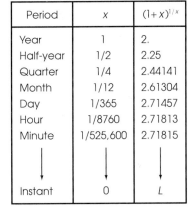

Period	x	$(1+x)^{1/x}$
Year	1	2.
Half-year	1/2	2.25
Quarter	1/4	2.44141
Month	1/12	2.61304
Day	1/365	2.71457
Hour	1/8760	2.71813
Minute	1/525,600	2.71815
↓	↓	↓
Instant	0	L

FIGURE 1

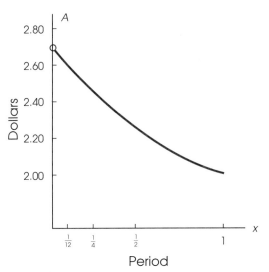

FIGURE 2

It turns out that the question we have posed has been studied in very great depth. The answer, an unending decimal traditionally denoted by the letter e, is known to thousands of decimal places. We write

$$\lim_{x \to 0} (1 + x)^{1/x} = e = 2.7182818 \ldots$$

Of course, to the nearest penny, the answer to our question is \$2.72.

The General Problem

Let c and L be fixed numbers and x a variable. Our aim is to give meaning to the statement

$$\lim_{x \to c} f(x) = L$$

which is read: *The limit as x approaches c of $f(x)$ is L*. Here is how we do it.

DEFINITION (LIMIT)
We say that $\lim_{x \to c} f(x) = L$ if $f(x)$ is arbitrarily close to L whenever x is sufficiently close to, but different from, c.

Note that we do not require anything to be true right at $x = c$. The function f does not even need a value at c; it had none in the compound interest example above, and it will have none in Example D below. All that matters is the behavior of f near $x = c$.

EXAMPLE A

Let $f(x) = 3x - 5$. Find $\lim_{x \to 2} f(x)$.

Solution The graph of $y = f(x)$ is shown in Figure 3. When x is close 2, $3x - 5$ is close to $3 \cdot 2 - 5 = 1$. Thus,

$$\lim_{x \to 2} f(x) = \lim_{x \to 2} (3x - 5) = 1 \qquad \blacksquare$$

EXAMPLE B

Let

$$g(x) = \begin{cases} 3x - 5 & \text{if } x \neq 2 \\ -2 & \text{if } x = 2 \end{cases}$$

Find $\lim_{x \to 2} g(x)$.

Solution The function g agrees with f in Example A except at $x = 2$ (Figure 4). At the point $x = 2$, the behavior of g is irrelevant. We conclude that

$$\lim_{x \to 2} g(x) = 1 \qquad \blacksquare$$

POINT OF INTEREST
Though the invention of calculus dates back to 1665–1666, its foundations lay in disarray until after 1800. A. Cauchy (1789–1857) set things right with a precise definition of limit. This Frenchman is credited with making calculus a coherent, logical subject.

FIGURE 3

FIGURE 4

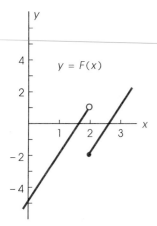

FIGURE 5

x	$\dfrac{3x^2 - 11x + 10}{x - 2}$
1	-2
1.5	$-.5$
1.9	$.7$
1.99	$.97$
↓	↓
2	?
↑	↑
2.01	1.03
2.1	1.3
2.5	1.5
3	4

FIGURE 6

RECALL
Factoring a three-term expression is usually a matter of trial and error. We must make sure that the product of the factors we obtain gives the correct middle term. Here is the check for our example

$$(3x - 5)(x - 2) = 3x^2 + (\ ) + 10$$
$$= 3x^2 - 11x + 10$$

EXAMPLE C

Let

$$F(x) = \begin{cases} 3x - 5 & \text{if } x < 2 \\ 3x - 8 & \text{if } x \geq 2 \end{cases}$$

Find $\lim_{x \to 2} F(x)$.

Solution Note the graph of F in Figure 5. As x approaches 2 from the left, $F(x)$ gets close to 1; but as x approaches 2 from the right, $F(x)$ gets close to -2. Though we have not emphasized it before, we now point out that our definition of limit requires that we get the same answer from both sides. Since in this case we do not, we conclude that $F(x)$ doesn't have a limit as $x \to 2$. However, we do have a way of describing the behavior of F. We write

$$\lim_{x \to 2^-} F(x) = 1$$

and

$$\lim_{x \to 2^+} F(x) = -2$$

The symbol $x \to 2^-$ is read "x approaches 2 from the left" while $x \to 2^+$ is read "x approaches 2 from the right." ∎

EXAMPLE D

Find $\lim_{x \to 2} \dfrac{3x^2 - 11x + 10}{x - 2}$.

Solution There is no immediately obvious way to find this limit, since both numerator and denominator are tending toward 0 as $x \to 2$. We remind you that $0/0$ is not a number; this symbol is meaningless.

An idea that would occur to a good scientist is to experiment. Try evaluating the quotient for numbers nearer and nearer to 2 and see what happens. A hand-held calculator produced the results shown in Figure 6. The evidence there suggests, but does not prove, that the limit is 1. In fact, it could be 1.01 or 0.998 for all that Figure 6 tells us.

We need a method more reliable than punching calculator keys; this method is provided by algebra—the algebra of factoring. Note that

$$3x^2 - 11x + 10 = (3x - 5)(x - 2)$$

Thus,

$$\lim_{x \to 2} \frac{3x^2 - 11x + 10}{x - 2} = \lim_{x \to 2} \frac{(3x - 5)(x - 2)}{x - 2}$$
$$= \lim_{x \to 2} (3x - 5) = 1 \qquad ∎$$

Perhaps you think we were just lucky in the example above. Not so. Whenever taking the limit as $x \to c$ of the quotient of two polynomials leads to the 0/0 form, you can be sure that both numerator and denominator have the common factor $x - c$. Here is another example.

EXAMPLE E

Find $\lim\limits_{x \to 3} \dfrac{x^2 - x - 6}{x^2 - 9}$.

Solution

$$\lim_{x \to 3} \frac{x^2 - x - 6}{x^2 - 9} = \lim_{x \to 3} \frac{(x - 3)(x + 2)}{(x - 3)(x + 3)}$$

$$= \lim_{x \to 3} \frac{x + 2}{x + 3}$$

Now the numerator tends toward 5 and the denominator toward 6. It seems clear that the quotient should approach $\frac{5}{6}$. That it does depends on one of the fundamental properties of limits, a matter we take up next. ∎

Properties of Limits

We need to know how sums, products, and quotients behave under the limit operations. For example, if $f(x) \to L$ and $g(x) \to M$ as $x \to c$, it seems reasonable that $f(x) + g(x) \to L + M$. This and the other properties of limits stated below can be proved and are proved in more advanced courses.

MAIN LIMIT THEOREM

Let n be a positive integer, k be a constant, and f and g be functions which have limits at c. Then

1. $\lim_{x \to c} k = k$
2. $\lim_{x \to c} x = c$
3. $\lim_{x \to c} k f(x) = k \lim_{x \to c} f(x)$
4. $\lim_{x \to c} [f(x) \pm g(x)] = \lim_{x \to c} f(x) \pm \lim_{x \to c} g(x)$
5. $\lim_{x \to c} [f(x) \cdot g(x)] = \lim_{x \to c} f(x) \cdot \lim_{x \to c} g(x)$
6. $\lim_{x \to c} \dfrac{f(x)}{g(x)} = \dfrac{\lim_{x \to c} f(x)}{\lim_{x \to c} g(x)}$ *provided* $\lim_{x \to c} g(x) \neq 0$
7. $\lim_{x \to c} [f(x)]^n = [\lim_{x \to c} f(x)]^n$
8. $\lim_{x \to c} \sqrt[n]{f(x)} = \sqrt[n]{\lim_{x \to c} f(x)}$ *provided* $\lim_{x \to c} f(x) > 0$ when n is even

---

Note that Property 4 states two results, one using the plus sign, the other using the minus sign.

We give two further examples to illustrate the use of the Main Limit Theorem. The numbers in circles refer to the corresponding parts of the theorem.

EXAMPLE F

Find $\lim_{x \to -2} (3x^4 - 2x)$, justifying each step by appealing to one of the properties of the Main Limit Theorem.

Solution

$$\lim_{x \to -2} (3x^4 - 2x) \overset{④}{=} \lim_{x \to -2} 3x^4 - \lim_{x \to -2} 2x$$

$$\overset{③}{=} 3 \lim_{x \to -2} x^4 - 2 \lim_{x \to -2} x$$

$$\overset{⑦}{=} 3\left(\lim_{x \to -2} x \right)^4 - 2 \lim_{x \to -2} x$$

$$\overset{②}{=} 3(-2)^4 - 2(-2) = 52 \qquad \blacksquare$$

EXAMPLE G

If $\lim_{x \to 3} f(x) = 2$ and $\lim_{x \to 3} g(x) = -4$, find

$$\lim_{x \to 3} \frac{2f(x) + 5}{3g(x)}$$

Solution

$$\lim_{x \to 3} \frac{2f(x) + 5}{3g(x)} \overset{⑥}{=} \frac{\lim_{x \to 3} [2f(x) + 5]}{\lim_{x \to 3} [3g(x)]}$$

$$\overset{④}{\underset{③}{=}} \frac{\lim_{x \to 3} 2f(x) + \lim_{x \to 3} 5}{3 \lim_{x \to 3} g(x)}$$

$$\overset{3,1}{=} \frac{2 \lim_{x \to 3} f(x) + 5}{3(-4)} = \frac{2 \cdot 2 + 5}{-12} = -\frac{3}{4} \qquad \blacksquare$$

Problem Set 2.1

In Problems 1 and 2, find each limit.

1. (a) $\lim_{x \to 3} (2x + 5)$ (b) $\lim_{x \to -4} (-3x + 10)$

   (c) $\lim_{x \to 2} (3x^2 - 2x + 1)$ (d) $\lim_{x \to -3} (x^2 - x + 5)$

2. (a) $\lim_{x \to -2} (-x + 9)$ (b) $\lim_{x \to 3} (7x - 8)$

   (c) $\lim_{x \to -1} (x^2 - 3x + 9)$ (d) $\lim_{x \to 4} (2x^2 - 2x + 1)$

Determine whether $\lim_{x \to 3} f(x)$ exists in Problems 3–8 and, if so, give its value.

3.

4.

5.

6.

7.

8.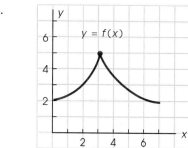

9. Let $f(x) = \begin{cases} 4x - 3 & \text{if } x \neq 3 \\ 5 & \text{if } x = 3 \end{cases}$

   (a) Sketch the graph of f.
   (b) Find $\lim_{x \to 3} f(x)$.

10. Let $g(x) = \begin{cases} -2x + 5 & \text{if } x \neq 4 \\ 2 & \text{if } x = 4 \end{cases}$

    (a) Sketch the graph of g.
    (b) Find $\lim_{x \to 4} g(x)$.

11. Sketch the graph of $F(x) = \begin{cases} x + 2 & \text{if } x \leq 1 \\ 2x & \text{if } x > 1 \end{cases}$.

    Then find (if possible) each of the following.
    (a) $F(1)$ (b) $\lim_{x \to 1} F(x)$
    (c) $\lim_{x \to 2} F(x)$ (d) $\lim_{x \to 1^+} F(x)$
    (e) $\lim_{x \to 1^-} F(x)$ (f) $\lim_{x \to -1} F(x)$

12. Sketch the graph of $G(x) = \begin{cases} 3x - 1 & \text{if } x < 2 \\ 2x - 2 & \text{if } x \geq 2 \end{cases}$.

Then find (if possible) each of the following.

(a) $G(2)$

(b) $\lim\limits_{x \to 1} G(x)$

(c) $\lim\limits_{x \to 3} G(x)$

(d) $\lim\limits_{x \to 2} G(x)$

(e) $\lim\limits_{x \to 2^-} G(x)$

(f) $\lim\limits_{x \to 2^+} G(x)$

In Problems 13–28, evaluate each limit (see Example D and E).

13. $\lim\limits_{x \to 4} \dfrac{x^2 - 4x}{x - 4}$

14. $\lim\limits_{x \to -3} \dfrac{-3x^2 - 9x}{x + 3}$

15. $\lim\limits_{x \to -3} \dfrac{x^2 + x - 6}{x + 3}$

16. $\lim\limits_{x \to 6} \dfrac{x^2 - 4x - 12}{x - 6}$

17. $\lim\limits_{x \to 3} \dfrac{3x^2 - 7x - 6}{x - 3}$

18. $\lim\limits_{x \to 5} \dfrac{2x^2 - 7x - 15}{x - 5}$

19. $\lim\limits_{x \to -2} \dfrac{x^2 - 4}{x^2 + 2x}$

20. $\lim\limits_{x \to 3} \dfrac{3x^2 - 9x}{x^2 - 9}$

21. $\lim\limits_{x \to 5} \dfrac{x^2 - 5x}{x^2 - 2x - 15}$

22. $\lim\limits_{x \to -3} \dfrac{x^2 - 9}{x^2 + 2x - 3}$

23. $\lim\limits_{x \to 4} \dfrac{x^2 + x - 12}{x^2 - 16}$

24. $\lim\limits_{x \to 0} \dfrac{x^3 - x^2}{x^2 + 2x}$

25. $\lim\limits_{x \to 0} \dfrac{x^4 - x^2}{3x^3 + 7x^2}$

26. $\lim\limits_{x \to 3} \dfrac{x^2 - 6x + 9}{2x^2 - 5x - 3}$

27. $\lim\limits_{x \to 2} \dfrac{(x - 1)(x^2 + 2x + 4)}{x^2 + 5x - 6}$

28. $\lim\limits_{x \to 1} \dfrac{(x - 1)(x^2 + 2x + 4)}{x^2 + 5x - 6}$

29. Find $\lim\limits_{x \to 2} (5x^3 - 4x^2 + 3x + 5)$ justifying each step by appealing to one of the properties in the Main Limit Theorem. See Example F.

30. Follow the instructions in Problem 29 for $\lim\limits_{x \to 3} (2x^4 - 7x^2 + 9x - 1)$.

Use the Main Limit Theorem to find the limits in Problems 31–42.

31. $\lim\limits_{x \to -2} (x^3 + 7)(x^2 + x - 1)$

32. $\lim\limits_{x \to 3} (2x^3 - 50)(2x^2 - 3x + 1)$

33. $\lim\limits_{x \to 5} (x^2 + 2x + 5)/(x + 3)$

34. $\lim\limits_{x \to -4} (2x^2 + 7x + 4)/(3x + 4)$

35. $\lim\limits_{x \to 3} (x^3 - 8x - 1)^6$

36. $\lim\limits_{x \to 6} (2x^2 - 10x - 9)^4$

37. $\lim\limits_{x \to 3} \sqrt[3]{x^4 - 10x^2 + 1}$

38. $\lim\limits_{x \to -4} \sqrt[3]{x^3 + 2x^2 + 5}$

39. $\lim\limits_{x \to 2} (\sqrt{x^2 + 5} + 2\sqrt[3]{2x^3 + 11})$

40. $\lim\limits_{x \to 3} (2x - 5)^4(x - \sqrt{x^2 - 5})$

41. $\lim\limits_{x \to 1} \dfrac{(x - 2)^3(x + 3)(x + 1)}{(x + 5)(x^2 + 2x - 2)}$

42. $\lim\limits_{x \to 2} \dfrac{(x^2 + x + 1)(2x - 1)(x + 3)}{(3x - 5)^4(x^2 + 1)}$

Find the limits in Problems 43 and 44, justifying each step by appealing to one of the properties in the Main Limit Theorem (see Example G).

43. If $\lim\limits_{x \to 2} f(x) = -3$ and $\lim\limits_{x \to 2} g(x) = 4$, find

$$\lim\limits_{x \to 2} \dfrac{3f(x) + 2x - 1}{2g(x)}$$

44. If $\lim\limits_{x \to 5} f(x) = 6$ and $\lim\limits_{x \to 5} g(x) = -2$, find

$$\lim\limits_{x \to 5} \dfrac{f(x) \cdot g(x) + 30}{f(x) - 2x + 3}$$

45. If $\lim\limits_{x \to 2} f(x) = 5$ and $\lim\limits_{x \to 2} g(x) = -4$ find each of the following limits.

(a) $\lim\limits_{x \to 2} [f(x) - x - 1]^4$

(b) $\lim\limits_{x \to 2} \left(\dfrac{f(x) - 2x}{g(x) + 6}\right)^4$

(c) $\lim\limits_{x \to 2} \sqrt[3]{7 - 5g(x)}$

(d) $\lim\limits_{x \to 2} [x^3 + x^2 + x + 1 - 2f(x)]^3$

46. If $\lim\limits_{x \to 3} f(x) = -4$ and $\lim\limits_{x \to 3} g(x) = 3$, find each of the following limits.

(a) $\lim\limits_{x \to 3} [g(x) - 2x + 1]^5$

(b) $\lim\limits_{x \to 3} \left[\dfrac{f(x) + x^2}{3g(x) + 1}\right]^3$

(c) $\lim\limits_{x \to 3} \sqrt{x^2 + 11 - 4f(x)}$

(d) $\lim\limits_{x \to 3} [4g(x) - x^2 - x + 2]^6$

In each of Problems 47–58 find the required limit or state that it doesn't exist.

47. $\lim\limits_{x \to 3} (2x - 4 + \sqrt{3x - 5})$

48. $\lim\limits_{x \to -4} (2x + 10)\sqrt{5 - 5x}$

49. $\lim\limits_{x \to 2} \dfrac{2x - 5}{(3x^2 - 4x + 3)^2}$

50. $\lim\limits_{x \to 5} \dfrac{(x - 5)^2}{x^2 + 4}$

51. $\lim\limits_{x \to 2} \dfrac{2x^3 - 16}{3x^2 - 12}$

52. $\lim\limits_{x \to -1} \dfrac{4x^3 + 4}{x^2 - 4x - 5}$

53. $\lim\limits_{x \to 3} \dfrac{2x - 5}{x^2 - 9}$

54. $\lim\limits_{x \to 4} \dfrac{3x + 1}{2x^2 - 32}$

55. $\lim\limits_{x \to 0} \dfrac{|x|}{x}$

56. $\lim\limits_{x \to 2^+} \dfrac{|x - 2|}{x - 2}$

57. $\lim\limits_{x \to 1} F(x)$, where $F(x) = \begin{cases} 2x + 1 & \text{if } x < 1 \\ 3x & \text{if } x \geq 1 \end{cases}$

58. $\lim\limits_{x \to 2} G(x)$, where $G(x) = \begin{cases} x^2 & \text{if } x \leq 2 \\ x + 4 & \text{if } x > 2 \end{cases}$

$\boxed{C}$ 59. By using successively $x = 0.1, 0.01$, and 0.001, estimate the following limits.
  (a) $\lim\limits_{x \to 0^+} x^x$
  (b) $\lim\limits_{x \to 0^+} \{[\ln (2 + x) - \ln 2]/x\}$

$\boxed{C}$ 60. Follow the instructions in Problem 59 for
  (a) $\lim\limits_{x \to 0^+} [\ln (x + 1)/(x^2 + 2x)]$
  (b) $\lim\limits_{x \to 0} [(\sqrt{2 + x} - \sqrt{2})/x]$

61. Suppose $\lim\limits_{x \to a} f(x) = 4$. Calculate each of the following or state that it doesn't exist.
  (a) $\lim\limits_{x \to a} [f(x) - 4]$
  (b) $\lim\limits_{x \to a} [f(x)/(x - a)]$
  (c) $\lim\limits_{x \to a} [\sqrt{f(x)} + f(x)]^2$
  (d) $\lim\limits_{x \to a} \{[f(x) - 4]/[f^2(x) - 16]\}$

2.2
THE DERIVATIVE

We have seen (Sections 1.5 and 1.6) that the symbol dy/dx can represent many different physical quantities (slope of a curve, rate of change, velocity, marginal profit, and so on). Good sense suggests that we study the underlying concept independently of these specialized vocabularies and diverse applications. Mathematicians have chosen the neutral name *derivative*.

DEFINITION (DERIVATIVE)

Let $y = f(x)$. Then dy/dx, called the **derivative** of y with respect to x, is given by

$$\frac{dy}{dx} = \lim_{\Delta x \to 0} \frac{f(x + \Delta x) - f(x)}{\Delta x} = \lim_{h \to 0} \frac{f(x + h) - f(x)}{h}$$

provided this limit exists.

We have used the definition above to introduce another bit of mathematical sleight of hand. Until now we have consistently used the symbol Δx to represent a small change in x. This double symbol will become increasingly cumbersome as our calculations get more complicated, and we shall often replace it by the single letter h. There's nothing mysterious here. It is analogous to calling Abraham Lincoln simply Abe.

While the symbol dy/dx is the one favored by most scientists, mathematicians are more likely to use another symbol that emphasizes the functional character of the derivative, namely, $f'(x)$. Thus, dy/dx and $f'(x)$ are symbols for the same thing. We will use both interchangeably.

We begin with a simple example that shows how the definition above is used to find derivatives.

EXAMPLE A

Use the definition of derivative to find $f'(x)$ if $f(x) = x^2 - 3x$.

Solution You know from the previous chapter that the answer has to be $2x - 3$. Here is the required demonstration.

$$f'(x) = \lim_{h \to 0} \frac{f(x + h) - f(x)}{h}$$

$$= \lim_{h \to 0} \frac{(x + h)^2 - 3(x + h) - (x^2 - 3x)}{h}$$

$$= \lim_{h \to 0} \frac{x^2 + 2xh + h^2 - 3x - 3h - x^2 + 3x}{h}$$

$$= \lim_{h \to 0} \frac{2xh + h^2 - 3h}{h}$$

$$= \lim_{h \to 0} \frac{h(2x + h - 3)}{h}$$

$$= \lim_{h \to 0} (2x + h - 3)$$

$$= 2x - 3$$

Two Key Examples

We use two examples to lead up to something very important.

EXAMPLE B

Find $f'(x)$ if $f(x) = 1/x$.

Solution

$$f'(x) = \lim_{h \to 0} \frac{f(x + h) - f(x)}{h} = \lim_{h \to 0} \frac{\dfrac{1}{x + h} - \dfrac{1}{x}}{h}$$

$$= \lim_{h \to 0} \frac{\dfrac{x - (x + h)}{(x + h)x}}{h} = \lim_{h \to 0} \frac{-h}{(x + h)x} \cdot \frac{1}{h}$$

$$= \lim_{h \to 0} \frac{-1}{(x + h)x}$$

All we have done so far is some rather straightforward algebra. Now we are ready to take the limit. Keep in mind that x is kept fixed; it is h that tends to 0. Using the properties of limits from Section 2.1, we obtain

$$f'(x) = \frac{\lim_{h \to 0} (-1)}{\lim_{h \to 0} (x + h) \cdot \lim_{h \to 0} x} = \frac{-1}{x \cdot x} = \frac{-1}{x^2}$$ ∎

EXAMPLE C

Find $g'(x)$ if $g(x) = \sqrt{x}$.

Solution

$$g'(x) = \lim_{h \to 0} \frac{g(x + h) - g(x)}{h} = \lim_{h \to 0} \frac{\sqrt{x + h} - \sqrt{x}}{h}$$

Here we face a problem in that this limit takes the nonsensical form 0/0, since both numerator and denominator tend to 0. This is typical; it also happened in Example A, though there we finessed the problem with some simple algebra that allowed us to cancel h between the numerator and denominator. Our goal is the same here, but the required algebra is fancier. We rationalize the numerator—a result that is achieved when we multiply numerator and denominator by $\sqrt{x + h} + \sqrt{x}$. See what happens.

$$g'(x) = \lim_{h \to 0} \frac{\sqrt{x + h} - \sqrt{x}}{h} \cdot \frac{\sqrt{x + h} + \sqrt{x}}{\sqrt{x + h} + \sqrt{x}}$$

$$= \lim_{h \to 0} \frac{x + h - x}{h(\sqrt{x + h} + \sqrt{x})} = \lim_{h \to 0} \frac{\not{h}}{\not{h}(\sqrt{x + h} + \sqrt{x})}$$

$$= \lim_{h \to 0} \frac{1}{\sqrt{x + h} + \sqrt{x}} = \frac{1}{\sqrt{x} + \sqrt{x}} = \frac{1}{2\sqrt{x}}$$ ∎

> **RECALL**
>
> $(a - b)(a + b) = a^2 - b^2$
>
> $(\sqrt{x + h} - \sqrt{x})(\sqrt{x + h} + \sqrt{x})$
>
> $= x + h - x$

If we rewrite the results just obtained using exponential notation (a subject we will review at the end of this section), we can observe something significant.

> **RECALL**
>
> $\dfrac{1}{x} = x^{-1}$
>
> $\sqrt{x} = x^{1/2}$

(i): $\dfrac{d(x^{-1})}{dx} = -1 \cdot x^{-2}$

(ii): $\dfrac{d(x^{1/2})}{dx} = \dfrac{1}{2} \cdot x^{-1/2}$

Both (i) and (ii) have the form

$$\frac{d(x^n)}{dx} = nx^{n-1}$$

a result familiar from Section 1.5 but demonstrated there only for n a positive integer. We are onto something truly marvelous.

The Power Rule Revisited

We are going to use the symbol r for any real number. It might be -1 or $\frac{1}{2}$ or even π or e.

POWER RULE

Let $y = f(x) = x^r$, where r is any real number. Then $f'(x) = rx^{r-1}$; that is,

$$\frac{d(x^r)}{dx} = rx^{r-1}$$

We emphasize again that r can be any real number. Thus,

$$\frac{d(x^\pi)}{dx} = \pi x^{\pi-1}$$

and

$$\frac{d(x^{2.0134})}{dx} = 2.0134 x^{1.0134}$$

Unfortunately we can't prove the Power Rule in the generality stated (see Purcell/Varberg: *Calculus with Analytic Geometry*, 5th Ed., Section 7.4 for this proof); we haven't developed enough tools. However, we will give more circumstantial evidence in Problems 29 and 30.

EXAMPLE D

If $f(x) = x^2\sqrt{x}$, find $f'(x)$.

Solution

$$f(x) = x^2 x^{1/2} = x^{2+1/2} = x^{5/2}$$

and so

$$f'(x) = \tfrac{5}{2}x^{3/2} \qquad\blacksquare$$

Note two special cases of the Power Rule that we will use constantly. Since $x^0 = 1$ and $x^1 = x$,

---

$$\frac{d(1)}{dx} = 0 \quad \text{and} \quad \frac{d(x)}{dx} = 1$$

Next we ask about constant multiples of powers. How do we find their derivatives? For example, what is the derivative of $\pi x^{5.4}$? The Constant Multiple Rule gives us the answer.

CONSTANT MULTIPLE RULE

Suppose that $f(x)$ has a derivative and that k is a constant. Then

$$\frac{d[k\,f(x)]}{dx} = k\,\frac{d[\,f(x)]}{dx}$$

Proof Let $F(x) = k\,f(x)$. Then

$$F'(x) = \lim_{h \to 0} \frac{F(x+h) - F(x)}{h} = \lim_{h \to 0} \frac{k\,f(x+h) - k\,f(x)}{h}$$

$$= \lim_{h \to 0} k\,\frac{f(x+h) - f(x)}{h} = k\,\lim_{h \to 0} \frac{f(x+h) - f(x)}{h}$$

$$= k\,f'(x)$$

The key step was moving k past the limit sign; this is legal because of part 3 of the Main Limit Theorem.   ■

EXAMPLE E

Find the derivatives of each of the following. (a) k, (b) $3x^{4/3}$, (c) $\pi x^{5.4}$.

Solution

(a) $\dfrac{d(k)}{dx} = \dfrac{d(k \cdot 1)}{dx} = k\dfrac{d(1)}{dx} = k \cdot 0 = 0$

(b) $\dfrac{d(3x^{4/3})}{dx} = 3\dfrac{d(x^{4/3})}{dx} = 3 \cdot \dfrac{4}{3}x^{1/3} = 4x^{1/3}$

(c) $\dfrac{d(\pi x^{5.4})}{dx} = \pi\dfrac{d(x^{5.4})}{dx} = \pi(5.4)x^{4.4} = 5.4\pi x^{4.4}$ ■

A Review of Exponents

We assume the reader is familiar with the basic definitions for exponents. In particular, if m and n are positive integers,

$$x^m = \underbrace{x \cdot x \cdot \ldots \cdot x}_{m \text{ factors}} \qquad x^{-m} = \frac{1}{x^m}$$

$$x^0 = 1 \qquad x^{m/n} = (\sqrt[n]{x})^m$$

For example,

$$5^{-2} = \frac{1}{5^2} = \frac{1}{25} \qquad 27^{2/3} = (\sqrt[3]{27})^2 = 3^2 = 9$$

The behavior of exponents is governed by five fundamental laws.

LAWS OF EXPONENTS

Let r and s be any real numbers. Then

1. $x^r x^s = x^{r+s}$ 2. $\dfrac{x^r}{x^s} = x^{r-s}$

3. $(x^r)^s = x^{rs}$ 4. $(xy)^r = x^r y^r$

5. $\left(\dfrac{x}{y}\right)^r = \dfrac{x^r}{y^r}$

EXAMPLE F

Simplify

(a) $(27x^6)^{1/3}$

(b) $\dfrac{\sqrt[3]{x}\,x^2}{x^{3/4}}$

(c) $\dfrac{x^{2/3}(8x)^{-2/3}}{x^{-2}}$

(d) $(x^{1/2} + x^{3/2})^2$

Solution

(a) $(27x^6)^{1/3} = (27)^{1/3}(x^6)^{1/3} = 3x^2$

(b) $\dfrac{\sqrt[3]{x}\,x^2}{x^{3/4}} = \dfrac{x^{1/3}x^2}{x^{3/4}} = x^{1/3+2-3/4} = x^{4/12+24/12-9/12} = x^{19/12}$

(c) $\dfrac{x^{2/3}(8x)^{-2/3}}{x^{-2}} = \dfrac{x^{2/3}8^{-2/3}x^{-2/3}}{x^{-2}} = \dfrac{x^0 8^{-2/3}}{x^{-2}}$

$\qquad\qquad = \dfrac{x^2}{8^{2/3}} = \dfrac{x^2}{(\sqrt[3]{8})^2} = \dfrac{x^2}{4}$

(d) $(x^{1/2} + x^{3/2})^2 = (x^{1/2})^2 + 2x^{1/2}x^{3/2} + (x^{3/2})^2$

$\qquad\qquad\qquad = x + 2x^2 + x^3$

Problem Set 2.2

In Problems 1–10, find $f'(x)$ by calculating $\lim_{h\to 0}$ $[f(x + h) - f(x)]/h$.

1. $f(x) = x^2 + 2x$
2. $f(x) = 3x^2 + 1$
3. $f(x) = x^3$
4. $f(x) = x^2 - x + 5$
5. $f(x) = 4/x$
6. $f(x) = 1/(x + 1)$
7. $f(x) = x/(x + 1)$
8. $f(x) = (2x + 1)/x$
9. $f(x) = \sqrt{x} + 2$
10. $f(x) = \sqrt{x} - 1$

In each of Problems 11–14, first write $f(x)$ as a single power of x and then find $f'(x)$ using the Power Rule (see Example D).

11. $f(x) = x\sqrt[3]{x}$
12. $x^2/\sqrt{x}$
13. $f(x) = x^3/x^{1.1}$
14. $f(x) = x^3 x^{1.5} x^{-2}$

Use the Power Rule and the Constant Multiple Rule to find $f'(x)$ in Problems 15–22 (see Example E).

15. $f(x) = 6\sqrt[3]{x}$

16. $f(x) = 6\sqrt{x}$

17. $f(x) = -2x^{5/4}$

18. $f(x) = \sqrt{2}x^{\sqrt{2}}$

19. $f(x) = 3x^{1.09}$

20. $f(x) = (2/\pi)x^{2\pi}$

21. $f(x) = -11x^{\sqrt{3}}$

22. $f(x) = 4x^{2.11}$

23. Simplify each of the following expressions.
    - (a) $(16x^8)^{1/4}$
    - (b) $x^2\sqrt{x^3}$
    - (c) $x^2/\sqrt{x^3}$
    - (d) $\sqrt[4]{x}\,x^3/x^{3/2}$
    - (e) $x^{3/4}(16x)^{-3/4}/x^{-3}$
    - (f) $(x^{3/2} - x^2)^2$

24. Simplify each of the following expressions.
    - (a) $(x^4/9)^{3/2}$
    - (b) $x^2\sqrt[4]{x}$
    - (c) $x^3/\sqrt{x}$
    - (d) $\sqrt{x}\,x^2/x^{2/3}$
    - (e) $x^{3/2}(9x)^{-3/2}/x^{-2}$
    - (f) $(x^{1/3} + x^{5/3})^2$

25. First simplify; then find dy/dx.
    - (a) $y = (9x^8)^{1/2}$
    - (b) $y = x^3\sqrt{4x^3}$
    - (c) $y = 2x^4/\sqrt{x^5}$
    - (d) $y = \sqrt[4]{x}\,x^5x^{-5/4}$

26. First simplify; then find dy/dx.
    - (a) $y = (x^6/16)^{3/2}$
    - (b) $y = x^5\sqrt[4]{16x}$
    - (c) $y = x^3/2\sqrt{x}$
    - (d) $y = \sqrt{9x}\,x^2/x^{3/2}$

---

27. Find dy/dx for each of the following using the Constant Multiple Rule and the Power Rule.
    - (a) $y = 14$
    - (b) $y = 4x^7$
    - (c) $y = 10x^{3.1}$
    - (d) $y = 12x^{7/6}$
    - (e) $y = -3x^{2\pi+1}$
    - (f) $y = \pi^{3.1}$

28. Find $f'(8)$ for each of the following functions.
    - (a) $f(x) = 2x^{4/3}$
    - (b) $f(x) = 3/\sqrt[3]{x}$
    - (c) $f(x) = 2^{4/3}$
    - (d) $f(x) = x^{2/3}x^{5/3}x^{-3}$

29. Use the definition of the derivative $f'(x) = \lim_{h\to 0}$ $[f(x + h) - f(x)]/h$ to find the derivative of $f(x) = x^{-2} = 1/x^2$.

30. Use the definition of the derivative to find $f'(x)$ if $f(x) = x^{1/3} = \sqrt[3]{x}$. *Hint:* Study Example C and recall that $(a - b)(a^2 + ab + b^2) = a^3 - b^3$.

31. The volume of a sphere is $\pi/6$ times the cube of the diameter. Find the rate of change in volume with respect to the diameter when the diameter is 10 centimeters.

32. Suppose that the size of a certain bacteria culture at time t (in minutes) is given by $N(t) = 3t^2\sqrt{t}$ (in milligrams). Find the rate of growth at $t = 1$ and at $t = 9$.

33. According to the Inverse Square Law for electric charges, the force of repulsion between charges of sizes q_1 and q_2 statcoulombs is $F = q_1q_2/r^2$ dynes when two charges are r centimeters apart. If $q_1 = 5$ and $q_2 = 80$, find the rate of change in force with respect to distance where $r = 10$.

34. It is estimated that a normal person can memorize $N = 30\sqrt{t}$ lines of a certain poem in t hours of concentrated study ($0 \le t \le 9$). What is the rate of learning when $t = 4$?

35. At what point on the graph of $y = 3x^{4/3}$ is the slope equal to 12?

36. Find the equation of the tangent line to $y = 4x^{3/2}$ at the point where $x = \frac{1}{4}$.

C 37. For common functions, $[f(4.001) - f(4)]/0.001$ and $f'(4)$ should be nearly equal. Why? Calculate both of these quantities for
    - (a) $f(x) = \sqrt{x}$
    - (b) $f(x) = 2/\sqrt{x}$

C 38. While common sense suggests that the electoral college method of electing presidents gives more power per voter to small states, an analysis of the swing power of the large states has led some political scientists to the opposite conclusion. They propose the "$\frac{3}{2}$ rule" according to which the effective voting strength of a state in a presidential election is proportional to its population to the $\frac{3}{2}$ power. Based on this model, the amount A a candidate spends in a state should vary with the population P according to the formula

$$A = kP^{3/2} \quad (k \text{ a constant})$$

- (a) Alfonso Hornblower plans to spend $1 million in his home state with a population of 4.5 million. Use this information to determine k.
- (b) How much should Hornblower plan to spend in New York (17.5 million people)?
- (c) Find the rate of change in spending with respect to population when $P = 4.5$ million.

2.3
RULES FOR DIFFERENTIATION

Our title introduces a new word—differentiation. The process of finding a derivative is called **differentiation.** If a function has a derivative, it is said to be **differentiable.** And the part of the calculus that deals with derivatives is called **differential calculus.**

We have already given two rules for differentiation—the Power Rule and the Constant Multiple Rule. This section develops several other rules that will greatly simplify our work.

The Sum Rule

How can we find the derivative of $6x^4 + 12x^{1/2} - 2x^{-3}$? The Sum Rule will help. Some would say it should be called the Sum and Difference Rule, but that name is too long.

SUM RULE

Let $u = u(x)$ and $v = v(x)$ be differentiable functions and let $F(x) = u(x) \pm v(x)$. Then $F'(x) = u'(x) \pm v'(x)$; that is,

$$\frac{d(u \pm y)}{dx} = \frac{du}{dx} \pm \frac{dv}{dx}$$

Proof   We give the proof for a sum; the case for a difference is completely similar.

$$F'(x) = \lim_{h \to 0} \frac{F(x + h) - F(x)}{h}$$

$$= \lim_{h \to 0} \frac{u(x + h) + v(x + h) - [u(x) + v(x)]}{h}$$

$$= \lim_{h \to 0} \left[\frac{u(x + h) - u(x)}{h} + \frac{v(x + h) - v(x)}{h} \right]$$

$$= \lim_{h \to 0} \frac{u(x + h) - u(x)}{h} + \lim_{h \to 0} \frac{v(x + h) - v(x)}{h}$$

$$= u'(x) + v'(x)$$

The key step was the next to last one; it is a consequence of part 4 of the Main Limit Theorem.   ■

EXAMPLE A

If $F(x) = 6x^4 + 12x^{1/2} - 2x^{-3}$, find $F'(x)$.

Solution

$$F'(x) = \frac{d(6x^4)}{dx} + \frac{d(12x^{1/2})}{dx} - \frac{d(2x^{-3})}{dx} \qquad \text{(Sum Rule)}$$

$$= 6\frac{d(x^4)}{dx} + 12\frac{d(x^{1/2})}{dx} - 2\frac{d(x^{-3})}{dx} \qquad \text{(Constant Multiple Rule)}$$

$$= 6(4x^3) + 12(\tfrac{1}{2}x^{-1/2}) - 2(-3x^{-4}) \qquad \text{(Power Rule)}$$

$$= 24x^3 + 6x^{-1/2} + 6x^{-4} \qquad \blacksquare$$

If you see how to write the answer in a problem like Example A without any intermediate steps—fine. We don't mean to create extra work for you.

EXAMPLE B

Find an equation of the tangent line to the graph of $y = 2\sqrt{x} + 5$ at $x = 4$.

Solution Let $f(x) = 2\sqrt{x} + 5 = 2x^{1/2} + 5$. Then

$$f'(x) = 2(\tfrac{1}{2})x^{-1/2} + 0 = 1/\sqrt{x}$$

and so

$$f'(4) = 1/\sqrt{4} = \tfrac{1}{2}$$

At $x = 4$, $y = 2\sqrt{4} + 5 = 9$. Thus the desired equation (in point-slope form) is

$$y - 9 = \tfrac{1}{2}(x - 4) \qquad \blacksquare$$

One of the advantages of the $f'(x)$ notation appears in the solution above. We can write $f'(4)$ for the value of the derivative at $x = 4$. In dy/dx notation, we would have to write something like

$$\left.\frac{dy}{dx}\right|_{x=4} \quad \text{or} \quad \left(\frac{dy}{dx}\right)_{x=4}$$

The Product Rule

Now we are in for a mild surprise. The derivative of a product is *not* equal to the product of the derivatives (see CAUTION). How then are we to find the derivative of something like

$$F(x) = (x^2 + 4)(x^3 + 3x + 2)$$

With the tools so far developed, we would have to first multiply the two fac-

---

CAUTION

A common error in Example B is to write

$$y - 9 = \frac{1}{\sqrt{x}}(x - 4)$$

This is *not* the equation of a line. The slope at $x = 4$ is

$$m = f'(4) = \tfrac{1}{2}$$

The equation of the tangent line at $x = 4$ is

$$y - 9 = \tfrac{1}{2}(x - 4)$$

CAUTION

$$\frac{d(uv)}{dx} \neq \frac{du}{dx}\frac{dv}{dx}$$

If $u = x^3$ and $v = x^4$,

$$\frac{du}{dx} = 3x^2 \quad \text{and} \quad \frac{dv}{dx} = 4x^3$$

but

$$\frac{d(uv)}{dx} = \frac{d(x^7)}{dx} = 7x^6$$

$$7x^6 \neq 3x^2 \cdot 4x^3$$

tors together and then differentiate. Thus,

$$F(x) = x^5 + 7x^3 + 2x^2 + 12x + 8$$

and

$$F'(x) = 5x^4 + 21x^2 + 4x + 12$$

At least we could do it, but how would you like to try

$$F(x) = (3x^{4/3} + 9x^{10/9} + x^3)(x^{-3} + 7x^{-1} + 5x)$$

Discouraging isn't it? The Product Rule bypasses the difficulty.

PRODUCT RULE

Let $u = u(x)$ and $v = v(x)$ be differentiable and let $F(x) = u(x)v(x)$.
Then $F'(x) = u(x)v'(x) + v(x)u'(x)$; that is,

$$\frac{d(uv)}{dx} = u\frac{dv}{dx} + v\frac{du}{dx}$$

Learn this rule in words. *The derivative of a product is the first times the derivative of the second plus the second times the derivative of the first.*

Proof

$$F'(x) = \lim_{h \to 0} \frac{F(x + h) - F(x)}{h}$$

$$= \lim_{h \to 0} \frac{u(x + h)v(x + h) - u(x)v(x)}{h}$$

$$= \lim_{h \to 0} \frac{u(x + h)v(x + h) - u(x + h)v(x) + u(x + h)v(x) - u(x)v(x)}{h}$$

$$= \lim_{h \to 0} \left[u(x + h)\frac{v(x + h) - v(x)}{h} + v(x)\frac{u(x + h) - u(x)}{h} \right]$$

$$= \lim_{h \to 0} u(x + h) \lim_{h \to 0} \frac{v(x + h) - v(x)}{h} + \lim_{h \to 0} v(x) \lim_{h \to 0} \frac{u(x + h) - u(x)}{h}$$

$$= u(x)v'(x) + v(x)u'(x) \qquad \blacksquare$$

SHARP CURVE
Subtracting $u(x + h)v(x)$ and then adding it right back in is a legal maneuver. It is this simple trick that makes the proof of the Product Rule work.

EXAMPLE C

Find the derivative of

$$F(x) = (x^2 + 4)(x^3 + 3x + 2)$$

Solution We use the Product Rule and then simplify.

$$F'(x) = (x^2 + 4)\frac{d}{dx}(x^3 + 3x + 2) + (x^3 + 3x + 2)\frac{d}{dx}(x^2 + 4)$$

$$= (x^2 + 4)(3x^2 + 3) + (x^3 + 3x + 2)(2x)$$

$$= 3x^4 + 15x^2 + 12 + 2x^4 + 6x^2 + 4x$$

$$= 5x^4 + 21x^2 + 4x + 12$$

Of course, our answer agrees with what we got earlier. ∎

EXAMPLE D

If $F(x) = (x^2 - 2x + 3)(x^3 + 3\sqrt{x} - 10)$, find $F'(1)$.

Solution

$$F'(x) = (x^2 - 2x + 3)(3x^2 + 1.5x^{-1/2}) + (x^3 + 3\sqrt{x} - 10)(2x - 2)$$

There is no need to simplify this; just substitute $x = 1$.

$$F'(1) = (1 - 2 + 3)(3 + 1.5) + (1 + 3 - 10)(2 - 2)$$

$$= (2)(4.5) + (-6)(0) = 9$$

∎

The Quotient Rule

Suppose we need the derivative of

$$F(x) = \frac{x^2 + 3x - 10}{x^2 + 9}$$

Then the next rule is very helpful.

QUOTIENT RULE

Let $u = u(x)$ and $v = v(x)$ be differentiable and let $F(x) = u(x)/v(x)$. Then

$$F'(x) = \frac{v(x)u'(x) - u(x)v'(x)}{[v(x)]^2}$$

or, equivalently,

$$\frac{d(u/v)}{dx} = \frac{v\frac{du}{dx} - u\frac{dv}{dx}}{v^2}$$

Learn this rule in words too. *The derivative of a quotient is equal to the denominator times the derivative of the numerator minus the numerator times the derivative of the denominator, all divided by the square of the denominator.*

EXAMPLE E

Find $F'(x)$ if $F(x) = (x^2 + 3x - 10)/(x^2 + 9)$.

Solution

$$F'(x) = \frac{(x^2 + 9)(2x + 3) - (x^2 + 3x - 10)(2x)}{(x^2 + 9)^2}$$

$$= \frac{2x^3 + 3x^2 + 18x + 27 - (2x^3 + 6x^2 - 20x)}{(x^2 + 9)^2}$$

$$= \frac{-3x^2 + 38x + 27}{(x^2 + 9)^2}$$

> **CAUTION**
> Be sure to notice in Example E that the minus sign applies to the whole second term
> $$2x^3 + 6x^2 - 20x$$

EXAMPLE F

If $F(x) = \dfrac{x^2 + 4}{x^{1/2}}$ find $F'(4)$ in two different ways.

Solution
Method 1. We use the Quotient Rule.

$$F'(x) = \frac{x^{1/2} \cdot 2x - (x^2 + 4) \cdot \frac{1}{2}x^{-1/2}}{x}$$

$$F'(4) = \frac{2 \cdot 8 - (16 + 4) \cdot \frac{1}{2} \cdot \frac{1}{2}}{4} = \frac{16 - 5}{4} = \frac{11}{4}$$

Method 2. We first simplify.

$$F(x) = \frac{x^2 + 4}{x^{1/2}} = \frac{x^2}{x^{1/2}} + \frac{4}{x^{1/2}} = x^{3/2} + 4x^{-1/2}$$

$$F'(x) = \tfrac{3}{2}x^{1/2} - 2x^{-3/2}$$

$$F'(4) = \frac{3}{2} \cdot 2 - 2 \cdot \frac{1}{4^{3/2}} = 3 - \frac{2}{8} = \frac{11}{4}$$

EXAMPLE G

Let n be a positive integer. Show using the Quotient Rule that

$$\frac{d}{dx}(x^{-n}) = -nx^{-n-1}$$

Solution

$$\frac{d(x^{-n})}{dx} = \frac{d\left(\dfrac{1}{x^n}\right)}{dx} = \frac{x^n \cdot 0 - 1 \cdot nx^{n-1}}{x^{2n}}$$

$$= \frac{-nx^{n-1}}{x^{2n}} = -nx^{n-1-2n} = -nx^{-n-1}$$

This example provides more evidence for the Power Rule in Section 2.2. ∎

Proof of the Quotient Rule

We do the proof in two stages. First let $G(x) = 1/v(x)$. Then

$$G'(x) = \lim_{h \to 0} \frac{\dfrac{1}{v(x+h)} - \dfrac{1}{v(x)}}{h} = \lim_{h \to 0} \frac{\dfrac{v(x) - v(x+h)}{v(x+h)v(x)}}{h}$$

$$= \lim_{h \to 0} \left[-\frac{v(x+h) - v(x)}{h} \cdot \frac{1}{v(x+h)v(x)} \right]$$

$$= -v'(x) \cdot \frac{1}{[v(x)]^2} = -\frac{v'(x)}{[v(x)]^2}$$

Next let $F(x) = \dfrac{1}{v(x)} \cdot u(x)$ and apply the Product Rule together with what we just learned.

$$F'(x) = \frac{1}{v(x)} \cdot u'(x) + u(x) \cdot \frac{-v'(x)}{[v(x)]^2}$$

$$= \frac{v(x)u'(x)}{[v(x)]^2} - \frac{u(x)v'(x)}{[v(x)]^2}$$

$$= \frac{v(x)u'(x) - u(x)v'(x)}{[v(x)]^2}$$

Problem Set 2.3

In Problems 1–8, find f '(x) as in Example A. In many cases you should rewrite f(x) before trying to find its derivative. For example, in Problem 7, write $f(x) = \sqrt{x}(x^3 - 2\sqrt{x} + 5) = \sqrt{x}x^3 - 2x + 5\sqrt{x} = x^{7/2} - 2x + 5x^{1/2}$.

1. $f(x) = 2x^3 - 3x^{2/3} + 5x^{-2}$
2. $f(x) = -6x^{5/3} + 2x^{-3} + 2x^{3/2}$
3. $f(x) = 2x^3 + 5\sqrt{x} - 4/x^5$
4. $f(x) = 3x^4 - 3\sqrt[4]{x} - 5/x^3$
5. $f(x) = 3\sqrt[3]{x} - 4/\sqrt{x}$

6. $f(x) = x\sqrt[3]{x} - 2/\sqrt[3]{x}$
7. $f(x) = \sqrt{x}(x^3 - 2\sqrt{x} + 5)$
8. $f(x) = \sqrt[3]{x}(2x + 4 - 5/\sqrt[3]{x})$

In Problems 9–12, find an equation for the tangent line to the given curve at the point with the indicated x-coordinate (see Example B).

9. $y = 2x^3 - 4x^2 - 3/x; \; x = -1$
10. $y = 3x^2 - 4x + 2\sqrt{x}; \; x = 4$
11. $y = 8\sqrt[4]{x} - 15 + 16/x; \; x = 16$
12. $y = 3x^{2/3} - 8/\sqrt[3]{x}; \; x = 8$
13. Let $F(x) = (x^2 - 9)(x^2 + 4)$. Find $F'(x)$ in two ways.
    (a) Use the Product Rule and then simplify.
    (b) First multiply out and then differentiate.
14. Let $G(t) = (t + 3)(t^2 - 3t + 1)$. Find $G'(t)$ in two ways as in Problem 13.

Use the Product Rule in Problems 15–18 to find $F'(x)$. Simplify your answer if you can.

15. $F(x) = (x^2 + 5)(2x^3 - 3x + 9)$
16. $F(x) = (3x^2 + 5x + 1)(2x^2 - 3)$
17. $F(x) = (\sqrt{x} + 4)(x^{3/2} + 2\sqrt{x} + 2)$
18. $F(x) = (3 + 5x^{-1})(7 + 4x^2)$

For the given function, find the indicated derivative (see Example D).

19. $F(x) = (x^2 + x - 4)(x + \sqrt[3]{x}); \; F'(1)$
20. $G(x) = (3x - 1)(2\sqrt[4]{x} + 3\sqrt{x}); \; G'(1)$
21. $f(t) = (t - 16/t)(3\sqrt{t} - t\sqrt{t}); \; f'(4)$
22. $g(t) = (\sqrt[3]{t} + 1)(t^2 - 8t); \; g'(8)$
23. Let $F(x) = (2x^2 + x - 3)/x$. Find $F'(x)$ in two ways.
    (a) Use the Quotient Rule and then simplify.
    (b) First divide and then differentiate.
24. Follow the directions in Problem 23 for $F(x) = (3x - 2)/\sqrt{x}$.

In Problems 25–28, use the Quotient Rule to find $f'(x)$.

25. $f(x) = \dfrac{2x^2 + 5}{2x - 1}$

26. $f(x) = \dfrac{4x^2}{x^3 - 2x + 1}$

27. $f(x) = \dfrac{4\sqrt{x}}{3 + \sqrt{x}}$

28. $f(x) = \dfrac{x^2 - 4x + 3}{x^2 + 2x}$

In Problems 29–34, find dy/dx by using the rules we have developed.

29. $y = 2x + 3 - 4\sqrt{x} + 5/x^3$
30. $y = (3x^2 + x - 7)(5 - 4x^3)$
31. $y = (x^{-3} + 4x)(2 + 3x^{-1})$
32. $y = (3x - 8)(x^2 + 7)$
33. $y = 2x^2 + 3x - x/(x - 4)$
34. $y = (x^2 - 3x + 1)(x^2 - 4) + x^3 - 3x^2 + 4x - 9$

35. Find an equation for the tangent line to the curve $y = 2x^2 - 3 + \sqrt[3]{x}$ at the point where $x = 1$.
36. Let $H(x) = [G(x)]^2 = G(x) \cdot G(x)$. Use the Product Rule to show that $H'(x) = 2G(x) \cdot G'(x)$.
37. Use the result from Problem 36 to find $H'(x)$ for each of the following.
    (a) $H(x) = (3x + 1)^2$
    (b) $H(x) = (x^2 - x + 4)^2$
    (c) $H(x) = (4\sqrt{x} + 5)^2$
    (d) $H(x) = (4x + 3/x^2)^2$
38. If the density of algae in a water tank t hours from now is expected to be $2\sqrt{t}/(\sqrt{t} + 10)$ algae per cubic centimeter, find the rate of change in density with respect to time 9 hours from now.
39. If an object moves along a line so that its position s in centimeters is given by $s = (2t^2 - 3\sqrt{t})/(t + 1)$ after t seconds, find its velocity after 4 seconds.
40. The number of units of oxygen per gallon of water in a lake t days after a sewage plant began dumping untreated sewage was found to be $N = 100\{2 - [t/(t + 8)]\}$. Find the rate of change in units of oxygen with respect to time at (a) $t = 2$, (b) $t = 92$.
41. A manufacturer of riding mowers expects to sell each mower for $720. The cost of making and selling x mowers per week is

    $$C(x) = 0.02x^2 + 500x + 3000$$

    (a) Find the profit $P(x)$ from producing and selling x mowers per week.
    (b) Find the marginal profit and evaluate it when $x = 1000$.
    (c) For what values of x is the marginal profit zero? What is the economic significance of this value of x?

[C] 42. Economists often talk about average cost, that is, cost per unit of output. If the total cost of making x units of something is $C(x)$, then the average cost is $C(x)/x$. Suppose that a manufacturer of jackets finds that the total cost in dollars of making x jackets per week is

    $$C(x) = 2000 + 12x - 8\sqrt{x} + 0.05x^{3/2}$$

If this manufacturer plans to make 500 jackets per week, what is

(a) The total cost?

(b) The average cost?

(c) The marginal cost?

(d) The marginal average cost?

2.4

THE GENERAL POWER RULE

Imagine trying to find the derivative of

$$F(x) = (x^2 + 4x - 9)^{50}$$

We would first multiply 50 factors of $x^2 + 4x - 9$ together, thus obtaining a polynomial with 101 terms. Then we would differentiate the polynominal. If we were very accurate and exceedingly lucky, we would get the right answer after about 5 hours of work. There is a better way.

The Derivative of u^n

Let $u = u(x)$ be a differentiable function. Our goal is to find a simple formula for $d(u^n)/dx$; our tool is the Product Rule. We begin with simple cases, looking for a pattern.

$$\frac{d(u^2)}{dx} = \frac{d(u \cdot u)}{dx} = uu' + uu' = 2uu'$$

$$\frac{d(u^3)}{dx} = \frac{d(u \cdot u^2)}{dx} = u \cdot 2uu' + u^2 \cdot u' = 3u^2u'$$

$$\frac{d(u^4)}{dx} = \frac{d(u \cdot u^3)}{dx} = u \cdot 3u^2u' + u^3 \cdot u' = 4u^3u'$$

The pattern seems clear and suggests that

$$\frac{d(u^n)}{dx} = nu^{n-1}\frac{du}{dx}$$

In fact, this formula is correct when n is any real number, but our proof will have to wait until Section 5.3.

GENERAL POWER RULE

Let $F(x) = [u(x)]^r$, where $u = u(x)$ is a differentiable function and r is a real number. Then

$$F'(x) = r[u(x)]^{r-1}u'(x)$$

Equivalently,

$$\frac{d(u^r)}{dx} = ru^{r-1}\frac{du}{dx}$$

Note that if $u(x) = x$, the General Power Rule becomes

$$\frac{d(x^r)}{dx} = rx^{r-1}\frac{dx}{dx} = rx^{r-1}(1) = rx^{r-1}$$

which is the ordinary power rule. The force of the General Power Rule is that we can easily find the derivative of a power of any function including the one in the introduction to this section.

EXAMPLE A

Find the derivative of

$$F(x) = (x^2 + 4x - 9)^{50}$$

Solution Let $u = x^2 + 4x - 9$. Then

$$F'(x) = \frac{d(u^{50})}{dx} = 50u^{49}\frac{du}{dx} = 50(x^2 + 4x - 9)^{49}(2x + 4) \qquad \blacksquare$$

EXAMPLE B

Find $F'(x)$ if $F(x) = \sqrt{3x + 7}$.

Solution Let $u = 3x + 7$. Then

$$F'(x) = \frac{d(u^{1/2})}{dx} = \frac{1}{2}u^{-1/2}\frac{du}{dx} = \frac{1}{2\sqrt{u}}\frac{du}{dx}$$

$$= \frac{1}{2\sqrt{3x + 7}}(3) = \frac{3}{2\sqrt{3x + 7}} \qquad \blacksquare$$

EXAMPLE C

> **CAUTION**
> A very common error in Example C is to forget the factor
> $$6x^2 + 9$$

If $y = \sqrt[3]{(2x^3 + 9x)^4}$, find dy/dx.

Solution Write $y = (2x^3 + 9x)^{4/3}$. Then

$$\frac{dy}{dx} = \frac{4}{3}(2x^3 + 9x)^{1/3}(6x^2 + 9)$$

If you can make the mental substitution $u = 2x^3 + 9x$ (as we did) without actually writing it down, that is fine. $\qquad \blacksquare$

EXAMPLE D

If $y = 2/(2x + 3)^4$, find dy/dx.

Solution We could use the Quotient Rule, but it is probably better to first write $y = 2(2x + 3)^{-4}$. Then from the Constant Multiple Rule followed by the General Power Rule, we obtain

$$\frac{dy}{dx} = 2 \cdot \frac{d(2x + 3)^{-4}}{dx} = 2 \cdot (-4)(2x + 3)^{-5}(2)$$

$$= -16(2x + 3)^{-5} = \frac{-16}{(2x + 3)^5} \qquad \blacksquare$$

Combining the Rules

We must be able to use the Sum Rule, Product Rule, Quotient Rule, and General Power Rule in all kinds of combinations. Often the key will be choosing the right rule to apply first.

EXAMPLE E

Find the derivative of $y = \left(\dfrac{2x + 3}{3x - 7}\right)^4$.

Solution At least mentally, let $u = (2x + 3)/(3x - 7)$. Then apply the General Power Rule followed by the Quotient Rule.

$$\overset{4 u^3}{} \qquad\qquad\qquad \overset{\dfrac{du}{dx}}{}$$

$$\frac{dy}{dx} = 4\left(\frac{2x + 3}{3x - 7}\right)^3 \cdot \frac{(3x - 7)2 - (2x + 3)3}{(3x - 7)^2}$$

$$= 4\left(\frac{2x + 3}{3x - 7}\right)^3 \frac{-23}{(3x - 7)^2}$$

$$= -92\frac{(2x + 3)^3}{(3x - 7)^5} \qquad \blacksquare$$

EXAMPLE F

Find the derivative of $y = 4x^3(5x^2 + 1)^{3/2}$.

Solution We begin with the Product Rule and then use the Power Rule.

$$\frac{dy}{dx} = 4x^3\frac{d}{dx}(5x^2 + 1)^{3/2} + (5x^2 + 1)^{3/2}\frac{d}{dx}(4x^3)$$

$$= 4x^3 \cdot \frac{3}{2}(5x^2 + 1)^{1/2}(10x) + (5x^2 + 1)^{3/2}(12x^2)$$

$$= 60x^4(5x^2 + 1)^{1/2} + 12x^2(5x^2 + 1)^{3/2}$$

It would not be wrong to leave the answer as shown. On the other hand, some old-fashioned algebra will improve the form of the result. Note that

$12x^2(5x^2 + 1)^{1/2}$ is a common factor of both terms. Thus,

$$\frac{dy}{dx} = 12x^2(5x^2 + 1)^{1/2}(5x^2 + 5x^2 + 1)$$

$$= 12x^2(5x^2 + 1)^{1/2}(10x^2 + 1) \qquad \blacksquare$$

EXAMPLE G

If $y = \dfrac{(x^2 + 1)^4}{(x + 5)^3}$, find $\dfrac{dy}{dx}$.

Solution We begin with the Quotient Rule.

$$\frac{dy}{dx} = \frac{(x + 5)^3 \dfrac{d}{dx}(x^2 + 1)^4 - (x^2 + 1)^4 \dfrac{d}{dx}(x + 5)^3}{(x + 5)^6}$$

$$= \frac{(x + 5)^3 \cdot 4(x^2 + 1)^3 \cdot 2x - (x^2 + 1)^4 \cdot 3(x + 5)^2}{(x + 5)^6}$$

It is tempting to leave the answer in this form. However, we can greatly simplify it if we factor out $(x + 5)^2(x^2 + 1)^3$ from both terms of the numerator. We get

$$\frac{dy}{dx} = \frac{(x + 5)^2(x^2 + 1)^3[8x(x + 5) - 3(x^2 + 1)]}{(x + 5)^6}$$

$$= \frac{(x + 5)^2(x^2 + 1)^3(8x^2 + 40x - 3x^2 - 3)}{(x + 5)^6}$$

$$= \frac{(x^2 + 1)^3(5x^2 + 40x - 3)}{(x + 5)^4} \qquad \blacksquare$$

Applications

Don't forget that the derivative allows us to calculate slopes and rates of change.

EXAMPLE H

Find the slope of the curve $y = x^2\sqrt{x^2 + 3x + 6}$ at the point where $x = 2$.

Solution Let $F(x) = x^2(x^2 + 3x + 6)^{1/2}$. Using first the Product Rule, then the General Power Rule, we obtain

$$F'(x) = x^2(\tfrac{1}{2})(x^2 + 3x + 6)^{-1/2}(2x + 3) + (x^2 + 3x + 6)^{1/2}(2x)$$

There is no need to simplify this before we evaluate at $x = 2$.

$$F'(2) = 4(\tfrac{1}{2})(16)^{-1/2}(7) + (16)^{1/2}(4)$$
$$= 4(\tfrac{1}{2})(\tfrac{1}{4})(7) + (4)(4) = \tfrac{39}{2}$$ ■

EXAMPLE I

Comtrol makes fancy programmable calculators at a cost of $150.00 each and is presently making 100 units per day. The president knows that if she increases production, she will have to lower the price. A market study she commissioned estimated that Comtrol can sell x units per day at a price of $p = 400/\sqrt{1 + 0.02x}$ dollars. Determine the marginal profit, evaluate it at $x = 100$, and interpret the result.

Solution Let $C(x)$, $R(x)$, and $P(x)$ denote the daily cost, daily revenue, and daily profit in producing x units. Then

$$C(x) = 150x$$

$$R(x) = px = \frac{400x}{\sqrt{1 + 0.02x}}$$

$$P(x) = \frac{400x}{\sqrt{1 + 0.2x}} - 150x$$

Thus, the marginal profit $P'(x)$ is

$$P'(x) = \frac{(1 + 0.02x)^{1/2}(400) - (400x)(\tfrac{1}{2})(1 + 0.02x)^{-1/2}(0.02)}{1 + 0.02x} - 150$$

and

$$P'(100) = \frac{(3)^{1/2}(400) - (400)(3)^{-1/2}}{3} - 150$$
$$\approx 153.96 - 150 = \$3.96$$

This means that increasing production by one unit will only increase the profit by about $3.96.

Note that at $x = 100$, Comtrol will receive $p = 400/\sqrt{3} \approx \$230.94$ for each unit at a cost of only $150.00 each, which means a profit of $80.94 each. How can you reconcile this figure with the $3.96 obtained above? ■

Problem Set 2.4

In Problems 1–16, find $f'(x)$.

1. $f(x) = (3x - 4)^{12}$
2. $f(x) = (7x - 8)^9$
3. $f(x) = (3x^2 + x + 1)^8$
4. $f(x) = (2x^2 + 5x - 6)^{14}$
5. $f(x) = (x + 2/x^2)^5$
6. $f(x) = (x^{2/3} - 2x)^6$
7. $f(x) = (x^2 - 2\sqrt{x})^{10}$
8. $f(x) = (2\sqrt{x} - 5/x)^4$

9. $f(x) = \sqrt{2 + 3x - x^2}$

10. $f(x) = \sqrt[3]{9x - 1}$

11. $f(x) = (2x^3 - 7x + 6)^{2/3}$

12. $f(x) = (x^4 - 2x^3 + 2)^{3/4}$

13. $f(x) = \sqrt[4]{(2x^3 + x)^3}$

14. $f(x) = (\sqrt{3x - x^5})^5$

15. $f(x) = 4/(3x - 1)^5$

16. $f(x) = 2/\sqrt{2x + 5}$

17. $f(x) = -2/\sqrt[3]{(2x - 1)^2}$

18. $f(x) = 4/(\sqrt{2x + 3})^3$

19. Find the slope of the curve $y = (t^2 + 3t - 1)^{-5/2}$ at $t = 2$.

20. Find an equation for the tangent line to the curve $y = (t^3 + t + 1)^{-3/4}$ at the point $(0, 1)$.

In Problems 21–32, find dy/dx. You will need to use various combinations of the rules we have learned (see Examples E, F, and G).

21. $y = \left(\dfrac{2x - 1}{3x + 2}\right)^3$

22. $y = \left(\dfrac{4x + 1}{3x - 2}\right)^4$

23. $y = \left(\dfrac{x^2 + 4}{2x - 5}\right)^5$

24. $y = \left(\dfrac{3x - 1}{x^2 + x - 4}\right)^3$

25. $y = (8x + 5)(2x - 1)^5$

26. $y = (3x - 1)^8(4x + 3)$

27. $y = 3x^2(2x^2 + 3)^{5/3}$

28. $y = 5x^3(x^2 + 4)^{5/2}$

29. $y = (x^2 + 1)^3(x^2 + 4)^2$

30. $y = (2x^3 + 1)^4(x^3 + x)^3$

31. $y = \dfrac{x^2}{(3x - 1)^4}$

32. $y = \dfrac{(4x + 1)^4}{(5 + 2x)^6}$

33. Calculate $F'(1)$ if $F(x) = (2x - 1)^8(6\sqrt[3]{x} + 1)$.

34. Calculate $g'(0)$ if $g(x) = (8x + 4)^{3/2}(2x - 1)^5$.

35. Find an equation for the tangent line to the curve $y = (\sqrt{x} - 2)^4/(2x + 4)$ at $x = 1$.

36. Find an equation for the tangent line to the curve $y = (x^2 + 3)(x^2 + 2x - 14)^{-2/5}$ at $x = 3$.

[C]37. An appliance manufacturer makes small refrigerators at a cost of $120 each. A study shows that x refrigerators can be sold per day if the price p is set at $p = 600/\sqrt{1 + 0.04x}$ dollars.
(a) Determine the marginal daily profit when $x = 100$.
(b) Interpret your answer to part (a).
(c) Calculate the average profit per unit when $x = 100$.

[C]38. An object moves along a line so that its position s in feet at time t seconds is given by

$$s = 16\sqrt{t^4 - 0.08t^2 + 9}$$

(a) Determine the position of the object at $t = 10$.

(b) Find the average velocity on the interval $t = 10$ to $t = 10.5$.

(c) Find the velocity at $t = 10$.

In problems 39–42, find $F'(4)$.

39. $F(x) = (4x - 10 - 8/x)^{3/2}$

40. $F(x) = \sqrt{x\sqrt{x + 1}} + \sqrt{x}$

41. $F(x) = [(\sqrt{x} + 3)(2x - 5)]^2$

42. $F(x) = (2x - 7)^5/\sqrt{2x + 1}$

43. At what value of x does the curve $y = (x^4 - 32x + 57)^{1/2}$ have a horizontal tangent?

44. Find an equation for the tangent line to the curve $y = (2x^3 - 4x^2 + 5x - 8)^5$ at the point having $x = 2$.

45. If $5000 is invested today at an interest rate of r compounded twice a year, it will grow to an amount A at the end of 4 years, where

$$A = 5000\left(1 + \frac{r}{2}\right)^8$$

(a) Find dA/dr, the rate at which A changes with respect to r.

[C](b) Calculate A and dA/dr when $r = 0.12$.

46. The population N of a certain city t years from now is projected to follow the formula

$$N = 24 - \frac{4}{t + 3} + \frac{36}{(t + 3)^2}$$

where N is measured in thousands.
(a) Calculate the population 7 years from now.
(b) Will the population be growing or declining at that time?

47. Based on a study of potato yields, a biologist has obtained the formula

$$y = \frac{1,000,000}{8.5r + 40,000}$$

where y is the yield per plant in kilograms and r is the density (number of plants per hectare).
(a) Find the formula for Y, the yield per hectare.
(b) Show that dy/dr is always negative but that dY/dr is always positive.
(c) Interpret the results from part (b). Do they make sense?

48. An object P is placed directly between a strong light S_1 and a weak light S_2, which are 20 feet apart (see Figure 7). When P is x feet from S_1, the intensity of

FIGURE 7

illumination at P is given by

$$I = \frac{4000}{x^2} + \frac{1000}{(20 - x)^2}$$

(a) Calculate I when $x = 10$ and $x = 15$.
(b) Calculate dI/dx when $x = 10$ and $x = 15$.
(c) If P is moved a bit toward S_2 from $x = 15$, will I increase or decrease?

2.5
HIGHER-ORDER DERIVATIVES

No object from everyday life is more familiar to Americans than the automobile. With this in mind, we use a car to introduce the next idea. Let us suppose that Sylvia drives from her home to the office along a straight road—a distance of exactly 6 miles. On one particular day, a device was attached to her car which recorded the distance s from her home at each time t. The results, shown as a graph, appear in Figure 8.

Let's try to interpret this graph. One obvious fact is that it took Sylvia 13 minutes to get to work. Of more interest are the linear segments between $t = 2$ and $t = 8$ and again between $t = 9$ and $t = 11$. On these intervals, the slope has a constant value of $\frac{1}{2}$, which means that the car was going at a constant speed of $\frac{1}{2}$ mile per minute (that is, 30 miles per hour).

Of still more interest are the curved parts of the graph. On the interval $0 \le t \le 2$, the slope (and hence the speed) was increasing. We say the car was accelerating, due, of course, to the fact that Sylvia was pressing down on the accelerator until she reached the desired speed of 30 miles per hour. At $t = 8$, she again began to accelerate, perhaps to pass a car, and then slowed down (decelerated) to 30 miles per hour. Finally at $t = 11$ she started slowing down to reach the desired speed of zero at her destination.

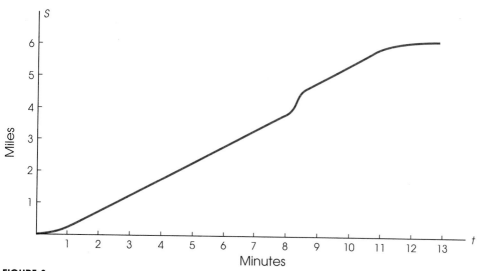

FIGURE 8

What then is acceleration? It is a measure of how speed is changing. More specifically, **acceleration** a is the rate of change in velocity v with respect to time t, that is,

$$a = \frac{dv}{dt} = \frac{d}{dt}\left(\frac{ds}{dt}\right)$$

When a is positive, velocity is increasing; when a is negative, velocity is decreasing. And the larger a is in magnitude, the more rapidly velocity is changing.

Higher Derivatives

Because time will play such an important role in this section, we use t rather than x as the independent variable. Suppose that $f(t) = t^5 + 4t^3 + 1$, so that $f'(t) = 5t^4 + 12t^2$. Notice that f' is itself a function. Its derivative, called the **second derivative of f** and denoted by f'', is given by $f''(t) = 20t^3 + 24t$. Now we see that the process can be repeated again and again. Here are the results.

$$f(t) = t^5 + 4t^3 + 1$$
$$f'(t) = 5t^4 + 12t^2$$
$$f''(t) = 20t^3 + 24t$$
$$f'''(t) = 60t^2 + 24$$
$$f^{(4)}(t) = 120t$$
$$f^{(5)}(t) = 120$$
$$f^{(6)}(t) = 0$$
$$f^{(7)}(t) = 0$$

Pay attention to the notation. More than three primes becomes clumsy, so we use a numeral enclosed in parentheses to indicate derivatives of higher order than 3.

Recall that if $y = f(t)$, we have two alternative symbols for the derivative, namely, $f'(t)$ and dy/dt. What corresponds to $f''(t)$? We could write

$$f''(t) = \frac{d}{dt}\left(\frac{dy}{dt}\right)$$

and we sometimes do. However, Gottfried Leibniz, one of the founders of the calculus, decreed that we might also write d^2y/dt^2. We have the following correspondence.

POINT OF INTEREST
Gottfried Wilhelm Leibniz (1646–1716) was a universal genius—an expert in law, religion, politics, history, and philosophy as well as mathematics. He gave us the notation dy/dx for the derivative and $\int$ for the integral.

Name	Prime Notation	Leibniz Notation
(First) derivative	$f'(t)$	$\dfrac{dy}{dt}$
Second derivative	$f''(t)$	$\dfrac{d^2y}{dt^2}$
Third derivative	$f'''(t)$	$\dfrac{d^3y}{dt^3}$
Fourth derivative	$f^{(4)}(t)$	$\dfrac{d^4y}{dt^4}$
Fifth derivative	$f^{(5)}(t)$	$\dfrac{d^5y}{dt^5}$
nth derivative	$f^{(n)}(t)$	$\dfrac{d^ny}{dt^n}$

EXAMPLE A

If $y = 3t^2 - 2t^{1/2}$, find d^3y/dt^3.

Solution

$$\frac{dy}{dt} = 6t - t^{-1/2}$$

$$\frac{d^2y}{dt^2} = 6 + \frac{1}{2}t^{-3/2}$$

$$\frac{d^3y}{dt^3} = -\frac{3}{4}t^{-5/2}$$

EXAMPLE B

If $f(t) = 1/\sqrt{2t + 3}$, find $f'''(3)$.

Solution

$$f(t) = (2t + 3)^{-1/2}$$
$$f'(t) = -\tfrac{1}{2}(2t + 3)^{-3/2}(2) = -(2t + 3)^{-3/2}$$
$$f''(t) = \tfrac{3}{2}(2t + 3)^{-5/2}(2) = 3(2t + 3)^{-5/2}$$
$$f'''(t) = 3(-\tfrac{5}{2})(2t + 3)^{-7/2}(2) = -15(2t + 3)^{-7/2}$$
$$f'''(3) = -15(9)^{-7/2} = -15/(\sqrt{9})^7 = -\tfrac{5}{729}$$

EXAMPLE C

If $y = (x^3 + 3)^4$, find d^2y/dx^2.

Solution

$$\frac{dy}{dx} = 4(x^3 + 3)^3 \cdot 3x^2 = 12x^2 \cdot (x^3 + 3)^3$$

$$\frac{d^2y}{dx^2} = 12x^2 \cdot 3(x^3 + 3)^2 \cdot 3x^2 + (x^3 + 3)^3 \cdot 24x$$

$$= 12x(x^3 + 3)^2[9x^3 + 2(x^3 + 3)]$$

$$= 12x(x^3 + 3)^2(11x^3 + 6) \qquad\blacksquare$$

Applications of the First and Second Derivatives

Differential calculus was developed to describe and predict motion. The first derivative measures velocity (rate of change in distance with respect to time); the second derivative measures acceleration (rate of change in velocity with respect to time). It is appropriate that we begin with a motion problem—the problem of a falling body.

If an object is thrown straight up from an initial height s_0 feet with an initial velocity v_0 feet per second, then its height s above the ground in feet after t seconds is

$$s = -16t^2 + v_0 t + s_0$$

This is what the physicists tell us, though they are quick to say that the formula is accurate only near the surface of the earth (not out in space) and for objects that are dense enough so air resistance is minimal (no feathers please).

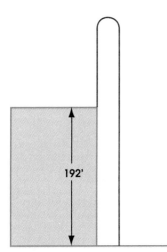

192'

FIGURE 9

EXAMPLE D

Suppose that a baseball was thrown upward from the roof of a building 192 feet high with an initial velocity of 64 feet per second (Figure 9).

(a) What was its velocity at $t = 1$? At $t = 5$?
(b) When did it reach maximum height?
(c) What was the maximum height?
(d) When did the ball hit the ground?
(e) With what velocity did it hit the ground?
(f) What was the ball's acceleration at $t = 1$? At $t = 5$?

Solution Here $s_0 = 192$ and $v_0 = 64$, so

$$s = -16t^2 + 64t + 192$$

$$v = \frac{ds}{dt} = -32t + 64$$

$$a = \frac{dv}{dt} = -32$$

(a) At $t = 1$, $v = -32 \cdot 1 + 64 = 32$ feet per second.
   At $t = 5$, $v = -32 \cdot 5 + 64 = -96$ feet per second.
   The negative sign means that the ball is falling at $t = 5$.
(b) The ball reached maximum height when the velocity was zero, that is, when $-32t + 64 = 0$. This occurred when $t = 2$.
(c) At $t = 2$,

$$s = -16(2)^2 + 64(2) + 192 = 256 \text{ feet}$$

(d) The ball hit the ground when $s = 0$.

$$-16t^2 + 64t + 192 = 0$$

$$t^2 - 4t - 12 = 0$$

$$(t - 6)(t + 2) = 0$$

$$t = 6, -2$$

Clearly, the solution $t = 6$ is the one we want.
(e) At $t = 6$, $v = -32(6) + 64 = -128$ feet per second.
(f) The acceleration is constant: -32 feet per second per second. This means that the velocity decreases by 32 feet per second every second. ∎

To introduce our next example we paraphrase an article from the business page of a large metropolitan newspaper (*St. Paul Dispatch*, Aug. 27, 1984).

A worldwide debt explosion is threatening the United States and other wealthy countries . . . . In the United States the ratio of government debt to national income remained unchanged at around 28 percent up to 1981 but then it began to increase sharply, reaching almost 36 percent in 1983 . . . . The IMF released a table showing that the speed of debt increase was greater in the United States than in Japan . . . .

EXAMPLE E

Interpret the facts in the article above in terms of derivatives. Sketch a graph reflecting these facts.

Solution Let R be the ratio of government debt to national income in the United States at time t.

(a) $R = 0.28$ $\qquad\qquad\qquad$ $(t < 1981)$

(b) $R = 0.36$ $\qquad\qquad\qquad$ $(t = 1983)$

(c) $\dfrac{dR}{dt} = 0$ $\qquad\qquad\qquad$ $(t < 1981)$

(d) $\dfrac{dR}{dt} > 0$ $\qquad\qquad\qquad$ $(1981 \leq t \leq 1983)$

(e) $\dfrac{d^2R}{dt^2} > 0$ $\qquad\qquad\qquad$ $(1981 \leq t \leq 1983)$

(f) $\left(\dfrac{d^2R}{dt^2}\right)_{\text{US}} > \left(\dfrac{d^2R}{dt^2}\right)_{\text{Japan}}$ $\qquad$ $(1981 \leq t \leq 1983)$

A graph that fits facts (a) through (e) is shown in Figure 10. ∎

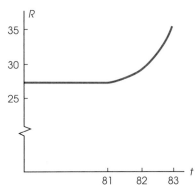

FIGURE 10

EXAMPLE F

GTS News reported in the spring of 1984 that unemployment in the United States was continuing to decrease but at a slower and slower rate. Interpret this statement in the language of derivatives.

Solution Let U denote the number of people unemployed at time t in the United States. Then in the spring of 1984, $dU/dt < 0$ and $d^2U/dt^2 > 0$. A graph of U against t would look like Figure 11. ∎

Occasionally, you will hear bankers and business people say, "We've got to watch that second derivative." Do you see why?

FIGURE 11

Spring 84

Problem Set 2.5

In Problems 1–6, find d^3y/dx^3.

1. $y = x^5 + 3x^3$
2. $y = 2x^3 - x^2 + 2$
3. $y = 4/x$
4. $y = 6\sqrt{x}$
5. $y = 2x^{3/2} - x^4$
6. $y = \frac{3}{20}x^5 + 3/x^2$

7. For $f(t) = \sqrt{3t + 7}$, find $f''(3)$.
8. For $g(t) = \sqrt[3]{2t + 6}$, find $g''(1)$.
9. For $h(s) = 3/\sqrt{s + 4}$, find $h'''(0)$.
10. For $F(u) = \sqrt{2u + 3}$, find $F'''(3)$.
11. Find d^2y/dx^2 for $y = (x^2 - 3x + 4)^3$.
12. Find d^2y/dx^2 for $y = (x^3 - x + 2)^2$.
13. Find $f''(4)$ if $f(t) = t/(t + 4)$.
14. Find $g''(0)$ if $g(u) = u^2/(u^2 + 4)$.

15. Suppose that a rock was thrown upward from the roof of a building 256 feet high with an initial velocity of 96 feet per second. See Example D.
    (a) Write a formula for its height s (in feet) above the ground after t seconds.
    (b) What was its height after 2 seconds?
    (c) What was its velocity after 2 seconds?
    (d) When did it reach its maximum height?
    (e) When did the rock hit the ground?
    (f) With what velocity did it hit the ground?
    (g) What was the rock's acceleration at $t = 2$? At $t = 6$?

16. Answer the same questions as in Problem 15 for the case where a rock is thrown downward from the roof of the same building with an initial speed of 64 feet per second.

17. A particle is moving in a straight line according to the equation $s = 2t^3 - 3t^2 - 12t + 8$ (s in centimeters, t in seconds).
    (a) Find the position, the velocity, and the acceleration of the particle when $t = 3$.
    (b) Find the velocity at the instant when the acceleration is zero.
    (c) Find the acceleration at the instant when $t > 0$ and the velocity is zero.
    (d) On what interval of time is the velocity negative?

18. Suppose that a particle is moving in a straight line so that its position s (in meters) at time t (in minutes) is given by
    $$s = 3t - 12\sqrt{t} + 2$$
    (a) Find the position, the velocity, and the acceleration of the particle when $t = 1$.
    (b) Where is the particle when the velocity is zero?
    (c) What value does the velocity approach as t gets larger and larger?
    (d) What value does the acceleration approach as t gets larger and larger?

19. The daily revenue R of Tony's Ice Cream Shop climbed during the period 1/1/72 to 12/31/77 and then began to fall. The rate of change in daily revenue increased from 1/1/72 to 6/30/74 but then began a decline.
    (a) When was $dR/dt > 0$?
    (b) When was $d^2R/dt^2 < 0$?
    (c) When was $dR/dt = 0$?
    (d) When was $d^2R/dt^2 = 0$?
    (e) Sketch a graph which reflects all the above facts.

20. The number $N(t)$ of bacteria in a culture was carefully monitored over a period of 10 hours. During all this time $N(t)$ increased—at a constant rate during the first 2 hours, then at a slower and slower rate after a drug was administered at $t = 2$, and finally at an increasing rate when the effect of the drug began to wear off 4 hours after it was administered.
    (a) When was $N'(t) > 0$?
    (b) When was $N''(t) = 0$?
    (c) When was $N''(t) < 0$?
    (d) Sketch a graph that reflects all the above facts.

21. Interpret each of the following statements in the language of derivatives.
    (a) Inflation I held steady during all of last year but is expected to increase more and more rapidly during the years ahead.
    (b) At the present time the price P of oil is declining rapidly, but this trend is expected to slow and then reverse directions in about 2 years.
    (c) John's temperature T is still rising, but the penicillin seems to be taking effect.
    (d) Nancy has the accelerator of her sports car pressed to the floor, but after another 2 minutes that won't do any good.

22. Let $T(n)$ be the time it takes a person to learn n French words. One experiment indicated that one learns words more and more rapidly until a vocabulary of 200 words is reached. Then the rate of learning remains constant until 400 words are reached, after which the rate of learning slows.
    (a) For what values of n is $T'(n) > 0$?
    (b) For what values of n is $T''(n) > 0$?
    (c) For what values of n is $T''(n) = 0$?

In Problems 23–26, find $f'''(1)$.

23. $f(x) = 6/\sqrt{2x - 1}$
24. $f(x) = 12x^{-2/3} + \sqrt{x}$
25. $f(x) = (x^2 + 1)^3$
26. $f(x) = x\sqrt{3 + x}$

27. Find the rate of change in the slope of $y = (4 + x)^3$ at the point $(1, 125)$.

28. At what point on the curve $y = 12\sqrt{x}$ is $d^2y/dx^2 = -\frac{1}{9}$?

29. A ball thrown upward from the top of a building 336 feet high hits the ground after 7 seconds.
    (a) What was the initial velocity? *Hint:* $s = -16t^2 + v_0t + 336$.
    (b) With what speed did the ball strike the ground?

30. In a small town of 400 people, the number $N(t)$ of people who had heard a certain rumor after t days is indicated by the graph in Figure 12.
    (a) When is the rate at which the rumor is spreading positive?
    (b) When is the rate of spread decreasing?
    (c) When is the rumor spreading most rapidly?

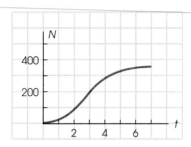

FIGURE 12

CHAPTER 2 REVIEW PROBLEM SET

1. Sketch the graph of $f(x) = \begin{cases} x + 2 & \text{if } x < 2 \\ x^2 - 4 & \text{if } x \geq 2 \end{cases}$

   Then find (if possible) each of the following.
   (a) $f(2)$ (b) $f(f(2))$ (c) $\lim\limits_{x \to 4} f(x)$
   (d) $\lim\limits_{x \to 2^-} f(x)$ (e) $\lim\limits_{x \to 2^+} f(x)$ (f) $\lim\limits_{x \to 2} f(x)$

2. Sketch the graph of $f(x) = \begin{cases} (x - 2)^2 & \text{if } x < 2 \\ 4 - x & \text{if } x \geq 2 \end{cases}$

   Find (if possible) each of the following.
   (a) $f(2)$ (b) $f(f(3))$ (c) $\lim\limits_{x \to 5} f(x)$
   (d) $\lim\limits_{x \to 2^-} f(x)$ (e) $\lim\limits_{x \to 2^+} f(x)$ (f) $\lim\limits_{x \to 2} f(x)$

Find each limit in Problems 3–30.

3. $\lim\limits_{x \to -2} (3x + 8)$

4. $\lim\limits_{x \to 3} (3x - 1)$

5. $\lim\limits_{x \to 5} (x^2 - 4x + 5)$

6. $\lim\limits_{x \to -3} [(x - 2)^2 + \sqrt{3}]$

7. $\lim\limits_{x \to 4} (x\sqrt{x} - 5)$

8. $\lim\limits_{x \to 2} x^2\sqrt{x + 7}$

9. $\lim\limits_{x \to 5} \dfrac{2x - 10}{x^2 - 5x}$

10. $\lim\limits_{x \to -3} \dfrac{3x + 9}{x^2 + 3x}$

11. $\lim\limits_{x \to 3} \dfrac{x^2 - 4x + 3}{x^2 - 3x}$

12. $\lim\limits_{x \to 5} \dfrac{x^2 - 3x - 10}{x^2 - 25}$

13. $\lim\limits_{x \to -4} \dfrac{(x + 4)^2}{x^2 + 2x - 8}$

14. $\lim\limits_{x \to 4} \dfrac{(x - 4)^3}{x^2 - x - 12}$

15. $\lim\limits_{x \to -2} \dfrac{x^2 - x - 6}{x^2 + 7x + 10}$

16. $\lim\limits_{x \to -6} \dfrac{x^2 + 5x - 6}{x^2 - x - 42}$

17. $\lim\limits_{x \to 2} \dfrac{x^3 - 8}{x^2 + 4x - 12}$

18. $\lim\limits_{x \to 5} \dfrac{x^3 - 125}{x^2 - 2x - 15}$

19. $\lim\limits_{x \to 3} \dfrac{x^3 - 8}{x^2 + 4x - 12}$

20. $\lim\limits_{x \to -5} \dfrac{x^3 - 125}{x^2 - 2x - 15}$

21. $\lim\limits_{x \to 3} \dfrac{2^x}{x}$

22. $\lim\limits_{x \to -1} \dfrac{2x + 3}{3^x}$

23. $\lim\limits_{x \to 2} (3x^2 - 11)\sqrt{x^2 + 5}$

24. $\lim\limits_{x \to 2} (2x^3 - 2x - 7)\sqrt{5x^2 + 7}$

25. $\lim\limits_{x \to -4} \dfrac{3x + 2}{x^2 - 10}$

26. $\lim\limits_{x \to 10} \dfrac{3x^2 - 2x - 5}{2x^2 + 5x + 25}$

27. $\lim\limits_{x \to 3} \dfrac{(2x - 5)^6(x - 3)}{(x^2 - 9)(x^2 + x - 11)}$

28. $\lim\limits_{x \to 1} \dfrac{(x^2 + x - 2)(4x - 3)^{12}}{(x^2 - 1)\sqrt{x^2 + 10x - 2}}$

29. $\lim\limits_{x \to 10} [\sqrt{4x + 9} - 2\sqrt[3]{x^2 - 36}]$

30. $\lim\limits_{x \to 5} \left[\dfrac{10}{\sqrt{4x + 5}} - 3\sqrt[3]{x^2 - 8x + 7} \right]$

In Problems 31–34, find $f'(x)$ by using the definition of the derivative, that is, by calculating

$$\lim\limits_{h \to 0} \dfrac{f(x + h) - f(x)}{h}$$

31. $f(x) = x^2 - 3x - 2$

32. $f(x) = 2x^2 + 3x + 1$

33. $f(x) = 2/\sqrt{x}$

34. $f(x) = \sqrt{x^2 + 1}$

In Problems 35–58, find $f'(x)$ using the rules for differentiation.

35. $f(x) = 3\sqrt[3]{x} = 3x^{1/3}$

36. $f(x) = 8\sqrt[4]{x}$

37. $f(x) = 6x^2 \sqrt[3]{x}$

38. $f(x) = 6x\sqrt[3]{x^2}$

39. $f(x) = x^{5/4}(16x)^{1/4}$

40. $f(x) = x^{3/2}\sqrt{25x^5}$

41. $f(x) = \sqrt{x\sqrt{x}}$

42. $f(x) = \sqrt[3]{x^4\sqrt{x}}$

43. $f(x) = (2x - 1)^6$

44. $f(x) = (3x + 1)^3$

45. $f(x) = \sqrt{x^2 - 2x + 5}$

46. $f(x) = \sqrt{x^2 + 8x - 10}$ 47. $f(x) = \left(2x - \dfrac{4}{\sqrt{x}}\right)^3$

48. $f(x) = (2x + 4/x)^4$

49. $f(x) = (x^3 - 3x)^{4/3}$ 50. $f(x) = (x^5 - 5x^2)^{7/5}$

51. $f(x) = \left(\dfrac{2x^2 + 1}{x^3 - 2x + 4}\right)^3$ 52. $f(x) = \left(\dfrac{3x - 2}{x^3 + 3}\right)^5$

53. $f(x) = (2x^2 + 5)\sqrt{2x + 5}$

54. $f(x) = (2x^2 + x + 1)\sqrt{4x + 1}$

55. $f(x) = (2x^2 + 5)\sqrt{2x^2 + 5}$

56. $f(x) = (2x^2 + x + 1)\sqrt{2x^2 + x + 1}$

57. $f(x) = [(x^2 + 2x)^4 + 3]^5$

58. $f(x) = [(2x + 1)^5 - 2]^4$

59. If $g(t) = 64\sqrt{t + 3}$, find $g'''(1)$.

60. If $h(s) = 54(2s + 1)^{1/3}$, find $h'''(-1)$.

61. Find an equation for the tangent line to the graph of $y = 5\sqrt[5]{x} - 6/x^2$ at the point with $x = 1$.

62. Find an equation of the tangent line to the graph of

$$y = 4/(2x - 1)^2 + 3(x^2 + 6x + 2)^{1/2}$$

at the point where $x = 1$.

63. At what point on the graph of $y = 4\sqrt{x}$ is the tangent line parallel to the line $x - 2y = 5$?

64. At what point of the graph of $y = 2x^{3/2}$ is the tangent line parallel to $6x - y = 5$?

65. If \$10,000 is invested today at an interest rate of x percent compounded 12 times a year, it will grow to an amount A by the end of 10 years, where

$$A = 10,000\left(1 + \dfrac{x}{1200}\right)^{120}$$

(a) Find the rate at which A changes with respect to x.

Ⓒ(b) Evaluate A and dA/dx when $x = 9$.

66. The temperature in degrees Fahrenheit during the period from $t = 0$ to $t = 6$ (hours) was given by

$$F(t) = 72 - 15t + 6/(t + 2)^2$$

How fast was the temperature falling at $t = 2$?

67. Suppose that a particle is moving along a coordinate line in such a way that its position s relative to the origin after t seconds is given by $s = 12t - 4t^{3/2} + 6$ (in feet).

(a) Find the position, velocity, and acceleration of the particle at $t = 1$.

(b) Find the acceleration at the instant when the velocity is zero.

(c) During what time interval is the velocity positive?

68. A particle is moving along a cordinate line so that its position s (in feet) relative to the origin after t seconds satisfies $s = 4t + 16t^{1/2} - 8t^{3/2} + 24$.

(a) Find its position, velocity, and acceleration at $t = 4$.

(b) Find its acceleration at the instant when its velocity is 0.

69. A flu epidemic hit a certain city. The number y of people who came down with the disease within t days of the outbreak is indicated in Figure 13.

(a) When is the rate of spread of the disease positive?

(b) When is the rate of spread increasing?

(c) When is the disease spreading most rapidly?

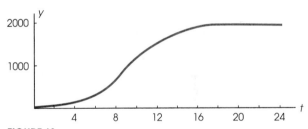

FIGURE 13

70. Suppose that a rumor spread through a city of 20,000 so that after 8 days almost everyone had heard it. During the first 4 days, the rate at which the rumor spread was increasing; after that the rate decreased. Sketch a graph of $y = N(t)$, where $N(t)$ represents the number of people who had heard the rumor after t days, given that $N(4) = 12,000$.

Do computers think? Is the human brain just a giant computer—a very complex jumble of electronic connections, switches, and diodes? These questions continue to fascinate scientists and they fascinated Norbert Wiener. A new branch of applied mathematics called artificial intelligence considers these questions.

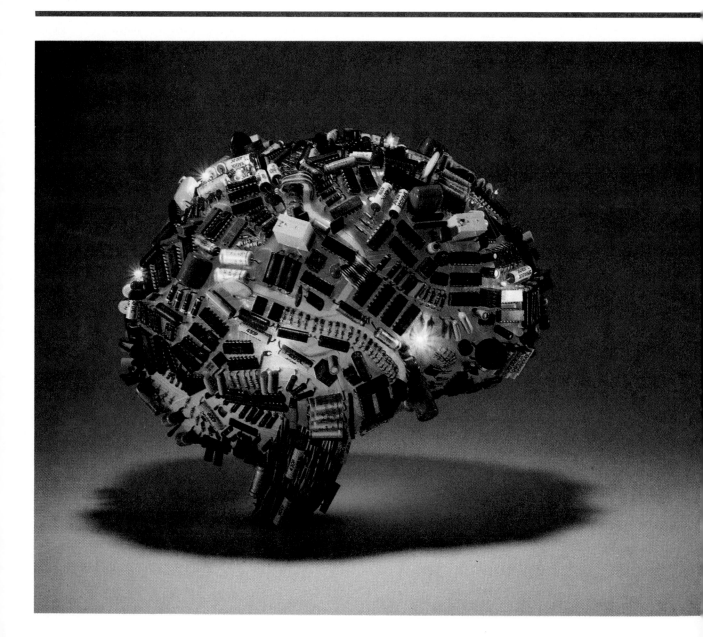

3

Applications of the Derivative to Graphing

Norbert Wiener
(1894–1964)

It is fitting that Norbert Wiener follows John von Neumann on our list of applied mathematicians, for the two were contemporaries with similar interests. Both made great contributions to pure mathematics, both were interested in physics, both helped develop the electronic computer, and both were fascinated by the analogy between the human brain and the computer. But while von Neumann eagerly worked on the atom bomb project, Wiener refused to do so and hoped the bomb would fail to work. In fact, Wiener was much concerned about the potential negative effects of modern technology and often articulated these dangers.

Born in Missouri, Norbert was the son of Harvard language professor, Leo Wiener. Leo quickly recognized the talents of his precocious son and drove him mercilessly. Thus, Norbert was reading and writing by age 3, had finished college at 14, and had obtained his Ph.D. from Harvard at 18. It was a regimen that left scars on the young man from which he never recovered. In spite of world recognition, he suffered from feelings of insecurity and social inferiority.

After postdoctoral study in England and Germany and some years of trying to find a suitable job, he was hired by the newly established Massachusetts Institute of Technology where he spent the rest of his life. He helped propel this institution to world renown, partly by his own contributions and also by his influence on many budding young scientists.

Wiener's early work dealt with the mathematics of Brownian motion and harmonic analysis and turned to the theory of information and communication. He can be called the founder of control theory, to which he gave the name *cybernetics*. With Arturo Rosenblueth, he made contributions to neurophysiology. His books include *Ex-prodigy, I Am a Mathematician, Cybernetics,* and *The Human Use of Human Beings.*

The authors of the present volume are two of Wiener's mathematical grandchildren. Both worked under Robert Cameron who had studied with Wiener. Both wrote their dissertations on the "Wiener integral."

3.1
CONTINUITY AND DIFFERENTIABILITY

The old maxim—one picture is worth a thousand words—holds also in calculus. The aim of a picture is to show the qualitative features of an object and thereby to reveal relationships that may be only dimly perceived in a mass of words. In calculus, the most important pictures are graphs of functions. This topic attracted our attention very early, especially in Section 1.2. In this chapter, we can probe much deeper because we have developed a powerful tool, the derivative. This tool acts as a microscope allowing us to examine the fine structure of a graph—to accurately locate its peaks, its valleys, and its turning points. All of this will become clear as we move through the chapter.

Our starting point is very basic, having to do with the fundamental nature of a curve. The subject is continuity.

Continuity

The words *continuity* and *continuous* are in our everyday vocabulary. We say that a manufacturing plant was in continuous production for 10 years, meaning that it operated throughout that whole period with no interruptions. We remark about the continuity of a certain author's writings, meaning that the ideas expressed follow a consistent path; if they change over time, they do so in a smooth, regular way.

Mathematicians use the words continuous and continuity in a very similar fashion. A function is continuous on a certain interval if its graph has no jumps or wild fluctuations. Put another way, a function is continuous if its graph can be drawn without lifting the pencil from the paper (or the chalk from the chalk board). These informal descriptions help us, but they are not adequate for what follows. A precise definition rests on the fundamental notion of limit.

DEFINITION
The function f is **continuous** at the point c if
$$\lim_{x \to c} f(x) = f(c)$$
and f is continuous on a set if it is continuous at each point of the set.

Perhaps we can understand this definition best by asking what it means for a function to be discontinuous at a point. The function f is discontinuous at c if one or more of the following are true.

1. $\lim_{x \to c} f(x)$ doesn't exist (Figure 1).
2. $f(c)$ is not defined (Figure 2).
3. $\lim_{x \to c} f(x)$ and $f(c)$ both exist but are unequal (Figure 3).

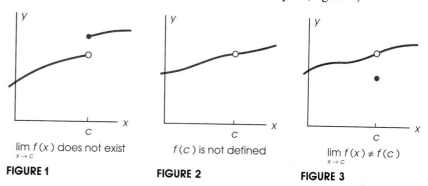

$\lim_{x \to c} f(x)$ does not exist

FIGURE 1

$f(c)$ is not defined

FIGURE 2

$\lim_{x \to c} f(x) \neq f(c)$

FIGURE 3

Figure 4 exhibits a function which is continuous at c. Note that the condition $\lim_{x \to c} f(x) = f(c)$ holds for this function.

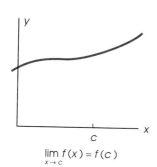

$\lim_{x \to c} f(x) = f(c)$

FIGURE 4

EXAMPLE A

Show that $f(x) = (x^2 - 4)/(x - 2)$ is continuous at $x = 1$ but is discontinuous at $x = 2$.

Solution By using the Main Limit Theorem (Section 2.1), we obtain

$$\lim_{x \to 1} f(x) = \frac{\lim_{x \to 1} (x^2 - 4)}{\lim_{x \to 1} (x - 2)} = \frac{-3}{-1} = 3$$

Since $f(1)$ also equals 3, the required condition for continuity holds.

At $x = 2$, the function f is undefined, so f is discontinuous there. Note the graph of f shown in Figure 5. ∎

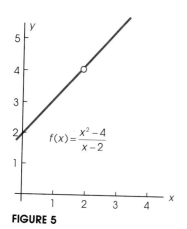

$f(x) = \dfrac{x^2 - 4}{x - 2}$

FIGURE 5

The function f in Example A is a rational function—a quotient of two polynomials. It is easy to show (using the Main Limit Theorem) that a polynomial is continuous everywhere and that a rational function is continuous except where its denominator is zero.

Sometimes a discontinuity can be removed by the simple device of defining (or redefining) the function at the point of discontinuity. This is the case with the function f in Example A. If we define $f(2)$ to be 4, the function f thereby becomes continuous at 2 (see Figure 5).

EXAMPLE B

Show that the discontinuity of

$$g(x) = \frac{x^2 - 4x - 5}{x + 1}$$

APPLICATIONS OF THE DERIVATIVE TO GRAPHING

at $x = -1$ can be removed by defining $g(-1)$ appropriately.

Solution

$$\lim_{x \to -1} g(x) = \lim_{x \to -1} \frac{(x - 5)(x + 1)}{x + 1} = \lim_{x \to -1} (x - 5) = -6$$

Thus if we define $g(-1) = -6$, the function g becomes continuous at -1. ∎

EXAMPLE C

The function $F(x) = (x^2 - x - 6)/(x^2 - 9)$ has discontinuities at -3 and 3. One of these discontinuities can be removed; the other cannot. Determine which can be removed and define F there to make F continuous.

Solution

$$\lim_{x \to -3} F(x) = \lim_{x \to -3} \frac{(x - 3)(x + 2)}{(x - 3)(x + 3)} = \lim_{x \to -3} \frac{x + 2}{x + 3}$$

Clearly this limit does not exist, since the numerator is approaching -1 and the denominator is approaching 0. Thus, the discontinuity at -3 cannot be removed.

In contrast,

$$\lim_{x \to 3} F(x) = \lim_{x \to 3} \frac{(x - 3)(x + 2)}{(x - 3)(x + 3)} = \lim_{x \to 3} \frac{x + 2}{x + 3} = \frac{5}{6}$$

Defining $F(3)$ to be $\frac{5}{6}$ will remove the discontinuity at 3. Figure 6 displays the graph of F and makes clear the nature of the two discontinuities. ∎

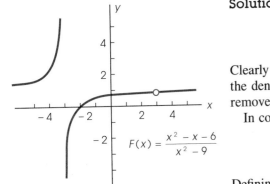

$F(x) = \dfrac{x^2 - x - 6}{x^2 - 9}$

FIGURE 6

EXAMPLE D

Discuss the continuity of

$$f(x) = \begin{cases} x - 1 & \text{if } x \leq 2 \\ -x^2 + 5 & \text{if } x > 2 \end{cases}$$

Solution The two parts of f are polynomials and therefore continuous everywhere. The only question is whether they have been pasted together correctly at 2 so that $\lim_{x \to 2} f(x) = f(2)$. The graph (Figure 7) suggests that the answer is yes, but it is also easy to give an algebraic argument. Note first that $f(2) = 2 - 1 = 1$. Also,

$$\lim_{x \to 2^-} f(x) = \lim_{x \to 2^-} (x - 1) = 1$$

and

$$\lim_{x \to 2^+} f(x) = \lim_{x \to 2^+} (-x^2 + 5) = 1$$

so $\lim_{x \to 2} f(x) = 1 = f(2)$. We conclude that f is continuous at 2 and therefore everywhere. ∎

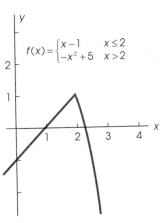

$f(x) = \begin{cases} x - 1 & x \leq 2 \\ -x^2 + 5 & x > 2 \end{cases}$

FIGURE 7

Differentiability

For a function to be differentiable means that it has a derivative, and this in turn means that the graph of the function has a slope. How can a function fail to have a slope, say, at the point c? There are several rather obvious ways.

1. The function fails to even be continuous at c (Figure 8).
2. The graph of the function has a corner at c and therefore a different slope from the right and from the left (Figure 9).
3. The graph of f has a vertical tangent at c (Figure 10).

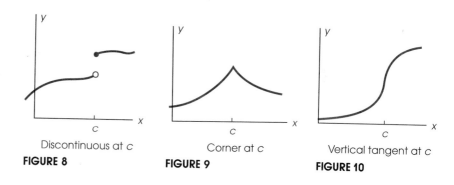

Discontinuous at c

FIGURE 8

Corner at c

FIGURE 9

Vertical tangent at c

FIGURE 10

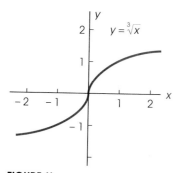

FIGURE 11

EXAMPLE E

At what points is $f(x) = \sqrt[3]{x} = x^{1/3}$ differentiable?

Solution $f'(x) = \frac{1}{3}x^{-2/3} = 1/3(\sqrt[3]{x})^2$, a formula valid for $x \neq 0$. Thus, f is differentiable for all $x \neq 0$. At $x = 0$, the formula for $f'(x)$ suggests that there may be a vertical tangent, and this is confirmed when we draw a graph (Figure 11). Note that f is continuous everywhere, even at $x = 0$. ∎

EXAMPLE F

At what points is $f(x) = |x|$ differentiable?

Solution The graph of f (Figure 12) makes the answer clear immediately. In fact, $f'(x) = -1$ if $x < 0$ and $f'(x) = 1$ if $x > 0$. At $x = 0$, there is no tangent line, hence no slope, and therefore no derivative. ∎

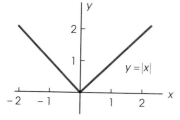

FIGURE 12

A moment's thought should convince you that a polynomial [for example, $P(x) = 3x^5 - 7x^4 + 11x - 16$] is differentiable everywhere and that a rational function [for example, $f(x) = (3x^3 - 2x + 11)/(x^2 - 9)$] is differentiable everywhere except where its denominator is zero. The reason is straightforward; we have formulas for their derivatives.

Refer again to Examples E and F. Note that in both cases, the given function was continuous at 0 but not differentiable there. Continuity does not imply differentiability. The converse, however, is true, as we now make clear.

> **THEOREM**
>
> If f is differentiable at c, then f is continuous at c.

Proof   For $h \neq 0$

$$f(c + h) = \frac{f(c + h) - f(c)}{h} \cdot h + f(c)$$

Thus by the Main Limit Theorem,

$$\lim_{h \to 0} f(c + h) = \lim_{h \to 0} \frac{f(c + h) - f(c)}{h} \cdot \lim_{h \to 0} h + \lim_{h \to 0} f(c)$$

$$= f'(c) \cdot 0 + f(c)$$

that is,

$$\lim_{h \to 0} f(c + h) = f(c)$$

This is equivalent to saying

$$\lim_{x \to c} f(x) = f(c)$$

the condition for continuity.

We summarize by noting two implications, but we emphasize that neither implication is reversible.

f is differentiable at c

$\Downarrow$

f is continuous at c

$\Downarrow$

$\lim f(x)$ exists at c

EXAMPLE G

Decide by studying Figure 13 which of the three conditions above holds at c_1, c_2, c_3, and c_4.

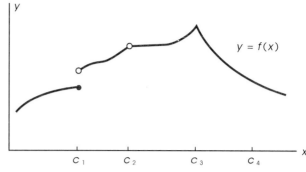

FIGURE 13

Solution

f is differentiable at c_4.

f is continuous at c_3 and c_4.

$\lim f(x)$ exists at c_2, c_3, and c_4.

None of the three conditions holds at c_1.

Problem Set 3.1

In Problems 1–8, indicate whether or not the given physical process is continuous for all $t > 0$.

1. The depth of water in a bathtub t hours after it was installed in a house.

2. The number of dollars in your checking account t hours from now.

3. The time showing on your digital wristwatch t hours after you bought it.

4. Your height t hours after you were born.

5. The number of hours you will spend studying mathematics during the next t hours.

6. The number of people inside the city limits of Chicago t hours from now.

7. The Celsius temperature of your body t hours from now as shown by a mercury thermometer.

8. The Celsius temperature that will be reported t hours from now on the electronic sign outside the local bank.

In Problems 9–12, show that the function f is continuous at $x = a$ but discontinuous at $x = b$ (see Example A).

9. $f(x) = \dfrac{x^2 - 4x}{x - 4}$; $a = 1$, $b = 4$

10. $f(x) = \dfrac{x^2 - x - 2}{x - 2}$; $a = 1$, $b = 2$

11. $f(x) = \dfrac{x^2 - 4}{3x + 6}$; $a = 4$, $b = -2$

12. $f(x) = \dfrac{2x^2 - 10x}{3x - 15}$; $a = 10$, $b = 5$

Each function in Problems 13–16 has a removable discontinuity at $x = a$. How should $f(a)$ be defined to remove this discontinuity? See Example B.

13. $f(x) = \dfrac{x^2 - 2x - 8}{3x - 12}$; $a = 4$

14. $f(x) = \dfrac{x^2 - 3x - 10}{2x + 4}$; $a = -2$

15. $f(x) = \dfrac{x^2 - 1}{2 - 2x}$; $a = 1$

16. $f(x) = \dfrac{x^2 - x - 6}{6 - 2x}$; $a = 3$

Each function in Problems 17–22 has discontinuities at two values of x. Find these values and, if the discontinuity is removable, indicate how f should be defined at that value to remove the discontinuity. See Example C.

17. $f(x) = \dfrac{x^2 - 4x - 12}{x^2 - 5x - 6}$

18. $f(x) = \dfrac{x^2 - x - 6}{x^2 - 2x - 3}$

19. $f(x) = \dfrac{x + 4}{x^2 - 5x + 4}$

20. $f(x) = \dfrac{x}{x^2 - x - 30}$

21. $f(x) = \dfrac{3x^2 - 12}{x^2 - 4}$

22. $f(x) = \dfrac{x^3 - x}{x^2 - 1}$

In Problems 23–32, determine all values of x where the given function is discontinuous. For the functions defined by multipart rules, see Example D.

23. $f(x) = x^5 - 5x + 2$

24. $f(x) = 4/(x^2 + 1)$

25. $f(x) = \sqrt{x^2 + 1}$

26. $f(x) = \sqrt{1 - x^2}$

27. $f(x) = 1/\sqrt{x}$

28. $f(x) = 1/\sqrt{x - 3}$

29. $f(x) = \begin{cases} 2x & \text{if } x \le 3 \\ x^2 - 3 & \text{if } x > 3 \end{cases}$

30. $f(x) = \begin{cases} \frac{1}{4}x^2 & \text{if } x \le 2 \\ 3 - x & \text{if } x > 2 \end{cases}$

31. $f(x) = \begin{cases} x^2 + 1 & \text{if } x \le 1 \\ 5x - \frac{2}{x} & \text{if } x > 1 \end{cases}$

32. $f(x) = \begin{cases} \frac{1}{3}x^3 & \text{if } x < 2 \\ 2x + 3 & \text{if } x \ge 2 \end{cases}$

In Problems 33–44, indicate the values of x (if any) where f fails to be differentiable (see Problems E and F).

33. $f(x) = x^{4/3}$

34. $f(x) = 2x^{1.2} + 5x$

35. $f(x) = 1/(x + 1)$

36. $f(x) = 1/(x^2 + 1)$

37. $f(x) = \sqrt[3]{x}(x + 1)$

38. $f(x) = \sqrt[5]{x}$

39. $f(x) = \sqrt[3]{2x + 1}$

40. $f(x) = \sqrt[3]{2x - 1}$

41. $f(x) = |x - 2|$

42. $f(x) = |x + 3|$

43. $f(x) = \begin{cases} x & \text{if } x \le 2 \\ 2 & \text{if } x > 2 \end{cases}$

44. $f(x) = \begin{cases} x^2 & \text{if } x \le 1 \\ x & \text{if } x > 1 \end{cases}$

Examine the graphs in Problems 45 and 46. As in Example G, decide at which of the points p, q, r, s, and t:

(a) *f has a value.*
(c) *f is continuous.*

(b) $\lim f(x)$ *exists.*
(d) *f is differentiable.*

45.

46.

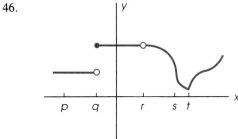

In Problems 47–54, decide whether or not the given function is continuous at x = 3. If it is continuous at x = 3, decide whether or not it is differentiable there.

47. $f(x) = (x - 3)^2 + \sqrt[3]{x - 3}$

48. $f(x) = |x - 3|$

49. $f(x) = |x - 3|/(x - 3)$

50. $f(x) = (2x - 6)/(x - 3)$

51. $f(x) = \dfrac{x^2 - 3x}{2x - 6}$

52. $f(x) = \begin{cases} \dfrac{x^2 - 3x}{2x - 6} & \text{if } x \ne 3 \\ 1.5 & \text{if } x = 3 \end{cases}$

53. $f(x) = \begin{cases} x^2 - 3x & \text{if } x \le 3 \\ x - 3 & \text{if } x > 3 \end{cases}$

54. $f(x) = \begin{cases} x + 3 & \text{if } x < 3 \\ x^2 - 3 & \text{if } x \ge 3 \end{cases}$

55. Let $[x]$ denote the *greatest integer in x*. For example, $[3.4] = 3$, $[3] = 3$, and $[-1.2] = -2$. Sketch the graph of $f(x) = [x]$. At what values of x is this function discontinuous?

56. Let $\langle x \rangle$ denote the *distance from x to the nearest integer*. For example, $\langle 3.4 \rangle = 0.4$, $\langle 3.7 \rangle = 0.3$, and

$\langle 3 \rangle = 0$. Sketch the graph of $f(x) = \langle x \rangle$. At what values of x is this function discontinuous? Not differentiable?

57. Sketch the graph of a function f defined on the interval $0 \le x \le 6$ and satisfying all of the following conditions.

(a) $f(0) = f(2) = f(4) = f(6) = 2$.

(b) f is continuous everywhere except at $x = 2$.

(c) f is differentiable on $0 < x < 6$ except at $x = 2$ and $x = 4$.

3.2
MONOTONICITY AND CONCAVITY

Several kinds of intervals will arise in this section, and we introduce special notation to describe them. The double inequality $a < x < b$ determines the **open interval** consisting of all numbers between a and b, not including a and b. This interval is denoted by (a, b). In contrast, the inequality $a \le x \le b$ determines the corresponding **closed interval,** which does include a and b. It is denoted by $[a, b]$. The two situations are illustrated in Figure 14.

$$(1, 6) = \{x : 1 < x < 6\}$$

$$[1, 6] = \{x : 1 \le x \le 6\}$$

FIGURE 14

Nine types of intervals can occur; they are shown in the accompanying table.

Inequality	Interval Notation	Graph
$a < x < b$	(a, b)	
$a \le x \le b$	$[a, b]$	
$a \le x < b$	$[a, b)$	
$a < x \le b$	$(a, b]$	
$-\infty < x \le b$	$(-\infty, b]$	
$-\infty < x < b$	$(-\infty, b)$	
$a \le x < \infty$	$[a, \infty)$	
$a < x < \infty$	(a, ∞)	
$-\infty < x < \infty$	$(-\infty, \infty)$	

Solving Inequalities

Solving inequalities is much like solving equations. Here are the allowable operations—operations that do not change the solution set.

1. We may add the same number to both sides of an inequality.
2. We may multiply both sides of an inequality by a positive number.
3. We may multiply both sides of an inequality by a negative number provided we reverse the direction of the inequality sign.

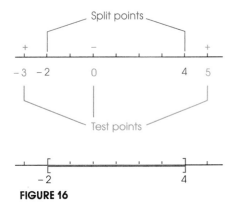

FIGURE 15

EXAMPLE A

Solve the inequality $3x - 8 < 5x + 2$ and write the solution set in interval notation.

Solution

$$3x - 8 < 5x + 2$$
$$3x < 5x + 10 \qquad \text{(adding 8)}$$
$$-2x < 10 \qquad \text{(adding } -5x)$$
$$x > 10/-2 \qquad \text{(multiplying by } -\tfrac{1}{2})$$
$$x > -5$$

The solution set is $(-5, \infty)$ (see Figure 15). ■

> **RECALL**
> If we cannot factor
> $$ax^2 + bx + c$$
> we can still find the split points by means of the **Quadratic Formula**
> $$x = \frac{-b \pm \sqrt{b^2 - 4ac}}{2a}$$

EXAMPLE B

Solve the quadratic inequality $x^2 - 2x \le 8$ and write the solution set in interval notation.

Solution As with quadratic equations, we bring all nonzero terms to one side and then factor (if possible).

$$x^2 - 2x \le 8$$
$$x^2 - 2x - 8 \le 0$$
$$(x - 4)(x + 2) \le 0$$

$f(x) = (x - 4)(x + 2)$
$f(-3) = (-7)(-1) = +7$
$f(0) = (-4)(2) = -8$
$f(5) = (1)(7) = +7$

The two numbers -2 and 4 where the left side is 0 are called **split points.** They split the real line into the three intervals $(-\infty, -2)$, $(-2, 4)$, and $(4, \infty)$. On each of them, $f(x) = (x - 4)(x + 2)$ has a constant sign, either positive or negative. We can determine which by picking any point on the interval as a **test point**; f will be positive or negative on the interval according as f at the test point is positive or negative. The whole process is displayed in Figure 16; it yields the solution set $[-2, 4]$. The endpoints -2 and 4 are included because the inequality $\le$ includes equality. ■

FIGURE 16

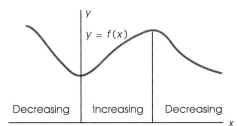

Decreasing   Increasing   Decreasing

FIGURE 17

Increasing or Decreasing

We say that f is **increasing** on an interval I if the graph of f rises to the right on I. Similarly f is **decreasing** on I if the graph of f falls to the right on I (see Figure 17). We use the adjective **monotonic** to describe a function which is either increasing or decreasing on an interval.

If the slope of a curve is positive, then the corresponding function is increasing; if the slope is negative, the function is decreasing. When stated in terms of derivatives, this gives a very important theorem.

MONOTONICITY THEOREM

Let f be differentiable on the open interval (a, b)

    (i) If $f'(x) > 0$ for all x in (a, b), then f is increasing on (a, b).

    (ii) If $f'(x) < 0$ for all x in (a, b), then f is decreasing on (a, b).

EXAMPLE C

Find the intervals on which $f(x) = 2x^2 - 3x + 11$ is increasing or decreasing.

Solution Since $f'(x) = 4x - 3$, our problem reduces to finding out where $4x - 3 < 0$ and where $4x - 3 > 0$. The first inequality is solved as follows.

$$4x - 3 < 0$$
$$4x < 3$$
$$x < \tfrac{3}{4}$$

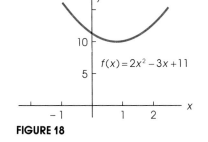

FIGURE 18

From this we conclude that f is decreasing on $(-\infty, \tfrac{3}{4})$. It is easy to see in a similar manner that f is increasing on $(\tfrac{3}{4}, \infty)$. Actually we could have anticipated a result like this, since we know that the graph of f is a parabola that turns up (Figure 18). ∎

EXAMPLE D

Find the intervals on which $g(x) = 2x^3 - 3x^2 - 12x$ is increasing or decreasing. Then sketch the graph of f.

Solution Since $g'(x) = 6x^2 - 6x - 12 = 6(x - 2)(x + 1)$, our problem is to find where $6(x - 2)(x + 1)$ is positive and where it is negative. The split points -1 and 2 split the real line into the three intervals $(-\infty, -1)$, $(-1, 2)$, and $(2, \infty)$. When tested (Figure 19), we find that g' is positive on the first and third of these intervals and negative on the second one. We conclude that g is increasing on $(-\infty, -1)$ and $(2, \infty)$, and decreasing on $(-1, 2)$. With this information plus the values of g at the split points,

FIGURE 19

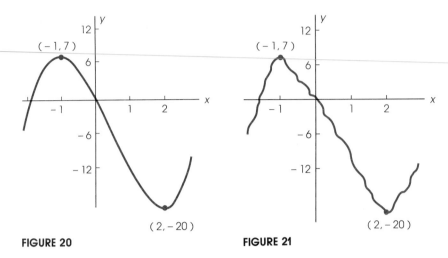

FIGURE 20   **FIGURE 21**

namely, $g(-1) = 7$ and $g(2) = -20$, we can quickly sketch a graph. But does it look like Figure 20 or Figure 21? Both increase and decrease on the right intervals and have their peak and pit in the right spots.

Since both we and nature have a predisposition toward simplicity, 99% of us would bet money on Figure 20. But to make certain, we need another concept called concavity. ■

Concave Up or Concave Down

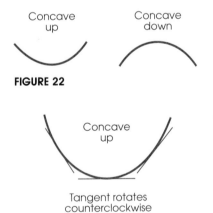

Concave up   Concave down

FIGURE 22

Concave up

Tangent rotates counterclockwise

FIGURE 23

We are all familiar with one manifestation of concavity. At some time, most of us have worn glasses, at least sunglasses. The inner surface of a typical glass lens is concave. For a graph to be concave means that it has this shape, but it may be **concave up** or **concave down** (Figure 22).

Of course, our aim is to relate concavity to the calculus, in particular, to the derivative. Let us agree first of all that what we really mean by saying that the graph of a function f is concave up is that the tangent line is rotating counterclockwise as it moves from left to right (Figure 23). This in turn means that the slope is increasing; that is, f' is an increasing function. But we can guarantee that f' is increasing by requiring that its derivative, namely, f'', be positive. We formulate the result as a theorem.

CONCAVITY THEOREM

Let f be twice differentiable on the open interval (a, b).

(i) If $f''(x) > 0$ for all x in (a, b), then the graph of f is concave up on (a, b).

(ii) If $f''(x) < 0$ for all x in (a, b), then the graph of f is concave down on (a, b).

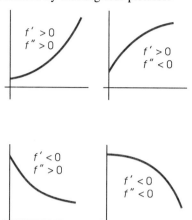

$f' > 0$
$f'' > 0$

$f' > 0$
$f'' < 0$

$f' < 0$
$f'' > 0$

$f' < 0$
$f'' < 0$

Consider again $f(x) = 2x^2 - 3x + 11$ in Example C. Note that $f'(x) = 4x - 3$ and $f''(x) = 4$. Since the second derivative is always positive, the graph is concave up on the whole real line, confirming that the graph we drew in Figure 18 is correct.

Of more interest is $g(x) = 2x^3 - 3x^2 - 12x$ in Example D. Since $g'(x) = 6x^2 - 6x - 12$, we obtain $g''(x) = 12x - 6 = 6(2x - 1)$. It is easy to see that $g''(x) < 0$ on $(-\infty, \frac{1}{2})$ and $g''(x) > 0$ on $(\frac{1}{2}, \infty)$. We conclude that the graph of g is concave down on the first interval, and concave up on the second interval. Now we can assert with confidence that the graph of g is much like Figure 20; it cannot wriggle as does the graph in Figure 21.

EXAMPLE E

Sketch the graph of a function f which satisfies all of the following conditions.

(i)   It is continuous everywhere.

(ii) $(-2, 2)$ and $(3, 7)$ are points on the graph.

(iii) $f'(-2) = 0$ and $f'(3) = 0$.

(iv) $f''(x) > 0$ on $(-\infty, 1)$ and $f''(x) < 0$ on $(1, \infty)$.

Solution We draw the graph in two stages. First, we plot the two points $(-2, 2)$ and $(3, 7)$ and show that the slope is 0 at these two points (shown in black in Figure 24). Second, condition (iv) means the graph must be concave up on $(-\infty, 1)$, concave down on $(1, \infty)$. This allows us to draw the colored curve.

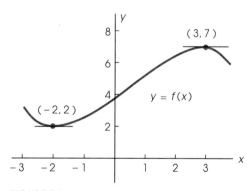

FIGURE 24

EXAMPLE F

Sketch the graph of a function f which satisfies all of the following conditions

(i)   It is continuous except at $x = 0$.

(ii) $\lim_{x \to 0^-} f(x) = 1$ and $\lim_{x \to 0^+} f(x) = 2$

(iii) $f'(x) = -1$ on $(-\infty, 0)$ and $f'(x) > 0$ on $(0, \infty)$

(iv) $f''(x) < 0$ on $(0, \infty)$

Solution The graph is shown in Figure 25. Clearly (i) and (ii) hold. Condition (iii) means that the graph is a line of slope -1 on $(-\infty, 0)$ and that the graph rises on $(0, \infty)$. Finally (iv) forces the graph to be concave down on $(0, \infty)$.

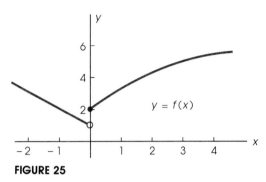

FIGURE 25

Problem Set 3.2

In Problems 1–6, show the indicated set on the real line. Recall that the symbol $\cup$ denotes union. Thus, $A \cup B$ is the set of points in A or B.

1. $[-3, 6)$
2. $(-2, 3]$
3. $[-2, 0) \cup [1, 4]$
4. $(-\infty, 1] \cup [2, 3]$
5. $\{x: -2 < x \le 5\}$
6. $\{x: 1 \le x < 4\}$

In Problems 7–12, solve the given inequality and write the solution set in interval notation (see Example A). In Problems 11 and 12, we suggest that you begin by clearing fractions. Thus, in Problem 11, begin by multiplying both sides by 6.

7. $4x - 11 < 7x + 7$
8. $6x + 13 \ge x - 17$
9. $2(x - 3) \ge 5x + 1$
10. $4 - 3x \le 5(x + 3)$
11. $\frac{1}{2}x - 2 < -\frac{2}{3}x + 5$
12. $\frac{3}{4}x + 3 > \frac{2}{5}x + 10$

Solve the quadratic inequalities in Problems 13–20 as illustrated in Example B. Then write the solution set in interval notation and show this set on the number line.

13. $(x + 7)(x - 3) \le 0$
14. $(x - 6)(x + 4) < 0$
15. $x^2 + 6x - 16 < 0$
16. $x^2 + 2x - 8 \le 0$
17. $x^2 - x \ge 6$
18. $x^2 - 5x > 6$
19. $2x^2 + 7x - 4 < 0$
20. $3x^2 + 13x - 10 \le 0$

In Problems 21–28, find the interval(s) on which f(x) is increasing and the interval(s) on which f(x) is decreasing (see Examples C and D).

21. $f(x) = 3x^2 - 6x + 11$
22. $f(x) = 2x^2 + 8x - 2$
23. $f(x) = -4x^2 + 16x - 9$
24. $f(x) = -x^2 + 8x - 7$
25. $f(x) = x^3 - 12x$
26. $f(x) = x^3 - 27x$
27. $f(x) = x^3 - 3x^2 - 12x$
28. $f(x) = 2x^3 + 3x^2 - 36x$

In Problems 29–34, determine the interval(s) on which the graph of f(x) is concave up and the interval(s) on which it is concave down (see the discussion preceding Example E).

29. $f(x) = x^3 - 12x$
30. $f(x) = x^3 - 27x$

31. $f(x) = x^3 - 3x^2 - 12x$

32. $f(x) = 2x^3 + 3x^2 - 36x$

33. $f(x) = x^4 - 24x^2 + 12x - 24$

34. $f(x) = x^4 - 6x^3 - 12x + 24$

35. Sketch the graph of $f(x) = x^3 - 12x$. Be sure to make use of what you learned in Problems 25 and 29.

36. Sketch the graph of $f(x) = x^3 - 27x$ (see Problems 26 and 30).

37. Sketch the graph of $f(x) = x^3 - 3x^2 - 12x$ (see Problems 27 and 31).

38. Sketch the graph of $f(x) = 2x^3 + 3x^2 - 36x$ (see Problems 28 and 32).

In each of Problems 39–42, we have shown the graph of a function $y = f(x)$. Indicate where (a) $f'(x) = 0$, (b) $f'(x) < 0$, (c) $f''(x) > 0$.

39.

40.

41.

42.

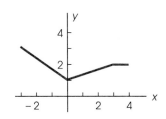

In Problems 43–50, sketch the graph of a function that satisfies all of the given conditions (see Examples D and E).

43. (i) f is continuous everywhere.
    (ii) $f(-2) = 4$ and $f(4) = -2$.
    (iii) $f'(-2) = f'(4) = 0$.
    (iv) $f''(x) < 0$ on $(-\infty, 1)$ and $f''(x) > 0$ on $(1, \infty)$.

44. (i) f is continuous everywhere.
    (ii) $f(0) = 2$ and $f(6) = 7$.
    (iii) $f'(0) = f'(6) = 0$.
    (iv) $f''(x) > 0$ on $(-\infty, 3)$ and $f''(x) < 0$ on $(3, \infty)$.

45. (i) f is continuous everywhere.
    (ii) $f(-1) = 2, f(3) = 5$, and $f(7) = 1$.
    (iii) $f'(-1) = f'(3) = f'(7) = 0$.
    (iv) $f'(x) > 0$ on $(-1, 3) \cup (7, \infty)$.
    (v) $f''(x) > 0$ on $(-\infty, 1) \cup (5, \infty)$ and $f''(x) < 0$ on $(1, 5)$.

46. (i) f is continuous everywhere.
    (ii) $f(-5) = 5, f(0) = 3$, and $f(5) = 10$.
    (iii) $f'(-5) = f'(0) = f'(5) = 0$.
    (iv) $f'(x) < 0$ on $(-5, 0) \cup (5, \infty)$.
    (v) $f''(x) > 0$ on $(-2, 3)$ and $f''(x) < 0$ on $(-\infty, -2) \cup (3, \infty)$.

47. (i) f is continuous everywhere except at $x = 6$.
    (ii) $\lim_{x \to 6^-} f(x) = 6$ and $\lim_{x \to 6^+} f(x) = 4$.
    (iii) $f'(x) = 0$ on $(-\infty, 2)$.
    (iv) $f''(x) < 0$ on $(2, 6)$ and $f''(x) > 0$ on $(6, \infty)$.

48. (i) f is continuous everywhere except at $x = 2$.
    (ii) $\lim_{x \to 2^-} f(x) = 2$ and $\lim_{x \to 2^+} f(x) = 0$.
    (iii) $f'(x) = 1$ on $(2, \infty)$.
    (iv) $f''(x) > 0$ on $(-\infty, -2)$ and $f''(x) < 0$ on $(-2, 2)$.

49. (i) f is continuous everywhere.
    (ii) $f'(x) = 1$ on $(-2, 1)$ and $f'(x) = -1$ on $(1, 3)$.
    (iii) $f''(x) > 0$ on $(-\infty, -2)$ and $f''(x) < 0$ on $(3, \infty)$.

50. (i) f is continuous everywhere.
    (ii) $f'(x) = 0$ on $(2, 5)$ and $f'(x) < 0$ on $(5, \infty)$.
    (iii) $f''(x) > 0$ on $(-\infty, 2)$ and $f''(x) < 0$ on $(5, \infty)$.

Solve the inequalities in Problems 51–58, writing your answers in interval notation.

51. $3(x - \frac{1}{2}) < 5(x + \frac{2}{3})$

52. $\frac{2}{3}(x + 2) > \frac{3}{4}(2x - 1)$

53. $x^2 - 10x \geq -16$

54. $x^2 - 7x \leq 18$

55. $(x + 3)(x - 2)^2(x - 4) < 0$

56. $(x + 1)^3(x - 1)(x - 2) > 0$

57. $(x + 3)/(x - 2) > 0$

58. $(x + 1)^2/(x - 3) < 0$

In each of Problems 59–62, interpret the given statement in terms of derivatives and sketch a possible graph.

59. From 1970 to 1990, unemployment $U = f(t)$ in a certain country continued to rise, but at a slower and slower rate.

60. The number $N = g(t)$ of cases of AIDS in the United States grew more and more rapidly from 1980 to 1990.

61. When given an antibiotic, John's temperature $T = f(t)$ fell rapidly at first but then more and more slowly and it finally leveled off at 99°F after 10 hours.

62. The number $I = g(t)$ of items produced per hour after t weeks by a new production worker tends to grow

rapidly at first but then more and more slowly, leveling off after about 15 weeks.

63. In the presence of an unlimited food supply, the population $y = g(t)$ of ants in a certain area at time t will tend to grow in such a way that $dy/dt = ky$, where k is a positive constant. Sketch a possible graph for y. *Hint:* Note that dy/dt is always positive, so y is increasing with t. What happens to dy/dt as t increases?

64. The amount $A = f(t)$ of the radioactive element radium in a sample at time t satisfies the equation $dA/dt = -kA$, where k is a positive constant. Sketch a possible graph for A.

65. In a city with population P, the number of people $N = f(t)$ who have heard a certain rumor t days after it was started tends to satisfy the equation $dN/dt = kN(P - N)$, where k is a positive constant. Sketch a possible graph for N.

3.3
EXTREME POINTS AND INFLECTION POINTS

We begin with the question: What are the most important points on a graph? A moment's reflection may suggest the correct answer: the points where the graph changes character. Chief among these are the *extreme points*, where the graph shifts from rising to falling (or vice versa), and the *inflection points*, where the graph changes its concavity from up to down (or vice versa). It is these points that play the major role in many applications of calculus to problems of the real world (see Chapter 4). Our task for this section is to locate precisely all extreme points and inflection points of a graph. Naturally, our tools are the first and second derivatives.

Theory of Extrema

Extreme values are of two types—maximum values and minimum values. These words are so important that we define them formally.

DEFINITION

Let I, the domain of f, contain the point c. We say that

   (i) $f(c)$ is the **maximum value** of f on I if $f(c) \geq f(x)$ for all x in I.

   (ii) $f(c)$ is the **minimum value** of f on I if $f(c) \leq f(x)$ for all x in I.

Since the maximum value is the largest value of f on the whole domain, it is sometimes called the **global maximum value.** This contrasts with a **local**

Domain I

FIGURE 26

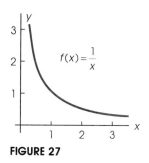

$f(x) = \dfrac{1}{x}$

FIGURE 27

POINT OF INTEREST

For f to be continuous on a closed interval $[a, b]$, it is sufficient that f be continuous at each point of (a, b) and that

$$\lim_{x \to a^+} f(x) = f(a)$$

and

$$\lim_{x \to b^-} f(x) = f(b)$$

maximum value; for it we only require that $f(c)$ be the largest value of f in a neighborhood (local region) about c. All of this is illustrated in Figure 26.

Not every function has a maximum value. Consider, for example, $f(x) = 1/x$ on the interval $(0, \infty)$ which clearly has no maximum value (Figure 27). Or consider the same function on the interval $(1, 2)$; it attains no maximum value even though it is bounded. Note that our definition requires that the function assume the maximum value. The function $f(x) = 1/x$ gets arbitrarily close to the value 1 on $(1, 2)$, but it never takes on this value. Actually, closed intervals are the natural domains to use in extrema theory because of the following theorem.

MAX-MIN EXISTENCE THEOREM

If f is continuous on a closed interval $[a, b]$, then f attains both a maximum value and a minimum value there.

Knowing that a function has a maximum value and a minimum value is helpful, but what we really want to know is how to locate the points where these values occur. A careful look at Figure 26 suggests the following result.

MAX-MIN LOCATION THEOREM

Let f be defined on an interval I containing the point c. If $f(c)$ is a maximum or minimum value (global or local), then c is one of the following:

(i)   an endpoint of I,
(ii) a point where the derivative is zero [$f'(c) = 0$],
(iii) a point where the derivative doesn't exist.

Putting the two theorems together, we can offer a simple procedure for finding the maximum and minimum values for a continuous function f on a closed interval $[a, b]$. Simply find all the **critical points** [points of the domain that are of type (i), (ii), or (iii) above]; then evaluate f at each of these points. The largest of these values is the maximum value, and the smallest is the minimum value.

EXAMPLE A

Find the maximum and minimum values of

$$f(x) = x^3 - 3x^2 + 1$$

on the closed interval $[-\frac{1}{2}, 4]$.

Solution The critical points are the endpoints $-\frac{1}{2}$ and 4 and the points where the derivative is zero. To locate the latter we find $f'(x)$, set it equal to zero, and solve.

$$f'(x) = 3x^2 - 6x$$

$$3x(x - 2) = 0$$

$$x = 0, 2$$

The values at the four critical points are $f(-\frac{1}{2}) = \frac{1}{8}, f(0) = 1, f(2) = -3$, and $f(4) = 17$. Thus the minimum value is -3 and the maximum value is 17. Figure 28 makes the situation completely clear.

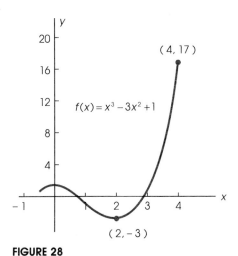

FIGURE 28

EXAMPLE B

Find the maximum and minimum values of $f(x) = x^{2/3}$ on $[-1, 8]$.

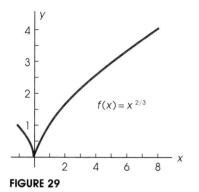

FIGURE 29

Solution $f'(x) = \frac{2}{3}x^{-1/3} = 2/(3\sqrt[3]{x})$ which is never 0. However, 0 is a critical point since $f'(0)$ fails to exist. In addition there are the two endpoints -1 and 8. Since $f(-1) = 1$, $f(0) = 0$, and $f(8) = 4$, the minimum value is 0 and the maximum value is 4. We don't need to go further but, to develop confidence in our methods, we show the graph in Figure 29. ∎

EXAMPLE C

Find the maximum and minimum values of

$$f(x) = \frac{x}{x^2 + 2}$$

on $[-1, 4]$.

Solution

$$f'(x) = \frac{(x^2 + 2)1 - x(2x)}{(x^2 + 2)^2} = \frac{-x^2 + 2}{(x^2 + 2)^2}$$

When we set this expression equal to 0 and solve, we get the two solutions $\pm\sqrt{2}$, but only $\sqrt{2}$ is in the interval $[-1, 4]$. The critical points are therefore -1, $\sqrt{2}$, and 4. Since $f(-1) = -\frac{1}{3} \approx -0.3333$, $f(\sqrt{2}) = \sqrt{2}/4 \approx 0.3536$, and $f(4) = \frac{4}{18} \approx 0.2222$, we conclude that the maximum value is $\sqrt{2}/4$ and the minimum value $-\frac{1}{3}$. ∎

Tests for Local Extrema

While it is global maxima and minima that are of most interest, we often wish to locate and evaluate local maxima and minima too. Fortunately, the location theorem applies to local extrema as well as global extrema; local extrema, too, can occur only at critical points. But once we have found the critical points, we would like a simple way of deciding whether they yield local maxima, local minima, or neither. The three graphs in Figure 30 suggest an important test.

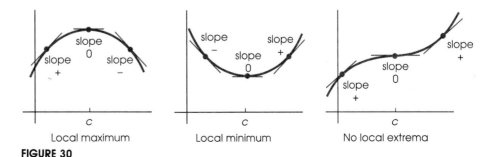

FIGURE 30

> **FIRST DERIVATIVE TEST FOR LOCAL EXTREMA**
>
> Let f be continuous on an interval containing the critical point c in its interior.
>
> (i) If $f'(x) > 0$ to the left of c and $f'(x) < 0$ to the right of c, then $f(c)$ is a local maximum value.
>
> (ii) If $f'(x) < 0$ to the left of c and $f'(x) > 0$ to the right of c, then $f(c)$ is a local minimum value.
>
> (iii) If $f'(x)$ has the same sign on both sides of c, then $f(c)$ is neither a local maximum nor a local minimum value.

EXAMPLE D

Find all local extrema of

$$f(x) = \tfrac{1}{3}x^3 - \tfrac{1}{2}x^2 - 2x + 1$$

on $(-\infty, \infty)$ and identify them correctly.

Solution

$$f'(x) = x^2 - x - 2 = (x - 2)(x + 1)$$

Sign of f'

FIGURE 31

The critical points -1 and 2 split the real line into the three intervals $(-\infty, -1)$, $(-1, 2)$, and $(2, \infty)$. On testing these intervals with the test points -2, 0, and 3, we find $f'(-2) > 0$, $f'(0) < 0$, and $f'(3) > 0$ (Figure 31). We conclude that $f(-1) \approx 2.17$ is a local maximum value and $f(2) \approx -2.33$ is a local minimum value. ■

Look at Figure 30 again and note that the graph of f is concave down at the local maximum and concave up at the local minimum. This suggests another test.

> **SECOND DERIVATIVE TEST FOR LOCAL EXTREMA**
>
> Let f' and f'' exist on an interval containing c in its interior and suppose $f'(c) = 0$.
>
> (i) If $f''(c) < 0$, then $f(c)$ is a local maximum value.
>
> (ii) If $f''(c) > 0$, then $f(c)$ is a local minimum value.
>
> (iii) If $f''(c) = 0$, the test fails.

Refer again to Example D for which

$$f(x) = \tfrac{1}{3}x^3 - \tfrac{1}{2}x^2 - 2x + 1$$
$$f'(x) = x^2 - x - 2 = (x - 2)(x + 1)$$
$$f''(x) = 2x - 1$$

At the critical points -1 and 2, $f''(-1) = -3 < 0$ and $f''(2) = 3 > 0$. The Second Derivative Test says that $f(-1)$ is a local maximum value and $f(2)$ is a local minimum value.

EXAMPLE E

Identify the critical points of

$$f(x) = \tfrac{1}{4}x^4 - x^3 - 5x^2 + 1$$

and use the Second Derivative Test to determine which yield local maximum values and which yield local minimum values.

Solution

$$f'(x) = x^3 - 3x^2 - 10x = x(x - 5)(x + 2)$$
$$f''(x) = 3x^2 - 6x - 10$$

The critical points are -2, 0, and 5. Since $f''(-2) = 14 > 0$, $f''(0) = -10 < 0$, and $f''(5) = 35 > 0$, we conclude that

$$f(-2) = -7 \text{ is a local minimum value.}$$
$$f(0) = 1 \text{ is a local maximum value.}$$
$$f(5) = -92.75 \text{ is a local minimum value.} \qquad \blacksquare$$

Inflection Points

A point on the graph of a continuous function f where the concavity changes from up to down, or vice versa, is called an **inflection point.** Candidates for inflection points are points where

(i) $f''(x) = 0$ (Figure 32), or
(ii) $f''(x)$ does not exist (Figure 33).

We can establish that a candidate is legal by checking that f'' changes sign there.

FIGURE 32

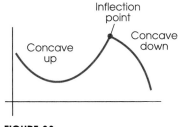

FIGURE 33

EXAMPLE F

Find the inflection points on the graph of

$$f(x) = \tfrac{1}{12}x^4 + \tfrac{1}{3}x^3 + 2x + 3$$

Solution

$$f'(x) = \tfrac{1}{3}x^3 + x^2 + 2$$

$$f''(x) = x^2 + 2x = x(x + 2)$$

The only candidates for inflection points are at $x = 0$ and $x = -2$ where $f''(x) = 0$. A quick check shows that the sign of f'' does change at these points (Figure 34). The inflection points with both coordinates given are $(-2, -2.3)$ and $(0, 3)$.   ■

Our final example wraps together most of the ideas of this section.

Sign of f''

FIGURE 34

EXAMPLE G

Sketch the graph of

$$f(x) = 3x^4 - 4x^3 + 2$$

showing all extreme points and inflection points.

Solution

$$f'(x) = 12x^3 - 12x^2 = 12x^2(x - 1)$$

$$f''(x) = 36x^2 - 24x = 12x(3x - 2)$$

The critical points (candidates for extrema) are $x = 0$ and $x = 1$. However f' does not change sign at 0, so there is no extremum there. At $x = 1$, there is a sign change from negative to positive, and so by the First Derivative Test there is a local minimum at $x = 1$.

The candidates for inflection points are at $x = 0$ and $x = \tfrac{2}{3}$, and since f'' does change sign at both places, we declare their legitimacy.

The graph is sketched in Figure 35 with both coordinates of the key points noted.

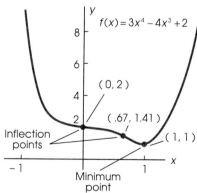

FIGURE 35   ■

Problem Set 3.3

Problems 1 and 2 show the graph of a function f on the interval [p, u] with critical values p, q, r, s, t, and u. Indicate which of these critical values will, when f is evaluated there, give the global maximum value and the global minimum value. Also indicate which of the others will give a local maximum value and a local minimum value.

1.

2.

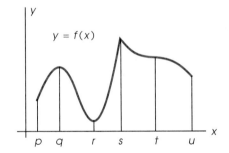

Problems 3–8 show the graph of a function f on the interval [0, ∞). Indicate whether f assumes a (global) maximum value and/or a (global) minimum value and, if so, give these value(s).

3.

4.

5.

6.

7.

8.

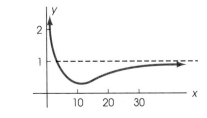

In Problems 9–26, find the maximum and minimum values of f(x) on the indicated closed interval. See Examples A, B, and C.

9. $f(x) = 2x^2 - 8x$; $[0, 3]$
10. $f(x) = 3x^2 + 6x$; $[-3, 2]$
11. $f(x) = x^2 - 6x + 3$; $[0, 5]$
12. $f(x) = 2x^2 - 8x + 3$; $[-1, 3]$
13. $f(x) = x^3 - 3x + 1$; $[-2, 3]$
14. $f(x) = x^3 - 12x + 3$; $[-3, 4]$
15. $f(x) = 2x^3 - 3x^2 - 12x + 2$; $[-3, 3]$
16. $f(x) = x^3 + 3x^2 - 24x + 10$; $[-5, 4]$
17. $f(x) = 2x^{1/3}$; $[-8, 1]$
18. $f(x) = x^{2/5}$; $[-1, 32]$
19. $f(x) = 3(x - 2)^{1/3}$; $[-6, 3]$
20. $f(x) = 2(x + 3)^{2/3}$; $[-4, 5]$
21. $f(x) = 3(x - 2)^{1/3} - x$; $[-6, 3]$
22. $f(x) = 2(x + 3)^{2/3} - \frac{4}{3}x$; $[-4, 5]$
23. $f(x) = 3x/(x^2 + 1)$; $[-2, 3]$
24. $f(x) = 2x/(x^2 + 4)$; $[-2, 3]$
25. $f(x) = x^2/(x^2 + 1)$; $[-2, 3]$
26. $f(x) = 2x^2/(x^2 + 4)$; $[-4, 2]$

In Problems 27–34, find all critical points and use the First Derivative Test to determine which yield local maximum values and which yield local minimum values (see Example D).

27. $f(x) = 2x^3 - 3x^2 + 4$
28. $f(x) = x^3 - 3x + 3$
29. $f(x) = -\frac{1}{4}x^3 + 3x - 2$
30. $f(x) = \frac{2}{27}x^3 - 2x + 1$
31. $f(x) = 2x^3 - 15x^2 + 24x + 10$
32. $f(x) = 2x^3 - 15x^2 + 36x - 20$
33. $f(x) = x + 4/x$
34. $f(x) = x^2 - 2/x$

In Problems 35–40, find all critical points and use the Second Derivative Test to decide which give local maximum values and which give local minimum values (see Example E).

35. $f(x) = x^3 - 6x^2 + 9x - 3$
36. $f(x) = x^3 + 3x^2 + 3x - 3$
37. $f(x) = x^4 - 8x^2 + 12$
38. $f(x) = 3x^4 + 8x^3 - 90x^2 + 4$
39. $f(x) = x - 4/x^2$
40. $f(x) = x^2 + 16/x$

In Problems 41–48, find all inflection points as in Example F.

41. $f(x) = x^3 - 12x$
42. $f(x) = x^3 - 27x$
43. $f(x) = x^3 - 6x^2 + 9x - 3$
44. $f(x) = x^3 + 3x^2 + 3x - 3$
45. $f(x) = x^4 - 4x^3 - 18x^2 + 10$
46. $f(x) = x^4 + 8x^3 + 24x - 20$
47. $f(x) = 3x^2 - x^{-2}$
48. $f(x) = 4(x^2 - x^{-1})$

Sketch the graph of f in each of Problems 49–52, showing the coordinates of all local extreme points and inflection points.

49. $f(x) = x^3 - 6x^2 + 9x - 3$ (Use the results from Problems 35 and 43.)
50. $f(x) = x^3 + 3x^2 + 3x - 3$ (Use the results from Problems 36 and 44.)
51. $f(x) = x^4 - 4x^3 + 10$
52. $f(x) = x^4 - 2x^3 - 5$

In Problems 53–56, find the maximum and minimum values of f on the indicated closed interval.

53. $f(x) = 2x + 8/x$; $[1, 4]$
54. $f(x) = x^2 + 4/\sqrt{x}$; $[\frac{1}{4}, 4]$
55. $f(x) = \begin{cases} x + 6 & \text{if } -1 \le x \le 0 \\ x^2 - 4x + 6 & \text{if } 0 < x \le 3 \end{cases}$
56. $f(x) = x^{4/3} - 4x^{1/3} + 2$; $[\frac{1}{8}, 8]$

In Problems 57–60, sketch the graph of f, showing the coordinates of all local extreme points and all inflection points.

57. $f(x) = x^3 - 3x^2 + 8$
58. $f(x) = x^5 - \frac{20}{3}x^3 - 10$
59. $f(x) = \begin{cases} -x^2 + 4 & \text{if } x \le 1 \\ x^2 - 4x + 6 & \text{if } x > 1 \end{cases}$
60. $f(x) = x + 9/x$
61. Sketch the graph of a continuous function f with domain $[0, 4)$ having its global minimum at $x = 0$, an inflection point at $x = 1$, and a global maximum at $x = 2$.

62. Sketch the graph of a continuous function f with domain [0, 4) which has a global maximum at $x = 0$, a local minimum at $x = 1$, a local maximum at $x = 3$, an inflection point at $x = 2$, and no global minimum.

Suppose that f has the graph shown in Problems 63 and 64. Sketch a possible graph for f '.

63.

64.

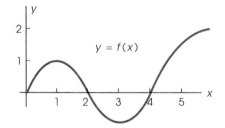

Suppose that f(0) = 0 and that f ' has the graph shown in Problems 65 and 66. Sketch a possible graph of f.

65.

66.

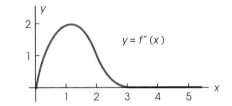

67. The concentration $f(t)$ of a certain chemical in the blood of a patient t hours after injection was given by $f(t) = 4t^2/(64 + t^4)$. Sketch the graph of f and determine when the concentration was greatest.

68. The cost C in dollars of producing x items of a certain product is given by the formula $C = 200 + 5x + 0.1x^2$. Find the value of x that will minimize the *average cost* of the product.

3.4
LIMITS INVOLVING INFINITY AND ASYMPTOTES

Our notation for infinite intervals (for example, $(4, \infty)$ and $(-\infty, 6]$) has alerted you to one use for the symbols ∞ and $-\infty$. They provide a shorthand way of indicating that an interval is unbounded to the right or to the left. Please note that we have not referred to ∞ and $-\infty$ as numbers. For example, we have never tried to add ∞ to a number or to divide ∞ by a number.

In this section, we develop other uses for the symbols ∞ and $-\infty$. We write

$$\lim_{x \to \infty} f(x) = L$$

and call it a **limit at infinity.** In writing $x \to \infty$, we simply mean that x gets larger and larger without bound. And in writing the limit statement above, we mean to say that $f(x)$ gets arbitrarily close to L as x gets larger and larger.

Similarly, in writing

$$\lim_{x \to a} f(x) = \infty$$

called an **infinite limit,** we are saying that as x approaches a, $f(x)$ gets larger and larger without bound. All of this will become clearer as we work a number of examples.

Limits at Infinity

It seems rather obvious that

$$\lim_{x \to \infty} \frac{1}{\sqrt{x}} = 0$$

After all, the numerator is fixed while the denominator grows without bound as $x \to \infty$. The table in Figure 36 confirms our reasoning. The same kind of thinking leads to the conclusion that if $p > 0$, then

$$\lim_{x \to \infty} \frac{1}{x^p} = 0$$

x	$1/\sqrt{x}$
200	.07091
2000	.02236
20000	.00707
200000	.00224
2000000	.00071
20000000	.00022

FIGURE 36

Here is another example with a similar theme.

EXAMPLE A

Find $\lim_{x \to \infty} (3x^2 + 2)/(2x^3 - 1)$.

Solution We first divide numerator and denominator by x^3, the highest power appearing in the denominator. Then we use the properties of limits stated in the Main Limit Theorem.

$$\lim_{x \to \infty} \frac{3x^2 + 2}{2x^3 - 1} = \lim_{x \to \infty} \frac{\dfrac{3x^2 + 2}{x^3}}{\dfrac{2x^3 - 1}{x^3}} = \lim_{x \to \infty} \frac{\dfrac{3}{x} + \dfrac{2}{x^3}}{2 - \dfrac{1}{x^3}}$$

$$= \frac{\lim\limits_{x \to \infty} \dfrac{3}{x} + \lim\limits_{x \to \infty} \dfrac{2}{x^3}}{\lim\limits_{x \to \infty} 2 - \lim\limits_{x \to \infty} \dfrac{1}{x^3}}$$

$$= \frac{0 + 0}{2 - 0} = 0 \qquad \blacksquare$$

The same procedure works in a problem where the limit is nonzero.

EXAMPLE B

Find $\lim_{x \to \infty} (2x^2 - x + 5)/(3x^2 + 2)$.

Solution This time we divide numerator and denominator by x^2, which amounts to dividing every term by x^2.

$$\lim_{x \to \infty} \frac{2x^2 - x + 5}{3x^2 + 2} = \lim_{x \to \infty} \frac{2 - \dfrac{1}{x} + \dfrac{5}{x^2}}{3 + \dfrac{2}{x^2}}$$

$$= \frac{2 - 0 + 0}{3 + 0} = \frac{2}{3} \qquad \blacksquare$$

EXAMPLE C

Find $\lim\limits_{x \to -\infty} \dfrac{\sqrt{4x^4 + 1}}{x^2 + \sqrt[3]{x}}$.

Solution

$$\lim_{x \to -\infty} \frac{\sqrt{4x^4 + 1}}{x^2 + \sqrt[3]{x}} = \lim_{x \to -\infty} \frac{\dfrac{\sqrt{4x^4 + 1}}{x^2}}{\dfrac{x^2 + x^{1/3}}{x^2}}$$

$$= \lim_{x \to -\infty} \frac{\dfrac{\sqrt{4x^4 + 1}}{\sqrt{x^4}}}{1 + \dfrac{1}{x^{5/3}}} = \lim_{x \to -\infty} \frac{\sqrt{\dfrac{4x^4 + 1}{x^4}}}{1 + \dfrac{1}{x^{5/3}}}$$

$$= \lim_{x \to -\infty} \frac{\sqrt{4 + \dfrac{1}{x^4}}}{1 + \dfrac{1}{x^{5/3}}} = \frac{\sqrt{4 + 0}}{1 + 0} = 2 \qquad ■$$

If $\lim_{x \to \infty} f(x) = L$ or if $\lim_{x \to -\infty} f(x) = L$, we call the line $y = L$ a **horizontal asymptote** for the graph of f. A horizontal asymptote provides a guideline for drawing the graph of f, as we demonstrate in our next example.

EXAMPLE D

Find the horizontal asymptotes (if any) for the graph of $f(x) = 2x^2/(1 + x^2)$ and show them as dotted lines. Then sketch the graph of f.

Solution

$$\lim_{x \to \infty} \frac{2x^2}{1 + x^2} = \lim_{x \to \infty} \frac{2}{\dfrac{1}{x^2} + 1} = \frac{2}{0 + 1} = 2$$

Similarly,

$$\lim_{x \to -\infty} \frac{2x^2}{1 + x^2} = 2$$

Thus the line $y = 2$ is a horizontal asymptote for the graph (in both directions). With this information and a brief table of values, we can sketch the graph (Figure 37). Note how the graph snuggles up to the line $y = 2$ as $x \to \pm\infty$.

x	$\dfrac{2x^2}{1+x^2}$
0	0
± 1	1
± 3	1.8
± 10	1.98

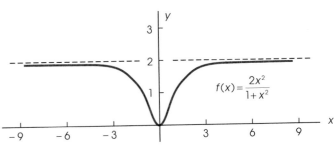

FIGURE 37

Infinite Limits

Consider the graph of $f(x) = 1/(x - 1)$ which is shown in Figure 38. It

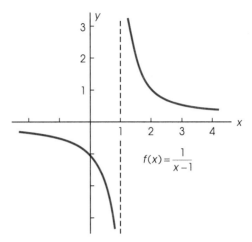

FIGURE 38

makes no sense to ask for $\lim_{x \to 1} 1/(x - 1)$, but it is reasonable to write

$$\lim_{x \to 1^-} \frac{1}{x - 1} = -\infty \qquad \lim_{x \to 1^+} \frac{1}{x - 1} = \infty$$

We say that the line $x = 1$ is a **vertical asymptote,** since the graph gets nearer and nearer to this line as $x \to 1^-$ and as $x \to 1^+$.

On the other hand, we write

$$\lim_{x \to 1} \frac{1}{(x - 1)^2} = \infty$$

since whether $x \to 1^-$ or $x \to 1^+$, $g(x) = 1/(x - 1)^2$ gets larger and larger in the positive direction. Again it is correct to say that the line $x = 1$ is a vertical asymptote (Figure 39). Note in both Figures 38 and 39 that the line $y = 0$ is a horizontal asymptote.

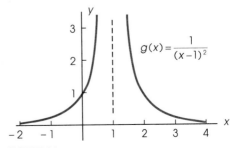

FIGURE 39

EXAMPLE E

Find the horizontal and vertical asymptotes of the graph of

$$f(x) = \frac{x^2}{x^2 - x - 6} = \frac{x^2}{(x-3)(x+2)}$$

Then sketch the graph.

Solution Since

$$\lim_{x \to \infty} \frac{x^2}{x^2 - x - 6} = 1 = \lim_{x \to -\infty} \frac{x^2}{x^2 - x - 6}$$

it follows that $y = 1$ is a horizontal asymptote (in both directions).

We expect vertical asymptotes where the denominator is zero, that is, at $x = -2$ and $x = 3$. After noting the sign of the denominator near -2 and 3 (Figure 40) and the fact that the numerator is positive, we deduce

$$\lim_{x \to -2^-} f(x) = \infty \qquad \lim_{x \to -2^+} f(x) = -\infty$$
$$\lim_{x \to 3^-} f(x) = -\infty \qquad \lim_{x \to 3^+} f(x) = \infty$$

Sign of $(x - 3)(x + 2)$

$$\begin{array}{ccccc} + & & - & & + \\ \hline & -2 & & 3 & \end{array}$$

FIGURE 40

All of this information together with a brief table of values allows us to sketch the graph shown in Figure 41.

x	$\dfrac{x^2}{x^2 - x - 6}$
-4	1.14
-3	1.5
-1	-0.25
0	0
1	-0.17
2	-1
4	2.67
5	1.79
6	1.5

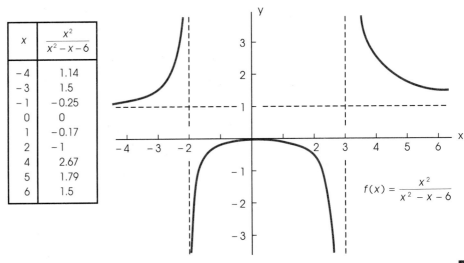

$$f(x) = \frac{x^2}{x^2 - x - 6}$$

FIGURE 41

Putting It All Together

In our study of graphs, we have developed a number of important concepts. Now it is time to summarize. We do this by offering a suggested graphing procedure.

1. Note any discontinuities.
2. Identify and draw any horizontal or vertical asymptotes.
3. Find $f'(x)$ and $f''(x)$. Use them to locate extreme points and inflection points and to determine intervals of monotonicity and concavity.
4. Make a table of values which includes the extreme points and inflection points.
5. Plot the points in Step 4 and sketch the graph.

EXAMPLE F

Sketch the graph of $f(x) = x^2/(x - 1)$.

Solution

1. There is a discontinuity at $x = 1$.
2. The line $x = 1$ is a vertical asymptote; there are no horizontal asymptotes.
3.

$$f'(x) = \frac{(x - 1)2x - x^2}{(x - 1)^2} = \frac{x^2 - 2x}{(x - 1)^2} = \frac{x(x - 2)}{(x - 1)^2}$$

$$f''(x) = \frac{(x - 1)^2(2x - 2) - (x^2 - 2x)2(x - 1)}{(x - 1)^4} = \frac{2}{(x - 1)^3}$$

The critical points are 0 and 2, and on checking the sign of f' (Figure 42) we conclude that there is a local maximum at $x = 0$ and a local minimum at $x = 2$. The sign of f'' is also shown in Figure 42 demonstrating that the graph of f is concave down on $(-\infty, 1)$ and concave up on $(1, \infty)$.

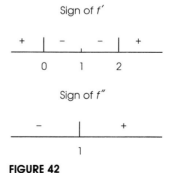

FIGURE 42

4.   See Figure 43.

5.   See Figure 43.

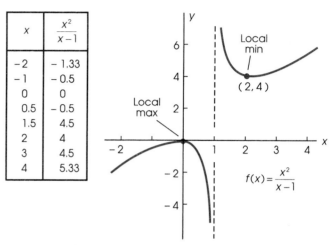

x	$\dfrac{x^2}{x-1}$
−2	−1.33
−1	−0.5
0	0
0.5	−0.5
1.5	4.5
2	4
3	4.5
4	5.33

FIGURE 43

Problem Set 3.4

Recall that the statement $\lim_{x \to \infty} f(x) = L$ means that f(x) gets closer and closer to L as x gets larger and larger. Using similar language, give meaning to each of the statements in Problems 1–6.

1. $\lim_{x \to -\infty} f(x) = 10$
2. $\lim_{x \to -3} f(x) = -\infty$
3. $\lim_{x \to \infty} f(x) = \infty$
4. $\lim_{x \to -4} f(x) = \infty$
5. $\lim_{x \to 2^+} f(x) = -\infty$
6. $\lim_{x \to 2^-} f(x) = \infty$

Find each of the limits in Problems 7–20.

7. $\lim_{x \to \infty} \dfrac{1}{\sqrt{x}}$
8. $\lim_{x \to \infty} \dfrac{2}{x^{4/3}}$
9. $\lim_{x \to 0} \dfrac{1}{x^2}$
10. $\lim_{x \to 0} \dfrac{1}{x}$
11. $\lim_{x \to 0^+} \dfrac{1}{\sqrt[3]{x}}$
12. $\lim_{x \to 0^-} \dfrac{1}{\sqrt[3]{x}}$

13. $\lim_{x \to 4^-} \dfrac{1}{x-4}$
14. $\lim_{x \to 4^+} \dfrac{1}{x-4}$
15. $\lim_{x \to 3} \dfrac{1}{(x-3)^4}$
16. $\lim_{x \to -2} \dfrac{1}{(x+2)^2}$
17. $\lim_{x \to \infty} \left(4 + \dfrac{3}{x}\right)$
18. $\lim_{x \to -\infty} \left(2 - \dfrac{2}{x^2}\right)$
19. $\lim_{x \to \infty} \dfrac{5 + \dfrac{3}{x}}{4 - \dfrac{2}{x^2}}$
20. $\lim_{x \to \infty} \dfrac{\dfrac{2}{x}}{1 + \dfrac{3}{x^2}}$

As in Examples A, B and C, find each of the limits in Problems 21–38.

21. $\lim_{x \to \infty} \dfrac{2x + 5}{3x^2 - 2}$
22. $\lim_{x \to \infty} \dfrac{3x - 2}{x^2 - 2x}$
23. $\lim_{x \to -\infty} \dfrac{3 - x}{5 - x^2}$
24. $\lim_{x \to -\infty} \dfrac{5 + 3x}{2x^2 - 4}$

25. $\lim\limits_{x\to\infty} \dfrac{2 - 4x}{6 + 3x}$

26. $\lim\limits_{x\to\infty} \dfrac{3x + 6}{8 - x}$

27. $\lim\limits_{x\to\infty} \dfrac{2x^2 - x + 11}{x^2 + 70}$

28. $\lim\limits_{x\to\infty} \dfrac{3x^2 + 25}{-x^2 + 5x}$

29. $\lim\limits_{x\to-\infty} \dfrac{2x^3 - 2x^2 + x + 7}{4x^3 - 10x + 2}$

30. $\lim\limits_{x\to-\infty} \dfrac{x^3 + 11x - 4}{5x^3 + 4x^2 + 11}$

31. $\lim\limits_{x\to\infty} \dfrac{x^4 + 10}{2x^6 - 4x + 1}$

32. $\lim\limits_{x\to\infty} \dfrac{2x^3 + x - 10}{x^4 + 3x^3 + 2x}$

33. $\lim\limits_{x\to\infty} \dfrac{5\sqrt{x} - 2}{\sqrt{x} + 1}$

34. $\lim\limits_{x\to\infty} \dfrac{2x^{3/2} + 1}{x^{3/2} + 5}$

35. $\lim\limits_{x\to-\infty} \dfrac{1 - \sqrt[3]{x}}{\sqrt[3]{x} + 2}$

36. $\lim\limits_{x\to-\infty} \dfrac{2\sqrt[3]{x}}{4 + \sqrt[3]{x}}$

37. $\lim\limits_{x\to-\infty} \dfrac{\sqrt{x^6 + 2x^2 + 7}}{2x^3 - x + 4}$

38. $\lim\limits_{x\to\infty} \dfrac{\sqrt{9x^4 + 3x + 7}}{(2x + 3)^2}$

In Problems 39–50, find the horizontal and vertical asymptotes (if any) for the graph of each function (see Examples D and E).

39. $f(x) = -\dfrac{3}{x^2}$

40. $f(x) = \dfrac{3x^2}{x^2 + 4}$

41. $f(x) = \dfrac{3 - x}{x + 2}$

42. $f(x) = \dfrac{2x + 7}{4x - 1}$

43. $f(x) = \dfrac{2 + 3x + 2x^2}{-x^2 + 5x}$

44. $f(x) = \dfrac{x^2 - 5x + 4}{2x^2 - 18}$

45. $f(x) = \dfrac{x^2 - 4}{x^2 - 2x - 3}$

46. $f(x) = \dfrac{4 - x^2}{2x^2}$

47. $f(x) = \dfrac{2x^3 + 7x - 3}{x^2 + x + 2}$

48. $f(x) = \dfrac{3x^2}{2x^2 - 2x + 1}$

49. $f(x) = \dfrac{x^2 + 2x - 3}{x^2 - 1}$

50. $f(x) = \dfrac{x^2 - x - 6}{x^2 - 2x - 3}$

In Problems 51–60, sketch the graph of f using the five-step procedure in Example F.

51. $f(x) = \dfrac{x}{x - 2}$

52. $f(x) = \dfrac{x - 3}{x - 2}$

53. $f(x) = \dfrac{x}{x^2 - 4}$

54. $f(x) = \dfrac{x + 1}{x^2 - 9}$

55. $f(x) = \dfrac{3x^2}{x^2 + 2}$

56. $f(x) = \dfrac{x^2 - 4}{x^2 + 4}$

57. $f(x) = \dfrac{x^2 + 1}{x}$

58. $f(x) = \dfrac{x^2 + 9}{2x}$

59. $f(x) = x - \dfrac{4}{x}$

60. $f(x) = x^2 - \dfrac{1}{x}$

61. Find the horizontal and vertical asymptotes (if any) for the graph of $f(x) = (2x^2 - 3)/[x(x^2 - 9)]$.

62. Find the horizontal and vertical asymptotes (if any) for the graph of $f(x) = 3x/(\sqrt{x} - 2)^2$.

63. Give the formula $f(x) = \ldots$ for a function that has a horizontal asymptote at $y = 4$ and vertical asymptotes at $x = -2$ and $x = 3$.

64. Give the formula $f(x) = \ldots$ for a function that has a horizontal asymptote at $y = -2$ and vertical asymptotes at $x = 0$, $x = 1$, and $x = 2$.

65. Find each of the following limits.

(a) $\lim\limits_{x\to\infty} \dfrac{\sqrt{x} + 2}{x}$

(b) $\lim\limits_{x\to\infty} \dfrac{2x - 3}{\sqrt{x^2 + 1}}$

66. Graph the function $f(x) = \dfrac{2x^2 - x - 1}{x^2 - 3x - 4}$

67. A manufacturer estimates that the total cost of producing x television sets per month is $C(x) = 42{,}000 + 250x$ dollars. Find the limiting value of the average cost per set, $C(x)/x$, as x gets larger and larger.

68. A manufacturer of freezers has fixed monthly costs of $28,000 and estimates the cost of labor and materials for each freezer to be $300. Assuming that it can make x such freezers each month and sell them at $475 each, find the limiting value of the average profit per freezer as x gets larger and larger.

69. Experience shows that an assembly line worker can perform a certain task in $12[1 + 1/(4\sqrt{n})]$ minutes after doing this task n times. How long will this task take in the long run?

70. A right triangle has one leg of length 10, the other leg of length x, and a hypotenuse of length h. Find (a) the limit of the ratio x/h as x gets larger and larger, (b) the limit of the ratio of the perimeter of the triangle to its area as x gets larger and larger.

71. For an equilateral triangle of side length x, find the limit of the ratio of its perimeter to its area as x tends to 0.

72. Show that the graph of $f(x) = (2x^2 + x + 1)/x$ has the line $y = 2x + 1$ as a slant asyinptote.

CHAPTER 3 REVIEW PROBLEM SET

1. Let $f(x) = (x^3 - 8)/(x^2 - 4)$.
   (a) Show that f is continuous at $x = 1$.
   (b) Is f continuous at $x = -2$?
   (c) How should $f(2)$ be defined to remove the discontinuity of f at $x = 2$?

2. Let $f(x) = (x^2 - x - 6)/(x^2 + x - 2)$.
   (a) Show that f is continuous at $x = 2$.
   (b) Is f continuous at $x = 1$?
   (c) How should $f(-2)$ be defined to remove the discontinuity at $x = -2$?

3. Solve each of the following inequalities, writing the solution set in interval notation.
   (a) $7x - 6 \le 12x + 1$
   (b) $2x^2 + 5x < 3$

4. Solve the following inequalities, writing the solution in interval notation.
   (a) $3(x - 2) \ge 2x + 14$
   (b) $3x^2 - 10x < 8$

In Problems 5–12, indicate whether or not the given function is continuous at $x = 1$. If it is continuous at $x = 1$, indicate whether or not it is differentiable there.

5. $f(x) = \dfrac{x^2 - 3x + 2}{x^2 + 4}$

6. $f(x) = \dfrac{x^2 + 4x - 5}{x^2 - x - 2}$

7. $f(x) = \dfrac{x^2 - 3x + 2}{x^2 + 2x - 3}$

8. $f(x) = \dfrac{x^2 + 3x - 4}{2x^2 + 3x - 5}$

9. $f(x) = |3x - 3|$

10. $f(x) = |6 - 3x|$

11. $f(x) = \begin{cases} x^3 & \text{if } x \le 1 \\ 2x^2 - x & \text{if } x > 1 \end{cases}$

12. $f(x) = \begin{cases} 2x^2 - x + 2 & \text{if } x \le 1 \\ 1.5x^2 + 1.5 & \text{if } x > 1 \end{cases}$

13. Find the global maximum and minimum values of
$$f(x) = \sqrt[3]{x} - \tfrac{1}{3}x + 10$$
on the interval $[0, 27]$.

14. Find the global maximum and minimum values of $f(x) = \sqrt[5]{x} - \tfrac{1}{3}x + 8$ on the interval $[-1, 32]$.

15. Find all critical points for $f(x) = x^3 + 3x^2 - 24x + 7$ and determine which yield local maximum values and which yield local minimum values.

16. Find all critical points for $f(x) = 4x^3 + 15x^2 - 18x - 200$ and determine which ones yield local maximum values and which ones yield local minimum values.

17. Find both coordinates for all inflection points of the graph of $f(x) = x^4 - 24x^2 + 36x - 10$.

18. Find both coordinates for all inflection points of the graph of $f(x) = x^4 - 12x^3 + 48x^2 + 4x - 200$.

19. Let $f(x) = x^4 - 4x^3 - 3$.
   (a) Where is f increasing? Decreasing?
   (b) Where is the graph of f concave up? Concave down?
   (c) Sketch the graph of f.

20. Let $f(x) = 3x^4 - 4x^3 + 8$.
   (a) Where is f increasing? Decreasing?
   (b) Where is the graph of f concave up? Concave down?
   (c) Sketch the graph of f.

Find each of the limits in Problems 21–28.

21. $\displaystyle\lim_{x \to 2^+} \dfrac{4}{x^2 - 4}$

22. $\displaystyle\lim_{x \to 1^+} \dfrac{2}{x^3 - 1}$

23. $\displaystyle\lim_{x \to 0^-} \dfrac{2}{\sqrt[3]{x}}$

24. $\displaystyle\lim_{x \to 2^-} \dfrac{5}{\sqrt[3]{x - 2}}$

25. $\displaystyle\lim_{x \to \infty} \dfrac{3x - 2}{x^2 + 2}$

26. $\displaystyle\lim_{x \to \infty} \dfrac{2x^3 + x - 1}{5x^2 + 4}$

27. $\displaystyle\lim_{x \to \infty} \dfrac{4 - \sqrt{x}}{\sqrt{x} + 2}$

28. $\displaystyle\lim_{x \to \infty} \dfrac{3 + 2x\sqrt{x}}{-x\sqrt{x} + 2}$

29. Consider $f(x) = x + (25/x)$ on $(0, \infty)$.
   (a) Where is f increasing? Decreasing?
   (b) Where is the graph of f concave up? Concave down?
   (c) Find any local maximum or local minimum values.
   (d) Sketch the graph of f, showing all asymptotes.

30. Consider $f(x) = x - 4/x^2$ on $(-\infty, 0) \cup (0, \infty)$.
    (a) Where is f increasing? Decreasing?
    (b) Where is the graph of f concave up? Concave down?
    (c) Determine any local maximum or local minimum values.
    (d) Sketch the graph of f, showing all asymptotes.

31. Sketch the graph of $f(x) = 2x^2/(x^2 - 9)$, showing all asymptotes and clearly labeling any local extreme points.

32. Sketch the graph of $f(x) = 3x^2/(x^2 - 1)$, showing all asymptotes and labeling any local extreme points.

33. Give a formula for $f(x)$ if f is a function whose graph has a horizontal asymptote at $y = 2$ and vertical asymptotes at $x = -1$, $x = 0$, and $x = 2$.

34. Give a formula for $f(x)$ if f is a function whose graph has a horizontal asymptote at $y = -2$ and vertical asymptotes at $x = -2$, $x = 0$, and $x = 3$.

35. Let $f(x) = \sqrt[3]{x}/(x + 2)$.
    (a) What is the domain of f?
    (b) Where is f discontinuous?
    (c) Find $\lim_{x \to \infty} f(x)$.
    (d) Where is f not differentiable?
    (e) Evaluate $f'(8)$.

36. Let $f(x) = \sqrt[3]{x} - 2/(x - 4)$.
    (a) What is the domain of f?
    (b) Where is f discontinuous?
    (c) Find $\lim_{x \to \infty} f(x)$.
    (d) Where does f fail to be differentiable?
    (e) Find $f'(3)$.

37. Evaluate

$$\lim_{h \to 0} \frac{\sqrt[3]{8 + h} - \sqrt[3]{8}}{h}$$

by recognizing this to be the value of a certain derivative.

38. Evaluate

$$\lim_{h \to 0} \frac{(16 + h)^{3/4} - 2(16 + h)^2 - [16^{3/4} - 2(16)^2]}{h}$$

by recognizing this to be the value of a certain derivative.

The chart below shows economic growth in the United States above and below the historical trend of $2\frac{1}{2}$% per year (represented by the zero line) and is typical of the many charts that appear in your economics textbook. Economists analyze such charts and propose mathematical models to explain them.

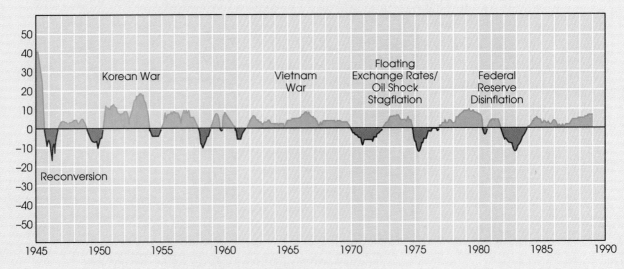

4

Other Applications of the Derivative

It has been said that Paul Samuelson is the world's best known economist. Literally millions of students have cut their economic teeth on his introductory text, *Economics,* first published in 1948 and still going strong in its umpteenth edition. In its influence, this book must rank with the nineteenth-century classics of John Stuart Mill and Alfred Marshall. Yet it represents just a small part of the output of this prolific scholar.

Born in Gary, Indiana, Samuelson went to the University of Chicago for his B.A. (1935) and then to Harvard for a Ph.D. (1941). His early essay on the interaction of the multiplier and the accelerator (originally composed as a term paper for a graduate course) demonstrated his originality and insight. His doctoral dissertation, largely written when he was 23 years old, became the momentous book *Foundations of Economic Analysis* and formed part of the basis for his award of the Nobel prize in economic science in 1970.

Moving to Massachusetts Institute of Technology in 1941, Samuelson established himself as a scholar supremely able to apply the tools of mathematics to economic problems. The topic of maximization subject to constraints (a topic treated in its simplest setting in this chapter) is a favorite theme in many of his writings.

Samuelson's view of making mathematical models to explain economic data is expressed in his own words near the beginning of his elementary text: "Even if we had more and better data, it would still be necessary—as in every science—to *simplify,* to *abstract* from the infinite mass of detail. No mind can comprehend a bundle of unrelated facts. All analysis involves abstraction. It is always necessary to *idealize,* to omit detail, to set up simple hypotheses and patterns by which the facts can be related, to set up the right questions before going out to look at the world. Every theory, whether in the physical or biological or social sciences, distorts reality in that it oversimplifies. But if it is good theory, what is omitted is outweighed by the beam of illumination and understanding thrown over the diverse empirical data."

Paul Anthony Samuelson
(1915–   )

OPTIMIZATION: GEOMETRIC PROBLEMS

We are all engaged in optimizing something. Students want to maximize their grades or perhaps minimize their study time. Professors aim to maximize their number of publications, the quality of their teaching, and the size of their salaries. These are complex problems and not easily subjected to mathematical analysis. There is, however, a host of practical optimization problems for which the tools of calculus are admirably suited. We begin with a particularly simple example.

EXAMPLE A

Mark Farmer has 100 feet of fencing with which he plans to enclose a rectangular garden plot. What are the dimensions of the rectangle that will make the area of his garden as large as possible?

Solution Several possible rectangular plots are shown in Figure 1. They lend support to our intuitive feeling that the rectangle should be a square. Calculus can turn this feeling into an irrefutable fact.

Consider a general rectangle with sides of length x and y, respectively (Figure 2.) We want to maximize the area $A = xy$, but unfortunately we as

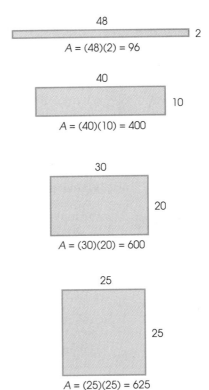

FIGURE 1

POINT OF INTEREST
In Example A, one could certainly argue that the allowable values for x are $0 < x < 50$, claiming that the values $x = 0$ and $x = 50$ do not yield rectangles. If we adopt this view, we have only one critical value, $x = 25$. To see that it gives the maximum area, we use the First Derivative Test, noting that $dA/dx > 0$ for $x < 25$ and $dA/dx < 0$ for $x > 25$. See Example F for a specific illustration of this point of view.

$$A = xy$$

FIGURE 2

yet have no theory for maximizing a function of two variables. However, the fact that Mark has 100 feet of fencing places a *constraint* on x and y, namely,

$$2x + 2y = 100$$

Happily, this allows us to solve for y in terms of x and thereby express A in terms of x alone.

$$x + y = 50$$
$$y = 50 - x$$
$$A = xy = x(50 - x) = 50x - x^2$$
$$\frac{dA}{dx} = 50 - 2x$$

The allowable values for x are $0 \le x \le 50$, so there are three critical points: the endpoints 0 and 50 and the point 25 where $dA/dx = 0$. At the endpoints, $A = 0$; at $x = 25$, $A = 50(25) - (25)^2 = 625$. Since a continuous function on a closed interval must attain a maximum (Max-Min Existence Theorem) and there are only the three indicated critical points, it must be that $x = 25$ yields the largest area. The rectangle of maximum area is a square. ∎

The example just considered suggests a procedure to follow in solving practical optimization problems.

1. If possible, draw a picture that represents the problem.
2. Assign symbols to the significant variables and express the quantity Q that is to be optimized in terms of them.
3. Determine the relations between the variables and use them to express Q in terms of a single variable.
4. Use the techniques in Section 3.3 to locate the point that gives the optimum value.

Variations on the Theme

When Mark Farmer thought about his problem, he realized that a bigger plot could be obtained by putting his garden next to a building.

FIGURE 3

EXAMPLE B

If Mark Farmer uses his 100 feet of fencing to fence a rectangular plot next to his long barn (Figure 3), what dimensions will make the garden as large as possible?

Solution The new feature is that the barn side of the garden does not need to be fenced. Following the suggested procedure, we have drawn a picture and assigned letters to the key variables. Again we want to maximize $A = xy$, but now the relation between x and y is $x + 2y = 100$. Thus,

$$y = 50 - \tfrac{1}{2}x$$
$$A = xy = x(50 - \tfrac{1}{2}x) = 50x - \tfrac{1}{2}x^2$$
$$\frac{dA}{dx} = 50 - x$$

The domain for x is $0 \le x \le 100$, and so the critical points are the endpoints 0 and 100 and the point 50 where $dA/dx = 0$. The latter gives the maximum value of $A = 50(50) - \tfrac{1}{2}(50)^2 = 1250$. The rectangle has dimensions 50 by 25 feet. ◼

EXAMPLE C

Could Mark Farmer do still better if he fitted his garden into the L-shaped corner next to his house (Figure 4)? What dimensions then give the maximum area and what is this area?

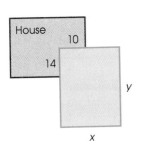

FIGURE 4

Solution All is as before except that now the relation between x and y is $x + y + (x - 10) + (y - 14) = 100$. Thus,

$$2x + 2y = 124$$
$$y = 62 - x$$
$$A = xy = x(62 - x) = 62x - x^2$$
$$\frac{dA}{dx} = 62 - 2x$$

Note that the smallest y can be is 14, which corresponds to $x = 48$. We conclude that the domain for x is $10 \le x \le 48$, and so the critical points are 10, 48, and 31. At these points, $A = 520$, 672, and 961, respectively. The dimensions of the plot of maximum area 961 are 31 by 31 feet. Farmer will do better by putting his garden next to the barn as in Example B. ◼

Best Containers

The next three examples illustrate how important it is to read a problem carefully before you try to solve it.

EXAMPLE D

An open box is to be made from a square sheet of tin 12 centimeters on a side by cutting small squares from each of the corners and turning up the edges (Figure 5). What are the dimensions of the resulting box if its volume is to be a maximum?

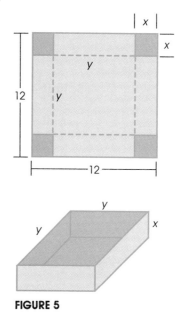

FIGURE 5

Solution The pictures have been drawn, and the significant variables labeled (Figure 5). We want to maximize the volume $V = xy^2$. Since $y = 12 - 2x$, we may write

$$V = xy^2 = x(12 - 2x)^2 = 144x - 48x^2 + 4x^3$$

$$\frac{dV}{dx} = 144 - 96x + 12x^2 = 12(12 - 8x + x^2) = 12(x - 6)(x - 2)$$

The domain for x is $0 \leq x \leq 6$, so we have the three critical points 0, 6, and 2. Both $x = 0$ and $x = 6$ make $V = 0$. The value $x = 2$ yields the maximum volume $V = 2(8)^2 = 128$ cubic centimeters. The box of maximum volume is 2 centimeters deep with a base 8 centimeters by 8 centimeters. ∎

EXAMPLE E

An open box with square base is to have a total of 144 square centimeters of tin in its sides and base. What are the dimensions of the box if its volume is to be a maximum?

FIGURE 6

Solution A picture with labels appears in Figure 6. We want to maximize $V = x^2y$ with the constraint that the surface area be 144. There are four sides of area xy, and the base has area x^2. Thus,

$$4xy + x^2 = 144$$

$$4xy = 144 - x^2$$

$$y = \frac{144 - x^2}{4x}$$

$$V = x^2y = \frac{x^2(144 - x^2)}{4x} = 36x - \frac{1}{4}x^3$$

$$\frac{dV}{dx} = 36 - \frac{3}{4}x^2$$

The domain for x is $0 \le x \le 12$. The endpoints 0 and 12 give volume 0. The other critical point is obtained by setting dV/dx to 0 and solving for x.

$$36 - \tfrac{3}{4}x^2 = 0$$

$$36 = \tfrac{3}{4}x^2$$

$$48 = x^2$$

$$x = \sqrt{48} = 4\sqrt{3} \approx 6.928$$

The maximum volume is

$$V = x^2y = \frac{(48)(144 - 48)}{4(4\sqrt{3})} = (48)(2\sqrt{3}) = 96\sqrt{3} \approx 166.3$$

and the required dimensions are depth $2\sqrt{3}$ centimeters and base a square of side $4\sqrt{3}$ centimeters.   ■

EXAMPLE F

A closed cylindrical can is to have a volume of 1000 cubic centimeters. Find the dimensions of the can that will minimize the amount of tin required to make it (assuming no waste).

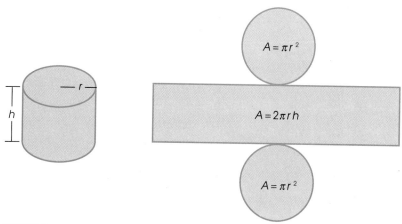

FIGURE 7

Solution Note the picture of the can together with the material needed to make it (Figure 7). We want to minimize the surface area $S = 2\pi r^2 + 2\pi rh$ given that $V = 1000 = \pi r^2 h$. We solve the latter equation for h and substitute in the expression for S.

$$S = 2\pi r^2 + 2\pi r \frac{1000}{\pi r^2} = 2\pi r^2 + 2000 r^{-1}$$

$$\frac{dS}{dr} = 4\pi r - \frac{2000}{r^2}$$

The domain for r is $0 < r < \infty$. The only critical point is obtained by setting dS/dr to 0.

$$4\pi r - \frac{2000}{r^2} = 0$$

$$4\pi r^3 - 2000 = 0$$

$$r^3 = \frac{2000}{4\pi}$$

$$r = \sqrt[3]{2000/4\pi} = \sqrt[3]{500/\pi} \approx 5.419$$

We find that $dS/dr < 0$ to the left of $\sqrt[3]{500/\pi}$ and $dS/dr > 0$ to the right of this point. This implies that $r = 5.419$ centimeters and $h = 1000/\pi r^2 \approx 10.839$ centimeters give the cylinder of smallest surface area. ∎

Best Routes

Our final examples are geometric in nature but also involve the familiar formula $D = RT$ (distance = rate × time) in its equivalent form $T = D/R$.

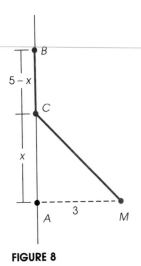

FIGURE 8

EXAMPLE G

Mary Drinkwater is in a boat 3 miles from the nearest point A on a straight shoreline. Down the shore 5 miles from A is her home at B. If Mary can row 6 miles per hour and run 10 miles per hour, where should she land her boat to arrive home in the least amount of time?

Solution We think the problem is to find x in Figure 8 so that the traveling time on the red path is a minimum. Note that $0 \le x \le 5$.

The distance MC is $\sqrt{9 + x^2}$, and the time required to row it is $\sqrt{9 + x^2}/6$. The time needed to run CB is $(5 - x)/10$. Mary's task is to minimize T, where

$$T = \frac{\sqrt{9 + x^2}}{6} + \frac{5 - x}{10}$$

Now

$$\frac{dT}{dx} = \frac{1}{6} \cdot \frac{1}{2}(9 + x^2)^{-1/2} \cdot 2x - \frac{1}{10}$$

$$= \frac{x}{6\sqrt{9 + x^2}} - \frac{1}{10}$$

When we set dT/dx to 0 and solve, we obtain

$$\frac{x}{6\sqrt{9 + x^2}} = \frac{1}{10}$$

$$10x = 6\sqrt{9 + x^2}$$

$$100x^2 = 36(9 + x^2)$$

$$64x^2 = 324$$

$$x = \pm\frac{18}{8} = \pm2.25$$

For the domain $0 \le x \le 5$, there are only the three critical points 0, 2.25, and 5. The corresponding values of T are 1, 0.9, and 0.972. Mary should head for the point 2.25 miles down the shore from A. ■

All problems considered so far have achieved their optimum value at a point where the derivative is 0. This is the typical but not the universal situation in applied problems. Our last example exhibits an exception.

EXAMPLE H

If Mary in Example G has a small motor which allows her boat to go 12 miles per hour, then where should she land?

Solution

$$T = \frac{\sqrt{9 + x^2}}{12} + \frac{5 - x}{10}$$

$$\frac{dT}{dx} = \frac{x}{12\sqrt{9 + x^2}} - \frac{1}{10}$$

When we try to solve $dT/dx = 0$, we discover it has no real solution. Thus, the only critical points are the endpoints 0 and 5 where T has values 0.75 and 0.486, respectively. Mary should head straight for B. ∎

> **CAUTION**
> We claim that the equation $dT/dx = 0$ in Example H has no real solution. Don't take our word for it.
>
> **CHECK IT OUT**
> You should always read a math book with paper and pencil handy for filling in the details of an argument.

Problem Set 4.1

In Problems 1–8, find the maximum value of z subject to the given condition. Assume throughout that $x \geq 0$ and $y \geq 0$. Note that in order to use the theory we have developed, you will first have to use the condition to express z in terms of a single variable. For example, in Problem 1, first note that $y = 200 - 2x$; then write $z = x(200 - 2x) = 200x - 2x^2$.

1. $z = xy; 2x + y = 200$
2. $z = xy; 3x + 2y = 450$
3. $z = 3x - 2y^{-2}; xy = 1$
4. $z = -4x - 3y; xy = 48$
5. $z = x^2y; x^2 + 4xy = 48$
6. $z = x^2y; 3x^2 + 4xy = 225$
7. $z = \pi x^2 y; 2x^2 + 2xy = 150$
8. $z = \pi x^2 y; x^2 + 2xy = 48$
9. The rectangle in Figure 9 has two sides along the coordinate axes and one vertex (x, y) in quadrant I on the line $x + 2y = 8$. What values of x and y make the area of the rectangle a maximum? What is the maximum area?

FIGURE 9

10. The rectangle in Figure 10 has two sides along the coordinate axes and one vertex (x, y) in quadrant I on the parabola $y = 9 - x^2$. Find the dimensions of the rectangle for which its area is a maximum. What is the maximum area?

FIGURE 10

For Problems 11–16, see Examples A, B, and C.

11. Mary Wilson has 300 meters of fencing with which to enclose two adjacent lots as shown in Figure 11. Find the dimensions x and y that will make the total area a maximum. What is the maximum area?

FIGURE 11

12. Suppose that Mary Wilson in Problem 11 chose to put the two lots along a river in such a way that the side labeled x in Figure 11 does not require a fence. Find the dimensions x and y that make the total area a maximum. What is the maximum area?

13. Suppose that Mary Wilson in Problem 11 chose to enclose three rather than two adjacent lots with her 300 meters of fencing. What dimensions x and y would then make the total area a maximum? What is the maximum area?

14. A 3000-foot race track encloses a rectangle and its adjoining semicircular ends (Figure 12). Find the dimensions x and y of the rectangle of maximum area so enclosed.

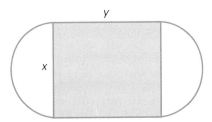

FIGURE 12

15. A gardener wishes to enclose two identical rectangular plots each of area 1200 square feet as in Figure 13. The outer boundary requires a heavy fence costing \$6 per foot, but the fence for the partition costs only \$4 per foot. For what dimensions x and y will the total cost of fencing be a minimum?

16. Christine Smith and Karen White have adjoining backyards. They have agreed to fence in adjoining garden plots (as in Figure 13) each of size 1000 square feet, sharing the cost of the partition fence on the common boundary. Determine the dimensions x and y that will minimize each woman's total cost. Assume that the same type of fence is used for the outer boundary and the partition.

FIGURE 13

For Problems 17–20, see Examples D, E, and F.

17. A tray is to be made from a 9-inch by 24-inch sheet of copper by cutting identical squares from the cor-

ners and folding up the flaps. Find the dimensions of the tray of maximum volume.

18. A closed cedar chest with a rectangular base twice as long as wide is to be built to have a volume of $\frac{64}{3}$ cubic feet. Determine the dimensions of the chest so as to minimize its surface area.

19. A cistern with a square base is to hold 16,000 cubic feet of water. The metal top costs three times as much per square foot as the concrete sides and base. What are the dimensions of the most economical cistern?

20. A closed box with a square base is to have a volume of 24 cubic feet. The material for the sides, top, and base costs 25¢, 50¢, and \$1 per square foot, respectively. Find the dimensions of the most economical box.

21. A powerhouse is located at P on one bank of a straight river that is 2000 feet wide. On the opposite bank and 10,000 feet downstream is a factory at F. A cable, costing twice as much for installation underwater as on land, is to be laid from P to B to F as shown in Figure 14. Determine x so as to minimize the cost of laying the cable.

FIGURE 14

22. Do Problem 21 again, this time assuming that underwater installation is three times as costly as on-land installation.

---

23. Mary has been asked to make a copper plaque that will contain 150 square inches of printed matter. It is to have 3-inch margins at the top and bottom and 2-inch margins at the sides. What are the outside dimensions of the plaque if it is to require the least amount of copper?

24. The Maxwell Company ships specialty fruit in rectangular boxes with square cross sections. The shipping firm requires that the sum of the length of the package and the perimeter of a cross section not exceed 100 inches. Find the dimensions of the box of largest volume that the company can use.

25. If the strength of a rectangular beam is proportional to the product of its width and the square of its depth ($S = kwd^2$), find the dimensions of the cross section of the strongest beam that can be cut from a circular log of radius 1 foot (see Figure 15).

FIGURE 15

26. Split Rock is located in a lake 3 miles from the nearest point P on a straight shoreline. Ole has been fishing all day anchored to this rock, but now it is time to head for home, located 12 miles down the shore from P. Ole will row at 4 miles per hour toward a point A on the shore and then jog at 6 miles per hour toward home. How far down the shore from P is A if Ole chooses the fastest way home?

27. A wire 40 centimeters in length is to be cut into two pieces. One piece is to be bent to form a circle and the other a square, thus enclosing two areas. How long should the shorter piece be if the sum of the two areas is to be (a) a minimum, (b) a maximum? (Allow the possibility of no cut.)

28. A window in the shape of a rectangle surmounted by an equilateral triangle is to be cut through a wall but is subject to the condition that its perimeter cannot exceed 24 feet. What are the dimensions of the rectangle if the window is to admit the maximum amount of light?

29. Find the dimensions of the right circular cylinder of maximum volume that can be inscribed in a sphere of radius 12 inches (see Figure 16).

FIGURE 16

30. Find the dimensions of the right circular cone of maximum volume that can be inscribed in a sphere of radius 12 inches.

31. Find the points on the graph of the parabola $y = x^2 - 2x + 1$ that are closest to the point $(1, \frac{5}{2})$. *Hint:* Let (x, y) be any point on the parabola. You are to minimize the distance between this point and the given point. An equivalent and easier problem is to minimize the square of this distance.

32. Consider a triangle formed by the coordinate axes and a line through the point $(4, 2)$ as in Figure 17. Find the x- and y-intercepts of the line so that the area of the triangle is a minimum.

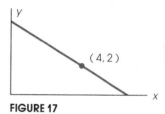

FIGURE 17

33. Find the x- and y-intercepts of the line in Problem 32 for which the hypotenuse has minimum length.

34. A paper company makes closed boxes from 5 feet by 8 feet sheets of cardboard (see Figure 18). To do this it cuts out the shaded parts shown in the figure and then folds on the dotted lines. Determine the dimensions x, y, and z which will give the box of maximum volume.

FIGURE 18

4.2
OPTIMIZATION: BUSINESS PROBLEMS

In Sections 1.4 and 1.6, we introduced a number of concepts used in business and economics. Of critical importance to a firm engaged in making and selling a product are three functions, always considered for some fixed time period (usually a quarter or a year).

$C(x)$: the cost of producing x units of the product

$R(x)$: the revenue obtained from selling these x units

$P(x) = R(x) - C(x)$: the profit generated by these x units.

In most cases, the product (refrigerators, bars of soap, reams of paper) will be in discrete units. Thus, in reality, the functions $C(x)$, $R(x)$, and $P(x)$ are defined only for nonnegative integers $x = 0, 1, 2, \ldots$ and consequently have graphs consisting of discrete points (Figure 19). However, it is conventional to connect these points with a smooth curve (Figure 20) and thereby pretend that $C(x)$, $R(x)$, and $P(x)$ are nice differentiable functions. The reason for doing this is quite obvious; it makes the powerful tools of the calculus available. Progress in analyzing real phenomena is often possible only when such simplifying assumptions are made. Making appropriate assumptions is a significant aspect of mathematical modeling.

Maximizing Profit

A firm may be concerned about many things—the pollution of its factories, the health of its employees, the reliability of its product. But in the final analysis, it will devote most of its energies to solving one problem—maximizing profit. Let's begin by looking at this problem from a general point of view.

We can expect revenue $R(x)$ to increase with x, though at a slower and slower rate. As the market becomes flooded, $R(x)$ may even decrease because of a lowering of the unit price that may be forced on the seller. As for cost $C(x)$, we can anticipate it to rise rapidly for small x (start-up costs, inefficiency of low production), continue to increase but more slowly as production becomes more efficient, and then rise rapidly as production nears capacity (overtime pay, breakdowns). A typical situation is shown graphically in Figure 21A and B.

> **POINT OF INTEREST**
> Because many real-world phenomena are discrete in nature, mathematicians are giving increasing attention to a field called Discrete Mathematics. Calculus, however, is part of Continuous Mathematics.

FIGURE 19

FIGURE 20

FIGURE 21 A

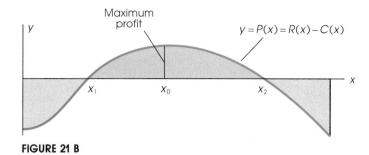

FIGURE 21 B

Note that the maximum profit occurs at the production level x_0 where $P'(x_0) = R'(x_0) - C'(x_0) = 0$, that is, where $R'(x_0) = C'(x_0)$. Thus, we have one of the fundamental principles of economics.

Maximum profit occurs when marginal revenue equals marginal cost.

To obtain appropriate cost and revenue functions for a particular firm is not a simple matter. Sometimes these functions are found by studying the history of the firm, sometimes they can be inferred from basic assumptions (see Example B), and occasionally they appear to be pulled out of a hat. Our first example is in the latter category.

EXAMPLE A

The Dresswell Company makes and sells specialty dresses. Its daily cost function and revenue function (in dollars) are

$$C(x) = 500 + 360x - 9x^2 + x^3$$
$$R(x) = 480x$$

What production schedule (value for x) yields the maximum daily profit? Assume that peak production is 14 dresses per day.

Solution

$$P(x) = R(x) - C(x)$$
$$= 480x - 500 - 360x + 9x^2 - x^3$$
$$= -x^3 + 9x^2 + 120x - 500$$

Thus,

$$P'(x) = -3x^2 + 18x + 120$$
$$= -3(x^2 - 6x - 40)$$
$$= -3(x - 10)(x + 4)$$

The only nonnegative solution to $P'(x) = 0$ is $x = 10$. This point together with the endpoints 0 and 14 must be checked for the maximum. Since $P(0) = -500$, $P(10) = 600$, and $P(14) = 200$, we conclude that a production schedule of 10 dresses per day will yield the maximum daily profit of $600. ∎

EXAMPLE B

A company estimates that it can sell 1000 units per week if it sets the unit price at $5.00, but that its weekly sales will rise by 100 units for each $0.10 decrease in price. The company has fixed costs (overhead) each week of $1050.00 and the costs in labor and materials to make a unit is $1.10.

(a) Find expressions for $R(x)$ and $C(x)$.
(b) Then determine the production level x that maximizes the profit $P(x)$ assuming $1000 \le x \le 3000$.

Solution

(a) Let $p(x)$ denote the unit price at production level x. Then

$$x = 1000 + 100 \frac{5 - p(x)}{0.10}$$

This equation (called the *demand equation*) can be solved for $p(x)$, yielding

$$p(x) = 6 - 0.001x$$

Revenue is, of course, just the number of units sold times the price received for each; that is,

$$R(x) = xp(x) = x(6 - 0.001x) = 6x - 0.001x^2$$

Total cost $C(x)$ is the sum of the *fixed cost* $1050.00 and the *variable cost* $1.10x$; that is,

$$C(x) = 1050 + 1.10x$$

(b) $P(x) = R(x) - C(x) = 6x - 0.001x^2 - 1050 - 1.10x$

$$= -0.001x^2 + 4.9x - 1050$$

and

$$P'(x) = -0.002x + 4.9$$

Setting $P'(x) = 0$ and solving gives the critical point $x = 2450$. Since $P(1000) = 2850$, $P(2450) = 4952.5$, and $P(3000) = 4650$, we conclude that a weekly production level of 2450 gives a maximum weekly profit of $4952.50. ∎

Minimizing Inventory Costs

To be successful, a retail store must pay attention to the size of its inventory. Overstocking leads to extra interest costs, excessive warehouse rental,

Lot
Size

Average
Inventory

1000

500

0

12 mo

500

250

0   6   12 mo

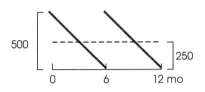

250

125

0   3   6   9   12 mo

x

$x/2$

FIGURE 22

and the danger of obsolescence or deterioration. Too small an inventory involves more paper work in reordering and extra delivery charges and increases the likelihood of running out of stock.

To illustrate, suppose the Sewbest Store expects to sell 1000 sewing machines each year. It could order in lots of size 1000, 500, 250 or in general lots of size x. On the average, the store will have $x/2$ sewing machines in stock on which it must pay inventory costs (Figure 22).

EXAMPLE C

Suppose that it costs the Sewbest Store \$20 to hold a sewing machine for a year. To reorder a lot of size x costs \$200 plus \$3 for each sewing machine. What lot size will result in the smallest inventory cost?

Solution Let $I(x)$ denote the yearly inventory cost corresponding to ordering in lots of size x.

$$I(x) = \text{carrying cost} + \text{reordering cost}$$

$$= \begin{pmatrix} \text{carrying cost} \\ \text{per unit} \end{pmatrix} \begin{pmatrix} \text{average} \\ \text{inventory} \end{pmatrix} + \begin{pmatrix} \text{cost of} \\ \text{each order} \end{pmatrix} \begin{pmatrix} \text{number} \\ \text{of orders} \end{pmatrix}$$

$$= (20)\left(\frac{x}{2}\right) + (200 + 3x)\left(\frac{1000}{x}\right)$$

$$= 10x + 200{,}000x^{-1} + 3000$$

$$I'(x) = 10 - 200{,}000x^{-2} = 10 - \frac{200{,}000}{x^2} = \frac{10(x^2 - 20{,}000)}{x^2}$$

When we set $I'(x) = 0$ and solve for x, we obtain $x = 141.42$. The First Derivative Test shows that this value of x minimizes I. Sewbest should order in lots of size about 141. ■

A Banking Problem

A bank naturally wants to increase its deposits and can do this by raising the interest it pays to savers. But raising the interest rate it pays lowers the net return it earns by reinvesting its deposits. Which is better: fewer deposits with a high net return or more deposits with a lower net return?

EXAMPLE D

The ABC Saving and Loan Association plans to encourage the opening of new time-deposit accounts by setting an attractive interest rate. It has heard that the amount of money people will deposit is proportional to the square of the interest rate they will receive. If the association can earn 16.5% simple interest on the money it raises, what simple interest rate should the association offer to savers in order to maximize its profit?

Solution Let x denote the interest rate the association will pay, and M the corresponding amount of money people will put in time deposits. Then

$$M = kx^2$$

with k being a positive constant. The association will earn on this money an amount

$$0.165M = 0.165kx^2$$

and will pay

$$xM = x(kx^2) = kx^3$$

resulting in an annual profit $P(x)$ given by

$$P(x) = 0.165kx^2 - kx^3$$

Setting $P'(x) = 0.33kx - 3kx^2 = 0$ gives the critical points $x = 0$ and $x = 0.33/3 = 0.11$. The association should offer an interest rate of 11%. ∎

Problem Set 4.2

In Problems 1–4, C(x) and R(x) represent the weekly cost function and the weekly revenue function (both in dollars) for manufacturing and selling x units of a certain product. Find (a) the weekly profit function P(x), (b) the marginal profit function P'(x), and (c) the value of x that maximizes the weekly profit.

1. $C(x) = 24x + 1000$, $R(x) = (4000x - x^2)/125$; $100 \le x \le 900$

2. $C(x) = 56x + 6000$, $R(x) = 140x - 0.1x^2$; $100 \le x \le 500$

3. $C(x) = 1000 + 20x - 0.05x^2$, $R(x) = 50x - 0.10x^2$; $50 \le x \le 350$

4. $C(x) = 400 + 3.5x - \frac{1}{300}x^2$, $R(x) = 7x - \frac{1}{200}x^2$; $200 \le x \le 1200$

5. In Problem 1, what is the unit price p for which the weekly profit is a maximum?

6. In Problem 2, what is the unit price p for which the weekly profit is a maximum?

7. The Radteck Company makes and sells radios. Its daily cost function and revenue function (both in dollars) are given by

$$C(x) = 80 + 8x - 1.65x^2 + 0.1x^3 \qquad R(x) = 32x$$

Assuming that the company can produce at most 30 radios per day, what production level x will yield the maximum daily profit?

8. Blizzard Blowers makes and sells snowblowers. At a weekly production level of x blowers ($0 \le x \le 30$), its weekly cost function and weekly revenue function (both in dollars) are given by

$$C(x) = 2000 + 40x - 33x^2 + 2x^3 \qquad R(x) = 400x$$

Determine x to give the maximum weekly profit.

9. A manufacturer of lamps estimates its daily production costs in producing x lamps as $C(x) = 600 + 8x + 0.01x^3$ dollars with $0 \le x \le 40$.
   (a) How many lamps (an integer) should be produced each day to minimize the average cost per lamp, $C(x)/x$?
   (b) What is this minimum average cost?

10. A furniture company estimates that the profit P in thousands of dollars resulting from spending x thousands of dollars per year in advertising is given by $P = 280 + 40x - \frac{1}{4}x^2$. How much would you recommend the company spend on advertising each year?

11. A manufacturer of pen and pencil sets can sell x sets per day if it sets the price per set at $p(x) = 4 - 0.002x$ dollars ($0 \le x \le 1200$).
   (a) What production level x and price p will maximize daily revenue?
   (b) If the corresponding cost function is $C(x) = 200 + 1.5x$ dollars, what production level x will maximize daily profit?

12. A maker of file cabinets with a capacity of 2500 units per month believes that it can sell x units if it sets the price at $p(x) = 75 - 0.025x$ dollars. Its monthly cost function in dollars is given by $C(x) = 3000 + 30x - 0.015x^2$. How many cabinets should it make each month to maximize its profit?

13. The ABC Company, a manufacturer of metric tool sets, has fixed monthly costs of $10,000 and direct costs of $120 for each set produced. The company estimates that 200 sets can be sold each month if the unit price is $200, and that 15 more units can be sold for each decrease of $10 in the price. Assume x, the number of units sold per month, satisfies $200 \leq x \leq 400$. ? See Example B.
    (a) Determine the monthly cost function $C(x)$.
    (b) Determine the monthly revenue function $R(x)$.
    (c) What number of tool sets should be produced to maximize monthly profit?

14. A manufacturer of hand-held calculators is confident that it can sell 3000 calculators per week at a price of $12 each and believes that reducing the price by 20¢ each will increase its weekly sales by 200.
    (a) Find the weekly revenue $R(x)$ if the manufacturer produces x calculators.
    (b) How many calculators should the manufacturer produce each week to maximize revenue? Assume a capacity of 10,000 calculators per week.

15. A stereo manufacturer estimates that in order to sell x stereos per year, the unit price must be $p(x) = 1200 - x$ dollars. The annual cost of producing x stereos is $C(x) = 30,000 + 100x$ dollars, and the annual capacity is 800 units.
    (a) Determine the annual revenue $R(x)$.
    (b) How many units should be produced to maximize the profit?
    (c) What is the maximum profit?
    (d) What price per unit maximizes the profit?
    (e) Find the value of the marginal revenue when $x = 200$.
    (f) Find (approximately) the added revenue gained by increasing production from 200 to 201.

16. Suppose the manufacturer in Problem 14 has weekly fixed costs of $2000 and unit costs of $2. How many calculators should it make each week to maximize its weekly profit?

17. The XYZ Company expects to sell 1200 power mowers per year. It costs the company $40 to hold a mower for a year. Reordering a lot of size x costs $350 plus $5 for each mower. What lot size will result in the smallest inventory cost for the year? See Example C.

18. Ace Sports Company can sell 24,000 bowling balls per year. It costs the company $5 to hold a bowling ball for a year, and reordering a lot of size x costs $250 plus $2 for each ball. In what lot size should

bowling balls be reordered to have the least annual inventory cost?

19. Eastwood Publishing Company expects to sell 100,000 copies of a new hardcover novel per year. If setting up each printing run costs $8000 and if storing a book for a year costs $0.40, what size printing run will minimize these costs?

20. Quick Start Battery Company expects to make and sell 20,000 Super A batteries per year. Setting up a production run costs $1200, and storing a battery for a year costs $3. What size run will minimize these costs?

21. To encourage time deposits, Peabody Savings and Loan Association plans to offer an attractive interest rate. Its research indicates that a simple interest rate x will result in deposits proportional to $x^{3/2}$. What interest rate should the association offer to maximize its profits if it can earn a 14% annual return on the money it raises? See Example D.

22. Suppose in Problem 21 that the amount people will deposit is proportional to $x^{5/2}$ but that the association can earn only a 12% return on the money it receives. What interest rate should be offered to maximize the association's profits?

23. A large state bank believes that the amount of money the public will deposit with it is proportional to the square of the increase in interest rate over 7%. If the bank expects to earn an 11% rate on the money it receives, what rate should it offer to investors to maximize profits? Assume that the rates are simple interest rates.

24. A research organization has told North Star Investments that it can expect deposits in proportion to the square of the increase in its interest rate beyond 4%. What rate should North Star offer to maximize profits if it can earn 10% on the money it receives?

---

25. An apartment complex has 80 units. When the rent is $400 per month, all of the units are rented. For each $10 increase in rent, one apartment unit becomes vacant. What rent should be charged to produce the maximum revenue?

26. A factory has 15 machines each capable of automatically assembling 40 toy trucks per hour. It costs $60.00 to set up each machine and $3.00 per hour to run it. One operator can oversee all 15 machines, and he earns $19.20 per hour. How many machines should be set up to produce 6000 toy trucks in order to minimize costs?

27. Susan Blackwood is raising a steer that now weighs 600 pounds and if sold today would bring 60¢ per pound. She estimates that the steer is gaining 30 pounds per week but that the price per pound will

likely drop by 2¢ a week over the next several weeks. After how many weeks should she market her steer to maximize the amount she will get for it?

28. Problem 27 took no account of the cost of feeding a steer. Suppose it costs $2.40 per week to feed a steer. After how many weeks should Susan sell her steer to maximize her profit?

29. Refer again to Problem 27. Suppose it costs 25¢ for the feed required to produce a 1-pound increase in the weight of the the steer. After how many weeks should Susan sell her steer if she wishes to maximize her profit?

30. John Adams estimates that his farm would yield 1000 bushels of sugar beets if harvested right now. Today's beet price is $2.50 per bushel, but it will drop $0.05 a bushel each day from now on. On the other hand, John guesses that the crop is still growing at 30 bushels per day. In how many days from now should John harvest and sell his crop to obtain maximum revenue (and presumably maximum profit)?

31. Northwood Independent Bank presently has $1 million in savings deposits on which it is paying the equivalent of 7% = 0.07 simple interest. To increase deposits, it plans to increase the rate it pays by an amount x and estimates that it will thereby gain new deposits M proportional to x^2 (that is, $M = kx^2$). To keep its present deposits, it will pay the new rate on all deposits—old as well as new. The bank can earn an annual rate of 11% on all this money.

(a) Show that the bank's annual profit $P(x)$ in millions of dollars on savings deposits under the new rate is given by

$$P(x) = 0.04 - x + 0.04kx^2 - kx^3$$

(b) One critical point is $x = 0$ (no increase). Show that the others are

$$x = \frac{0.08 \pm \sqrt{0.0064 - 12/k}}{6}.$$

[C](c) Assuming that $k = 10,000$, determine the interest rate that maximizes $P(x)$.

[C](d) What is the maximum value of P in dollars?

4.3*

MORE APPLICATIONS

Differential calculus achieved its earliest successes in the fields of astronomy and physics. It is, in fact, in the physical sciences that the concepts of calculus find their most obvious uses. Yet, more and more, the life sciences and the social sciences are turning to mathematics (including calculus) to find tools with which to solve their problems. We saw in the previous section how economists use calculus to minimize costs and maximize profits. Here we offer a broad spectrum of other applications.

Optimum Density

It is easy to be too greedy. Plant too many corn stalks on an acre of ground and you will actually reduce your yield; move tables too close together in a restaurant and diners will stay away in droves; put too many secretaries in the same office and their output will decrease. Calculus can help us analyze the problem of crowding.

EXAMPLE A

An apple grower estimates that each tree will produce 35 baskets of apples when there are 40 trees per acre, but that for each additional tree planted the yield per tree will go down by three-fourths of a basket. What density of trees will result in the largest apple harvest?

Solution The total yield y in baskets per acre can be expressed in terms of t, the number of trees per acre, by the formula

$$\text{total yield} = (\text{number of trees})(\text{yield per tree})$$

$$y = t[35 - \tfrac{3}{4}(t - 40)]$$

$$= t(35 - \tfrac{3}{4}t + 30)$$

$$= 65t - \tfrac{3}{4}t^2$$

Thus

$$\frac{dy}{dt} = 65 - \tfrac{3}{2}t$$

and when set equal to 0, this gives

$$t = (65)(\tfrac{2}{3}) \approx 43.33$$

It is a simple matter to check that this value of t makes y as large as possible. Our apple grower should plant about 43 trees per acre. ∎

Best Truck Speed

Here is a problem faced by all firms that deliver their products by truck. Drivers are paid by the hour, which suggests that they should be asked to drive as fast as possible. On the other hand, truck operating costs tend to increase with speed; in particular, gas mileage goes down as speed increases. At what speed should a truck be driven to minimize the total cost?

POINT OF INTEREST
In connection with Example B, someone is sure to complain that 69 miles per hour exceeds the speed limit. Fair enough. Study the graph of cost C against speed s shown below. Conclude that to minimize cost a trucker should drive at the speed limit provided this limit is 69 miles per hour or less.

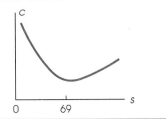

EXAMPLE B

The cost of operating a certain truck (gas, oil, and so on) is $40 + s/4$ cents per mile when driven at s miles per hour. The driver gets \$12 per hour. At what speed should the truck be driven to minimize the total cost of making a delivery to a distant city?

Solution Consider a trip of k miles. The total cost C (in cents) is given by

$$\text{total cost} = \text{truck cost} + \text{driver cost}$$

$$C = k\left(40 + \frac{s}{4}\right) + \frac{k}{s}(1200)$$

$$= 40k + \frac{ks}{4} + \frac{1200k}{s}$$

$$\frac{dC}{ds} = 0 + \frac{k}{4} - \frac{1200k}{s^2}$$

When we set dC/ds equal to 0 and solve for s, we obtain

$$\frac{k}{4} = \frac{1200k}{s^2}$$

$$s^2 = 1200k\left(\frac{4}{k}\right) = (1200)(4)$$

$$s \approx 69 \text{ miles per hour}$$

Note that the constant k canceled out at the end. The answer of 69 miles per hour is independent of the length of the trip. ∎

Drug Dosage

Doctors need to understand the characteristics of the drugs they prescribe. For a given drug, one important item is the maximum allowable dosage C. Another is the so-called **sensitivity** of the body to the drug. If x measures the dosage and $R(x)$ the strength of the reaction (change in body temperature or blood pressure or some other measurable body function), then $R'(x)$ measures sensitivity. It tells the doctor (approximately) how much change to expect in R as the result of a unit change in the dosage. Generally, doctors like to prescribe a dosage of maximum sensitivity.

EXAMPLE C

A common mathematical model for drug reactions is

$$R(x) = x^2\left(\frac{C}{2} - \frac{x}{3}\right) = \frac{Cx^2}{2} - \frac{x^3}{3} \qquad 0 \le x \le C$$

Show that the dosage of maximum sensitivity is $C/2$.

Solution Note that we want to maximize $R'(x)$ which we do by finding where $R''(x) = 0$.

$$R'(x) = Cx - x^2$$
$$R''(x) = C - 2x$$
$$R'''(x) = -2$$

Setting $R''(x) = 0$ yields $x = C/2$. Since $R'''(x)$ is negative, we can be certain that $R'(C/2)$ is at least a local maximum value for $R'(x)$. To check that it is the global maximum value, we calculate R' at the three critical points: 0, $C/2$, and C. We obtain $R'(0) = R'(C) = 0$, while $R'(C/2) = C^2/4$. ∎

Maximum Sustainable Harvest

We pointed out the danger of being too greedy earlier in this section. If one is in the business of harvesting fish, fur-bearing animals, or trees, restraint is a prerequisite to long-term success. Overharvesting will obviously interfere with the reproductive capabilities of the population.

To model this situation, let us suppose that a population of size x will grow to size $f(x)$ in 1 year if unmolested by human beings. Then to keep from depleting the population, the yearly harvest should not exceed $h(x) = f(x) - x$. If our goal is to secure the largest possible harvest year after year, we should allow the population to grow to the size which makes $h(x)$ a maximum.

EXAMPLE D

Each year, many states allow hunters to shoot deer during a limited open season whose length is carefully chosen to ensure a harvest that is sustainable year after year. For one state, the yearly growth curve for the deer population is estimated to be

$$f(x) = 1.4x - 0.0004x^2$$

where x is measured in thousands. What is the optimal population size and what is the yearly kill that it will sustain?

Solution

$$h(x) = f(x) - x = 0.4x - 0.0004x^2$$
$$h'(x) = 0.4 - 0.0008x$$
$$h''(x) = -0.0008 < 0$$

The maximum value of h is attained when $h'(x) = 0$, that is, at $x = 0.4/0.0008 = 500$. Thus, the state authorities should allow the deer population to climb to 500,000 which will then sustain a yearly harvest of

$$h(500) = 0.4(500) - 0.0004(500)^2 = 100$$

that is, 100,000 deer per year. ■

Learning Theory

There are many phenomena in real life whose behavior over time is modeled by a curve of the type $y = P(t)$ shown in Figure 23. Here are some examples.

(a) $P(t)$ is the percentage of the population infected by a contagious disease during the first t days of an epidemic.
(b) $P(t)$ is the number of people who have heard a rumor t days after the rumor was started.
(c) $P(t)$ measures the amount of learning of a skill that takes place during the first t hours of practice.

In each case P grows slowly at first, then more and more rapidly, but eventually slower and slower. The time of most interest to doctors in case (a), to

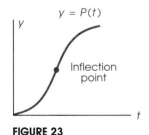

$y = P(t)$

Inflection point

t

FIGURE 23

sociologists in case (b), and to psychologists in case (c) is the moment of fastest growth, that is, the moment when $P'(t)$ is largest. This is the time when $P''(t) = 0$ and corresponds to the inflection point on the curve in Figure 23.

EXAMPLE E

Typing skill is usually measured by typing speed in words per minute. For a class of 12-year-old boys, a psychologist discovered that on the average they could achieve a speed of $P(t)$ words per minute after t hours of practice where

$$P(t) = 0.015t^2 - 0.0001t^3 \qquad (0 \le t \le 100)$$

After how many hours of practice are the boys learning most rapidly?

Solution

$$P'(t) = 0.03t - 0.0003t^2$$
$$P''(t) = 0.03 - 0.0006t$$

Setting $P''(t) = 0$ gives $t = 50$. The boys are learning most rapidly after 50 hours of practice.   ∎

Problem Set 4.3

1. A commercial orange grower used to plant 30 trees per acre and normally harvested about 20 bushels of oranges per tree. A research study indicates that planting trees closer together may improve the total harvest per acre; in particular, the study estimates that for each additional tree (up to 15) beyond 30 that is planted on an acre, the yield per tree is reduced only by $\frac{1}{2}$ bushel. How many trees per acre should be planted to maximize the orange yield per acre?

2. A nut farmer estimates that a walnut tree will yield 60 pounds of nuts when there are 40 trees per acre and that for each additional tree per acre the yield per tree is reduced by 1 pound. If so, how many trees per acre should be planted to maximize yield?

3. Experience convinced the farmer in Problem 2 that the yield is reduced only $\frac{3}{4}$ pound for each additional tree beyond 40 (up to the maximum of 55 trees per acre permitted by government regulations).
   (a) Determine the yield per acre when there are 52 trees per acre.
   (b) What density of trees will maximize the per acre yield?

4. Acme Car Rental is able to rent out 120 cars per day when it charges $12 per day (there is an additional mileage charge that just covers expenses). For each 25¢ increase in the daily rate, Acme thinks one less car would be rented.
   (a) Find the net daily income when the rate is set at $15.
   (b) What daily rate would maximize the net daily income?

5. A downtown movie theater has had an average attendance of 150 people with an admission charge of $2.50. Assume that it will consider only a charge that is a multiple of $0.10 and that it believes it will lose three customers for each $0.10 increase in the admission charge. What admission charge will maximize revenue and what revenue does this produce for each showing?

6. The Mark Twain Showboat advertises excursions to groups of 300 at $15 per ticket but promises a reduction of 20¢ on all tickets for each group of 10 passengers in excess of 300 (assume that a group of 9 additional passengers produces no reduction in the ticket price). Exactly what (whole) number of passengers produces the maximum revenue for the showboat and what is this revenue?

7. Portage Transportation Company estimates the cost of operating a moving van is $40 + \frac{1}{4}x$ cents per mile

when driven at x miles per hour. In addition, the driver gets $10 per hour.

(a) At what speed should the van be driven to minimize the total cost of a 400-mile run?

(b) Find the minimum total cost.

8. Suppose that a truck averages $400/(x + 2)$ miles per gallon when driven at a speed of x miles per hour $(40 \leq x \leq 65)$. Suppose further that gasoline costs $1.20 per gallon, that the driver gets $10.00 per hour, and that other costs are ignored. What speed will minimize the cost of a 200-mile run?

9. A long-distance trucker finds that when he maintains a speed of x miles per hour, the truck consumes $V = 3x^{-1} + 0.0015x$ gallons of diesel fuel per mile.

(a) If diesel fuel costs $1.30 per gallon, what speed will result in the lowest fuel cost for a 400-mile trip?

(b) Find the minimum fuel cost.

10. Suppose the driver in Problem 9 gets $12 per hour of driving time. Ignoring costs other than fuel and wages, what speed will minimize the cost of the 400-mile trip?

11. Suppose that the human reaction R to a certain drug is measured as

$$R(x) = \frac{Cx^2}{3} - \frac{x^3}{4}$$

where x is the dosage and C is the maximum allowable dosage.

(a) Find the dosage that maximizes the reaction $R(x)$.

(b) Find the dosage that maximizes the sensitivity $R'(x)$.

12. The reaction R to an experimental drug is given by

$$R(x) = \frac{Cx^2}{a} - \frac{x^3}{a}$$

where a is a constant and C is the maximum allowable dosage. Show that the dosage that maximizes sensitivity to the drug is one-half the dosage that maximizes the reaction.

13. Blackhawk County annually attracts a large number of hunters for its fall hunting season and would like to attract even more. It is claimed that if there were no hunting season in the county, the pheasant population would grow from size x (in thousands) to size $f(x) = 2.5x - 0.02x^2$ in 1 year. To what size should the county let the pheasant population grow to be able to maintain the largest yearly kill? What would this yearly kill then be?

14. In one of our western states, it was estimated that, if no hunting were allowed, an elk population of size x (in thousands) would grow in 1 year to a population

of size $f(x) = 1.6x - 0.0005x^2$. What population size allows for the maximum yearly kill year after year? What is the maximum kill?

15. A psychologist has found that if a 17-year-old girl is allowed to study t hours for a certain test, her score S can be predicted from the formula $S(t) = 30t^2 - 3t^3$ $(0 \leq t \leq 7)$.

(a) How many hours of study produce the highest score?

(b) At what time is learning occurring most rapidly?

16. A government study deals with the spread of a certain strain of flu. The study says that in a city of 10,000, the number of people N who will contract the disease within the first t weeks after the first reported case can be predicted from the formula

$$N(t) = 54t^2 - 2t^3 \qquad (0 \leq t \leq 18)$$

(a) How many people from the city will get the disease?

(b) When will the disease be spreading most rapidly?

---

17. Arne Carlson, a fruit grower in British Columbia, knows from experience that when he plants 24 cherry trees per acre, he can expect a yield per tree of 44 pounds. The local expert from the agricultural college claims that for each additional tree that Arne plants per acre, his yield per tree will drop by only 1 pound. Assume that Arne believes the expert.

(a) How many additional trees (beyond 24) should Arne plant on each acre to achieve maximum yield per acre?

(b) What is the maximum yield per acre?

18. A certain toll bridge averages 80,000 cars a day when the toll is $1.50. The Bridge Authority is considering increasing the toll but, because there is a free bridge only 10 miles away, estimates that 1000 fewer cars will use its bridge for each $0.05 increase in the toll. What toll (a multiple of $0.05) will maximize the daily revenue?

19. At $80 a night per room, the 120-room Royal Plaza Hotel is full to capacity. Suppose that each 80¢ increase in the room rate results in an additional vacant room and that it costs $5 to service a rented room. What should the management charge per room to maximize its profit from room rentals? Assume that the number of rented rooms must be an integer.

20. Lake Serene has to be treated periodically to control the growth of harmful bacteria. If t days after a treatment the concentration of bacteria per cubic centimeter is given by

$$C(t) = 18t^2 - 216t + 900 \qquad (0 \leq t \leq 12)$$

when would you recommend that your children go

swimming and what is the concentration of bacteria at this time?

21. When the treatment facility overflows and raw sewage is dumped into a lake, the oxygen content decreases until it reaches a minimum and then begins to increase. Suppose that t days after raw sewage was dumped into Lake Wobegon, the concentration C of oxygen (measured in appropriate units) in the lake was given by

$$C(t) = 600 \left[1 - \frac{8}{t + 4} + \frac{64}{(t + 4)^2} \right]$$

(a) After how many days will the oxygen concentration be lowest?
(b) When will the oxygen concentration be increasing most rapidly?

22. Mark Farmer has found that when he uses x bags of fertilizer per acre, his soybean yield in bushels per acre is

$$V(x) = \frac{32x}{x + 8} + 50$$

If soybeans sell at $10 per bushel and fertilizer costs $30 per bag, how many bags per acre should he use to maximize his profit?

C 23. A school is to be built along a road between factories A and B which are 8 miles apart. Both factories emit air pollutants. At a point x miles from factory A, the concentration of pollutants is measured as

$$p(x) = \frac{27c}{x} + \frac{8c}{8 - x} \qquad (0 < x < 8)$$

where c is a constant. How far from factory A should the school be built to minimize the air pollution problem?

24. The Hoodoo University Dragons are anxious to maximize the revenue from their football games. For several years they have charged $5.00 per ticket and have averaged 60,000 at their games. A typical fan spends $2.25 on concessions. A consulting firm says that for each $1.00 increase in the admission charge, there will be a loss of 6000 fans and on this basis recommends a new ticket price. What was their recommendation?

4.4

IMPLICIT DIFFERENTIATION AND RELATED RATES

The terms *explicit* and *implicit* are pretty standard English words. For example, we speak of explicit and implicit assumptions. An explicit assumption is openly stated and clearly formulated. In contrast, an implicit assumption may be hidden or vaguely expressed, but it is an assumption just the same. An important task for politicians, philosophers, and ordinary people is to make implicit assumptions explicit.

In mathematics, we use these two words to indicate two ways of determining a function. A function (with domain variable x) is described explicitly if it is determined by an equation of the type $y = f(x)$. For example, the equation

$$y = \frac{2 + x}{3 - x}$$

gives an explicit description of a function. In contrast, the equation

$$3y - xy = 2 + x$$

also determines a function (actually the same function), but now this fact is somewhat disguised. We must do some work (solve for y) to make the description explicit.

Sometimes it is easy to turn an implicit description into an explicit one; sometimes it is difficult; and sometimes it is impossible.

The equation $3y - xy = 2 + x$ is readily solved for y. First factor y from the left side; then divide by $3 - x$:

$$3y - xy = 2 + x$$
$$y(3 - x) = 2 + x$$
$$y = \frac{2 + x}{3 - x}$$

Not many of us would want to try solving the equation

$$y^9 + 2xy + \log xy = 3y$$

for y; we believe this to be impossible. Here is an example of intermediate difficulty.

EXAMPLE A

Sove $x^2 + y^2 + 2y = 3$ for y.

Solution We begin by adding 1 to both sides, thus completing the square on the y's.

$$x^2 + y^2 + 2y + 1 = 3 + 1$$
$$x^2 + (y + 1)^2 = 4$$
$$(y + 1)^2 = 4 - x^2$$
$$y + 1 = \pm\sqrt{4 - x^2}$$
$$y = -1 \pm \sqrt{4 - x^2}$$

The interesting fact discovered here is that our original equation determines two functions:

$$y = f(x) = -1 + \sqrt{4 - x^2} \quad \text{and} \quad y = g(x) = -1 - \sqrt{4 - x^2} \quad \blacksquare$$

Implicit Differentiation

Now we ask a question. Is it possible to find the derivative dy/dx without first having y expressed explicitly in terms of x? The surprising answer is yes. Consider the equation

$$3y - xy = 2 + x$$

with which we started our discussion. Take the derivative of both sides with respect to x and equate the results. Keep in mind that y is a function of x;

CAUTION

A very common error is to leave out the parentheses around

$$x\frac{dy}{dx} + y$$

in the first step. This is fatal because it results in the wrong sign on y.

thus, we will have to use the Product Rule on the xy term. We obtain

$$3\frac{dy}{dx} - \left(x\frac{dy}{dx} + y \cdot 1\right) = 0 + 1$$

$$3\frac{dy}{dx} - x\frac{dy}{dx} - y = 1$$

Next move y to the right side; then factor dy/dx from the left side.

$$\frac{dy}{dx}(3 - x) = 1 + y$$

$$\frac{dy}{dx} = \frac{1 + y}{3 - x}$$

The only troublesome feature of our answer is that it involves both y and x (we would prefer that it involved just x), but in many situations this is a minor problem.

An inquisitive reader is sure to ask whether the answer we just got for dy/dx is equivalent to the one we would have gotten if we had first solved the original equation for y and then taken the derivative. The answer is yes, though we shall not take the space to demonstrate this.

The method of finding the derivative dy/dx without first solving the given equation explicitly for y in terms of x is called the method of **implicit differentiation.**

EXAMPLE B

Find the slope of the tangent line to the circle

$$x^2 + y^2 = 25$$

at the point $(3, -4)$ using the method of implicit differentiation (see Figure 24).

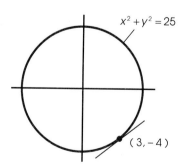

FIGURE 24

Solution We take the derivative of both sides with respect to x and equate the results. Keep in mind that y is a function of x; thus, we must use the General Power Rule on the term y^2.

$$2x + 2y \frac{dy}{dx} = 0$$

$$2y \frac{dy}{dx} = -2x$$

$$\frac{dy}{dx} = \frac{-2x}{2y} = \frac{-x}{y}$$

At $(3, -4)$, $dy/dx = -3/-4 = \frac{3}{4}$. The slope of the tangent line at this point is $\frac{3}{4}$. ∎

EXAMPLE C

Find the equation of the tangent line to

$$xy^2 + y^3 = 12x$$

at the point $(1, 2)$.

Solution You should first check to see that $(1, 2)$ really does satisfy the equation. Then take the derivative of both sides with respect to x. On the term xy^2, you will have to use the Product Rule together with the General Power Rule; on the y^3 term, you will need just the General Power Rule.

$$x \cdot 2y \frac{dy}{dx} + y^2 \cdot 1 + 3y^2 \frac{dy}{dx} = 12$$

$$\frac{dy}{dx} (2xy + 3y^2) = 12 - y^2$$

$$\frac{dy}{dx} = \frac{12 - y^2}{2xy + 3y^2}$$

At $(1, 2)$, dy/dx has the value $(12 - 4)/(4 + 12) = \frac{1}{2}$. We conclude that the tangent line has the equation

$$y - 2 = \tfrac{1}{2}(x - 1)$$ ∎

Related Rates

Here is a common problem arising in the world of experience. Two variables, say x and y, are both functions of a third variable t though these functions are not known explicitly. An equation connecting x and y is given. Can we find dy/dt if we know dx/dt? An example will clarify the question and provide a positive answer.

FIGURE 25

RECALL

In a right triangle,

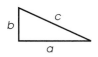

the three sides are related by the equation

$$a^2 + b^2 = c^2$$

This result is known as the Pythagorean Theorem.

EXAMPLE D

A 20-foot ladder leans against a vertical wall (Figure 25). If the bottom of the ladder is pulled away from the wall at 3 feet per second, how fast is the top of the ladder sliding down the wall when it is 12 feet above the ground?

Solution Variables x and y have been assigned in Figure 25. Note that both x and y are functions of time t. We are given that $dx/dt = 3$, and the Pythagorean Theorem implies that

$$x^2 + y^2 = 400$$

We are asked to find dy/dt when $y = 12$.

We use the method of implicit differentiation; that is, we take the derivative of both sides with respect to t, equating the results. Since both x and y are functions of t, we must apply the General Power Rule to both the x^2 and y^2 terms. We obtain

$$2x \frac{dx}{dt} + 2y \frac{dy}{dt} = 0$$

or since dx/dt is given to be 3,

$$2x(3) + 2y \frac{dy}{dt} = 0$$

$$\frac{dy}{dt} = -\frac{6x}{2y} = -\frac{3x}{y}$$

This expression for dy/dt is to be evaluated when $y = 12$. Unfortunately, we also need to know x. Once again the Pythagorean Theorem comes to our aid.

$$x^2 + (12)^2 = 400$$

$$x^2 = 256$$

$$x = 16$$

We conclude that when $y = 12$, $x = 16$ and

$$\frac{dy}{dt} = -\frac{3x}{y} = -\frac{48}{12} = -4 \text{ feet per second}$$

The negative sign is no surprise, since y is decreasing with time. ∎

<table>
<tr><td>

CAUTION

In Example E, we are asked for the rate of increase in profit with respect to time, that is, for dP/dt. Be sure to distinguish this from marginal profit which is dP/dx, the rate of change in profit with respect to the number of units produced.

</td></tr>
</table>

EXAMPLE E

A manufacturer of hand-held calculators has weekly cost and revenue functions given by

$$C = 6000 + 5x$$

$$R = 10x - 0.0001x^2$$

where x is the number of calculators produced each week. If x is increasing at the rate of 400 calculators per week, find the rate of increase in profit P when weekly production reaches 4000.

Solution We are given that $dx/dt = 400$ and want to find dP/dt when $x = 4000$. Now

$$P = R - C = 10x - 0.0001x^2 - 6000 - 5x$$

that is,

$$P = -0.0001x^2 + 5x - 6000$$

We differentiate both sides of this equation with respect to t, keeping in mind that x depends on t. We obtain

$$\frac{dP}{dt} = (-0.0001)(2x)\frac{dx}{dt} + 5\frac{dx}{dt} - 0$$

When $x = 4000$ and $dx/dt = 400$,

$$\frac{dP}{dt} = (-0.0001)(8000)(400) + 5(400)$$

$$= -320 + 2000 = 1680$$

Profit is increasing at $1680 per week. ∎

Problem Set 4.4

The equations in Problems 1–12 determine y in terms of x but do so implicitly. Solve for y, thereby giving y explicitly in terms of x (see Example A and the discussion preceding it).

1. $x^2 + 2y = 3$
2. $2xy - 5x = 6$
3. $4y + 3xy = 2 - x$
4. $x^2y - 3y = 2x + 1$
5. $x^2 + xy = y + x + 3$
6. $x^2y + 9 = 5x + 2y$
7. $(y - 4)(x^2 + x) = 3x + 2$
8. $(y - 2x)(x + 2) = 1 - x^2$
9. $y^2 + 2x = x^2 + 4$
10. $xy^2 + x^3 = 4$
11. $y^2 - x^2 - 8y + 11 = 0$

12. $y^2 + 2y - 2x + 4 = 0$
13. Find dy/dx for Problem 5 in two ways (by differentiating the explicit relation you found there and by implicit differentiation). Show that your answers are equivalent.
14. Find dy/dx for Problem 6 in two different ways.

In Problems 15–24, find dy/dx by implicit differentiation (see the discussion preceding Example B as well as Examples B and C).

15. $2x^2y = 7$
16. $3x^3y = 4x + 1$

17. $x^2y + 5 = x^2 - 4y$

18. $3xy - 2x + 4 = x^2 + 2y$

19. $y^3 + 3xy = 27$　　　　20. $2y^2 - 3x^2y = 15$

21. $x^2 + 2xy + 3y^2 = 36$　　22. $xy^2 - yx^2 = 4y + 1$

23. $(2x^2 + y^2)^4 = 20$　　　24. $(3 + y^2)^5 = 100x^2$

In Problems 25–30, find the equation of the tangent line to the given curve at the specified point (see Examples B and C). Don't forget that you must evaluate dy/dx to get the slope.

25. $x^2 + y^2 = 169$; $(-12, 5)$

26. $x^2 - y^2 = 16$; $(5, 3)$

27. $x^2 + 4y^2 = 4y + 2x$; $(2, 1)$

28. $x^2 - y^3 = 34 + y$; $(2, -3)$

29. $x^2 - 2xy + 5y^2 = 13$; $(4, 1)$

30. $y^3 - 2xy = 5$; $(-2, 1)$

31. Suppose that the point P is moving along the curve $y = x^2$ in such a way that dx/dt is always 3 units per second. Find dy/dt when $x = 2$. See Example D.

32. Suppose that the point P moves along the curve $x = y^3$ so that dx/dt is always 4 units per second. Find dy/dt when $y = 2$.

33. A 50-foot ladder leaning against a vertical wall is beginning to slide. At the instant when the top of the ladder is 30 feet from the ground, the top is sliding down at the rate of 2 feet per second. How fast is the bottom of the ladder moving away from the wall at that instant?

34. A man who is 6 feet tall is walking directly away from a street light that is 15 feet above the ground. If he walks at 5 feet per second, how fast is the tip of his shadow moving when he is 100 feet from the base of the light post? *Hint:* In Figure 26 you are to find dx/dt when $y = 100$.

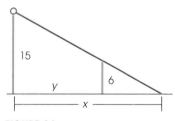

FIGURE 26

35. Suppose the point P in Figure 27 is moving to the right along the line $y = 3$ at the rate of 4 units per second.

(a) How fast is the distance $\overline{OP}$ increasing at the instant when $\overline{AP} = 6$?

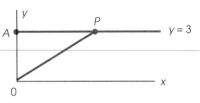

FIGURE 27

(b) How fast is the area of the triangle OAP increasing at this instant?

36. The wind causes a girl's kite to move in a horizontal plane 300 feet above the ground at 30 feet per second. If she stands still, holding her end of the cord 5 feet above the ground, how fast is the cord slipping through her fingers when 500 feet of cord are out?

37. A manufacturer of bulldozers has a weekly revenue of R thousand dollars when x bulldozers are produced and sold per week, where

$$R = 200\sqrt{x} + 180x - x^2$$

If 100 bulldozers are now being produced per week and production is being increased at the rate of 3 units per week, at what rate is the weekly revenue changing?

38. A wholesale fruit company supplies a chain of grocery stores with x thousand crates of peaches per day at p dollars per crate. Demand x is thought to be related to price p (in dollars) by the equation

$$px + 8p - 3x - 240 = 0$$

Today's demand is 9000 crates but is expected to decrease at the rate of 180 crates per day. At what rate is the price changing right now?

39. In a discount supermarket, the weekly demand for a certain brand of tea satisfies the equation $px + 12p = 600$, where x is the weekly demand in pounds and p is the price in dollars per pound. Right now the price is \$6 per pound but is decreasing at the rate of 8¢ per week.

(a) At what rate is the demand changing?

(b) At what rate is the weekly revenue changing?

40. A fast-food restaurant has determined the cost C (in dollars) of producing x hamburgers per month is $C = 3000 + 0.48x$ and that the monthly demand satisfies $p = (40,000 - x)/15,000$, where p is the price (in dollars). Right now the restaurant charges \$1.50 for a hamburger but plans to steadily increase the price at the rate of \$0.02 per month.

(a) Find the rate of change in the demand x.

(b) Find the rate of change in monthly revenue when the price of a hamburger reaches \$1.75.

(c) Find the rate at which the monthly profit is changing when the price of a hamburger reaches \$1.75.

41. Find the equation of the tangent line to $(y^2 + x - 1)^4 = 3x^2 + 4$ at the point $(2, 1)$.

42. A squirrel is running along a telephone wire which has the shape of a parabola with equation $y = 15 + 0.01x^2$ $(-20 \le x \le 20)$, where x and y are in feet. Suppose that the horizontal component of the speed dx/dt is 4 feet per second. Find the vertical component of the speed dy/dt at the instant when the squirrel is at the point $(10, 16)$.

43. Suppose that the legs $\overline{AB}$ and $\overline{BC}$ of the right triangle in Figure 28 are increasing at the constant rates of 6 centimeters per minute and 2 centimeters per minute, respectively. Find each of the following rates at the instant when $x = 12$ centimeters and $y = 5$ centimeters.
    (a) The rate at which $\overline{AC}$ is increasing.
    (b) The rate at which the perimeter is increasing.
    (c) The rate at which the area is increasing.

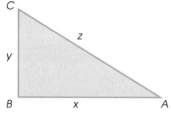

FIGURE 28

44. Suppose the radius of the cylinder in Figure 29 is decreasing at the rate of 0.5 centimeters per minute and the height h is increasing at 1.5 centimeters per minute. Find the rate of change in volume with respect to time at the instant when $r = 6$ centimeters and $h = 12$ centimeters.

FIGURE 29

45. The elephant at the local zoo has a (spherical) tumor whose radius is growing at the alarming rate of 0.2 centimeters per day. Today its radius measures 8 centimeters. How fast is the volume of this tumor growing today?

46. Water is being poured into the hemispherical tank in Figure 30 at the constant rate of 30 cubic feet per minute. Find the rate at which the height of the water in the tank is rising when the height is 4 feet. *Hint:* The volume of the spherical segment of height h is $V = \pi h^2(24 - h)/3$.

FIGURE 30

47. A manufacturer of 10-speed bicycles estimates that the yearly demand x (in hundred thousands) for its bicycles will decrease in the future in such a way that t years from now

$$x = \frac{50}{t + 4} + 3$$

On the other hand, its annual profit P (in hundred thousands of dollars) is related to the number of bicycles it sells by the equation

$$3P - 2x^2 = 10$$

At what rate will the annual profit be changing 6 years from now?

48. A manufacturer of home computers estimates that the demand x (in thousands of units per year) for its computers is related to the price p in dollars by the equation $x = 4500 - 1.2p$. At present the price is \$1800 but is falling at the rate of \$150 per year. At what rate is the total annual revenue changing?

4.5*

APPROXIMATIONS AND THE DIFFERENTIAL

As usual, let y be a function of x; that is, $y = f(x)$. We are interested in what happens to y when we change x by a small amount Δx (the change in x is also denoted by dx). We denote the corresponding change in y by the

FIGURE 31

symbol Δy, a quantity that is often difficult to calculate. We propose to approximate Δy by another quantity dy; called the differential of y. The **differential** of y is the corresponding change on the tangent line and is therefore defined by

$$dy = f'(x)\, \Delta x = f'(x)\, dx$$

To understand what we have just said, refer to Figure 31. Note that dy is a rather good approximation to Δy and that this approximation will get better and better the smaller we make Δx.

We emphasize that Δx and dx are used interchangeably to indicate a small change in x, but that Δy and dy denote different quantities: Δy is the exact change in y, whereas dy is the corresponding change on the tangent line (refer again to Figure 31).

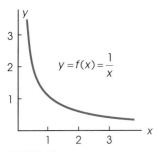

FIGURE 32

EXAMPLE A

Let $f(x) = 1/x$. Suppose x changes from 2 to 2.15, so that $dx = \Delta x = 0.15$. Find the corresponding exact change Δy and its differential approximation dy (see Figure 32).

Solution

$$\Delta y = f(2.15) - f(2) = \frac{1}{2.15} - \frac{1}{2} = -0.03488372$$

To calculate dy, we first find $f'(x) = -1/x^2$ so that $f'(2) = -\frac{1}{4}$. Thus, dy (evaluated when $x = 2$ and $\Delta x = 0.15$) is given by

$$dy = f'(x)\, \Delta x = (-\tfrac{1}{4})(0.15) = -0.0375$$

Clearly, dy is not equal to Δy, but it does approximate it.   ■

EXAMPLE B

Suppose we need to know $\sqrt{81.3}$ and don't have a calculator handy. Use a differential approximation to find this number.

Solution We know that $\sqrt{81} = 9$. Let $y = f(x) = \sqrt{x}$ and consider what happens to y when x changes from 81 to 81.3. The corresponding approximate change in y is given by the differential dy evaluated when $x = 81$ and $\Delta x = 0.3$. Now $f'(x) = \frac{1}{2}x^{-1/2} = 1/2\sqrt{x}$, so $f'(81) = 1/2\sqrt{81} = \frac{1}{18}$. Thus,

$$dy = f'(x)\, \Delta x = (\tfrac{1}{18})(0.3) = 0.016666667$$

We conclude that

$$\sqrt{81.3} \approx \sqrt{81} + dy = 9 + 0.016666667 = 9.016666667$$

For comparison purposes, we mention that a direct calculation of $\sqrt{81.3}$ on our calculator gives 9.016651263.   ■

Estimating Errors Due to Inaccurate Measurement

Whenever we try to measure something (radius of a cylinder, strength of an electrical current, acidity of a solution), we are faced with a troublesome fact of life. Measurements cannot be made exactly. Thus, all scientists report measurements with an indication of the accuracy. For example, a chemist reports the mass of a sample as 2.0136 ± 0.0003 grams, an anthropologist reports the age of a bone specimen as 1230 ± 25 years, and a sociologist states that the percentage of one-parent homes in a certain state is 36 ± 2 percent. Any further calculation based on an inaccurate measurement of the basic data will be contaminated by these measurement errors. We pose the problem: Can we give a good estimate of the possible error in a calculated result if we have an estimate of the error in the data? The answer is yes, and it is based on the use of the differential approximation discussed above. We illustrate with a simple example.

EXAMPLE C

A scientist reported that she had measured the radius of a sphere as 5.34 ± 0.13 centimeters. Calculate the volume of this sphere together with an estimate of the possible error in the answer.

Solution The formula for the volume of a sphere is $V = f(r) = \frac{4}{3}\pi r^3$. If 5.34 were the exact radius, we would calculate the volume as

$$V = \frac{4}{3}\pi (5.34)^3 = 637.8409242$$

The possible error of 0.13 in the measurement of the radius leads to an error in the calculated volume which we approximate by dV. Now,

$$dV = f'(r)\,\Delta r = 4\pi r^2\,\Delta r = 4\pi (5.34)^2(0.13) \approx 46.58$$

It would be reasonable to report the volume of the sphere as 638 ± 47 cubic centimeters. ∎

Relative Errors

An error of 1 foot in measuring the radius of the moon is of much less significance than is an error of 1 foot in measuring the height of a man. Often, it is the size of the error relative to the size of the quantity being measured that is of interest. Thus, we define **relative error:**

$$\text{relative error} = \frac{\text{size of error}}{\text{size of quantity being measured}}$$

Sometimes, the relative error is expressed as a percentage and then it is called the **percentage error.**

In Example C above, the relative error in the measurement of the radius is $0.13/5.34 \approx 0.024$, while the (approximate) relative error in the calculated volume is $47/638 \approx 0.074$. In alternative language, the percentage error in the measurement of the radius is 2.4%, while the (approximate) percentage error in the volume is 7.4%.

EXAMPLE D

The side of a square is measured as 15.6 ± 0.2 feet. Use differentials to find the (approximate) error and relative error in the calculated area of the square.

Solution The formula for the area of a square is $A = f(x) = x^2$. Thus,

$$dA = f'(x)\,\Delta x = 2x\,\Delta x$$

The calculated area is

$$A = (15.6)^2 = 243.36$$

with an error of

$$dA = 2(15.6)(0.2) = 6.24$$

A sensible way to report the answer would be $A = 243.36 \pm 6.24$. The relative error is $6.24/243.36 \approx 0.026$ or 2.6%. ∎

The Economic Concept of Elasticity

A major concern of a manufacturer or a business person is to know how a change in the price of goods will affect the total revenue. To analyze the responsiveness of changes in demand (and of revenue) to changes in price, economists have introduced the concept of elasticity.

To begin, we suppose a manufacturer or business person is engaged in selling one type of goods. The **demand function** $x = f(p)$ relates the number of units x that can be sold to the price p charged and is always a decreasing function (the higher the price, the fewer units people will buy; see Figure 33). As a measure of responsiveness, it seems reasonable to calculate

FIGURE 33

the ratio of the relative change in demand to the relative change in price but (in a calculus course) we should do this as the change in price tends to 0. Thus, consider

$$\lim_{\Delta p \to 0} \frac{\Delta x/x}{\Delta p/p} = \lim_{\Delta p \to 0} \frac{p \, \Delta x/\Delta p}{x} = \frac{p}{x} \lim_{\Delta p \to 0} \frac{\Delta x}{\Delta p} = \frac{p \, dx/dp}{x} = \frac{pf'(p)}{f(p)}$$

Now $f'(p)$ will be negative, whereas p and x are positive, so the measure above is negative. Since economists prefer to work with positive numbers, they define the **elasticity of demand** $E(p)$ to be the negative of the above expression.

DEFINITION (ELASTICITY OF DEMAND)

Let $x = f(p)$ be the demand function; that is, let x denote the quantity demanded at price p per unit. Then the **elasticity of demand** $E(p)$ is given by

$$E(p) = -\frac{p \, dx/dp}{x} = -\frac{pf'(p)}{f(p)}$$

To appreciate the significance of this concept, let us consider how the revenue responds to changes in price. Recall that

$$\text{revenue} = (\text{quantity sold})(\text{price per unit})$$

that is,

$$R(p) = x \cdot p = f(p) \cdot p$$

By the Product Rule for differentiation,

$$\frac{dR}{dp} = \frac{d}{dp}[f(p) \cdot p] = f(p) \cdot 1 + p \cdot f'(p)$$

$$= f(p)\left[1 + \frac{pf'(p)}{f(p)}\right]$$

$$= f(p)[1 - E(p)]$$

Suppose that $E(p) > 1$ (the so-called **elastic** case); then dR/dp is negative and therefore R is decreasing with p. This means that an increase in price will decrease revenue or, alternatively, that a cut in price will increase revenue. Similarly, if $E(p) < 1$ (the so-called **inelastic** case), then dR/dp is positive and R is increasing with p. This means that an increase in price will increase revenue (alternatively, that a cut in price will decrease revenue). In summary,

a cut in price will increase revenue in the elastic case, but an increase in price will increase revenue in the inelastic case.

EXAMPLE E

A sales expert has estimated the monthly sales x of a certain book to be related to its price according to the formula $x = f(p) = 10{,}000/\sqrt{p} = 10{,}000p^{-1/2}$, where p is in dollars (see Figure 34). Presently the book is selling at $9. Determine whether the demand is elastic or inelastic at that price level and advise the sales manager as to whether the price should be increased or decreased (to increase revenue).

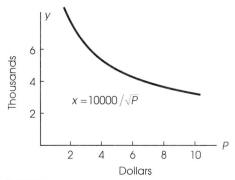

FIGURE 34

Solution First note that $f(9) = 10{,}000/3 \approx 3{,}333$. Also, $f'(p) = -5000p^{-3/2}$, and so $f'(9) = -5000(9)^{-3/2} \approx -185$. Thus,

$$E(9) = \frac{-9f'(9)}{f(9)} \approx \frac{-9(-185)}{3333} \approx 0.5$$

Demand is inelastic; the sales manager is advised to increase the price. ■

A good question to ask is, What value of the elasticity $E(p)$ will make the revenue a maximum? The maximum value of R occurs when its derivative dR/dp is 0. Referring to the formula $dR/dp = f(p)[1 - E(p)]$ derived above, we see that this happens when $E(p) = 1$, the so-called **unit elasticity** case.

Problem Set 4.5

In Problems 1–6, find the differential dy in terms of x and dx.

1. $y = 2x^{3/2}$
2. $y = 3/x^2$
3. $y = 4\sqrt{x} + 5x^{-1}$
4. $y = 2/\sqrt{x} + x$
5. $y = (4 - 3x)^5$
6. $y = \sqrt{x^2 + 3x + 2}$

[C]*In Problems 7–12, find the exact change Δy and the differential approximation dy corresponding to the change in x from x_1 to x_2 (see Example A).*

7. $y = 3x^2 - x + 3$; $x_1 = 2$, $x_2 = 2.1$
8. $y = 4 + 2x - x^2$; $x_1 = 1$, $x_2 = 1.2$

9. $y = \sqrt{x}$; $x_1 = 4$, $x_2 = 4.02$
10. $y = 4/x^2$; $x_1 = 2$, $x_2 = 2.03$
11. $y = \sqrt{2x^3 + 9}$; $x_1 = 2$, $x_2 = 1.95$
12. $y = (3 + 2\sqrt{x})^3$; $x_1 = 1$, $x_2 = 0.98$

In Problems 13–18, use a differential approximation to find N (see Example B).

13. $N = \sqrt{9.04}$
14. $N = \sqrt{25.06}$
15. $N = \sqrt[3]{26.75}$
16. $N = \sqrt[3]{63.8}$
17. $N = 4(1.98)^3 - 3/1.98$
18. $N = [2\sqrt{1.01} + 3(1.01)^2]^3$

19. A large cube of ice melted so that the length of each edge decreased from 25 inches to 24.5 inches. Approximate the corresponding change in (a) the volume, (b) the surface area.

20. One model for growing children says that the pulse rate R in beats per minute is related to height x in inches by $R = 610/\sqrt{x}$. About how much will the pulse rate of a child decrease as his or her height increases from 49 to 51 inches?

$\boxed{C}$ 21. One leg of a right triangle is exactly twice as long as the other. The shorter leg is measured as 48.62 ± 0.21. Calculate the length of the hypotenuse and give an estimate for the possible error in the answer. See Example C.

$\boxed{C}$ 22. The edge of a cube is measured as 24.6 ± 0.4. Calculate the volume and surface area of the cube, giving an estimate of the possible error in each case.

23. The height of a right circular cone is exactly four times the radius of the base. Recall that the volume of such a cone is $V = \pi r^2 h/3 = 4\pi r^3/3$. Suppose that the radius of the base can be measured with a possible error of 0.3 inch. Find (approximately) the corresponding relative error in the calculated volume if the radius is measured as (a) 8.4 inches, (b) 80.4 inches. Give your answers in percent.

24. Give (in percent) the size of the relative error in the calculated values in volume and surface area in Problem 22.

25. The radius of a sphere was measured as 10.5 inches with a percentage error of 1%. Determine (approximately) the percentage error in the calculated value of the volume. *Hint:* $V = 4\pi r^3/3$.

26. Recall that for a sphere the surface area is given by the formula $S = 4\pi r^2$. Deduce that $dS/S = 2\ dr/r$. Now use the latter fact to decide how big a percentage error we can allow in measuring r if we insist that the percentage error in S be less than 6%.

27. A manufacturer is presently selling its line of women's earrings at $12 per pair and has reason to think that weekly demand x is related to price by the equation $x = f(p) = 500 - 1.5p$.
    (a) Is the demand elastic or inelastic at this price?
    (b) Should the price be increased or decreased to increase revenue?
    (c) If the price is increased from $12.00 to $12.10, about how much will the weekly revenue change. *Hint:* From the formula $dR/dp = f(p)[1 - E(p)]$, which was derived just before Example E, we obtain $dR = f(p)[1 - E(p)]\ dp$.

28. A distributer of deluxe ballpoint pens is presently charging $3.25 for each pen. A research study suggests that it could sell x pens per week if it set the price at $p = 4 - 0.002x$. Note that this is equivalent to $x = f(p) = (4 - p)/0.002 = 2000 - 500p$.
    (a) Is the demand elastic or inelastic at this price?

(b) Should the distributer increase or decrease the price to increase revenue?
(c) If the price per pen is increased from $3.25 to $3.35, approximate the corresponding change in weekly revenue.

29. A maker of hand-held calculators knows that it can sell 3000 calculators per week at $12 each and estimates that, for each 20¢ decrease in price, sales each week would increase by 200 calculators.
    (a) Write the demand equation relating the weekly demand x to the price p.
    (b) Determine the elasticity of demand $E(p)$.
    (c) For what interval of p values is the demand elastic?

30. Suppose the demand equation is $x = f(p) = 5625 - 3p^2$.
    (a) Find the elasticity of demand $E(p)$.
    (b) Determine the values of p for which the demand is elastic.
    (c) If the present price is $10, what percentage change in demand is caused by a 10% increase in price?

---

31. Find the differential dy in each case.
    (a) $y = 3/x + x^2$ (b) $y = \sqrt{x^2 + 9}$
    (c) $y = (3t + 1)^4$ (d) $y = (2s + 4)/(s - 2)$

32. Suppose that $y = (x^2 + 2x - 2)^3$ and that x increases from 3 to 3.1.
    (a) Find the corresponding approximate change dy in y.
    (b) Find the corresponding exact change Δy in y.

33. If $f(x) = \sqrt[3]{x} + 5x^2 - 280$, use differentials to approximate $f(8.01)$.

34. A cubical block of cheese with sides 18 inches long is to be covered with a coat of wax 0.15 inch thick. Use differentials to approximate the volume of wax needed.

35. The demand equation for a certain product (in units per month) is given by $x = 400(30 - \sqrt{p})$.
    (a) Find the elasticity of demand.
    (b) For what values of the price is the demand elastic?
    (c) If the price is increased from $625 to $630, find the approximate change in the monthly revenue.

36. A stereo distributor can sell x stereos annually when the unit price is $p = 1200 - x$ dollars. Its cost is estimated at $C(x) = 3000 + 400x$.
    (a) Use differentials to find the approximate change in the annual profit if x is increased from 400 to 410.
    (b) Find the approximate change in the average cost per unit, $C(x)/x$, when x is increased from 400 to 410.
    (c) For what values of p is the demand elastic?

CHAPTER 4 REVIEW PROBLEM SET

1. Find the maximum value of $z = x^2y$ subject to the constraints $5x + 2y = 120$ and $x \geq 0$.

2. Find the maximum value of $z = xy + \frac{1}{2}x^2$ subject to the constraints $3x + 4y = 60$ and $x \geq 0$.

3. Find the points on the parabola $x = y^2$ that are closest to the point $(10, 0)$. *Hint:* Minimize the square of a distance.

4. Let x and y be positive numbers satisfying $x + 2y = 12$. Determine these numbers if their product xy is a maximum.

5. A window is to have the shape of a rectangle topped by a semicircle as shown in Figure 35. If the perimeter is to be 20 feet, what are the dimensions x and y of the window admitting the most light?

FIGURE 35

6. Suppose that the semicircular part of the window in problem 5 admits one-half as much light per square foot as does the rectangular part. What are the dimensions x and y of the window admitting the most light?

7. A cistern with a rectangular base twice as long as wide is to hold 24,000 cubic feet of water. If the metal top costs twice as much per square foot as the concrete sides and bottom, what are the dimensions of the most economical cistern?

8. A rectangular handbill is to contain 54 square inches of printed matter with 3-inch margins at the top and bottom and 2-inch margins on each side. What dimensions for the handbill would use the least paper?

9. A manufacturer of picnic table sets estimates the weekly cost of producing and selling x sets to be

$$C(x) = 1000 + 50x + 0.07x^2$$

with $0 \leq x \leq 200$.
   (a) How many sets (an integer) should be produced each week to minimize the average cost per set, $C(x)/x$?
   (b) What is the minimum average cost?

10. Wonder Builders makes bird houses. The company estimates the weekly cost in dollars of producing x bird houses to be $C(x) = 1200 + 60x + 0.08x^2$.
   (a) How many houses (an integer) should be produced each week to minimize the average cost per house?
   (b) What is this minimum average cost per house?

11. Suppose that the manufacturer in Problem 9 can sell picnic tables at a price of $100 each.
   (a) How many sets (an integer) should be produced to maximize weekly profit?
   (b) What is the maximum weekly profit?

12. Suppose the company of Problem 10 can sell its bird houses at $105 each.
   (a) How many houses (an integer) should be produced each week to maximize profit?
   (b) What is this maximum weekly profit?

13. An apartment complex with 220 rental units can fill all units when the monthly rent is set at $480. It is estimated that for each $15 per month increase in rent, 5 units will become vacant. The complex has fixed monthly costs of $60,000 and a maintenance cost of $60 for each rented apartment. What monthly rent should be charged to maximize the total profit of the complex?

14. A grapefruit orchard yields an average of 32 bushels of grapefruit per tree when 24 trees are planted on an acre of land. For each additional tree planted per acre, the yield of a tree is decreased by 0.75 bushels. How many trees per acre should be planted to get the highest yield?

15. The Royal Appliance Company expects to sell 6000 clock radios per year. It costs the company $2 to hold a radio in stock for a year, and the cost of reordering a lot of size x is $100 plus $1 for each radio. In what lot size should radios be reordered to minimize inventory costs?

16. The Hi-Life Sports Company expects to sell 8000 tennis rackets per year. It cost the company $1.50 to keep a racket in stock for a year and the cost of reordering a lot of size x is $125 plus $1.50 for each racket. In what lot size should tennis rackets be reordered to minimize inventory costs?

17. Martha Johnson, an apple grower, has 600 bushels of apples which she will sell to a single wholesaler. Today she could get $10.00 per bushel, and the price is going up $0.25 per day. On the other hand, she can count on spoilage of about 8 bushels per day. When

should she sell the apples in order to maximize the total revenue?

18. A local club has 120 members, the annual dues per member being $36. A recent survey suggests that for each dollar decrease in the dues, 2 additional members will join the club. At what level should the dues be set to maximize the total revenue from dues?

19. Long Haul Movers estimates the vehicle cost of operating a moving van at $50 + \frac{1}{3}x$ cents per mile when driven at x miles per hour. It pays its drivers $12.50 per hour. At what speed should a van be driven to minimize the total operating cost per mile of one of its vans?

20. Suppose that a milk truck averages $420/(x + 3)$ miles per gallon when driven at an average speed of x miles per hour, $40 \le x \le 70$. If gasoline costs $1.25 per gallon and the driver gets $12.50 per hour, what speed will minimize the cost of a 300-mile run? Calculate this minimum cost.

21. Find an equation for the tangent line to the curve $xy^3 + x^2y = 3x + 7$ at the point $(1, 2)$.

22. Find an equation for the tangent line to the curve $2xy^2 - x^3 = 4y - 11$ at the point in the first quadrant where $x = 3$.

23. A tank has the shape of a right circular cone 12 feet deep and 12 feet across at the top (Figure 36). Water is running into the tank at the rate of 120 cubic feet per minute. How fast is the level of water rising when this level reaches (a) 3 feet, (b) 9 feet? Recall that $V = \frac{1}{3}\pi r^2 h$ for a cone.

24. An airplane flying west at 400 miles per hour goes over a certain town at 12:00 noon. A second plane at the same altitude, flying south at 560 miles per hour, goes over the town at 12:30 P.M. How fast are the planes separating at 1:30 P.M.?

FIGURE 36

25. Refer to Figure 36 and assume that the height of the cone is exactly twice the radius of the top. Suppose the radius of the top is measured to be 6 feet with a possible error of 0.02 feet. Calculate the volume of

the cone and use differentials to give an estimate of the possible error in this value.

26. A point is moving along the curve $y = x^2 + x + 4$ so that $dx/dt = 3$. Find the rate at which y is changing with respect to t when the point is at $(2, 10)$.

27. A wholesale store is presently charging $6.50 for a pen and pencil set. It estimates that it can sell x sets per week when the price per set is $p(x) = 8 - 0.005x$ dollars.
    (a) Is the demand elastic or inelastic?
    (b) Should the store increase or decrease the price to increase total revenue?
    (c) Approximate the change in weekly revenue when the price is increased from $6.50 to $7.00.

28. It is claimed in Boothill County that if there were no hunting season, the grouse population would grow from size x (in thousands) to size $f(x) = 2.8x - 0.03x^2$ in one year. To what size should the grouse population be allowed to grow to maintain the largest yearly kill? What would this yearly kill be?

29. Consider $f(x) = x^2 + 4x^{-1}$.
    (a) Find the slope of the tangent line at $x = 1$ and at $x = 2$.
    (b) Show that the graph of f is concave up on $(0, \infty)$.
    (c) Find the global minimum value of $f(x)$ on $(0, \infty)$.
    (d) Find the global maximum value of $f(x)$ on $[\frac{1}{2}, \frac{5}{2}]$.
    (e) Sketch the graph of $y = f(x)$ on $(0, 4]$.

30. Consider $f(x) = (x + 2)^2/x$ on the domain $\{x: x > 0\}$.
    (a) Find the slope of the tangent line at $x = 1/2$.
    (b) Where is the graph of f concave up?
    (c) Show that $y = x + 4$ is an asymptote of the graph.
    (d) Find the global minimum value of f.
    (e) Sketch the graph of f.

31. Sketch the graph of a function with domain $[0, 8]$ having all of the following properties.
    (a) f is continuous everywhere on $[0, 8]$.
    (b) f fails to have a derivative at $x = 4$.
    (c) $f'(x) > 0$ on $(2, 4) \cup (4, 8)$.
    (d) $f''(x) > 0$ on $(0, 4)$ and $f''(x) < 0$ on $(4, 8)$.

32. Sketch the graph of a function with domain $[0, 8]$ having all of the following properties.
    (a) $f(0) = f(3) = 3$ and $f(x) = 5$ on $[6, 8]$
    (b) f is continuous on $[0, 8]$ except at $x = 3$.
    (c) $f'(x) > 0$ on $(0, 3) \cup (3, 6)$.
    (d) $f''(x) < 0$ on $(0, 3)$ and $f''(x) > 0$ on $(3, 6)$.

33. Evaluate $\lim_{x \to \sqrt{2}} \dfrac{(2x - 3)(x - \sqrt{2})}{x^2 - 2}$.

34. Evaluate $\lim_{x \to 2} \dfrac{(x - 2)^2(2x - 1)}{(x^3 - 8)^2}$.

35. Evaluate $f'(2)$, given that $f(x) = (3x^2 - \sqrt{x + 2})^4$.

36. Evaluate $f'(2)$ if $f(x) = \left[2\sqrt{x + 7} - \left(\dfrac{2}{x - 1}\right)^2\right]^4$.

No picture better symbolizes the modern field of genetics than the double-helix model of DNA (deoxyribonucleic acid). DNA carries the genetic code; a tiny amount in a fertilized egg cell determines the physical characteristics of the fully developed animal. Unraveling this code is a complex mathematical task.

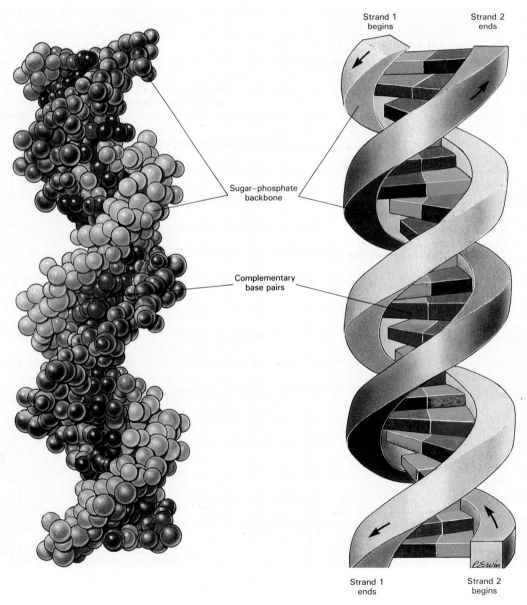

(a) Molecular model of DNA

(b) Ribbon model of DNA

5

Exponential and Logarithmic Functions

Sir Ronald Aylmer Fisher
(1890–1962)

Ronald Fisher was the creator of much of modern statistical theory and was a pioneer in the mathematical theory of genetics. Born in London, he was educated at Cambridge University where he specialized in mathematics and physics. He suffered from extreme myopia, but this seems only to have strengthened his innate geometrical sense and forced him to solve problems entirely mentally.

In 1919, Fisher was appointed research statistician at the Rothamsted Experimental Station in Hertfordshire. His job was to analyze and reinterpret the records of 66 years of agricultural experiments at this station. Here he began the work that was to revolutionize statistical techniques. His 1925 book, *Statistical Methods for Research Workers*, became the bible of a whole generation of statisticians and still wields considerable influence. Its central theme is hypothesis testing and includes such notions as use of controls, randomization, and analysis of variance.

Fisher's second book, *The Genetical Theory of Natural Selection* (1930), is a classic in population genetics. It is a book of considerable mathematical depth and attempts to reconcile Mendel's and Darwin's theories of genetics. Two more books, *The Design of Experiments* (1935) and *Statistical Tables for Biological, Agricultural and Medical Research* (1938), have gone through many editions and had a profound influence.

In 1933, Fisher accepted the position of Galton Professor of Genetics at University College, London, and later the Balfour Chair of Genetics at Cambridge University. He was knighted in 1952.

Fisher's association with this chapter is appropriate, since both the exponential and logarithmic functions appear prominently in his work. The most important distribution in statistical theory is the standard normal distribution whose density function is given by

$$f(x) = (2\pi)^{-1/2} e^{-x^2/2}$$

EXPONENTIAL AND LOGARITHMIC FUNCTIONS

It is quite clear by now that the concept of function is basic to all that we do in calculus. However, if you review what we have done, you will realize that we have restricted our study to a fairly narrow class of functions, namely, the class of **algebraic functions.** These are the functions whose rule of correspondence can be obtained by using the real numbers, the symbol x, and the five algebraic operations of addition, subtraction, multiplication, division, and root taking. Examples are

$$f(x) = 3xxx - 5xx + 7x - 9 = 3x^3 - 5x^2 + 7x - 9$$

$$g(x) = \frac{3x^3 - 5x^2 + 7x - 9}{4x^5 - 3}$$

$$h(x) = 2\sqrt{x} + \frac{3}{\sqrt{x}} = 2x^{1/2} + 3x^{-1/2}$$

$$k(x) = (3x^2 - 1)^{2/3}(5x + 1)^{-5/6}$$

Note that though we have used exponents a great deal, in each case the exponent was a constant. Never have we allowed an exponent to be a variable. Removing this restriction is the key notion in this chapter. And doing this will open the door to a whole new world, the world of so-called **transcendental functions,** which is just a fancy term for nonalgebraic functions. Among these are the exponential functions, logarithmic functions, and trigonometric functions.

The Exponential Functions

By an exponential function, we mean a function of the form

$$f(x) = b^x$$

where b is a positive constant, called the **base** of the exponential function. While such functions have not previously appeared in this book, we feel confident that you have met them in earlier mathematics courses. To refresh your memory, we begin with a simple example.

EXAMPLE A

Sketch the graph of $f(x) = 2^x$.

Solution A table of values (partially obtained on a hand-held calculator) and the corresponding curve are shown in Figure 1. Note that the domain for this function consists of all real numbers, while its range consists of the positive real numbers. Perhaps the most significant feature of this function is its growth, which becomes more and more rapid as x increases. ∎

x	$y = 2^x$
-4	0.0625
-3	0.125
-2	0.25
-1	0.5
0	1
0.5	1.4142
1	2
1.3	2.4623
2	4
3	8
4	16

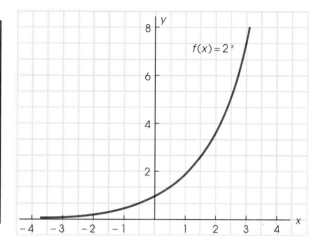

FIGURE 1

The graphs of a number of other exponential functions are shown in Figure 2. Note that if the base $b > 1$, then the graph rises to the right—more and more steeply the larger the base. In contrast, if $0 < b < 1$, the graph falls to the right. Also, the graphs of $f(x) = b^x$ and $g(x) = (1/b)^x = b^{-x}$ are symmetric to each other in the y-axis.

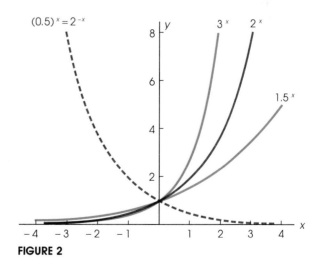

FIGURE 2

The properties of exponents were stated at the end of Section 2.2 and now are restated as they apply to exponential functions.

PROPERTIES OF EXPONENTS

For any real x and y and any $b > 0$,

1. $b^x b^y = b^{x+y}$
2. $b^x/b^y = b^{x-y}$
3. $(b^x)^y = b^{xy}$

EXAMPLE B

Simplify each of the following using the properties of exponents.

(a) $\left(\dfrac{b^3 b^4}{b^5}\right)^3$

(b) $\left(\dfrac{b^5 b^{-2}}{b^3 b^{-1/2}}\right)^{1/3}$

Solution

(a) $\left(\dfrac{b^3 b^4}{b^5}\right)^3 = \left(\dfrac{b^7}{b^5}\right)^3 = (b^2)^3 = b^6$

(b) $\left(\dfrac{b^5 b^{-2}}{b^3 b^{-1/2}}\right)^{1/3} = \left(\dfrac{b^3}{b^{5/2}}\right)^{1/3} = (b^{1/2})^{1/3} = b^{1/6}$ ∎

The Logarithmic Functions

The functions to be introduced next are closely related to the exponential functions. We begin with a definition.

DEFINITION

Assume that $b > 0$ and $b \neq 1$. The logarithm of x to the base b (written $\log_b x$) is the exponent to which b must be raised to yield x, that is,

$$b^{\log_b x} = x$$

Another way to state the above definition is to say that

$$y = \log_b x \quad \text{if and only if} \quad b^y = x$$

Let's see if we can apply this definition in an example.

EXAMPLE C

Find each of the following logarithms.

(a) $\log_2 16$, (b) $\log_2 \frac{1}{8}$, (c) $\log_4 8$, (d) $\log_4 1$, (e) $\log_{10} 0.0001$

Solution

(a) $\log_2 16 = 4$ since $2^4 = 16$
(b) $\log_2 \frac{1}{8} = -3$ since $2^{-3} = 1/2^3 = \frac{1}{8}$
(c) $\log_4 8 = \frac{3}{2}$ since $4^{3/2} = (\sqrt{44})^3 = 8$
(d) $\log_4 1 = 0$ since $4^0 = 1$
(e) $\log_{10} 0.0001 = -4$ since $10^{-4} = 0.0001$ ∎

The properties of logarithms follow from those for exponents, as we shall show in a moment.

PROPERTIES OF LOGARITHMS

For any positive numbers u and v and any real number r,

1. $\log_b (uv) = \log_b u + \log_b v$
2. $\log_b (u/v) = \log_b u - \log_b v$
3. $\log_b (u^r) = r \log_b u$

To see why Property 1 is true, let $\log_b u = x$ and $\log_b v = y$. Then $b^x = u$ and $b^y = v$. It follows from the first property of exponents that

$$uv = b^x b^y = b^{x+y}$$

But this is precisely what it means to say that

$$\log_b(uv) = x + y = \log_b u + \log_b v$$

The other two properties of logarithms follow in a similar way.

EXAMPLE D

CAUTION
Perhaps the most common error in the use of logarithms is to confuse $\log (x/y)$ and $\log x/\log y$ (see Example D).

Given that $\log_b 3 = 0.477$ and $\log_b 5 = 0.699$, find each of the following.

(a) $\log_b (3b^2/125)$ (b) $(\log_b 3b^2)/\log_b 125$

Solution

(a) $\log_b \dfrac{3b^2}{125} = \log_b 3b^2 - \log_b 125$

$= \log_b 3 + \log_b b^2 - \log_b 5^3$

$= \log_b 3 + 2 \log_b b - 3 \log_b 5$

$= 0.477 + 2(1) - 3(0.699) = 0.380$

(b) Be sure to distinguish this problem from the one just completed.

$$\frac{\log_b 3b^2}{\log_b 125} = \frac{\log_b 3 + 2\log_b b}{3\log_b 5}$$

$$= (0.477 + 2)/3(0.699)$$

$$= 2.477/2.097 = 1.181 \qquad \blacksquare$$

EXAMPLE E

Write each of the following as a single logarithm by using the properties of logarithms.

(a) $\log_b x + \log_b y - \log_b z$
(b) $3\log_b x - \frac{1}{2}\log_b y$

Solution

(a) $\log_b x + \log_b y - \log_b z = \log_b xy - \log_b z = \log_b \dfrac{xy}{z}$

(b) $3\log_b x - \frac{1}{2}\log_b y = \log_b x^3 - \log_b y^{1/2}$

$$= \log_b \frac{x^3}{y^{1/2}} = \log_b \frac{x^3}{\sqrt{y}} \qquad \blacksquare$$

Graphs of Logarithmic Functions

Let us begin by noting that the functions $f(x) = b^x$ and $g(x) = \log_b x$ are **inverse functions.** By this we mean that they undo each other; that is,

$$f(g(x)) = x \quad \text{and} \quad g(f(x)) = x$$

This amounts to saying that

$$b^{\log_b x} = x \quad \text{and} \quad \log_b b^x = x$$

both of which are direct consequences of the definition of a logarithm. From this it follows that the graphs of $y = b^x$ and $y = \log_b x$ are reflections of each other across the line $y = x$. We illustrate this fact for the case $b = 2$ in Figure 3.

Figure 3 illustrates, for the case $b = 2$, several important facts about the exponential function $f(x) = b^x$ and its inverse, $g(x) = \log_b x$. The domain of f is the set of all real numbers; its range is the set of positive real numbers. For the logarithmic function g, exactly the opposite is true. Its domain is the set of positive real numbers, and its range is the set of all real numbers. In particular, *negative numbers and 0 do not have logarithms,* but every real number is the logarithm of something. Finally, note from Figure 3 how rapidly 2^x grows as x becomes large, but conversely how slowly $\log_2 x$ grows. These opposite behaviors are characteristic of b^x and $\log_b x$ provided $b > 1$.

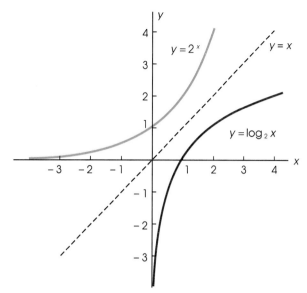

FIGURE 3

Problem Set 5.1

Evaluate each of the numerical expressions in Problems 1–8.

1. $16^{3/2}$
2. $27^{2/3}$
3. $5^2 5^{-1}$
4. $3^{4/3} 3^{2/3} 3^{-2}$
5. $(2^{-3} 2^4)^{-1}$
6. $(4^{-1/4} 4^{1/3})^{12}$
7. $(1.7)^{2.3}$ [C]
8. $\pi^{1.234}$ [C]
9. $(5.12)^{-2.3}(1.23)^2$ [C]
10. $(2.23)^{4.12}/(3.21)^3$ [C]

In Problems 11–14, sketch the graphs of each set of functions using the same axes (see Example A).

11. $f(x) = 4^x$, $g(x) = (\frac{1}{4})^x$
12. $f(x) = 5^x$, $g(x) = 5^{-x}$
13. $f(x) = (0.8)^x$, $g(x) = (1.25)^x$, $h(x) = (2.25)^x$ [C]
14. $f(x) = (1.1)^x$, $g(x) = (1.8)^x$, $h(x) = (2.3)^x$ [C]

Simplify the expressions in Problems 15–24, leaving no negative exponents in your answers (see Example B).

15. $(b^{-3} b^{7/2})^2$
16. $(b^{-2/3} b^2)^3$
17. $(c^3 c^{-4} c^{1/2})^4$
18. $(c^{-2/3} c^2 c^{-1})^3$
19. $[b^4/(b^{-2} b^5)]^3$
20. $(b^{-7} b^3/b^{-8})^2$
21. $(\pi^8 \pi^{-4})^{-2}$
22. $(\pi^{4/3} \pi^0 \pi^{2/3})^{-3/2}$
23. $(a + 1)^{2/3}(a + 1)^{-1/2}$
24. $(b - 1)^{-3/4}(b - 1)^{5/3}$

Evaluate each of the logarithms in Problems 25–38 (see Example C). Note that on many calculators the $\log_{10}$ key is denoted simply by log.

25. $\log_3 27$
26. $\log_4 16$
27. $\log_{10} 100{,}000$
28. $\log_{10} 0.0001$
29. $\log_9 27$
30. $\log_{16} 64$
31. $\log_b b^{-3/2}$
32. $\log_b (1/b^2)$
33. $\log_{10} 0.4321$ [C]
34. $\log_{10} 43.21$ [C]
35. $\log_{10} (5.14/6.32)$ [C]
36. $(\log_{10} 5.14)/(\log_{10} 6.32)$ [C]
37. $3^{\log_3 (4.32)}$
38. $\log_{2.2}(2.2^{3.3})$

Find each of the logarithms in Problems 39–46 given that $\log_a 2 = 0.693$, $\log_a 3 = 1.099$, and $\log_a 5 = 1.609$. For example, in Problem 39 use the fact that $12 = 2^2 3$ and proceed as in Example D.

39. $\log_a 12$
40. $\log_a \frac{6}{5}$
41. $\log_a \frac{27}{4}$
42. $\log_a (5 \sqrt{2})$
43. $\log_a (40/a)$
44. $\log_a (30a^2)$
45. $[\log_a (5a^2)]/\log_a 81$
46. $\log_a (5a^2/81)$

In problems 47–52, write each expression as a single logarithm by using the properties of logarithms (see Example E).

47. $\log_b x - 2 \log_b 8$
48. $\log_b x + \log_b y - \log_b z$
49. $5 \log_b x - 4 \log_b y$
50. $2 \log_b (x + 3) + 4 \log_b (x - 3)$
51. $3[\log_b x - \log_b y + 2 \log_b z]$
52. $\frac{1}{2}[\log_b x - 2(\log_b x - \log_b z)]$

In Problems 53–56, sketch the graphs of each pair of functions using the same coordinate axes.

53. $f(x) = 3^x,\ g(x) = \log_3 x$
54. $f(x) = \log_2 x,\ g(x) = \log_2 x^3$
55. $f(x) = \log_3 x,\ g(x) = \log_3 (x - 2)$
56. $f(x) = \log_2 x,\ g(x) = \log_2 (x + 3)$

Solve for x in the equations in Problems 57–66.

57. $\log_2 16 = x$
58. $\log_3 x = -2$
59. $\log_x \frac{1}{8} = -3$
60. $\log_3 3^\pi = x$
61. $5^{\log_5 9} = x$
62. $3^{3\log_3 2} = x$
63. $\log_5 (x + 2) = 2$
64. $\log_4 (x - 1) = 1.5$
65. $3^{2x-1} = 27$
66. $2^{3x+2} = 64$

67. Which is larger, 10^{1000} or $10,000^{200}$?
68. Which is larger, $\log_5 117$ or $\log_6 117$?
69. [C] Which is larger, 10^{10000} or 13^{9000}? *Hint:* Take logs.
70. [C] Think of a large sheet of paper being folded 40 times (this is possible only in our imagination). Calculate the height of the resulting stack of paper assuming 1000 thicknesses of paper to the inch. Express your answer in miles.
71. [C] One of the authors (Walter Fleming) remembers seeing the following problem on a streetcar transfer (long ago when American cities had streetcars and ordinary people knew about horses). "A rich man went to the blacksmith to have his horse shoed. It takes 28 nails to shoe a horse (7 for each shoe). The man was charged 1¢ for the first nail, 2¢ for the second nail, 4¢ for the third, 8¢ for the fourth, and so on. How much did he have to pay?"
72. [C] An old legend says that an Eastern prince was so fond of playing chess that he offered the inventor of the game a reward of his choice. The man called for a chessboard and asked for one grain of rice for the first square, two grains for the second square, four grains for the third, and so on. The prince exploded with laughter at such a modest request. Show that the inventor had the last laugh by calculating the number of kilograms of rice that would be needed. You will need to know that a chess board has 64 squares and that a grain of rice weighs about 0.02 gram.

5.2

THE NATURAL EXPONENTIAL AND LOGARITHMIC FUNCTIONS

Historically, the most common base used for the exponential and logarithmic functions was base 10. However, in calculus and all of advanced mathematics another base plays a fundamental role. This base is the number e, a number we first met back in the introduction to Section 2.1. The number e is defined by a limit as follows.

$$e = \lim_{u \to 0} (1 + u)^{1/u}$$

Figure 4 gives the results of some calculations done on a hand-held calculator and suggests an approximate value for e.

$$e \approx 2.71828$$

u	$(1+u)^{1/u}$
.1	2.59374
.01	2.70481
.001	2.71692
.0001	2.71815
.00001	2.71827
.000001	2.718280
.0000001	2.718282
.00000001	2.718282

FIGURE 4

Actually, e is an irrational number whose decimal expansion is known to hundreds of thousands of places; however, the approximation above is good enough for any calculations that we will need to make.

The functions $f(x) = e^x$ and $g(x) = \log_e x$ are called the **natural exponential function** and the **natural logarithmic function,** respectively. We use the abbreviation $\ln x$ in place of the more cumbersome $\log_e x$. The two functions e^x and $\ln x$ are inverse functions, and so

$$e^{\ln x} = x \quad \text{and} \quad \ln e^x = x$$

Thus, the graphs of $y = e^x$ and $y = \ln x$ are reflections of each other across the line $y = x$ as we show in Figure 5.

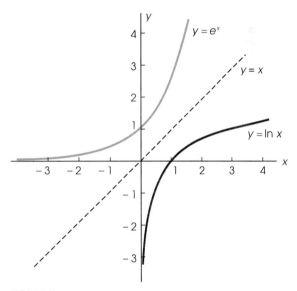

FIGURE 5

Note that the domain of e^x is the set of all real numbers, whereas for $\ln x$ it is the set of positive real numbers. Our graphs suggest that both of these functions should have derivatives. We will investigate that subject soon.

Some Calculations with ln x and e^x

The appendix includes tables of $\ln x$ and e^x. However, most of our readers will own a scientific calculator which allows calculations with these functions at the touch of a button. (On some calculators e^x is obtained as inv $\ln x$.) To gain experience in using these functions, we suggest you work through Example A on your calculator.

EXAMPLE A

Calculate each of the following using a scientific calculator. (a) ln (2.0123), (b) ln (0.00542), (c) ln (−1.2373), (d) $e^{0.01234}$, (e) $e^{3.765}$, (f) $e^{-2.3672}$, (g) $[\ln (2.367 + e^{-5.43})]/e^{0.154}$

Solution

CAUTION

Only positive numbers have logarithms. For example, ln (−1.2373) is undefined. The reason is simple. There is no number x for which

$$e^x = -1.2373$$

(a) ln (2.0123) = 0.6992783465
(b) ln (0.00542) = −5.217659464
(c) ln (−1.2373) = error; negative numbers do not have logarithms.
(d) $e^{0.01234}$ = 1.012416452
(e) $e^{3.765}$ = 43.16370586
(f) $e^{-2.3672}$ = 0.0937428391
(g) $[\ln (2.367 + e^{-5.43})]/e^{0.154}$ = 0.7402315607 ∎

EXAMPLE B

Use natural logarithms to solve the equation

$$3^{2x + 1} = 4.9$$

RECALL

For natural logarithms, the three properties of logarithms take the form:

1. $\ln uv = \ln u + \ln v$

2. $\ln \dfrac{u}{v} = \ln u - \ln v$

3. $\ln u^r = r \ln u$

Solution When the unknown appears in an exponent, it is a good idea to begin by taking logarithms of both sides (we will use natural logarithms, though logarithms to other bases could be used). Note our use of the third property of logarithms.

$$\ln (3^{2x+1}) = \ln 4.9$$

$$(2x + 1) \ln 3 = \ln 4.9$$

$$2x + 1 = (\ln 4.9)/(\ln 3) = 1.446584224$$

$$2x = 0.446584224$$

$$x = 0.223292112$$ ∎

The Derivative of $f(x) = \ln x$

Our approach is very basic; we go back to the definition of the derivative. In the derivation that follows, we use the properties of logarithms and then at a key step we replace h/x by u (note that if $h \to 0$, then $u \to 0$).

$$f'(x) = \lim_{h \to 0} \frac{f(x + h) - f(x)}{h} = \lim_{h \to 0} \frac{\ln (x + h) - \ln x}{h}$$

$$= \lim_{h \to 0} \frac{1}{h} \ln \left(\frac{x + h}{x}\right) = \lim_{h \to 0} \frac{1}{x} \frac{x}{h} \ln \left(1 + \frac{h}{x}\right)$$

$$= \frac{1}{x} \lim_{h \to 0} \ln \left(1 + \frac{h}{x}\right)^{x/h} = \frac{1}{x} \lim_{u \to 0} \ln (1 + u)^{1/u}$$

$$= \frac{1}{x} \ln e = \frac{1}{x} \log_e e = \frac{1}{x}$$

We have shown that

$$\frac{d(\ln x)}{dx} = \frac{1}{x}$$

This is a fundamental result. When combined with the fact that $d^2(\ln x)/dx^2 = -1/x^2$, it yields the conclusion that the graph of $\ln x$ is always rising and concave down, agreeing with the picture in Figure 5. Now let us use the boxed result in two simple examples.

EXAMPLE C

Find the equation of the tangent line to $y = \ln x$ at the point $(1, 0)$.

Solution At $x = 1$, the derivative $1/x$ has a value of 1, which means that the slope of the tangent line is 1. The equation of this line is $y - 0 = 1(x - 1)$; that is, $y = x - 1$. ■

EXAMPLE D

Find $f'(x)$ if (a) $f(x) = x^3 \ln x$, (b) $f(x) = (\ln x)^5$.

Solution

(a) We use the Product Rule.

$$f'(x) = (x^3)\frac{1}{x} + (\ln x)(3x^2) = x^2 + 3x^2 \ln x = x^2(1 + 3\ln x)$$

(b) We use the General Power Rule.

$$f'(x) = 5(\ln x)^4 \frac{1}{x}$$

■

The Derivative of $f(x) = e^x$

Once more, we go back to the definition of the derivative.

$$f'(x) = \lim_{h \to 0} \frac{f(x + h) - f(x)}{h} = \lim_{h \to 0} \frac{e^{x+h} - e^x}{h}$$

$$= \lim_{h \to 0} \frac{e^x(e^h - 1)}{h} = e^x \lim_{h \to 0} \frac{e^h - 1}{h}$$

Our problem thus reduces to evaluating the last limit. The calculations in Figure 6 were done on a hand-held calculator. They suggest that

$$\lim_{h \to 0} \frac{e^h - 1}{h} = 1$$

h	$(e^h - 1)/h$
.1	1.05171
.01	1.00502
.001	1.00050
.0001	1.00005
.00001	1.00000

FIGURE 6

Taking this for granted, we conclude that

$$\frac{d(e^x)}{dx} = e^x$$

another fundamental fact. It says that the derivative of e^x is itself. You have a right to be a little skeptical of the demonstration above, since it rests on guessing at a limit based on the numbers in Figure 6. However, we promise another and very different demonstration of the boxed result in the next section.

From the fact that $d(e^x)/dx = e^x$, we infer that $d^2(e^x)/dx^2 = e^x$, which means that both the first and second derivatives of e^x are always positive. This implies that the graph of $y = e^x$ is always rising and concave up, confirming the shape of the graph in Figure 5.

EXAMPLE E

Find $g'(x)$ if (a) $g(x) = e^x(x^2 + 3x)$ (b) $g(x) = xe^{-x} = x/e^x$
(c) $g(x) = \sqrt{e^x + 5x} = (e^x + 5x)^{1/2}$

Solution

(a) We use the Product Rule.

$$g'(x) = e^x(2x + 3) + (x^2 + 3x)e^x = e^x(x^2 + 5x + 3)$$

(b) We use the Quotient Rule.

$$g'(x) = \frac{(e^x)(1) - (x)(e^x)}{(e^x)^2} = \frac{e^x(1 - x)}{(e^x)^2} = \frac{1 - x}{e^x}$$

(c) This calls for the General Power Rule.

$$g'(x) = \tfrac{1}{2}(e^x + 5x)^{-1/2}(e^x + 5)$$ ∎

EXAMPLE F

Find the maximum value of $f(x) = x/e^x$ on the interval $[0, \infty)$. Also determine where the graph of $f(x)$ is concave up and where it is concave down. Finally, sketch the graph of $f(x)$.

Solution From Example E, we know that $f'(x) = (1 - x)/e^x$. Thus, the critical points are the endpoint 0 and the point 1, where the derivative is 0. Since $f(0) = 0$ while $f(1) = 1/e$, we tentatively accept $f(1) = 1/e \approx 0.36788$ as the maximum value. To confirm this, we note that $f'(x) > 0$ when $x < 1$, and $f'(x) < 0$ for $x > 1$.

FIGURE 7

To analyze concavity, we find the second derivative.

$$f''(x) = \frac{e^x(-1) - (1 - x)e^x}{(e^x)^2} = \frac{-2e^x + xe^x}{e^{2x}}$$

$$= \frac{e^x(x - 2)}{e^{2x}} = \frac{x - 2}{e^x}$$

From this we see that $f''(x) < 0$ for $x < 2$, and $f''(x) > 0$ for $x > 2$. We conclude that the graph of $f(x)$ is concave down for $x < 2$ and concave up for $x > 2$. The graph is shown in Figure 7. ∎

Problem Set 5.2

[C]*Use your calculator to evaluate each of the expressions in Problems 1–14.*

1. ln 14.6
2. ln 3961
3. ln 0.0096
3. ln 0.109
5. ln $(9.82)^3$
6. 3 ln 9.82
7. ln $\sqrt[3]{61.9}$
8. ln $\sqrt[4]{4.63}$
9. $e^{3.18}$
10. $e^{0.023}$
11. $e^{-\pi}$
12. π^e
13. $(2 \ln 14.1 + 19)/(e^{2.1} - 14)$
14. $e^{3.2} (\ln 61 + e^{-1.2})^3$

[C]*Use natural logarithms and your calculator to solve the equations in Problems 18–24 (see Example B).*

15. $3^x = 15$
16. $5^x = 0.41$
17. $2^{3x-1} = 20.5$
18. $4^{x-3} = 9.82$
19. $(\pi^2 + 1)^x = 2$
20. $(\sqrt{3} + 2)^x = 144$
21. $4^{\sqrt{x}} = 10$
22. $10^{1/x} = 4.214$
23. $2^{x^2-x} = 5$
24. $3^{x^2+x} = 7$

25. Find the slope of the tangent line to the curve $y = \ln x$ at $x = 0.5, 1, 3$, and 100.

26. Find the slope of the tangent line to the curve $y = e^x$ at $x = 0.5, 1, 3$, and 100.

In Problems 27–30, find the equation of the tangent line to the given curve at the indicated value of x (see Example C).

27. $y = 3 \ln x$ at $x = e$
28. $y = 2 \ln x + x$ at $x = 1$
29. $y = (3 + \ln x)^2$ at $x = 1/e$
30. $y = (x^2 + \ln x)^3$ at $x = 1$

31. At what point on the curve $y = 4 + 3 \ln x$ is the tangent line parallel to the line $5y - x = 10$?

32. For what values of x is the slope of the curve $y = x - 2 \ln x$ less than 0.3?

Find $f'(x)$ for each of the functions given in Problems 33–46 (see Examples D and E).

33. $f(x) = x^2 \ln x$
34. $f(x) = (\ln x)/x^3$
35. $f(x) = 3(\ln x)^2$
36. $f(x) = \sqrt{2 + \ln x}$
37. $f(x) = 4/(\ln x)^3$
38. $f(x) = (x^2 + 2 \ln x)/x$
39. $f(x) = (x^2 + 2 \ln x)^3$
40. $f(x) = (3x + \ln x)^{1.2}$
41. $f(x) = (2x + 1)e^x$
42. $f(x) = 3x^2e^x$
43. $f(x) = x^2/(e^x + 3)$
44. $f(x) = 2e^x/(3x - 1)$
45. $f(x) = \sqrt[5]{5e^x + x^5}$
46. $f(x) = (e^x + 2x)^3$

47. Consider (as in Example F) the function $f(x) = xe^x$.
    (a) Where is this function increasing and where is it decreasing?
    (b) Find its minimum value on $(-\infty, \infty)$.
    (c) Determine where the graph of f is concave up and where it is concave down.
    (d) Sketch the graph of f.

48. Follow the instructions in Problem 47 for the function $f(x) = x^2/e^x$ except that in part (b) find the maximum value.

[C]49. Calculate.
    (a) $(e^{3.1} + \ln 43.2)^{1/3}$
    (b) $[\ln (\pi^3 + \sqrt{e})]/[e^{\sqrt{\pi}} - \ln 2]$

[C]50. Solve for x.
    (a) $(3.2)^x = (2.8)^{x+1}$
    (b) $e^{x^2-x-6} = 1$
    (c) $\ln (3x + 1) = 2.966$
    (d) $\ln (x^2 + x) = 1$

51. Write $2 \ln 3x + \ln (x^2 - 1) - \ln (x - 1) - \ln x$ as a single logarithm and simplify.

52. Simplify each of the following.
    (a) $(e^{\ln x})^2$ (b) $e^{3\ln 2}$

    (c) $e^{(0.5)\ln(x^2)}$ (d) $\dfrac{\ln(x^4)}{\ln(x^2)}$

53. Find $g'(0)$ for each of the following.
    (a) $g(x) = (3e^x + 5x)^2$
    (b) $g(x) = (x^2 + 1)/\sqrt{4 + 5e^x}$

54. Find $g''(x)$ for each of the following. Simplify your answer.
    (a) $g(x) = x^3 e^x$ (b) $g(x) = x(x + \ln x)^3$

55. Consider the function $f(x) = e^x/x$ on $(0, \infty)$.
    (a) Where is this function increasing and where is it decreasing?
    (b) Determine its minimum value.
    (c) Determine where its graph is concave up and where it is concave down.
    (d) Sketch its graph.

5.3
THE CHAIN RULE

In the previous section, we learned how to find the derivatives of two important functions, namely, e^x and $\ln x$. Now we want to consider how to find the derivative of some complicated combinations of these functions, functions like

$$F(x) = e^{x^3+4x-9}$$

and

$$G(x) = \ln \frac{x^2 + 5}{x - 1}$$

Both of the displayed functions can be thought of as *composite functions* (see Section 1.1). For example,

$$F(x) = f(g(x)) \quad \text{where} \quad f(x) = e^x \quad \text{and} \quad g(x) = x^3 + 4x - 9$$

$$G(x) = h(k(x)) \quad \text{where} \quad h(x) = \ln x \quad \text{and} \quad k(x) = \frac{x^2 + 5}{x - 1}$$

Our problem boils down to this. How can we find the derivative of a composite function given that we know the derivatives of the constituent parts? The answer is given by the Chain Rule, certainly one of the most useful results in calculus.

The Chain Rule Stated

We begin by asking the question about the derivative of a composite function in a slightly different way. Suppose that

$$y = f(u) \quad \text{where} \quad u = g(x)$$

We ask the question: How can we find dy/dx given that we know dy/du and du/dx? Stated this way, there is a simple intuitive answer. Remember that derivatives represent rates of change. If y changes r times as fast as u, and u changes s times as fast as x, then y ought to change rs times as fast as x. For example, if Renee can type three times as fast as Susan, and Susan can type twice as fast as Tammy, then Renee can type $(3)(2) = 6$ times as fast as Tammy. This suggests the following pattern.

$$\begin{array}{ccc} \text{Left} & \text{Middle} & \text{Right} \\ \text{variable} & \text{variable} & \text{variable} \\ \downarrow & \nearrow \searrow & \downarrow \\ y = f(u) & \text{and} \quad u = & g(x) \end{array}$$

$$\begin{bmatrix} \text{Derivative of} \\ \text{left variable} \\ \text{with respect to} \\ \text{right variable} \end{bmatrix} = \begin{bmatrix} \text{Derivative of} \\ \text{left variable} \\ \text{with respect to} \\ \text{middle variable} \end{bmatrix} \cdot \begin{bmatrix} \text{Derivative of} \\ \text{middle variable} \\ \text{with respect to} \\ \text{right variable} \end{bmatrix}$$

$$\frac{dy}{dx} \quad = \quad \frac{dy}{du} \quad \cdot \quad \frac{du}{dx}$$

Here is a formal statement of the Chain Rule. We postpone its proof until the end of the section.

CHAIN RULE

Let $y = f(u)$ and $u = g(x)$ determine the composite function $y = f(g(x))$. If both dy/du and du/dx exist, then dy/dx exists and is given by the formula

$$\frac{dy}{dx} = \frac{dy}{du} \cdot \frac{du}{dx}$$

The Chain Rule Illustrated

We illustrate first with the two functions mentioned in the introduction to this section.

EXAMPLE A

Find dy/dx for each of the following.

(a) $y = e^{x^3 + 4x - 9}$ (b) $y = \ln{[(x^2 + 5)/(x - 1)]}$

Solution

(a) Think of this as $y = e^u$, where $u = x^3 + 4x - 9$. Then by the Chain Rule,

$$\frac{dy}{dx} = \frac{dy}{du}\frac{du}{dx} = e^u(3x^2 + 4) = e^{x^3 + 4x - 9}(3x^2 + 4)$$

Note that the final answer is stated in terms of x, the variable in the original problem.

(b) Think of this as $y = \ln u$, where $u = (x^2 + 5)/(x - 1)$. Then by the Chain Rule followed by the Quotient Rule,

$$\frac{dy}{dx} = \frac{dy}{du}\frac{du}{dx} = \frac{1}{u}\frac{(x - 1)(2x) - (x^2 + 5)(1)}{(x - 1)^2}$$

$$= \frac{x - 1}{x^2 + 5}\frac{x^2 - 2x - 5}{(x - 1)^2} = \frac{x^2 - 2x - 5}{(x^2 + 5)(x - 1)} \qquad \blacksquare$$

EXAMPLE B

Find $f'(x)$ if (a) $f(x) = xe^{-x}$, (b) $f(x) = [\ln (x^2 + 1)]/x$.

Solution

(a) We apply the Product Rule first.

$$f'(x) = x\frac{d(e^{-x})}{dx} + e^{-x}(1)$$

To find the derivative of e^{-x}, we let $u = -x$ and apply the Chain Rule, obtaining $e^u(-1) = e^{-x}(-1)$. We conclude that

$$f'(x) = x(-e^{-x}) + e^{-x} = e^{-x}(-x + 1)$$

(b) We begin with the Quotient Rule and then apply the Chain Rule to the derivative of $\ln (x^2 + 1)$.

$$f'(x) = \frac{x\dfrac{d[\ln (x^2 + 1)]}{dx} - \ln (x^2 + 1)\dfrac{d(x)}{dx}}{x^2}$$

$$= \frac{x\dfrac{1}{x^2 + 1}2x - \ln (x^2 + 1)}{x^2}$$

$$= \frac{2x^2 - (x^2 + 1)\ln (x^2 + 1)}{x^2(x^2 + 1)} \qquad \blacksquare$$

EXAMPLE C

For $y = \ln (e^{-x} + x^2)$, find dy/dx.

Solution Initially, we let $u = e^{-x} + x^2$ and apply the Chain Rule to obtain

$$\frac{dy}{dx} = \frac{dy}{du}\frac{du}{dx} = \frac{1}{u}\frac{du}{dx} = \frac{1}{e^{-x} + x^2}\frac{du}{dx}$$

Calculating du/dx will involve a second use of the Chain Rule in finding the derivative of e^{-x} (we saw how this was done in Example B). We get $du/dx = -e^{-x} + 2x$, which yields the final result

$$\frac{dy}{dx} = \frac{-e^{-x} + 2x}{e^{-x} + x^2}$$

∎

The General Power Rule Revisited

We have been using one instance of the Chain Rule all along. Recall that if $u = u(x)$ and if $y = u^r$, then $dy/dx = ru^{r-1}(du/dx)$. In other words, to find the derivative of $y = u^r$, where u is a function of x, we first take the derivative of y with respect to u (which is ru^{r-1}) and multiply the result by du/dx. This means that when we prove the Chain Rule (coming shortly), we will have automatically proven the General Power Rule, a matter we left dangling back in Chapter 2.

The three uses of the Chain Rule so far observed can be summarized as follows.

1. $\dfrac{d(u^r)}{dx} = ru^{r-1}\dfrac{du}{dx}$

2. $\dfrac{d(e^u)}{dx} = e^u\dfrac{du}{dx}$

3. $\dfrac{d(\ln u)}{dx} = \dfrac{1}{u}\dfrac{du}{dx}$

We must develop skill in using these three results, so much so that we can use them mechanically. Here is another three-part example.

EXAMPLE D

Find dy/dx for (a) $y = [\ln (3x + 1)]^3$, (b) $y = (e^{\sqrt{x}} + 2)^4$, (c) $y = \ln [\ln (3x^2)]$.

Solution Each of these involves a double use of the Chain Rule. See if you can make the mental substitutions required to obtain the following answers.

(a) $\dfrac{dy}{dx} = 3[\ln (3x + 1)]^2\dfrac{1}{3x + 1}3$

(b) $\dfrac{dy}{dx} = 4(e^{\sqrt{x}} + 2)^3(e^{\sqrt{x}})\dfrac{1}{2}x^{-1/2}$

(c) $\dfrac{dy}{dx} = \dfrac{1}{\ln (3x^2)}\dfrac{1}{3x^2}6x$

∎

Our next example will play an important role in Chapter 7.

EXAMPLE E

Find dy/dx if $y = \ln |x|$.

Solution We suppose that $x \neq 0$, since $\ln 0$ is undefined. For $x > 0$, $|x| = x$, and so in this case $\dfrac{d(\ln |x|)}{dx} = 1/x$. For $x < 0$, $|x| = -x$, and so (using the Chain Rule),

$$\frac{d(\ln |x|)}{dx} = \frac{d[\ln (-x)]}{dx} = \frac{1}{-x}(-1) = \frac{1}{x}$$

We conclude that whether $x > 0$ or $x < 0$,

$$\frac{d(\ln |x|)}{dx} = \frac{1}{x} \qquad x \neq 0$$

∎

The Chain Rule Proved

Refer to the statement of the Chain Rule near the beginning of this section. Let

$$\Delta u = g(x + \Delta x) - g(x) \quad \text{and} \quad \Delta y = f(u + \Delta u) - f(u)$$

Note that if $\Delta x \to 0$, then $\Delta u \to 0$. Thus,

$$\frac{dy}{dx} = \lim_{\Delta x \to 0} \frac{\Delta y}{\Delta x} = \lim_{\Delta x \to 0} \frac{\Delta y}{\Delta u} \frac{\Delta u}{\Delta x}$$

$$= \lim_{\Delta u \to 0} \frac{\Delta y}{\Delta u} \lim_{\Delta x \to 0} \frac{\Delta u}{\Delta x} = \frac{dy}{du} \cdot \frac{du}{dx}$$

SHARP CURVE
Our proof of the Chain Rule ultimately collapses under the weight of the maxim—Thou shalt not divide by zero. To construct a rigorous proof requires a complicated maneuver best left for more advanced books.

This proof seems almost too simple and, in fact, there is a subtle flaw. The division and multiplication by Δu in the first step would be illegal if $\Delta u = 0$. This would happen, for example, if g were a constant function. The Chain Rule turns out to be valid even in this case, though our proof breaks down. For a complete proof of the Chain Rule which covers all possible cases, see *Calculus With Analytic Geometry,* 5th edition, by Purcell and Varberg, p. 827.

Another Demonstration That $d(e^x)/dx = e^x$

Whenever we know the derivative of a function, we can use the Chain Rule to find the derivative of its inverse function. We illustrate with the familiar function $\ln x$. Assume we know that $d(\ln x)/dx = 1/x$. Let $u = e^x$ so that $\ln u = x$. In this last equation, take the derivative of both sides with respect to x, keeping in mind that u is a function of x. On the left we use the Chain Rule. We obtain the following in straightforward steps.

$$\ln u = x$$

$$\frac{1}{u}\frac{du}{dx} = 1$$

$$\frac{du}{dx} = u$$

$$\frac{d(e^x)}{dx} = e^x$$

Problem Set 5.3

In Problems 1–16, find dy/dx.

1. $y = 4e^{3x}$
2. $y = 2e^{-3x}$
3. $y = e^{x^2} = e^{(x^2)}$
4. $y = 3e^{x^3}$
5. $y = 4 \ln (2x + 5)$
6. $y = 3 \ln (1 - x^2)$
7. $y = \ln (5x + 1/x)$
8. $y = \ln (\sqrt{x} + x^{-2})$
9. $y = 3e^{4\sqrt{x}}$
10. $y = -2e^{1/x}$
11. $y = \ln [(3x + 1)/(x^2 + 2)]$
12. $y = \ln [x^2/(2x - 1)]$
13. $y = e^{2x/(x+3)}$
14. $y = -e^{(1-x)/x^2}$
15. $y = \ln [(2x^2 + 5)(3x - 2)]$
16. $y = \ln [(5x - 3)(x^3 + 2x)]$

Find f'(x) in Problems 17–24 (see Example B).

17. $f(x) = x^2 e^{3x}$
18. $f(x) = (2x + 1) \ln 5x$
19. $f(x) = (2x + 1)/e^{2x}$
20. $f(x) = [\ln (4x + 1)]/3x$
21. $f(x) = (1 + 3/x) \ln (x^2 + 4)$
22. $f(x) = 5\sqrt{x}e^{2/x}$
23. $f(x) = e^{x^2} \ln (x^2)$
24. $f(x) = [\ln (2x + 1)]/e^{2x+1}$

Find g'(t) in Problems 25–32 (see Example C).

25. $g(t) = \ln (2t + e^{5t})$
26. $g(t) = 3 \ln (t^2 - e^{2t})$
27. $g(t) = e^{t - \ln (t^2+1)}$
28. $g(t) = e^{3t + \ln (t^3 - 2)}$
29. $g(t) = \ln [\ln (t)]$
30. $g(t) = e^{e^t}$
31. $g(t) = \ln [\ln (\ln t)]$
32. $g(t) = \ln [\ln (t^2 - t)]$

In Problems 33–44, find dy/dx (see Example D).

33. $y = (3e^{2x} - 1)^3$
34. $y = (2x + e^{2x})^{1/2}$
35. $y = [\ln (x^2 + 4)]^4$
36. $y = (2x + 3 \ln x)^5$
37. $y = 4(e^x + \ln x)^{-2}$
38. $y = (2xe^x + 4)^{3/2}$
39. $y = x^2(e^x + 2)^3$
40. $y = (x + \ln x)^2/e^x$
41. $y = e^{(2x+1)^2}$
42. $y = (e^{2x+1})^2$
43. $y = [\ln (x + 4 \ln x)]^3$
44. $y = [\ln (x \ln x)]^3$

45. Find dy/dt.
    (a) $y = 2e^{3t} - 4e^{-t}$
    (b) $y = 3 \ln (t^2 - 4)$
    (c) $y = t^2(e^t - \ln t)$
    (d) $y = e^{2t}/(t^2 - 5)$
    (e) $y = e^{(2t+1)^2}$
    (f) $y = \ln [(2t + \sqrt{t})^4]$
    (g) $y = [\ln (2t + \sqrt{t})]^4$
    (h) $y = (4t - e^{2/t})^6$

46. Find $g'(s)$.
    (a) $g(s) = s^2 - 5e^{2s}$
    (b) $g(s) = 2 \ln (s + s^2 + s^3)$
    (c) $g(s) = (e^s - s)/e^{2s}$
    (d) $g(s) = 3s \ln (s^2 + 1)$
    (e) $g(s) = \ln (\sqrt{3 + s^2} + 2)$
    (f) $g(s) = 3e^{(2-s)^4}$
    (g) $g(s) = [(5/s) + e^{3s}]^4$
    (h) $g(s) = [s^2 \ln (1 + 5s)]^3$

47. If $f(x) = x^2 e^{-x^2}$, find $f''(x)$.

48. Find the equation of the tangent line to the curve $y = [2 + \ln (1 + x)]^3$ at the point where $x = 1$.

49. Find the equation of the tangent line to the curve $y = \sqrt{x^2 + 1} \ln (e^{2x} + x)$ at the point where $x = 0$.

C50. Suppose that after t weeks of practice, a typical typing student is able to type $f(t) = 80(1 - e^{-0.25t})$ words per minute.
    (a) How fast can such a student type after 4 weeks?
    (b) When will the student be able to type 75 words per minute?
    (c) How fast is the student's typing speed increasing after 3 weeks?

C51. If P dollars is invested at rate r *compounded continuously*, its accumulated value after t years is given by

$A = Pe^{rt}$ (a subject to be discussed fully in Section 6.3). Taking this for granted, answer the following questions under the assumption that $r = 9\% = 0.09$ and $P = 10,000$.

(a) What is the accumulated value of this investment after 5.5 years?

(b) At what rate in dollars per year is this investment growing at the end of 10 years?

(c) How long will it take this investment to double in value?

C52. To test retention, a group of calculus students was given a final exam and then equivalent forms of the exam at 1-month intervals thereafter. Students were told not to study between exams. The average score on these exams after t months was found to be $S(t) = 72 - 15 \ln (t + 1)$.

(a) Determine the average score after 12 months.

(b) At what rate was the average score changing after 4 months?

(c) When will the average score drop to three-fourths its original value?

C53. A company believes that it will be able to sell $n = n(x)$ units of its product each year if it spends x thousand dollars on advertising, where $n = 2500 + 400 \ln x$. Further, its annual profit P (ignoring advertising costs) in thousands of dollars is related to the number of units sold by the formula $P = -30 + 0.22n - 0.000002n^2$.

(a) Determine the profit when \$40,000 is spent on advertising.

(b) Calculate the rate of change in profit with respect to advertising when \$40,000 is spent on advertising.

(c) Interpret your answer to part (b).

54. Sketch the graph of $y = e^{-(x-1)^2}$, showing the coordinates of the maximum point and the inflection points.

5.4*

GENERAL LOGARITHMIC AND EXPONENTIAL FUNCTIONS

We have said that it is the natural logarithmic and exponential functions (ln x and e^x) that are important in calculus and its applications. Even the most superficial examination of the literature will support this statement. It is also true that any problem involving other logarithmic or exponential functions can easily be transformed into a problem about the corresponding natural functions. We demonstrate this now.

We begin with the equation that actually defines the general logarithm $\log_b x$, namely,

$$b^{\log_b x} = x$$

When we take natural logarithms of both sides of this equation and equate the results, we get

$$(\log_b x)(\ln b) = \ln x$$

or

$$\log_b x = \frac{\ln x}{\ln b}$$

This makes it obvious that any question involving $\log_b x$ can be rephrased as a question about $\ln x$.

In a similar manner, recall that for any u, $e^{\ln u} = u$. In particular

$$e^{\ln b^x} = b^x$$

or by Property 3 of logarithms,

$$e^{x \ln b} = b^x$$

We may rewrite this as

$$b^x = e^{kx} \qquad \text{where } k = \ln b$$

EXAMPLE A

Using the $\ln x$ and e^x keys on a calculator, evaluate (a) $\log_3 14$, (b)$3^{2.24}$

Solution

(a) $\log_3 14 = (\ln 14)/(\ln 3) = 2.63905733/1.098612289$
$$= 2.402173503$$
(b) $3^{2.24} = e^{(\ln 3)(2.24)} = e^{(1.098612289)(2.24)}$
$$= e^{2.460891527} = 11.71525135$$

Of course $3^{2.24}$ can be obtained on your calculator using the y^x key, but internally the calculator probably made the calculation as we have shown it. ∎

The Derivatives of $\log_b x$ and b^x

Using the boxed results above, it is a simple matter to find formulas for the required derivatives. After noting that $\ln b$ is a constant, we write

$$\frac{d(\log_b x)}{dx} = \frac{1}{\ln b} \frac{d(\ln x)}{dx} = \frac{1}{\ln b} \cdot \frac{1}{x}$$

That is,

$$\frac{d(\log_b x)}{dx} = \frac{1}{x \ln b}$$

Similarly, a simple use of the Chain Rule gives

$$\frac{d(b^x)}{dx} = \frac{d(e^{kx})}{dx} = ke^{kx} = (\ln b)e^{kx}$$

That is,

$$\frac{d(b^x)}{dx} = b^x \ln b$$

EXAMPLE B

Find dy/dx if (a) $y = \log_3 x$, (b) $y = \log_2 (x^2 + 3)$, (c) $y = 2^x$,
(d) $y = 3^{e^x - e^{-x}}$

Solution

(a) $\dfrac{dy}{dx} = \dfrac{1}{x \ln 3}$

(b) We apply the Chain Rule with $u = x^2 + 3$ to obtain

$$\frac{dy}{dx} = \frac{1}{u \ln 2} \frac{du}{dx} = \frac{2x}{(x^2 + 3) \ln 2}.$$

(c) $\dfrac{dy}{dx} = 2^x \ln 2$

(d) $\dfrac{dy}{dx} = 3^{e^x - e^{-x}} (\ln 3)(e^x + e^{-x})$ ∎

The Derivative of $(g(x))^{f(x)}$

We now know how to find the derivative of $[f(x)]^c$ and $c^{f(x)}$, c being a constant. The first is handled by the General Power Rule; the second is handled by the method illustrated in part (d) of Example B above. Thus,

$$\frac{d[(x^2 - 5)^6]}{dx} = 6(x^2 - 5)^5(2x)$$

while

$$\frac{d(6^{x^2-5})}{dx} = 6^{x^2-5}(\ln 6)(2x)$$

But what about the derivative of a function raised to an exponent which is itself a function? For example, how does one find the derivative of $(x^2 - 5)^{3x}$? Rather than state a new rule, we are going to suggest a procedure that goes by the name **logarithmic differentiation.** It amounts to this: Before differentiating $y = f(x)$, take natural logarithms of both sides.

> **CAUTION**
> Be sure to distinguish between the four functions x^e, e^x, x^x, and e^e. Their derivatives are:
>
> $\dfrac{d(x^e)}{dx} = ex^{e-1}$
>
> $\dfrac{d(e^x)}{dx} = e^x$
>
> $\dfrac{d(x^x)}{dx} = x^x(1 + \ln x)$
>
> $\dfrac{d(e^e)}{dx} = 0$

EXAMPLE C

Find dy/dx for $y = (x^2 + 5)^{3x}$.

Solution As suggested, we begin by taking natural logarithms, obtaining

$$\ln y = 3x \ln(x^2 + 5)$$

Then we take the derivative of both sides. On the left, we must keep in mind that y is a function of x; on the right side, we use the Product Rule.

We get

$$\frac{1}{y}\frac{dy}{dx} = 3x\frac{d[\ln(x^2 + 5)]}{dx} + [\ln(x^2 + 5)](3)$$

$$= 3x\frac{1}{x^2 + 5}2x + 3\ln(x^2 + 5)$$

$$= \frac{6x^2}{x^2 + 5} + 3\ln(x^2 + 5)$$

Next, we solve for dy/dx (multiplying by y). We get

$$\frac{dy}{dx} = y\left[\frac{6x^2}{x^2 + 5} + 3\ln(x^2 + 5)\right]$$

$$= (x^2 + 5)^{3x}\left[\frac{6x^2}{x^2 + 5} + 3\ln(x^2 + 5)\right]$$ ∎

The method of logarithmic differentiation is sometimes helpful in finding the derivative of complicated expressions which involve products and quotients of powers. We illustrate.

EXAMPLE D

Find dy/dx if

$$y = \frac{(2x - 3)^4(3x + 5)^5}{(5x + 4)^6}$$

Solution First, we take logarithms of both sides and simplify using the properties of logarithms; then we take the derivative of both sides with respect to x; finally, we solve for dy/dx.

$$\ln y = 4\ln(2x - 3) + 5\ln(3x + 5) - 6\ln(5x + 4)$$

$$\frac{1}{y}\frac{dy}{dx} = 4\frac{2}{2x - 3} + 5\frac{3}{3x + 5} - 6\frac{5}{5x + 4}$$

$$\frac{dy}{dx} = \frac{(2x - 3)^4(3x + 5)^5}{(5x + 4)^6}\left(\frac{8}{2x - 3} + \frac{15}{3x + 5} - \frac{30}{5x + 4}\right)$$ ∎

Exponential Functions Versus Power Functions

Once again, we emphasize the difference between the exponential function $f(x) = 2^x$ and the power function $g(x) = x^2$. Both functions are increasing for $x > 0$ (Figure 8), but eventually $f(x)$ grows incomparably faster. When trying to model a real-life situation, one often is faced with choosing between an exponential model and a power model. How can one make this decision in a sensible manner?

FIGURE 8

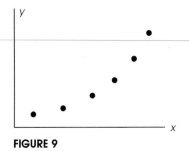

FIGURE 9

Let us state the problem more generally. Suppose you have a set of (x, y) data where y increases rather rapidly as x increases as in Figure 9. You would like to fit a curve to these data. Consider for $c > 0$ and $b > 1$ the following two models.

Exponential model	Power model
$y = cb^x$	$y = cx^b$

Is there a definite way of distinguishing between them and of choosing c and b? Yes, there is, though we shall not go into all the details. The basic idea is to take logarithms to obtain, respectively,

$$\ln y = \ln c + x \ln b \qquad \ln y = \ln c + b \ln x$$

$$y = C + Bx \qquad\qquad Y = C + bX$$

Here we have made the substitutions $Y = \ln y$, $C = \ln c$, $B = \ln b$, and $X = \ln x$. The important thing to notice is that in both cases, we get a linear equation (with a straight-line graph). But in the first case, we get the straight-line graph when we plot $\ln y$ against x; in the second, we get the straight line when we plot $\ln y$ against $\ln x$. Two kinds of graph paper are available to make this procedure easy. One is called semilog graph paper and has a log scale on the vertical axis; the other is called log-log graph paper and has logarithmic scales on both axes. If plotting data on semilog paper leads to a straight line, the exponential model is appropriate; on the other hand, if plotting data on log-log paper gives a straight line, the power model is indicated.

Problem Set 5.4

[c] *In Problems 1–8, first write each expression in terms of e and ln. Then evaluate the result on your calculator. See Example A.*

1. $\log_3 29$
2. $\log_5 14$
3. $4^{2.63}$
4. $3^{1.98}$
5. $\log_6 0.432$
6. $\log_5 0.098$
7. $(3.2)^{2.4}$
8. $(5.06)^{0.31}$
[c] 9. Write $(4.71)^t$ in the form e^{kt}.
[c] 10. Write $(0.987)^t$ in the form e^{kt}.
[c] 11. Write $(8.12)^{-2t}$ in the form e^{kt}.
[c] 12. Write $\log_3(t)$ in the form $k \ln t$.
[c] 13. Write $\log_6(t^3)$ in the form $k \ln t$.
[c] 14. Write $\log_\pi(t^{\sqrt{2}})$ in the form $k \ln t$.

Find dy/dx in Problems 15–28 (see Example B).

15. $y = \log_5 x$
16. $y = \log_9 x$
17. $y = 4^x$
18. $y = 10^x$
19. $y = \log_3(2x + 1)$
20. $y = \log_5(1 - 3x)$
21. $y = 6^{4 - 5x^2}$
22. $y = 3^{x^2 + 2x}$
23. $y = 3 \log_4(3x^2 + 1)$
24. $y = 10 \log_2(x^3 - 2x^2)$
25. $y = 5^{4e^x + 3}$
26. $y = 2^{x + e^{1/x}}$
27. $y = 2^x 3^{-2x}$
28. $y = 3^{2x} \log_2(3x)$

In Problems 29–36, use logarithmic differentiation to find dy/dx (see Example C).

29. $y = (2x + 1)^x$
30. $y = (3x + 2)^x$
31. $y = (x^2 + 5x)^{2x}$
32. $y = (3x - 2x^2)^{3x}$
33. $y = (5x)^{x^2 - x - 1}$
34. $y = (2x + 4)^{4e^{-x}}$
35. $y = (e^x + 4)^{\ln x}$
36. $y = (3 \ln x + 2)^{e^x}$

Use logarithmic differentiation to find dy/dx in Problems 37–44 (see Example D).

37. $y = x^{10}(2x - 1)^6$
38. $y = (3x + 1)^4 / x^8$

39. $y = (x + 1)^3(x + 2)^4/(x + 3)^5$
40. $y = (2x - 1)^5(x + 3)^6(2 - x)^3$
41. $y = (x^2 + 1)^8(2x + 5)^5$
42. $y = (3x + 5)^4/(x^2 - x + 9)^3$
43. $y = \sqrt{x^2 + 2}(x + 2)^4/\sqrt[3]{3x - 5}$
44. $y = (3x + 2)^{5/3}(x - 1)^9/e^{2x}$

C45. Calculate $\log_b 100$ for $b = 2$, e, 10, and 200.
C46. Write each of the following in the form e^{kt}: 10^{-t}, 3^{-t}, 1.5^t, and 9^t.
47. Find dy/dx.
   (a) $\log_3 (\pi^2 + e^3)$ (b) $\log_3 (e^{2x} + x)$
   (c) $y = 5^{3e^{2x}}$ (d) $y = (4 + 3x^2)^{x^2}$
48. Find the equation of the tangent line to the curve $y = 4^{x^2-3}$ at the point where $x = 2$.
49. Use logarithmic differentiation to find $f'(0)$, given that
$$f(x) = (\sqrt{x + 1} + 1)^4[2(x + 1)^{-1} - 1]^{10}$$
50. Show that $f(x) = \log_5(\log_2 x)$ can be rewritten as $f(x) = [\ln (\ln x) - \ln (\ln 2)]/\ln 5$ and use this to find $f(e)$ and $f'(e)$.
51. Transform the equation $xy^2 = e^{-5}$ to a linear equation by an appropriate change in variables. Then find the slope and the "y-intercept" of the resulting equation.
C52. For the following data sets, decide whether $y = cb^x$ or $y = cx^b$ is the better model. One way to do this (suggested in the text) is to plot $\ln y$ against x and $\ln y$ against $\ln x$ to see which more closely approximates a line.

(a)

x	1	2	3	4
y	96	145	216	325

(b)

x	0	1	2	4
y	243	162	108	48

(c)

x	1	2	3	5
y	12	190	975	7490

(d)

x	1	4	9
y	16	128	432

C53. Suppose that the number of bacteria in a certain culture t hours from now will be $400(1.6)^t$.
   (a) Determine the bacteria count 4.5 hours from now.
   (b) When will the count reach 4000?
54. The magnitude R of an earthquake on the *Richter scale* is given by $R = \log_{10} (I/I_0)$, where I is the actual intensity of the earthquake and I_0 is a standard intensity used for comparison. A recent quake measured 4.2 on this scale, whereas the 1906 San Francisco quake measured 8.2. How much more intense was the San Francisco quake than the recent one?
55. The Weber-Fechner law in psychology says that people react to external stimuli on a logarithmic scale. Because of this, the loudness L of sounds is usually measured in decibels using the formula $L = 10 \log_{10} (I/I_0)$. Here I is the actual intensity of the sound in watts per square meter and I_0 is a constant (the least intense sound that can be heard). A certain conversation seemed to be five times louder than a whisper. If so, how did the actual intensity of the sound compare with that of a whisper?

CHAPTER 5 REVIEW PROBLEM SET

1. Evaluate each of the following without the help of a calculator.
   (a) $9^{-3/2}$ (b) $(5^{-3/2}5^{2/3}5^{3/4})^{12}$
   (c) $\log_5 125$ (d) $\log_8 2$
   (e) $\log_8 16$ (f) $\log_5 5^{\pi+1}$
2. Evaluate each of the following without using a calculator.
   (a) $8^{-4/3}$ (b) $(3^{-5/2}3^{4/3} 3^{11/12})^{12}$
   (c) $\log_{10} 10,000$ (d) $\log_{16} 4$
   (e) $\log_4 4^{3.2}$ (f) $\log_4 32$
C 3. Use your calculator to evaluate each of the following.
   (a) $(3\pi)^{2.13}$ (b) $(4.11)^{2.1}/(1.79)^{1.4}$
   (c) $\ln [(14.2)^3 - 962]$ (d) $e^{\pi^2/2}$
   (e) $(e^{-2.5} + 2 \ln 1.08) \ln 142$
   (f) $\log_7 72.45$
C 4. Evaluate using a calculator.
   (a) $(2e + 1)^5$ (b) $e^{\sqrt{3} + 2}$
   (c) $\ln(6.17^2 + 12.3^3)$ (d) $e^{(3-\sqrt{2})^2} \ln 13.68$
   (e) $(14.2^2 - 5 \ln 14.6)/\sqrt[4]{98.1}$
   (f) $\log_8 (e^3 + 192.4)$
C 5. Sketch the graphs of $f(x) = e^{-0.1x}$ and $g(x) = e^{0.4x}$ using the same axes.

$\boxed{c}$ 6. Sketch the graphs of $f(x) = e^{0.6x}$ and $g(x) = e^{-0.3x}$ using the same axes.

7. If $\log_a 2 = 0.356$, $\log_a 3 = 0.565$, and $\log_a 5 = 0.827$, find each value.
   (a) $\log_a 40$ (b) $\log_a \frac{12}{25}$ (c) $\log_a (\sqrt{15}/a)$

8. If $\log_a 2 = 0.431$, $\log_a 3 = 0.683$, and $\log_a 7 = 1.209$, find each value.
   (a) $\log_a 54$ (b) $\log_a (\sqrt{7}/6)$ (c) $\log_a (1.5a\sqrt{a})$

9. Express $2 \ln x + 3 \ln 5 - 5 \ln y$ as a single logarithm.

10. Express $3 \ln x - 2 \ln y + \frac{1}{2} \ln 15$ as a single logarithm.

$\boxed{c}$11. Solve for x.
   (a) $\ln (3x + 1) = 4$ (b) $e^{2+3x} = \pi^x$
   (c) $\ln (x^2 - x) - \ln 3 = 1$ (d) $e^{x^2-3} = \ln 140$

12. Solve for x.
   (a) $\ln(5x - 2) = 3$ (b) $3^{2-x} = \pi^2$
   (c) $\ln(2x + 1) - \ln(3 + 4x) = 2$
   (d) $e^{2x^2+1} = \ln 2000$

13. Find $f'(x)$ in each case.
   (a) $f(x) = 2e^{x^2-3}$ (b) $f(x) = 5 \ln (4 + 3/x)$
   (c) $f(x) = e^{x/(2x+1)}$
   (d) $f(x) = \ln [(x^2 + 3)^2 + 1]$
   (e) $f(x) = (x^2 + 4) \ln (2x + 1)$
   (f) $f(x) = e^{2x}/(x + 2)$
   (g) $f(x) = (3 + 2x \ln x)^4$
   (h) $f(x) = (xe^{-2x} - 1)^{3/2}$

14. Find $f'(x)$ in each case.
   (a) $f(x) = \frac{1}{2}e^{x^2-2x+10}$ (b) $f(x) = 12 \ln\left(x - \frac{4}{x}\right)$
   (c) $f(x) = e^{(3x-2)(3x+2)}$
   (d) $f(x) = \ln(\sqrt{x^2 + 5} - 3)$
   (e) $f(x) = (2x - 1)^5 \ln(4 - 2x)$
   (f) $f(x) = e^{2+\sqrt{x}}/(4 - 3x)$
   (g) $f(x) = (x^2 + 2 \ln x)^5$
   (h) $f(x) = (x^2 e^{4x} + 4)^{5/2}$

15. Consider $f(x) = (x - 2)e^x$.
   (a) Where is f increasing? Decreasing?
   (b) Where is the graph of f concave up? Concave down?
   (c) Determine the minimum value of f.
   (d) Sketch the graph of f.

16. Consider $g(x) = (x - 3)e^{-x}$.
   (a) Where is g increasing? Decreasing?
   (b) Where is the graph of g concave up? Concave down?
   (c) Determine the maximum value of g.
   (d) Sketch the graph of g.

17. Where does the graph of $f(x) = e^{-x^2/2}$ change the direction of its concavity?

18. Where does the graph of $y = \ln(x^2 + 4)$ change its concavity?

$\boxed{c}$19. To test retention of European history, a class was given equivalent forms of a final exam at 1-month intervals after the class had ended. A careful study of the results showed that the average exam score $S(t)$ at the end of t months could be modeled by the formula

$$S(t) = 76 - 16 \ln (t + 1).$$

   (a) Determine the average score at the end of 4 months.
   (b) At what rate was the average score changing at the end of 4 months?
   (c) When did the average score first fall below 40?

20. The number N of bacteria in a certain culture t hours from now is predicted to be

$$N = 1000 + 20,000 \ln(1 + 0.8t)$$

   (a) What is the size of the present population?
   (b) At what rate will the population be increasing 10 hours from now?
   (c) When will the population reach 16,000?

$\boxed{c}$21. Write $(2\pi)^t$ in the form e^{kt}.

$\boxed{c}$22. Write $(\sqrt{3} + 2)^t$ in the form e^{kt}.

$\boxed{c}$23. Write $\log_5 t$ in the form $k \ln t$.

$\boxed{c}$24. Write $\log_8 t^3$ in the form $k \ln t$.

25. Find dy/dx.
   (a) $y = 10^{4\sqrt{x}}$ (b) $y = 6 \log_4(3x^2 + 5)$

26. Find du/dt.
   (a) $u = 7^{\sqrt{4+t^2}}$ (b) $u = \log_5(t^2 + 2t - 4)$

27. Find $f'(1)$ using logarithmic differentiation.
   (a) $f(x) = (x^2 + 1)^{x^3+1}$
   (b) $f(x) = (3x^2 - 2)^6(2x - 1)^8/\sqrt{2x + 1}$

28. Find $g'(2)$ using logarithmic differentiation.
   (a) $g(x) = (2x - 3)^{\sqrt{2x-3}}$
   (b) $g(x) = (3x^2 - 11)^5(3 - x)^4/\sqrt{x + 7}$

29. Transform $y = 4e^{-3x+4}$ to a linear equation by an appropriate change in variables. Then find the slope and "y-intercept" of the resulting equation.

30. Transform $y = 3e^{\pi x+2}$ to a linear equation by a suitable change of variables. Then find the slope and "y-intercept" of the resulting equation.

31. Find the equation of the tangent line to the curve $x \ln y + y^2 = 5x - 9$ at the point $(2, 1)$.

32. Find the equation of the tangent line to the curve

$$y^2 \ln x + 2ye^{x-1} = 3y - 3$$

at the point $(1, 3)$.

33. A point is moving along the curve $y = 6e^{x^2/4}$ in such a way that the horizontal component dx/dt of its velocity is always 3. Find dy/dt at the instant when $x = 2$.

34. A point is moving along the curve $y = 3e^{2(x-1)}$ in such a way that the horizontal component of its velocity is always 4.
    (a) Find dy/dt at the instant when $x = 1$.
    (b) Find the rate (with respect to time) at which the distance from the origin is increasing when $x = 1$.

35. Evaluate

$$\lim_{h \to 0} \frac{[\ln (5 + h)]^2 - [\ln 5]^2}{h}$$

by recognizing it as a value of a certain derivative.

36. Evaluate $\lim\limits_{h \to 0} \dfrac{e^{3(2+h)^2} - e^{12}}{h}$ by recognizing it as a value of a certain derivative.

37. Let $f(x) = x^{\pi} - \pi^x$.
    (a) Is $f(e)$ positive or negative?
    (b) Is $f'(e)$ positive or negative?

38. Let $f(x) = \ln(\ln(\ln x))$.
    (a) Determine the domain of f.
    (b) Evaluate $f[e^{(e^e)}]$.
    (c) Evaluate $f'(e^4)$.

39. Let $f(x) = (\ln x)^2 - \ln x$ on $(0, \infty)$.
    (a) Where is $f(x) > 0$?
    (b) Where is $f'(x) > 0$?
    (c) Where is $f''(x) > 0$?

40. Let $f(x) = \ln x - \sqrt{\ln x}$ on $(1, \infty)$.
    (a) Where is $f(x) > 0$?
    (b) Where is $f'(x) > 0$?
    (c) Evaluate $f''(e)$.

Differential equations provide mathematical models for complicated physical phenomena such as the flow of particles past airplane wings. Solving such equations is a tremendous challenge and often can only be accomplished by numerical methods using supercomputers.

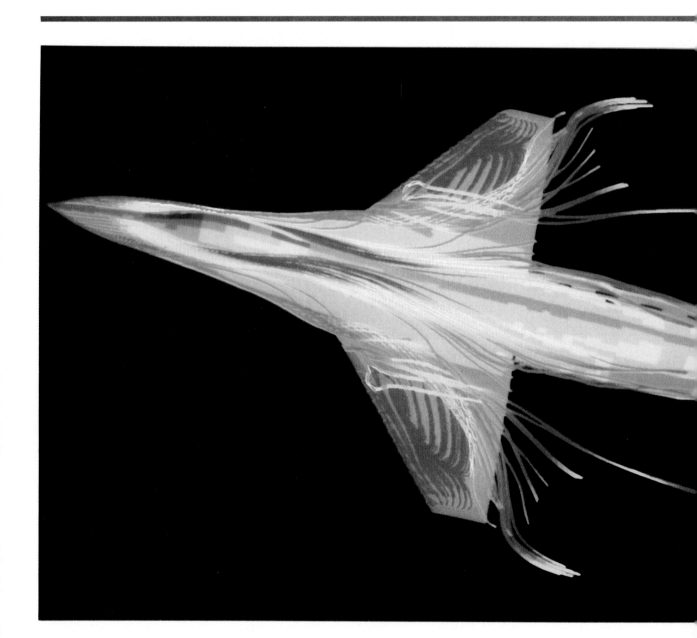

6

Applications of Exponential and Logarithmic Functions

6.1 Exponential Growth and Decay

6.2 Other Growth Models

6.3 Mathematics of Finance

6.4 More on the Mathematics of Finance

Ol'ga Arsen'evna Oleinik
(1925–   )

Ol'ga Arsen'evna Oleinik, a superb Soviet mathematician and head of the Department of Differential Equations at Moscow State University, was born in 1925 near Kiev, Russia. After completing her undergraduate work with honors in mathematics, she pursued graduate study under I. G. Petrovskii and in 1954 successfully defended her doctoral dissertation. It was on a topic in differential equations, a subject in which she had already obtained important new results as a student and to which she was to be a major contributor throughout her career.

Her rise through the ranks at Moscow State University has been rapid, culminating in her appointment as head of the Department of Differential Equations in 1973. Besides administering this department, she regularly teaches a course on differential equations in which she uses and constantly revises her own textbook, has supervised over 40 dissertations, serves as editor-in-chief of *Transactions of the Moscow Mathematical Society,* is a member of the editorial board of several other journals, participates in international congresses and conferences, and is invited to give lectures at prestigious institutions around the world. But what is still more amazing is that by the time she celebrated her sixtieth birthday in 1985, she had published 233 research papers. One can only wonder at such energy, such productivity, such scholarship.

This is not the place to attempt a description of her research except to say that it covers three main areas: differential equations (a subject introduced in this chapter), algebraic geometry, and mathematical physics. In terms of applications, she has studied filtration of fluids in porous media, heat transfer, boundary layer problems, inelasticity theory, and the mechanics of inhomogeneous media.

Oleinik's work has been recognized in many ways. She has won the Chebotarev and the Lomonosov prizes in her own country, as well as the medals of the College de France and of Charles University. She has been elected to several foreign mathematical societies and holds an honorary doctorate from the University of Rome. Her contributions to encouraging world peace resulted in a medal from the World Peace Council. She is said to be a person of integrity, kindliness, and personal charm. Ol'ga Arsen'evna Oleinik is a truly remarkable woman.

EXPONENTIAL GROWTH AND DECAY

We are interested in a whole class of real-life problems where we are given the time rate of change in a quantity y and want to determine explicitly how y depends on time. In particular in this section, we treat the case where the time rate of change in y is proportional to the amount present, that is, where $dy/dt = ky$, k being a constant. This is an example of a **differential equation,** an equation involving derivatives in which the unknown is a function. An important task is to solve such equations, that is, to determine the unknown function. Methods for solving differential equations are discussed in Section 7.4. Here we merely state the result we need and apply it in practical problems.

THEOREM

The solution $y = f(t)$ to the differential equation

$$\frac{dy}{dt} = ky$$

subject to $y = y_0$ at $t = 0$ is

$$y = y_0 e^{kt}$$

It is easy to verify that $y = y_0 e^{kt}$ is a solution to the given differential equation, a matter we ask you to investigate in Problems 1 and 2. What is not so clear is how this solution was found or why it is the only solution. These are matters that will be clarified in Section 7.4.

The significance of the theorem is that any physical phenomenon satisfying $dy/dt = ky$ will have an exponential growth (or decay) curve. It will grow exponentially if the constant k is positive; it will decay exponentially if k is negative (Figure 1).

Exponential growth

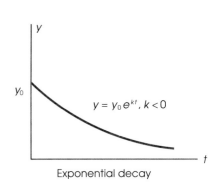

Exponential decay

FIGURE 1

EXAMPLE A

Solve the differential equation $dy/dt = 0.04y$, given that $y_0 = 15$, and determine y when $t = 6$.

Solution We simply appeal to the theorem. The solution is

$$y = 15e^{0.04t}$$

When $t = 6$,

$$y = 15e^{0.04(6)} \approx 19.0687$$

Biological Growth

Consider a biological population, such as a bacteria culture, which is growing in the presence of adequate food supplies. It seems reasonable to conjecture that the rate of growth of such a population is proportional to the size of the population. (The number of offspring per time period is proportional to the size of the population.) Experience confirms this conjecture.

EXAMPLE B

The cells in a certain bacterial culture divide on the average every 2.5 hours. If there are 500 cells initially, how many cells would we expect to find after 12 hours?

Solution Let y denote the number of cells present at time t. The appropriate differential equation is $dy/dt = ky$ with $y_0 = 500$, and this has the solution

$$y = 500e^{kt}$$

To determine k, we note that the population will double in size every 2.5 hours, and so

$$1000 = 500e^{k(2.5)}$$

or on dividing by 500,

$$2 = e^{2.5k}$$

Since the unknown k is in the exponent, we solve by taking natural logarithms of both sides.

$$\ln 2 = 2.5k$$

$$k = (\ln 2)/2.5 \approx 0.277259$$

We conclude that

$$y = 500e^{0.277259t}$$

a result valid for all t. Finding y when $t = 12$ is simply a matter of substituting $t = 12$ in this equation.

$$y = 500e^{0.277259(12)} \approx 13,929$$

It would probably be better to report an answer of 13,900 or even 14,000, since our differential equation is surely only an approximate model for the real situation. ∎

The differential equation $dy/dt = ky$ has been used in describing the world population of human beings. History indicates that the population y of the world (during the last 200 years) has been growing approximately according to this model with growth constant $k = 0.0198$.

EXAMPLE C

The population of the world was about 4 billion on January 1, 1975. Assuming that the exponential growth model with $k = 0.0198$ is appropriate, when will the world's population reach 10 billion?

Solution Let t denote the number of years after January 1, 1975, and y the population in billions at time t. Then $y_0 = 4$, and so

$$y = 4e^{0.0198t}$$

We wish to determine when $y = 10$; that is, we wish to solve the equation

$$10 = 4e^{0.0198t}$$

for t. Dividing by 4 and taking natural logarithms yields

$$\ln 2.5 = 0.0198t$$

$$t = (\ln 2.5)/0.0198 \approx 46.28$$

According to this model, the world's population will reach 10 billion in 1975 + 46.28 or during the year 2021. ∎

It seems clear that the world's population cannot continue to grow indefinitely according to the model above. Food supplies cannot continue to keep up with this kind of population growth. We will consider other and perhaps more realistic models for human population growth in the next section.

Radioactive Decay

It is an experimental fact that radioactive elements (uranium, radium, and so on) decay according to the differential equation $dy/dt = ky$ with k being a negative constant associated with the corresponding element. The decay constant k is usually not specified. Rather, scientists give the **half-life** of the element, that is, the time required for a given amount of the element to decay to half its size.

EXAMPLE D

Radium has a half-life of 1690 years. Determine the decay constant k for radium and then calculate how much of 10 grams of radium will be left after 2400 years.

Solution From the given information, we conclude that 2 units of radium will decay to 1 unit in 1690 years. Thus,

$$1 = 2e^{k(1690)}$$

or

$$0.5 = e^{k(1690)}$$

Following our standard procedure, we take natural logarithms of both sides and solve for k.

$$\ln 0.5 = k(1690)$$

$$k = (\ln 0.5)/1690 = -0.000410146$$

We conclude that the amount y of radium present after t years is $y = y_0 e^{-0.000410146t}$, where y_0 is the initial amount. When $t = 2400$, then 10 grams of radium will have decayed to

$$y = 10e^{-0.000410146(2400)} \approx 3.737 \text{ grams} \quad \blacksquare$$

Carbon Dating

The ratio of the radioactive isotope carbon 14 to the stable isotope carbon 12 in the atmosphere remains constant, since the amount of carbon 14 that decays is exactly balanced by the formation of new carbon 14 as a result of cosmic rays hitting the atmosphere. All living things contain these two isotopes of carbon in the same ratio as in the atmosphere. However, when a plant or animal dies, it ceases to absorb carbon from the atmosphere, and so the proportion of carbon 14 in dead organic material decreases over time as a result of radioactive decay. This allows scientists to date old organic objects, as we now illustrate.

EXAMPLE E

Human hair from a grave in Africa proved to have only 74% of the carbon 14 found in living tissue. Given that carbon 14 has a half-life of 5570 years, determine when the body died.

Solution We first determine the decay constant k from the equation

$$0.5 = e^{k(5570)}$$

$$\ln 0.5 = k(5570)$$

$$k = (\ln 0.5)/5570 = -0.000124443$$

Thus, if y denotes the fraction of the original amount of carbon 14 present at time t,

$$y = e^{-0.000124443t}$$

and so we must solve the equation

$$0.74 = e^{-0.000124443t}$$

for t. This we do by first taking natural logarithms.

$$\ln 0.74 = -0.000124443t$$

$$t = (\ln 0.74)/(-0.000124443) \approx 2420$$

The body died about 2420 years ago.   ■

Problem Set 6.1

1. In each case, differentiate the given y to show that it satisfies the accompanying differential equation.
   (a) $y = 5e^{0.03t}$; $dy/dt = 0.03y$
   (b) $y = 20e^{-0.6t}$; $dy/dt = -0.6y$
2. Differentiate $y = y_0 e^{kt}$ to show that it satisfies the differential equation $dy/dt = ky$.

[C] *In Problems 3–8, solve the given differential equation with the given value of y_0. Then determine the value of y when $t = 8$.*

3. $dy/dt = 0.8y$; $y_0 = 16$
4. $dy/dt = 0.09y$; $y_0 = 12$
5. $dy/dt = -0.05y$; $y_0 = 20$
6. $dy/dt = -0.006y$; $y_0 = 40$
7. $dy/dt = 0.193y$; $y_0 = 3.21$
8. $dy/dt = -2.1y$; $y_0 = 612$
9. A colony of bacteria has 100 members now and will double every 70 minutes. How many members will it have
   (a) after 210 minutes?
   [C] (b) after t minutes?
   [C] (c) after 120 minutes?
10. A population of bacteria numbers 600 now and will double every 3.25 hours. How many bacteria will there be
    (a) after 13 hours?
    [C] (b) after t hours?
    [C] (c) after 9.7 hours?
[C] 11. An exponentially growing culture had 16,000 cells initially and grew to 60,000 cells in 12 hours.
    (a) How many cells were there after t hours?
    (b) How long did it take to reach 80,000 cells?

[C] 12. Suppose that initially there were 800 bacteria in a certain culture and that now, after 6 hours, there are 3500.
    (a) How many bacteria will be present 4 hours from now?
    (b) How many hours from now will the number of bacteria reach 10,000?
[C] 13. Assuming that the growth model for world population given in Example C is correct ($k = 0.0198$), how long does it take the world's population to double?
[C] 14. Population figures for the world are unreliable, and some statisticians give a considerably lower figure for k than that reported in Example C. How long does it take the world's population to double if $k = 0.0183$?
[C] 15. The population of a certain country was 10.5 million on January 1, 1978, and 14.6 million 10 years later. Assume that the exponential growth model is appropriate.
    (a) Find the formula for the population t years after January 1, 1978.
    (b) What will the population be on January 1, 2000?
[C] 16. The population of the world was 4 billion on January 1, 1975, and a newspaper reported it to be 5 billion on April 1, 1985 (approximate figures). This information allows a determination of the value for k in the exponential growth model. Find this k.
[C] 17. Uranium 238 has a half-life of 4.51 billion years. Determine the decay constant k and then calculate how much of 20 grams of this element will be left after 1 billion years. See Example D.
[C] 18. The dangerously radioactive element, radon 220, has a half-life of 3.92 seconds. How much of 5 grams of radon 220 will be left after 2.5 seconds?

19. Polonium 214 disintegrates according to the differential equation $dy/dt = -4621y$, where y is the number of grams present after t seconds. Find the half-life of this element.

20. Uranium 237 satisfies the differential equation $dy/dt = -0.1022y$, where y is the number of grams present after t days. Find the half-life of this element.

21. A skeleton found in a farmer's field has only 62% of the amount of carbon 14 found in living human bones. How old is the skeleton? See Example E.

22. A fragment of wood found during the excavation of an old building site has 51% of the amount of carbon 14 found in living wood of the same type. About how many years ago was the building erected?

23. An anthropologist claimed that a human skull unearthed in southern Europe was 24,500 years old. If carbon dating was used to make this claim, what percentage of the amount of carbon 14 found in living human bones was in that skeleton?

24. When an ancient burial site was excavated, the amount of carbon 14 present in the skeletons ranged from 71 to 75% of that in living bones. About how many years had this burial site been used?

---

25. A certain population is growing exponentially so that it doubles every 25 years. If the population was 10,000 in 1980, what will it be in 2080?

26. A radioactive element decays with a half-life of 20 days. If I have 64 grams of this element now, how much will I have after 120 days?

27. A fossilized leaf contains 25% of the carbon 14 of a living leaf. How old is this fossil?

28. Twenty-five percent of a radioactive substance remained after 10 years. What is its half-life?

29. Solve the differential equation $dy/dt = ky$, given that $y = 10$ when $t = 3$ and $y = 24$ when $t = 4$. *Hint:* The solution is $y = y_0 e^{kt}$, but you must determine both y_0 and k.

30. The rabbit population in a certain area is said to be growing exponentially. This population was estimated at 48,000 in 1975 and at 62,000 in 1985. How large will this population be in 2005?

31. To diagnose a certain disease, a tracer dye is injected into the pancreas. A healthy person will secrete 5% of the dye remaining in the pancreas each minute. Thirty minutes after injecting 0.3 gram of this dye into a patient, 0.1 gram remained. How much would have remained in a healthy individual?

32. Atmospheric pressure p (in millibars) is related to altitude h above sea level (in meters) by the differential equation $dp/dh = kp$. At sea level, $p = 1013.5$, and at $h = 1500$ meters, $p = 845.6$.
    (a) What is the pressure at the top of Mount Everest which is 8848 meters above sea level?
    (b) How high is Mount A if the pressure at its peak measures 627.3 millibars?

33. Colony A and colony B of bacteria grow according to the differential equations $dN/dt = 0.4N$ and $dN/dt = 1.2N$, respectively, where t is measured in hours. Right now colony A is 10 times as large as colony B. When will the two colonies have the same population?

34. The wolf population in a certain state grew exponentially from 1950 to 1960 with growth constant $k = 0.035$ but then began a long exponential decline with decay constant $k = -0.015$. How many wolves did the state have in 1985 if there were 6400 in 1950?

6.2
OTHER GROWTH MODELS

There are many situations where the exponential growth model in the last section is inappropriate. The equation $y = y_0 e^{kt}$, with $k > 0$, describes processes which continue to grow—and at a faster and faster rate (Figure 2). Clearly, the world population of humans cannot continue to grow in this way indefinitely, since the amount of food the earth can produce is limited. There may be an upper bound L beyond which the population of the world cannot grow.

To take another, simpler, example, consider what happens when a cold object is put in an oven whose temperature is L. The temperature of the ob-

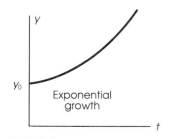

FIGURE 2

ject will increase, but not indefinitely. It will never exceed the upper bound L. Our task through most of this section is to consider growth models where there is an upper limit beyond which growth is impossible.

Simple Bounded Growth

Many processes in nature can be modeled by the differential equation

$$\frac{dy}{dt} = k(L - y) \qquad (k > 0)$$

This says that the rate of increase in y is large when y is small, but that this rate tends to zero as y gets closer and closer to L. We will show how to solve this differential equation in a later section (Section 7.4), but for now we ask you to accept the following result.

THEOREM

The solution to the differential equation

$$\frac{dy}{dt} = k(L - y)$$

subject to $y = y_0$ at $t = 0$, is

$$y = L - (L - y_0)e^{-kt} = L(1 - e^{-kt}) + y_0 e^{-kt}$$

Note in the solution that as $t \to \infty$, $e^{-kt} \to 0$, and so $y \to L$. The graph of the solution is shown in Figure 3.

FIGURE 3

EXAMPLE A

Solve the differential equation

$$\frac{dy}{dt} = 0.4(10 - y)$$

given that $y_0 = 3$. Then calculate y when $t = 6$.

Solution Since $k = 0.4$, $L = 10$, and $y_0 = 3$, the theorem says that

$$y = 10 - (10 - 3)e^{-0.4t} = 10 - 7e^{-0.4t}$$

At $t = 6$,

$$y = 10 - 7e^{-0.4(6)} \approx 9.36$$ ∎

Sociologists have found that important news tends to be diffused in a population according to the model above. At first the news spreads very rapidly, then more and more slowly as the number of people who have heard it nears the total population.

EXAMPLE B (SPREAD OF INFORMATION)

The news that the mayor of a certain city had been killed was announced at noon, and in 3 hours it was thought that 75% of the people in the city had heard it. How long will it take for 99% of the people to hear it?

Solution Let y denote the percentage (written as a decimal) of the people who hear the news within t hours and note that $y_0 = 0$ and that the bound L is 1 (corresponding to 100%). Using these results and appealing to the theorem (second form of the solution), we obtain

$$y = 1 - e^{-kt}$$

We can determine k from the fact that $y = 0.75$ at $t = 3$; that is,

$$0.75 = 1 - e^{-k(3)}$$

or

$$e^{-3k} = 0.25$$

Taking natural logarithms yields

$$-3k = \ln 0.25$$
$$k = (\ln 0.25)/(-3) \approx 0.462098$$

The value for y at any time t is thus

$$y = 1 - e^{-0.462098t}$$

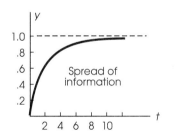

FIGURE 4

an equation graphed in Figure 4. We want to know t when $y = 0.99$; that is, we want to solve

$$0.99 = 1 - e^{-0.462098t}$$

for t. Simplifying and taking logarithms gives

$$e^{-0.462098t} = 0.01$$
$$-0.462098t = \ln 0.01$$
$$t = (\ln 0.01)/(-0.462098) \approx 9.97$$

In just under 10 hours, 99% of the people in the city will have heard the news. ∎

EXAMPLE C (VON BERTALANFFY FISH GROWTH MODEL)

The classical model for fish growth assumes that the rate of change in fish length is proportional to the difference between a theoretical maximum length L and the actual length y. A certain variety of fish is hatched at length 0.5 inches and never grows beyond 12 inches. If a typical such fish has a length of 6 inches in 20 weeks, how long will it be after 50 weeks?

Solution If y denotes length at time t, then $dy/dt = k(12 - y)$ and $y_0 = 0.5$. Thus, by our theorem,

$$y = 12 - (12 - 0.5)e^{-kt} = 12 - 11.5e^{-kt}$$

Substituting $y = 6$ when $t = 20$ gives

$$6 = 12 - 11.5e^{-k(20)}$$

or

$$11.5e^{-20k} = 6$$

$$e^{-20k} = 6/11.5 \approx 0.521739$$

$$-20k = \ln 0.521739$$

$$k = (\ln 0.521739)/(-20) \approx 0.032529$$

We conclude that

$$y = 12 - 11.5e^{-0.032529t}$$

At $t = 50$,

$$y = 12 - 11.5e^{-0.032529(50)} \approx 9.74 \text{ inches}$$ ▪

Logistic Growth

If the food supply (or space) is limited, a better model for population growth than either exponential growth or bounded growth is a compromise between them called **logistic growth.** We suppose that the rate of change in y is proportional to both y and $L - y$; that is,

$$\frac{dy}{dt} = ky(L - y) \qquad (k > 0)$$

A population growing according to this model will appear to exhibit exponential growth initially (when y is small and food is plentiful) and then bounded growth for large t (when y nears L). The relevant theorem, proved in Section 7.4, is

---

THEOREM

The solution to the differential equation

$$\frac{dy}{dt} = ky(L - y)$$

subject to $y = y_0$ at $t = 0$, is

$$y = \frac{Ly_0}{y_0 + (L - y_0)e^{-Lkt}}$$

$$= \frac{L}{1 + Be^{-Lkt}} \qquad [B = (L - y_0)/y_0]$$

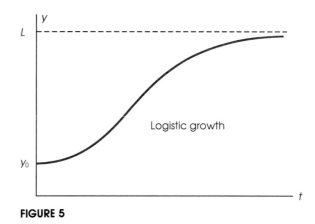

FIGURE 5

The graph of a typical logistic growth curve is the S-shaped curve shown in Figure 5.

EXAMPLE D

A fruit fly population, initially numbering 20, is growing in a laboratory container. After 3 days, there are 56 flies, and experience indicates that not more than 240 flies can live in such a confined space. Assuming the logistic growth model is appropriate, how many flies can we expect after 10 days?

Solution Let y denote the number of fruit flies at time t. Then, $y_0 = 20$ and $L = 240$, so $B = (240 - 20)/20 = 11$. Thus,

$$y = \frac{L}{1 + Be^{-Lkt}} = \frac{240}{1 + 11e^{-240kt}}$$

Substituting $y = 56$ when $t = 3$ gives

$$56 = \frac{240}{1 + 11e^{-720k}}$$

or

$$1 + 11e^{-720k} = 240/56 = 4.28571$$

$$e^{-720k} = 3.28571/11 = 0.29870$$

$$-720k = \ln 0.29870$$

$$k = (\ln 0.29870)/(-720) = 0.0016782$$

Thus,

$$y = \frac{240}{1 + 11e^{-240(0.0016782)t}} = \frac{240}{1 + 11e^{-0.40277t}}$$

When we substitute $t = 10$ in the last equation, we get $y = 200.67$. There will be approximately 200 flies after 10 days. ∎

C Problem Set 6.2

In Problems 1–4, solve the given differential equation with the specified initial value y_0. Then calculate y for the indicated value of t.

1. $dy/dt = 0.09(8 - y)$; $y_0 = 2$; $t = 6$
2. $dy/dt = 0.3(10 - y)$; $y_0 = 4$; $t = 8$
3. $dy/dt = 0.45(20 - y)$; $y_0 = 12$; $t = 4$
4. $dy/dt = 0.08(25 - t)$; $y_0 = 12$; $t = 5$
5. It was announced on the 9:00 A.M. news that movie star Golden Eddie was coming to visit his home town of Blooming Meadow. By 12:00 noon of that day, 60% of the people in the town had heard the exciting news. See Example B.
   (a) What percentage of the townspeople had heard the news by 3:00 P.M.?
   (b) How long did it take for 98% of the people to hear it?
6. Central Auto Parts recently began an advertising campaign on a local TV station offering huge discounts in a closeout sale. After 10 days, a poll showed that 40% of the viewership of 24,000 had seen the advertisement.
   (a) How many viewers will see the advertisement within the first t days?
   (b) How long must the store continue the advertising to reach 80% of the station's viewers?
7. Newton's Law of Cooling says that an object at temperature T when placed in a medium of temperature T_m with $T_m < T$, will cool at a time rate that is proportional to the difference $T - T_m$.
   (a) Write the differential equation that models this situation.
   (b) A kettle of soup at a temperature of 180°F is placed on a table in a room where the temperature is 70°F. If the temperature drops to 140°F after 10 minutes, what will its temperature be after another 10 minutes?
8. A ham at room temperature 27°C was placed in a refrigerator set at 3°C. In 20 minutes, the temperature of the ham dropped to 15°C. How long will it take to drop to 5°C?
9. Let $W = W(t)$ denote the number of words per minute that an average person can type after t weeks of practice. Suppose that W satisfies the differential equation $dW/dt = 0.6(72 - W)$.
   (a) Find the formula for W.
   (b) What is the maximum typing speed an average person can hope to attain?
   (c) How long does it take an average person to achieve a typing speed of 68 words per minute?
10. Based on one model, the yield y (in bushels) of corn per acre should increase in relation to the number x (in pounds) of fertilizer used per acre according to the differential equation $dy/dx = k(240 - y)$. When no fertilizer was used, the yield was 80 bushels per acre, and when 100 pounds of fertilizer per acre was applied, the yield was 140 bushels per acre. What yield can one expect if 150 pounds per acre are applied?
11. A variety of bass, hatched at a length of 2 centimeters, never grows beyond a length of 100 centimeters. Assuming that the Von Bertalanffy model applies, how long will a bass be after 20 weeks that was 8 centimeters long after 7 weeks? See Example C.
12. A type of worm increases in length according to the Bertalanffy model and never grows beyond 20 centimeters long. A worm with an initial length of 1 centimeter grew to 6 centimeters in 1 month.
    (a) What was its length after 4 months?
    (b) How long did it take to reach 18 centimeters?
13. An initial bacteria population of 2000 was growing in a container that can support a maximum population of 36,000. After 1.5 hours, this population numbered 10,000. Assuming the logistic model for growth, how large was it after 5 hours? See Example D.
14. Suppose that the bacteria in a certain culture had an initial density of 2×10^7 cells per milliliter and that the limiting density is 2×10^8 cells per milliliter. If this density increased to 5×10^7 after 3 hours, what will it be after 6 hours? Assume the logistic growth model.
15. A very contagious type of Asian flu is spreading through a city of 120,000 at a rate proportional to both the number already infected and the number not infected. When first diagnosed, 30 people had the disease, and 10 days later 3000 were infected. Assuming that everyone is susceptible to the disease, how long will it take until 100,000 have been infected? Use the logistic growth model.
16. A newly created lake has been stocked with 500 adult (1 pound and bigger) walleyed pike. It is estimated that the lake can support only 40,000 adult walleyes. No fishing is to be allowed until the count of adult walleyes reaches 10,000. How long will this take if a survey at the end of 1 year indicated a count of 2000 walleyes? Use the logistic growth model.

---

17. Suppose that the total revenue R (in thousands of dollars) from sales of a certain model of home freezer during the first t years it has been on the market satisfies the differential equation $dR/dt = k(3500 - R)$ and that the revenue during the first year was $1500 thousand. When will the revenue reach $3000 thousand?

18. Let $P = P(x)$ be the number of Tinker Builder sets that a new worker on a production line can assemble on the xth day on the job and suppose that P can be modeled by the differential equation $dP/dx = k(40 - P)$. If a worker assembled 25 sets on the fifth day, how many sets can we expect this worker to assemble on the eighth day?

19. An experimental psychologist has discovered that the number of new French words a typical language student learns in x hours of a special tutoring course fits the bounded growth model. If learning 1500 vocabulary words is the most that can be expected and if it took Samantha 4 hours to learn 100 words, how many hours of special tutoring would she require to reach the 1000-word level?

20. Several years ago a service organization began a drive to increase its membership, with a goal of reaching 600 members. Two years ago, the membership was 80, and today it is 250. Assuming the logistic growth model is appropriate, what can the organization expect its membership to be 4 years from now?

21. A suburb of a certain city has enough space for 20,000 single-family homes. When incorporated, there were only 100 such homes. Five years later, there were 800 of these homes. No multiple-family dwellings are planned for this suburb. Assuming the logistic growth model for this suburb and that (on the average) each home will have 1.2 school-age children, how large a school system will be needed after 20 years?

22. Refer to Problem 21. If the average home in this suburb has 4.2 members and this is expected to continue to be true, what will the rate of increase in population be at the end of 20 years?

23. Let us suppose that the yield y of tomatoes in bushels per acre increases, in relation to the number x pounds of fertilizer applied, according to the differential equation $dy/dx = k(480 - y)$. When no fertilizer is applied, the yield is 100 bushels per acre, and when 200 pounds per acre is applied, the yield is 280 bushels per acre. If tomatoes sell for $6.00 per bushel and fertilizer costs $0.80 per pound, how many pounds of fertilizer per acre should be applied to yield the maximum profit per acre?

24. Refer to Problem 23. Let us make the model a little more realistic by supposing that it costs 50¢ a bushel to pick and market a bushel of tomatoes. What level of fertilizer application will then yield the maximum profit per acre?

6.3
MATHEMATICS OF FINANCE

In the western world, people expect money that is invested to grow in value. The simplest and surest way of achieving such growth is to put money in a savings account at a bank. Suppose that a **principal** of P dollars is put in a savings account today and left there for t years. Then at periodic times, the bank will add amounts called **interest** to the original amount P, thus causing the value of the account to grow. There are a number of different rules for specifying how the interest is figured, and it is important that we understand them. We do not exaggerate in saying that this understanding is fundamental to success in the business world.

If only the original principal P draws interest, we say that the bank is paying **simple interest.** Suppose, for example, that Mary puts $100 in the bank today in an account that pays 9% simple interest annually. What will the account be worth at the end of 6 years? At the end of each year, the bank will add 9% of $100 [that is, $0.09(100) = \$9$] to the account. In 6 years, the bank will add a total of $6(9) = \$54$ to the account, bringing its value to $154.

In general, if P dollars is invested in an account drawing simple interest at rate r (written as a decimal), then after t years this account will be worth

$$F = P + Prt$$

We use the letter F to indicate that it represents the final or future value of the original P dollars.

EXAMPLE A

If \$500 is invested at 12% simple interest, what will its value be after 9 years?

Solution

$$F = P + Prt = 500 + 500(0.12)(9) = \$1040 \qquad \blacksquare$$

If we graph $F = P + Prt$ (note that P and r are fixed), we get a line with slope Pr (Figure 6). Actually, this graph is not a true reflection of what is in the account except at year ends, since it is only at these times that the bank adds interest to the account. However, a straight-line graph is a good conceptual model for thinking about simple interest and is to be contrasted with the curved graphs used to model compound interest, the subject we consider next.

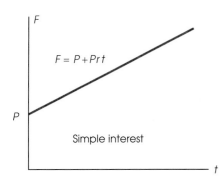

FIGURE 6

Compound Interest (With Annual Compounding)

Of far more significance than simple interest in today's world is compound interest, since most banks and investment firms offer this type of interest. If at periodic intervals interest is converted to principal so that the interest also earns interest, then we say that the account is earning **compound interest.** Suppose that Mary puts her \$100.00 in an account earning 9% compound interest with interest compounded (converted to principal) annually. Then how much will she have at the end of 6 years? At the end of 1 year, the bank will add \$9.00 to the account just as in the simple interest problem treated earlier. But now \$109.00 becomes the principal during the second year and will therefore earn $0.09(109) = \$9.81$ in interest during that year. The principal for the third year is thus $109 + 9.81 = \$118.81$. And so on.

Rather than carry out this six-step process completely, let us see if we can develop a formula that will simplify our work. Note that the value at the end of 1 year is

$$100 + 0.09(100) = 100(1 + 0.09) = 100(1.09)$$

Thus, to get the value at the end of a year, we multiply the principal at the beginning of the year by 1.09. To get the value at the end of the second year, we take the principal at the beginning of that year, namely, $100(1.09)$, and multiply it by 1.09, obtaining

$$100(1.09)(1.09) = 100(1.09)^2$$

You can check, using your calculator, that this is 118.81, the number we got earlier. At the end of the third year, the value will be

$$100(1.09)^2(1.09) = 100(1.09)^3 = \$129.50$$

and at the end of 6 years, it will be

$$100(1.09)^6 = \$167.71$$

As expected, this is considerably more than the $154.00 Mary received at 9% simple interest.

The general situation is similar. If P dollars is invested at a rate r (written as a decimal) with interest compounded annually, then after t years the value will be

$$F = P(1 + r)^t$$

$F = P(1+r)^t$

Compound interest

FIGURE 7

The graph of this equation is a curve, an exponential curve (Figure 7). It rises more and more rapidly as t gets bigger. Money invested at compound interest eventually grows very rapidly in value, a fact you should never forget.

EXAMPLE B

If $500 is invested at 12% interest compounded annually, what will it be worth after 9 years?

Solution

$$F = P(1 + r)^t = 500(1.12)^9 = \$1386.54 \qquad \blacksquare$$

Compounding k Times Per Year

Interest rates are always stated as annual rates (also called **nominal rates**). However, compounding may be done annually, quarterly, monthly, or at any stated time intervals (called periods) of equal length. Suppose interest is compounded k times per year. Then a nominal rate r corresponds to a rate of $i = r/k$ per period. For example, a nominal rate of 12% corresponds to a period rate of $i = 0.12/12 = 0.01$ if interest is compounded monthly. If P dollars is invested at rate r compounded k times per year, then its value at the end of one period is $P(1 + i)$, at the end of two periods $P(1 + i)^2$, at the end of three periods $P(1 + i)^3$, and so on. In t years, there is a total of $n = kt$ periods, so at the end of t years the value is

$$F = P(1 + i)^{kt} = P(1 + i)^n$$

Here $i = r/k$ is the interest rate per period, and n is the total number of periods.

EXAMPLE C

Find the value of $500 at the end of 9 years if it is invested at 12% interest compounded (a) annually, (b) quarterly, (c) monthly, (d) daily.

Solution

(a) We worked this out in Example B; the answer is $1386.54.

(b) $i = 0.12/4 = 0.03$ and $n = 4(9) = 36$, so

$$F = 500(1.03)^{36} = \$1449.14$$

(c) $i = 0.12/12 = 0.01$ and $n = 12(9) = 108$, so

$$F = 500(1.01)^{108} = \$1464.46$$

(d) $i = 0.12/365 = 0.00032876712$ and $n = 365(9) = 3285$, so

$$F = 500(1.00032876712)^{3285} = \$1472.08$$

Several things are to be noted about these calculations. First, we give our answers to the nearest penny. This requires that we take a rather accurate value for i (using as many digits as our calculator allows). Second, we ignore leap years in making the calculation in part (d). Third, the final value F increases as we compound more and more frequently. You expected this didn't you?   ■

Continuous Compounding

In recent years, most banks have offered to compound interest daily, at least on some of their savings accounts. A few banks have gone even further; they offer continuous compounding of interest. What does this mean? Imagine what would happen to F if interest were compounded every day, every minute, every second, and so on. That is, consider what happens to F as we let the period of compounding grow shorter and shorter. Appealing to the formula above, we are asking for an evaluation of

$$\lim_{k \to \infty} P\left(1 + \frac{r}{k}\right)^{kt}$$

With a little algebra, we can rewrite this limit in a more manageable form.

$$\lim_{k \to \infty} P\left(1 + \frac{r}{k}\right)^{kt} = \lim_{k \to \infty} P\left[\left(1 + \frac{r}{k}\right)^{k/r}\right]^{rt}$$

$$= P\left[\lim_{k \to \infty} \left(1 + \frac{r}{k}\right)^{k/r}\right]^{rt}$$

$$= P\left[\lim_{u \to 0} (1 + u)^{1/u}\right]^{rt} \qquad \text{(replacing } r/k \text{ by } u\text{)}$$

$$= Pe^{rt} \qquad \text{(using Section 5.2)}$$

Here is what we have learned. If P dollars is invested at rate r compounded continuously, then after t years its value will be

$$F = Pe^{rt}$$

EXAMPLE D

If \$500 is invested at 12% interest compounded continuously what will its value be at the end of 9 years?

Solution Comparing with Example C, we expect the answer to be larger than that obtained by daily compounding, that is, larger than \$1472.08.

$$F = Pe^{rt} = 500e^{0.12(9)} = \$1472.34$$

is larger, but not by much. The difference between daily compounding and continuous compounding is a matter of pennies unless applied to very large principals. ∎

Effective Rates

Which is the better interest rate—9% compounded daily or 9.1% compounded monthly? Because most people do not know how to answer this question, the government now requires banks to state not only their nominal interest rate but also something called the effective rate. For a given compound rate, the corresponding **effective rate** of interest is the rate which compounded annually yields the same return as the given rate. Here is a simpler way to say the same thing. The effective rate is just the interest earned on \$1 during 1 year at the given rate. Thus, the effective rate r_e corresponding to the nominal rate r compounded k times per year is

$$r_e = \left(1 + \frac{r}{k}\right)^k - 1$$

EXAMPLE E

Which yields the better return: (a) 9% compounded daily or (b) 9.1% compounded monthly?

Solution To answer, we calculate the corresponding effective rates.

(a) $r_e = (1 + 0.09/365)^{365} - 1 \approx 0.094162$
(b) $r_e = (1 + 0.091/12)^{12} - 1 \approx 0.094893$

The second rate is somewhat better. ∎

Doubling Times

How long will it take money to double in value at a given interest rate? The answer, called the **doubling time** for the interest rate, is of considerable interest to investors.

EXAMPLE F

How long does it take money to double if the interest rate is 10% compounded monthly?

Solution Let n be the number of months required for \$1 to grow to \$2. Then

$$2 = 1(1 + 0.10/12)^n = (1.008333333)^n$$

an equation we wish to solve for n. Since the unknown is in the exponent, our first step is to take natural logarithms of both sides. We obtain

$$\ln 2 = n \ln 1.008333333$$

$$n = (\ln 2)/(\ln 1.008333333) = 83.52$$

The doubling time is about 84 months, or 7 years. ∎

---

[c] **Problem Set 6.3**

In Problems 1–6, find the value F after t years if P dollars is invested at the rate r simple interest.

1. $P = 1000$, $r = 0.08$, $t = 4$
2. $P = 800$, $r = 0.09$, $t = 3$
3. $P = 1200$, $r = 0.095$, $t = 4.5$
4. $P = 1600$, $r = 0.075$, $t = 5.25$
5. $P = 1463$, $r = 0.093$, $t = 16.75$
6. $P = 2182$, $r = 0.0821$, $t = 14.5$
7. If \$1000 is invested at the rate $r = 8\% = 0.08$ compounded annually, what will it be worth at the end of 4 years?
8. If \$800 is invested at 9% compounded annually, what will it be worth after 3 years?
9. If \$1250 is invested at 9.5% compounded annually, what will it be worth after 11 years?
10. If \$1800 is invested at 7.75% compounded annually, what will it be worth after 8 years?
11. Find the value of \$1000 at the end of 10 years if it is invested at 9% compounded (a) annually, (b) semiannually, (c) quarterly, (d) monthly, and (e) daily.

12. Do Problem 11 assuming the interest rate is 8%.
13. Henry Baker borrowed \$3650 from Jimmy Hawks at 11% compounded quarterly and paid off his debt in a lump sum 6 years and 6 months later. What was this lump sum?
14. On January 1, 1990, Linda borrowed \$4200 from her mother, promising to repay this sum together with interest at 8% compounded semiannually on July 1, 1995. How much will she have to pay her mother then?
15. If \$1000 is invested at 9% interest compounded continuously, what will it be worth at the end of 10 years? Be sure to compare this value with the results for Problem 11.
16. If \$1000 is invested at 8% interest compounded continuously, what will it be worth at the end of 10 years?
17. Find the accumulated value at the end of 16 years and 9 months if \$1463 is invested at 9.3% compounded continuously.
18. If \$2182 is invested at 8.21% compounded continuously, what will its value be at the end of 14 years and 6 months?

In Problems 19–22, find the effective rate of interest for the given compound rate and period of compounding.

19. 10%, quarterly   20. 9%, monthly
21. 8.9%, daily   22. 12%, daily

In Problems 23–26, determine which of the two rates gives the better return (see Example E).

23. Eight percent compounded monthly or 8.2% compounded semiannually.
24. Ten percent compounded quarterly or 9.9% compounded monthly.
25. Ten percent compounded annually or 9.6% compounded monthly.
26. Ten percent compounded annually or 9.51% compounded daily.
27. How long does it take money to double in value at 11% compounded monthly?
28. How long does it take money to double in value at 10% compounded daily? See Example F.
29. How long does it take money to double in value at 9.5% compounded continuously?
30. How long does it take money to triple in value at 6.36% compounded daily?

31. If $2400 is invested today, find its value at the end of 8 years if the interest rate is 9.6% (a) simple interest, (b) compounded monthly, (c) compounded daily, (d) compounded continuously.

32. How much money should be invested today at 12% compounded monthly in order to have $4000 at the end of 5 years? *Hint:* Solve $4000 = P(1.01)^{60}$ for P.
33. What principal P invested at 8.4% compounded quarterly will accumulate to a value of $4800 at the end of 12 years?
34. What principal P invested at 8.4% simple interest will accumulate to a value of $4800 at the end of 12 years?
35. How long will it take $1000 to grow to $4800 if it is invested at 10% compounded daily?
36. How long will it take $1000 to grow to $4800 if it is invested at 10% compounded continuously?
37. At what rate of interest compounded annually will an investment of $2100 accumulate to $3400 by the end of 6 years?
38. At what rate of interest compounded monthly will an investment of $1000 grow to $3000 in 15 years?
39. At what rate of interest compounded continuously will an investment of $1000 grow to $3000 in 15 years?
40. Jack McDuff borrowed $5000 from Jim Slye, promising to pay back $7000 at the end of 2 years. What rate compounded monthly was Jack paying?
41. Amy Wise invested $2500 in a savings account. Find its value at the end of 8 years if it earned 7.3% compounded daily during the first 4 years and 7.9% compounded daily during the last 4 years.
42. When Lisa was born, her father put $1000 in a savings account for her with an initial interest rate of 10% compounded monthly. At the end of each 4-year period, the interest went up by $\frac{1}{2}$%. How much was in her account when she went away to college at age 18?

6.4
MORE ON THE MATHEMATICS OF FINANCE

The basic principles on which the mathematics of finance rests were enunciated in the previous section. Here our goal is to flesh out some details and to take into account the important factor of inflation, that is, the continuing decrease in the purchasing power of money.

We begin with the fundamental notion of a time line, a line extending indefinitely into both the future and the past. Because money earns interest, a given amount of money with value P right now will have increased value at a future time but had less value at a time in the past. We imagine that associated with each point on the time line is the value of P at that time. For example, suppose that Mary has $100 right now and that interest is at a rate

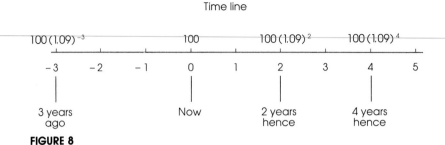

FIGURE 8

of 9% compounded annually. The value of the $100 at various times is shown on the time line in Figure 8.

More generally, if Mary has P dollars right now and interest is compounded at a rate of i per period, then the value of P at a time n periods in the future will be $P(1 + i)^n$ and its value at a time m periods into the past was $P(1 + i)^{-m}$. We **accumulate** money n periods into the future by multiplying by $(1 + i)^n$; we **discount** money m periods into the past by multiplying by $(1 + i)^{-m}$. The corresponding time line diagram is shown in Figure 9.

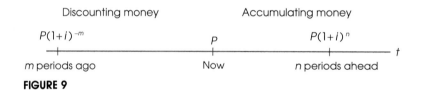

FIGURE 9

EXAMPLE A

Zero-coupon U.S. Treasury bonds are popular with conservative investors. These bonds carry no interest coupons but rather are bought at huge discounts. How much should Celia be willing to pay for a zero-coupon bond with a face value of $10,000 due in 15 years if interest is at 13% compounded annually?

Solution This is simply a matter of discounting $10,000 back 15 years. The answer is

$$10,000(1.13)^{-15} = \$1598.91$$

■

Equivalent Values

A common financial problem is to calculate the value of a set of money payments at some specified time.

EXAMPLE B

Chung Tai expects to receive royalty payments of $2500 one year from now, $3000 two years from now, and $3500 three years from now. What is the total value of these payments right now? That is, what is their total present value? Assume that interest is 12% compounded monthly.

Solution We must discount the three payments by the appropriate number of periods, namely, 12 months, 24 months, and 36 months, respectively. The interest rate per period is $i = 0.12/12 = 0.01$. We conclude that the total present value T of the three payments is

$$T = 2500(1.01)^{-12} + 3000(1.01)^{-24} + 3500(1.01)^{-36}$$
$$= 2218.62 + 2362.70 + 2446.24 = \$7027.56 \qquad \blacksquare$$

EXAMPLE C

Marcia has an obligation of $5000 which is due 1 year from now. She is negotiating with her creditor to make instead three equal payments at the end of 2, 3, and 4 years. If interest is assumed to be at 14% compounded annually, what should be the size of the three payments?

FIGURE 10

Solution In terms of time line diagrams, Marcia wants to replace the top diagram in Figure 10 by the bottom one. We choose to determine the values of the payments shown in the two diagrams at the common date of 1 year from now (any date can be chosen as the comparison date in stating the equivalence of two sets of values). We obtain

$$5000 = x(1.14)^{-1} + x(1.14)^{-2} + x(1.14)^{-3}$$
$$= x[(1.14)^{-1} + (1.14)^{-2} + (1.14)^{-3}]$$
$$= x(2.321632027)$$
$$x = 5000/2.321632027 = \$2153.66 \qquad \blacksquare$$

The Effect of Inflation

While it is true that money invested in a savings account will earn interest and so grow in value, part of this increased value will be eroded away be-

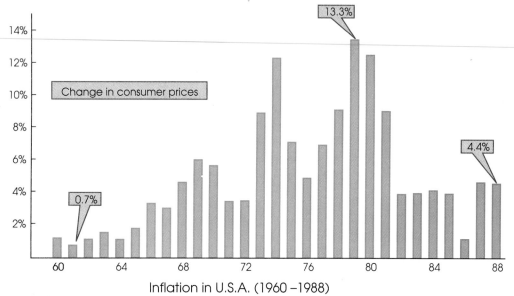

FIGURE 11

cause (at least in recent years) the purchasing power of the dollar is continually decreasing. Since 1960, the so-called **inflation rate** has varied from a low of 0.7% in 1961 to a high of 13.3% in 1979 (Figure 11). If we want to know what is happening to money in terms of its true purchasing power, we must correct for inflation.

For purposes of illustration, let us suppose that the interest rate per period is i and that the inflation rate per period is j. It is natural to guess that the true growth rate g of money is $i - j$, but it turns out that this is wrong though it is approximately correct. Let us find the exact value of g. Because it is earning interest at rate i, P dollars will grow to $P(1 + i)^n$ after n periods. But this value must be discounted because of the inflation rate j to find its true value (that is, its value in constant dollars). The true value is

$$P(1 + i)^n(1 + j)^{-n}$$

We have called the true growth rate g. Thus,

$$P(1 + i)^n(1 + j)^{-n} = P(1 + g)^n$$

In other words,

$$\frac{(1 + i)^n}{(1 + j)^n} = (1 + g)^n$$

or

$$\frac{1 + i}{1 + j} = 1 + g$$

POINT OF INTEREST

If both interest and inflation are assumed to be compounded continuously at rates i and j, respectively, then the true value of P dollars after t years is

$$Pe^{it}e^{-jt} = Pe^{(i-j)t}$$

This means that the true growth rate g considered as a continuous rate is $g = i - j$.

or

$$g = \frac{1 + i}{1 + j} - 1 = \frac{i - j}{1 + j}$$

EXAMPLE D

In 1984, inflation was running at an annual rate of about 4%, while one could expect to earn interest in a savings account at an annual rate of abov 8.5%. At that time, Rodriguez had $2500 in a savings account. He wanteα to know how much he would have in 10 years in constant value dollars. Assuming the rates of interest and inflation to remain constant (a huge assumption), give Rodriguez an answer.

Solution The true growth rate of money under the given assumptions is

$$g = (0.085 - 0.04)/1.04 \approx 0.0432692$$

In 10 years, Rodriguez will have

$$2500(1.0432692)^{10} = \$3818.60$$

measured in constant value dollars.   ■

Depreciation

The Internal Revenue Service keeps changing the rules for figuring depreciation of machines, equipment, and real property. However, there are two basic theoretical models for depreciation. The first, straight-line depreciation, was discussed in Chapter 1. The second, exponential depreciation, allows for more rapid depreciation during the early years of service. One version of the latter is the *double-declining-balance* method. If this is used over N years, the original value C is depreciated each year by $100(2/N)$ percent of its value at the beginning of that year. Thus, the value V after n years is given by

$$V = C\left(1 - \frac{2}{N}\right)^n$$

EXAMPLE E

A piece of equipment costing $10,000 is to be depreciated by the double-declining-balance method over 15 years. What is its value after (a) 5 years, (b) 10 years, (c) 15 years?

Solution $1 - \dfrac{2}{N} = 1 - \dfrac{2}{15} = 0.8666666667.$

(a) $V = 10,000(0.8666666667)^5 = \$4889.46$
(b) $V = 10,000(0.8666666667)^{10} = \$2390.68$
(c) $V = 10,000(0.8666666667)^{15} = \$1168.91$ ∎

Problem Set 6.4

1. Accumulate $2500 for 6 years at 8% compounded monthly.

2. Accumulate $4200 for 8 years at 9% compounded quarterly.

3. Discount $2500 for 10 years at 8% compounded monthly.

4. Discount $4200 for 18 years at 9% compounded annually.

5. How much should a person be willing to pay for a zero-coupon bond with a face value of $5000 due in 12 years if the interest rate is 8.5% compounded annually? See Example A.

6. Do Problem 5 assuming the interest rate is 10% compounded annually.

7. Helen intends to buy a zero-coupon bond with a face value of $10,000 due in 14 years. What will she have to pay if the interest rate is 9% compounded monthly?

8. John is planning to buy a zero-coupon bond with a face value of $5000 due 8 years from now. What should he be willing to pay if the prevailing interest rate is 8.4% compounded quarterly?

In Problems 9–12, find the total present value of the given set of payments if the interest rate is 9% compounded annually (see Example B).

9. $2000 two years from now and $3500 six years from now.

10. $5000 three years from now and $3200 four years from now.

11. $2000 at the end of each of the next 4 years.

12. $2400 at the end of 2 years, 4 years, and 6 years.

13. Do Problem 9 assuming interest is compounded monthly.

14. Do Problem 12 assuming interest is compounded monthly.

15. Romulo has an obligation of $6000 due 4 years from now but has negotiated instead to pay $3000 three years from now and then a payment of x dollars 5

years after that. Determine x if the interest rate is 8% compounded annually. See Example C.

16. To settle a debt, Mary had promised to pay Phyllis $4500 three years from now and another $2500 four years after that. Mary wants to renegotiate this contract and has asked to be allowed to make a single payment 8 years from now. How large should this single payment be if Phyllis insists on earning interest at 9% compounded quarterly?

17. Walter has an obligation to pay Margaret $10,000 four years from now but has asked instead to make two equal payments at the ends of 3 and 6 years. If interest is figured at 9% compounded annually, what should be the size of these payments?

18. Tony, having borrowed money to go to college, owes his alma mater $8000 right now. He has agreed to pay off this debt by making one payment 2 years from now and a second payment twice as large 4 years from now. Determine the size of the first payment if interest is at 6% compounded quarterly.

19. Maria has $10,000 in a savings account earning 6.4% compounded annually, and the annual inflation rate is 3.5%. Assuming these rates hold steady, what will the account be worth in constant value dollars 5 years from now? What is the true growth rate of her money? See Example D.

20. Mark Farmer anticipates that he will need a new tractor 6 years from now. Assuming that the interest rate and the inflation rate will remain steady at 9% and 5%, respectively, how much money needs to be set aside to purchase a model that sells for $60,000 today?

21. Which is better from the point of view of an investor: 8% interest with 4% inflation or 14% interest with 10% inflation? Explain.

22. Show that for the true growth rate to be at least one-half of the annual interest rate i, it is necessary that the annual inflation rate j satisfy the inequality $j < i/(i + 2)$.

23. A $50,000 tractor is to be depreciated by the double-declining-balance method over 10 years. What is its value after (a) 3 years, (b) 6 years, (c) 10 years? See Example E.

24. A bulldozer costing $80,000 is to be depreciated over 12 years by the double-declining-balance method. What is its value after (a) 4 years, (b) 8 years, (c) 12 years?

25. A building valued at $100,000 is to be depreciated by the double-declining-balance method over 18 years. How long does it take to lose half of its value?

26. How long did it take the bulldozer in Problem 24 to lose half of its value?

---

27. An obligation consisting of payments of $2000 in 2 years and $4000 in 5 years is to be replaced by two equal payments at the ends of 3 and 6 years, respectively. Determine the size of the latter payments if interest is at 9% compounded monthly.

28. John is supposed to pay his friend David $2000 two years from now but has asked instead to make equal payments at the end of each of the next four years. How large should these payments be if interest is at 8% compounded quarterly?

29. To pay for a new car that she expects to cost $16,000 three years from now, Betty determines to make three equal deposits in her savings account—now and at the end of the next 2 years. How large must these deposits be if the bank pays 7% compounded annually?

30. As a result of a divorce settlement, Alice is to receive payments of $10,000 at the end of each of the next 4 years. To raise the down payment for a house, she has decided to sell these payments to a bank that has of-

fered to buy them paying 6% interest compounded annually. How much will she receive as a lump sum?

31. On Monday morning, Rene bought a zero-coupon bond with a face value of $20,000. It matures in 15 years and pays interest at 8% compounded annually.
    (a) How much did she pay for the bond?
    (b) On Monday afternoon of the same day, she had to sell the bond because of an emergency, but in that short time the prevailing interest rate had risen to 8.5% compounded annually. How much did she lose not counting broker fees?

32. Roberta bought a zero-coupon bond with a face value of $10,000. It matures in 12 years and carries an interest rate of 9% compounded annually. One year later when Roberta sold the bond the prevailing interest rate had dropped to 7% compounded annually. How much more did Roberta get for the bond than she paid for it?

33. Howard bought a zero-coupon bond with a face value of $25,000 for $10,000. The bond matures in 10 years. What interest rate is he getting based on annual compounding?

34. In a certain Latin American country, the annual inflation rate is 80%. What annual interest rate would a lender need in order to achieve a true annual growth rate of 5%?

35. A road-paving company makes a practice of selling each of its pieces of heavy equipment when its value depreciates to one-fourth its original value. After how many years does it sell equipment if it uses the double-declining-balance method of depreciation over 12 years?

CHAPTER 6 REVIEW PROBLEM SET

C 1. Solve the differential equation $dy/dt = -0.029y$, given that $y_0 = 80$. Then determine the value of y when $t = 21$.

C 2. Solve the differential equation $dy/dt = 0.013y$ given that $y_0 = 1200$. Then determine the value of y when $t = 15$.

C 3. Solve the differential equation $dy/dt = 0.07(12 - y)$, given that $y_0 = 3$. Then evaluate y when $t = 5$.

C 4. Solve the differential equation $dy/dt = 0.045(20 - y)$, given that $y_0 = 8$. Then find the value of y when $t = 11$.

C 5. Solve the differential equation $dy/dt = 0.05y(20 - y)$, given that $y_0 = 8$. Then evaluate y when $t = 3.5$.

C 6. Solve the differential equation $dy/dt = 0.04y(30 - y)$, given that $y_0 = 10$. Then evaluate y when $t = 6$.

C 7. Suppose that a bacteria population is growing exponentially, that it numbers 2400 now, and that it will number 12,000 after 12 hours.
    (a) Write a formula for the number y of bacteria t hours from now.
    (b) How many bacteria will there be at the end of 5 hours?
    (c) At what rate is the population growing at the end of 1 hour?
    (d) After how many hours will the population reach 16,000?

C 8. Suppose that the population of a certain country was 26 million on January 1, 1980, and 34.5 million 8

years later. Assume that the exponential growth model is appropriate.
(a) Find the formula for the population t years after January 1, 1980.
(b) What will the population be on January 1, 1996?

C 9. Recall that carbon 14 has a half-life of 5570 years. The thigh bone of a prehistoric animal contains 6% of the carbon 14 found in living bones. How old is this fossil?

C 10. Polonium 214 disintegrates according to the differential equation $dy/dt = -4621y$, where y is the number of grams present after t seconds. Determine the half-life of this element.

C 11. The news that Morningvale's favorite son, Henry Fastman, had drowned in a boating accident in Sweden was first heard by 1000 people on the 8:00 A.M. news. By 11:00 A.M., two-thirds of Morningvale's population of 48,000 had learned the bad news. Assuming the bounded growth model applies, how many people will have heard the news by 3:00 P.M.?

C 12. The news that 10-year-old Lisa had been kidnapped was reported on the 9:00 A.M. news. Only 1500 of Porter City's population of 84,000 heard this report, but by noon the news had spread to three-fourths of the population. Assuming the bounded growth model is valid, determine how many people will have heard the news by 4:00 P.M.

C 13. When a certain national park opened, it had a deer population of 4000. Three years later, this population had risen to 4600. It is estimated that the park cannot support more than 9000 deer. How many deer will there be 20 years after the park opening, assuming the logistic growth model is appropriate?

C 14. A flu epidemic is spreading through a city of 50,000 at a rate proportional to both the number already infected and the number not yet infected. When the disease was first recognized, 60 people had it, and within 5 days a total of 4000 had been infected. Assuming everyone is subject to the disease, how long will it take until 40,000 have been infected?

C 15. Find the accumulated value at the end of 10 years of $3000 invested today at:
(a) 8% simple interest,
(b) 8% compounded annually,
(c) 8% compounded monthly,
(d) 8% compounded continuously.

C 16. If $4500 is invested today, find its accumulated value at the end of 8 years if interest is at:
(a) 9% simple interest;
(b) 9% compounded annually;
(c) 9% compounded quarterly;
(d) 9% compounded continuously.

C 17. How long does it take money to double at 8.5% compounded (a) semiannually, (b) continuously?

C 18. How long does it take an investment of $3200 to accumulate to $5000 if interest is at 9.5% compounded: (a) monthly, (b) continuously?

C 19. Find the effective rate corresponding to a rate of 9.4% compounded monthly.

C 20. Which rate of interest gives a lender a better return on his money: 9.2% compounded monthly or 9.15% compounded continuously?

C 21. What is the present value of a set of four equal annual payments of $560 each, the first one occurring 1 year from now, if the interest is at the rate of 10% compounded annually?

C 22. Joanne will discharge a debt by making 6 annual payments of $1000, the first one occurring one year from now. If interest is figured at 8.5% compounded annually, what is her present indebtedness?

C 23. Luanne Fisher bought a zero-coupon bond with a face value of $20,000 for $11,000. The bond matures at the end of 8 years. What interest rate is she getting based on annual compounding?

C 24. John paid $4500 for a zero-coupon bond which has a face value of $10,000 and matures at the end of 10 years. What interest rate is John realizing on his investment based on semiannual compounding?

C 25. Maurice has an obligation of $10,000 due 3 years from now. He has negotiated with his creditor to make instead one payment at the end of 2 years and another twice as large at the end of 5 years. Determine the size of these payments if the interest is at 7.5% compounded annually.

C 26. Felicia has promised to pay off her debt to her alma mater by making two payments of $5000 each, one at the end of 2 years, the other at the end of 5 years. She wishes to renegotiate her debt so that she will make one payment at the end of 1 year and another three times as large at the end of 6 years. Determine the size of these payments if interest is at 7.6% compounded annually.

C 27. A bus costing $120,000 is to be depreciated over 10 years by the double-declining-balance method.
(a) Determine its value after 5 years.
(b) When will its value first fall below $60,000?

C 28. How much money should be invested today at 9% compounded quarterly to have $4800 at the end of $8\frac{1}{2}$ years?

C 29. The wolf population in a certain state is believed to have grown exponentially from 1950 to 1980. If the population was 12,000 in 1970 and 13,500 in 1980, what was it in 1950?

C 30. A certain radioactive element decays exponentially so that 67% of a given amount will remain after 10 years. What percent of an initial amount will remain after 500 years?

C 31. A moving company knows that it must buy a new van 5 years from now. The interest rate and the inflation rate are expected to remain steady at 8.4% and 5.6%, respectively, each compounded annually. How much money should be set aside now to take care of this purchase if the required type of van sells for $60,000 now?

C 32. Mary invested $5000 in a fund that earned interest at the rate of 7.5% compounded annually during the first 8 years. Then the interest rate jumped to a higher rate for the next 6 years, again compounded annually. What was this higher interest rate if the accumulated amount at the end of the 14 years was $15,375?

33. Show by direct substitution that $y = L - (L - y_0)e^{-kt}$ satisfies the differential equation $dy/dt = k(L - y)$.

34. Show by substitution that $y = 100/(1 + e^{-100kt})$ satisfies the differential equation $dy/dt = ky(100 - y)$.

35. Show that $y = (A + Bt)e^{2t}$ satisfies the differential equation $y'' - 4y' + 4y = 0$.

36. Show that $y = A + (B + Ct)e^{3t}$ satisfies the differential equation $y''' - 6y'' + 9y' = 0$.

37. A population y is said to be growing so that the rate of growth is proportional to the time elapsed and to the number present. Write down the corresponding differential equation. In the long run, will this growth be faster or slower than exponential growth?

38. An object is moving along a straight line so that its acceleration is inversely proportional to the square root of its velocity.
    (a) What differential equation does the velocity v satisfy?
    (b) Find the formula for v in terms of t, given that $v = 4$ when $t = 0$ and $v = 9$ when $t = 2$.

39. An object is moving along a line so that its acceleration is proportional to the square of its velocity. If y is its distance from a fixed point, what differential equation does y satisfy?

C 40. Bacteria populations A and B grow exponentially, doubling in size in 2 hours and 3 hours, respectively. If the present population of A is 400 and that of B is 600, after how many hours will the two populations be equal in size?

Are there definite stages in a child's mental development? Do teachers sometimes introduce abstract concepts too early? Does manipulation of concrete objects improve geometric understanding? Would children learn mathematics better if teachers understood and used Piaget's learning theory model? The answers to all these questions appear to be yes.

7

The Antiderivative

Jean Piaget was born in 1896 in Neuchâtel, Switzerland. At age 10, he published his first article about an albino sparrow he had seen in a park and, while still in high school, developed sufficient reputation as a zoologist to be offered a position as curator of the mollusk collection at the Geneva Museum. Wisely, he turned down this heady offer and chose rather to study biology at the University of Neuchâtel where he earned his Ph.D. in 1918. There followed further study in Zurich and at the Sorbonne where he pursued the philosophy of science, logic, and the emerging field of psychology. More important than any formal learning was his intense personal study over many years of the history and significance of the principal concepts of biology, physics, and mathematics. Asked which scientist he admired most, he answered, "Einstein."

Piaget's international reputation rests most securely on his work in developmental psychology. Careful observation of the learning patterns of his own children and later of large numbers of others led him to propose that children go through four definite stages of development: sensorimotor period (birth to 2 years), preoperational period (2 to 7 years), concrete operational period (7 to 11 years), and formal operational period (11 to 15 years). It is in the latter stage that children begin to think abstractly and logically—to do true mathematical thinking. More than any other person, Piaget has investigated how children learn the concepts and operations of arithmetic, geometry, algebra, and logic.

Piaget's psychological writings are studded with mathematical ideas and terminology. He is the author of a book on algebraic logic (1949) and another on groups and lattices (1952). When asked why mathematics is important in the study of the development of knowledge, he replied, "Because, along with formal logic, mathematics is the only entirely deductive discipline."

Jean Piaget
(1896–1980)

ANTIDERIVATIVES

Most of the operations that we perform in mathematics can be undone. The operation of adding 10 to a number can be undone by subtracting 10 from the result of the first operation. We call addition and subtraction inverse operations. Other examples of inverse operations are multiplication and division, squaring and taking square roots, and taking the natural logarithm of a number and exponentiating with the base e. Our task in this section is to study the operation that is inverse to differentiation (finding the derivative). We call this operation *antidifferentiation,* and the object obtained by this process is an *antiderivative*. Here is a formal definition.

DEFINITION

We call $F(x)$ an **antiderivative** of $f(x)$ on an interval if $F'(x) = f(x)$ for all x in the interval, that is, if

$$\frac{d[F(x)]}{dx} = f(x)$$

EXAMPLE A

Find an antiderivative of $f(x) = x^2$.

Solution The function $F(x) = x^3$ occurs to us, but we immediately reject it because $d(x^3)/dx = 3x^2$. However, this does suggest $F(x) = x^3/3$, which is correct since

$$\frac{d(x^3/3)}{dx} = \frac{1}{3}\frac{d(x^3)}{dx} = \frac{1}{3}3x^2 = x^2$$

A little more thought suggests other possibilities. The functions $\frac{1}{3}x^3 + 7$ and $\frac{1}{3}x^3 - 129$ also have the right derivative. In fact, any function of the form $F(x) = \frac{1}{3}x^3 + C$ is an antiderivative of $f(x) = x^2$. ∎

The General Antiderivative of a Function

Example A above indicates that there are many antiderivatives for any given function. If you find one antiderivative, you may add any constant to it and have another antiderivative. Are all antiderivatives of a given function obtained this way? The answer is yes, and we now indicate why.

We begin with the intuitive fact that the only type of function whose derivative is identically 0 is a constant function. For if a function is not constant, then it will increase or decrease somewhere and this will make its derivative nonzero.

FIGURE 1

Next consider two functions $F(x)$ and $G(x)$ which have the same derivative everywhere on an interval (see Figure 1). Then the function $H(x) = F(x) - G(x)$ has a derivative which is identically 0, which means that it is a constant function; that is, $H(x) = C$ for some constant C. But this means that $F(x) - G(x) = C$ or, equivalently, $F(x) = G(x) + C$. We have shown that if two functions have the same derivative, then one can be obtained from the other by the addition of a constant. In the language of antiderivatives, we have shown that if we find one antiderivative of a given function $f(x)$, we can get any other antiderivative from it by the addition of a constant. The whole family of antiderivatives so obtained is called the **general antiderivative of** $f(x)$. In Example A, we would call $F(x) = \frac{1}{3}x^3 + C$ the general antiderivative of $f(x) = x^2$.

EXAMPLE B

Find the general antiderivative of $f(x) = x^4$.

Solution After experimenting a little, we discover that $\frac{1}{5}x^5$ is one antiderivative. Thus $F(x) = \frac{1}{5}x^5 + C$ is the general antiderivative of $f(x) = x^4$. ∎

From now on, when we ask you to find the antiderivative of a function, we mean that you should find the general antiderivative. Thus, your answers and ours will involve an arbitrary constant, traditionally denoted by the letter C.

We need a notation for antiderivatives and a systematic way of finding them. These are matters we take up next.

Some Rules for Antiderivatives

First, we introduce a symbol that goes way back to one of the founders of the calculus, Gottfried Leibniz. To denote the antiderivative of $f(x)$, he used the symbol $\int f(x)\, dx$. The elongated ess, $\int$, is called the **integral sign,** the function $f(x)$ is called the **integrand,** and for now dx is just a symbol that indicates that we are antidifferentiating with respect to x. In this notation, we indicate the results of Examples A and B as follows.

$$\int x^2\, dx = \tfrac{1}{3}x^3 + C \quad \text{and} \quad \int x^4\, dx = \tfrac{1}{5}x^5 + C$$

We remark that an antiderivative is also sometimes called an **indefinite integral.**

These two results suggest a general rule; it is called the Power Rule (for antiderivatives).

POWER RULE

If r is any real number except -1, then

$$\int x^r \, dx = \frac{x^{r+1}}{r+1} + C$$

The proof of this rule is simple. Just take the derivative of the right side and note that it is the integrand on the left side. This is how you should remember this rule: *To find the antiderivative of a power, increase the exponent by 1 and divide by the new exponent.*

Be sure to note that the rule above has one exception, namely, when $r = -1$. However, we take care of this exception by using Example E from Section 5.3.

$$\int x^{-1} \, dx = \int \frac{1}{x} \, dx = \ln|x| + C \qquad x \neq 0$$

EXAMPLE C

Find (a) $\displaystyle\int x^{15} \, dx$, (b) $\displaystyle\int \sqrt{x} \, dx$.

Solution

(a) $\displaystyle\int x^{15} \, dx = \frac{x^{16}}{16} + C$

(b) $\displaystyle\int \sqrt{x} \, dx = \int x^{1/2} \, dx = \frac{x^{3/2}}{\frac{3}{2}} + C = \frac{2}{3}x^{3/2} + C$ ∎

As with derivatives, the antiderivative of a sum is the sum of the antiderivatives, and a constant may be moved past an integral sign. In symbols,

---

$$\int [f(x) + g(x)] \, dx = \int f(x) \, dx + \int g(x) \, dx$$

$$\int cf(x) \, dx = c \int f(x) \, dx$$

With these two properties in hand, we can find many more antiderivatives.

RECALL

It is clear that

$$\int 1 \, dx = x + C$$

since the derivative of $x + C$ is 1. This is consistent with the Power Rule. Just remember that $1 = x^0$, apply the Power Rule, and you will get the boxed result.

EXAMPLE D

Find the antiderivative of $4x^2 - 6x + 5$.

Solution

$$\int [4x^2 - 6x + 5] \, dx = \int 4x^2 \, dx + \int (-6)x \, dx + \int 5(1) \, dx$$

$$= 4 \int x^2 \, dx - 6 \int x \, dx + 5 \int 1 \, dx$$

$$= \frac{4x^3}{3} - \frac{6x^2}{2} + 5x + C$$

$$= \tfrac{4}{3}x^3 - 3x^2 + 5x + C$$

EXAMPLE E

Find the antiderivative of $x^\pi + 3x^{-1/3} + 5x^{-1}$.

Solution

$$\int [x^\pi + 3x^{-1/3} + 5x^{-1}] \, dx = \int x^\pi \, dx + 3 \int x^{-1/3} \, dx + 5 \int x^{-1} \, dx$$

$$= \frac{x^{\pi+1}}{\pi + 1} + \frac{3x^{2/3}}{\frac{2}{3}} + 5 \ln |x| + C$$

$$= \frac{x^{\pi+1}}{\pi + 1} + \frac{9}{2}x^{2/3} + 5 \ln |x| + C$$

We mention one more rule that will be used continuously from now on, the rule for antidifferentiating e^x.

$$\int e^x \, dx = e^x + C$$

As usual, the proof is simple. Just take the derivative of the right side and see that you get the integrand on the left.

EXAMPLE F

Find the antiderivative of $2e^x - 3$.

Solution

$$\int (2e^x - 3) \, dx = 2 \int e^x \, dx - 3 \int 1 \, dx = 2e^x - 3x + C$$

Antiderivatives with Side Conditions

We have emphasized the fact that the general antiderivative of a given function involves an arbitrary constant and thus is a whole family of functions. Thus the general antiderivative of the function $f(x) = 2x$ is the family of functions $F(x) = x^2 + C$. This family is displayed in Figure 2.

Sometimes we are given a condition which picks out from a family of antiderivatives a particular function which satisfies that condition.

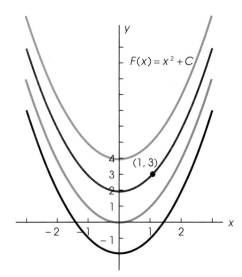

FIGURE 2

EXAMPLE G

Find the antiderivative of $f(x) = 2x$ which satisfies the side condition $F(1) = 3$, that is, whose graph goes through the point $(1, 3)$.

Solution The general antiderivative of $f(x) = 2x$ is $F(x) = x^2 + C$. The condition $F(1) = 3$ means that

$$3 = F(1) = 1^2 + C$$

which requires that $C = 2$. We conclude that $F(x) = x^2 + 2$. Geometrically, the side condition has picked out the blue curve from the whole family of curves (Figure 2). ∎

Problem Set 7.1

Fill in the blanks in Problems 1–6.

1. If $G'(x) = g(x)$, we call $G(x)$ an ＿＿＿＿ of $g(x)$.

2. If $G'(x) = H'(x)$ for all x, then $G(x) - H(x)$ is ＿＿＿＿.

3. The derivative of the antiderivative of $f(x)$ is ———————.

4. The general antiderivative of $f'(x)$ is ———————.

5. x^4 is an antiderivative of ———————.

6. The general antiderivative of x^4 is ———————.

In Problems 7–18, find the general antiderivative of the given function.

7. $f(x) = x^3$

8. $f(x) = x^5$

9. $f(x) = x^{-3}$

10. $f(x) = x^{-5}$

11. $f(x) = x^{5/2}$

12. $f(x) = x^{3/4}$

13. $f(x) = x^{\pi}$

14. $f(x) = x^{\sqrt{2}}$

15. $f(x) = 1/x^4$

16. $f(x) = 1/x$

17. $f(x) = x^{-1/2}/x^{1/2}$

18. $f(x) = x^4/\sqrt{x}$

In Problems 19–40, find the indicated antiderivatives.

19. $\displaystyle\int (3x + 2)\, dx$

20. $\displaystyle\int (x^2 + 4x)\, dx$

21. $\displaystyle\int (5x^4 - 6x)\, dx$

22. $\displaystyle\int (3x^3 + 2x - 3)\, dx$

23. $\displaystyle\int (5x^{3/2} - 3x^{1/2})\, dx$

24. $\displaystyle\int (4x^{1/3} + 3)\, dx$

25. $\displaystyle\int (3\sqrt{x} + 5/x)\, dx$

26. $\displaystyle\int (4/x^3 + 2\sqrt[3]{x})\, dx$

27. $\displaystyle\int (5/x^2 - 2/x^3)\, dx$

28. $\displaystyle\int (2/x^5 - 3)\, dx$

29. $\displaystyle\int (5e^x - 4/x)\, dx$

30. $\displaystyle\int (x^{-2} + 4e^x)\, dx$

31. $\displaystyle\int (3\sqrt[3]{t} + 2t)\, dt$

32. $\displaystyle\int (4\sqrt[3]{s} - 1/s)\, ds$

33. $\displaystyle\int (2u^3 - 4/u^2)\, du$

34. $\displaystyle\int (5/v^6 + 2v^5)\, dv$

35. $\displaystyle\int (2x - 1)(x + 3)\, dx$

36. $\displaystyle\int (x + 2)(3x - 1)\, dx$

37. $\displaystyle\int (2t + 3)^2\, dt$

38. $\displaystyle\int (3u^2 - 1)^2\, du$

39. $\displaystyle\int (x^2 + 4x + 2)/x^2\, dx$

40. $\displaystyle\int (3x^2 - 5)/x^{3/2}\, dx$

In Problems 41–48, find the particular antiderivative $F(x)$ of $f(x)$ that satisfies the given side condition (see Example G).

41. $f(x) = 4x - 2;\ F(1) = 3$

42. $f(x) = 3 - 2x;\ F(2) = 7$

43. $f(x) = x^2 - 2x + 5;\ F(2) = \frac{23}{3}$

44. $f(x) = 3x^2 + x - 4;\ F(-1) = \frac{13}{2}$

45. $f(x) = 4e^x;\ F(0) = 2$

46. $f(x) = e^x - 2x;\ F(0) = 5$

47. $f(x) = 3/x + 6x^2;\ F(1) = 5$

48. $f(x) = x^2 - 2/x;\ F(1) = \frac{10}{3}$

49. Find the following antiderivatives.

(a) $\displaystyle\int (2x - 4/x)\, dx$

(b) $\displaystyle\int (x + \sqrt{x})^2\, dx$

(c) $\displaystyle\int (\sqrt{x} + 1)(3/\sqrt{x})\, dx$

(d) $\displaystyle\int (3x^2 - 4)/x^3\, dx$

(e) $\displaystyle\int (6 + 4e^s)\, ds$

(f) $\displaystyle\int (e^t + t^e)\, dt$

50. Suppose that the slope of the graph of f at x is given by $3x^2 - e^x + 5$ and that the graph passes through $(0, 4)$. Find $f(x)$.

51. Let $R(x)$ be the revenue in dollars from producing and selling x camping trailers per year and note that $R(0) = 0$. If the marginal revenue is $3000 - 0.9x$, find the revenue in a year in which 1500 trailers were produced and sold.

52. Suppose that $g''(t) = 6/\sqrt{t}$, $g'(4) = 0$, and $g(1) = 10$. Find $g(t)$. *Hint:* Antidifferentiate twice.

53. Suppose that an object moves along a straight line in such a way that its velocity v after t seconds is given by $v = ds/dt = 6t^2 - 4t + 2$ feet per second. Find the position $s = s(t)$ of the object at time t, given that $s(1) = 14$.

54. Suppose that a car starting from rest accelerates at 8 feet per second. Find its velocity and how far it has traveled by the end of 5 seconds.

55. A rapidly growing city had a population of 10,000 in 1985. If its growth rate t years after 1985 is estimated as $dP/dt = 250 + 300\sqrt{t}$, what will the population be in 2001?

☐C 56. During the first 13 days of chemotherapy, the mass $M(t)$ of a malignant tumor decreases at the rate $M'(t) = -0.2t + 0.015t^2$ grams per day. What is the mass of such a tumor after 10 days of treatment if it started out weighing 180 grams?

☐C 57. Let $C(x)$ denote the number of dollars it costs a publishing company to produce x copies of a certain calculus book. If the marginal cost is $400/\sqrt{x}$ and if the set-up and preparation costs are $150,000, find (a) the total cost of producing 25,600 books and (b) the average cost per book at this production level.

58. A nursery sells a certain kind of pine tree when such a tree reaches a height of 2 feet and claims that it will have a growth rate t years later given by $dh/dt = 4t^{-1/3}$, where h is the height in feet. How tall will such a tree be 8 years later?

7.2
THE SUBSTITUTION RULE

The Chain Rule plays a fundamental role in finding derivatives. The analog of this rule for antiderivatives is the Substitution Rule. It is equally important and will be used in almost every section from here on.

Before we can state this important rule, we need to review an idea introduced for another purpose in Section 4.5. Let $u = g(x)$ and let dx be a symbol called the **differential** of x. You may think of dx as representing a small change in x if you wish. Then du, the differential of u, is defined by

$$du = g'(x) \, dx$$

If you studied Section 4.5, you know that du gives an approximation to the corresponding change in u, an approximation that gets better the smaller dx is chosen. For right now, all that is important is that you remember the boxed formula. With it, we can state the Substitution Rule in a particularly convenient form.

> **THE SUBSTITUTION RULE**
> If $u = g(x)$, then $du = g'(x) \, dx$ and
> $$\int f(g(x))g'(x) \, dx = \int f(u) \, du$$

To see why this rule is correct, let $F(u)$ be an antiderivative of $f(u)$, so that

$$F'(u) = f(u) \quad \text{and} \quad \int f(u) \, du = F(u) + C$$

Now by the Chain Rule,

$$\frac{d[F(u)]}{dx} = \frac{d[F(u)]}{du} \frac{du}{dx} = F'(u)g'(x)$$

That is,

$$\frac{d[F(g(x))]}{dx} = F'(g(x))g'(x) = f(g(x))g'(x)$$

In terms of antiderivatives, this says

$$F(u) + C = F(g(x)) + C = \int f(g(x))g'(x) \, dx$$

or, equivalently,

$$\int f(u) \, du = \int f(g(x))g'(x) \, dx$$

The Substitution Rule Illustrated

In applying the Substitution Rule to an integral, we must be able to recognize the integrand as having the form $f(g(x))g'(x) \, dx$ so that the substitution $u = g(x)$ will change this expression to the form $f(u) \, du$. This requires insight and practice. Our first example can be done two ways—with or without the Substitution Rule.

EXAMPLE A

Find $\int (2x^4)^3 8x^3 \, dx$.

Solution Method 1 (using the Substitution Rule). Let $u = 2x^4$, so that $du = 8x^3 \, dx$. Then

$$\int (2x^4)^3 8x^3 \, dx = \int u^3 \, du = \frac{u^4}{4} + C = \frac{(2x^4)^4}{4} + C = 4x^{16} + C$$

Method 2 (by first multiplying out the integrand).

$$\int (2x^4)^3 8x^3 \, dx = \int 8x^{12} 8x^3 \, dx = \int 64x^{15} \, dx$$

$$= \frac{64x^{16}}{16} + C = 4x^{16} + C$$

Naturally, we get the same answer both ways. ■

EXAMPLE B

Find $\int (x^2 + 3)^3 2x \, dx$.

Solution This example could be done by cubing $x^2 + 3$, multiplying by $2x$, and then finding the antiderivative; but that is far too much work. Rather, we make the substitution $u = x^2 + 3$, so $du = 2x \, dx$. This gives

$$\int (x^2 + 3)^3 2x \, dx = \int u^3 \, du = \frac{u^4}{4} + C = \frac{(x^2 + 3)^4}{4} + C$$ ■

SHARP CURVE
Two methods of doing the same integration may lead to different-looking answers.

$$\int (2x + 3)2 \, dx$$

$$= \frac{1}{2}(2x + 3)^2 + C$$

$$= 2x^2 + 6x + \frac{9}{2} + C$$

$$\int (2x + 3)2 \, dx = \int (4x + 6) \, dx$$

$$= 2x^2 + 6x + K$$

However, these answers are equivalent since $\frac{9}{2} + C$ represents the same arbitrary constant in the first case that K does in the second case.

EXAMPLE C

Find $\int (x^4 - 5)^6 x^3 \, dx$.

Solution The substitution $u = x^4 - 5$ occurs to us, but unfortunately, $du = 4x^3 \, dx$, differing from what we have by a factor of 4. However, if we remember that constants can be shifted past an integral sign, we see what to do. Multiply and divide by 4.

$$\int (x^4 - 5)^6 x^3 \, dx = \int (x^4 - 5)^6 \frac{1}{4} 4x^3 \, dx$$

$$= \frac{1}{4} \int u^6 \, du = \frac{1}{4} \frac{u^7}{7} + C$$

$$= \frac{1}{28} (x^4 - 5)^7 + C \qquad \blacksquare$$

EXAMPLE D

Find $\int e^{x^2} x \, dx$.

Solution We try the substitution $u = x^2$, noting that $du = 2x \, dx$. The factor of 2 is missing in the integrand, but we can introduce this factor provided we compensate with a factor of $\frac{1}{2}$.

$$\int e^{x^2} x \, dx = \int e^{x^2} \frac{1}{2} 2x \, dx = \frac{1}{2} \int e^u \, du$$

$$= \frac{1}{2} e^u + C = \frac{1}{2} e^{x^2} + C \qquad \blacksquare$$

EXAMPLE E

Find $\int \frac{x^2 + 2x}{x^3 + 3x^2 + 4} \, dx$.

Solution We make the substitution $u = x^3 + 3x^2 + 4$ so $du = (3x^2 + 6x) \, dx$. Clearly, we need an extra factor of 3 in the integrand. We may introduce this factor provided we compensate for it with a factor of $\frac{1}{3}$, a factor which may be passed in front of the integral sign.

$$\int \frac{x^2 + 2x}{x^3 + 3x^2 + 4} \, dx = \frac{1}{3} \int \frac{3(x^2 + 2x)}{x^3 + 3x^2 + 4} \, dx = \frac{1}{3} \int \frac{1}{u} \, du$$

$$= \frac{1}{3} \ln |u| + C = \frac{1}{3} \ln |x^3 + 3x^2 + 4| + C \qquad \blacksquare$$

EXAMPLE F

Find $\displaystyle\int \frac{(\ln x)^4}{x}\,dx$

Solution Let $u = \ln x$ so $du = (1/x)\,dx$. Then

$$\int \frac{(\ln x)^4}{x}\,dx = \int u^4\,du = \frac{u^5}{5} + C = \frac{(\ln x)^5}{5} + C \qquad \blacksquare$$

A Miscellany of Examples

The examples so far considered are rather standard; they all use basically the same idea. However, we must be alert to a number of variations.

EXAMPLE G

Find $\displaystyle\int (x + 3)^{50} x\,dx$.

Solution The substitution $u = x + 3$ occurs to us but does not seem promising since $du = dx$. What shall we do with the extra factor x that appears in the integrand? Note that if $u = x + 3$, then $x = u - 3$. If we replace all factors involving x by their correspondents in terms of u, we obtain

$$\int (x + 3)^{50} x\,dx = \int u^{50}(u - 3)\,du$$

which still looks rather complicated. However, we have achieved something, because we can now multiply the factors of the integrand together. We get

$$\int u^{50}(u - 3)\,du = \int (u^{51} - 3u^{50})\,du$$

$$= \frac{u^{52}}{52} - \frac{3u^{51}}{51} + C$$

$$= \frac{(x + 3)^{52}}{52} - \frac{(x + 3)^{51}}{17} + C \qquad \blacksquare$$

EXAMPLE H

Find $\displaystyle\int (x^3 - 3)^2\,dx$.

Solution No substitution will work here. However, the simplest possible idea does work: Expand the integrand.

$$\int (x^3 - 3)^2\,dx = \int (x^6 - 6x^3 + 9)\,dx$$

$$= \frac{x^7}{7} - \frac{6x^4}{4} + 9x + C \qquad \blacksquare$$

EXAMPLE I

Find $\int (2x^4 - 3x^2 + 5)/5x\ dx$.

Solution No substitution will work. Here the trick is to break the integrand into three terms.

$$\int \frac{2x^4 - 3x^2 + 5}{5x}\ dx = \int \left[\frac{2}{5}x^3 - \frac{3}{5}x + \frac{1}{x}\right] dx$$

$$= \frac{2x^4}{20} - \frac{3x^2}{10} + \ln|x| + C$$

$$= \frac{x^4}{10} - \frac{3x^2}{10} + \ln|x| + C \qquad \blacksquare$$

EXAMPLE J

Find $\int [x^3/(x^2 + 1)]\ dx$ and $\int (x^3 - 6x + 8)/(x - 2)\ dx$.

Solution Both integrands are examples of what are called improper fractions—rational functions in which the degree of the numerator is greater than or equal to that of the denominator. Our first step is to perform long division (see RECALL). In the first case, this gives

$$\frac{x^3}{x^2 + 1} = x - \frac{x}{x^2 + 1} = x - \frac{1}{2}\frac{2x}{x^2 + 1}$$

This combined with the mental substitution $u = x^2 + 1$ gives

$$\int \frac{x^3}{x^2 + 1}\ dx = \int x\ dx - \frac{1}{2}\int \frac{2x}{x^2 + 1}\ dx$$

$$= \frac{x^2}{2} - \frac{1}{2}\ln (x^2 + 1) + C$$

In the second case, we obtain

$$\int \frac{x^3 - 6x + 8}{x - 2}\ dx = \int (x^2 + 2x - 2)\ dx + 4\int \frac{1}{x - 2}\ dx$$

$$= \frac{x^3}{3} + x^2 - 2x + 4\ln|x - 2| + C$$

Here, we made the mental substitution $u = x - 2$ in the last step. $\qquad \blacksquare$

RECALL

Here we illustrate what we hope is a familiar process—long division.

$$\begin{array}{r} x^2 + 2x - 2 \\ x - 2\overline{)x^3 \qquad\ \ - 6x + 8} \\ \underline{x^3 - 2x^2} \\ 2x^2 - 6x \\ \underline{2x^2 - 4x} \\ - 2x + 8 \\ \underline{- 2x + 4} \\ 4 \end{array}$$

Thus,

$$\frac{x^3 - 6x + 8}{x - 2}$$

$$= x^2 + 2x - 2 + \frac{4}{x - 2}$$

Problem Set 7.2

In Problems 1–4, find the general antiderivative in two ways: by a substitution and by first multiplying out the integrand (see Examples A and B).

1. $\int (2x^3)^4 6x^2\ dx$

2. $\int (2x^5)^3 10x^4\ dx$

3. $\int (x^2 + 3)^2 2x\ dx$

4. $\int (2x^2 - 1)^2 4x\ dx$

In Problems 5–14, use an appropriate substitution to find the indicated antiderivative (see Examples B and C).

5. $\int (x^3 - 4)^5 3x^2 \, dx$

6. $\int (2x^3 + 1)^3 6x^2 \, dx$

7. $\int (x^4 + 5)^6 x^3 \, dx$

8. $\int (3x^2 - 4)^4 x \, dx$

9. $\int (x^2 + 6x + 4)^4 (2x + 6) \, dx$

10. $\int (2x^2 + 8x - 2)^3 (4x + 8) \, dx$

11. $\int (x^3 + 3x - 2)^{10} (x^2 + 1) \, dx$

12. $\int (x^3 + 3x^2 - 11)^7 (x^2 + 2x) \, dx$

13. $\int (\sqrt{x} + 1)^4 (1/\sqrt{x}) \, dx$

14. $\int (3x^{-1} + 1)^3 (x^{-2}) \, dx$

Find each antiderivative in Problems 15–50 (see Examples D–J).

15. $\int e^{3x} \, dx$

16. $\int e^{-2x} \, dx$

17. $\int 3e^{2t+1} \, dt$

18. $\int 4e^{3u-4} \, du$

19. $\int e^{2x^2+1} x \, dx$

20. $\int e^{4-3x^2} x \, dx$

21. $\int e^{\sqrt{x}} (1/\sqrt{x}) \, dx$

22. $\int e^{4/x} (1/x^2) \, dx$

23. $\int \dfrac{2x + 5}{x^2 + 5x + 4} \, dx$

24. $\int \dfrac{3x^2 - 1}{x^3 - x + 7} \, dx$

25. $\int \dfrac{x^3 + x}{x^4 + 2x^2 - 2} \, dx$

26. $\int \dfrac{x^3 - 1}{2x^4 - 8x + 5} \, dx$

27. $\int \dfrac{e^x}{3e^x + 1} \, dx$

28. $\int \dfrac{e^x + 2}{4e^x + 8x} \, dx$

29. $\int (4 + \ln x)^{3/2} (1/x) \, dx$

30. $\int \sqrt{2 + 3 \ln x} \, (1/x) \, dx$

31. $\int [\ln (2x + 1)]^4 / (2x + 1) \, dx$

32. $\int \int (x \ln x)^3 (1 + \ln x) \, dx$

33. $\int [\ln (x^2 + 1)][x/(x^2 + 1)] \, dx$

34. $\int [\ln (\ln x)]^4 [1/(x \ln x)] \, dx$

35. $\int (x - 4)^{14} x \, dx$

36. $\int (x - 2)^{10} x \, dx$

37. $\int (2x + 1)^8 4x \, dx$

38. $\int (1 - 3x)^7 x \, dx$

39. $\int (x^2 + 5)^2 \, dx$

40. $\int (2x^2 - 3)^2 \, dx$

41. $\int (\sqrt{x} + 3)^2 \, dx$

42. $\int (3/x + 2)^2 \, dx$

43. $\int \dfrac{3x^3 + x - 7}{x^2} \, dx$

44. $\int \dfrac{5x^4 - 2x^2 + 7}{x^3} \, dx$

45. $\int \dfrac{2\sqrt{x} + 5}{\sqrt{x}} \, dx$

46. $\int \dfrac{x^{5/2} + 2x^{3/2} + 1}{x^{3/2}} \, dx$

47. $\int \dfrac{x^2 + x + 1}{x - 1} \, dx$

48. $\int \dfrac{x^2 + 5x - 4}{x + 2} \, dx$

49. $\int \dfrac{x^3 + 2x}{x^2 - 1} \, dx$

50. $\int \dfrac{x^3 + 6x}{x^2 + 2} \, dx$

Find the antiderivatives in Problems 51–62.

51. $\int (2u^2 + 1)^3 \, du$

52. $\int (2t^2 + 1)^{11} t \, dt$

53. $\int x^2 e^{x^3} \, dx$

54. $\int (e^{2x} - 3)^2 \, dx$

55. $\int (x + 1)^{10} x \, dx$

56. $\int (\ln x)^3 x^{-1} \, dx$

57. $\int \dfrac{t^2 + 1}{t^3 + 3t - 4} \, dt$

58. $\int \left(t^2 + \dfrac{4}{t} \right)^2 dt$

59. $\int \dfrac{\sqrt{t} + 2t + 3}{t} \, dt$

60. $\int \dfrac{t^2 + 9t}{t + 3} \, dt$

61. $\int \dfrac{e^t}{e^t + 2} \ln (e^t + 2) \, dt$

62. $\int \dfrac{(t + 1)^3}{(t + 2)^5} \, dt$

63. Find the equation of the curve that passes through the point (0, 3) and whose tangent line at any point (x, y) on the curve has slope $(2x + 3)/(x^2 + 3x + 1)$.

C 64. A commercial truck, purchased for $60,000, depreciates at a rate of $28{,}000e^{-t/2}$ dollars per year t years after purchase (that is, $dV/dt = -28{,}000e^{-t/2}$). Find its value at the end of 10 years.

C 65. A manufacturer estimates that the marginal cost of producing x picnic tables per week is $C'(x) =$

$50(1 + e^{-0.02x})$ and that the fixed costs per week are $2500.

(a) How much will it cost to produce 80 tables per week?

(b) What is the average cost per table in this case?

C 66. A recreational lake is treated periodically with a chemical during the summer to reduce the bacteria level. The rate at which the concentration of bacteria is changing t days after treatment is given by $dB/dt = -1500t^2/(1 + t^3)$, where $B = B(t)$ is the number of bacteria per milliliter of water. If the ini-

tial concentration of bacteria was 6000 per liter, find the bacteria concentration after 8 days.

C 67. A particle moves along a straight line in such a way that its velocity v after t minutes is $v = t(3 - 0.2t)^{1/2}$ feet per second $(0 \le t \le 15)$. If $s = s(t)$ denotes the coordinate of the particle at time t and $s(10) = 10$, find (a) $s(t)$ and (b) the distance traveled during the interval $0 \le t \le 15$. *Hint:* Let $u = (3 - 0.2t)^{1/2}$, so that $u^2 = 3 - 0.2t$ and $2u\, du = -0.2\, dt$.

7.3
SOME INTEGRATION TECHNIQUES

We have not used the word *integration* before because the word *antidifferentiation* seems more descriptive. However, from now on, the two words will be used interchangeably. In like manner, the expression *integral* (technically *indefinite integral*) will often replace *antiderivative*.

Our main tool for integration is the *Substitution Rule* introduced in the previous section. There is another general rule that is sometimes useful. It is traditionally known as **integration by parts,** but the phrase *double substitution rule* is more descriptive.

Integration by Parts

The Product Rule for derivatives says that

$$\frac{d[u(x)v(x)]}{dx} = u(x)v'(x) + v(x)u'(x)$$

When we integrate both sides of this equation, we obtain

$$u(x)v(x) = \int u(x)v'(x)\, dx + \int v(x)u'(x)\, dx$$

> **RECALL**
> Keep in mind that antidifferentiation and differentiation are inverse operations. Thus
> $$\int \frac{d}{dx} f(x)\, dx = f(x) + C$$
> In particular,
> $$\int \frac{d}{dx} [u(x)v(x)]\, dx = u(x)v(x) + C$$
> a fact that we have used in the derivation at the right. Why were we able to suppress the constant C in this derivation?

or, equivalently,

$$\int u(x)v'(x)\, dx = u(x)v(x) - \int v(x)u'(x)\, dx$$

In differential symbols, this takes the following form called Integration by Parts.

> **INTEGRATION BY PARTS**
> Let $u = u(x)$ and $v = v(x)$ be differentiable functions. Then,
> $$\int u\, dv = uv - \int v\, du$$

For this result to be of use in finding an integral $\int f(x)\ dx$, we must be able to write $f(x)\ dx$ in the form $u\ dv$. Even then, it has little value unless the integral $\int v\ du$ is something we know how to handle. An example will clarify what we mean.

EXAMPLE A

Find $\displaystyle\int xe^x\ dx$.

Solution First, you should convince yourself that the ordinary Substitution Rule is of no help. Turning to integration by parts, make the double substitution

$$u = x \qquad dv = e^x\ dx$$

Then

$$du = dx \qquad v = \int e^x\ dx = e^x$$

and so,

$$\int xe^x\ dx = \int u\ dv = uv - \int v\ du$$

$$= xe^x - \int e^x\ dx$$

$$= xe^x - e^x + C \qquad \blacksquare$$

EXAMPLE B

Find $\displaystyle\int xe^{x^2}\ dx$.

Solution Don't jump to integration by parts. Always begin by trying an ordinary substitution. This integral was found by the substitution $u = x^2$ in the previous section (Example D). $\qquad \blacksquare$

EXAMPLE C

Find $\displaystyle\int x^2 e^x\ dx$.

Solution After a bit of thought, we decide that an ordinary substitution will not be fruitful so we try integration by parts. Let

$$u = x^2 \qquad dv = e^x\ dx$$

Then,

$$du = 2x \, dx \qquad v = e^x$$

and so,

$$\int \overbrace{x^2}^{u} \overbrace{e^x \, dx}^{dv} = \overbrace{x^2 e^x}^{u \, v} - \int \overbrace{e^x}^{v} \overbrace{2x \, dx}^{du} = x^2 e^x - 2 \int x e^x \, dx$$

To find the last integral, we appeal to Example A (otherwise, we would have to give up or, if especially clever, apply integration by parts again). We obtain

$$\int x^2 e^x \, dx = x^2 e^x - 2(x e^x - e^x + C)$$

$$= x^2 e^x - 2 x e^x + 2 e^x + K \qquad \blacksquare$$

EXAMPLE D

Find $\int x^2 \ln x \, dx$.

Solution This is an integration by parts problem, but what shall we substitute for u and dv? If we let $dv = \ln x \, dx$, we are in trouble because we don't know how to integrate $\ln x$. As an alternative, let us try

$$u = \ln x \qquad dv = x^2 \, dx$$

Then

$$du = \frac{1}{x} \, dx \qquad v = \frac{x^3}{3}$$

and

$$\int x^2 \ln x \, dx = \frac{(\ln x) x^3}{3} - \int \frac{x^3}{3} \frac{1}{x} \, dx$$

$$= \frac{1}{3}(x^3 \ln x) - \frac{1}{3} \int x^2 \, dx$$

$$= \frac{1}{3}(x^3 \ln x) - \frac{1}{3} \frac{x^3}{3} + C$$

$$= \frac{1}{3}(x^3 \ln x) - \frac{1}{9} x^3 + C \qquad \blacksquare$$

Integration Using Partial Fractions

It is a standard algebraic exercise to combine two or more simple fractions into one fraction. For example,

$$\frac{3}{x-1} - \frac{2}{x+2} = \frac{3x+6-2x+2}{(x-1)(x+2)} = \frac{x+8}{(x-1)(x+2)} = \frac{x+8}{x^2+x-2}$$

In integration problems, it will be to our advantage to go in the opposite direction since, while we do not know how to integrate the expression at the right, we are able to integrate the one at the left. We illustrate.

EXAMPLE E

Find $\int \dfrac{x+8}{x^2+x-2}\, dx$.

Solution

$$\int \frac{x+8}{x^2+x-2}\, dx$$

$$= \int \left(\frac{3}{x-1} - \frac{2}{x+2}\right) dx = 3\int \frac{1}{x-1}\, dx - 2\int \frac{1}{x+2}\, dx$$

The first of these integrals is found by the substitution $u = x - 1$, and the second by the substitution $u = x + 2$. Note that in both cases $du = dx$. We obtain as our final answer

$$3 \ln |x-1| - 2 \ln |x+2| + C \qquad \blacksquare$$

We pose now this question: Can a complicated fraction always be written as a sum of simpler fractions? More precisely, can a fraction of the form

$$\frac{f(x)}{(x-a)(x-b)(x-c)\cdots(x-p)} \qquad (a, b, c, \ldots, p \text{ all different})$$

be expressed as a sum

$$\frac{A}{x-a} + \frac{B}{x-b} + \frac{C}{x-c} + \cdots + \frac{P}{x-p}$$

The answer is yes provided the numerator $f(x)$ is a polynomial with degree less than that of the denominator. The process is called decomposing into partial fractions.

EXAMPLE F

Decompose $4/(x^2 + 5x + 6)$ into partial fractions.

Solution First, we note that the denominator can be factored as the product $(x + 3)(x + 2)$. Thus, we expect a decomposition of the form

$$\frac{4}{x^2+5x+6} = \frac{A}{x+3} + \frac{B}{x+2}$$

After multiplying both sides by $(x + 3)(x + 2)$, this takes the form

$$4 = A(x + 2) + B(x + 3)$$

We want the latter to be true for all x, in particular for $x = -3$ and $x = -2$. Substituting these values gives

$$4 = A(-3 + 2)$$

and

$$4 = B(-2 + 3)$$

We conclude that $A = -4$ and $B = 4$. Thus,

$$\frac{4}{x^2 + 5x + 6} = \frac{-4}{x + 3} + \frac{4}{x + 2}$$ ■

EXAMPLE G

Find $\displaystyle\int 4/(x^2 + 5x + 6)\ dx$.

Solution We use the result from Example F to write the above integral as

$$-4 \int \frac{1}{x + 3}\ dx + 4 \int \frac{1}{x + 2}\ dx$$

When we make the substitution $u = x + 3$ in the first integral and $u = x + 2$ in the second, we obtain

$$-4 \ln|x + 3| + 4 \ln|x + 2| + C$$ ■

Tables of Integrals

Faced with an integration problem, you now have a number of techniques to try. Suppose none of them work. What then? One possibility is to turn to a *table of integrals*. There are whole books consisting of integral formulas, carefully arranged by category. You can probably find one or more of them in your college library. A small such table of integrals is shown at the back of this book, and we urge you to familiarize yourself with it. Here is an example of its use.

POINT OF INTEREST
Some things are hard; others are impossible. It is impossible to represent π as a quotient of two integers. But π can be approximated by 22/7 or even better by 31416/10000. Similarly $\int e^{x^2}\ dx$ cannot be expressed in terms of elementary functions. It can, however, be approximated by such functions. This is a huge topic and one that we cannot delve into at this point.

EXAMPLE H

Find $\displaystyle\int \frac{1}{x^2 \sqrt{9 - x^2}}\ dx$

Solution Apply Formula 58 from the table of integrals at the back of the book to obtain

$$\int \frac{1}{x^2 \sqrt{9 - x^2}}\ dx = -\frac{\sqrt{9 - x^2}}{9x} + C$$ ■

We conclude by correcting a wrong impression that we may have given. Even with all of the known techniques and with the biggest table of integrals that has been devised, we are unable to perform many indefinite integrations. Some integrations are difficult, some are extremely difficult, and some are impossible. Yes, we do mean impossible. It has been shown that the indefinite integrals of many functions cannot be expressed in terms of the elementary functions of calculus. For example, $\int e^{x^2}\,dx$ is known to be such an indefinite integral.

Problem Set 7.3

Find the integrals in Problems 1–12 using integration by parts.

1. $\int xe^{2x}\,dx$

2. $\int xe^{-3x}\,dx$

3. $\int x\ln x\,dx$

4. $\int \sqrt{x}\ln x\,dx$

5. $\int x^{-2}\ln x\,dx$

6. $\int x^{7/2}\ln x\,dx$

7. $\int x(1+x)^6\,dx$

8. $\int x(2x+1)^6\,dx$

9. $\int x^2 e^{-2x}\,dx$

10. $\int x^2 e^{4x}\,dx$

11. $\int x\ln(x+2)\,dx$

12. $\int \ln(x-2)\,dx$

Decompose each of the expressions in Problems 13–20 into partial fractions as in Example F.

13. $3/[(x+1)(x+2)]$ 14. $24/[(x-3)(x+3)]$

15. $(5x-9)/(x^2-3x)$ 16. $(x+11)/(x^2+x-2)$

17. $(6x+24)/[(x+1)(x+2)(x+3)]$

18. $(6x^2-8x-4)/[(x-2)(x-1)(x+2)]$

19. $(3x^2-11x-30)/(x^3+x^2-6x)$

20. $(x^2-26x+9)/[(x^2-9)(x-1)]$

21–28. Use the results obtained above to find the integrals of the expressions in Problems 13–20 as in Examples E and G.

Find the integrals in Problems 29–40.

29. $\int 4e^{2x}\,dx$

30. $\int 6e^{-3x}\,dx$

31. $\int (x+2)e^{-5x}\,dx$

32. $\int (x-3)e^{4x}\,dx$

33. $\int xe^{4x^2}\,dx$

34. $\int x^3 e^{-x^2}\,dx$

35. $\int (\ln x)/\sqrt{x}\,dx$

36. $\int (\ln x)^4/x\,dx$

37. $\int \dfrac{2x-1}{x^2-x-2}\,dx$

38. $\int \dfrac{2x+3}{x^2+3x-10}\,dx$

39. $\int \dfrac{2x+7}{x^2+3x-10}\,dx$

40. $\int \dfrac{3x-9}{x^2-x-2}\,dx$

41. Find $\int \dfrac{7x^2-7x-9}{(x+1)^2(x-4)}\,dx$

Hint: The integrand can be decomposed in the form
$$\frac{A}{x+1}+\frac{B}{(x+1)^2}+\frac{C}{x-4}.$$

42. See the hint in Problem 41 to help you find
$$\int \frac{-10x^2+36x-17}{(x-2)^2(x+1)}\,dx$$

43. An object is moving along a line in such a way that its velocity at time t seconds is $2te^{2t}$ feet per second. How far will the object travel between $t=0$ and $t=8$?

44. Find the function f such that the tangent line to the graph of f at the point $[x, f(x)]$ has slope $x^2\ln x$, given that $f(1)=2$.

C 45. Let $P(t)$ denote the number of milligrams of tetracycline assimilated into the bloodstream of a person t minutes after taking a pill containing the drug. Suppose the rate of assimilation is $3te^{-0.4t}$ milligrams per minute at time t. How much of the drug will be assimilated in 15 minutes?

C 46. The rate of growth of a certain city is $10(12-t)e^{t/4}$ people per year, for $0 \le t \le 12$, where t denotes the number of years after 1990. What will the population of this city be in 2000 if it was 13,640 in 1990?

DIFFERENTIAL EQUATIONS

The phrase *differential equation* occurred earlier in the book (Sections 6.1 and 6.2). A **differential equation** is an equation involving derivatives (or differentials) in which the unknown is a function. A typical example is the differential equation of exponential growth and decay, namely, $dy/dt = ky$. In the earlier sections, we gave the solutions to three important differential equations, but we did it without indicating how these solutions were found. For example, we said that $dy/dt = ky$ has the solution $y = y_0 e^{kt}$. We propose in this section to introduce a fairly general method for solving simple differential equations, but we warn our readers that we will only scratch the surface of an immense subject.

Our discussion of antiderivatives (indefinite integrals) makes the solution of one kind of differential equation obvious. The solution to $dy/dx = f(x)$ is $y = \int f(x)\, dx$. For example, the solution to $dy/dx = 3x^2$ is $y = x^3 + C$. Actually, this is a whole family of solutions, the family shown graphically in Figure 3. In this section, we refer to this family of solutions as the **general solution** to the differential equation. Sometimes, we will pick a **particular solution** from this family by giving an initial condition. Thus, we may say: Solve the differential equation $dy/dx = 3x^2$ subject to the initial condition that $y_0 = 2$. This picks from the family the particular solution $y = x^3 + 2$ (see again Figure 3).

Separation of Variables

We now formulate a new way of solving the differential equation $dy/dx = 3x^2$. Interpret dy and dx in this equation as differentials and multiply by dx to obtain

$$dy = 3x^2\, dx$$

Then integrate both sides; that is, write successively

$$\int dy = \int 3x^2\, dx$$

$$y + C_1 = x^3 + C_2$$

$$y = x^3 + (C_2 - C_1)$$

Nothing is lost by replacing the constant $C_2 - C_1$ by C. Accordingly, from now on, we shall add a constant of integration to one side only. In this case, we obtain the same solution we got earlier, namely, $y = x^3 + C$.

What is significant about the method just illustrated is that it has wide applicability. It works whenever a simple algebraic manipulation of the differential equation results in a **separation of variables,** that is, a grouping of all the y's and dy's on one side of the equation and all the x's and dx's on the other side.

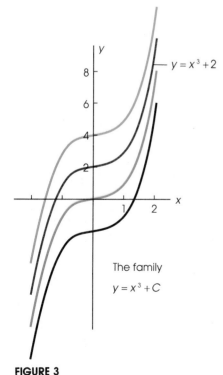

$y = x^3 + 2$

The family

$y = x^3 + C$

FIGURE 3

EXAMPLE A

Solve the differential equation

$$\frac{dy}{dx} = \frac{2x}{3y}$$

Solution We begin by multiplying both sides by $3y\,dx$ to separate variables and then integrate both sides.

$$3y\,dy = 2x\,dx$$

$$\int 3y\,dy = \int 2x\,dx$$

$$\frac{3y^2}{2} = x^2 + C$$

We could now solve this last equation for y explicitly, but since it would involve both a plus and minus and a messy square root, we accept $3y^2/2 = x^2 + C$ as an adequate representation of the general solution. ∎

EXAMPLE B

Solve the differential equation

$$\frac{dy}{dx} = 3\sqrt{xy}$$

subject to the condition that $y = 256$ when $x = 4$.

Solution We separate variables (multiply by dx and divide by $\sqrt{y}$) and then integrate.

$$y^{-1/2}\,dy = 3x^{1/2}\,dx$$

$$\int y^{-1/2}\,dy = \int 3x^{1/2}\,dx$$

$$2y^{1/2} = 2x^{3/2} + C$$

Substituting the initial data yields

$$2\sqrt{256} = 2(4)^{3/2} + C$$

$$32 = 16 + C$$

$$16 = C$$

The required particular solution therefore satisfies

$$2y^{1/2} = 2x^{3/2} + 16$$

or

$$y^{1/2} = x^{3/2} + 8$$

or

$$y = (x^{3/2} + 8)^2 = x^3 + 16x^{3/2} + 64$$

■

Three Examples Related to Growth Problems

We consider now the three differential equations which were used to model various types of growth in Sections 6.1 and 6.2. In each case, y represents a positive quantity whose rate of growth (or decay) dy/dt has a known form. The results obtained here justify the theorems of the earlier sections.

EXAMPLE C (EXPONENTIAL GROWTH MODEL)

Solve the differential equation $dy/dt = ky$ subject to the condition that $y = y_0$ when $t = 0$ (k is a constant and $y_0 > 0$).

Solution To separate variables, we multiply by dt and divide by y; then we integrate.

$$\frac{1}{y} \, dy = k \, dt$$

$$\int \frac{1}{y} \, dy = \int k \, dt$$

$$\ln y = kt + C$$

Changing the latter equation to its equivalent exponential form yields

$$y = e^{kt+C} = e^C e^{kt}$$

Substituting $y = y_0$ when $t = 0$ gives

$$y_0 = e^C$$

Thus, the desired solution is

$$y = y_0 e^{kt}$$

■

EXAMPLE D (BOUNDED GROWTH MODEL)

Solve the differential equation $dy/dt = k(L - y)$ subject to $y = y_0$ when $t = 0$ (k and L are constants and $0 \le y_0 < L$).

Solution We separate variables and integrate.

$$\frac{dy}{L - y} = k \, dt$$

$$\int \frac{dy}{L - y} = \int k \, dt$$

In the left integral, we make the substitution $u = L - y$ and $du = -dy$, leading to

$$-\int \frac{du}{u} = \int k \, dt$$

$$-\ln u = kt + C$$

$$\ln u = -kt - C$$

$$u = e^{-kt-C} = e^{-C}e^{-kt}$$

$$L - y = e^{-C}e^{-kt}$$

$$y = L - e^{-C}e^{-kt}$$

Substituting $y = y_0$ at $t = 0$ gives

$$y_0 = L - e^{-C} \quad \text{or} \quad e^{-C} = L - y_0$$

We conclude that

$$y = L - (L - y_0)e^{-kt}$$

Our derivation assumed that $y < L$ (so that $u = L - y > 0$), an assumption consistent with the bounded growth model. ■

EXAMPLE E (LOGISTIC GROWTH MODEL)

Solve the differential equation $dy/dt = ky(L - y)$ subject to $y = y_0$ when $t = 0$ (k and L are constants and $0 < y_0 < L$).

Solution We rewrite the differential equation as

$$\frac{1}{y(L - y)} \, dy = k \, dt$$

and then as

$$\left(\frac{1}{L} \frac{1}{y} + \frac{1}{L} \frac{1}{L - y} \right) dy = k \, dt$$

Thus,

$$\frac{1}{L} \int \frac{1}{y} \, dy + \frac{1}{L} \int \frac{1}{L - y} \, dy = \int k \, dt$$

> **SHARP CURVE**
> Writing $1/y(L - y)$ as the sum of two simple fractions is an example of decomposing a rational expression into partial fractions, a subject treated in Section 7.3

As in Example D, these integrations lead to

$$\frac{1}{L} \ln y - \frac{1}{L} \ln (L - y) = kt + C$$

From here on, the details are somewhat messy, but they are of a straightforward algebraic nature (Problem 35). The final result is

$$y = \frac{Ly_0}{y_0 + (L - y_0)e^{-Lkt}}$$ ■

An Economic Application

In Section 4.5, we introduced the important economic concept of *elasticity*. If $x = f(p)$ represents the demand function for a certain manufactured item (that is, x is the number of units that can be sold at price p per unit in some specified length of time), then the elasticity of demand E is given by

$$E = -\frac{p}{x}\frac{dx}{dp}$$

EXAMPLE F

If the elasticity of demand for a certain item is constant with value $\frac{1}{2}$, determine the demand function $x = f(p)$.

Solution We must solve the differential equation

$$-\frac{p}{x}\frac{dx}{dp} = \frac{1}{2}$$

On separation of variables, this takes the form

$$-2\frac{dx}{x} = \frac{dp}{p}$$

or after integration of both sides

$$-2 \ln x = \ln p + C$$

We can rewrite this successively as

$$\ln p + 2 \ln x = -C$$
$$\ln (px^2) = -C$$
$$px^2 = e^{-C}$$
$$x^2 = \frac{e^{-C}}{p}$$
$$x = \frac{K}{\sqrt{p}}$$

Here K is a constant that cannot be determined without more information. ∎

Problem Set 7.4

Find the general solution of the differential equations in Problems 1–12.

1. $\dfrac{dy}{dx} = \dfrac{x}{y^2}$

2. $\dfrac{dy}{dx} = \dfrac{1}{xy}$

3. $\dfrac{dy}{dx} = \dfrac{x^2}{1 - 3y^2}$

4. $\dfrac{dy}{dx} = \dfrac{1 + x^2}{y^{3/2}}$

5. $\dfrac{dy}{dx} = \dfrac{xy^2}{1 + x^2}$

6. $\dfrac{dy}{dx} = (1 + 2x)e^{-y}$

7. $\dfrac{dy}{dx} = \dfrac{x^3(1 + y^2)}{y(1 + x^4)}$

8. $\dfrac{dy}{dx} = \left(\dfrac{y + 4}{2 - x}\right)^2$

9. $\dfrac{dy}{dx} = \dfrac{2y}{x^2 - x}$

10. $\dfrac{dy}{dx} = \dfrac{7x - 1}{(x^2 - 1)y}$

11. $\dfrac{dy}{dx} = \dfrac{3x^2}{ye^y}$

12. $\dfrac{dy}{dx} = \dfrac{\sqrt{x}\ln x}{\sqrt{y}}$

In Problems 13–24, find the particular solution of the indicated differential equation that satisfies the given condition (see Examples B–E).

13. $\dfrac{dy}{dx} = \dfrac{x}{2y}$; $y = 4$ when $x = 2$

14. $\dfrac{dy}{dx} = \sqrt[3]{x/y}$; $y = 8$ when $x = 1$

15. $\dfrac{dy}{dt} = -y^2 t(t^2 + 2)^4$; $y = 1$ when $t = 0$

16. $\dfrac{dy}{dt} = \sqrt[3]{2t + 1}$; $y = 0$ when $t = 0$

17. $\dfrac{dy}{dt} = 0.02y$; $y = 4$ when $t = 0$

18. $\dfrac{dy}{dt} = -0.18y$; $y = 10$ when $t = 0$

19. $\dfrac{dy}{dt} = -0.36y$; $y = 5$ when $t = 1$

20. $\dfrac{dy}{dt} = 0.0125y$; $y = 20$ when $t = 4$

21. $\dfrac{dy}{dt} = 4(12 - y)$; $y = 8$ when $t = 0$

22. $\dfrac{dy}{dt} = 0.9(20 - y)$; $y = 13$ when $t = 0$

23. $\dfrac{dy}{dt} = 0.03y(12 - y)$; $y = 10$ when $t = 0$

24. $\dfrac{dy}{dt} = 0.04y(6 - y)$; $y = 5$ when $t = 0$

25. Suppose the elasticity of demand, $-(p/x)(dx/dp)$, is always equal to the constant $\frac{3}{4}$ and that $x = 10$ when $p = 16$. See Example F.
    (a) Express x in terms of p.
    (b) Find x when $p = 81$.

26. Suppose that the elasticity of demand (see Problem 25) is equal to x and that $x = 1000$ when $p = 1$. Find x when $p = 1.2$.

In Problems 27–34, find the general solution of the given differential equation.

27. $\dfrac{dy}{dt} = -4y$

28. $\dfrac{dy}{dx} = x^2 y^3$

29. $\dfrac{dy}{dx} = \dfrac{e^x}{y}$

30. $\dfrac{dy}{dx} = \dfrac{y \ln x}{x}$

31. $\dfrac{dy}{dx} = xy^2 e^{x^2}$

32. $\dfrac{dy}{dx} = xye^{-x}$

33. $\dfrac{dy}{dx} = \dfrac{8y}{x(x - 2)}$

34. $\dfrac{dy}{dx} = (y - 5)(10 - y)$

35. Fill in the missing steps in the solution to Example E; that is, show how to get from the line

$$\frac{1}{L}\ln y - \frac{1}{L}\ln(L - y) = kt + C$$

to the final result

$$y = \frac{Ly_0}{y_0 + (L - y_0)e^{-Lkt}}$$

36. Show that if the elasticity of demand for a certain item is always equal to 1, then the product of the price and the demand is a constant.

C 37. Suppose that the rabbit population $P = P(t)$ in a certain area t months after January 1, 1980, was growing at a rate equal to 9% of P (in rabbits per month). If $P(0)$ was 1200, what was the increase in population during the month of January 1981?

C 38. A company is advertising a new brand of bath soap on local television. Suppose that the rate at which new consumers become aware of the product is proportional to the number who are not yet aware of it. Let $p(t)$ denote the percentage of the total number of possible viewers who are aware of the product after t weeks.
    (a) Write the differential equation that $p(t)$ must satisfy.
    (b) If no one knew about the new soap when the advertising campaign began and 20% were aware of it after 1 week, what percentage were aware of it at the end of 3 weeks?

C 39. A city of 5000 has been hit by a flu epidemic. The rate at which the epidemic is spreading is proportional to the product of y, the number of people already infected, and the number not yet infected.
    (a) Write the differential equation that y must satisfy.
    (b) If 5 people were infected initially and 30 were infected after 1 week, how many were infected at the end of 4 weeks?

C 40. Let $P = P(t)$ denote the population of Wonderland t years after 1980. The internal growth rate (that is, neglecting migration) in the population of this country is equal to 2% of P, but there is also a net migration into the country of 20,000 per year.
(a) Write the differential equation for P.
(b) If the population was 12 million in 1980, what will it be in the year 2000?

C 41. Suppose that in a certain country the population $P = P(t)$, t years after 1990, satisfies the differential equation $P'(t) = 0.02P(t)$; and suppose also that the gross national product $G(t)$ satisfies the differential

equation $G'(t) = 0.015G(t)$. If the 1990 per capita income G/P was $920, what will the per capita income be in the year 2000?

C 42. In a certain wilderness region, the bear population ranges from a low of 500 to a high of 2500 and presently numbers 1000. A biologist has suggested that the growth rate of this population is equal to $0.0002(P - 500)(2500 - P)$, where $P = P(t)$ is the population t years from now. Estimate the population 15 years from now.

CHAPTER 7 REVIEW PROBLEM SET

1. Determine the general antiderivative of each of the following functions.
(a) $f(x) = 8x^7 - 4x^3 + \pi$
(b) $f(x) = 2\pi x^{\pi-1} - 10x^{3/2}$
(c) $f(x) = -\dfrac{4}{x} + 2e^{4x}$
(d) $f(x) = 3x^2 e^{x^3} - 2$

2. Determine the general antiderivative of each function.
(a) $f(x) = 15x^4 - 8x^3 + 2x - \sqrt{3}$
(b) $f(x) = 3\sqrt{2}x^{\sqrt{2}-1} - 14x^{5/2}$
(c) $f(x) = 4e^{2x} - \sqrt{x} + \dfrac{5}{x}$
(d) $f(x) = \dfrac{3x^2 + 1}{x^3 + x - 4} + \dfrac{1}{x^2}$

3. Find each indefinite integral.
(a) $\displaystyle\int (3x^2 - 12x + 5)\, dx$
(b) $\displaystyle\int \left(3\sqrt{x} + \dfrac{5}{x^2}\right) dx$
(c) $\displaystyle\int \dfrac{(2x + 1)^2}{x}\, dx$
(d) $\displaystyle\int \left(4e^{3x} - \dfrac{6}{\sqrt[3]{x}}\right) dx$

4. Find each indefinite integral.
(a) $\displaystyle\int (4x^3 - 6x^2 - 3)\, dx$
(b) $\displaystyle\int \dfrac{(2x^3 + 1)^2}{x^4}\, dx$
(c) $\displaystyle\int \left(5t\sqrt{t} + \dfrac{4}{\sqrt{t}}\right) dt$
(d) $\displaystyle\int (e^{2x} + 3x^e)\, dx$

5. Determine $F(x)$ if $F'(x) = 4/x^2 + 4 + x^2$ and $F(2) = 3$.

6. Find $F(x)$ if $F'(x) = x^3 + 6x + \dfrac{12}{x} + \dfrac{8}{x^3} + 4$ and $F(1) = 4$.

7. If $f''(t) = 12/t^5$, $f'(1) = 1$, and $f(1) = -7$, find $f(t)$.

8. If $f''(t) = 15\sqrt{t}$, $f'(1) = 12$, and $f(1) = 3$, find $f(t)$.

9. The Deluxe Apparel Company estimates the marginal cost of making and selling x tuxedos per month to be $280 - 12x + 0.6x^2$ dollars. Find the formula for $C(x)$, the cost of making and selling x tuxedos per month, given that $C(10) = \$3100$.

10. The Good Time Company estimates the marginal cost of making and selling x alarm clocks per month to be $5 - 0.08x + 0.12x^2$ dollars. Find the formula for $C(x)$, the cost of making and selling x alarm clocks per month, given that $C(100) = \$42,100$.

11. A proposed model for the deer population P in a certain state is that its rate of growth satisfies $dP/dt = 2000/\sqrt{t}$, where t is the number of years after January 1, 1980. If this model is a good one and if the deer population was 18,500 on January 1, 1980, what deer population is predicted for January 1, 2005?

12. Suppose that the population P of a certain city, t years after January 1, 1990, satisfies the differential equation $dP/dt = 150e^{-0.015t}$. If the population on January 1, 1990, was 25,000, find the population on January 1, 2000.

13. Find each indefinite integral.
(a) $\displaystyle\int (x^3 + 2x)^4(3x^2 + 2)\, dx$
(b) $\displaystyle\int (e^{2x} + 4)^3 e^{2x}\, dx$
(c) $\displaystyle\int \dfrac{2x + 1}{x^2 + x - 7}\, dx$
(d) $\displaystyle\int \dfrac{x^3 + x}{(x^4 + 2x^2 - 7)^3}\, dx$
(e) $\displaystyle\int \dfrac{1}{x(\ln x)^2}\, dx$
(f) $\displaystyle\int x(x + 3)^9\, dx$
(g) $\displaystyle\int \dfrac{x^2 + x - 15}{x - 3}\, dx$
(h) $\displaystyle\int (x^3 + 2)^2 x\, dx$

14. Find each indefinite integral.

(a) $\displaystyle\int (2x^3 - 5x + 4)^3(6x^2 - 5)\, dx$

(b) $\displaystyle\int (\ln t + 5)^4(1/t)\, dt$

(c) $\displaystyle\int \frac{4x - 1}{2x^2 - x + 8}\, dx$

(d) $\displaystyle\int \frac{6t^2 + 6}{(2t^3 + 6t - 5)^4}\, dt$

(e) $\displaystyle\int \frac{3\sqrt{\ln x}}{x}\, dx$

(f) $\displaystyle\int (x - 4)^{18}x\, dx$

(g) $\displaystyle\int \frac{4x^3 - 5x + 7}{2x - 1}\, dx$

(h) $\displaystyle\int (y^4 + 3)^2 y^2\, dy$

15. Find using integration by parts.

(a) $\displaystyle\int xe^{4x}\, dx$

(b) $\displaystyle\int \frac{\ln x}{x^{3/2}}\, dx$

16. Find by using integration by parts.

(a) $\displaystyle\int xe^{-3x}\, dx$

(b) $\displaystyle\int 49x^{5/2} \ln x\, dx$

17. Use a partial fraction decomposition to find each integral.

(a) $\displaystyle\int \frac{3x + 2}{x^2 - x - 12}\, dx$

(b) $\displaystyle\int \frac{2x^2 - 9x - 12}{x^3 + x^2 - 12x}\, dx$

18. Use a partial fraction decomposition to find each integral.

(a) $\displaystyle\int \frac{x - 16}{x^2 + 3x - 10}\, dx$

(b) $\displaystyle\int \frac{x^2 - 15x + 18}{x(x^2 - 9)}\, dx$

19. Find the general solution of the following differential equations.

(a) $\displaystyle\frac{dy}{dx} = 8x^3 y^2$

(b) $\displaystyle t\frac{dy}{dt} = \sqrt{y}\ \ln t$

20. Find the general solution to each differential equation.

(a) $\displaystyle\frac{dy}{dx} = 6x^2\sqrt{y}$

(b) $\displaystyle (t + 1)e^s \frac{ds}{dt} = (e^s + 2)t$

21. Find the particular solution to the given differential equation with the associated condition.

(a) $\displaystyle\frac{dy}{dx} = \frac{1}{2}y^3 e^x;\ y = \frac{1}{3}$ when $x = 0$.

(b) $\displaystyle\frac{dy}{dx} = -4y^2 x\sqrt{5 + x^2};\ y = \frac{1}{2}$ when $x = 2$.

22. Determine the indicated particular solution.

(a) $\displaystyle\frac{dy}{dx} = -y^2 x^2(x^3 + 2)^3;\ y = \frac{1}{8}$ when $x = 1$.

(b) $\displaystyle (e^x + 2)\frac{dy}{dx} = e^{x-y};\ y = 0$ when $x = 0$.

23. Find the equation of the curve that passes through the point $(1, 2)$ and whose slope at any point (x, y) on the curve is $(1 - 2\sqrt{x})^3/\sqrt{x}$.

24. Find the equation of the curve that passes through the point $(1, 5)$ and whose slope at any point (x, y) on the curve is $(9x^2\sqrt{x^3 + 3})/y$.

25. A strain of Asian flu has hit a city of 25,000. The rate at which the flu is spreading is proportional to y, the number of people already infected, and to the number not yet infected.
    (a) Write the differential equation that y satisfies.
    (b) If 50 people were infected initially and a total of 750 by the end of 2 weeks, how many were infected by the end of 5 weeks?

26. Ten people started the rumor of a scandal in city government in Abbeville, a city of 50,000 people. Assume that the rate at which the rumor spread was proportional to y, the number who had heard the rumor, and to the number who had not heard the rumor.
    (a) Write the differential equation satisfied by y.
    (b) If 5,000 people had heard the rumor by the end of 5 days, how many had heard it by the end of 12 days?

27. Let f have domain $(-3, \infty)$ and satisfy $f''(x) = -10/(x + 3)^3$, $f'(0) = \frac{5}{9}$, and $f(0) = \frac{-2}{3}$.
    (a) Where is the graph of f concave up?
    (b) Where is f increasing?
    (c) Where is $f(x) > 0$?
    (d) Find $\lim\limits_{x \to \infty} f(x)$.
    (e) Sketch the graph of f.

28. Let f have domain $(-1, \infty)$ and satisfy $f''(x) = -10/(x + 1)^2$, $f'(0) = 1$, and $f(0) = 1$.
    (a) Where is the graph of f concave up?
    (b) Where is f increasing?
    (c) Where is the slope of the graph equal to $1/3$?
    (d) Where is $f(x) > 0$?
    (e) Sketch the graph of f.

Take a closed wire loop of arbitrary shape, dip it in liquid soap, and a soap film of smallest area spanning the loop will form. Richard Courant analyzed these minimal surfaces and was fascinated by them. Below, soap films are used to find minimal paths connecting four points and five points. Intersections always form at 120° angles.

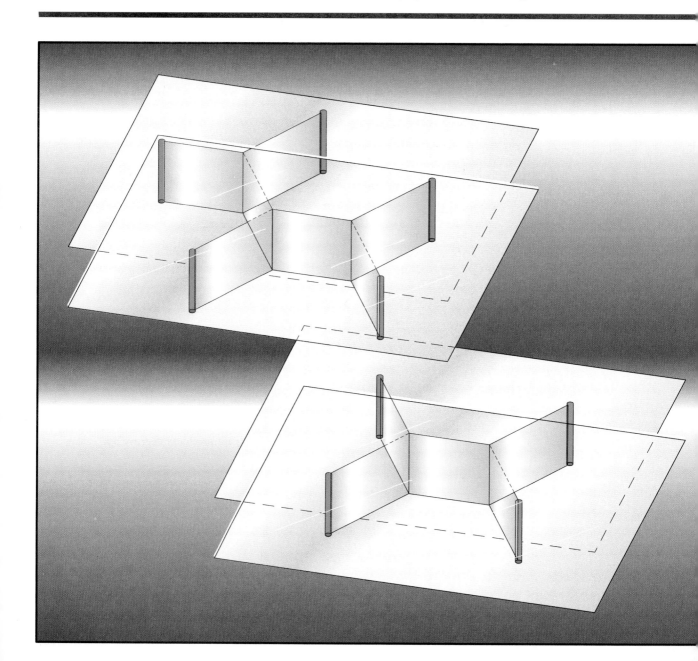

8

The Definite Integral

Richard Courant represents one of the richest traditions in applied mathematics. After his early training in mathematics, physics, and philosophy in Breslau, Courant went to Göttingen in central Germany to obtain his Ph.D. Since the days of the great Karl Gauss (c. 1800), the university in this small city had enjoyed the reputation of being the center of the world in mathematics and physics. It was *the* place to study if one had mathematical talent. There Courant came under the influence of the greatest mathematician of the day, David Hilbert. Hilbert recognized both the mathematical and administrative gifts of the young Courant, made him his assistant, and later helped to secure him the position of director of the mathematics institute.

As director, Courant was determined to continue and increase the influence of the institute, and he was successful in attracting a majority of the world's greatest mathematicians to either permanent or visiting positions. While Courant valued pure mathematics, he was ever concerned that mathematical research should both grow out of and help solve the problems of the real world. He felt that what distinguished the institute at Göttingen was its integration of mathematics and science.

Within months after he came to power, Adolf Hitler began to destroy what Courant and his predecessors had built because many of the brightest stars at the institute (including Courant) were Jewish. Fortunately, most of them were able to come to the United States. Before long, Courant was beginning to build at New York University a mathematical center (now called the Courant Institute of Mathematical Sciences) modeled after the one at Göttingen. Here the great tradition of Göttingen lives on.

While Courant is said to have been egotistical, self-serving, and often controversial, all agree that he was a shrewd judge of mathematical talent, an exceptional administrator, an able mathematician, and a superb mathematical writer. His books, always with an applied flavor, include *Differential and Integral Calculus*, *What Is Mathematics* (with Herbert Robbins), and *Methods of Mathematical Physics* (with David Hilbert).

Richard Courant
(1888–1972)

AREA AND THE DEFINITE INTEGRAL

We used the geometric notion of *slope of a curve* to introduce the first of the two most important concepts in calculus, the derivative. Another geometric idea, *area of a curved region,* will lead to the second of these concepts, the definite integral.

What is area? This is not a simple question, and the definition for curved regions must finally be based on the notion of limit just as was that of slope. We will get at this question in Section 8.4. However, for now we ask you to trust your intuition and accept the following properties of area.

1. The area of a plane region is a nonnegative number.
2. Congruent regions have the same area.
3. The area of the union of two regions whose interiors do not overlap is the sum of the areas of the two regions; that is, if R_1 and R_2 are as in Figure 1, then

$$A(R_1 \cup R_2) = A(R_1) + A(R_2)$$

4. The area of a rectangle is the product of its length and breadth.

FIGURE 1

We remind you of some area formulas that you probably learned in grade school (Figure 2).

Rectangle
b
$A = bh$

Parallelogram
h
b
$A = bh$

Triangle
h
b
$A = \frac{1}{2} bh$

Trapezoid
a
h
b
$A = \frac{1}{2}(a+b)h$

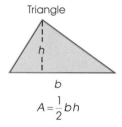

Circle
r
$A = \pi r^2$

FIGURE 2

EXAMPLE A

Find the area of the region $R = R_1 \cup R_2 \cup R_3$ shown in Figure 3.

FIGURE 3

Solution

$$A(R) = A(R_1) + A(R_2) + A(R_3)$$
$$= (\tfrac{1}{2})(8)(6) + (8)(10) + (\tfrac{1}{2})(\pi)(4^2)$$
$$= 24 + 80 + 8\pi \approx 129.13 \qquad \blacksquare$$

The Definite Integral

Consider a region R in the xy-plane bounded by a curve $y = f(x)$ and the x-axis and lying between $x = a$ and $x = b$ (Figure 4). In general R will consist of two parts: R_u (the part above the x-axis and referred to as R up) and R_d (the part below the x-axis and referred to as R down). We define the definite integral from a to b in terms of the areas of these two regions as follows.

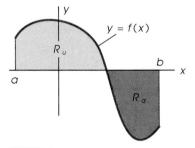

FIGURE 4

DEFINITION

The symbol $\int_a^b f(x)\,dx$, called the **definite integral of $f(x)$ from a to b,** is defined to be the area of R_u minus the area of R_d. That is,

$$\int_a^b f(x)\,dx = A(R_u) - A(R_d)$$

This innocent-sounding definition has far-reaching consequences that we will be exploring in both this chapter and the next. Our first job is to make sure that we are able to calculate some simple definite integrals.

EXAMPLE B

Let $f(x) = -\frac{1}{2}x$ on the interval -2 to 4. Sketch its graph and then calculate $\int_{-2}^{4} f(x)\, dx$.

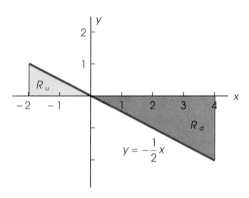

FIGURE 5

Solution The graph is shown on the left in Figure 5. The diagram on the right of this figure shows R_u and R_d.

$$\int_{-2}^{4} f(x)\, dx = A(R_u) - A(R_d) = 1 - 4 = -3 \qquad \blacksquare$$

EXAMPLE C

Let the graph of $f(x)$ be as shown in the left part of Figure 6. Calculate $\int_{-1}^{3} f(x)\, dx$.

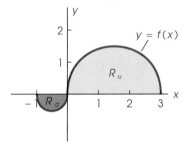

FIGURE 6

Solution The regions R_u and R_d are shown in the right part of Figure 6; both are semicircles. Thus,

$$\int_{-1}^{3} f(x)\, dx = A(R_u) - A(R_d) = \frac{\pi (1.5)^2}{2} - \frac{\pi (0.5)^2}{2} \approx 3.14 \qquad \blacksquare$$

EXAMPLE D

Sketch the graph and calculate $\int_{-1}^{2} f(x)\, dx$ for

$$f(x) = \begin{cases} 3x & -1 \le x \le 1 \\ -x + 1 & 1 < x \le 2 \end{cases}$$

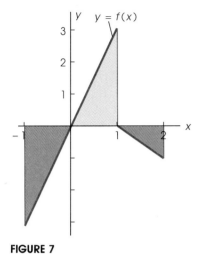

FIGURE 7

Solution The graph of f and the regions R_u and R_d are shown in Figure 7. Note that between -1 and 1, the part below the x-axis and the part above the x-axis are the same size, and so their contribution to the integral is 0. The contribution on the interval 1 to 2 is $-\frac{1}{2}$. We conclude that

$$\int_{-1}^{2} f(x)\ dx = -\frac{1}{2}$$ ∎

Use of Symmetry

Example D suggests a way to use symmetry as an aid in evaluating definite integrals. We explore this further now. Recall from Section 1.2 that if f is an odd function [meaning that $f(-x) = -f(x)$], then its graph is symmetric with respect to the origin. Similarly, if f is an even function [meaning that $f(-x) = f(x)$], then its graph is symmetric with respect to the y-axis. For example, $f(x) = x^3$ is an odd function, while $g(x) = x^2$ is an even function (Figure 8). On an interval $[-a, a]$ that is symmetric about the origin, the

FIGURE 8

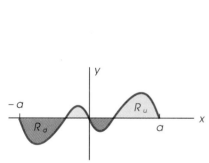

Odd function: $A(R_u) = A(R_d)$

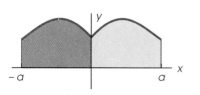

Even function: $A(R_l) = A(R_r)$

FIGURE 9

graph of an odd function and the x-axis will enclose a region with the same amount of area above the x-axis as below this axis (Figure 9). On the other hand, the region bounded by the graph of an even function and the x-axis on the interval $[-a, a]$ has exactly the same area to the left of the y-axis as to the right of this axis. These facts lead to the following conclusions.

1. If f is an odd function, $\int_{-a}^{a} f(x)\ dx = 0$.
2. If f is an even function, $\int_{-a}^{a} f(x)\ dx = 2 \int_{0}^{a} f(x)\ dx$.

EXAMPLE E

In the next section, we will learn that

$$\int_0^a x^n \, dx = \frac{a^{n+1}}{n+1} \qquad n \neq -1$$

Use this fact and symmetry considerations to calculate
(a) $\int_{-4}^4 x^2 \, dx$, (b) $\int_{-2}^2 5x^3 \, dx$, (c) $\int_{-1}^1 x^6 \, dx$

Solution

(a) $\int_{-4}^4 x^2 \, dx = 2 \int_0^4 x^2 \, dx = \frac{2(4^3)}{3} = \frac{128}{3}$

(b) $\int_{-2}^2 5x^3 \, dx = 0$, since $f(x) = 5x^3$ is an odd function

(c) $\int_{-1}^1 x^6 \, dx = 2 \int_0^1 x^6 \, dx = \frac{2(1^7)}{7} = \frac{2}{7}$ ∎

Problem Set 8.1

In Problems 1–10, make use of standard area formulas to find the area of each shaded region.

1.

2.

3.

4.

5.

6.

7.

8.

9.

10.

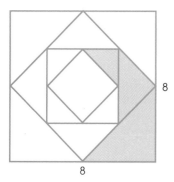

In Problems 11–22, sketch the graph of f on the interval from a to b and then calculate $\int_a^b f(x)\,dx$. In Problems 17–22, a and b are understood to be the left and right end points of the given domain.

11. $f(x) = 2x;\ a = -1,\ b = 2$
12. $f(x) = 1.5x;\ a = -4,\ b = 1$
13. $f(x) = -0.5x + 2;\ a = 0,\ b = 6$
14. $f(x) = 2x - 3;\ a = 0,\ b = 4$

15. $f(x) = 3x - 6;\ a = 0,\ b = 2$
16. $f(x) = -3x + 6;\ a = 2,\ b = 6$
17. $f(x) = \begin{cases} 2x & -1 \le x \le 1 \\ 2 & 1 < x \le 3 \end{cases}$
18. $f(x) = \begin{cases} x + 2 & -2 \le x < 0 \\ -x + 2 & 0 \le x \le 4 \end{cases}$
19. $f(x) = \begin{cases} |x| & -2 \le x < 2 \\ 4 - x & 2 \le x \le 5 \end{cases}$
20. $f(x) = \begin{cases} |x - 2| & 0 \le x \le 4 \\ 8 - 2x & 4 < x \le 7 \end{cases}$
21. $f(x) = \begin{cases} x - 1 & 0 \le x \le 2 \\ 1 & 2 < x \le 3 \\ -x + 3 & 3 < x \le 5 \end{cases}$
22. $f(x) = \begin{cases} -x & -2 \le x < 2 \\ x & 2 \le x \le 4 \\ 5 - x & 4 < x \le 5 \end{cases}$

23. Calculate $\int_{-2}^{8} f(x)\,dx$, where f is the function whose graph is shown below.

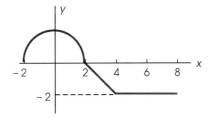

24. Calculate $\int_{-3}^{3} f(x)\,dx$ for the function f whose graph is shown below.

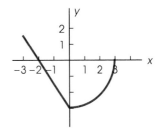

25. Calculate $\int_{-2}^{5} f(x)\,dx$ for the function f whose graph is shown below.

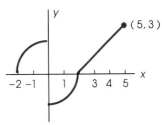

26. Calculate $\int_{-1}^{8} f(x)\, dx$ for the function f whose graph is shown below.

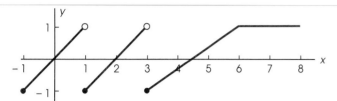

Use the formula $\int_{0}^{a} x^n\, dx = a^{n+1}/(n+1)$ to evaluate each of the definite integrals in Problems 27–32.

27. $\int_{0}^{3} x^2\, dx$

28. $\int_{0}^{2} x^3\, dx$

29. $\int_{0}^{\pi} x^4\, dx$

30. $\int_{0}^{\sqrt{2}} x^3\, dx$

31. $\int_{0}^{4} \sqrt{x}\, dx$

32. $\int_{0}^{8} x^{2/3}\, dx$

33. Decide whether the given function is even, odd, or neither.
(a) $f(x) = 2x^3 - x$ (b) $f(x) = 2x^2 + 5$
(c) $f(x) = x^3 + 2x^2 - 1$
(d) $f(x) = (x^2 + 4)/(x^7 + 4x)$

34. Determine whether the given function is even, odd, or neither.
(a) $f(x) = x^4 - 6$ (b) $f(x) = x^5 - 4x$
(c) $f(x) = |x|$ (d) $f(x) = (x^3 + x)(x^5 - 3x)$

In Problems 35–42, use symmetry and the formula mentioned before Problem 27 to evaluate each definite integral (see Example E).

35. $\int_{-2}^{2} 2x^3\, dx$

36. $\int_{-2}^{2} x^2\, dx$

37. $\int_{-1}^{1} x^4\, dx$

38. $\int_{-3}^{3} (x^5 - 4x^3)\, dx$

39. $\int_{-2}^{2} |x|\, dx$

40. $\int_{-4}^{4} \sqrt{|x|}\, dx$

41. $\int_{-5}^{5} 3x/(x^2 + 1)\, dx$

42. $\int_{-\pi}^{\pi} (e^x - e^{-x})\, dx$

In Problems 43–50, sketch the graph of the integrand and then evaluate the definite integral using standard area formulas. Recall that [x] denotes the greatest integer in x.

43. $\int_{0}^{4} (2x - 3)\, dx$

44. $\int_{-2}^{2} (-3x + 1)\, dx$

45. $\int_{0}^{3} |x - 2|\, dx$

46. $\int_{-1}^{2} |2x - 1|\, dx$

47. $\int_{-3}^{3} \sqrt{9 - x^2}\, dx$

48. $\int_{0}^{4} \sqrt{16 - x^2}\, dx$

49. $\int_{0}^{4} [x]\, dx$

50. $\int_{0}^{4} (x - [x])\, dx$

If $v(t)$ is the velocity at time t of an object moving along a straight line, then the total distance traveled by the object between time $t = a$ and time $t = b$ is given by $\int_{a}^{b} |v(t)|\, dt$. Use this fact to find the total distance traveled by an object having the velocity specified in Problems 51–54. Note that in each case, the velocity is sometimes negative and sometimes positive, which means that the object does some doubling back.

51. $v(t) = t; a = -2, b = 3$
52. $v(t) = t^3; a = -2, b = 2$
53. $v(t) = 2 - t; a = 0, b = 5$
54. $v(t) = 5 - 2t; a = 0, b = 4$

8.2
THE FUNDAMENTAL THEOREM OF CALCULUS

You know from the title that something very important is coming. It is the theorem that will allow us to evaluate a wide class of definite integrals in an almost trivial manner.

> **THE FUNDAMENTAL THEOREM OF CALCULUS**
>
> Let $f(x)$ be continuous on the interval $[a, b]$ and let $F(x)$ be any antiderivative of $f(x)$ there. Then
>
> $$\int_a^b f(x)\, dx = F(b) - F(a)$$

Before offering a proof of this remarkable theorem, we present one example to show you its power.

EXAMPLE A

Evaluate $\displaystyle\int_1^4 x^2\, dx$.

Solution We know from Chapter 7 that $F(x) = x^3/3$ is an antiderivative of $f(x) = x^2$. Thus, by the Fundamental Theorem,

$$\int_1^4 x^2\, dx = F(4) - F(1) = \frac{4^3}{3} - \frac{1^3}{3} = \frac{63}{3} = 21 \qquad \blacksquare$$

Proof of the Fundamental Theorem

To prepare the way, let us first note that the variable x in the symbol $\int_a^b f(x)\, dx$ is a *dummy variable*. By this we mean that any other variable would serve as well. The result of a definite integration is a number, and this number does not depend on what variable is used in the integrand.

$$\int_a^b f(x)\, dx = \int_a^b f(t)\, dt = \int_a^b f(u)\, du$$

We mention this because we now want to let x act as the upper limit on the integral sign, and so we will want to use a different variable in the integrand.

Let a be a fixed number and let $x \geq a$. Consider the function G determined by

$$G(x) = \int_a^x f(t)\, dt$$

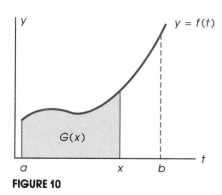

FIGURE 10

Figure 10 gives a geometric interpretation of the function G. If f is a nonnegative function, $G(x)$ is the area under the curve $y = f(t)$ between a and x. Note that $G(a) = 0$ and $G(b)$ is the definite integral we want to evaluate.

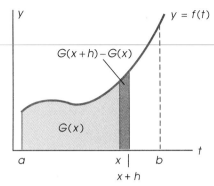

FIGURE 11

For reasons that will be clear shortly, we want to find $G'(x)$. From the definition of the derivative, we may write

$$G'(x) = \lim_{h\to 0} \frac{G(x + h) - G(x)}{h}$$

Referring to Figure 11, we see that we can interpret the numerator of this expression (at least if f is nonnegative and h is positive) as the area under the curve between x and $x + h$. But since f is continuous, the value of this area will be very close to $f(x)h$. In fact, it will equal $f(\bar{x})h$, where $\bar{x}$ is some point between x and $x + h$. We conclude that

$$G'(x) = \lim_{h\to 0} \frac{f(\bar{x})h}{h} = \lim_{h\to 0} f(\bar{x}) = f(x)$$

This is just another way of saying that $G(x)$ is an antiderivative of $f(x)$. Now we have two antiderivatives of $f(x)$ in view, namely, $G(x)$ and $F(x)$. Since we know that we can get any antiderivative from another by the addition of a constant, we conclude that there must be a constant C such that $G(x) = F(x) + C$. On the other hand, we noted earlier that $G(a) = 0$, so $F(a) + C = 0$ and consequently $C = -F(a)$. Thus,

$$G(x) = F(x) - F(a)$$

or in integral symbols,

$$\int_a^x f(t)\,dt = F(x) - F(a)$$

When we let $x = b$ in the latter equation, we have the result we wanted.

It should be pointed out that our argument did assume that $f(x)$ was nonnegative and that h was positive. A slight modification will establish the result when these assumptions are dropped.

The Theorem Illustrated

To increase our confidence in the result, we apply it to the integral of Example B in the previous section where we know the answer should be -3.

EXAMPLE B

Evaluate $\int_{-2}^4 f(x)\,dx$ where $f(x) = -\frac{1}{2}x$.

Solution Since $F(x) = -\frac{1}{2}x^2/2 = -\frac{1}{4}x^2$ is an antiderivative of $f(x)$, we obtain

$$\int_{-2}^4 -\frac{1}{2}x\,dx = F(4) - F(-2) = \left(-\frac{1}{4}\right)4^2 - \left(-\frac{1}{4}\right)(-2)^2$$

$$= -4 + 1 = -3 \qquad \blacksquare$$

EXAMPLE C

Evaluate $\displaystyle\int_0^2 3e^x\, dx$.

Solution Since $F(x) = 3e^x$ is an antiderivative of $f(x) = 3e^x$, we see that

$$\int_0^2 3e^x\, dx = F(2) - F(0) = 3e^2 - 3e^0 \approx 19.167 \qquad \blacksquare$$

Because the expression $F(b) - F(a)$ comes up in every integration problem, we introduce a special notation for it, namely, $[F(x)]_a^b$. Thus,

$$\left[x^2 - 3\right]_2^4 = 4^2 - 3 - (2^2 - 3) = 13 - 1 = 12$$

EXAMPLE D

Evaluate $\displaystyle\int_1^2 (4x^3 + 3x)\, dx$.

Solution

$$\int_1^2 (4x^3 + 3x)\, dx = \left[x^4 + \frac{3x^2}{2}\right]_1^2 = 16 + 6 - \left(1 + \frac{3}{2}\right) = \frac{39}{2} \qquad \blacksquare$$

EXAMPLE E

Find the area of the region under the curve $y = \sqrt{x}$ between $x = 1$ and $x = 4$ (Figure 12).

Solution We calculate this area by means of an integral.

$$\int_1^4 \sqrt{x}\, dx = \int_1^4 x^{1/2}\, dx = \left[\frac{2}{3}x^{3/2}\right]_1^4 = \frac{16}{3} - \frac{2}{3} = \frac{14}{3} \qquad \blacksquare$$

FIGURE 12

Some Properties of Definite Integrals

Here are three properties that will often be used.

1. $\displaystyle\int_a^b cf(x)\, dx = c \int_a^b f(x)\, dx$, c a constant.
2. $\displaystyle\int_a^b [f(x) \pm g(x)]\, dx = \int_a^b f(x)\, dx \pm \int_a^b g(x)\, dx$.
3. $\displaystyle\int_a^b f(x)\, dx = \int_a^c f(x)\, dx + \int_c^b f(x)\, dx$.

The first two properties are inherited from the corresponding properties for antiderivatives. You can see why the third property is true by studying Figure 13, at least for the case where $a < c < b$. Actually, Property 3 is valid no matter how a, b, and c are arranged provided we make the following agreement.

FIGURE 13

$$\int_b^a f(x)\, dx = -\int_a^b f(x)\, dx$$

EXAMPLE F

If $\displaystyle\int_1^3 f(x)\,dx = 4$ and $\displaystyle\int_1^3 g(x)\,dx = -2$, evaluate $\displaystyle\int_1^3 [3f(x) + 5g(x)]\,dx$

Solution We use Property 2 and then Property 1 to write

$$\int_1^3 [3f(x) + 5g(x)]\,dx = \int_1^3 3f(x)\,dx + \int_1^3 5g(x)\,dx$$

$$= 3\int_1^3 f(x)\,dx + 5\int_1^3 g(x)\,dx$$

$$= 3(4) + 5(-2) = 2 \qquad ■$$

EXAMPLE G

Evaluate $\displaystyle\int_{-2}^4 f(x)\,dx$ where

$$f(x) = \begin{cases} -2x & -2 \le x \le 0 \\ x^2 - 2 & 0 < x \le 2 \\ -2x + 8 & 2 < x \le 4 \end{cases}$$

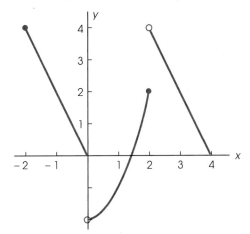

FIGURE 14

Solution The graph of f is shown in Figure 14. To evaluate the integral we will need to break it into three parts by using Property 3.

$$\int_{-2}^4 f(x)\,dx = \int_{-2}^0 (-2x)\,dx + \int_0^2 (x^2 - 2)\,dx + \int_2^4 (-2x + 8)\,dx$$

$$= \left[-x^2\right]_{-2}^0 + \left[\frac{x^3}{3} - 2x\right]_0^2 + \left[-x^2 + 8x\right]_2^4$$

$$= [0 - (-4)] + \left[\frac{8}{3} - 4 - (0 - 0)\right] + [-16 + 32 - (-4 + 16)]$$

$$= \frac{20}{3} \qquad ■$$

The Average Value of a Function

We are all familiar with the notion of the *average* of a set of scores. It is simply the sum of the scores divided by the number of scores. Thus, if S_1, S_2, S_3, . . . , S_n are the scores on a test taken by a class of n students, the average score is

$$S_{ave} = \frac{1}{n}(S_1 + S_2 + S_3 + \cdots + S_n)$$

But what shall we mean by the average value of a function? It won't make sense to add up all its values and divide by the number of values, because a function has infinitely many values.

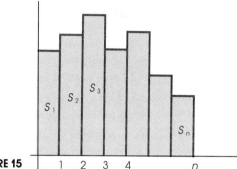

FIGURE 15

We can reinterpret the formula for S_{ave} displayed above in a geometric way. Refer to Figure 15 and note that S_1 is simply the area of the first rectangle, S_2 is the area of the second rectangle, and so on. Moreover, S_{ave} is just the sum of the areas of all the rectangles divided by the length of the interval $[0, n]$. This suggests a definition for the average value of a function. The **average value** of the function f on the interval $[a, b]$ is defined by

$$f_{ave} = \frac{1}{b - a} \int_a^b f(x)\, dx$$

We will give a more adequate basis for this definition in Section 8.4.

EXAMPLE H

Find the average value of $f(x) = -x^2 + 4x + 5$ on the interval $[1, 5]$.

Solution Refer to Figure 16 which should suggest to you that the average value of f is something like 6 or 7. By our definition

$$f_{ave} = \frac{1}{5 - 1} \int_1^5 (-x^2 + 4x + 5)\, dx = \frac{1}{4}\left[-\frac{1}{3}x^3 + 2x^2 + 5x\right]_1^5$$

$$= \frac{1}{4}\left(-\frac{125}{3} + 50 + 25\right) - \frac{1}{4}\left(-\frac{1}{3} + 2 + 5\right) \approx 6.66667 \quad \blacksquare$$

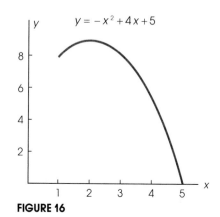

$y = -x^2 + 4x + 5$

FIGURE 16

Problem Set 8.2

In problems 1–4, evaluate the given definite integral in two ways: (i) by drawing a sketch and then calculating $A(R_u) - A(R_d)$, and (ii) by using the Fundamental Theorem of Calculus.

1. $\int_{-1}^{3} 2x \, dx$

2. $\int_{-3}^{1} (x + 2) \, dx$

3. $\int_{0}^{3} (-x + 1) \, dx$

4. $\int_{0}^{3} (-2x + 4) \, dx$

Calculate the definite integrals in Problems 5–20 by using the Fundamental Theorem of Calculus.

5. $\int_{-1}^{2} -\frac{3}{4}x \, dx$

6. $\int_{-2}^{3} -\frac{2}{3}x \, dx$

7. $\int_{-2}^{3} 6x^2 \, dx$

8. $\int_{1}^{2} 4x^3 \, dx$

9. $\int_{1}^{3} (3x^2 + 2x - 8) \, dx$

10. $\int_{0}^{2} (x^2 + 4x - 3) \, dx$

11. $\int_{0}^{2} (2x^3 - x - 3) \, dx$

12. $\int_{-1}^{2} (5x^4 - 3x) \, dx$

13. $\int_{-2}^{2} (x - 2)^2 \, dx$

14. $\int_{-1}^{1} (x + 1)^3 \, dx$

15. $\int_{1}^{8} 4\sqrt[3]{x} \, dx$

16. $\int_{1}^{32} 14x^{2/5} \, dx$

17. $\int_{1}^{e} \frac{1}{x} \, dx$

18. $\int_{1}^{3} -2x^{-2} \, dx$

19. $\int_{0}^{1} (5e^x - 2x) \, dx$

20. $\int_{0}^{1} (4e^{-x} + 1) \, dx$

In Problems 21–24, sketch the graph of the given function on the interval [a, b] and then find the area of the region under the graph (see Example E).

21. $f(x) = x^2 + 4$; $a = -1$, $b = 3$

22. $f(x) = x^2 + 4x + 4$; $a = -2$, $b = 1$

23. $f(t) = t^{2/3}$; $a = 0$, $b = 8$

24. $f(t) = t^{3/2}$; $a = 0$, $b = 4$

Suppose that $\int_{-1}^{3} f(x) \, dx = -8$, $\int_{3}^{5} f(x) \, dx = 4$, $\int_{-1}^{3} g(x) \, dx = 5$, and $\int_{3}^{5} g(x) \, dx = 2$. Use the properties of the definite integral to calculate the definite integrals in Problems 25–36. See Example F and the discussion preceding it.

25. $\int_{3}^{-1} f(x) \, dx$

26. $\int_{5}^{3} g(x) \, dx$

27. $\int_{-1}^{3} -2f(x) \, dx$

28. $\int_{-1}^{3} 0.4 \, g(x) \, dx$

29. $\int_{3}^{5} [2f(x) + 3g(x)] \, dx$

30. $\int_{3}^{5} [2g(x) - f(x)] \, dx$

31. $\int_{-1}^{5} 3f(x) \, dx$

32. $\int_{-1}^{5} -2g(x) \, dx$

33. $\int_{-1}^{5} [2f(x) + 3] \, dx$

34. $\int_{-1}^{3} [4 - 2g(x)] \, dx$

35. $\int_{-1}^{7} g(x) \, dx + \int_{7}^{5} g(x) \, dx$

36. $\int_{-1}^{7} f(x) \, dx - \int_{5}^{7} f(x) \, dx$

In Problems 37–40, evaluate $\int_{-2}^{3} f(x) \, dx$ as in Example G.

37. $f(x) = \begin{cases} x^2 + 4x + 3 & -2 \le x < 0 \\ -x + 3 & 0 \le x \le 3 \end{cases}$

38. $f(x) = \begin{cases} x + 1 & -2 \le x < 1 \\ x^2 + x - 2 & 1 \le x \le 3 \end{cases}$

39. $f(x) = \begin{cases} x^2 & -2 \le x < 0 \\ x & 0 \le x < 2 \\ 4 - x^2 & 2 \le x \le 3 \end{cases}$

40. $f(x) = \begin{cases} -x/2 & -2 \le x < 0 \\ x - 1 & 0 \le x < 1 \\ x^2 - 4 & 1 \le x \le 3 \end{cases}$

Find the average value of f on the given interval in Problems 41–46. Make a guess first (see Example H).

41. $f(x) = x^3 + 1$ on [0, 2]

42. $f(x) = 3x^2 - 4x$ on [0, 3]

43. $f(x) = x^3 + 1$ on [−1, 2]

44. $f(x) = 3x^2 - 4x$ on [0, 2]

45. $f(x) = 4 - x^2$ on [−1, 2]

46. $f(x) = e^x$ on [−1, 2]

In Problems 47–58, perform the indicated calculation.

47. $\int_{1}^{4} (5x^{3/2} - 3\sqrt{x}) \, dx$

48. $\int_{1}^{4} (x^{-2} + 2x^{-1}) \, dx$

49. $\int_{0}^{1} (e^x + x^e) \, dx$

50. $\int_{-1}^{1} (x^3 - 2)^2 \, dx$

51. $\int_0^3 2|t - 2| \, dt$

52. $\int_{-5}^0 4|u + 3| \, du$

53. $\int_0^3 f'(x) \, dx$ where $f(x) = 12/(3 + x^2)$

54. $\int_0^4 g'(t) \, dt$ where $g(t) = e^{\sqrt{t}}$

55. $G'(0)$ where $G(x) = \int_{-1}^x e^t \, dt$

56. $H'(1)$ where $H(x) = \int_0^x \dfrac{4}{9 + t^2} \, dt$

57. $\int_0^1 2x \, e^{x^2} \, dx$. *Hint:* Make the substitution $u = x^2$.

58. $\int_0^1 3x^2(1 + x^3)^4 \, dx$

59. The base of a solid is a rectangle 3 inches by 10 inches. The solid is rectangular but with a curved top. The height of the solid x inches from the left end is e^x. Find the volume of this solid if the left end is (a) the short end, (b) the long end.

60. Find the area of the region under the curve $y = xe^x$ on the interval $0 \leq x \leq 2$.

61. The temperature T (in degrees Fahrenheit) in an auditorium during a 12-hour period beginning at 6:00 A.M. was given by $68 + 0.5t + 0.03t^2$, t being the number of hours after 6:00 A.M.
    (a) What was the temperature at 2:00 P.M.?
    (b) What was the average temperature over the 12-hour period?

62. The speed v of a motorist t seconds after she applied the brakes was given by $v = 96 - 2t^2$ in feet per second.
    (a) How long did it take her to come to a full stop?
    (b) How far did she travel during this time interval?
    (c) What was her average speed during this time interval?

63. Suppose that the weight of a steer t months after birth is given by $W = 40 + 50t - 0.6t^2$ pounds for $0 \leq t \leq 20$. Find the average weight of the steer during this 20-month period.

8.3
CALCULATION OF DEFINITE INTEGRALS

We now want to consider the calculation of more complicated definite integrals, integrals in which a substitution will be required. As a first example, consider the problem of evaluating

$$\int_0^1 (2x + 1)^5 \, dx$$

One way to handle this problem is by a two-step process. First, we find the corresponding indefinite integral; then we apply the Fundamental Theorem of Calculus. This is how it works. From the Substitution Rule (Section 7.2), we may let $u = 2x + 1$ and $du = 2 \, dx$ to obtain

$$\int (2x + 1)^5 \, dx = \frac{1}{2} \int (2x + 1)^5 \, 2 \, dx = \frac{1}{2} \int u^5 \, du$$

$$= \frac{1}{2} \frac{u^6}{6} + C = \frac{1}{12} (2x + 1)^6 + C$$

Then by the Fundamental Theorem of Calculus, we may write

$$\int_0^1 (2x + 1)^5 \, dx = \left[\frac{1}{12} (2x + 1)^6 \right]_0^1$$

$$= \frac{3^6}{12} - \frac{1^6}{12} = \frac{728}{12} \approx 60.67$$

Is there a way to short-cut this two-step process? The answer is yes. It is called **substitution in definite integrals** and is useful when we cannot antidifferentiate the integrand in our heads.

Substitution in Definite Integrals

Substitution in definite integrals is just like that in indefinite integrals, with the exception that when we change from x to u, we must make the corresponding change in the limits on the integral sign. We illustrate by redoing the example above.

EXAMPLE A

Evaluate $\displaystyle\int_0^1 (2x + 1)^5 \, dx$.

Solution Let $u = 2x + 1$, so $du = 2 \, dx$, but now also note that when $x = 0$, $u = 1$ and when $x = 1$, $u = 3$. Thus,

$$\int_0^1 (2x + 1)^5 \, dx = \frac{1}{2} \int_1^3 u^5 \, du = \left[\frac{u^6}{12}\right]_1^3 = \frac{3^6}{12} - \frac{1^6}{12}$$

$$= \frac{728}{12} \approx 60.67$$

Note that the solution can be written as one chain of equalities; it is simple and elegant. ∎

EXAMPLE B

Evaluate $\displaystyle\int_3^6 [3/(x - 2)] \, dx$.

Solution Let $u = x - 2$ so $du = dx$ and note that $u = 1$ when $x = 3$ and $u = 4$ when $x = 6$. Thus,

$$\int_3^6 \frac{3}{x - 2} \, dx = 3 \int_1^4 \frac{1}{u} \, du = 3\left[\ln |u|\right]_1^4 = 3(\ln 4 - \ln 1) = 3 \ln 4 \approx 4.16$$

∎

EXAMPLE C

Evaluate $\displaystyle\int_0^2 e^{-3t} \, dt$.

Solution Let $u = -3t$ so $du = -3dt$ and note that $u = 0$ when $t = 0$ and $u = -6$ when $t = 2$. We conclude that

$$\int_0^2 e^{-3t} \, dt = -\frac{1}{3} \int_0^2 e^{-3t}(-3 \, dt) = -\frac{1}{3} \int_0^{-6} e^u \, du$$

$$= -\frac{1}{3}\left[e^u\right]_0^{-6} = -\frac{1}{3}(e^{-6} - e^0) \approx 0.3325$$

∎

EXAMPLE D

Evaluate $\displaystyle\int_{\frac{1}{4}}^{1} (\ln 4x)^3 \frac{1}{x}\, dx$.

Solution Let $u = \ln 4x$. Then by the Chain Rule, $du = (1/4x)(4)\, dx = (1/x)\, dx$. Also, $u = \ln 1 = 0$ when $x = \frac{1}{4}$ and $u = \ln 4$ when $x = 1$. Thus,

$$\int_{\frac{1}{4}}^{1} (\ln 4x)^3 \frac{1}{x}\, dx = \int_{0}^{\ln 4} u^3\, du = \left[\frac{u^4}{4}\right]_{0}^{\ln 4}$$

$$= \frac{(\ln 4)^4}{4} - \frac{0^4}{4} \approx 0.9233 \qquad \blacksquare$$

Integration by Parts in Definite Integrals

The topic of integration by parts for indefinite integrals was treated in Section 7.3, but that was an optional section so we begin at the beginning. The idea is this. Faced with an integral of the form $\int u(x)v'(x)\, dx$, abbreviated $\int u\, dv$, we may always transform this to an expression involving an integral of the form $\int v\, du$. The latter may be much easier to evaluate than the original integral and, if so, the method gives us a tool for evaluation of integrals. Here is a formal statement of the method as it applies to definite integrals.

INTEGRATION BY PARTS IN DEFINITE INTEGRALS

$$\int_{a}^{b} u(x)v'(x)\, dx = \left[u(x)v(x)\right]_{a}^{b} - \int_{a}^{b} v(x)u'(x)\, dx$$

or in abbreviated form,

$$\int_{a}^{b} u\, dv = \left[uv\right]_{a}^{b} - \int_{a}^{b} v\, du$$

To use integration by parts amounts to making a double substitution, one for u and the other for dv, as we now illustrate.

EXAMPLE E

Evaluate $\displaystyle\int_{1}^{2} x \ln x\, dx$.

Solution We always begin by looking for an ordinary substitution. In this case, none seems to work, so we turn to integration by parts. We first try the double substitution $u = x$ and $dv = \ln x\, dx$ but are stymied because we don't know how to integrate $\ln x$ to obtain v. As an alternative, we try letting $u = \ln x$ and $dv = x\,dx$. Then, $du = (1/x)\,dx$ and $v = x^2/2$. The integration by parts formula gives

$$\int_1^2 x \ln x \, dx = \left[\frac{x^2}{2} \ln x\right]_1^2 - \int_1^2 \frac{x^2}{2} \frac{1}{x} \, dx$$

We are no better off unless we can evaluate the latter integral. Fortunately, in this case we can, since the integrand simplifies to $x/2$, which integrates to $x^2/4$. Thus,

$$\int_1^2 x \ln x \, dx = \left[\frac{x^2}{2} \ln x\right]_1^2 - \left[\frac{x^2}{4}\right]_1^2$$

$$= 2 \ln 2 - 0 - \left(\frac{4}{4} - \frac{1}{4}\right)$$

$$= 2 \ln 2 - \frac{3}{4} \approx 0.6363 \qquad \blacksquare$$

EXAMPLE F

Evaluate $\displaystyle\int_1^e \ln x \, dx$.

Solution At first glance, this integral would seem to be intractable even to integration by parts. However, the double substitution

$$u = \ln x \qquad dv = 1\,dx$$

$$du = \frac{1}{x}\,dx \qquad v = x$$

leads to

$$\int_1^e \ln x \, dx = \left[(\ln x)(x)\right]_1^e - \int_1^e x \frac{1}{x} \, dx$$

$$= (e \ln e - 1 \ln 1) - \int_1^e 1 \, dx$$

$$= e - [x]_1^e = e - (e - 1) = 1 \qquad \blacksquare$$

Tables of Integrals

We mentioned in Section 7.3 how tables of integrals can help us in finding indefinite integrals. Of course, the same is true for definite integrals. Here is an example which illustrates use of the small table of integrals at the back of the book.

EXAMPLE G

Find $\int_0^4 \sqrt{x^2 + 9}\, dx$.

Solution Use Integral Formula 44 from the back of the book.

$$\int_0^4 \sqrt{x^2 + 9}\, dx = \left[\frac{x}{2}\sqrt{x^2 + 9} + \frac{9}{2} \ln \left| x + \sqrt{x^2 + 9} \right| \right]_0^4$$

$$= 2(5) + \frac{9}{2} \ln (4 + 5) - \frac{9}{2} \ln (0 + 3)$$

$$= 10 + \frac{9}{2}(\ln 9 - \ln 3)$$

$$= 10 + 4.5 \ln \frac{9}{3} = 10 + 4.5 \ln 3 \approx 14.9438 \quad \blacksquare$$

Problem Set 8.3

Calculate the definite integrals in Problems 1–4 in two ways: (i) by multiplying out the integrand and then integrating, and (ii) by using a substitution.

1. $\int_0^1 (2x - 1)^2\, 2\, dx$

2. $\int_{-1}^2 (3x + 2)^2\, 3\, dx$

3. $\int_0^2 (x^2 - 3)^2\, x\, dx$

4. $\int_0^1 (x^2 + 4)^2 x\, dx$

Evaluate the definite integrals in Problems 5–24. You can either antidifferentiate the integrand in your head and apply the Fundamental Theorem of Calculus or use the method illustrated in Examples A–D.

5. $\int_0^1 (2x - 1)^4\, dx$

6. $\int_0^1 (3x - 2)^3\, dx$

7. $\int_0^4 \sqrt{x^2 + 9}\, x\, dx$

8. $\int_0^3 x/\sqrt{x^2 + 16}\, dx$

9. $\int_5^8 5/(x - 4)\, dx$

10. $\int_{-1}^3 3/(x + 2)\, dx$

11. $\int_0^2 x/(2x^2 + 1)\, dx$

12. $\int_3^5 x/(x^2 - 8)\, dx$

13. $\int_0^1 3e^{2x}\, dx$

14. $\int_0^2 4e^{-2x}\, dx$

15. $\int_0^2 xe^{-x^2}\, dx$

16. $\int_1^4 e^{\sqrt{x}}/\sqrt{x}\, dx$

17. $\int_0^3 (x^2 - 2x)^3(x - 1)\, dx$

18. $\int_0^2 (x^3 - 6x)^4(x^2 - 2)\, dx$

19. $\int_0^1 (\sqrt{x} + 1)^5/\sqrt{x}\, dx$

20. $\int_1^8 (\sqrt[3]{x} - 1)^4 x^{-2/3}\, dx$

21. $\int_0^3 [\ln (1 + 3x)]^2/(1 + 3x)\, dx$

22. $\int_0^1 [\ln (e^x + 1)]^3[e^x/(e^x + 1)]\, dx$

23. $\int_{-2}^1 e^{(x+2)^2}(x + 2)\, dx$

24. $\int_0^2 (x + 2e^{x/2})^3(1 + e^{x/2})\, dx$

Use integration by parts to evaluate the definite integrals in Problems 25–30.

25. $\int_0^2 xe^{2x}\, dx$

26. $\int_0^3 xe^{-x}\, dx$

27. $\int_1^e x^{3/2} \ln x\, dx$

28. $\int_e^{e^2} \sqrt{x} \ln x\, dx$

29. $\int_0^1 x(1 - x)^4 \, dx$

30. $\int_0^2 (x - 2)^6 \, x \, dx$

Use the table of integrals at the end of the book to evaluate the definite integrals in Problems 31–36.

31. $\int_4^5 \sqrt{x^2 - 16} \, dx$

32. $\int_3^4 \frac{\sqrt{x^2 + 16}}{x} \, dx$

33. $\int_0^3 \frac{x^2}{\sqrt{x^2 + 16}} \, dx$

34. $\int_3^4 \frac{1}{x^2\sqrt{25 - x^2}} \, dx$

35. $\int_{-2}^1 \frac{x}{2x + 5} \, dx$

36. $\int_0^1 x\sqrt{3x + 1} \, dx$

Calculate each integral in Problems 37–42.

37. $\int_5^{\sqrt{34}} \frac{1}{\sqrt{u^2 - 9}} \, du$

38. $\int_5^{\sqrt{34}} \frac{u}{\sqrt{u^2 - 9}} \, du$

39. $\int_5^{\sqrt{34}} \frac{u^2}{\sqrt{u^2 - 9}} \, du$

40. $\int_2^4 e^{(x-2)^2}(x - 2) \, dx$

41. $\int_2^4 (x - 2)e^{x-2} \, dx$

42. $\int_0^2 e^{2\ln(x+1)} \, dx$

43. Find the area under the curve $y = x \ln x$ between $x = 1$ and $x = e^3$.

44. Find the area under the curve $y = xe^{-x}$ between $x = 0$ and $x = \ln 4$.

45. Find the average value of $f(x) = (x^3 - 1)^4 x^2$ on the interval $1 \le x \le \sqrt[3]{2}$.

46. How far did an object travel along a line during the time interval $1 \le t \le e$ if its velocity at time t seconds was given by $v = (\ln t)^3/t$ centimeters per second.

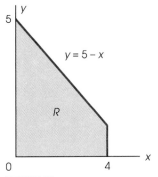

FIGURE 17

8.4
THE DEFINITE INTEGRAL AS A LIMIT

Our definition of the definite integral was in terms of area. But what is area? Earlier we made little attempt to answer this question and, in fact, it is a deep question. We do intend to give an answer now.

To give the flavor of our approach, consider the region R bounded by the line $y = 5 - x$ and the x-axis, between $x = 0$ and $x = 4$ (Figure 17). Partition the interval $[0, 4]$ into 5, then 10, then 20, and in general n equal subintervals and construct the corresponding inscribed rectangular polygons R_5, R_{10}, R_{20}, and R_n as shown in Figure 18. As we increase the number of division points in the partition, the areas of the corresponding rectangular

FIGURE 18

polygons approach nearer and nearer to what we wish to call the area of R, that is,

$$A(R) = \lim_{n\to\infty} A(R_n)$$

Later in this section, we will actually carry out this limit process and show you that the result is the same number you would get by applying the familiar formula for the area of a trapezoid. The significant observation to make now is that the process above can be carried out also for a curved region. In fact, we will use this process in a moment to give the limit definition of the definite integral. It is essentially this definition that is given in all rigorous treatments of calculus. Moreover, though this definition is a bit formidable, it is important because it shows us how to use the definite integral in applications, a subject we take up in Chapter 9.

A Formal Definition of the Definite Integral

Let f be a continuous function defined on the interval $[a, b]$. Partition this interval into n equal subintervals by means of points.

$$a = x_0 < x_1 < x_2 < x_3 < \cdots < x_{n-1} < x_n = b$$

(see Figure 19) and let $\Delta x = x_i - x_{i-1} = (b - a)/n$. We define the **definite integral of $f(x)$ from a to b** as follows.

$$\int_a^b f(x)\, dx = \lim_{n\to\infty} [\, f(x_1)\, \Delta x + f(x_2)\, \Delta x + f(x_3)\, \Delta x + \cdots + f(x_n)\, \Delta x]$$

$$= \lim_{n\to\infty} [\, f(x_1) + f(x_2) + f(x_3) + \cdots + f(x_n)]\, \Delta x$$

Note that each term $f(x_i)\, \Delta x$ is the area of a rectangle or the negative of such an area (depending on whether $f(x_i)$ is positive or negative).

We can write the above definition more compactly after we introduce a special notation for sums, called sigma notation.

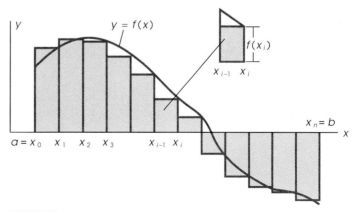

FIGURE 19

Sigma Notation

The Greek letter Σ is used throughout mathematics to denote summation. In particular,

$$\sum_{i=1}^{n} x_i = x_1 + x_2 + x_3 + \cdots + x_{n-1} + x_n$$

and

$$\sum_{i=1}^{n} i = 1 + 2 + 3 + \cdots + (n - 1) + n$$

The symbol $\displaystyle\sum_{i=1}^{n}$ is read *sum as i runs from 1 to n* and means that we are to add together all terms of the indicated form as i runs through the integers 1, 2, 3, . . . , n.

EXAMPLE A

Evaluate each of the following sums.

(a) $\displaystyle\sum_{i=1}^{6} i^2$, (b) $\displaystyle\sum_{i=2}^{9} (2i - 3)$

Solution

(a) $\displaystyle\sum_{i=1}^{6} i^2 = 1^2 + 2^2 + 3^2 + 4^2 + 5^2 + 6^2$

$$= 1 + 4 + 9 + 16 + 25 + 36 = 91$$

(b) $\displaystyle\sum_{i=2}^{9} (2i - 3) = 1 + 3 + 5 + 7 + 9 + 11 + 13 + 15 = 64$ ▪

EXAMPLE B

Write each of the following sums in sigma notation.

(a) $1 + \frac{1}{2} + \frac{1}{3} + \frac{1}{4} + \cdots + \frac{1}{99} + \frac{1}{100}$
(b) $b_2 + b_3 + b_4 + \cdots + b_{19} + b_{20}$
(c) $f(x_1) \, \Delta x + f(x_2) \, \Delta x + f(x_3) \, \Delta x + \cdots + f(x_n) \, \Delta x$

Solution

(a) $\displaystyle\sum_{i=1}^{100} (1/i)$, (b) $\displaystyle\sum_{i=2}^{20} b_i$, (c) $\displaystyle\sum_{i=1}^{n} f(x_i) \, \Delta x$ ▪

We mention two important properties that Σ shares with $\int$, the sum property and the constant-multiple property.

1. $\displaystyle\sum_{i=1}^{n} (a_i \pm b_i) = \sum_{i=1}^{n} a_i \pm \sum_{i=1}^{n} b_i$

2. $\displaystyle\sum_{i=1}^{n} ca_i = c \sum_{i=1}^{n} a_i$

Rather than give the simple proofs of these properties, we illustrate them.

EXAMPLE C

Given that $\sum_{i=1}^{5} a_i = 20$ and $\sum_{i=1}^{5} b_i = 5$, calculate $\sum_{i=1}^{5} (3a_i - 2b_i)$.

Solution

$$\sum_{i=1}^{5} (3a_i - 2b_i) = 3 \sum_{i=1}^{5} a_i - 2 \sum_{i=1}^{5} b_i = 3(20) - 2(5) = 50 \qquad \blacksquare$$

Using Σ notation, we can write the definition of the definite integral in a particularly compact and suggestive way.

$$\int_a^b f(x) \, dx = \lim_{n \to \infty} \sum_{i=1}^{n} f(x_i) \, \Delta x$$

Thus, we see that the definite integral is the limit of a sum. And if f is a continuous function (or even a function with a finite number of jump discontinuities), the above limit will always exist. For a proof of this latter fact, you will have to go to an advanced calculus book.

Some Calculations Using the Limit Definition

We are going to need two formulas that you have probably seen at some point in your precalculus training. If not, don't worry too much about where they come from, though you might check that they are correct for a few small values of n.

(i) $\quad 1 + 2 + 3 + \cdots + n = \sum_{i=1}^{n} i = \dfrac{n(n + 1)}{2}$

(ii) $\quad 1^2 + 2^2 + 3^2 + \cdots + n^2 = \sum_{i=1}^{n} i^2 = \dfrac{n(n + 1)(2n + 1)}{6}$

EXAMPLE D

Use the limit definition to calculate $\int_0^4 (5 - x) \, dx$. Note that this corresponds to the area of the trapezoid in Figure 17.

Solution We partition the interval $[0, 4]$ into n subintervals each of length $\Delta x = (b - a)/n = 4/n$. Also, the division points of the interval are

$x_1 = 4/n$, $x_2 = 2(4/n)$, $x_3 = 3(4/n)$, and so on. Since $f(x) = 5 - x$, we obtain

$$\sum_{i=1}^{n} f(x_i)\, \Delta x$$

$$= [\,f(x_1) + f(x_2) + f(x_3) + \cdots + f(x_n)\,]\, \Delta x$$

$$= \left[\left(5 - \frac{4}{n}\right) + \left(5 - 2\frac{4}{n}\right) + \left(5 - 3\frac{4}{n}\right) + \cdots + \left(5 - n\frac{4}{n}\right)\right]\frac{4}{n}$$

$$= \left[5n - \frac{4}{n}(1 + 2 + 3 + \cdots + n)\right]\frac{4}{n}$$

$$= \left[5n - \frac{4}{n}\frac{n(n+1)}{2}\right]\frac{4}{n}$$

$$= 20 - \frac{8}{n}(n + 1) = 20 - 8 - \frac{8}{n} = 12 - \frac{8}{n}$$

We conclude that

$$\int_{0}^{4} (5 - x)\, dx = \lim_{n \to \infty} \sum_{i=1}^{n} f(x_i)\, \Delta x = \lim_{n \to \infty}\left(12 - \frac{8}{n}\right) = 12$$

Naturally, this is the answer you get if you use the formula for the area of a trapezoid. ∎

EXAMPLE E

Use the limit process to evaluate $\displaystyle\int_{0}^{2} x^2\, dx$.

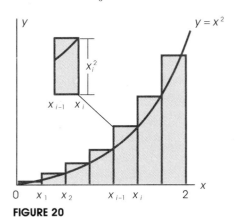

FIGURE 20

Solution We partition the interval $[0, 2]$ into n equal subintervals each of length $\Delta x = 2/n$ (see Figure 20). The partition points have coordinates $x_1 = 2/n$, $x_2 = 2(2/n)$, $x_3 = 3(2/n)$, and in general $x_i = i(2/n)$. Since $f(x) = x^2$, we see that

$$\sum_{i=1}^{n} f(x_i)\, \Delta x = \sum_{i=1}^{n} \left[i\left(\frac{2}{n}\right)\right]^2 \left[\frac{2}{n}\right] = \sum_{i=1}^{n} i^2 \left[\frac{8}{n^3}\right]$$

$$= \frac{8}{n^3} \sum_{i=1}^{n} i^2 \qquad \text{(by Property 2 of } \Sigma\text{)}$$

$$= \frac{8}{n^3} \frac{n(n+1)(2n+1)}{6} \qquad \text{[by Formula (ii), p. 267]}$$

$$= \frac{4}{3} \frac{(n+1)(2n+1)}{n^2}$$

Thus,

$$\int_{0}^{2} x^2\, dx = \lim_{n \to \infty} \frac{4}{3} \frac{(n+1)(2n+1)}{n^2} = \frac{4}{3} \lim_{n \to \infty} \frac{2n^2 + 3n + 1}{n^2} = \frac{8}{3}$$

Naturally, this is the same answer we would get by using the Fundamental Theorem of Calculus. ∎

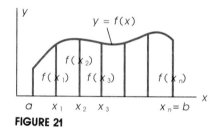

FIGURE 21

The Average Value of a Function

We can now give the more adequate discussion of the average value of a function that we promised in Section 8.2. Consider a function f defined on the interval $[a, b]$ (see Figure 21). Divide the interval $[a, b]$ into n equal pieces of length $\Delta x = (b - a)/n$ and let $x_i = a + i\,\Delta x$. A reasonable approximation to the average value of f ought to be obtained by averaging the values of f at the sample points $x_1, x_2, \ldots, x_n$, especially if n is large.

$$f_{\text{ave}} \approx \frac{1}{n}[f(x_1) + f(x_2) + \cdots + f(x_n)]$$

$$= \frac{1}{b - a} \frac{b - a}{n}[f(x_1) + f(x_2) + \cdots + f(x_n)]$$

$$= \frac{1}{b - a}[f(x_1)\,\Delta x + f(x_2)\,\Delta x + \cdots + f(x_n)\,\Delta x]$$

$$= \frac{1}{b - a} \sum_{i=1}^{n} f(x_i)\,\Delta x$$

SHARP CURVE
The basis of our derivation of the formula for the average value of a function was to sample more and more of the infinitely many values of f. But we also used a mathematical trick, dividing and multiplying by $b - a$. It is often surprising what a large reward a little trick can bring.

This approximation will get better and better as we sample f at more and more points, that is, as n gets larger and larger. Thus, we define the **average value** of f on the interval $[a, b]$ to be the limit of the above expression as $n \to \infty$. This gives

$$f_{\text{ave}} = \frac{1}{b - a} \int_{a}^{b} f(x)\, dx$$

Problem Set 8.4

Calculate each of the sums in Problems 1–8 (see Example A).

1. $\sum_{i=1}^{6} (i/5)$

2. $\sum_{i=1}^{5} (i - 1)^2$

3. $\sum_{i=3}^{6} (3i - 4)$

4. $\sum_{i=2}^{5} (5i - 8)$

5. $\sum_{i=1}^{3} \frac{1}{i(i + 1)}$

6. $\sum_{i=4}^{6} \frac{12}{i(i - 2)}$

7. $\sum_{i=0}^{4} 2^i$

8. $\sum_{i=1}^{3} (4^i - 5^{i-1})$

In Problems 9–20, write each sum in sigma notation (see Example B).

9. $1^3 + 2^3 + 3^3 + \cdots + 10^3$

10. $2 + 4 + 6 + 8 + \cdots + 100$

11. $\frac{3}{2} + \frac{3}{2^2} + \frac{3}{2^3} + \cdots + \frac{3}{2^{100}}$

12. $5 + \frac{5}{2} + \frac{5}{3} + \frac{5}{4} + \cdots + \frac{5}{70}$

13. $x_4 + x_5 + x_6 + \cdots + x_{21}$

14. $t_1 + t_2 + t_3 + \cdots + t_n$

15. $g(t_1) \Delta t + g(t_2) \Delta t + g(t_3) \Delta t + \cdots + g(t_n) \Delta t$

16. $f(x_1) \Delta x + f(x_2) \Delta x + f(x_3) \Delta x + \cdots + f(x_m) \Delta x$

17. $3x_1 \Delta x + 3x_2 \Delta x + 3x_3 \Delta x + \cdots + 3x_n \Delta x$

18. $x_1^2 \Delta x + x_2^2 \Delta x + x_3^2 \Delta x + \cdots x_n^2 \Delta x$

19. $(x_1^3 + 4) \Delta x + (x_2^3 + 4) \Delta x + (x_3^3 + 4) \Delta x + \cdots + (x_n^3 + 4) \Delta x$

20. $(x_1^2 + 5x_1) \Delta x + (x_2^2 + 5x_2) \Delta x + (x_3^2 + 5x_3) \Delta x + \cdots + (x_n^2 + 5x_n) \Delta x$

Calculate each of the sums in Problems 21–28, given that $\sum_{i=1}^{10} x_i = 30$ and $\sum_{i=1}^{10} y_i = 4$ (see Example C). In Problem 25, you will want to use the fact that the constant 2 appears in every term and so contributes $10(2) = 20$ to the sum. A similar remark applies in Problems 26–28.

21. $\sum_{i=1}^{10} 4x_i$

22. $\sum_{i=1}^{10} (-0.5)y_i$

23. $\sum_{i=1}^{10} (3x_i - 20y_i)$

24. $\sum_{i=1}^{10} (2x_i + 11y_i)$

25. $\sum_{i=1}^{10} (x_i + 2)$

26. $\sum_{i=1}^{10} (y_i + 3)$

27. $\sum_{i=1}^{10} (2x_i - 3y_i - 3)$

28. $\sum_{i=1}^{10} (x_i - 3y_i + 1)$

In Problem 29–38, use the limit process to evaluate each definite integral (see Examples D and E). Check your answer by using the Fundamental Theorem of Calculus.

29. $\int_0^3 (4 - x) \, dx$

30. $\int_0^4 (6 - x) \, dx$

31. $\int_0^2 (2x + 1) \, dx$

32. $\int_0^3 (3x + 5) \, dx$

33. $\int_0^3 (x^2 + 1) \, dx$

34. $\int_0^4 (x^2 - 2) \, dx$

35. $\int_0^1 (x^2 - 2x) \, dx$

36. $\int_0^1 (2x^2 + x) \, dx$

37. $\int_1^3 (x + 4) \, dx$
   Hint: $x_i = 1 + i(2/n)$

38. $\int_2^5 (2x + 1) \, dx$
   Hint: $x_i = 2 + i(3/n)$

Calculate each of the sums in Problems 39–46. For some of them, you will need to use the formulas (i) and (ii) on page 267.

39. $1 + 2 + 3 + \cdots + 499$

40. $1^2 + 2^2 + 3^2 + \cdots + 20^2$

41. $10 + 11 + 12 + \cdots + 99$

42. $41^2 + 42^2 + 43^2 + \cdots + 100^2$

43. $\sum_{i=1}^{100} (i^2 - 3i)$

44. $\sum_{j=1}^{100} (j^2 + 2j)$

45. $\sum_{k=1}^{100} (2 - 3k + k^2)$

46. $\sum_{i=1}^{100} \left(\frac{1}{i} - \frac{1}{i + 1} \right)$ Hint: Write out this sum before trying to calculate it.

Write the expressions in Problems 47 and 48 as definite integrals.

47. $\lim_{n\to\infty} \sum_{i=1}^{n} \left[i^2\left(\frac{4}{n}\right)^2 + 3i\frac{4}{n} \right] \frac{4}{n}$

48. $\lim_{n\to\infty} \sum_{i=1}^{n} \left[3i^2\left(\frac{2}{n}\right)^2 + i\frac{2}{n} - 2 \right] \frac{2}{n}$

49. Calculate $\sum_{i=1}^{100} \ln\left(\frac{i}{i+1}\right)$

50. Calculate $\sum_{i=2}^{20} \ln\left[1 - \frac{1}{i} \right]$

51. (Statistics) For the set of data $\{x_1, x_2, \cdots, x_n\}$, we define the mean $\bar{x}$ and variance s^2 by $\bar{x} = \frac{1}{n}\sum_{i=1}^{n} x_i$ and $s^2 = \frac{1}{n}\sum_{i=1}^{n} (x_i - \bar{x})^2$. Find the mean and variance for the set of data $\{3, 4, 5, 5, 6, 6, 6, 7, 8, 10\}$.

CHAPTER 8 REVIEW PROBLEM SET

1. Calculate $\int_{-3}^{7} f(x)\,dx$, where f is the function whose graph is shown.

(a)

(b)

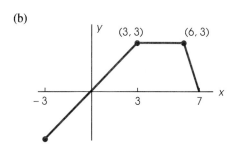

2. Sketch the graph of f on the interval $[-2, 4]$ and then evaluate $\int_{-2}^{4} f(x)\,dx$.

(a) $f(x) = \begin{cases} -x & -2 \le x \le 2 \\ -2 & 2 < x \le 4 \end{cases}$

(b) $f(x) = \begin{cases} -x - 2 & -2 \le x \le 0 \\ (4 - x)/2 & 0 < x \le 4 \end{cases}$

3. Sketch the graph of f on the interval $[a, b]$ and then calculate $\int_a^b f(x)\,dx$.
(a) $f(x) = x - 2;\ a = 0,\ b = 5$
(b) $f(x) = \sqrt{4 - x^2};\ a = -2,\ b = 2$
(c) $f(x) = \begin{cases} |x + 1| & x \le 1 \\ -2x + 4 & 1 < x \le 3 \\ -2 & x > 3 \end{cases}$
$a = -3,\ b = 6$

4. Sketch the graph of f on the interval $[-2, 5]$ and then calculate $\int_{-2}^{5} f(x)\,dx$.

(a) $f(x) = \begin{cases} |x| & -2 \le x \le 2 \\ 2 & 2 \le x \le 5 \end{cases}$

(b) $f(x) = \begin{cases} \sqrt{4 - x^2} & -2 \le x \le 2 \\ (1 - x)/2 & 2 < x \le 5 \end{cases}$

5. Use symmetry and the formula $\int_0^a x^n\,dx = a^{n+1}/n + 1$ to evaluate each integral.

(a) $\int_{-1}^{1} 5x^6\,dx$

(b) $\int_{-8}^{8} \pi x^{101}\,dx$

(c) $\int_{-1}^{1} (x^7 - 2x^4)\,dx$

(d) $\int_{-3}^{3} \frac{2x^7}{\sqrt{x^2 + 4}}\,dx$

6. Suppose that f is an odd function, that g is an even function, and that $\int_0^2 f(x)\,dx = \int_0^2 g(x)\,dx = 3$. Use symmetry to evaluate each of the following.

(a) $\int_{-2}^{2} f(x)\,dx$

(b) $\int_{-2}^{2} |f(x)|\,dx$

(c) $\int_{-2}^{2} g(x)\,dx$

(d) $\int_{-2}^{2} x^3 g(x)\,dx$

7. An object moves along a straight line in such a way that after t seconds its velocity is given by $v(t) = 2t - 3$ feet per second. Find the total distance (backward and forward) traveled by the object between $t = 0$ and $t = 5$.

8. An object moves along a straight line so that after t seconds its velocity is $v = 3t - 2$ feet per second. Find the distance between its position at $t = 0$ and its position at $t = 5$.

Evaluate the definite integrals in Problems 9–30.

9. $\int_1^9 3\sqrt{x}\,dx$

10. $\int_1^8 2\sqrt[3]{x}\,dx$

11. $\int_0^6 \left(\frac{1}{2}x^2 - 4\right)dx$

12. $\int_0^2 \left(3x^2 - \frac{1}{2}x + 4\right)dx$

13. $\int_1^3 \frac{x+4}{x^2 + 8x - 3}\,dx$

14. $\int_0^3 t(t^2 + 16)^{1/2}\,dt$

15. $\int_{1/4}^{1/2} \frac{e^{1/x}}{x^2}\,dx$

16. $\int_3^5 \frac{t-2}{3t^2 - 12t + 10}\,dt$

17. $\int_0^3 \frac{2x+1}{\sqrt{x^2 + x + 4}}\,dx$

18. $\int_2^5 \frac{4x+1}{(2x^2 + x - 6)^{3/2}}\,dx$

19. $\int_0^2 [\ln(x^2 + 1)]^3 \frac{x}{x^2 + 1}\,dx$

20. $\int_1^4 \frac{e^{\sqrt{x}}}{\sqrt{x}}\,dx$

21. $\int_0^5 4|t - 3|\,dt$

22. $\int_0^{\ln 11} \frac{d}{dx}\ln(e^x + 4)\,dx$

23. $\int_0^{\sqrt{e-1}} \frac{d}{dx}[\ln(1 + x^2)]\,dx$

24. $\int_0^{(e-1)/2} \frac{\ln(2x + 1)^4}{2x + 1}\,dx$

25. $\int_{-3}^5 f(x)\,dx$ where $f(x) = \begin{cases} e^{-x} & -3 \le x \le 0 \\ x + 1 & 0 < x \le 2 \\ 3 & 2 < x \le 5 \end{cases}$

26. $\int_0^{e^3} f(x)\,dx$ where

$$f(x) = \begin{cases} 2e^x & 0 \le x \le 1 \\ 2x - 2 & 1 < x < e \\ (\ln x)^2/x & e \le x \le e^3 \end{cases}$$

27. $\int_1^3 (x - 3)^5 x\,dx$ *Hint:* Let $u = x - 3$.

28. $\int_2^4 (x - 4)^5 x\,dx$

29. $\int_0^{\sqrt{5}} (9 - x^2)^{-3/2}\,dx$.   *Hint:* Use table of integrals.

30. $\int_0^4 (9 + x^2)^{3/2}\,dx$ *Hint:* Use a table of integrals.

31. Sketch the graph of $y = 2\sqrt[3]{x} + 1$ on the interval $[1, 8]$ and then find the area of the region under this graph.

32. Sketch the graph of $y = 4(x + 1)/\sqrt{x}$ on the interval $[1, 4]$ and then find the area of the region under the graph.

33. Suppose that the height h (in inches) of a tree t months after it was planted was given by

$$h = 52 + 3.2t - 0.02t^2 \qquad 0 \le t \le 30$$

Find the average height of the tree during the 30-month period.

34. After x years, a certain company was making snow-blowers at the rate of $140 + 2x + 3x^2$ units per year. On the average, how many units did the company make per year during the first 10 years?

35. Calculate the sum $\sum_{i=1}^4 (3^i - 2i)$.

36. Calculate $\sum_{k=0}^4 (k^3 - 2^k)$.

37. Write in sigma notation.
(a) $1 \cdot 2 + 2 \cdot 3 + 3 \cdot 4 + \cdots + 97 \cdot 98$
(b) $(2x_1 + 3)\,\Delta x + (2x_2 + 3)\,\Delta x + \cdots + (2x_n + 3)\,\Delta x$

38. Write the sigma notation.
(a) $1^2 + 2^2 + 3^2 + \cdots + 100^2$
(b) $t_1{}^3 \Delta t + t_2{}^3 \Delta t + t_3{}^3 \Delta t + \cdots + t_n{}^3 \Delta t$

39. Calculate $\sum_{i=1}^8 (4x_i - 6y_i)$, given that $\sum_{i=1}^8 x_i = 36$ and $\sum_{i=1}^8 y_i = 15$.

40. Calculate $\sum_{k=1}^{12} (3x_k - 5y_k + 1)$, given that $\sum_{k=1}^{12} x_k = 40$ and that $\sum_{k=1}^{12} y_k = 20$.

41. Calculate $\sum_{i=1}^{100} (3i^3 + 4i)$ by using the two formulas
$$\sum_{i=1}^n i = \frac{n(n+1)}{2} \text{ and } \sum_{i=1}^n i^3 = \left[\frac{n(n+1)}{2}\right]^2.$$

42. Calculate $\sum_{i=1}^{10} (i^3 - 2i^2 - 3i)$ given the formulas in Problem 41 and the additional formula $\sum_{i=1}^n i^2 = \frac{n(n+1)(2n+1)}{6}$.

43. Evaluate $\int_1^e f''(t)\,dt$, given that $f(t) = t \ln t$.

44. Find $\frac{d}{dx}(x \ln x)$. Then evaluate $\int_1^e (1 + \ln x)\,dx$.

45. Use the limit process to evaluate $\int_0^2 (4x + 3)\,dx$. Then think of two ways to check your answer.

46. Use the limit process to find $\int_0^3 (x^2 + 1)\,dx$. Check your answer by using the Fundamental Theorem of Calculus. You will need a formula from Problem 42.

47. Evaluate

$$\lim_{n \to \infty} \frac{4}{n} \sum_{i=1}^n \frac{4i/n}{1 + (4i/n)^2}$$

by recognizing this limit as a definite integral and then evaluating the definite integral.

48. Evaluate $\lim_{n \to \infty} \frac{2}{n} \sum_{i=1}^n \left[3\left(\frac{2i}{n}\right)^2 + \frac{2i}{n}\right]$.

49. Consider the function $f(x) = 1/x^2$, $x > 0$.
   (a) Sketch the graph of f.
   (b) Find the point on the graph where the tangent line has slope $-1/4$.
   (c) If a point moves along the graph of $y = 1/x^2$ so that $dx/dt = 2.5$, find dy/dt at $x = 10$.
   (d) Find the area under the graph of f from $x = 1$ to $x = 1000$.
   (e) Find $\lim\limits_{t \to \infty} \int_1^t \dfrac{1}{x^2}\, dx$.

50. Consider the function $g(x) = x/\sqrt{x^2 + 16}$.
   (a) Determine the domain of g.
   (b) Evaluate $\lim\limits_{x \to \infty} g(x)$.
   (c) Where is g increasing?
   (d) Find the point on the graph of g where the tangent line has slope $1/4$.
   (e) Find the area under the graph between $x = 0$ and $x = 3$.

Trying to predict weather is a daunting problem involving the collection of masses of data and the construction of complicated probability models. To collect data on temperature, pressure, and humidity everywhere is impossible; rather it must be done at selected points such as the intersection points of the grid shown below.

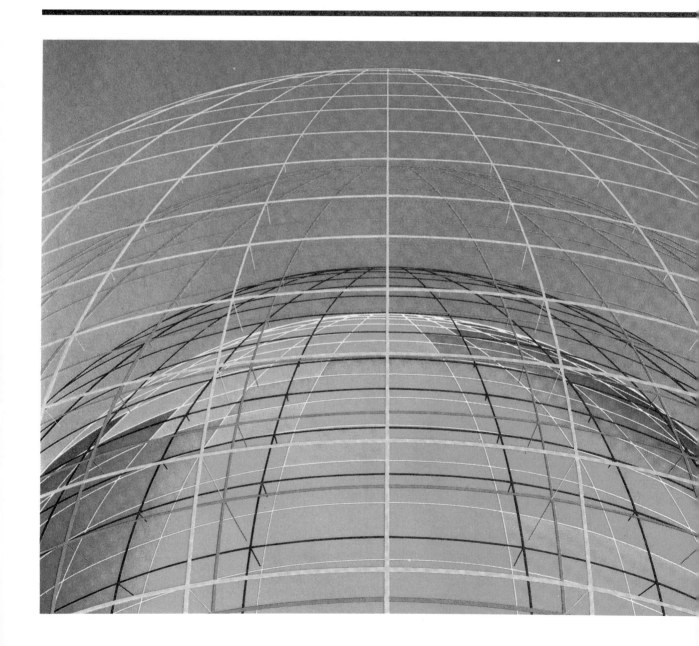

9

Applications of the Definite Integral

David Blackwell is internationally recognized as a distinguished mathematician and statistician. He has made important contributions to Bayesian statistics, probability theory, game theory, dynamic programming, and information theory. Few would have predicted that a young lad with black skin, born in the backwaters of Southern Illinois, and of parents with little education, could attain such stature. That Blackwell did is a tribute to his ability and determination and to the encouragement of some enlightened teachers.

After graduation from high school at age 16, Blackwell entered the University of Illinois with the goal of becoming an elementary teacher. When he learned that his father was borrowing money to keep him in college, he insisted on supporting himself by working as a waiter, dishwasher, and lab cleaner. Still, he managed to obtain a bachelor's degree in three years. Mathematics was easy for him, but it was not until he studied G. H. Hardy's *Pure Mathematics* that he began to see the real beauty of the subject. This started a love affair that continues to the present. Soon the young man was setting his sights higher—still teaching but maybe at the college level. This, of course, would require an advanced degree.

Fortunately, Blackwell's talent was recognized with a fellowship. And even more fortunately, one of Illinois's mathematical stars, Joseph Doob, agreed to supervise his Ph.D. work. That degree was awarded in 1941, three short years after his A.B. There followed a year of study at the Institute for Advanced Study at Princeton.

The times did not favor black mathematicians. Blackwell fully expected that he would have to teach at an under-financed and unheralded institution and, in fact, he did teach at several such schools for a number of years. Finally in 1954, he was invited to join the statistics faculty at the University of California at Berkeley, and when the famed Jerzy Neyman gave up chairing this group, it was Blackwell who was asked to take over. There he is part of what is generally recognized as the most prestigious statistics department in the world.

David Blackwell
(1919–   )

AREA BETWEEN TWO CURVES

The concept of area was used to motivate and actually define the definite integral. In this section, we are going to reverse direction and use the definite integral to calculate areas. We begin by recalling that if the graph of $y = f(x)$ is above the x-axis, then the area of the region bounded by this graph and the x-axis between $x = a$ and $x = b$ is just $\int_a^b f(x)\,dx$. We refer to this area as *the area under the curve $y = f(x)$ between a and b.*

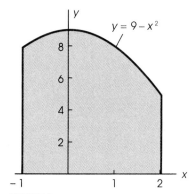
FIGURE 1

EXAMPLE A

Find the area under the curve $y = 9 - x^2$ between -1 and 2 (see Figure 1).

Solution This is simply a matter of evaluating a definite integral

$$\int_{-1}^{2} (9 - x^2)\,dx = \left[9x - \frac{x^3}{3} \right]_{-1}^{2} = 18 - \frac{8}{3} - \left[-9 - \frac{(-1)^3}{3} \right] = 24 \quad \blacksquare$$

A Helpful Way of Thinking

Consider any area problem of the type illustrated in Example A. If you studied Section 8.4, you know that this area can be approximated by the sum of the areas of a collection of thin rectangles (Figure 2). Concentrate now on a typical thin rectangle (Figure 3) whose right edge has x-coordinate x, whose height is then $f(x)$, and whose width is dx. It will have area $f(x)\,dx$. We want to add up the areas of such rectangles and then take the limit as the number of these rectangles tends to infinity. You may think of $\int_a^b$ as standing for the operation of adding up and taking the limit. This limiting process turns the approximation into an exact value for the desired area. Thus, the required area is $\int_a^b f(x)\,dx$.

From now on, we shall consider the task of finding the area of a region to be a three-step process.

1. Sketch the region and imagine slicing it into thin strips. Show a typical strip as in Figure 3.
2. Approximate the area of this strip by pretending it is a rectangle. Label the right coordinate of the rectangle with x, the height with $f(x)$,

FIGURE 2

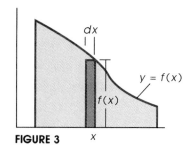
FIGURE 3

and the width with dx. This gives $f(x)\, dx$ for the area of the rectangle.

3. Add up the areas of these rectangles and take the limit, which is another way of saying "integrate."

We shall describe this three-step process by using the three words *slice, approximate, integrate*. We illustrate with another example.

EXAMPLE B

Find the area under the curve $y = 3e^{-x}$ between 0 and 2.

Solution Step 1 and 2 (*slice* and *approximate*) are accomplished by drawing a picture and labeling it appropriately (Figure 4). This picture in turn suggests the correct integral to write down, namely, $\int_0^2 3e^{-x}\, dx$. Finally, we carry out the integration (using the substitution $u = -x$ and $du = -dx$).

$$\int_0^2 3e^{-x}\, dx = -\int_0^2 3e^{-x}(-dx) = -3 \int_0^{-2} e^u\, du = -3\left[e^u \right]_0^{-2}$$

$$= -3(e^{-2} - e^0) \approx 2.5940$$ ■

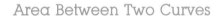

3
2   dx
1        $y = 3e^{-x}$

x 1   2

FIGURE 4

Area Between Two Curves

You will see that the process *slice, approximate, integrate* is the basic notion behind all applications of the definite integral. We next illustrate it for the area of a region between two curves.

EXAMPLE C

Find the area of the region between the curves $y = f(x) = x^2 + 1$ and $y = g(x) = x - 1$ on the interval $[0, 3]$.

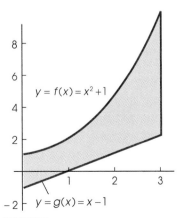

$y = f(x) = x^2 + 1$

$y = g(x) = x - 1$

FIGURE 5

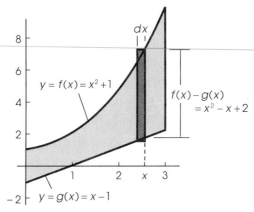

FIGURE 6

Solution The region of interest is shown in Figure 5, but it is Figure 6 that represents the steps *slice* and *approximate*. The height of the rectangle at x is $f(x) - g(x)$; that is, it is $x^2 + 1 - (x - 1)$. Note that this expression for the height is correct also in that part of the region to the left of $x = 1$, for there $x - 1$ is negative and subtracting it is really adding just the right positive number. Thus, we see that the area of a typical rectangle is always $(x^2 - x + 2) \, dx$ and the area of the whole region is

$$\int_0^3 (x^2 - x + 2) \, dx = \left[\frac{x^3}{3} - \frac{x^2}{2} + 2x \right]_0^3$$

$$= 9 - \frac{9}{2} + 6 - (0) = \frac{21}{2} \qquad \blacksquare$$

EXAMPLE D

Find the area of the region between the curves $y = x^2 - 4$ and $y = -x + 2$.

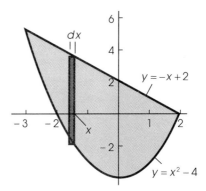

FIGURE 7

Solution The appropriate sketch with a typical rectangle is shown in Figure 7. The height of this rectangle is $(-x + 2) - (x^2 - 4)$, and so its area is $(-x^2 - x + 6)\ dx$. But now we face another problem. The limits that go on the integral sign (namely, the smallest and largest x-values in the region) are not obvious. They are the x-coordinates of the intersection points of the two curves and are obtained by solving the two equations simultaneously. To find them, we equate the two expressions for y and solve as follows.

$$x^2 - 4 = -x + 2$$

$$x^2 + x - 6 = 0$$

$$(x + 3)(x - 2) = 0$$

$$x = -3, 2$$

We complete the solution by calculating the required integral.

$$\int_{-3}^{2} (-x^2 - x + 6)\ dx = \left[-\frac{x^3}{3} - \frac{x^2}{2} + 6x \right]_{-3}^{2}$$

$$= \left(-\frac{8}{3} - 2 + 12 \right) - \left(9 - \frac{9}{2} - 18 \right)$$

$$\approx 20.8333 \qquad\blacksquare$$

EXAMPLE E

Find the area of the region bounded by the two curves $y = x^2$ and $y = x^3 + 2x^2 - 2x$.

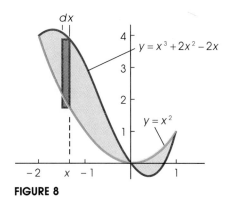

FIGURE 8

Solution A sketch of the required region together with a typical rectangle is shown in Figure 8. Note that the region is in two parts and that the curves exchange positions as to which is on top for the two parts. For the left part, the height of the rectangle is $(x^3 + 2x^2 - 2x) - x^2$, whereas for the right part, the height is the negative of this expression. The areas of the corresponding rectangles are $(x^3 + x^2 - 2x)\ dx$ and $(-x^3 - x^2 + 2x)\ dx$. To find the intersection points of the two curves, we solve the equation obtained by setting the two expressions for y equal to each other.

$$x^3 + 2x^2 - 2x = x^2$$

$$x^3 + x^2 - 2x = 0$$

$$x(x + 2)(x - 1) = 0$$

$$x = -2, 0, 1$$

The required area A is given by

$$A = \int_{-2}^{0} (x^3 + x^2 - 2x)\,dx + \int_{0}^{1} (-x^3 - x^2 + 2x)\,dx$$

$$= \left[\frac{x^4}{4} + \frac{x^3}{3} - x^2\right]_{-2}^{0} + \left[-\frac{x^4}{4} - \frac{x^3}{3} + x^2\right]_{0}^{1}$$

$$= 0 - \left(4 - \frac{8}{3} - 4\right) + \left(-\frac{1}{4} - \frac{1}{3} + 1\right) - 0$$

$$= \frac{37}{12} \approx 3.0833 \qquad\blacksquare$$

> **POINT OF INTEREST**
> We can always represent the area A of the region between two curves $y = f(x)$ and $y = g(x)$ by a single integral, namely,
>
> $$A = \int_{a}^{b} |f(x) - g(x)|\,dx$$
>
> However, to evaluate this integral may require us to rewrite it as two or more integrals as in Example E.

Horizontal Slicing

It will sometimes happen that an area problem is much easier to handle by horizontal slicing than by the vertical slicing that we have so far illustrated. Consider for example the region R shown in Figure 9. Note that vertical slices to the left of b always go from the curve to the line, whereas those to the right of b go from the lower branch of the curve to the upper branch of the curve. Thus to find the area of R by vertical slicing would require that we cut R into two pieces (the parts to the left and right of b) and do two separate integrations. A much better choice is to slice horizontally so that a slice always goes from the line on the left to the curve on the right. Then one integration will give the area. We illustrate in an example.

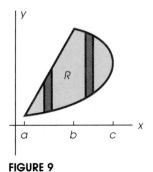

FIGURE 9

EXAMPLE F

Find the area of the region between the curves $x = y^2$ and $x = y + 2$.

FIGURE 10

Solution The region with a typical horizontal slice and the corresponding approximating rectangle are shown in Figure 10. The width of this rectangle is denoted by dy (a small change in y), and its length at y is the x-coordinate on the line minus the x-coordinate on the curve; that is, the length is $y + 2 - y^2$. Thus, the area of the rectangle is $(-y^2 + y + 2) \, dy$; this is the expression we must integrate.

Next, we must choose the limits of integration. They are, respectively, the smallest and largest values of y on the region. We obtain them by solving the equations $x = y^2$ and $x = y + 2$ simultaneously, that is, by solving $y^2 = y + 2$. This gives $y = -1$ and $y = 2$.

Finally, we perform the required integration.

$$A(R) = \int_{-1}^{2} (-y^2 + y + 2) \, dy = \left[-\frac{1}{3}y^3 + \frac{1}{2}y^2 + 2y \right]_{-1}^{2} = 4.5 \quad \blacksquare$$

Problem Set 9.1

In Problems 1–12, find the area of the region under the given curve between a and b (see Example B).

1. $f(x) = x^2$; $a = 1, b = 4$
2. $f(x) = 4 - x^2$; $a = -1, b = 2$
3. $f(x) = (x + 1)^2$; $a = 1, b = 3$
4. $f(x) = (x - 3)^2$; $a = 0, b = 5$
5. $f(x) = x^3 - 4x$; $a = -2, b = -1$
6. $f(x) = 9x - x^3$; $a = 1, b = 3$
7. $f(x) = 4e^x$; $a = 0, b = 2$
8. $f(x) = 3e^{x/2}$; $a = -1, b = 0$
9. $f(x) = 1/(x + 3)$; $a = -2, b = 0$
10. $f(x) = 12/x^2$; $a = 2, b = 4$
11. $f(x) = x/(x^2 + 1)$; $a = 0, b = 2$
12. $f(x) = x\sqrt{4 - x^2}$; $a = 0, b = 2$

13. In the figure below, we show the graph of a region R with a typical slice, properly labeled. Find the area of R.

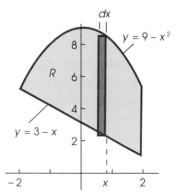

14. Find the area of the region R shown in the figure below.

In Problems 15–22, we give the equations of two curves and an interval. Find the area of the region between the two curves on the given interval. As in Example C, be sure to show an appropriate figure properly labeled (see Figure 6).

15. $y = x^2 + 2, y = -2x^2$; [0, 2]
16. $y = x^2 - 1, y = 8 - x^2$; [-1, 2]
17. $y = x^3 - 1, y = x + 2$; [-2, 1]
18. $y = 2x^3, y = 4 - x^2$; [-2, 1]
19. $y = e^x, y = 2x + 3$; [-1, 1]
20. $y = 2e^{-x}, y = (x - 2)^2$; [3, 5]
21. $y = 6/(x + 3), y = -x + 1$; [-2, 2]
22. $y = (\ln x)/x, y = x^2 + x$; [1, 3]

Each of Problems 23–32 gives the equations of two curves, curves that bound a region. Follow the method illustrated in Example D (including a good figure properly labeled) to find the area of this region.

23. $y = x^2 + 2, y = 2x + 5$
24. $y = x^2 - 1, y = 2x + 2$
25. $y = 5 - x^2, y = -x - 1$
26. $y = 9 - x^2, y = -x - 3$
27. $y = x^2 - 3x + 2, y = x - 1$
28. $y = x^2 + 6x, y = 5x + 2$
29. $y = 2\sqrt{x}, y = (2x + 4)/3$

30. $y = \sqrt{x}, y = (x + 3)/4$
31. $y = x^2 - 12, y = 6 - x^2$
32. $y = x^2 + 3, y = 11 - x^2$

The graphs of each pair of equations in Problems 33–36 bound a region (which may be in two parts). Find the total area of this region as in Example E.

33. $y = x^3, y = 4x$
34. $y = 2x^3, y = 2x$
35. $y = 2x^3, y = -x^3 + x^2 + 2x$
36. $y = x^2, y = x^3 - 2x$

In Problems 37–42, use horizontal slicing to find the area of the region bounded by the given curves (see Example F).

37. $x = y^2, x = 4$
38. $x = -y^2, x = -3$
39. $x = 4y - y^2, x = y$
40. $x = y^2 - 6y, x = -y$
41. $x = y^2 - 4y + 3, x = y - 1$
42. $x = y^2 - 2y - 3, x = 2y - 3$

In Problems 43–50, find the area of the region bounded by the given curves.

43. $y = x^2 - 9, y = 0$
44. $x = -y^2 - 2y + 3, x = 0$
45. $y = x(x + 2)(x - 3), y = 0$
46. $y = e^x, y = e^2, x = 0$
47. $x = 1/y, x = 0, y = 1, y = 3$
48. $y = x, y = 6 - x, y = x/2$
49. $y = x + 6, y = x^3, 2y + x = 0$
50. $y = x^2 - x - 6, y = x^3 + x^2 - 5x - 6$
51. $y = x/(x^2 + 1), y = x/2$
52. $y = e^x, x = (y - 1) \ln 2$
53. Find the area of the triangle with vertices at $(-1, 0)$, $(4, 5)$, and $(0, 9)$.
54. Find the area of the parallelogram with vertices at $(0, 9)$, $(5, 3)$, $(9, 7)$, and $(4, 4)$.

9.2
VOLUMES OF SOLIDS OF REVOLUTION

It is not surprising that the definite integral can be used to calculate areas; it was invented for that purpose. What is somewhat surprising is the number of physical and geometrical quantities that can be calculated by means of definite integrals. In fact, almost any quantity that can be thought of as arising from slicing something into small pieces, approximating each piece, adding up, and taking the limit will finally lead to a definite integral. In particular, this is true of volumes of solids of revolution.

But what is a solid of revolution? Take a plane region that lies entirely on one side of a fixed line called the **axis.** When this region is revolved about the axis, it generates a **solid of revolution** (Figure 11). We wish to calculate the volume of such a solid.

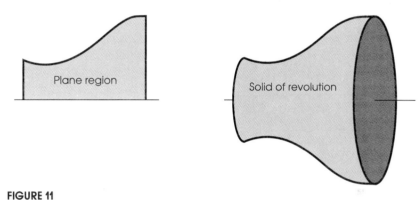

FIGURE 11

Imagine slicing such a solid into thin pieces of equal thickness by means of planes perpendicular to the axis (Figure 12) and then approximating each piece by a disk (Figure 13). If we can find the volume of a disk, add up the volumes of all the disks, and then take the limit as the number of pieces becomes infinite, we should get a definite integral.

FIGURE 12   **FIGURE 13**

$V = \pi[f(x)]^2 dx$

FIGURE 14

The process we have described for calculating volumes of solids of revolution is called the **method of disks.** Let us see how it works in some examples.

The Method of Disks

Before we can actually do an example, we have to answer one more preliminary question. How do you find the volume of a disk? A disk is just a thin right circular cylinder. We hope you recall that the volume of a right circular cylinder is $\pi r^2 h$; that is, it is the area of the base times the height. For the thin disks we will be considering, this formula will become $\pi[f(x)]^2 dx$ (Figure 14).

EXAMPLE A

Find the volume of the solid formed by revolving the region bounded by $y = \sqrt{x}$, $y = 0$, and $x = 4$ about the x-axis.

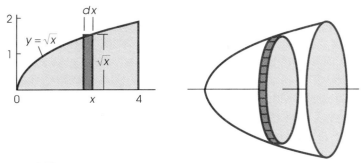

FIGURE 15

Solution The region together with a typical rectangle is shown in the left part of Figure 15. The corresponding solid of revolution and the disk generated by the rectangle are shown at the right in Figure 15. Note that the volume of a typical disk is

$$\pi (\sqrt{x})^2 \, dx = \pi x \, dx$$

We conclude that the volume of the solid is

$$\int_0^4 \pi x \, dx = \left[\frac{\pi x^2}{2} \right]_0^4 = \pi (8) - 0 = 8\pi \qquad \blacksquare$$

Many of us have trouble drawing three-dimensional pictures like the one at the right in Figure 15, but all of us can draw two-dimensional pictures like the one at the left. It is enough to visualize the solid mentally; all the information needed to write down the correct integral appears in the plane figure. We suggest that you concentrate on drawing good, properly labeled two-dimensional figures.

EXAMPLE B

Find the volume of the solid formed by revolving the plane region bounded by $y = \sqrt{9 - x^2}$ and $y = 0$ about the x-axis.

FIGURE 16

Solution The equation $y = \sqrt{9 - x^2}$ represents a semicircle (Figure 16), and so the solid generated is a solid sphere. We know that the volume of a sphere is $\frac{4}{3}\pi r^3$, so we will have a good chance to check that our integration method is correct. The rectangle in Figure 16 will generate a disk of volume $\pi(\sqrt{9 - x^2})^2 \, dx$, and so the volume of the whole sphere is

$$\int_{-3}^{3} \pi(9 - x^2)\,dx = 2\int_{0}^{3} \pi(9 - x^2)\,dx$$

$$= 2\pi\left[9x - \frac{x^3}{3}\right]_0^3$$

$$= 2\pi(27 - 9) = 36\pi$$

In the first step above, we used the symmetry of the region about the y-axis. Note that the answer 36π is $\frac{4}{3}\pi(3)^3$ as expected.   ■

The Method of Washers

Sometimes slicing a solid of revolution produces disks with holes in the middle; we call them **washers.** A typical washer is shown in Figure 17. Its

$$V = \pi(R^2 - r^2)h$$

FIGURE 17

volume is obtained by calculating the volume of a solid disk and then sub-tracting the volume of the hole. Thus, the volume formula for a washer of thickness h is

$$V = \pi R^2 h - \pi r^2 h = \pi (R^2 - r^2)h$$

where R is the radius of the disk and r is the radius of the hole.

EXAMPLE C

Find the volume of the solid formed when the region bounded by $y = x^2 + 2$ and $y = 1$ between $x = 0$ and $x = 2$ is revolved about the x-axis.

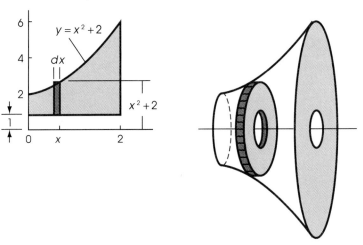

FIGURE 18

Solution The plane region (with a typical rectangle) and the corresponding solid (with a typical disk) are shown in Figure 18. The volume of this typical disk is

$$\pi[(x^2 + 2)^2 - (1)^2] \, dx$$

and so the volume of the solid is

$$\int_0^2 \pi[(x^2 + 2)^2 - 1] \, dx = \pi \int_0^2 (x^4 + 4x^2 + 3) \, dx$$

$$= \pi \left[\frac{x^5}{5} + \frac{4x^3}{3} + 3x \right]_0^2$$

$$= \pi \left(\frac{32}{5} + \frac{32}{3} + 6 \right) \approx 72.4661 \qquad \blacksquare$$

Horizontal Slicing

Of course, we can generate a solid of revolution by revolving a region about the y-axis or some other vertical line too. This will lead to a y-integration, but otherwise all is as before.

EXAMPLE D

Find the volume of the solid formed by revolving the region bounded by $y = x^3$, $y = 8$, and $x = 0$ about the line $x = -1$.

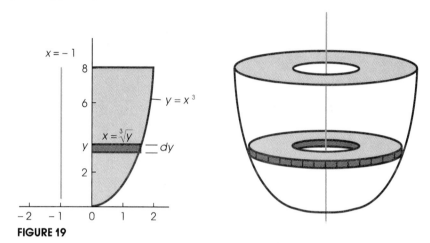

FIGURE 19

Solution Now we must slice horizontally, and this slicing results in horizontal washers. The plane region (with a typical rectangle) and the corresponding solid (with a typical washer) are shown in Figure 19. Clearly $y = x^3$ means that $x = \sqrt[3]{y}$. Note that $x + 1$ is the outer radius of the washer, whereas 1 is the radius of the hole. Thus, the volume of a typical washer is

Outer radius   Radius of hole

$$\pi[(\sqrt[3]{y} + 1)^2 - (1)^2] \, dy$$

We conclude that the volume of the whole solid is

$$\int_0^8 \pi[(\sqrt[3]{y} + 1)^2 - 1)] \, dy = \pi \int_0^8 (y^{2/3} + 2y^{1/3}) \, dy$$

$$= \pi \left[\frac{3}{5} y^{5/3} + \frac{3}{2} y^{4/3} \right]_0^8$$

$$= \pi \left(\frac{3}{5} 8^{5/3} + \frac{3}{2} 8^{4/3} \right)$$

$$= \pi \left(\frac{96}{5} + 24 \right) \approx 135.7168 \quad \blacksquare$$

EXAMPLE E

A certain lake has the shape of the solid of revolution generated by revolving the region bounded by

$$y = 30\left[\left(\frac{x}{1000}\right)^2 - 2\right]$$

$y = 0$, and $x = 0$ about the y-axis, distances being measured in meters. If a full-grown walleyed pike needs 10,000 cubic meters of water to live comfortably, how many pike can the lake support?

Solution The problem is simply one of finding the total volume of the lake and dividing the result by 10,000. A picture of the region with a typical rectangle is shown in Figure 20. When we solve the given equation for x, we obtain

$$x = 1000\left(2 + \frac{y}{30}\right)^{1/2}$$

Thus, the volume of the disk generated by the displayed rectangle is

$$\pi\left\{\left[1000\left(2 + \frac{y}{30}\right)\right]^{1/2}\right\}^2 dy$$

We conclude that the volume of the lake is

$$\int_{-60}^{0} \pi\left[1,000,000\left(2 + \frac{y}{30}\right)\right] dy = 1,000,000\pi \int_{-60}^{0}\left(2 + \frac{y}{30}\right) dy$$

$$= 1,000,000\pi\left[2y + \frac{y^2}{60}\right]_{-60}^{0}$$

$$= 1,000,000\pi[0 - (-120 + 60)]$$

$$= 60,000,000\pi \approx 188,500,000$$

The lake will support about 18,850 full-grown pike. ■

FIGURE 20

Problem Set 9.2

In Problems 1–6, find the volume of the solid generated when the indicated region is revolved about the x-axis (see Example A). Begin by showing a typical rectangle properly labeled, as we have already done in Problems 1 and 2.

1.

2.

3.

4.

5.

6.

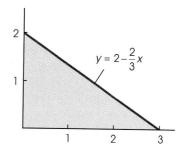

In Problems 7–12, sketch the region R bounded by the graphs of the given equations, showing a typical vertical rectangle. Then find the volume of the solid generated when R is revolved about the x-axis (see Example B).

7. $y = 4 - x^2$, $x = 0$, $y = 0$, $x \geq 0$
8. $y = 4x - x^2$, $y = 0$
9. $y = 4/x$, $x = 1$, $x = 4$, $y = 0$
10. $y = 4/x^2$, $x = 1$, $x = 2$, $y = 0$
11. $y = 2e^{2x}$, $x = 0$, $x = 2$, $y = 0$
12. $y = e^{-x}$, $x = -2$, $x = 0$, $y = 0$

In Problems 13–18, sketch the region R bounded by the graphs of the given equations. Then use the method of washers to find the volume of the solid generated when R is revolved about the x-axis (see Example C).

13. $y = 5x^2$, $y = 5$
14. $y = \sqrt{x}$, $y = 2$, $x = 0$
15. $y = x^2$, $y = x$
16. $y = 2x^2$, $y = x + 1$
17. $y = 8 - x^2$, $y = x^2$
18. $y = x^3$, $y = 4x$

In Problems 19–26, sketch the region R bounded by the graphs of the given equations. Then use horizontal slicing to find the volume of the solid generated when R is revolved about the y-axis (see Example D).

19. $x = 4y - y^2$, $x = 0$
20. $x = y^{1/3}$, $x = 0$, $y = 8$
21. $2x + y = 6$, $x = 0$, $y = 0$
22. $2x + 3y = 6$, $x = 0$, $y = 0$
23. $y = x^3$, $y = 0$, $x = 2$
24. $y = x^2$, $y = 0$, $x = 3$
25. $y = 2x^2$, $y = 2x$
26. $y = 4 - x^2$, $y = -2x + 4$

In Problems 27–36, find the volume of the solid generated when the given region is revolved about the given line.

27. Region in Problem 1 about $y = -2$
28. Region in Problem 2 about $y = -1$
29. Region in Problem 3 about $y = 9$
30. Region in Problem 4 about $y = 8$
31. Region in Problem 5 about $x = -3$
32. Region in Problem 6 about $x = -2$
33. Region in Problem 13 about $y = 5$
34. Region in Problem 14 about $y = 2$
35. Region in Problem 25 about $x = 4$
36. Region in Problem 26 about $x = 3$

37. Use the ideas of this section to find the volume of the solid that remains when a hole of radius 3 is drilled through the center of a ball of radius 5.

38. The base of a solid is the region bounded by $y = e^x$, $y = 0$, $x = 0$, and $x = 1$. Cross sections of the solid perpendicular to the x-axis are squares. Find the volume of this solid. *Hint:* Slicing through this solid with planes perpendicular to the x-axis produces thin slabs (like slices of cheese) whose volumes are easy to approximate.

39. The interior surface of a tank has the shape that is obtained by revolving the curve $y = e^{x^2}$ between $x = 0$ and $x = 2$ about the y-axis. Determine the capacity of the tank.

9.3
SOCIAL SCIENCE APPLICATIONS

The most impressive applications of the definite integral are in the physical sciences—to such concepts as work, energy, fluid force, center of mass, and moment of inertia. But there are also many applications in the social sciences; it is to the latter that we address our attention in this section.

Recovering a Quantity from Its Margin

A standard application is to determine an economic quantity $Q(x)$ when we know its marginal rate $Q'(x)$. From the Fundamental Theorem of Calculus,

$$\int_0^b Q'(x)\, dx = Q(b) - Q(0)$$

In other words,

$$Q(b) = Q(0) + \int_0^b Q'(x)\, dx$$

Example A shows how we use this fact.

POINT OF INTEREST

Since

$$\int_a^b Q'(x)\, dx = Q(b) - Q(a)$$

we can always determine the net change in a quantity Q on the interval $a \le x \le b$ by simply integrating its rate of change over that interval.

EXAMPLE A

Suppose that the marginal cost (in dollars per item) of manufacturing x units of a product is given by

$$C'(x) = 200e^{-x/100}$$

In simple terms, you can think of $C'(x)$ as representing the cost (approximately) of manufacturing the xth unit. For example, it costs about $200e^{-50/100} = \$121.31$ to make the 50th unit. Note that the cost of making the xth unit decreases as x increases, presumably as a result of economy of scale. Find the cost of making 300 of these units, assuming that there are $10,000 of fixed costs (that is, costs that occur even if no units are made). What is the average cost per unit?

Solution Note that $C(0)$ corresponds to the fixed costs; that is, $C(0) = 10,000$. Thus, the total cost of making 300 units is

$$C(300) = C(0) + \int_0^{300} C'(x)\, dx$$

$$= 10,000 + \int_0^{300} 200e^{-x/100}\, dx$$

$$= 10,000 + 200\int_0^{300} e^{-x/100}\, dx$$

In this integral, we make the substitution $u = -x/100$, and $du = -(\frac{1}{100})\, dx$. We obtain

$$C(300) = 10,000 + 200(-100)\int_0^{-3} e^u\, du$$

$$= 10,000 - 20,000\left[e^u\right]_0^{-3}$$

$$= 10,000 - 20,000[e^{-3} - e^0] = \$29,004.26$$

The average cost is $29,004.26/300 = \$96.68$. ■

Consumer and Producer Surplus

In our earlier discussions of demand functions, we usually assumed that x, the quantity demanded, is expressed in terms of p, the price charged, that is, that $x = f(p)$. Actually, economists are more likely to think of the **demand function** as relating the price p to the quantity x that people will buy, that is, $p = D(x)$. It is really just a matter of perspective; for if $x = -\frac{1}{2}p + 40$, then $p = -2x + 80$, and vice versa. Similarly, the **supply function**, $p = S(x)$, relates the price p to the supply x that is available. Graphs of typical demand and supply functions are shown in Figure 21. Note that the demand function is decreasing (the price must come down for people to buy more) and the supply function is increasing (producers will supply more as prices increase). The **equilibrium price** is the price at which the two curves intersect. This, say economists, is the only price that can persist over the long run.

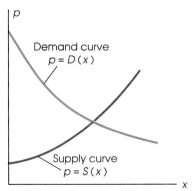

FIGURE 21

EXAMPLE B

Find the equilibrium price if the demand and supply functions (in dollars) for a certain product are given by $p = 500 - 5x$ and $p = \frac{1}{2}x^2 + 200$, respectively (Figure 22a).

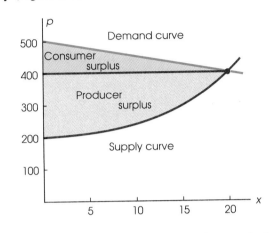

FIGURE 22

Solution We solve the two equations simultaneously by equating the two expressions for p.

$$\tfrac{1}{2}x^2 + 200 = 500 - 5x$$

$$\tfrac{1}{2}x^2 + 5x - 300 = 0$$

$$x^2 + 10x - 600 = 0$$

$$(x + 30)(x - 20) = 0$$

$$x = -30 \qquad x = 20$$

Only the positive answer $x = 20$ makes sense. The equilibrium price is $p = 500 - 5(20) = \$400$. ∎

In Figure 22b, we have indicated two quantities of economic significance, namely, consumer surplus and producer surplus. The **consumer surplus** is the total amount saved by consumers who are willing to pay a price higher than the equilibrium price and corresponds to the area of the region between the demand curve and the horizontal line through the equilibrium point. Similarly, the **producer surplus** is the increased revenue realized by producers who are willing to sell at a price lower than the equilibrium price. It is the area of the region between the same horizontal line and the supply curve.

EXAMPLE C

Find the consumer surplus CS and the producer surplus PS for the situation described in Example B.

Solution

$$CS = \int_0^{20} [(500 - 5x) - 400]\, dx$$

$$= \left[100x - \frac{5x^2}{2} \right]_0^{20} = \$1000$$

$$PS = \int_0^{20} \left[400 - \left(\frac{x^2}{2} + 200 \right) \right] dx$$

$$= \left[200x - \frac{x^3}{6} \right]_0^{20} \approx \$2667$$

Distribution of Income

While no country has been able to achieve equal distribution of income, a desirable goal in the eyes of many political scientists is to reduce the inequalities that exist. The table and corresponding graph (Figure 23) indicate how incomes in the United States were distributed in 1973 (*Source:* Paul Samuelson, *Economics* (10th Ed.), New York: McGraw-Hill, 1976). The left column of the table assumes that people have been ranked according to income.

Note in the table that the lowest paid 40% of the population receives only 18% of the income, whereas if income were absolutely equally distributed, they would get 40% of the income. In fact, if income were equally distributed, the data would lie along the 45-degree line. In actuality, the data always lie along a curve below this line, a curve called a **Lorenz curve.** The further this curve moves from the 45-degree line, the more unequal the in-

Cumulative percentage of people	Cumulative percentage of income
0	0
20	6
40	18
60	35
80	59
100	100

Income distribution: 1973

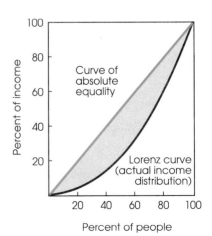

FIGURE 23

come distribution. As a measure of inequality, we use the so-called **Gini coefficient of inequality** g defined by

$$g = \frac{\text{area between the Lorenz curve and the line } y = x}{\text{area under the line } y = x}$$

The numerator is a definite integral whose value is between 0 and $\frac{1}{2}$; the denominator has a value of exactly $\frac{1}{2}$. Thus, g is between 0 (absolute equality) and 1 (absolute inequality). If $y = f(x)$ is the equation of the Lorenz curve, the formula for g takes the simple form

$$g = 2 \int_0^1 [x - f(x)] \, dx$$

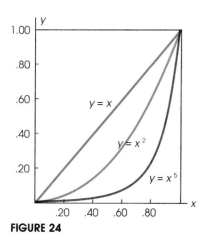

FIGURE 24

EXAMPLE D

Sketch the Lorenz curves $y = x^2$ and $y = x^5$ on the same graph and determine the corresponding Gini coefficients.

Solution The graphs are shown in Figure 24. The required Gini coefficients are

$$g = 2 \int_0^1 [x - x^2] \, dx = 2\left[\frac{x^2}{2} - \frac{x^3}{3}\right]_0^1 = \frac{1}{3}$$

and

$$g = 2 \int_0^1 [x - x^5] \, dx = 2\left[\frac{x^2}{2} - \frac{x^6}{6}\right]_0^1 = \frac{2}{3}$$

■

Problem Set 9.3

In doing Problems 1–4, keep in mind that $\int_a^b Q'(x) \, dx = Q(b) - Q(a)$. Thus, to find the net change in Q between a and b, we simply integrate its derivative between a and b.

1. The population (in hundreds) of a bacteria colony is changing at the rate of $6t^2 + 30t^{1/4} + 100$ per hour at time t. How much will the population increase between $t = 1$ and $t = 16$?

[c]2. Suppose that the value of a piece of property is growing at a rate of $1000t + 100e^{-t/4}$ dollars per year at time t. How much will its value increase between $t = 3$ and $t = 6$?

3. Hermantown has a population of 50,000 now, and it is predicted that x years from now its population will be increasing at the rate of $450 + 35x^{4/3}$ people per year. What is its expected population 8 years from now?

4. A research group has advised a midwestern stock exchange that over the next t months the number of shares traded will change at the rate of $2t + 3\sqrt{t} - 4$ million shares per month. Find the number of shares the stock exchange can expect to trade during the next 16 months.

5. A manufacturer's marginal cost (in dollars per unit) to make x units of a certain product is given by

$C'(x) = 30 - 0.012x$. If there are fixed costs of $2000, find (a) the total cost of making 150 units, (b) the average cost per unit of making 150 units.

[C]6. Suppose that the marginal cost (in dollars per unit) to manufacture x units of a product is given by $C'(x) = 300e^{-x/90}$ and that fixed costs are $8000. Find (a) the total cost of making 180 units, (b) the average cost per unit of making 180 units.

7. Suppose that the marginal revenue (in dollars per unit) for the manufacturer in Problem 5 from selling x units is given by $R'(x) = 50 + 0.36x - 0.003x^2$.
   (a) Find the total revenue from selling 150 units. [Note that $R(0) = 0$, since you get zero revenue from selling zero units.]
   (b) Find the total profit from making and selling 150 units.

[C]8. Suppose that the marginal revenue (in dollars per unit) for the manufacturer in Problem 6 from selling x units is given by $R'(x) = 500e^{-x/80}$.
   (a) Find the total revenue from selling 180 units.
   (b) Find the total profit from making and selling 180 units.

[C]9. Based on the information in Problems 5 and 7, what number of units will produce the maximum profit? *Hint:* We can expect the maximum for $P(x)$ to occur at a point where $P'(x) = 0$.

[C]10. Based on the information in Problems 6 and 8, what number of units will produce the maximum profit?

In Problems 11–16, find the equilibrium price corresponding to the given demand and supply functions $D(x)$ and $S(x)$ as in Example B.

11. $D(x) = 8 - \frac{1}{3}x$, $S(x) = \frac{1}{3}x$
12. $D(x) = 9 - \frac{1}{4}x$, $S(x) = \frac{1}{2}x$
13. $D(x) = 400 - \frac{1}{3}x^2$, $S(x) = \frac{1}{9}x^2$
14. $D(x) = 120 - \frac{1}{5}x^2$, $S(x) = \frac{1}{10}x^2$
15. $D(x) = 240 - 30x - 20x^2$, $S(x) = 20x^2 + 10x$
16. $D(x) = 720 - 180x - 10x^2$, $S(x) = 20x^2 + 120x$

17–22. For each of Problems 11–16, sketch the graphs of the demand and supply functions, showing the regions whose areas correspond to the consumer surplus and producer surplus. Then calculate the consumer surplus CS and the producer surplus PS (see Example C).

23. Suppose that the demand function for a certain commodity is $p = D(x) = 100 - 6\sqrt{x}$. How many units are demanded when the price is set at $28? Determine the consumer surplus if the equilibrium price is $28.

24. If the demand function for a certain product is $D(x) = 45e^{-x/30}$, find the consumer surplus when the equilibrium price is $15.

25. If the supply function for a certain type of jacket is $S(x) = 10e^{0.02x}$ and the equilibrium price is $40, determine the producer surplus.

26. The supply function for ABD jogging shoes is $S(x) = 12x^{1/3} + 30$, and the equilibrium price is $66. Find the producer surplus.

In Problems 27–32, we give the equation of a Lorenz curve. Sketch this curve on the interval $0 \leq x \leq 1$ and then calculate the corresponding Gini coefficient.

27. $y = x^3$
28. $y = x^4$
29. $y = (x^2 + x)/2$
30. $y = (2x^2 + x)/3$
31. $y = (e^x - 1)/(e - 1)$
32. $y = (e^{2x} - 1)/(e^2 - 1)$

33. It is estimated that the rate of repair costs in dollars per mile for a city bus after it has been driven x miles is $0.12 + 0.01\sqrt[3]{x}$. Find the total repair costs for driving a bus an additional 61,000 miles if its odometer shows 64,000 miles now.

[C]34. A certain country is using oil at the rate $0.05e^{0.015t}$ billion barrels per year, where t is the number of years after January 1, 1985. At this rate, how many billion barrels of oil will it use during the decade of the 1990s?

35. If the total attendance at a racetrack during the first x days of the season was $\int_0^x (10,500 + 50t)\, dt$, on what day did the total attendance first surpass 150,000?

36. A factory is expected to discharge pollutants into a lake at the rate of $100t\sqrt{t^2 + 9}$ tons per year, t years after it goes into operation. How much will it discharge during the fifth year of operation?

37. A manufacturer produces small recreational vehicles. The demand function is $D(x) = 20,000 - 100x$, and the supply function is $S(x) = 8000 + 500x$, where x is the number of vehicles produced per month. Find (a) the equilibrium price and (b) the consumer surplus.

38. The manager of a paper company estimates that after x years of use a certain machine produces revenue at the rate of $40,000 - 60x^2$ dollars per year, but that the cost of running the machine (labor, depreciation, maintenance) is $9675 + 24x^2$ dollars per year.
   (a) How long will it be profitable to use the machine?
   (b) Find the total profit produced by this machine during its useful life.

9.4

CONTINUOUS MONEY STREAMS

We introduced the notion of continuous compounding of interest in Section 6.3. Recall that if P dollars is invested at annual rate r compounded continuously, then its value F at the end of t years is given by $F = Pe^{rt}$. Solving this equation for P in terms of F gives the equation

$$P = Fe^{-rt}$$

Written in this form, the equation tells us the present value P of F dollars that will be received t years from now, assuming that interest is compounded continuously at the annual rate r.

EXAMPLE A

Mary knows that she will need to replace her car in 3 years. How much would she have to put in the bank today at 8% interest compounded continuously in order to have the $12,000 she expects to need 3 years from now?

Solution We apply the boxed formula, remembering that the interest rate of 8% must be written in its decimal form.

$$P = 12,000e^{-0.08(3)} = 12,000e^{-0.24} = \$9439.53 \qquad \blacksquare$$

Continuous Money Streams

There are many situations in business and industry where it is useful to think of money as flowing continuously into an account. For example, the ABC Company plans to buy an expensive machine which it estimates will increase the company's net income by $10,000 per year. But this income won't come at the end of each year, it will come in dribbles throughout the year. As a model for what will happen, it is convenient to think of the machine as if it will produce income continuously. Our next example raises an important question for the company to answer.

EXAMPLE B

If the ABC Company expects the machine described above to last for 8 years and if money is considered to be worth 9% compounded continuously, what is the present value PV of the income stream that the machine will produce over its lifetime?

Solution Recall our basic advice: *slice, approximate, integrate.* Divide the time interval $0 \le t \le 8$ into little pieces with a typical piece at time t being of length dt (Figure 25). The income produced by the machine during this

FIGURE 25

short time interval is approximately $10,000 dt, and the present value of this piece of income is approximately

$$10,000 \, dt \, (e^{-0.09t})$$

Integrating this amount over the period $0 \leq t \leq 8$ should give us the present value of the whole income stream; that is,

$$PV = \int_0^8 10,000e^{-0.09t} \, dt = -\frac{10,000}{0.09} \int_0^8 e^{-0.09t} \, (-0.09) \, dt$$

$$= -\frac{10,000}{0.09} \left[e^{-0.09t} \right]_0^8 = -\frac{10,000}{0.09} (e^{-0.72} - 1) \approx \$57,027.53$$

The company can afford to pay anything less than this amount for the machine. ∎

Let us generalize this result. Suppose that the annual rate of income of an income stream is $R(t)$ at time t [in Example B, $R(t)$ was the constant 10,000]. Suppose further that this income stream will last over the period $0 \leq t \leq T$. If interest is at rate r compounded continuously, then the present value PV of this income stream is given by

$$PV = \int_0^T R(t)e^{-rt} \, dt$$

EXAMPLE C

Suppose the ABC Company in Example B takes a more optimistic view and estimates that the new machine will produce income at the rate $R(t) = 10,000 + 200t$. What is the present value of this income stream, again assuming interest at 9% compounded continuously and a lifetime of 8 years?

Solution

$$PV = \int_0^8 (10,000 + 200t)e^{-0.09t} \, dt$$

$$= \int_0^8 10,000e^{-0.09t} \, dt + \int_0^8 200te^{-0.09t} \, dt$$

We evaluated the first integral in Example B; the second can be evaluated using integration by parts with $u = 200t$ and $dv = e^{-0.09t} \, dt$. We obtain

$$PV = 57,027.53 + \left[-\frac{200}{0.09} te^{-0.09t} \right]_0^8 + \frac{200}{0.09} \int_0^8 e^{-0.09t} \, dt$$

$$= 57,027.53 - 8653.37 - \frac{200}{0.0081} \left[e^{-0.09t} \right]_0^8 = \$61,046.94 \quad ∎$$

Let us modify the problem in Example B in another way by assuming that the machine will last indefinitely. More generally, let us ask for the present value PV of a perpetual (meaning infinitely long) income stream which produces income at an annual rate $R(t)$, assuming that interest is at an annual rate r compounded continuously. The result in this case is

$$PV = \int_0^\infty R(t)e^{-rt}\, dt$$

Here, we face a new kind of integral, one with an infinite limit. Such integrals are called **improper integrals** and must be given a clear definition. We do this by defining

$$\int_0^\infty R(t)e^{-rt}\, dt = \lim_{T\to\infty} \int_0^T R(t)e^{-rt}\, dt$$

EXAMPLE D

Redo Example B assuming the machine will last indefinitely. In other words, find the present value PV of a perpetual income stream with an annual rate of \$10,000 per year, assuming interest is at 9% compounded continuously.

Solution

$$PV = \int_0^\infty 10{,}000e^{-0.09t}\, dt = \lim_{T\to\infty} \int_0^T 10{,}000e^{-0.09t}\, dt$$

$$= \lim_{T\to\infty} -\frac{10{,}000}{0.09}\left[e^{-0.09t}\right]_0^T = \lim_{T\to\infty} \frac{10{,}000}{0.09}(1 - e^{-0.09T})$$

$$= \frac{10{,}000}{0.09} = \$111{,}111.11 \qquad \blacksquare$$

Improper Integrals

Improper integrals play an important role in mathematics and its applications. For this reason, we now give a general discussion and illustrate with several examples. First, we give the appropriate definitions.

1. $\displaystyle\int_a^\infty f(x)\, dx = \lim_{b\to\infty} \int_a^b f(x)\, dx$

2. $\displaystyle\int_{-\infty}^b f(x)\, dx = \lim_{a\to-\infty} \int_a^b f(x)\, dx$

3. $\displaystyle\int_{-\infty}^\infty f(x)\, dx = \int_{-\infty}^0 f(x)\, dx + \int_0^\infty f(x)\, dx$

If these limits exist and are finite, we say the corresponding improper integrals **converge** and have the indicated values; otherwise, we say the given improper integrals **diverge.**

EXAMPLE E

Show that $\int_0^\infty e^{-3x}\,dx$ converges and has a value of $\frac{1}{3}$.

Solution

$$\int_0^\infty e^{-3x}\,dx = \lim_{b\to\infty} \int_0^b e^{-3x}\,dx = \lim_{b\to\infty}\left[-\frac{1}{3}e^{-3x}\right]_0^b$$

$$= \lim_{b\to\infty}\left(-\frac{1}{3}e^{-3b} + \frac{1}{3}\right) = \frac{1}{3}$$
∎

EXAMPLE F

Show that $\int_1^\infty \frac{1}{x^2}\,dx$ converges, but that $\int_1^\infty \frac{1}{x}\,dx$ diverges.

Solution

$$\int_1^\infty \frac{1}{x^2}\,dx = \lim_{b\to\infty}\int_1^b x^{-2}\,dx = \lim_{b\to\infty}\left[-\frac{1}{x}\right]_1^b = \lim_{b\to\infty}\left[-\frac{1}{b} + 1\right] = 1$$

Also,

$$\int_1^\infty \frac{1}{x}\,dx = \lim_{b\to\infty}\int_1^b \frac{1}{x}\,dx = \lim_{b\to\infty}[\ln b] = \infty,$$

which implies that $\int_1^\infty \frac{1}{x}\,dx$ diverges.
∎

EXAMPLE G

Show that the present value of a perpetual money stream at the rate of P dollars per year is P/r, assuming that interest is compounded continuously at an annual rate of r.

Solution

$$PV = \int_0^\infty Pe^{-rt}\,dt = \lim_{b\to\infty}\int_0^b Pe^{-rt}\,dt = \lim_{b\to\infty}\left[-\frac{P}{r}e^{-rt}\right]_0^b$$

$$= \lim_{b\to\infty}\left(-\frac{P}{r}e^{-rb} + \frac{P}{r}\right) = \frac{P}{r}$$
∎

EXAMPLE H

Harry (a business executive) is going through a divorce. His income from investments and salary is estimated at $100,000 per year and is expected to continue at this rate throughout his lifetime. The court wants to know the present value of this income stream in order to make a fair distribution of assets. A consultant recommended assuming Harry's income to be a perpetual money stream at 8% interest compounded continuously. What present value does this lead to?

Solution From Example G, we obtain

$$PV = 100,000/0.08 = \$1,250,000$$ ∎

Problem Set 9.4

In Problems 1–4, find the present value (the value at 0) of each payment (or set of payments) due at the indicated times, assuming that interest is compounded continuously at the given rate r. See Example A.

[c]1.
$r = 0.075$

				$2000	
0	1 yr	2	3	4	5

[c]2.
$r = 0.084$

					$9000
0	1 yr	2	3	4	5

[c]3.
$r = 0.091$

		$2400			$6200
0	1 yr	2	3	4	5

[c]4.
$r = 0.076$

	$3000			$5400	
0	1 yr	2	3	4	5

[c]5. To settle an estate, Janet has promised to pay her sister $8500 eight years from now. If she were to pay off this obligation right now, what should she pay, assuming interest is at 8.25% compounded continously?

[c]6. Karen is to pay me $34,000 eight years from now, and Robert is to pay me $40,000 ten years from now. Which person owes me more and how much more, assuming interest is at 8.5% compounded continuously?

[c]7. The XYZ Company plans to buy a new computer which will, according to estimates, produce additional income at the rate of $25,000 a year for 10 years. Find the present value of the income stream generated by the computer over 10 years, if interest is at 9.5% compounded continuously. See Example B.

[c]8. A farmer has decided to erect two new silos at a total cost of $80,000. He figures that the new silos will produce an income stream at the rate of $15,000 per year for the next 20 years. Determine the present value of this income stream to see if he has made a wise investment. Assume that interest is at 10.6% compounded continuously.

[c]9. Suppose that the XYZ Company in Problem 7 estimates that the new computer will produce additional income at the rate $R(t) = 25,000 + 350t$ dollars per year, t being the number of years from now. Assuming again an interest rate of 9.5% compounded continuously, determine the present value of the income stream the computer will generate over the next 10 years. See Example C.

[c]10. Suppose that the farmer in Problem 8 predicts more optimistically that the new silos will produce income at the rate $R(t) = 15,000 + 300t$ dollars per year, t years after the silos are built. Determine the present value of this income stream over the next 20 years, assuming again an interest rate of 10.6% compounded continuously.

Evaluate the improper integrals in Problems 11–16 (see Examples D–F). In Problems 15 and 16, you will need the fact that $\lim_{x \to \infty} xe^{-kx} = 0$ provided $k > 0$.

11. $\displaystyle\int_1^\infty t^{-3} \, dt$ 12. $\displaystyle\int_1^\infty 3t^{-4} \, dt$

13. $\displaystyle\int_0^\infty e^{-0.09t}\,dt$

14. $\displaystyle\int_0^\infty e^{-0.075t}\,dt$

15. $\displaystyle\int_0^\infty te^{-0.09t}\,dt$

16. $\displaystyle\int_0^\infty te^{-0.075t}\,dt$

The results for Problems 13–16 should prove helpful in doing problems 17–20.

17. A restaurant is remodeling its facilities based on the assumption that this remodeling will produce additional income at the rate of $8000 per year for an indefinitely long period. Determine the present value of this income stream if interest is at 9% compounded continuously. (See Examples D and G.)

18. A small department store is installing a new energy-efficient heating and air conditioning system which will, according to estimates, produce savings at the rate of $1200 per year for an indefinitely long period of time. If interest is at 7.5% compounded continuously, what is the present value of the savings effected by the new system.

C 19. Redo Problem 17 assuming that the additional income will be at the rate of $R(t) = 8000 + 100t$ dollars per year, t being the number of years after remodeling.

C 20. Redo Problem 18, assuming the savings will be at the rate $R(t) = 1200 + 40t$ dollars per year, t being the number of years after installation of the new equipment.

___

C 21. How much money should Charles Moyer invest today at 10% compounded continuously to have the $16,500 that he will need for a new car 6 years from now?

C 22. How much does Sarah actually owe today if she must make payments of $800 at the end of each of the next 4 years? Assume interest is at 10.8% compounded continuously.

C 23. Determine the present value of a continuous money stream which will produce income at the rate of $500 per year for the next 10 years. Assume interest is at 8% compounded continuously.

C 24. Redo Problem 23 assuming income is produced at the rate of $500 + 20t$ dollars per year for the next 10 years, t being the number of years from now.

C 25. Redo Problem 23 assuming income will be produced at the rate of $500 per year forever.

C 26. Redo Problem 23 assuming income will be produced at the rate of $500 + 20t$ dollars per year forever.

In Problems 27–32, determine whether the given improper integral converges or diverges and, if convergent, determine its value.

27. $\displaystyle\int_0^\infty e^{0.03x}\,dx$

28. $\displaystyle\int_1^\infty \frac{\ln x}{x}\,dx$

29. $\displaystyle\int_0^\infty xe^{-x^2}\,dx$

30. $\displaystyle\int_4^\infty \frac{4}{(x-3)^5}\,dx$

31. $\displaystyle\int_0^\infty \frac{x}{1+x^2}\,dx$

32. $\displaystyle\int_0^\infty \frac{x}{(1+x^2)^2}\,dx$

33. Show that $\displaystyle\int_1^\infty x^{-p}\,dx$ converges if $p > 1$ and find its value.

34. Show that $\displaystyle\int_0^\infty \frac{x}{(1+x^2)^p}\,dx$ converges if $p > 1$ and find its value.

35. J. P. Gorgan has made provision for a perpetual flow of money to Yoohoo University at an annual rate of $750,000 per year. How much money did he have to set aside to fund this flow if he was able to set up a trust paying 7.4% compounded continuously?

9.5*

CONTINUOUS PROBABILITY

We assume that the reader is familiar with the basic notion of **probability.** In particular, the probability of an event is a number between 0 and 1 which measures the chance of the event occurring in an experiment. Roughly speaking, it gives the percentage of times we expect the event to occur in a large number of trials of the experiment. We are interested in the properties of a continuous **random variable X,** by which we mean a variable that takes values (in a random manner) on an interval of the real line. For example, the weight **X** of people is a random variable which may lie anywhere on

FIGURE 26

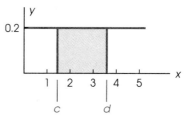

FIGURE 27

the interval, say, from 1 to 1000 pounds. We will be interested in calculating the probabilities of certain events, for example, prob $\{70 \leq \mathbf{X} \leq 270\}$.

As a particular example, consider the spinner device shown in Figure 26 which has the interval $[0, 5)$ wrapped around its circumference. Our experiment will consist of spinning the spinner, and $\mathbf{X}$ will be the coordinate of the stopping point. The probability of $\mathbf{X}$ taking values in two intervals of equal length should be the same, and the probability of its stopping somewhere in $[\,0, 5)$ is 1. This leads to the conclusion that if $0 \leq c < d < 5$, then prob $\{c \leq \mathbf{X} \leq d\} = (d - c)/5$. If we let $f(x) = \frac{1}{5}$ for $0 \leq x < 5$ and $f(x) = 0$ elsewhere, then (see Figure 27)

$$\text{prob } \{c \leq \mathbf{X} \leq d\} = \int_c^d f(x)\, dx$$

This example suggests some very general ideas.

Probability Density Functions

Associated with each continuous random variable $\mathbf{X}$ is a function f defined on $(-\infty, \infty)$ satisfying two properties:

1. $f(x)$ is nonnegative; that is, $f(x) \geq 0$ for all x.
2. The area under the graph of $y = f(x)$ is 1; that is,

$$\int_{-\infty}^{\infty} f(x)\, dx = 1$$

Moreover,

$$\text{prob } \{c \leq \mathbf{X} \leq d\} = \int_c^d f(x)\, dx$$

The function f is called the **probability density function** associated with the random variable $\mathbf{X}$. If $f(x)$ is a positive constant on an interval (and 0 otherwise) as in the spinner example above, then we say that $\mathbf{X}$ is **uniformly distributed** on that interval.

EXAMPLE A

Suppose that $\mathbf{X}$ is uniformly distributed on the interval $[1, 9]$. Determine the probability density function f and calculate prob $\{2.3 \leq \mathbf{X} \leq 3.9\}$.

FIGURE 28

Solution In order for the area under $y = f(x)$ to be 1, it must be that $f(x) = \frac{1}{8}$ on $[1, 9]$ and $f(x) = 0$ elsewhere (Figure 28). Thus,

$$\text{prob } \{2.3 \leq \mathbf{X} \leq 3.9\} = \int_{2.3}^{3.9} \frac{1}{8}\, dx = \frac{1}{8}(3.9 - 2.3) = 0.2 \qquad \blacksquare$$

EXAMPLE B

Let $f(x) = kx$ for x in $[1, 3]$ and $f(x) = 0$ elsewhere. Determine k so that f is a probability density function. Then calculate prob $\{2.1 \leq \mathbf{X} \leq 2.5\}$.

Solution The area under the graph of $y = f(x)$ must be 1, so

$$1 = \int_1^3 kx\, dx = \left[\frac{1}{2}kx^2\right]_1^3 = \frac{9}{2}k - \frac{1}{2}k = 4k$$

Thus, $k = \frac{1}{4}$ and

$$\text{prob } \{2.1 \leq \mathbf{X} \leq 2.5\} = \int_{2.1}^{2.5} \frac{1}{4}x\, dx = \left[\frac{1}{8}x^2\right]_{2.1}^{2.5}$$

$$= \frac{1}{8}[2.5^2 - 2.1^2] = 0.23 \qquad \blacksquare$$

EXAMPLE C

Papa's Ice Cream Shop guarantees that no customer will wait more than 3 minutes for service and that the time X one must wait for service has the probability density function $f(x) = \frac{1}{18}(9 - x^2)$ for x on $[0, 3]$ and $f(x) = 0$ elsewhere. Sketch the graph of f and show that it is a probability density function. Then calculate the probability that a customer will wait more than 2 minutes; that is, calculate prob $\{X \geq 2\}$.

FIGURE 29

Solution The graph of $y = f(x)$ is shown in Figure 29. We must demonstrate that the area under this curve is 1.

$$\int_0^3 \frac{1}{18}(9 - x^2)\, dx = \frac{1}{18}\left[9x - \frac{1}{3}x^3\right]_0^3 = \frac{1}{18}(27 - 9) = 1$$

Also,

$$\text{prob } \{X \geq 2\} = \int_2^3 \frac{1}{18}(9 - x^2)\, dx = \frac{1}{18}\left[9x - \frac{1}{3}x^3\right]_2^3$$

$$= \frac{1}{18}\left[27 - 9 - \left(18 - \frac{8}{3}\right)\right] = \frac{4}{27}$$ ∎

EXAMPLE D

A transportation engineer claims that the distance X (in feet) between cars on a certain freeway during rush hour has probability density function $f(x) = 0.01e^{-0.01x}$ on $(0, \infty)$ and $f(x) = 0$ otherwise. Sketch the graph of f, show that it is a density function, and calculate prob $\{X \leq 200\}$.

FIGURE 30

Solution The graph of f is shown in Figure 30. Note that this model allows for the distance (in feet) between cars to be any positive number but makes it likely that it is a small number. Now

$$\int_0^\infty 0.01e^{-0.01x}\,dx = \lim_{b\to\infty} \int_0^b 0.01e^{-0.01x}\,dx = \lim_{b\to\infty}\left[-e^{-0.01x}\right]_0^b$$

$$= \lim_{b\to\infty}\left[-e^{-0.01b}+1\right] = 1$$

which shows that f really is a density function. Finally,

$$\text{prob}\{X \le 200\} = \int_0^{200} 0.01e^{-0.01x}\,dx = \left[-e^{-0.01x}\right]_0^{200}$$

$$= -e^{-2}+1 \approx 0.86 \qquad\blacksquare$$

The Mean of a Random Variable

If we were to choose one number as most representative of the values of a continuous random variable X with density function f, what would it be? The most commonly used such number is the mean μ defined by

$$\mu = \int_{-\infty}^\infty x\,f(x)\,dx$$

Geometrically, the mean is the balance point of the distribution. To be specific, if a rigid wire of total mass 1 is constructed so that its density (mass per unit length) at the point x is $f(x)$, then the wire will balance on a fulcrum at μ (Figure 31). Thus, the mean of a uniform distribution is the midpoint of this distribution (for example, the mean of the random variable in Example A is 5, as we will soon check). On the other hand, one look at Figure 29 should convince you that the mean in Example C is to the left of 1.5. Our next example gives the exact results for each of the random variables so far considered.

FIGURE 31

EXAMPLE E

Find the mean of the random variable in (a) Example A, (b) Example B, (c) Example C, and (d) Example D.

Solution

(a) $\mu = \displaystyle\int_1^9 x\frac{1}{8}\,dx = \left[\frac{x^2}{16}\right]_1^9 = \frac{81}{16} - \frac{1}{16} = 5$

(b) $\mu = \displaystyle\int_1^3 x\left(\frac{1}{4}x\right)\,dx = \left[\frac{1}{12}x^3\right]_1^3 = \frac{27}{12} - \frac{1}{12} = \frac{13}{6}$

(c) $\mu = \displaystyle\int_0^3 x\left[\frac{1}{18}(9 - x^2)\right]\,dx = \frac{1}{18}\left[\frac{9}{2}x^2 - \frac{1}{4}x^4\right]_0^3 = \frac{9}{8}$

(d) $\mu = \displaystyle\int_0^\infty x(0.01e^{-0.01x})\,dx = \lim_{b \to \infty}\int_0^b x(0.01e^{-0.01x})\,dx$

To evaluate the latter integral, we use integration by parts with $u = x$ and $dv = 0.01e^{-0.01x}\,dx$. We obtain

$$\mu = \lim_{b \to \infty}\left[-xe^{-0.01x}\right]_0^b + \lim_{b \to \infty}\int_0^b e^{-0.01x}\,dx$$

$$= 0 + \lim_{b \to \infty}\left[-\frac{1}{0.01}e^{-0.01x}\right]_0^b = \frac{1}{0.01} = 100 \qquad ∎$$

Problem Set 9.5

1. Suppose that the random variable $\mathbf{X}$ is uniformly distributed on $[1, 5]$. See Example A.
   (a) Determine the probability density function f corresponding to $\mathbf{X}$.
   (b) Calculate prob $\{2.9 \le \mathbf{X} \le 4.1\}$.

2. Let $\mathbf{X}$ denote the result of picking a number at random from the interval $[0, 10]$. *Note:* The phrase "at random," when it is not qualified, is interpreted to mean that the corresponding random variable is uniformly distributed.
   (a) Determine the probability density function f for $\mathbf{X}$.
   (b) Calculate prob $\{|\mathbf{X} - 3| \le 2\}$.

3. A certain traffic light stays red for 30 seconds. You arrive at the light at random and find it red. Calculate the probability that you will have to wait at least 20 seconds before the light turns green.

4. Let $\mathbf{X}$ be the result of selecting a number at random on the interval $[-a, a]$. Calculate prob $\{|\mathbf{X}| > a/4\}$.

5. Let $f(x) = k(x + 1)$ for $0 \le x \le 4$ and $f(x) = 0$ elsewhere. See Example B.
   (a) Determine k so that f is a probability density function.
   (b) Calculate prob $\{\mathbf{X} > 3\}$.

6. Let $f(x) = k\sqrt{x}$ for $1 \le x \le 4$ and $f(x) = 0$ elsewhere.
   (a) Determine k so that f is a probability density function.
   (b) Calculate prob $\{2 \le \mathbf{X} \le 3\}$.

7. Let $f(x) = k/x$ on $[1, e^2]$ and $f(x) = 0$ otherwise.

(a) Show that $k = \frac{1}{2}$ makes a density function.

(b) Find prob $\{X \le 2\}$.

(c) Determine a so that prob $\{X > a\} = \frac{1}{2}$.

8. Let $f(x) = k/(x + 1)^2$ on $[0, 2]$ and $f(x) = 0$ elsewhere.

(a) Show that $k = \frac{3}{2}$ makes a density function.

(b) Calculate prob $\{X > 1\}$.

(c) Determine b so that prob $\{X \le b\} = \frac{1}{8}$.

9. A chow mein takeout shop claims that the time X one must wait for service has density function $f(x) = (36 - x^2)/144$ for $0 \le x \le 6$ and $f(x) = 0$ otherwise. Note that this implies that no customer will wait more than 6 minutes for service. See Example C.

(a) Show that f is a probability density function.

(b) Sketch the graph of f.

(c) Calculate the probability that a customer will wait at least 1 minute for service.

(d) Use your graph to guess at the mean waiting time and then calculate it exactly.

10. Under new management, the chow mein shop in Problem 9 was able to change the probability density function for X to $f(x) = 3(4 - x^2)/16$ for $0 \le x \le 2$ and $f(x) = 0$ otherwise.

(a) Now what is the maximum time any customer has to wait for service?

(b) Calculate the probability that a customer will wait at least 1 minute for service.

(c) Determine the mean waiting time.

11. A restaurant chain has determined that the fraction X of its waffle shops opened during any month that show a profit during the first 2 years of operation has probability density function $f(x) = 20x(1 - x)^3$ for $0 \le x \le 1$ and $f(x) = 0$ otherwise. Calculate the probability that a new shop just opening has at least a 60% chance of showing a profit during the first 2 years.

12. The cells in a certain cell population divide in a maximum of 6 days. The age X in days of a cell chosen at random from this population has probability density function $f(x) = 2ke^{-kx}$ for $0 \le x \le 6$ and $f(x) = 0$ elsewhere.

(a) Show that $k = (\ln 2)/6$.

(b) Calculate the probability that a cell is at least 4 days old.

13. A highway engineer claims that the number of minutes T between automobile accidents on a certain 10-mile stretch of highway has probability density function $f(t) = 0.02e^{-0.02t}$ for $t > 0$ and $f(t) = 0$ otherwise. See Example D.

(a) Calculate the probability that after an accident has occurred there won't be another for at least 40 minutes.

(b) Find the mean length of time between accidents.

14. The number of minutes T between successive telephone calls received at the emergency ward of a large hospital has probability density function $f(t) = 0.08e^{-0.08t}$ for $t > 0$ and $f(t) = 0$ otherwise.

(a) Calculate the probability that the time between calls is at least 1 minute.

(b) Find the mean length of time between calls.

15. Let $f(x) = kx(2 - x)$ on $[0, 2]$ and $f(x) = 0$ elsewhere.

(a) Determine k so that f is a probability density function.

(b) Calculate prob $\{X \le 0.3\}$.

(c) Find the mean of X.

16. Let $f(x) = k/x^3$ on $[1, \infty)$ and $f(x) = 0$ elsewhere.

(a) Determine k so that f is a density function.

(b) Find the mean.

17. A gasoline station can sell X thousand gallons of gasoline per week, where X has probability density function $f(x) = k/(x + 1)^2$ on $[2, \infty)$ and $f(x) = 0$ otherwise.

(a) Determine k.

(b) If the manager orders 4000 gallons at the beginning of a week, what is the probability that he will run out before the week is over?

18. Suppose that X is uniformly distributed on $[0, b]$ and that prob $\{X > 4\} = \frac{1}{3}$. What is b?

19. Let $f(x) = 6x^2$ on $[-a, a]$ and $f(x) = 0$ elsewhere. Determine a so that f is a density function.

20. Determine which of the following improper integrals converge and which diverge and, if convergent, give its value.

(a) $\displaystyle\int_{2}^{\infty} \frac{1}{\sqrt{x - 1}}\, dx$

(b) $\displaystyle\int_{e}^{\infty} \frac{1}{x(\ln x)^2}\, dx$

(c) $\displaystyle\int_{0}^{\infty} t^2 e^{-t^3}\, dt$

(d) $\displaystyle\int_{0}^{\infty} \frac{t}{2 + t^2}\, dt$

21. Let $f(x) = ke^{-kx}$ for $x > 0$ and $f(x) = 0$ elsewhere. Show that f is a probability density function for every $k > 0$ and determine the corresponding mean.

22. A dart is thrown at random at a circular dart board of radius 1. This means that the probability it will land in a region R is proportional to the area of R. Let X denote the distance from the center to where the dart lands.

(a) For $0 \le x \le 1$, determine prob $\{X \le x\}$.

(b) Determine the density function for X.

CHAPTER 9 REVIEW PROBLEM SET

1. Find the area of the region under the curve

$$y = 3x^2 + 2/(x + 1)$$

   between $x = 0$ and $x = 2$.

2. Find the area of the region under the curve $y = 4x^3 + 2x + e^{x/2}$ between $x = 0$ and $x = 2$.

3. Find the area of the shaded region in Figure 32.

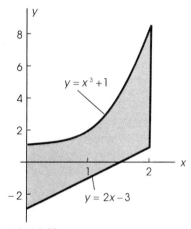

FIGURE 32

4. Find the area of the part of the shaded region in Figure 32 that is above the x-axis.

5. Find the area of the region bounded by the curve $y = 3 - x^2$ and the lines $y = 4 - x$, $x = -1$, and $x = 2$. Begin by sketching this region.

6. Sketch the region bounded by $y = e^x$ and the lines $y = 4$ and $x = 1$. Then find its area.

7. Find the area of the region bounded by $x = y^2 - y$ and $x = y$.

8. Sketch the region bounded by the parabola $x = 4y - y^2$ and the line $x + 2y = 5$. Then find its area.

9. Sketch the region R bounded by the parabola $y = x^2$ and the line $y = 4$. Then find the volume of the solid generated when R is revolved about (a) the x-axis, (b) the y-axis.

10. Sketch the region R bounded by the parabola $y = \sqrt{x}$ and the lines $y = 0$ and $x = 4$. Then find the volume of the solid generated when R is revolved about:
    (a) the x-axis   (b) the y-axis

11. Find the volume of the solid generated when the region bounded by the parabolas $x = y^2$ and $x = 4y - y^2$ is revolved about the y-axis.

12. Find the volume of the solid generated when the region bounded by the parabola $x = (y - 2)^2$ and the line $x + y = 4$ is revolved about the y-axis.

13. Grassway Company makes and sells lawnmowers. Its weekly marginal cost and weekly marginal revenue functions (in making and selling x lawnmowers) are given, respectively, by

$$C'(x) = 270 - 2x + 0.039x^2$$

$$R'(x) = 450$$

   both in dollars per mower. If the company has weekly fixed costs of $640, find:
   (a) the cost of making and selling 100 mowers per week.
   (b) the profit realized in making and selling 100 mowers per week.

14. A manufacturer of bicycles estimates the weekly marginal cost function and weekly marginal revenue function in making and selling x bicycles to be given by

$$C'(x) = 200 - 3x + 0.06x^2$$

$$R'(x) = 400 - 0.03x$$

   both in dollars per bicycle. If the weekly fixed costs are $450, find:
   (a) the cost of making and selling 80 bicycles per week;
   (b) the corresponding weekly profit.

15. Find the equilibrium price if the demand and supply functions (in thousands of dollars) for a certain product are given by $p = 15 + 19/(x + 1)$ and $p = \frac{1}{3}x + 10$, respectively.

16. Suppose that the demand and supply functions (in thousands of dollars) for a certain product are given by $y = 9 - x^2$ and $p = x^2 + 2$, respectively. Find the equilibrium price.

17. For the data in Problem 15, show graphically the regions whose areas correspond to the consumer surplus CS and the producer surplus PS. Then calculate the values of these quantities.

18. Graph the equations in Problem 16, indicate the regions corresponding to the consumer surplus and the producer surplus, and calculate these latter quantities.

19. Find the Gini coefficient for the Lorenz curve

$$y = \frac{1}{2}x^3 + \frac{1}{4}x + \frac{1}{4}x^{1/5}$$

20. Calculate the Gini coefficient for the Lorenz curve $y = 0.25x^4 + 0.75x^2$.

21. Flowingwell has a population of 10,000 right now,

and it has been predicted that x years from now its population will be increasing at the rate of $100 + 5\sqrt{x}$ people per year.
(a) What is its projected population 9 years from now?
(b) Find the average yearly increase in population over the 9-year period.

22. Let $P(t)$ denote the number of individuals in a bacteria colony t hours from now. Suppose the present population is 12,000 and it is growing at the rate $P'(t) = 9200t + 15,000$.
(a) Find the formula for $P(t)$.
(b) Calculate the size of the population 5 hours from now.
(c) Determine the average population during the initial 5-hour period.

C 23. To satisfy his creditors, John has agreed to make a payment of $3600 three years from now and a second payment of $2800 six years from now. What is John's present indebtedness, assuming that interest is at 8.8% compounded continuously?

C 24. Helen owes $8400 to her brother who charges her interest at the rate of 8.5% compounded semiannually. She has agreed to discharge her debt by making a payment of $4000 3 years from now and a second payment 7 years from now. Find the size of her second payment.

25. How much money must the McMaster Foundation put in trust now if it wishes to provide a perpetual money stream for Morningstar Children's Home at the rate of $50,000 per year and if interest is at 7.6% compounded continuously?

26. A foundation wishes to provide a perpetual money stream to the Golden Years Nursing Home at the rate of $75,000 per year. How much money must be put in trust now to provide this money stream if interest is projected at 8.3% compounded continuously?

27. Suppose that $f(x) = k\sqrt[3]{x}$ on $[0, 8]$ and $f(x) = 0$ elsewhere determines the probability density function for a random variable $\mathbf{X}$. Find:
(a) the value of k (b) prob $\{0 \le \mathbf{X} \le 1\}$
(c) the mean of $\mathbf{X}$

28. Let the probability density function for a random variable $\mathbf{X}$ be given by $f(x) = k(5x + 3)\sqrt{x}$ on $[0, 4]$ and $f(x) = 0$ elsewhere. Find: (a) k, (b) prob $\{1 \le \mathbf{X} \le 4\}$, and (c) the mean of $\mathbf{X}$.

29. The useful life $\mathbf{T}$ in years of a Maxipower vacuum cleaner is a random variable with density function $f(t) = \frac{1}{8}e^{-t/8}$ for $t > 0$ and $f(t) = 0$ otherwise.
(a) What fraction of these cleaners are usable after 8 years?
(b) Determine the mean of $\mathbf{T}$.

30. The time $\mathbf{T}$ in seconds for a rat to learn a certain be-

havior is a random variable with probability density function $f(t) = 0.03e^{-0.03t}$ for $t > 0$. Calculate the probability that a rat will learn this behavior in 2 minutes or less.

31. Consider the shaded region R shown in Figure 33. Write a definite integral which represents
(a) the area of R.
(b) the volume of the solid generated when R is revolved about the x-axis.

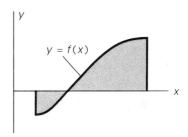

FIGURE 33

32. Sketch the region R bounded by the curve $y = 2\sqrt[3]{x} - 3$ and the x-axis between $x = 2$ and $x = 11$.
(a) Determine the area of R.
(b) Find the volume of the solid generated when R is revolved about the x-axis.

33. The function $f(x) = 27x^2/(1 + x^3)^2$ on $[a, \infty)$ and $f(x) = 0$ elsewhere is a probability density function. Determine a.

34. Determine k so that $f(x) = kx/(1 + x^2)^2$ is the probability density function for a random variable $\mathbf{X}$ on $[0, \infty)$. Then find the *median value* of $\mathbf{X}$, that is, find a so that prob $\{0 \le \mathbf{X} \le a\} = 0.5$.

35. The work done by a constant force of F pounds in moving an object along a line through a distance d feet is defined to be Fd foot-pounds. Suppose that the variable force required to move an object is $F(x)$ pounds at the point with coordinate x. Determine the formula for the work done in moving the object along a coordinate line from $x = a$ to $x = b$. *Hint:* Slice, approximate, integrate.

36. Suppose that the formula for $F(x)$ in Problem 35 is $F(x) = 5x^{2/3}$. Determine the work done in moving the object from $x = 1$ to $x = 8$.

37. $\displaystyle \lim_{n\to\infty} \frac{3}{n} \sum_{i=1}^{n} \left[1 + f\left(\frac{3i}{n}\right) \right]^2$ corresponds to what definite integral?

38. Evaluate $\displaystyle \lim_{n\to\infty} \frac{2}{n} \sum_{i=1}^{n} \sqrt{5 + \left(\frac{2i}{n}\right)^2} \left(\frac{2i}{n}\right)$ by recognizing this limit as a definite integral and then evaluating this integral.

George Dantzig solved the problem of finding an optimum way of allocating resources by constructing a convex polytope model. His simplex algorithm involves moving from vertex to vertex (ABC . . . M) improving the allocation at each step until an optimum is achieved. The model shown is 3-dimensional; real applications may require thousands of dimensions.

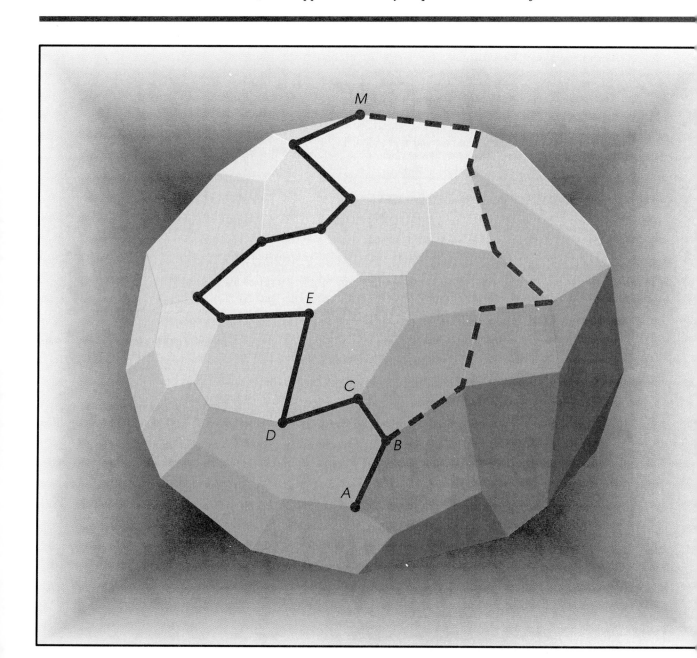

10

Multivariate Calculus

All workers in the fields of management science and operations research know the name George Dantzig. In fact, for all practical purposes, Dantzig founded these fields. The story is one of the most exciting in recent mathematical history.

George, with good mathematical genes inherited from his father, Tobias Dantzig, and a fresh Ph.D. from the University of California at Berkeley, was working for the the planning division of the U.S. Air Force. As anyone who has thought about it realizes, it takes fantastic coordination of supplies, equipment, and personnel to make an operation as big as the Air Force run efficiently. Dantzig formulated the Air Force's planning problem as a *linear program*, that is, as the problem of optimizing a linear function subject to linear constraints. Unfortunately, the problem involved thousands of variables and therefore seemed to be intractable. But Dantzig proceeded to invent an algorithm, the so-called *simplex method*, which offered a clear step-by-step procedure for solving the problem. This method is hardly feasible for hand calculations, but another exciting event occurred almost simultaneously—the invention of the modern high-speed computer. With Dantzig's simplex method and a computer, linear programs with as many as 15,000 variables could be handled with relative ease.

Of course, all the large businesses in the United States face problems similar to the Air Force. Soon they were applying Dantzig's methods to their problems and thereby saving millions of dollars. Perhaps no other mathematical process has so captured the fancy of the corporate world.

But the story is not finished. Many problems of interest are nonlinear in character. Dantzig's early success has attracted a host of workers to these harder problems, and notable advances have been made often with Dantzig himself playing a key role. In 1984, a young Indian-born mathematician, Nerendra Karmarkar, provided the newest breakthrough, a method even better than the simplex method for linear programs with a large number of variables.

George Bernard Dantzig
(1914–   )

FUNCTIONS OF TWO OR MORE VARIABLES

So far, our interest has focused on functions of one variable, that is, functions whose rule of correspondence has the form $y = f(x)$. We have evaluated such functions, graphed them, differentiated them, and integrated them. The restriction to functions of one variable is very limiting, since most problems that arise in the social sciences involve many variables. In particular, problems in economics often involve hundreds or even thousands of variables. It is time for us to lift our sights. Fortunately, all of calculus can be studied in the context of functions of several variables.

Examples of Functions of Several Variables

Our first task is to clarify the language and notation. By a function f of two variables x and y, we mean a rule that associates with each ordered pair (x, y) of numbers in some set (called the domain) a definite number $f(x, y)$. The rule is usually given by an equation. For example, the equation $f(x, y) = x^2 + y$ associates with the ordered pair (x, y) the number $x^2 + y$. Thus,

$$f(2, 3) = 2^2 + 3 = 7$$

$$f(-4, 5) = (-4)^2 + 5 = 21$$

and

$$f\left(\frac{1}{a}, b^2\right) = \left(\frac{1}{a}\right)^2 + b^2 = \frac{1 + a^2b^2}{a^2}.$$

There is no conceptual reason for limiting our discussion to functions $f(x, y)$ of two variables. We can and occasionally will consider a function $f(x, y, z)$ of three variables or, more generally, a function $f(x_1, x_2, \ldots, x_n)$ of n variables. All that we do is valid in the most general context. However, it will simplify life for everyone if we center most of our discussion around cases with two and three variables.

EXAMPLE A

Let $f(x, y, z) = x^2y/z$. Find and simplify:

(a) $f(3, 2, 12)$,   (b) $f(-2, -2, -2)$,
(c) $f(x, x, 1/x)$,   (d) $f(x + h, y, z) - f(x, y, z)$

Solution

(a) $f(3, 2, 12) = (3)^2(2)/12 = 18/12 = 3/2$
(b) $f(-2, -2, -2) = (-2)^2(-2)/(-2) = 4$
(c) $f\left(x, x, \dfrac{1}{x}\right) = \dfrac{x^2x}{1/x} = x^4$

(d) $f(x + h, y, z) - f(x, y, z) = \dfrac{(x + h)^2 y}{z} - \dfrac{x^2 y}{z}$

$$= \dfrac{(x^2 + 2xh + h^2)y - x^2 y}{z}$$

$$= \dfrac{2xhy + h^2 y}{z} \qquad \blacksquare$$

EXAMPLE B

The material for the top and bottom of a cylindrical can costs 0.05¢ per square inch, while the material for the curved sides costs 0.03¢ per square inch. Let $C(r, h)$ denote the cost of materials for a can of radius r inches and height h inches. Find a formula for $C(r, h)$ and evaluate it when $r = 2$ and $h = 6$.

Solution The top and bottom are circles of area πr^2, and the curved portion has area $2\pi rh$ (Figure 1). Thus,

$$C(r, h) = 2\pi r^2 (0.05) + 2\pi rh (0.03)$$

$$\approx 0.314159 r^2 + 0.188496 rh$$

and

$$C(2, 6) = 2\pi (4)(0.05) + 2\pi (2)(6)(0.03) \approx 3.52¢ \qquad \blacksquare$$

πr^2

$2\pi rh$

πr^2

FIGURE 1

The costs of a manufacturing process are divided between labor costs L and capital costs K (raw materials, equipment, and so on). In many cases, economists find that a **Cobb-Douglas production function** $F(L, K) = cL^a K^{1-a}$ provides a good model for the number of units that will be produced (c and a are constants, $0 < a < 1$).

EXAMPLE C

A certain manufacturing process is believed to have the production function

$$F(L, K) = 300 L^{1/3} K^{2/3}$$

(a) How many units of output will result from inputs of 64 units of labor and 27 units of capital?

(b) Which would increase output more, doubling the number of units of labor or doubling the number of units of capital?

Solution

(a) $F(64, 27) = 300(64)^{1/3}(27)^{2/3} = 300(4)(9) = 10{,}800$

(b) $F(128, 27) = 300(128)^{1/3}(27)^{2/3} = 13{,}607$
$F(64, 54) = 300(64)^{1/3}(54)^{2/3} = 17{,}144$

The calculation shows that output is increased more by doubling the input of capital than by doubling the input of labor. This would be true at any production level, since doubling capital multiplies output by $2^{2/3} \approx 1.5874$, whereas doubling labor multiplies the output by only $2^{1/3} \approx 1.2599$.   ■

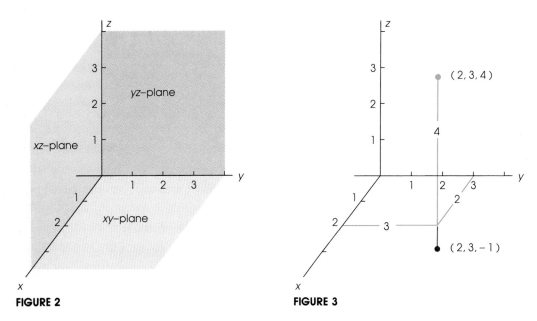

FIGURE 2   **FIGURE 3**

Three-Dimensional Space

To graph $y = f(x)$ required the Cartesian plane (two-dimensional space); to graph $z = f(x, y)$ will require Cartesian three-dimensional space. In space, introduce three mutually perpendicular axes, each scaled with the real numbers and intersecting at their common zero point called the origin. We will almost always do this so the x- and y-axes are in the horizontal plane, with the x-axis pointing forward, the y-axis pointing to the right, and the z-axis pointing straight up (Figure 2). The three axes determine the three coordinate planes (the xy-, xz- and yz-planes). Then every point in space gets a three-number label (x, y, z) called its coordinates and measuring its directed distances from these three planes (Figure 3).

Graphs of Functions of Two Variables

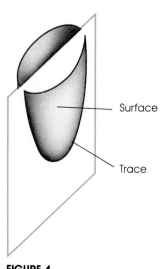

Surface

Trace

FIGURE 4

The graph of the function $f(x)$ of one variable consists of all points (x, y) in the plane satisfying $y = f(x)$; in general, it is a curve. The graph of the function $f(x, y)$ of two variables consists of all points (x, y, z) in space satisfying $z = f(x, y)$; it is a surface. To sketch the graph of a surface, we usually begin by finding its **traces** in various planes, especially the three coordinated planes. A trace is just the intersection of the surface with a plane and is therefore a curve (Figure 4).

FIGURE 5

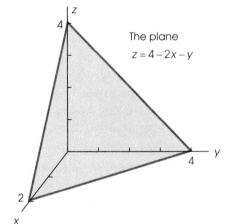

FIGURE 6

EXAMPLE D

Sketch the graph of $z = f(x, y) = 4 - 2x - y$.

Solution Since the variables all occur to the first power, you might guess that the graph will be a plane, the natural generalization of a line. This guess is correct. To sketch the graph, we first sketch the traces in the three coordinate planes. For example, to find the trace in the xz-plane, set $y = 0$ in the given equation, obtaining the linear equation $z = 4 - 2x$. The traces in the yz-plane and xy-plane are obtained in similar fashion by setting $x = 0$ and $z = 0$, respectively. These three traces already suggest the required plane (Figure 5); those with artistic talents can provide a more vivid picture by the use of shading (Figure 6). Of course, we can show only a small part of the plane, since it stretches infinitely far. We choose to show the part of the plane in the first octant (where all coordinates are positive). ∎

EXAMPLE E

Sketch the graph of $z = f(x, y) = x^2 + y^2$.

Solution The traces in the xz-plane and yz-plane are the graphs of $z = x^2$ and $z = y^2$, respectively. The trace in the xy-plane $(z = 0)$ reduces to a single point, the origin. The trace in the plane $z = 4$ is the circle $x^2 + y^2 = 4$. All this is shown in Figures 7 and 8. Again, we show only the first octant part of the surface, a surface called a circular paraboloid. The complete surface stretches back behind the xz- and yz-planes and infinitely high up in the z direction. ∎

FIGURE 7

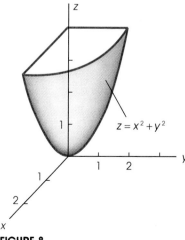

FIGURE 8

Computer Graphics

With a computer and appropriate software, it is possible to display excellent graphs of some very complicated functions. Figure 9 shows two computer-drawn surfaces. Since we do not assume that all our readers have the required equipment at their fingertips, we will place our emphasis on hand-drawn graphs. It will help to have a little artistic ability.

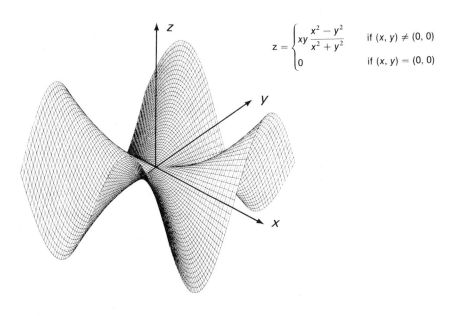

$$z = \begin{cases} xy\,\dfrac{x^2 - y^2}{x^2 + y^2} & \text{if } (x, y) \neq (0, 0) \\ 0 & \text{if } (x, y) = (0, 0) \end{cases}$$

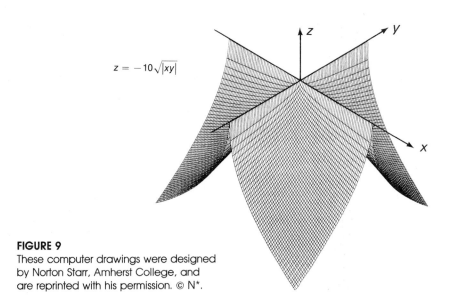

$$z = -10\sqrt{|xy|}$$

FIGURE 9
These computer drawings were designed
by Norton Starr, Amherst College, and
are reprinted with his permission. © N*.

Problem Set 10.1

In Problems 1–6, find and simplify (a) $f(1, 4)$, (b) $f(2, 2)$, (c) $f(x, 2x)$, (d) $f(x + h, y) - f(x, y)$. *See Example A.*

1. $f(x, y) = x^2 - xy + y^2$
2. $f(x, y) = x^2 + 4\sqrt{y}$
3. $f(x, y) = x/y^4$
4. $f(x, y) = (x + 2y)/(x - 3y)$
5. $f(x, y) = (3x + 1)e^{y-4}$
6. $f(x, y) = x^2 \ln (xy - 3) + y^2 - 5$

In Problems 7–12, find and simplify (a) $f(1, 0, 1)$, (b) $f(3, 2, 6)$ (c) $f(-2z, z, z)$, (d) $f(z, y, x)$

7. $f(x, y, z) = 2xy + yz - 3xz$
8. $f(x, y, z) = x^2(y + z) - 2xyz$
9. $f(x, y, z) = (xy - 2z)/(yz - 2x)$
10. $f(x, y, z) = \sqrt{x^2 + y^2 + z^2}$
11. $f(x, y, z) = (x + 2y)(y + 2z)(z + 2x)$
12. $f(x, y, z) = (y^2 + z)e^{x-2y+4z}$

13. For a rectangle of dimensions x and y, let $f(x, y)$ denote its perimeter and $g(x, y)$ its area.
    (a) Write explicit expressions for $f(x, y)$ and $g(x, y)$.
    (b) Calculate $f(6, 2)$ and $g(6, 2)$.
    (c) If the perimeter and area of a rectangle have the same numerical value, how is y related to x?

14. The picture in Figure 10 shows a garden with dimensions x and y, surrounded by a sidewalk of width z.
    (a) Write explicit expressions for $f(x, y, z)$, the combined area of the garden and the walk, and $g(x, y, z)$, the area of the walk.
    (b) Calculate $f(12, 8, 2)$ and $g(12, 8, 2)$.
    (c) Suppose that $x = 24$ and $y = 10$. Find z so that the garden and the walk have the same area.

FIGURE 10

15. Let the dimensions of a closed rectangular box with a square base be x by x by y (all in feet).
    (a) Write explicit expressions for its volume $V(x, y)$ and surface area $S(x, y)$.
    (b) What happens to $V(x, y)$ if we double x and halve y?

(c) If the box is made of thin material costing 18¢ per square foot for the sides and 30¢ per square foot for the top and bottom, find an explicit expression for $C(x, y)$, the total cost (in dollars) of material for the box.

16. Consider a rectangular box with base x by y and height z (all in centimeters).
    (a) Write explicit expressions for its volume $V(x, y, z)$ and surface area $S(x, y, z)$.
    (b) Suppose the box is made of thin material costing 30¢ per square centimeter for the top, 25¢ per square centimeter for the sides, and 50¢ per square centimeter for the bottom. Write an explicit expression $C(x, y, z)$ for the cost (in dollars) of materials needed to make the box.

17. Let $U = F(L, K) = 420L^{1/4}K^{3/4}$ be a production function.
    (a) Calculate $F(16, 81)$ and $F(81, 16)$.
    (b) What is the effect on U of doubling L? Of doubling K? Of doubling both L and K?

☐18. If $F(L, K) = 285L^{0.78}K^{0.22}$, calculate
    (a) $F(200, 200)$ (b) $F(100, 300)$ (c) $F(300, 100)$.

19. Consider the plane whose equation is

$$z = f(x, y) = 6 - 3x - 2y$$

    (a) Which of the points $(4, 1, -8)$, $(3, 0, 3)$, $(2, -6, 12)$ are in this plane?
    (b) Find the equations of the traces in the three coordinate planes.
    (c) Sketch the graphs of the three traces found in (b) and then the graph of the corresponding plane.

20. Consider the plane with equation $z = 18 - 6x - 2y$.
    (a) Find the equations of the traces in the three coordinate planes.
    (b) Sketch the graphs of these three traces and then the graph of the given plane.

To find the x-intercept of the graph of the equation $z = f(x, y)$, set $y = z = 0$ and solve for x in the equation. Similiarly, to find the y-intercept, set $x = z = 0$ and solve for y. Thus the x, y, and z-intercepts for the plane $z = 12 + 4x - 3y$ are $x = -3$, $y = 4$, and $z = 12$. The graph of the plane $z = 4x - 8$ has no y-intercept, since setting $x = z = 0$ gives $0 = -8$. Thus this plane never crosses the y-axis but is parallel to it. Finding intercepts can be a great aid in graphing, especially for planes. In Problems 21–26, find the intercepts for the given plane and use this information to help you sketch its graph.

21. $z = 5 - x - y$

22. $z = 6 - 3x - y$

23. $z = 8 - 2x + y$

24. $z = 6 - 6y + x$

25. $z = 8 - 4x$

26. $z = 10 - 2y$

In Problems 27–36, sketch the graph of the given equation. First find and sketch the traces (if they exist) in the three coordinate planes. In some cases, it may be wise to find traces in other planes. For example, the graph in Problem 29 has no trace in the xy-plane ($z = 0$), so you might consider its traces in the planes $z = 4$ and $z = 9$ which are circles. See Example E.

27. $z = x^2 + y^2 - 9$

28. $z = x^2 + y^2 - 16$

29. $z = x^2 + y^2 + 4$

30. $z = x^2 + y^2 + 1$

31. $4z = x^2 + y^2$

32. $z = 4(x^2 + y^2)$

33. $z = 9 - x^2 - y^2$

34. $z = -x^2 - y^2 - 1$

35. $z = 4x^2 + y^2 - 4$

36. $z = x^2 + \frac{1}{9}y^2$

37. A shoe manufacturer produces shoes and boots for men. Shoes sell at \$60 a pair and boots at \$75 a pair. Labor and materials cost \$30 per pair for shoes and \$45 per pair for boots, and there are fixed weekly costs of \$2000. Write explicit expressions for the total weekly profit $P(x, y)$ from producing x pairs of shoes and y pairs of boots.

38. A candy store sells milk chocolate at p dollars per pound and fudge at q dollars per pound. The corresponding demand equations for chocolate and fudge in pounds per day are $x = 300 - 2p + 5q$ and $y = 400 + 6p - 2q$, respectively.
    (a) Write an explicit expression for the daily revenue function $R(p, q)$.
    (b) Calculate $R(6, 7.5)$ and $R(5, 8)$.

39. The stopping distance $S = S(w, r)$ for a car after the brakes are applied is directly proportional to its weight w and the square of its speed r (that is, $S = kwr^2$ for some constant k). A car weighing 1800 pounds and traveling 40 miles per hour will stop in 86.4 feet. Determine an explicit expression for $S(w, r)$ and use it to find the stopping distance for a car weighing 2400 pounds and traveling 60 miles per hour.

40. A silo has the shape of a right circular cylinder topped with a hemisphere. Suppose that the cylinder has radius r amd height h. Write an explicit expression for the volume $V = V(r, h)$ of the silo.

41. Sketch the graphs of the following equations.
    (a) $x = 8 - 2y - 4z$ (b) $x = y^2 + z^2$
    (c) $z = \sqrt{25 - x^2 - y^2}$

42. Each of the following equations has one of the variables x, y, or z missing. This means that the graph in three-dimensional space will be parallel to the axis of the missing variable. Sketch the graph of each equation in three-space.
    (a) $3x + y = 9$ (b) $x + 4z = 8$
    (c) $z = x^2$ (d) $y = (x - 1)^2$

10.2
PARTIAL DERIVATIVES

Let $y = f(x)$ determine a function of one variable. Then the derivative dy/dx measures the rate of change in y with respect to x. Let $z = f(x, y)$ determine a function of two variables. Then $\partial z/\partial x$ denotes the rate of change in z with respect to x keeping y fixed, and $\partial z/\partial y$ denotes the rate of change in z with respect to y keeping x fixed. These new derivatives, $\partial z/\partial x$ and $\partial z/\partial y$, are called the **partial derivative of z with respect to x** and the **partial derivative of z with respect to y,** respectively. Note the use of ∂ rather than d to distinguish partial derivatives from ordinary derivatives. Fortunately, calculation of partial derivatives involves no new ideas. Simply treat y as a constant in the calculation of $\partial z/\partial x$, and treat x as a constant in the calculation of $\partial z/\partial y$.

EXAMPLE A

Let $z = 2xy^3 + 5x^4 - 3y^5$. Find $\partial z/\partial x$ and $\partial z/\partial y$.

Solution In calculating $\partial z/\partial x$, we treat y as a constant and obtain (see SHARP CURVE)

$$\frac{\partial z}{\partial x} = 2y^3 + 20x^3 - 0$$

Similarly, in calculating $\partial z/\partial y$, we treat x as a constant and get

$$\frac{\partial z}{\partial y} = 6xy^2 + 0 - 15y^4$$

EXAMPLE B

For $w = e^{xy} + xyz^2$, find $\partial w/\partial x$, $\partial w/\partial y$, and $\partial w/\partial z$.

Solution Note that w is a function of the three variables x, y, and z. Thus, $\partial w/\partial x$ denotes the derivative of w with respect to x holding both y and z fixed, that is, treating y and z as constants. Similar understandings apply to $\partial w/\partial y$ and $\partial w/\partial z$. We obtain (using the Chain Rule on e^{xy})

$$\frac{\partial w}{\partial x} = e^{xy}\frac{\partial}{\partial x}(xy) + yz^2 = ye^{xy} + yz^2$$

$$\frac{\partial w}{\partial y} = e^{xy}\frac{\partial}{\partial y}(xy) + xz^2 = xe^{xy} + xz^2$$

$$\frac{\partial w}{\partial z} = 0 + 2xyz$$

EXAMPLE C

For $z = x^2 \ln(x/y)$, find $\partial z/\partial x$ and $\partial z/\partial y$.

Solution First, we write $z = x^2 \ln x - x^2 \ln y$. Then for $\partial z/\partial x$, we use the Product Rule on the first term.

$$\frac{\partial z}{\partial x} = x^2\frac{1}{x} + 2x \ln x - 2x \ln y$$

$$= x + 2x \ln\frac{x}{y}$$

$$\frac{\partial z}{\partial y} = 0 - x^2\frac{1}{y} = -\frac{x^2}{y}$$

MULTIVARIATE CALCULUS

More Notation

Recall that for a function f of one variable x, we have two basic notations for the derivative, namely, df/dx and $f'(x)$. For a function f of two variables x and y, the prime notation $f'(x, y)$ would be confusing, since it does not indicate which of x and y is to be treated as a variable and which as a constant. Rather, we use $f_x(x, y)$ for $\partial f/\partial x$ and $f_y(x, y)$ for $\partial f/\partial y$. Thus, we have the following analagous notations.

Functions of one variable	Functions of two variables
$f'(x)$ or $\dfrac{df}{dx}$	$f_x(x, y)$ or $\dfrac{\partial f}{\partial x}$
$f'(x_0)$ or $\left(\dfrac{df}{dx}\right)_{x=x_0}$	$f_x(x, y_0)$ or $\left(\dfrac{\partial f}{\partial x}\right)_{y=y_0}$
	$f_x(x_0, y_0)$ or $\left(\dfrac{\partial f}{\partial x}\right)_{x=x_0,\, y=y_0}$

EXAMPLE D

For $f(x, y) = xy^2 + xe^{xy} + x^2/2$, find (a) $f_x(x, y)$, (b) $f_x(x, 0)$, (c) $f_x(0, 2)$, (d) $f_y(3, 1)$.

Solution

(a) $f_x(x, y) = y^2 + [x(ye^{xy}) + e^{xy}(1)] + x = y^2 + xye^{xy} + e^{xy} + x$
(b) $f_x(x, 0) = 0 + 0 + e^0 + x = 1 + x$
(c) $f_x(0, 2) = 4 + 0 + e^0 + 0 = 5$
(d) $f_y(x, y) = 2xy + x(xe^{xy}) + 0 = 2xy + x^2e^{xy}$, and so
$\qquad f_y(3, 1) = 6 + 9e^3$ ∎

Geometric Interpretation

You will recall that $f'(x_0)$ is the slope of the tangent line to the curve $y = f(x)$ at the point x_0 (Figure 11). In a similar fashion $f_x(x_0, y_0)$ is the

FIGURE 11

FIGURE 12

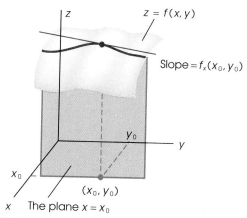

FIGURE 13

slope at (x_0, y_0) of the tangent line to the trace of the surface $z = f(x, y)$ in the plane $y = y_0$ (Figure 12). Likewise, $f_y(x_0, y_0)$ is the slope at (x_0, y_0) of the tangent line to the trace of the surface in the plane $x = x_0$ (Figure 13).

EXAMPLE E

Jack and Jill ran up the hill $z = f(x, y) = e^{-x^2-y^2}$, running directly above the x-axis (Figure 14). What was the grade of their path at the point $(1, 0)$? Distances are in miles.

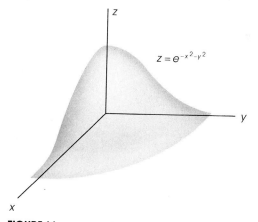

FIGURE 14

Solution The grade is just the absolute value of the slope but is given as a percentage (see end of Section 1.3). Since

$$f_x(x, y) = -2xe^{-x^2-y^2}$$

we conclude that the slope $m = f_x(1, 0) = -2e^{-1} \approx -0.73576$. Thus, the grade is about 74%. That is a very steep grade, and it is no wonder that Jack fell down and broke his crown.

Rate of Change Interpretation

Suppose that $w = f(x, y)$ and that $f_x(5, 6) = 10$ and $f_y(5, 6) = 2$. Then at the point $(5, 6)$, w is changing 10 times as fast as x but only 2 times as fast as y. Rephrased another way, if x is increased by 1 unit, w will increase by about 10 units; if y is increased by 1 unit, w will increase by about 2 units.

EXAMPLE F

Let $W = f(L, K) = 90L^{2/3}K^{1/3}$ be the production function for a manufacturing process, with W, L, and K denoting the number of units of goods produced, labor used, and capital required, respectively. Calculate $f_L(8, 27)$ and $f_K(8, 27)$ and interpret the meaning of these numbers.

Solution $f_L(L, K) = 60L^{-1/3}K^{1/3}$ and $f_K(L, K) = 30L^{2/3}K^{-2/3}$, so $f_L(8, 27) = (60)(\frac{1}{2})(3) = 90$ and $f_K(8, 27) = (30)(4)(\frac{1}{9}) \approx 13.3$. Thus, an increase in labor of 1 unit (from 8 to 9 while keeping K fixed at 27) will increase production by about 90 units. On the other hand, an increase in capital of 1 unit (from 27 to 28 while keeping L fixed at 8) will increase production by about 13.3 units. The partial derivatives f_L and f_K are called the *marginal productivity of labor* and the *marginal productivity of capital*, respectively. ∎

Higher-Order Derivatives

The partial derivative $f_x(x, y)$ is itself a function, and so we may take its partial derivative with respect to either x or y. We use the following notation.

Second-Order Partial Derivatives

Second partial of f with respect to x	$\dfrac{\partial^2 f}{\partial x^2} = \dfrac{\partial}{\partial x}\left(\dfrac{\partial f}{\partial x}\right)$	or	$f_{xx} = (f_x)_x$
Second partial of f with respect to y	$\dfrac{\partial^2 f}{\partial y^2} = \dfrac{\partial}{\partial y}\left(\dfrac{\partial f}{\partial y}\right)$	or	$f_{yy} = (f_y)_y$
Mixed second partials of f	$\dfrac{\partial^2 f}{\partial y\,\partial x} = \dfrac{\partial}{\partial y}\left(\dfrac{\partial f}{\partial x}\right)$	or	$f_{xy} = (f_x)_y$
	$\dfrac{\partial^2 f}{\partial x\,\partial y} = \dfrac{\partial}{\partial x}\left(\dfrac{\partial f}{\partial y}\right)$	or	$f_{yx} = (f_y)_x$

Higher-order partial derivatives like f_{xxy} and f_{xyy} are defined in an analagous way, but we shall have little use for them.

EXAMPLE G

For $f(x, y) = xy^3 + yxe^x$, find (a) f_{xx}, (b) f_{xy}, (c) f_{yx}, (d) f_{yy}.

Solution First note that

$$f_x(x, y) = y^3 + y(xe^x + e^x) = y^3 + yxe^x + ye^x$$

and

$$f_y(x, y) = 3xy^2 + xe^x$$

(a) $f_{xx}(x, y) = 0 + y(xe^x + e^x) + ye^x = yxe^x + 2ye^x$
(b) $f_{xy}(x, y) = 3y^2 + xe^x + e^x$
(c) $f_{yx}(x, y) = 3y^2 + xe^x + e^x$
(d) $f_{yy}(x, y) = 6xy$ ∎

We hope you noted that $f_{xy} = f_{yx}$ in Example G, a circumstance that was not accidental. In fact, these mixed partial derivatives are always equal provided they are continuous.

An Application to Economics

Economists classify pairs of consumer products by using three words: *competitive, complementary,* and *independent*. Fords and Chevrolets are competitive products. An increase in sales of one is likely to produce a decrease in sales of the other (unless total car sales are expanding rapidly). Cars and gasoline are complementary products. Increases in the sale of one tend to be accompanied by increases in the sale of the other. Cars and rat poison would seem to be independent products. Sales of one are unlikely to have any effect on sales of the other. Let us see what this means in terms of partial derivatives.

Let the two products labeled A and B be selling at prices per unit p_A and p_B, respectively. We suppose that the corresponding demands x_A and x_B depend in some way on both p_A and p_B, that is,

$$x_A = f(p_A, p_B) \quad \text{and} \quad x_B = g(p_A, p_B)$$

Consider first the competitive situation. Keep p_A fixed but allow p_B to increase. This will cause the demand for product A to increase, that is, $\partial x_A/\partial p_B > 0$, a circumstance shown graphically in Figure 15. Similarly, an increase in p_A with p_B fixed will cause an increase in the demand for product B, that is, $\partial x_B/\partial p_A > 0$. Thus, competition between the two products is characterized by both of these partial derivatives being positive.

FIGURE 15

$$\frac{\partial x_A}{\partial P_B} > 0$$

x_A

P_B

P_A Competitive products

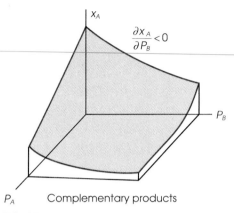

FIGURE 16

$$\frac{\partial x_A}{\partial P_B} < 0$$

x_A

P_B

P_A Complementary products

In the complementary situation, an increase in p_B lowers the demand for product B and therefore also for product A, that is, $\partial x_A/\partial p_B < 0$ (Figure 16). Similarly, $\partial x_B/\partial p_A < 0$. We summarize in the following chart.

Competitive Products	Complementary Products
$\dfrac{\partial x_A}{\partial p_B} > 0$ and $\dfrac{\partial x_B}{\partial p_A} > 0$	$\dfrac{\partial x_A}{\partial p_B} < 0$ and $\dfrac{\partial x_B}{\partial p_A} < 0$

EXAMPLE H

Suppose that a company makes two products, A and B, with unit prices p_A and p_B, respectively. If the corresponding demand equations are

$$x_A = -200p_A + 250p_B$$

$$x_B = 400 + 100p_A - 150p_B$$

decide whether the two products are competitive, complementary, or neither.

Solution $\partial x_A/\partial p_B = 250$ and $\partial x_B/\partial p_A = 100$. Both are positive; the products are competitive. ∎

Problem Set 10.2

In Problems 1–16, find $\partial z/\partial x$ and $\partial z/\partial y$ (see Examples A and C).

1. $z = x^2y^3$
2. $z = (3x + 2y)^2$
3. $z = xy^2 - 4x^2 + 3y + 5$
4. $z = 3x^2y - 6y^2 + 4x + 2$

5. $z = 2x\sqrt{y} + 4/y^2$
6. $z = x^2/\sqrt{y} - 3x^{2/3}$
7. $z = 4/xy^2$
8. $z = 12/y\sqrt{x}$
9. $z = x^2ye^{2x}$
10. $z = y^2 \ln(2x + 3y)$
11. $z = (x^2 + 2y)^3 \ln y$
12. $z = y^2e^{x^2/y}$
13. $z = \sqrt{2xy + y^2}\, e^{2y}$

14. $z = (y^2/x) \ln (x + 3)$
15. $z = \ln [x^3y^2/(2x + y)^3]$ 16. $z = \ln [e^{xy} \ln (xy^2)]$

In Problems 17–20, find $\partial w/\partial z$ (see Example B).

17. $w = 2xyz + 3yz^2$
18. $w = (xyz^3 - z)^4$
19. $w = xye^{z+2} + xz/y$
20. $w = z^2 \ln (x^2y + z)$

In Problems 21–26, find $f_x(2, 1)$ and $f_y(2, -1)$ as in Example D. To do this, you will first want to find $f_x(x, y)$ and $f_y(x, y)$.

21. $f(x, y) = x^2 + 2xy^2 - y$
22. $f(x, y) = 3xy - 4x^2 + 5y$
23. $f(x, y) = y^2 \ln (2x + 3y)$
24. $f(x, y) = x^2e^{(x-2)y}$
25. $f(x, y) = (x + 2y)e^{x(y^2-1)}$
26. $f(x, y) = (x + 2y)^2/(y^2 + 1)$

In Problems 27–30, find the slope of the tangent line at (x_0, y_0) to the trace of $z = f(x, y)$ in the plane $y = y_0$.

27. $f(x, y) = 9 - x^2 - y^2; x_0 = 2, y_0 = 1$
28. $f(x, y) = x^2 + y^2 - 4; x_0 = 3, y_0 = -1$
29. $f(x, y) = \sqrt{25 - x^2 - y^2}; x_0 = 3, y_0 = 0$
30. $f(x, y) = e^{x^2+2y^2}; x_0 = 1, y_0 = -1$

31. Suppose that Jack and Jill in Example E ran down the hill $z = f(x, y) = e^{-x^2-y^2}$ directly above the line $y = 2$ (in the xy-plane). What was the grade of their path at the point $(3, 2)$?

32. If Jack and Jill in Example E ran down the given hill directly above the line $x = 3$, what was the grade of their path at the point $(3, 2)$?

33. Suppose that you traveled down the hill represented by the surface $z = f(x, y) = 4/\ln (2 + 2x^2 + 2y^2)$ on a path directly above the x-axis. What was the grade of your path at the point $(1, 0)$?

34. If in Problem 33 you traveled down the hill directly above the y-axis, what was the grade of your path at $(0, 2)$?

35. Recall that the volume of a right circular cone with radius of base r and height h is given by $V = \pi r^2h/3$. Find the rate of change in volume
    (a) with respect to the radius when $r = 6$ and $h = 8$.
    (b) with respect to the height when $r = 6$ and $h = 8$.

36. The resistance R to blood flow in a human blood vessel depends on the length L and radius r of the vessel according to the formula $R = kL/r^4$, where k is a constant. Find the rate of change in resistance
    (a) with respect to length.
    (b) with respect to radius.

C 37. Empirical evidence suggests that the surface area A (in square centimeters) of an average person's body is related to his or her weight W (in kilograms) and height H (in centimeters) according to the formula $A = 700W^{0.42}H^{0.73}$.
    (a) Determine $\partial A/\partial W$ and $\partial A/\partial H$.
    (b) About how much will A change if a person's weight increases from 70 to 71 kilograms while he or she stays 180 centimeters in height?
    (c) About how much will A change if a person grows from 180 to 181 centimeters in height while maintaining a weight of 70 kilograms?

38. Suppose that the stopping distance in feet for a car weighing w pounds traveling v miles per hour is given by $S = S(w, v) = 0.000012wv^2$.
    (a) Evaluate $S(2000, 70)$, $S_w(2000, 70)$, and $S_v(2000, 70)$.
    (b) About how much will the stopping distance increase for a car of weight 2000 pounds if its velocity is increased from 70 to 71 miles per hour?

In Problems 39–44, find (a) $f_{xx}(x, y)$, (b) $f_{xy}(x, y)$, and (c) $f_{yx}(x, y)$ as in Example G.

39. $f(x, y) = 2xy^3 - 2x^2y$
40. $f(x, y) = y^4 - 2xy^2 + 5x^2$
41. $f(x, y) = e^{xy^2}$
42. $f(x, y) = 2y \ln (x^2 + y)$
43. $f(x, y) = (x^2 + 2y)^3$ 44. $f(x, y) = (y^2 + e^x)^4$

In Problems 45–48, p_A and p_B represent unit prices of two products A and B, while x_A and x_B represent the corresponding demands. Determine as in Example H whether the products A and B are competitive, complementary, or neither.

45. $x_A = -300p_A - 250p_B + 300;$
    $x_B = -420p_A - 180p_B + 640$
46. $x_A = -45p_A - 60p_B + 120;$
    $x_B = -50p_A - 35p_B + 80$
47. $x_A = -100p_A + 84p_B + 2000;$
    $x_B = 45p_A - 60p_B + 1200$
48. $x_A = -55p_A + 48p_B + 420;$
    $x_B = -60p_A - 90p_B + 180$

49. A corporation estimates that, with annual expenditures of x millions of dollars on research and y millions of dollars on advertising, its revenue in millions of dollars will be given by $R = R(x, y) = 42 + 3x^2 + 2xy$. Evaluate $R_x(20, 30)$ and $R_y(20, 30)$ and interpret the meaning of these answers.

C 50. A ski manufacturer has the weekly production function $W = F(L, K) = 250L^{0.63}K^{0.37}$, where W is the number of pairs of skis produced, L the number of units of labor used, and K the number of units of capital required. Evaluate $F_L(10, 20)$ and $F_K(10, 20)$ and interpret your answers.

51. The temperature T in Celsius degrees in a thin rectangular plate at a point x feet from the left edge and y feet from the bottom edge satisfies $T = T(x, y) = 64 - 0.5x^2 - 1.5y^2$. Find $T_x(4, 2)$ and $T_y(4, 2)$ and interpret your answers.

c 52. In the winter, some weather stations report the temperature T in degrees Celsius, the wind velocity v in miles per hour, and the wind chill W in degrees Celsius using a formula like

$$W = W(v, T)$$

$$= 30 - 0.05(11 + 9\sqrt{v} - 0.09v)(30 - T)$$

Presumably W measures how cold it feels.
(a) Calculate $W(30, -20)$ and $W_v(30, -20)$.
(b) What significance do you attach to $W_v(30, -20)$?

53. A factory employs x workers at $6 per hour, y workers at $9 per hour, and z workers at $12 per hour.
(a) Write an expression for the average hourly wage $A(x, y, z)$ of the $x + y + z$ workers.
(b) Calculate $A_x(10, 20, 10)$.
(c) Interpret your answer to part (b).

c 54. When P dollars are invested at the rate r compounded continuously, the accumulated amount A at the end of

t years is given by $A = A(P, r, t) = Pe^{rt}$. Evaluate $A_r(1000, 0.07, 10)$ and interpret its meaning.

Recall for functions of one variable $y = f(x)$ that the actual change Δy when x goes from x_0 to $x_0 + \Delta x$ can be approximated by $dy = f'(x_0) \Delta x$. In a similar fashion, the change Δz in $z = f(x, y)$ when (x, y) goes from (x_0, y_0) to $(x_0 + \Delta x, y_0 + \Delta y)$ is approximately equal to $dz = f_x(x_0, y_0) \Delta x + f_y(x_0, y_0) \Delta y$. Use this fact in Problems 55–58.

55. Find approximately the change in $z = 2x^2\sqrt{y}$ when x changes from 10 to 10.2 and y changes from 16 to 16.1.

56. Approximate the change in $z = 12x/y^2$ when x changes from 20 to 20.5 and y changes from 10 to 10.2.

c 57. About how much did the volume of a right circular cylinder increase when its radius changed from 10 to 10.3 and its height from 30 to 30.4 (all in centimeters)?

c 58. P dollars invested for 1 year at rate r compounded continuously will accumulate to Pe^r at the end of the year. About how much extra will John have at the end of a year if he can increase his investment from $1000 to $1050 and get a rate of 0.075 rather than 0.07?

10.3
OPTIMIZATION

If optimization was important for functions of a single variable, it is equally important for functions of two or more variables. The behavior of functions of several variables can be quite complicated. Figure 17 illustrates some of the possibilities for a function of two variables.

The language we will use is quite clear from Figure 17, but for completeness, we give a formal definition.

DEFINITION

Let D, the domain of $f(x, y)$, contain the point (a, b). We say that

(i) $f(a, b)$ is the **maximum value** of $f(x, y)$ on D if $f(a, b) \geq f(x, y)$ for all (x, y) in D.

(ii) $f(a, b)$ is the **minimum value** of $f(x, y)$ on D if $f(a, b) \leq f(x, y)$ for all (x, y) in D.

The same definition applies to **local maximum value** and **local minimum value,** except that we require the inequalities to hold only in a neighborhood of (a, b)

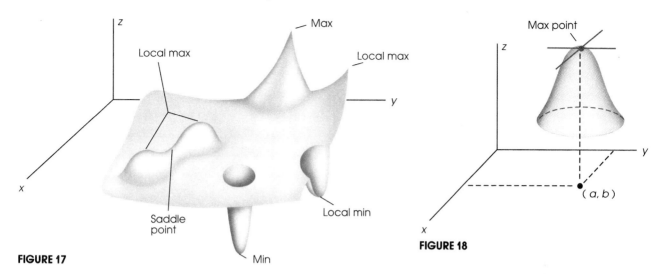

FIGURE 17

FIGURE 18

We use the phrase **extreme value** to refer to either a maximum value or a minimum value. Just as in the one-variable case, an extreme value can occur only at a critical point. A **critical point** is (i) a boundary point of D, (ii) a point where the first partial derivatives are 0, or (iii) a point where the function has some type of pathological behavior (its graph has a break, has a corner, or fluctuates wildly). Our interest is mainly in critical points of type (ii). Three types of behavior are typical of this case. They are illustrated in Figures 18, 19, and 20.

Note that in both Figures 18 and 19, the partial derivatives are 0 at (a, b); that is,

$$f_x(a, b) = 0 = f_y(a, b)$$

However, this relation does not guarantee an extremum at (a,b), as the **saddle point** in Figure 20 illustrates.

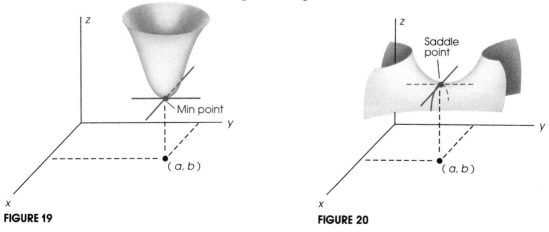

FIGURE 19

FIGURE 20

Second Partials Test for Extrema

As was true of functions of one variable, we have a test for locating maxima, minima, and saddle points and for distinguishing between them. We state this test as a theorem, a theorem whose proof goes far beyond the level of this course.

THEOREM (SECOND PARTIALS TEST)

Suppose that $f(x, y)$ has continuous second partial derivatives in a neighborhood of (a, b) and that

$$f_x(a, b) = 0 = f_y(a, b)$$

Let

$$T(a, b) = f_{xx}(a, b)f_{yy}(a, b) - [f_{xy}(a, b)]^2$$

Then

(i) If $T(a, b) > 0$ and $f_{xx}(a, b) < 0$, $f(a, b)$ is a local maximum value.

(ii) If $T(a, b) > 0$ and $f_{xx}(a, b) > 0$, $f(a, b)$ is a local minimum value.

(iii) If $T(a, b) < 0$, there is a saddle point at (a, b).

(iv) If $T(a, b) = 0$, the test is inconclusive.

EXAMPLE A

Find the local extrema, if any, of

$$f(x, y) = x^2 + 3y^3 + 4x - 9y + 5$$

Solution Since $f_x(x, y) = 2x + 4$ and $f_y(x, y) = 9y^2 - 9$, we find the critical points by solving simultaneously the simple pair of equations

$$2x + 4 = 0$$
$$9y^2 - 9 = 0$$

There are two solutions, namely, $(-2, 1)$ and $(-2, -1)$.

Now $f_{xx}(x, y) = 2$, $f_{xy}(x, y) = 0$, and $f_{yy}(x, y) = 18y$. Thus, at $(-2, 1)$,

$$f_{xx} > 0 \quad \text{and} \quad T = (2)(18) - 0 = 36 > 0$$

We conclude $f(-2, 1) = -5$ is a local minimum value for $f(x, y)$. However, at $(-2, -1)$,

$$T = (2)(-18) - 0 = -36 < 0$$

and therefore there is a saddle point at $(-2, -1)$ (not an extremum).   ∎

A Geometric Application

Geometric applications are nice because our intuition guides us along.

FIGURE 21

EXAMPLE B

Determine the dimensions of the open rectangular box of volume 32 cubic feet that has the least (outer) surface area.

Solution The box, appropriately labeled, is shown in Figure 21. Its surface area is $A = xy + 2xz + 2yz$. Since the volume is 32, we have $xyz = 32$ or, equivalently, $z = 32x^{-1}y^{-1}$. This we substitute in the expression for A to obtain

$$A = xy + 2x(32x^{-1}y^{-1}) + 2y(32x^{-1}y^{-1})$$
$$= xy + 64y^{-1} + 64x^{-1}$$

Note that $0 < x < \infty$, $0 < y < \infty$ is the appropriate domain, x and y being dimensions of the base. To find the critical values, we calculate the partial derivatives A_x and A_y, set them equal to 0, and solve the resulting pair of equations simultaneously.

$$A_x = y - 64x^{-2} = 0 \qquad A_y = x - 64y^{-2} = 0$$
$$y = 64x^{-2} \qquad\qquad x = 64y^{-2}$$

To solve this pair of equations, we substitute the expression for x from the second equation into the first and then solve the resulting equation for y.

$$y = 64x^{-2} = 64(64y^{-2})^{-2} = \frac{y^4}{64}$$
$$64y = y^4$$
$$y^4 - 64y = 0$$
$$y(y^3 - 64) = 0$$
$$y = 0, 4$$

Only $y = 4$ is in the domain, and substituting this value gives

$$x = 64y^{-2} = 64(4)^{-2} = 4$$

We conclude that $(4, 4)$ is the only critical point for A. Now we could use the Second Partials Test to establish that $(4, 4)$ gives a local minimum for A (we leave this to the reader). There is finally the subtle question of whether this local minimum is actually the (global) minimum. Physical considerations suggest that it is. Finally, we calculate z from the equation $z = 32x^{-1}y^{-1}$, obtaining $z = 2$. The dimensions of the box of minimum surface area are 4 by 4 by 2. ∎

A Business Application

We propose a maximum revenue problem.

EXAMPLE C

A company sells two competing products A and B at unit prices p and q dollars, respectively. It estimates that the demands x for product A and y for product B satisfy

$$x = 50 - 4p + q$$
$$y = 60 + p - 2q$$

Determine the prices p and q that will maximize the total revenue from these two products.

Solution The total revenue R is given by

$$R = px + qy = p(50 - 4p + q) + q(60 + p - 2q)$$
$$= -4p^2 + 2pq - 2q^2 + 50p + 60q$$

Thus, the partial derivatives are

$$R_p = -8p + 2q + 50 \quad \text{and} \quad R_q = -4q + 2p + 60$$

We set these two partial derivatives equal to 0 to obtain the pair of equations

$$8p - 2q = 50$$
$$-2p + 4q = 60$$

This pair of equations can be solved by the method explained in Example D in Section 1.3. The result is $p = \$11.43$, $q = \$20.71$. Since these values for p and q determine the only critical point, we feel sure that they give the maximum revenue. We can, of course, use the Second Partials Test to establish that they at least give a local maximum. Finally, we mention that the total revenue corresponding to these choices of p and q is $\$907.14$. ∎

Problem Set 10.3

In Problems 1–14, find all critical points. Indicate which of these gives a local maximum, a local minimum, or a saddle point (see Example A).

1. $f(x, y) = x^2 + y^2 - 4x + 6y - 4$
2. $f(x, y) = x^2 + y^2 + 8y - 2x + 9$
3. $f(x, y) = (x + 1)^2 + (y - 4)^2 + 4x - 2y$
4. $f(x, y) = (x - 3)^2 + y^2 - 6x + 4y - 10$
5. $f(x, y) = 5 + 6x + 2y - 2x^2 + xy - y^2$

6. $f(x, y) = 3x^2 - xy + y^2 + 5x + y - 10$
7. $f(x, y) = 4y + 2xy - x^2 - 2y^2 + 4$
8. $f(x, y) = 2y - 6x - x^2 - 2y^2 + 3xy$
9. $f(x, y) = x^3 + 2x^2y + 4x - 8y$
10. $f(x, y) = 6xy - x^2 - y^3 + 4$
11. $f(x, y) = x^3 + y^3 - 12xy - 10$
12. $f(x, y) = x^3 + y^3 - 3x - 12y + 24$
13. $f(x, y) = ye^x - 2x - y + 10$
14. $f(x, y) = 2xy + 4 \ln x + y^2 - 4$

15. A rectangular tank open at the top is to hold 256 cubic feet of water. Determine the dimensions of the tank so that it will require the least amount of sheet metal to build. See Example B.

16. A rectangular storage shed will be built with a capacity of 24,000 cubic feet. The cost of the construction is $8 per square foot for the flat roof, $6 per square foot for the sides, and $12 per square foot for the floor. Find the dimensions of the shed that minimize the cost of construction.

17. Mailing instructions for a rectangular box require that the length plus the girth shall not exceed 84 inches (girth is the distance around, as illustrated by the dotted rectangle in Figure 22). Find the dimensions of the box that give the largest possible volume.

FIGURE 22

18. An open rectangular bin with two partitions is to be built to have a total capacity of 8 cubic feet (Figure 23). Find the dimensions of the bin requiring the least amount of material in its construction.

FIGURE 23

19. Suppose that a company sells two products, A and B, at corresponding unit prices p and q and with corresponding demands

$$x = -100p + 84q + 2000$$
$$y = 45p - 60q + 1200$$

respectively. Determine the prices p and q for which the total revenue for the two products is a maximum. See Example C.

20. A department store sells two kinds of women's blazers at prices p and q with corresponding weekly demands

$$x = 40 - 0.3p + 0.2q$$
$$y = 50 + 0.2p - 0.2q$$

respectively. Determine p and q so as to maximize the total revenue.

21. A manufacturer makes two kinds of children's tricycles. Making x units of the first kind and y units of the second kind per day will allow it to set the corresponding prices at $80 - 2.5x$ dollars and $120 - 4y$ dollars, respectively. The total production cost for these tricycles is estimated at $C = 30x + 40y - 4xy$. Determine x and y to maximize profit.

22. The ABC Company can make candy A for 60¢ per pound and candy B for 70¢ per pound. The weekly demands x and y for the two kinds of candy if priced at p cents and q cents per pound, respectively, are

$$x = 300q - 400p$$
$$y = 250(100 + p - q)$$

both measured in pounds. If the ABC Company wants to maximize its profit, how should it set the prices p and q?

---

23. Find all critical points for the function $f(x, y) = y^4 - 4xy + 0.5x^2 - 20$. Then identify each as to whether it gives a local maximum, a local minimum, or a saddle point.

24. Find three positive numbers x, y, and z whose product is 24, such that $2x + 2y + z$ is as small as possible.

25. A large firm estimates that if it spends x million dollars on research and y million dollars on advertising per year, then its annual profit will be $P(x, y) = -3x^2 + 3xy - y^2 + 22x - 8y + 20$ million dollars. Determine x and y so that the annual profit will be a maximum.

26. A campaign committee to reelect the mayor knows that the mayor can count on a certain number of votes with no advertising, that some advertising will increase the vote total, but that too much advertising may actually backfire. It estimates that spending x thousand dollars on TV advertising and y thousand dollars on radio advertising will produce z thousand extra votes for the mayor, where $z = 12xy - 2x^3 - y^2$. Find x and y so as to maximize the mayor's vote total.

27. The reaction (in appropriate units) of a person to x units of a certain drug t hours after it is administered

is given by

$$R(x, t) = 4x(10 - x)te^{-2t}$$

Find the maximum reaction.

28. When x units of drug A and y units of drug B are used in the treatment of one type of cancer, the reac- tion of the patient (in appropriate units) 12 hours after their administration is given by $R(x, y) = 10xy(20 - x - 2y)$. Find x and y so as to maximize this reaction.

10.4*
CONSTRAINED OPTIMIZATION AND LAGRANGE'S METHOD

Consider the following two minimization problems.

> Problem 1. Minimize $z = x^2 + y^2 + 2$.
> Problem 2. Minimize $z = x^2 + y^2 + 2$ subject to $x + y = 8$.

The first problem is a free optimization problem; the second is a constrained optimization problem. The geometric interpretation of the two problems is shown in Figures 24 and 25.

We can turn the second problem into one of the first type by solving the constraint for y and substituting in the expression for z. Problem 2 then takes the form:

> Rephrased Problem 2. Minimize $z = x^2 + (8 - x)^2 + 2$.

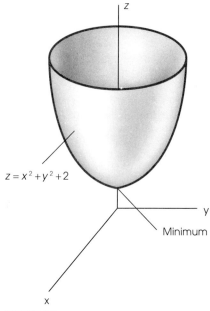

$z = x^2 + y^2 + 2$

Minimum

FIGURE 24

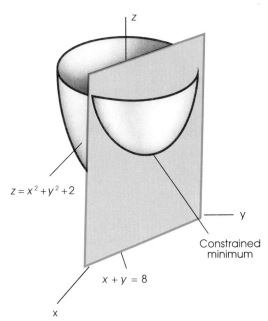

$z = x^2 + y^2 + 2$

Constrained minimum

$x + y = 8$

FIGURE 25

POINT OF INTEREST
Joseph-Louis Lagrange (1736–1813) called the loss of his father's fortune a lucky disaster, for otherwise "I should probably not have cast my lot with mathematics." In addition to his many theoretical contributions, Lagrange helped perfect the metric system of weights and measures.

In this form the problem is simple enough and can actually be handled by the methods in Section 3.3. The minimum occurs at $x = 4$, which in turn makes $y = 4$ and $z = 34$.

But what if the problem had read:

Minimize $z = x^2 + y^2 + 2$ subject to the constraint $xy^5 + x^5y^2 = 8$.

We could not have solved the constraint for y in terms of x (or vice versa); the standard method breaks down. The eighteenth-century French mathematician, Joseph-Louis Lagrange suggested an alternative. Often it works better than the standard method even when that method works.

The Method of Lagrange Multipliers

We consider the general problem of optimizing $f(x, y)$ subject to the constraint $g(x, y) = 0$. Here is Lagrange's method.

LAGRANGE'S METHOD
For the problem of optimizing $z = f(x, y)$ subject to the constraint $g(x, y) = 0$, let

$$F(x, y, \lambda) = f(x, y) + \lambda g(x, y)$$

The local extreme points (x, y) for this optimization problem will be found among the points (x, y, λ) that satisfy the three equations

$$F_x(x, y, \lambda) = 0$$
$$F_y(x, y, \lambda) = 0$$
$$F_\lambda(x, y, \lambda) = 0$$

It would take us too far afield to prove this result, but we will illustrate it with several examples. Our first example was discussed at the beginning of this section.

EXAMPLE A

Minimize $z = f(x, y) = x^2 + y^2 + 2$ subject to the constraint $g(x, y) = x + y - 8 = 0$.

Solution Note that we have expressed the constraint in the form required by Lagrange's method. Let

$$F(x, y, \lambda) = x^2 + y^2 + 2 + \lambda(x + y - 8)$$

Lagrange's three equations take the form

$$F_x = 2x + \lambda = 0$$

$$F_y = 2y + \lambda = 0$$

$$F_\lambda = x + y - 8 = 0$$

To solve them simultaneously, eliminate λ between the first two equations, obtaining $2x - 2y = 0$ or $x = y$. Substitute this result in the third equation to get $2x = 8$ or $x = 4$. Then $y = 4$ (and though we are not interested in it, $\lambda = -8$). The only critical point is $(x, y) = (4, 4)$, the point we found earlier by the standard method. The minimum value for z is $f(4, 4) = 34$. ∎

A Three-Variable Problem

Lagrange's method generalizes to three or more variables in the obvious way.

EXAMPLE B

An open box with two partitions is to be made of cardboard as shown in Figure 26. Maximize the total volume subject to the constraint that only 216 square inches of cardboard can be used.

FIGURE 26

Solution Using the variables x, y, and z as indicated in Figure 26, our job is to maximize $V = xyz$ subject to $xy + 2xz + 4yz = 216$, that is, subject to

$$g(x, y, z) = xy + 2xz + 4yz - 216 = 0$$

Let

$$F(x, y, z, \lambda) = xyz + \lambda(xy + 2xz + 4yz - 216)$$

and consider the four equations:

(i) $F_x = yz + \lambda(y + 2z) = 0$
(ii) $F_y = xz + \lambda(x + 4z) = 0$
(iii) $F_z = xy + \lambda(2x + 4y) = 0$
(iv) $F_\lambda = xy + 2xz + 4yz - 216 = 0$

Solving the first two equations for λ gives

$$\lambda = \frac{-yz}{y + 2z} \qquad \lambda = \frac{-xz}{x + 4z}$$

$$\frac{-yz}{y + 2z} = \frac{-xz}{x + 4z}$$

$$-xyz - 4yz^2 = -xyz - 2xz^2$$

$$4yz^2 = 2xz^2$$

$$2y = x \qquad \text{(we can assume } z \neq 0)$$

Similarly, from the second and third equations, we obtain

> **CAUTION**
>
> Multiple-variable problems are both messy and difficult, as Examples B and C illustrate. Moreover, we have glossed over important points. How can we be sure in Example B that (12, 6, 3) gives the maximum and not a saddle point? At this stage, we are not going to give a definitive answer. Rather, we ask you to use your physical intuition and note that (12, 6, 3) are reasonable dimensions for the maximum volume box.

$$\lambda = \frac{-xz}{x + 4z} \qquad \lambda = \frac{-xy}{2x + 4y}$$

$$\frac{-xz}{x + 4z} = \frac{-xy}{2x + 4y}$$

$$-2x^2z - 4xyz = -x^2y - 4xyz$$

$$2x^2z = x^2y$$

$$z = \frac{y}{2} \qquad \text{(we can assume } x \neq 0)$$

Substituting $x = 2y$ and $z = y/2$ in the fourth equation gives

$$(2y)y + 2(2y)(y/2) + 4y(y/2) = 216$$

$$6y^2 = 216$$

$$y^2 = 36$$

We conclude that $y = 6$, $x = 12$, and $z = 3$, and $V = xyz = 216$ cubic inches. ∎

An Economic Application

We introduced the concept of a Cobb-Douglas production function in Section 10.1.

EXAMPLE C

Suppose the Cobb-Douglas production function for a certain manufacturing process is given by

$$z = f(x, y) = 200x^{3/4}y^{1/4}$$

where x denotes the number of units of labor and y the number of units of capital. If a unit of labor costs \$250 and a unit of capital is \$400 and if total expenditures are limited to \$120,000, find the maximum production level.

Solution We want to maximize $f(x, y)$ subject to the constraint $250x + 400y = 120,000$, that is, $120,000 - 250x - 400y = 0$. Let

$$F(x, y, \lambda) = 200x^{3/4}y^{1/4} + \lambda(120,000 - 250x - 400y)$$

and consider the three equations:

$$F_x = 150x^{-1/4}y^{1/4} - 250\lambda = 0$$

$$F_y = 50x^{3/4}y^{-3/4} - 400\lambda = 0$$

$$F_\lambda = 120,000 - 250x - 400y = 0$$

If we solve for λ in the first two equations, we obtain

$$\lambda = \tfrac{3}{5}x^{-1/4}y^{1/4}$$

and

$$\lambda = \tfrac{1}{8}x^{3/4}y^{-3/4}$$

and therefore

$$\tfrac{3}{5}x^{-1/4}y^{1/4} = \tfrac{1}{8}x^{3/4}y^{-3/4}$$

Multiplying both sides by $x^{1/4}y^{3/4}$ yields

$$\tfrac{3}{5}y = \tfrac{1}{8}x$$

or

$$x = \tfrac{24}{5}y$$

This expression for x can be substituted into the third equation, giving

$$120,000 - 250(\tfrac{24}{5})y - 400y = 0$$

$$120,000 = 1600y$$

$$75 = y$$

It follows that $x = 360$, and the maximum number of units of production is

$$f(360, 75) = 200(360)^{3/4}(75)^{1/4} \approx 48,643 \quad \blacksquare$$

In the example just completed,

$$\lambda = (\tfrac{3}{5})(360)^{-1/4}(75)^{1/4} \approx 0.405$$

This value of the Lagrange multiplier λ has a nice economic interpretation. It can be shown that the magnitude of λ gives the **marginal productivity of money.** One additional dollar spent on production will yield 0.405 additional units of product.

Problem Set 10.4

In Problems 1–4, find the minimum value of f(x, y) subject to the constraint g(x, y) = 0 (see Example A).

1. $f(x, y) = x^2 + 2y^2 + 4$; $g(x, y) = 2x + y - 18$
2. $f(x, y) = 2x^2 + 3y^2 + 12$; $g(x, y) = x + 3y - 14$
3. $f(x, y) = 2x^2 - xy + 4y^2 + 100$;
   $g(x, y) = 2x + 5y - 76$
4. $f(x, y) = x^2 + 3xy + 4y^2 - 35$;
   $g(x, y) = 3x + y - 16$
5. Find the maximum and minimum values of $f(x, y) = xy + 10$ subject to the constraint $x^2 + 4y^2 = 32$.
6. Find the maximum and minimum values of $f(x, y) = x^2y$ subject to the constraint $x^2 + y^2 = 20$.
7. A farmer has 300 meters of fence with which to make a rectangular three-pen enclosure as shown in Figure 27. Find the dimensions of the enclosure for which the total area is a maximum.

FIGURE 27

8. Suppose that the long side of the enclosure in Problem 7 is along a building and so does not require fencing. Find the dimensions of the enclosure that give the largest total area.

In Problems 9–12, find the minimum value of f(x, y, z) subject to the constraint g(x, y, z) = 0 as illustrated in Example B.

9. $f(x, y, z) = x^2 + y^2 + z^2$;
   $g(x, y, z) = 2x + y + 3z - 14$
10. $f(x, y, z) = x^2 + 2y^2 + 3z^2 + 20$;
    $g(x, y, z) = x + y + z - 44$
11. $f(x, y, z) = 2x + y + 4z$;
    $g(x, y, z) = x^2 + y^2 + z^2 - 84$
12. $f(x, y, z) = xyz$; $g(x, y, z) = x^2 + y^2 + z^2 - 3$
13. A rectangular box with a lid is to be constructed from 36 square feet of material. If the bottom requires a double thickness while the top and sides require only a single thickness, find the dimensions of the box of maximum volume.

14. A rectangular box is to have a volume of 252 cubic feet. The cost of construction is $2 per square foot for the lid, $3 per square foot for the sides, and $5 per square foot for the bottom. Find the dimensions of the box which minimize the cost of construction.
15. A manufacturing process has Cobb-Douglas production function

$$z = f(x, y) = 300x^{2/3}y^{1/3}$$

where x denotes the number of units of labor and y the number of units of capital. A unit of labor costs $400, a unit of capital is $500, and total expenditures for labor and capital are limited to $180,000. Find the maximum production level and the marginal productivity of money. See Example C.

16. Follow the instructions for Problem 15 for the production function

$$z = f(x, y) = 600x^{3/4}y^{1/4}$$

given that a unit of labor costs $600 and a unit of capital is $400 and the total expenditures are limited to $240,000.

17. A standard sheet of paper measures 8.5 inches by 11 inches and so has a perimeter of 39 inches. What dimensions, subject to having a perimeter of 39 inches, would give the maximum area?
18. A can is to hold 1 liter (1000 cubic centimeters) of oil. Determine the dimensions of the can so that the surface area is a minimum.
19. Find the least distance between the origin and the plane $2x + 4y - 6z = 14$. *Hint:* Minimize the square of the distance, namely, $x^2 + y^2 + z^2$.
20. Consider a rectangular box with three of its faces in the coordinate planes, one vertex at $(0, 0, 0)$, and the diagonally opposite vertex having positive coordinates and lying in the plane $x/a + y/b + z/c = 1$ (a, b, and c all positive). Find the maximum volume of the box.
21. A manufacturer makes three types of jackets for which the profits per jacket are $12, $14, and $20, respectively. Suppose that the daily production levels, x, y, and z, for the three jackets must satisfy the constraint $x^2 + y^2 + z^2 = 740$. Find x, y, and z for maximum daily profit and calculate this profit.
22. A firm with $92,000 available for advertising each month will apportion x thousand dollars to television, y thousand dollars to radio, and z thousand dollars to

newspapers. It estimates the corresponding monthly profit P thousand dollars to be given by

$$P = P(x, y, z)$$

$$= x^2 + y^2 + z^2 - 2x - 4y - 4z + 20$$

Find x, y, and z that will maximize profit.

23. If the temperature T in degrees Celsius at any point (x, y, z) in space is given by the formula $T =$ $xy + xz$, find the hottest point on the sphere $x^2 + y^2 + z^2 = 25$.

24. Let the Cobb-Douglas production function for a manufacturing process be given by $z = f(x, y) = kx^a y^b$, where $a + b = 1$ ($a > 0$, $b > 0$), with the associated cost constraint $cx + dy = M$. Show that z is maximized when $x = aM/c$ and $y = bM/d$.

10.5 *
APPLICATION: THE LEAST SQUARES LINE

Surely you have wondered how social scientists obtain the mathematical models they use. How, for example, can an economist claim that the relation between the demand x for a certain product and its unit price p is given by

$$x = -0.345p + 400$$

A serious economist will have looked at the available data (or perhaps gathered new data) on the sales of the product at various prices. When plotted in the px-plane, the data typically look something like Figure 28. Often there is a tendency for them to lie along a line (as there is in Figure 28). The economist's problem reduces to finding the line that best fits the data (Figure 29). The problem of finding a line (or, more generally, a curve of some special type) that best fits a set of data is one of the most important problems in the sciences.

But what do we mean by *best*? What is the criterion for saying that a line best fits a set of data? There is general agreement that we should choose the line that minimizes the sum of the squares of the vertical deviations of the data from the line, hence the name **least squares line.**

FIGURE 28

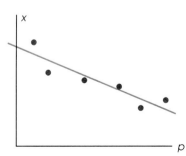

FIGURE 29

A Simple Example

To get a handle on the mathematics involved in least squares, we begin with an example that has only three data points: $(1, 2)$, $(2, 5)$, and $(3, 4)$. Our task is to choose m and b so as to minimize $S = d_1^2 + d_2^2 + d_3^2$, where d_1, d_2, and d_3 are the vertical deviations of the data from the line $y = mx + b$ (Figure 30).

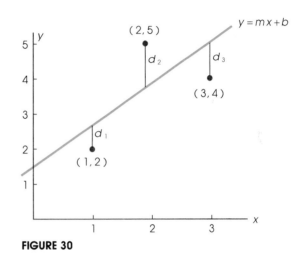

FIGURE 30

The point on the line with x-coordinate 1 has y-coordinate $m(1) + b$, and so

$$d_1^2 = (m + b - 2)^2$$

Similarly,

$$d_2^2 = (2m + b - 5)^2$$

and

$$d_3^2 = (3m + b - 4)^2$$

Thus,

$$S = (m + b - 2)^2 + (2m + b - 5)^2 + (3m + b - 4)^2$$

We want to minimize S, a function of the two variables m and b. The theory in Section 10.3 applies: Take partial derivatives and set them equal to 0.

$$S_m = 2(m + b - 2) + 4(2m + b - 5) + 6(3m + b - 4) = 0$$

$$S_b = 2(m + b - 2) + 2(2m + b - 5) + 2(3m + b - 4) = 0$$

This pair of equations simplifies to

$$28m + 12b = 48$$

$$12m + 6b = 22$$

These equations are easily solved, giving $m = 1$ and $b = \frac{5}{3}$. Since there is only one critical point, we are confident that it produces the minimum S. The least squares line is $y = x + \frac{5}{3}$.

The General Case

Consider now a set of n data points (x_1, y_1), (x_2, y_2), (x_3, y_3), . . . , (x_n, y_n). The corresponding sum of squares S to be minimized is

$$S = d_1^2 + d_2^2 + d_3^2 + \cdots + d_n^2$$

where

$$d_i = mx_i + b - y_i$$

To simplify our work, we are going to use the summation notation introduced in Section 8.4, namely, Σ. And since all of our sums will have i running from 1 to n, we write simply Σ rather than the more complete $\Sigma_{i=1}^n$. We will also make use of the properties of Σ discussed in Section 8.4. Get ready for a bit of fancy mathematical footwork.

$$S = \sum d_i^2 = \sum (mx_i + b - y_i)^2$$

Thus,

$$S_m = \sum 2x_i(mx_i + b - y_i)$$

$$= \sum (2mx_i^2 + 2bx_i - 2x_i y_i)$$

$$= 2m \sum x_i^2 + 2b \sum x_i - 2 \sum x_i y_i$$

Also,

$$S_b = \sum 2(mx_i + b - y_i)$$

$$= \sum (2mx_i + 2b - 2y_i)$$

$$= 2m \sum x_i + 2b(n) - 2 \sum y_i$$

Setting S_m and S_b equal to 0 and dividing by 2 gives the pair of equations

$$\left(\sum x_i^2 \right) m + \left(\sum x_i \right) b = \sum x_i y_i$$

$$\left(\sum x_i \right) m + nb = \sum y_i$$

When these equations are solved for m and b, we obtain the following boxed formulas. To use them, m must be calculated first, since the formula for b involves m.

POINT OF INTEREST
The important task of finding least squares lines is routinely done with computer software packages such as Minitab and SPSS. Note that the formulas we give for m and b are easy to implement on a computer and a computer doesn't much care whether there are 10 or 10,000 pairs of data.

$$m = \frac{(\Sigma \, x_i)(\Sigma \, y_i) - n(\Sigma \, x_i y_i)}{(\Sigma \, x_i)^2 - n(\Sigma \, x_i^2)} \qquad b = \frac{\Sigma \, y_i - m(\Sigma \, x_i)}{n}$$

We summarize. The least squares line for a set of data (x_1, y_1), (x_2, y_2), $\ldots$, (x_n, y_n) is $y = mx + b$, where m and b are given by the formulas in the box. We mention that the least squares line is also called the **regression line of y on x.**

Predicting

A common use of the least squares line is in predicting the future based on data collected in the past. To cut down on the arithmetic, we illustrate with an example that uses only 10 pairs of data, but you can imagine that n is 1000 rather than $n = 10$.

EXAMPLE A

Suppose that we have collected the data shown in the left two columns in Figure 31. The value of x_i indicates high school percentile rank in graduating class. The corresponding y_i is the grade point average in the first term of college (A = 4, B = 3, C = 2, D = 1, and F = 0). Use these data to find the least squares line, and use it to make your best prediction for the college calculus grade of a student who had a high school percentile rank of 76.

	x_i	y_i	x_i^2	$x_i y_i$
	80	3.0	6400	240.0
	72	2.6	5184	187.2
	51	2.2	2601	112.2
	90	3.3	8100	297.0
	95	3.5	9025	332.5
	99	3.1	9801	306.9
	82	2.9	6724	237.8
	85	3.0	7225	255.0
	77	2.5	5929	192.5
	64	2.0	4096	128.0
Σ	795	28.1	65,085	2289.1

FIGURE 31

Solution In the table in Figure 31, we have arranged our data in a form convenient for using the boxed formulas for m and b. Note the calculations of Σx_i, Σy_i, Σx_i^2, and $\Sigma x_i y_i$ obtained by adding the corresponding columns. We conclude that

$$m = \frac{(795)(28.1) - (10)(2289.1)}{(795)^2 - (10)(65085)} \approx 0.029$$

and

$$b = \frac{(28.1) - (0.029)(795)}{10} \approx 0.505$$

Thus, the least squares line is $y = 0.029x + 0.505$, and a student with a high school percentile rank of 76 can expect a grade point average of

$$y = (0.029)(76) + 0.505 \approx 2.7 \qquad \blacksquare$$

You can see from Example A that finding m and b involves a large amount of calculation. Fortunately, computers can easily be programmed to do these calculations, and even some hand-held calculators have buttons that give m and b automatically as soon as the xy-data are entered.

Problem Set 10.5

In Problems 1–4, find the least squares line for the given data by going through the following steps.

    (i) Determine the sum S of the squares of the deviations of the given data points from the line $y = mx + b$. For example, in Problem 1,

$$S = (m + b - 3)^2 + (4m + b - 2)^2$$
$$+ (5m + b - 2)^2$$

    (ii) Find the partial derivatives S_m and S_b and set them equal to 0.

    (iii) Solve the resulting pair of equations simultaneously.

1. (1, 3), (4, 2), (5, 2)
2. (1, 3), (4, 5), (6, 8)
3. (−1, 1) (1, 2), (2, 2), (5, 3)
4. (−1, 2), (1, 2), (2, 1), (5, 0)

In Problems 5–8, use the formulas for m and b given in the text to find the equation of the least squares line for the given set of data.

5. (1, 1), (3, 2), (4, 5), (5, 5)
6. (1, 3), (3, 2), (4, 1), (6, 2)
7. (2, 3), (3, 4), (4, 3), (5, 4), (6, 7)
8. (2, 5), (4, 4), (4, 2), (5, 1), (7, 0)
9. Table 1 gives the per capita yearly income y in thousands of dollars for a certain country in the years 1960, 1965, and so on. Note that x represents the number of years after 1960.
   (a) Find the least squares line for this set of data.
   (b) Use this equation to estimate the per capita income in 1978.
   (c) Predict the per capita income in the year 2000.

TABLE 1

Year	x	y
1960	0	1.8
1965	5	2.1
1970	10	2.6
1975	15	3.0
1980	20	4.2
1985	25	5.6

10. Table 2 gives the amount y in billions of dollars that a certain country spent on new factories and equipment during the years 1970, 1975, and so on. Note that $x = 0$ corresponds to 1970, and $x = 1$ to 1975 (what value of x corresponds to 1971?). Find the least squares line for these data and use it to predict the expenditures for new factories and equipment during 2000 and 2002.

TABLE 2

Year	x	y
1970	0	1.8
1975	1	2.1
1980	2	2.4
1985	3	3.2
1990	4	3.8

11. Table 3 shows the appraised value in thousands of dollars of a house in various years. Find a least squares line for these data and use it to predict the value of the house in the year 2000. *Hint:* See Problems 9 and 10 for two possible ways of introducing variables x and y.

TABLE 3

Year	Value
1975	180
1980	195
1985	215
1990	225

12. Table 4 reports the number of new cases of flu in a certain city. Use these data and a least squares line to predict the number of new cases of flu on day 9.

TABLE 4

x (day number)	1	2	3	4	5	6	7
y (new cases)	10	15	18	21	24	33	35

C 13. A company wishes to determine the demand equation for one of its products. At four outlets, A, B, C, and D, located in comparable communities, the number x of units sold in 1 month with the product priced at p dollars was shown in Table 5. Find the linear demand equation which best fits these data in the least squares sense and use the result to predict the demand if the price is set at $12.

TABLE 5

Price p	Demand x
10.00	1200
10.50	1100
11.50	880
13.50	720

C 14. A stage show has run for 4 weeks with the attendance as shown in Table 6. Find the equation of the least squares line $y = mx + b$ for these data, with x being the week number and y the attendance for the week. Then predict the week in which the attendance will fall below 2500.

TABLE 6

Week no.	Attendance
1	6400
2	5400
3	4600
4	4100

C 15. Find the curve $y = ae^{mx}$ that fits the data in Table 7 best in the least squares sense. *Hint:* The equation $y = ae^{mx}$ is equivalent to $\ln y = \ln a + mx$. The substitutions $Y = \ln y$ and $b = \ln a$ transform this equation to the linear equation $Y = mx + b$. Use the method of least squares to find the values of m and b that give the best fit to the data pairs $(x, Y) = (x, \ln y)$. Then determine a and write the corresponding equation $y = ae^{mx}$.

TABLE 7

x	y
1	3.1
2	6.3
3	11.9
4	24.1
5	47.8

C 16. A certain city spent for services (police, fire, administration) the amount y (in millions of dollars) shown in Table 8 for the years 1970, 1975, and so on. Find the equation of the form $y = ae^{mx}$ that best fits these data and use it to predict expenditures in the year 2000.

TABLE 8

Year	x (year no.)	y (expenditure)
1970	0	40
1975	1	44
1980	2	49
1985	3	54
1990	4	60

CHAPTER 10 REVIEW PROBLEM SET

1. If $f(x, y, z) = x^2 y^3 z^4$, find and simplify:
   (a) $f(3, -1, -2)$ (b) $f(x, 2x, x/2)$
   (c) $f(2a, 2b, c/2)/f(a, b, c)$

2. If $f(x, y, z) = x^3 y z^{1/3}$, find and simplify:
   (a) $f(2, 4, \frac{1}{8})$ (b) $f(y^2, y, y^6)$
   (c) $f(2a, 2b, 2c)/f(a, b, c)$

3. Let $f(r, h)$ denote the total surface area (inner plus outer) of a spherical shell of inner radius r and thickness h. Write a formula for $f(r, h)$ and calculate $f(3, 0.5)$. *Hint:* The formula for the surface area of a sphere is $S = 4\pi r^2$.

4. Let $g(r, h)$ denote the volume of the spherical shell of Problem 3. Write a formula for $g(r, h)$ and calculate $g(4, 0.3)$. *Hint:* The formula for the volume of a solid sphere of radius r is $V = \frac{4}{3}\pi r^3$.

5. Write the formula for the volume $g(a, b)$ of a prism of length a if each cross section is an equilateral triangle of side b.

6. Write the formula for the volume of a prism of height h if each cross section is a trapezoid whose parallel sides have lengths a and b and whose altitude is d.

7. Sketch the graph of each of the following equations, showing the intercepts and appropriate traces.
   (a) $4x + y - 2z = 8$ (b) $x^2 + y^2 + z^2 = 16$
   (c) $x + z = 4$ (d) $z = 9 - x^2 - y^2$

8. Sketch the graph of each equation, showing the intercepts and appropriate traces.
   (a) $3x - 2y + 4z = 12$
   (b) $x^2 + y^2 + (z - 2)^2 = 4$
   (c) $4x + 5z = 20$ (d) $z = x^2 + y^2 - 9$

9. For each of the following, find $f_x(x, y)$ and $f_y(1, 0)$.
   (a) $f(x, y) = 2x^2 - 3xy + y^3$
   (b) $f(x, y) = x^2 \ln(2x + 3y)$
   (c) $f(x, y) = e^{x^2 y}$
   (d) $f(x, y) = \left(3x - \dfrac{2y}{x}\right)^4$

10. For each of the following, find $f_y(x, y)$ and $f_x(0, 2)$.
    (a) $f(x, y) = (x - 2y)^3$ (b) $f(x, y) = y^2 e^{2xy}$
    (c) $f(x, y) = y + 3\ln(x^2 + y^2 - x - 1)$
    (d) $f(x, y) = \left(xy + \dfrac{8}{y}\right)^4$

11. Find the slope of the tangent line at $(2, 3)$ to the trace of $z = 25 - 4x^2 - y^2$ in the plane $x = 2$.

12. Find the slope of the tangent line at $(3, 1)$ to the trace of the surface $z = x^2 - 3xy + 4y^2 + 12$ in the plane $y = 1$.

13. Suppose you ran down the hill represented by the surface $z = 12/(x^2 + y^2 + 1)^2$ directly above the line $y = 1$. What was the slope of your path at $(2, 1)$?

14. The plane $x = 2$ intersects the surface $z = 24e^{9-x^2-y^2}$ in a curve. What is the slope of this curve at $(2, 2)$?

C 15. If P dollars is invested for 20 years at the rate r compounded annually, the accumulated amount at the end of that time will be $f(P, r) = P(1 + r)^{20}$. Calculate $f_P(2000, 0.08)$ and $f_r(2000, 0.08)$ and give interpretations of these numbers.

16. Suppose that the maximum safe load $S(x, y, z)$ for a horizontal beam supported at both ends is given by $S(x, y, z) = 1200xy^2/z$ where x and y are the width and depth in inches, z is the length in feet, and S is in pounds. Calculate $S_x(2, 4, 10)$ and $S_y(2, 4, 10)$ and interpret these numbers.

17. Find all critical points of $f(x, y) = x^2 + y^3 - 6xy$ and indicate which of them gives a local maximum, a local minimum, or a saddle point.

18. Find all critical points of $f(x, y) = x^3 + y^3 - 9xy$ and indicate which of them gives a local maximum, a local minimum, or a saddle point.

19. Find three positive numbers, x, y, and z, satisfying the condition $2x + y + z = 2$ for which the product xyz is a maximum.

20. Find the dimensions of the box with maximum volume for which the length plus twice the width plus twice the height equals 24.

21. A manufacturing company sells two kinds of riding mowers at prices p dollars and q dollars with corresponding weekly demands

$$x = 300 - 0.5p + 0.3q$$

$$y = 400 + 0.2p - 0.4q$$

Determine p and q so as to maximize weekly revenue.

22. Find the coordinates of the point on the surface $z = 4 + x^2 + y^2$ that is nearest to the point $(1, 2, 0)$. *Hint:* Minimize $d^2 = (x - 1)^2 + (y - 2)^2 + z^2$ with (x, y, z) on the surface.

23. Use Lagrange's method to find the maximum value of

$$f(x, y) = xy$$

subject to the constraints $4x^2 + 9y^2 = 36$, $x > 0$, $y > 0$.

24. Use Lagrange's method to find the maximum value of $f(x, y) = x^2 y$ subject to the constraint $x^2 + y^2 = 27$.

25. Jane Kelly recently opened an ice cream parlor. The weekly revenue (in hundreds of dollars) for the first 6 weeks was as shown in the accompanying table.

x (week number)	1	2	3	4	5	6
y (revenue)	32	36	35	44	46	50

(a) Find the equation of the least squares line for these data.
(b) Use this equation to predict revenue for the tenth week.

26. For the first six months after Jim MacMillan opened an automotive center, the monthly profit (in hundreds of dollars) was as shown.

x (month number)	1	2	3	4	5	6
y (profit)	42	48	50	54	57	62

(a) Find the equation of the least squares line.
(b) Use this equation to predict the profit for the 12th month.

27. The maximum safe load L that can be supported by a beam of width w, depth d, and length s is given by $L = k(wd^2/s)$, where k is a constant (Figure 32).
   (a) Determine k if a beam 2 inches wide, 1 foot deep, and 20 feet long, can safely support a load of 1400 pounds.
   (b) Use a partial derivative to determine approximately what changing the dimensions from 2, 1, 20 to 2, 1, 21 does to the safe load.

FIGURE 32

28. How would you define $f_x(a, b)$ as a limit?

From Michelangelo to Yang, from art to physics, the concept of right and left is significant.
Mathematicians, too, must be aware of this distinction. In graphing $z = f(x, y)$, we always
choose a right-handed coordinate system. A cupped right hand curling from the positive
x-axis to the positive y-axis has its thumb pointing in the direction of the positive z-axis.

11

Double Integrals

11.1 **Double Integrals over Rectangles**

11.2 **Double Integrals over General Regions**

11.3 **Applications of Double Integrals**

Chen Ning Yang, son of a mathematician, was born in 1922 in Hofei, Anhwei, China. He received his education through the master's degree in China and briefly taught high school mathematics there. At the end of World War II, he came to the United States on a fellowship to study physics at the University of Chicago. There he was greatly influenced by Enrico Fermi, though he actually wrote his dissertation under Edward Teller. In 1949, he went to the Institute for Advanced Study in Princeton, New Jersey where he became a professor in 1955.

Yang's professional life has been intimately associated with a slightly younger Chinese colleague, Tsung Dao Lee, who also came to the University of Chicago and later joined him at the Institute for Advanced Study. Together they produced 32 joint papers. Yang and Lee dared to question a revered concept—that physics knows no distinction between right and left. Another Chinese physicist, Chien-Shiung Wu, carried out the crucial experiment that confirmed their conjecture. Some particles are left-handed, some are right-handed. For this insight, Yang and Lee shared the 1957 Nobel Prize.

Robert Oppenheimer, director of the institute, was especially proud of the two young physicists that he had lured to Princeton; here was collaboration that really worked. But Yang and Lee gradually drifted apart and today are at separate institutions—Yang at the State University of New York at Stony Brook, Lee at Columbia University. No longer do they write papers together or even see each other. The drive for individual recognition is very strong among scientists.

Yang is an eminent physicist with powerful mathematical intuition. Occasionally his papers are published in mathematics rather than physics journals. The Yang-Mills theory of non-Abelian gauge fields is today an area of extensive research by both mathematicians and physicists.

Yang has received many honors (U.S. Chamber of Commerce outstanding young man of 1957, the Albert Einstein Commemorative Award, an honorary doctorate from Princeton University) in addition to his Nobel Prize. He says that his main job is to think, and he laments the lack of respect in the United States for "learning and learnedness."

Chen Ning Yang
(1922–   )

11.1
DOUBLE INTEGRALS OVER RECTANGLES

Our goal for this chapter is to extend integration theory to functions $f(x, y)$ of two variables. It would be well to review Chapter 8 ("The Definite Integral"), for it is that material we wish to generalize. Recall that for $f(x) \geq 0$, $\int_a^b f(x)\,dx$ was defined to be the area $A(R)$ of the region R under the curve $y = f(x)$ between a and b. Moreover, for a function that is sometimes positive and sometimes negative,

$$\int_a^b f(x)\,dx = A(R_u) - A(R_d)$$

Here R_u and R_d denote the parts above and below the x-axis of the region R trapped between the curve and the axis (Figure 1).

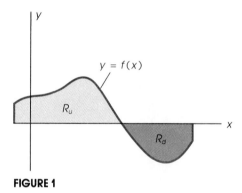

FIGURE 1

Consider now a function $f(x, y)$ defined on a domain D in the xy-plane. The graph of $z = f(x, y)$ is a surface which traps a solid region S between it and the xy-plane. If $f(x, y) \geq 0$, we define the **double integral** $\int_D\int f(x, y)\,dy\,dx$ to be the volume $V(R)$ of this solid. More generally,

$$\int_D\int f(x, y)\,dy\,dx = V(S_u) - V(S_d)$$

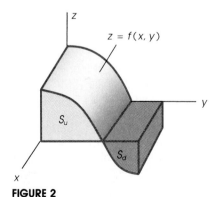

FIGURE 2

where S_u and S_d are the parts of the solid region S that are above and below the xy-plane (Figure 2). In making this definition, we are taking volume as an intuitively understood concept. In particular, we assume that the volume of a rectangular box is given by length times width times height and that the volume of a general cylinder is the area of the base times the height (Figure 3).

$$V = lwh$$

$$V = Ah$$

FIGURE 3

Initially, we shall assume that the domain D is a rectangle R which may be partitioned into smaller rectangles R_k of area $\Delta x\, \Delta y$, $k = 1, 2, 3, \ldots,$ n. In analogy with Section 8.4, the double integral can be written as a limit, namely,

$$\int\!\!\int_R f(x, y)\, dy\, dx = \lim_{n\to\infty} \sum_{k=1}^{n} f(x_k, y_k)\, \Delta y\, \Delta x$$

The partition process leading to this interpretation of the double integral is illustrated in Figure 4.

None of what we have said so far is of much help in evaluating double integrals. For this we need the Fundamental Theorem of Calculus, but we also need two new ideas, namely, partial integrals and iterated integrals. We digress to discuss these ideas.

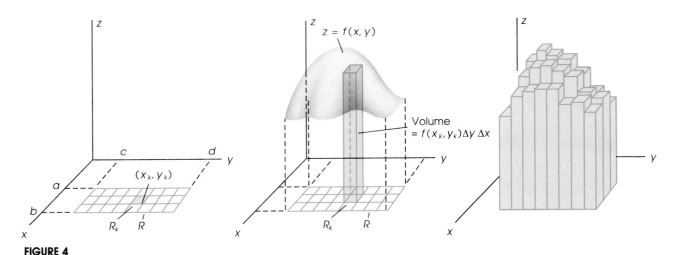

FIGURE 4

Partial Integrals and Iterated Integrals

Consider the symbol $\int_c^d f(x, y)\, dy$. We call it a **partial integral** because, though the integrand is a function of two variables, only one integration is indicated. In analogy with partial derivatives, we are to treat x as a constant and integrate with respect to y (the presence of dy tell us that y is the variable of integration).

EXAMPLE A

Evaluate the partial integral $\displaystyle\int_1^2 3xy^2\, dy$.

Solution We treat x as a constant so $3x$ can be passed across the integral sign.

$$\int_1^2 3xy^2\, dy = 3x \int_1^2 y^2\, dy = 3x\left[\frac{y^3}{3}\right]_1^2 = 3x\left(\frac{8}{3} - \frac{1}{3}\right) = 7x$$

We say that we have integrated out the variable y, leaving an answer that involves x.   ■

Now we are ready for the concept of an **iterated integral**. The symbol we use is

$$\int_a^b \left[\int_c^d f(x, y)\, dy\right] dx$$

We do the partial integration with respect to y first, this resulting in a function of x. Then we integrate this function of x between the constants a and b, which gives us a numerical answer.

EXAMPLE B

Evaluate $\displaystyle\int_0^1 \left[\int_1^2 3xy^2\, dy\right] dx$.

Solution We evaluated the inner partial integral in Example A. Thus,

$$\int_0^1 \left[\int_1^2 3xy^2\, dy\right] dx = \int_0^1 7x\, dx = \left[\frac{7x^2}{2}\right]_0^1 = \frac{7}{2}$$ ■

There is no reason why y has to be the first variable of integration; it can just as well be x, as we now illustrate.

EXAMPLE C

Evaluate $\displaystyle\int_1^2 \left[\int_0^1 3xy^2\, dx\right] dy$.

Solution We work on the inner partial integral first.

$$\int_0^1 3xy^2\, dx = \left[\frac{3y^2x^2}{2}\right]_0^1 = \frac{3y^2}{2}$$

Thus,

$$\int_1^2 \left[\int_0^1 3xy^2\, dx\right] dy = \int_1^2 \frac{3y^2}{2}\, dy = \left[\frac{y^3}{2}\right]_1^2 = \frac{7}{2}$$

Note that the answers we got in Examples B and C are the same. This is no accident; it always happens. ■

With the digression on partial integrals and iterated integrals finished, we turn to our main job—evaluation of double integrals.

Evaluation of Double Integrals over Rectangles

We want to evaluate the double integral

$$\iint_R f(x,\ y)\ dy\ dx$$

where R is the rectangle $\{(x,\ y): a \le x \le b, c \le y \le d\}$. Assume temporarily that $f(x,\ y) \ge 0$. Then this double integral is the volume of the solid S under the surface $z = f(x,\ y)$ and above the rectangle R in the xy-plane (Figure 5). But now we point out that this volume can also be represented as an iterated integral. Note that if x is fixed, $\int_c^d f(x,\ y)\ dy$ is the area of the cross section determined by the plane at x perpendicular to the x-axis (see Figure 6). If we multiply this area by dx, we get (approximately) the volume of a slab, and if we add the volumes of these slabs and take the limit (integrate), we get the volume of the solid S. Thus, the double integral can be expressed as an iterated integral; that is,

$$\iint_R f(x,\ y)\ dy\ dx = \int_a^b \left[\int_c^d f(x,\ y)\ dy\right] dx$$

FIGURE 5

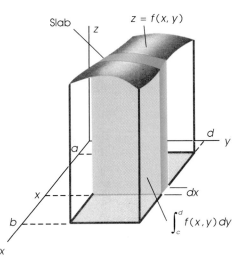

FIGURE 6

Moreover, this result is valid even if $f(x, y)$ is sometimes positive and sometimes negative.

EXAMPLE D

Evaluate the double integral $\int_R\!\int 6x^3y^2\, dy\, dx$, where R is the rectangle $\{(x, y): 1 \leq x \leq 3, 0 \leq y \leq 2\}$.

Solution We rewrite this double integral as an iterated integral and then evaluate the latter.

$$
\int_R\!\int 6x^3y^2\, dy\, dx = \int_1^3 \left[\int_0^2 6x^3y^2\, dy \right] dx
$$

$$
= \int_1^3 \left[2x^3y^3 \right]_0^2 dx = \int_1^3 16x^3\, dx
$$

$$
= \left[4x^4 \right]_1^3 = 4(81) - 4(1) = 320 \qquad \blacksquare
$$

EXAMPLE E

Evaluate the double integral $\int_R\!\int (3x - 2ye^{2x})\, dy\, dx$, where $R = \{(x, y): -1 \leq x \leq 1, -1 \leq y \leq 0\}$.

Solution We evaluate the corresponding iterated integral.

$$
\int_{-1}^1 \left[\int_{-1}^0 (3x - 2ye^{2x})\, dy \right] dx = \int_{-1}^1 \left[3xy - y^2e^{2x} \right]_{-1}^0 dx
$$

$$
= \int_{-1}^1 (3x + e^{2x})\, dx = \left[\frac{3}{2}x^2 + \frac{1}{2}e^{2x} \right]_{-1}^1
$$

$$
= \frac{1}{2}e^2 - \frac{1}{2}e^{-2} \approx 3.62686 \qquad \blacksquare
$$

EXAMPLE F

Evaluate the double integral $\int_R\!\int 2y\, e^{2xy}\, dy\, dx$, where

$$
R = \{(x, y): 0 \leq x \leq 1, 0 \leq y \leq \ln 2\}
$$

Solution We learned earlier that we can write a double integral as an iterated integral in two ways corresponding to the two different orders of integration. Doing the y-integration first confronts us with $\int_0^{\ln 2} 2y\, e^{2xy}\, dy$, which is a nontrivial integration-by-parts problem. But doing the x-integration first is straightforward, since the substitution $u = 2xy$ means that $du = 2y\, dx$, exactly what we have.

$$\int_0^{\ln 2} \left[\int_0^1 2y\, e^{2xy}\, dx \right] dy = \int_0^{\ln 2} \left[e^{2xy} \right]_0^1 dy = \int_0^{\ln 2} (e^{2y} - 1)\, dy$$

$$= \left[\frac{1}{2} e^{2y} - y \right]_0^{\ln 2} = \frac{1}{2} e^{2\ln 2} - \ln 2 - \frac{1}{2}$$

$$= \frac{1}{2} e^{\ln 4} - \ln 2 - \frac{1}{2} = \frac{1}{2}(4) - \frac{1}{2} - \ln 2$$

$$= \frac{3}{2} - \ln 2 \qquad\qquad \blacksquare$$

Problem Set 11.1

Evaluate the partial integrals in Problems 1–4.

1. $\displaystyle \int_0^3 (x^2 + 3y^2)\, dy$

2. $\displaystyle \int_1^4 3x\sqrt{y}\, dy$

3. $\displaystyle \int_2^5 \frac{y^2}{x}\, dx$

4. $\displaystyle \int_0^1 ye^{2x}\, dx$

Evaluate the iterated integrals in Problems 5–8 using your answers to Problems 1–4.

5. $\displaystyle \int_0^1 \left[\int_0^3 (x^2 + 3y^2)\, dy \right] dx$

6. $\displaystyle \int_0^2 \left[\int_1^4 3x\sqrt{y}\, dy \right] dx$

7. $\displaystyle \int_0^3 \left[\int_2^5 \frac{y^2}{x}\, dx \right] dy$

8. $\displaystyle \int_1^7 \left[\int_0^1 ye^{2x}\, dx \right] dy$

9. Reverse the order of integration in Problem 5 (that is, integrate first with respect to x) and then evaluate the resulting iterated integral.

10. Reverse the order of integration in Problem 6 and evaluate the resulting iterated integral.

11. Reverse the order of integration in Problem 7 and evaluate the resulting iterated integral.

12. Reverse the order of integration in Problem 8 and evaluate the resulting iterated integral.

In Problems 13–22, evaluate the double integral over the indicated rectangle R. See Examples D, E, and F.

13. $\displaystyle \iint_R (x^2 + 2xy)\, dy\, dx;\ R = \{(x, y): 0 \le x \le 2,\ 1 \le y \le 2\}$

14. $\displaystyle \iint_R (x^2/\sqrt{y})\, dy\, dx;\ R = \{(x, y): 0 \le x \le 3,\ 1 \le y \le 4\}$

15. $\displaystyle \iint_R xe^y\, dy\, dx;\ R = \{(x, y): 0 \le x \le 2,\ 0 \le y \le \ln 3\}$

16. $\displaystyle \iint_R (2y/x)\, dy\, dx;\ R = \{(x, y): 1 \le x \le e,\ 0 \le y \le 3\}$

17. $\displaystyle \iint_R xe^{xy}\, dy\, dx;\ R = \{(x, y): 0 \le x \le 1,\ 0 \le y \le 2\}$

18. $\displaystyle \iint_R xye^{xy^2}\, dy\, dx;\ R = \{(x, y): 1 \le x \le 2,\ 0 \le y \le 1\}$

19. $\displaystyle \iint_R \frac{1}{(x + y)^2}\, dy\, dx;\ R = \{(x, y): 1 \le x \le 2,\ 0 \le y \le 1\}$

20. $\displaystyle \iint_R 3\sqrt{x + 2y}\, dy\, dx;\ R = \{(x, y): 0 \le x \le 2,\ 0 \le y \le 1\}$

21. $\displaystyle \iint_R 4x^3(x^2y + 1)^3\, dy\, dx;\ R = \{(x, y): 0 \le x \le 1,\ 0 \le y \le 1\}$

22. $\displaystyle \iint_R 4x(x^2 + y)^3\, dy\, dx;\ R = \{(x, y): 0 \le x \le \sqrt{2},\ 0 \le y \le 1\}$

Evaluate the integrals in Problems 23–34.

23. $\displaystyle\int_0^4 e^{x^2 y}\, dy$

24. $\displaystyle\int_0^3 xe^{2x^2 y}\, dx$

25. $\displaystyle\int_0^1 \left[\int_1^4 (x^2 + 2xy + 3\sqrt{y})\, dy\right] dx$

26. $\displaystyle\int_0^3 \left[\int_1^e \frac{x^2 + 4x}{y}\, dy\right] dx$

27. $\displaystyle\int_2^3 \left[\int_2^4 \frac{2x + 1}{(x^2 + x - 2)y^2}\, dy\right] dx$

28. $\displaystyle\int_0^1 \left[\int_0^4 x(y + 1)(y^2 + 2y)^3\, dx\right] dy$

29. $\displaystyle\int_0^1 \left[\int_0^1 6x(x + y)^5\, dx\right] dy$

30. $\displaystyle\int_0^1 \left[\int_0^2 xe^{x^2 + y}\, dx\right] dy$

31. $\displaystyle\int_0^2 \left[\int_0^2 r\theta e^\theta\, d\theta\right] dr$

32. $\displaystyle\int_0^1 \left[\int_0^2 xe^{x + y}\, dx\right] dy$

33. $\displaystyle\int_0^2 \left\{\int_0^1 \left[\int_4^9 \sqrt{x}\, y^2 z^3\, dx\right] dy\right\} dz$

34. $\displaystyle\int_1^2 \left\{\int_0^1 \left[\int_1^e \frac{ze^{yz}}{x}\, dx\right] dy\right\} dz$

11.2
DOUBLE INTEGRALS OVER GENERAL REGIONS

We want now to consider the double integral

$$\iint_D f(x,\, y)\, dy\, dx$$

where D is a set in the plane with a curved boundary. We do not allow D to be completely arbitrary but rather restrict attention to sets that are either y-simple or x-simple. A set D is **y-simple** (Figure 7) if there are functions g and h on $[a, b]$ such that

$$D = \{(x,\, y)\colon g(x) \le y \le h(x),\ a \le x \le b\}$$

and D is **x-simple** (Figure 8) if there are functions p and q on $[c, d]$ such that

$$D = \{(x,\, y)\colon p(y) \le x \le q(y),\ c \le y \le d\}$$

Figure 9 shows a set D that is neither y-simple nor x-simple.

FIGURE 7

FIGURE 8

FIGURE 9

FIGURE 10

FIGURE 11

Double Integrals to Iterated Integrals

Suppose now that we want to evaluate $\iint_D f(x, y)\, dy\, dx$, where D is the y-simple set shown in Figure 10. Pretending for the moment that $f(x, y) \geq 0$, we may interpret this double integral as the volume of the solid S under the surface $z = f(x, y)$ and above D. But this volume can also be interpreted as an iterated integral. If x is fixed, then the partial integral $\int_{g(x)}^{h(x)} f(x, y)\, dy$ is the area of a cross section of the solid S determined by the plane at x perpendicular to the x-axis (Figure 11). If we multiply this area by dx, we get the volume of a slab. If we add the volumes of these slabs and take the limit (integrate from a to b), we get the volume of S. Thus,

$$\iint_D f(x, y)\, dy\, dx = \int_a^b \left[\int_{g(x)}^{h(x)} f(x, y)\, dy \right] dx$$

There is a corresponding formula where the x-integration is the inner integration. Moreover, these important relationships are valid whether f is always nonnegative or not.

EXAMPLE A

Evaluate $\iint_D 3xy^2\, dy\, dx$, where D is the region bounded by the curves $y = x^2$ and $y = 2x$ (see Figure 12).

Solution The region D is y-simple, so we will write our integral as an iterated integral with y as the first variable of integration. In this first integration, x is held fixed. Thus the integration is along the line shown in Figure

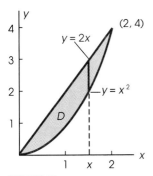

FIGURE 12

12 from the smallest value of y to the largest, that is, from $y = x^2$ to $y = 2x$.

$$\iint_D 3xy^2\, dy\, dx = \int_0^2 \left[\int_{x^2}^{2x} 3xy^2\, dy \right] dx$$

We first evaluate the inner partial integral, keeping in mind that x is treated as a constant.

$$\int_{x^2}^{2x} 3xy^2\, dy = \left[xy^3 \right]_{x^2}^{2x} = 8x^4 - x^7$$

Using this, we evaluate the iterated integral.

$$\int_0^2 \left[\int_{x^2}^{2x} 3xy^2\, dy \right] dx = \int_0^2 (8x^4 - x^7)\, dx$$

$$= \left[\frac{8}{5}x^5 - \frac{1}{8}x^8 \right]_0^2 = \frac{256}{5} - 32 = \frac{96}{5} \qquad \blacksquare$$

The region D in Example A is also x-simple, so it is possible to do the integrations in the opposite order and get the same answer. We make this our next example.

EXAMPLE B

Redo Example A but with the integrations in the opposite order.

Solution In this case, y is held fixed in the first integration, and so we integrate along the line in Figure 13 from the smallest value of x to the largest, that is, from $x = y/2$ to $x = \sqrt{y}$.

$$\int_0^4 \left[\int_{y/2}^{\sqrt{y}} 3xy^2\, dx \right] dy = \int_0^4 \left[\frac{3}{2}x^2y^2 \right]_{y/2}^{\sqrt{y}} dy = \int_0^4 \left(\frac{3}{2}y^3 - \frac{3}{8}y^4 \right) dy$$

$$= \left[\frac{3}{8}y^4 - \frac{3}{40}y^5 \right]_0^4 = \frac{3}{2}(64) - \frac{3}{10}(256) = \frac{96}{5}$$

We do get the same answer as before. $\qquad \blacksquare$

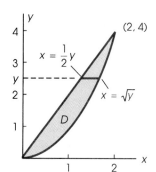

FIGURE 13

EXAMPLE C

Evaluate $\iint_D 2y\, dy\, dx$, where D is the region shown in Figure 14.

Solution We note that D is y-simple (but not x-simple). We will need the equation of the line which forms the upper boundary of D. Since this line has slope 1 and goes through $(-1, 1)$, we determine its equation to be $y - 1 = 1(x + 1)$ or $y = x + 2$. Thus, the given double integral can be rewritten as the following iterated integral and then evaluated.

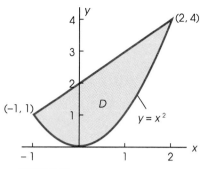

FIGURE 14

$$\int_{-1}^{2}\left[\int_{x^2}^{x+2} 2y\, dy\right] dx = \int_{-1}^{2}\left[y^2\right]_{x^2}^{x+2} dx = \int_{-1}^{2}[(x+2)^2 - x^4]\, dx$$

$$= \left[\frac{1}{3}(x+2)^3 - \frac{1}{5}x^5\right]_{-1}^{2} = \frac{72}{5} \qquad \blacksquare$$

EXAMPLE D

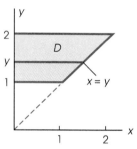

FIGURE 15

Evaluate $\int_D\int (xy + 4x^3)\, dy\, dx$, where D is the region in the xy-plane bounded by $x = y$ and $x = 0$ between $y = 1$ and $y = 2$ (Figure 15).

Solution Since the region D is x-simple but not y-simple, we are forced to do the x-integration first. In this integration y is fixed, so the integration is along the line in Figure 15 from $x = 0$ to $x = y$.

$$\int_{1}^{2}\left[\int_{0}^{y} (xy + 4x^3)\, dx\right] dy = \int_{1}^{2}\left[\frac{1}{2}x^2y + x^4\right]_{0}^{y} dy$$

$$= \int_{1}^{2}\left(\frac{1}{2}y^3 + y^4\right) dy = \left[\frac{1}{8}y^4 + \frac{1}{5}y^5\right]_{1}^{2}$$

$$= 2 + \frac{32}{5} - \left(\frac{1}{8} + \frac{1}{5}\right) = \frac{323}{40} \qquad \blacksquare$$

EXAMPLE E

By changing the order of integration, evaluate the iterated integral

$$\int_{0}^{4}\left[\int_{x/2}^{2} e^{y^2}\, dy\right] dx$$

Solution Note that we cannot do the integration as it stands, since we do not know any antiderivative of e^{y^2}. However, this iterated integral is equal to the double integral $\int_D\int e^{y^2}\, dy\, dx$, where

$$D = \{(x, y): \tfrac{1}{2}x \leq y \leq 2, 0 \leq x \leq 4\}$$

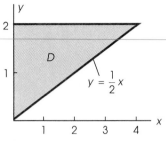

FIGURE 16

that is, D is the set shown in Figure 16. We now write this double integral as an iterated integral in which the x-integration is done first. We obtain

$$\int_0^2 \left[\int_0^{2y} e^{y^2} \, dx \right] dy = \int_0^2 \left[x e^{y^2} \right]_0^{2y} = \int_0^2 2y e^{y^2} \, dy$$

$$= \left[e^{y^2} \right]_0^2 = e^4 - 1 \qquad \blacksquare$$

What if we are asked to integrate over a set D that is neither x-simple nor y-simple? Is there anything we can do? Yes, provided we can break D into parts that are simple.

EXAMPLE F

Evaluate $\int_D\int 2y \, dy \, dx$, where D is the set in Figure 17 consisting of the two parts D_1 and D_2.

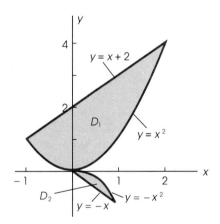

FIGURE 17

Solution The double integral over D_1 was treated in Example C; its value is $\frac{72}{5}$. The set D_2 is y-simple, and so the double integral over it can be written as an iterated integral and evaluated as follows.

$$\int_0^1 \left[\int_{-x}^{-x^2} 2y \, dy \right] dx = \int_0^1 \left[y^2 \right]_{-x}^{-x^2} dx = \int_0^1 (x^4 - x^2) \, dx$$

$$= \left[\frac{1}{5} x^5 - \frac{1}{3} x^3 \right]_0^1 = -\frac{2}{15}$$

Thus,

$$\int_D\int 2y \, dy \, dx = \frac{72}{5} - \frac{2}{15} = \frac{214}{15} \qquad \blacksquare$$

Problem Set 11.2

Evaluate the iterated integrals in Problems 1–6.

1. $\int_{1}^{2} \left[\int_{0}^{x^2} 2x \, dy \right] dx$ 2. $\int_{0}^{3} \left[\int_{0}^{2x} (y + 2) \, dy \right] dx$

3. $\int_{0}^{2} \left[\int_{1}^{2y} (3x^2 + 2xy) \, dx \right] dy$

4. $\int_{0}^{1} \left[\int_{0}^{y} (y^2 - x^2) \, dx \right] dy$

5. $\int_{0}^{1} \left[\int_{0}^{3x} 2ye^{x^3} \, dy \right] dx$ 6. $\int_{0}^{1} \left[\int_{1}^{e^x} (e^{x^2}/y) \, dy \right] dx$

In Problems 7–10, evaluate $\int_{D} \int f(x, y) \, dy \, dx$, where f is the given function and D is the indicated region. See Examples A and B.

7. $f(x, y) = x$

8. $f(x, y) = y$

9. $f(x, y) = 2x$

10. $f(x, y) = 2x$

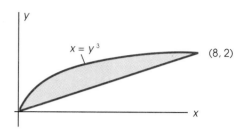

In Problems 11–22, sketch the region D. Then evaluate $\int_{D} \int f(x, y) \, dy \, dx$ as in Examples C and D.

11. $D = \{(x, y): 0 \leq y \leq \sqrt{x}, 0 \leq x \leq 2\}$;
    $f(x, y) = 4xy^3$

12. $D = \{(x, y): 0 \leq y \leq \sqrt{4 - x^2}, 0 \leq x \leq 4\}$;
    $f(x, y) = 2xy$

13. $D = \{(x, y): 4x^2 \leq y \leq 4x, 0 \leq x \leq 1\}$;
    $f(x, y) = 3y^2$

14. $D = \{(x, y): x^3 \leq y \leq 8, 0 \leq x \leq 2\}$;
    $f(x, y) = 2x + 3y$

15. $D = \{(x, y): x^2 \leq y \leq x + 2, -1 \leq x \leq 1\}$;
    $f(x, y) = 4y$

16. $D = \{(x, y): x \leq y \leq 4x - x^2, 1 \leq x \leq 2\}$;
    $f(x, y) = x$

17. $D = \{(x, y): 0 \leq x \leq \sqrt{y}, 1 \leq y \leq 4\}$;
    $f(x, y) = 3x^2 + 2y$

18. $D = \{(x, y): 2 \leq x \leq 2 + \sqrt{y}, 1 \leq y \leq 4\}$;
    $f(x, y) = 2x/\sqrt{y}$

19. D is the region bounded by $y = 2x^2$ and $y = 8$;
    $f(x, y) = 2y$

20. D is the region bounded by $y = \sqrt{x}/2$ and
    $y = x^2/16$; $f(x, y) = 16x$

21. D is the region bounded by $y = 2x$, $x = 1$, and
    $y = 4$; $f(x, y) = 2y/x$

22. D is the region bounded by $y = x$ and $y = 3x - x^2$;
    $f(x, y) = 2x$

Write the interated integrals in Problems 23–26 with the order of integration reversed. (See Example E.)

23. $\int_{0}^{2} \left[\int_{y^3}^{4y} f(x, y) \, dx \right] dy$

24. $\int_{0}^{1} \left[\int_{x^3}^{x} f(x, y) \, dy \right] dx$

25. $\int_{1}^{e} \left[\int_{0}^{\ln x} f(x, y) \, dy \right] dx$

26. $\displaystyle\int_1^{e^e}\left[\int_{\ln y}^e f(x,\, y)\, dx\right] dy$

27. Sketch the graph of the two-part region D bounded by $y = 2x^3$, $y = 0$, $x = -1$, and $x = 2$. Then evaluate $\int_D\int x\, dy\, dx$. See Example F.

28. Let $f(x) = 4x + x^2$ if $-4 \le x \le 0$ and $f(x) = x^2/4$ if $0 < x \le 4$. Sketch the region D bounded by $y = f(x)$, $y = 0$, and $x = 4$ and then evaluate $\int_D\int (5 - 2x)\, dy\, dx$.

The volume of a cylinder of height 1 has the same numerical value as the area of its base. It follows that $\int_D\int 1\, dy\, dx$ is the area of D. Using this fact, write the area of the region in Problems 29–32 as an iterated integral and then evaluate this integral.

29.

30.

31.

32.

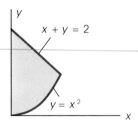

33. Calculate $\int_D\int (x + 2)\, dy\, dx$, where D is the region bounded by $y = e^x$, $y = 0$, $x = 0$, and $x = 1$.

34. Calculate $\int_D\int (1/y)\, dy\, dx$, where D is the region bounded by $y = e^{-x}$, $y = e^{-x/2}$, and $x = 1$.

In Problems 35 and 36, evaluate $\int_D\int x^2 y\, dy\, dx$, where D is the indicated region.

35.

36.

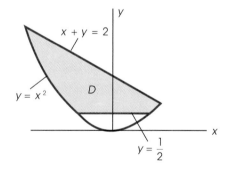

11.3
APPLICATIONS OF DOUBLE INTEGRALS

You will not be surprised when we say that a major application of double integrals is the calculation of volumes. After all, we originally defined the double integral of a nonnegative function $f(x, y)$ over a set D to be the volume of the solid trapped between the surface $z = f(x, y)$ and D. Other applications treated in this section are the mass of a thin lamina of variable density, the population of a city of known population density, and the average value of a function $f(x, y)$. Another important application (to bivariate probability distributions) will appear in Section 16.4.

Volume of a Solid

We begin with a problem where we know the answer from elementary geometry. Consider the tetrahedron S bounded by the three coordinate planes and the plane $2x + y + z = 6$ (see Figure 18). The base of this tetrahedron is a triangle of area $\frac{1}{2}(3)(6) = 9$ and the height is 6. Since the volume of a tetrahedron is one-third the area of the base times the height, our tetrahedron has volume 18.

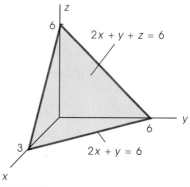

FIGURE 18

EXAMPLE A

Find the volume of the tetrahedron S described above using a double integral.

Solution First we note that the plane $2x + y + z = 6$ intersects the xy-plane $(z = 0)$ in the line $2x + y = 6$ or $y = 6 - 2x$. Let $D = \{(x, y): 0 \le y \le 6 - 2x, 0 \le x \le 3\}$. Then the volume of S is equal to $\int\int_D (6 - 2x - y)\, dy\, dx$. We evaluate this integral in the standard way as an iterated integral,

$$\int_0^3 \left[\int_0^{6-2x} (6 - 2x - y)\, dy \right] dx = \int_0^3 \left[6y - 2xy - \frac{1}{2}y^2 \right]_0^{6-2x} dx$$

$$= \int_0^3 (2x^2 - 12x + 18) \, dx$$

$$= \left[\frac{2}{3}x^3 - 6x^2 + 18x \right]_0^3 = 18 \qquad \blacksquare$$

EXAMPLE B

Find the volume $V(S)$ of the solid S under the surface $z = (9 - y^2)^{3/2}$ and above the triangle D shown in Figure 19.

Solution Sketching this solid would be tricky, and it is not necessary. Since $(9 - y^2)^{3/2} \geq 0$ on D, we know that the required volume is given by the double integral $\int_D\int (9 - y^2)^{3/2} \, dy \, dx$. Moreover, this double integral can be written as an iterated integral and in two ways. An inner y-integration appears to be impossible, so we will write the iterated integral with x as the first variable of integration.

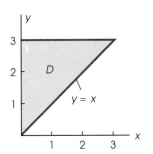

FIGURE 19

$$V(S) = \int_0^3 \left[\int_0^y (9 - y^2)^{3/2} \, dx \right] dy = \int_0^3 \left[x(9 - y^2)^{3/2} \right]_0^y \, dy$$

$$= \int_0^3 y(9 - y^2)^{3/2} \, dy = -\frac{1}{2} \int_0^3 -2y(9 - y^2)^{3/2} \, dy$$

$$= -\frac{1}{2} \left[\frac{2}{5}(9 - y^2)^{5/2} \right]_0^3 = -\frac{1}{5}(0 - 9^{5/2}) = \frac{243}{5} \qquad \blacksquare$$

Mass of a Lamina

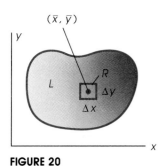

FIGURE 20

A lamina is a thin sheet, so thin that we can consider it to be two-dimensional. Consider a lamina L of constant mass density k (say in grams per square centimeter). Its total mass is just $kA(L)$, that is, k times the area of L. For example, a lamina of area 25 square centimeters and density 3 grams per square centimeter has a total mass of 75 grams. Simple arithmetic suffices for this problem. However, calculus is needed as soon as we consider a lamina of variable density. Imagine that the lamina L in Figure 20 has density $k(x, y)$ at the point (x, y). Then the little rectangular piece R will have a mass of approximately $k(\bar{x}, \bar{y}) \, \Delta y \, \Delta x$, where $(\bar{x}, \bar{y})$ is the midpoint of R. When we add the masses of all the pieces and take the limit as the pieces get smaller and smaller, we obtain a double integral, namely, $\int_L\int k(x, y) \, dy \, dx$.

EXAMPLE C

The lamina L in Figure 21 has density at (x, y) given by $k(x, y) = 2xy$. Assuming that distances are in centimeters and density is in grams per square centimeter, find the total mass $M(L)$ of this lamina.

FIGURE 21

FIGURE 22

Solution

$$M(L) = \iint_L 2xy \, dy \, dx = \int_0^4 \left[\int_0^{\sqrt{x}} 2xy \, dy \right] dx = \int_0^4 \left[xy^2 \right]_0^{\sqrt{x}} dx$$

$$= \int_0^4 x^2 \, dx = \left[\frac{1}{3} x^3 \right]_0^4 = \frac{64}{3} \text{ grams} \qquad \blacksquare$$

EXAMPLE D

A city having the shape shown in Figure 22 has a population density in people per square mile given by

$$k(x, y) = 10{,}000 e^{-|x|} e^{-y}$$

Find the total population P of the city.

Solution Note that the density is a maximum at the origin and diminishes as we move away from this point. Also, the density function is symmetric about the y-axis, so we may find the total population by doubling the population of the right half D of the city. We will use the fact that on D, $|x| = x$.

$$P = 2 \iint_D 10{,}000 e^{-x} e^{-y} \, dy \, dx = 20{,}000 \int_0^2 \left[\int_0^2 e^{-x} e^{-y} \, dy \right] dx$$

$$= 20{,}000 \int_0^2 \left[-e^{-x} e^{-y} \right]_0^2 dx = 20{,}000 \int_0^2 -e^{-x}(e^{-2} - 1) \, dx$$

$$= 20{,}000(e^{-2} - 1) \left[e^{-x} \right]_0^2 = 20{,}000(e^{-2} - 1)^2 \approx 14{,}950 \qquad \blacksquare$$

The Average Value of a Function

By analogy with functions of a single variable, we define the **average value of a function $f(x, y)$ over a region D** by

$$\text{Average value} = \frac{1}{\text{area of } D} \iint_D f(x, y) \, dy \, dx$$

EXAMPLE E

Find the average value of $f(x, y) = 2x/y$ over the rectangle D shown in Figure 23.

Solution We first compute the required double integral which can, of course, be written as an iterated integral.

$$\int_0^2 \left[\int_1^2 \frac{2x}{y} \, dy \right] dx = \int_0^2 \left[2x \ln y \right]_1^2 dx = \int_0^2 2x \ln 2 \, dx$$

$$= \left[x^2 \ln 2 \right]_0^2 = 4 \ln 2$$

Since the area of D is 2, the required average value is 2 ln 2. $\qquad \blacksquare$

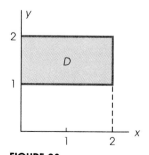

FIGURE 23

EXAMPLE F

The elevation $E(x, y)$ in kilometers at the point with coordinates (x, y) is given by $E(x, y) = (9 - x^2)e^{-y}$. Sketch the graph of the corresponding surface over the rectangle $D = \{(x, y): 0 \le x \le 2, 0 \le y \le 3\}$ and find its average elevation.

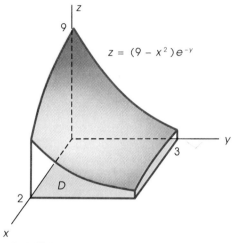

FIGURE 24

Solution The surface $z = E(x, y)$ is sketched in Figure 24. Also,

$$\int_0^2 \left[\int_0^3 (9 - x^2)e^{-y} \, dy \right] dx = \int_0^2 \left[(9 - x^2)(-e^{-y}) \right]_0^3 dx$$

$$= (1 - e^{-3}) \int_0^2 (9 - x^2) \, dx$$

$$= (1 - e^{-3}) \left[9x - \frac{1}{3}x^3 \right]_0^2$$

$$= (1 - e^{-3}) \left(18 - \frac{8}{3} \right) \approx 14.57$$

Thus, the average elevation is $\frac{1}{6}(14.57) \approx 2.43$ kilometers.   ∎

Problem Set 11.3

In Problems 1–4, sketch the tetrahedron bounded by the given plane and the three coordinate planes. (Finding the three intercepts is the key.) Then use integration to calculate the volume of this tetrahedron. Fi- *nally, check your answer by using a formula from geometry. See Example A.*

1. $2x + 2y + z = 6$
2. $x + 2y + 4z = 8$
3. $2x - y + 2z = 6$
4. $x - 3y - 2z = -6$

In Problems 5–8, find the volume of the solid under the surface z = f(x, y) and above the region D. See Example B.

5. $z = x^2 + 2y$; $D = \{(x, y): 0 \le y \le 2 - x, 0 \le x \le 2\}$
6. $z = 4 - x^2$; $D = \{(x, y): 0 \le y \le x, 0 \le x \le 2\}$
7. $z = x^2y$; $D = \{(x, y): x^2 \le y \le 4, 0 \le x \le 2\}$
8. $z = 4y + x$; $D = \{(x, y): 0 \le y \le x^2, 0 \le x \le 3\}$

In Problems 9–14, find the total mass of the lamina with the indicated shape L and density k(x, y). See Example C.

9. $k(x, y) = x + 2y$

10. $k(x, y) = 2x^2y$

11. $k(x, y) = xe^y$

12. $k(x, y) = 10\sqrt{x}\, y$

13. $k(x, y) = 2x^2y$

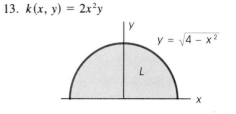

14. $k(x, y) = x^2\sqrt{y}$

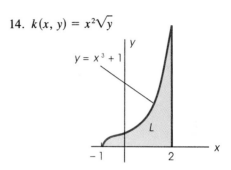

15. A city has the shape of the square shown in Figure 25. Its population density in people per square mile at (x, y) is given by $k(x, y) = 16{,}000e^{-|x|/2}e^{-|y|}$. Find the population of the city. See Example D.

FIGURE 25

16. A city situated along a straight river is within the rectangle shown in Figure 26. If the population density in people per square mile at (x, y) is given by $k(x, y) = 20{,}000(4 - y^2)$, find the population of the city.

FIGURE 26

Find the average value of $f(x, y)$ over the region D in Problems 17–20. See Example E.

17. $f(x, y) = (2x + 1)/y$; $D = \{(x, y): 1 \leq y \leq 3,$
$$0 \leq x \leq 2\}$$

18. $f(x, y) = x^{1/2}y^{3/2}$; $D = \{(x, y): 1 \leq y \leq 4,$
$$0 \leq x \leq 4\}$$

19. $f(x, y) = 2x^2y$; $D = \{(x, y): (x - 2)^2 + y^2 \leq 4,$
$$2 \leq x \leq 4\}$$

20. $f(x, y) = y/x$; $D = \{(x, y): 0 \leq y \leq \ln x,$
$$1 \leq x \leq e^2\}$$

21. The elevation in kilometers at the point (x, y) is $x^2 + y^2$. Find the average elevation over the rectangle $D = \{(x, y): 0 \leq y \leq 2, 0 \leq x \leq 3\}$.

22. Redo Problem 21 assuming the elevation is $25 - x^2 - 2y$.

23. The temperature in degrees Celsius at the point (x, y) on the triangular plate shown in Figure 27 is $T(x, y) = 40 - 3x$. Find the average temperature of the plate.

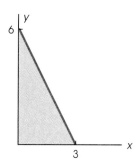

FIGURE 27

24. If the temperature in degrees Celsius at the point (x, y) on the square plate shown in Figure 28 is $T(x, y) = 45 - x^2 - y^2$, find the average temperature of the plate.

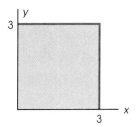

FIGURE 28

25. Sketch the solid bounded by the planes $z = \frac{1}{3}(6 - 2x)$, $x = 0$, $y = 0$, $z = 0$, and $y = 6$. Then find its volume.

26. Find the volume of the solid under the surface $z = e^{x+2y}$ and above the region
$$D = \{(x, y): 0 \leq y \leq 2, 0 \leq x \leq 3\}.$$

27. A lamina L has the shape shown in Figure 29. If its density at (x, y) is $k(x, y) = ye^x$, find the mass of the lamina.

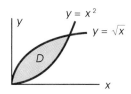

FIGURE 29

28. Find the average value of $f(x, y) = 2x^{3/2}y$ over the region D shown in Figure 30.

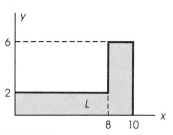

FIGURE 30

*Let L be a lamina with mass $M(L)$ and density $k(x, y)$ at the point (x, y). Its **center of mass** has coordinates $(\bar{x}, \bar{y})$, where*

$$\bar{x} = \frac{\iint\limits_L x\, k(x, y)\, dy\, dx}{M(L)} \qquad \bar{y} = \frac{\iint\limits_L y\, k(x, y)\, dy\, dx}{M(L)}$$

Find the coordinates of the center of mass for the laminas in Problems 29 and 30.

29. $L = \{(x, y): 0 \leq y \leq \frac{1}{2}x, 0 \leq x \leq 4\}$;
$$k(x, y) = 2xy$$

30. $L = \{(x, y): 3x \leq y \leq 6, 0 \leq x \leq 2\}$;
$$k(x, y) = 2x + 4y$$

Chapter 11 Review Problem Set

In Problems 1–8, evaluate each iterated integral.

1. $\displaystyle\int_1^3 \left[\int_0^2 (2x^2y + 3y^2)\, dy \right] dx$

2. $\displaystyle\int_0^2 \left[\int_1^2 (2xy^3 + 3x^2)\, dy \right] dx$

3. $\displaystyle\int_0^{\ln 3} \left[\int_0^{\ln 2} e^{2x+y}\, dy \right] dx$

4. $\displaystyle\int_0^2 \left[\int_0^3 \frac{x^2 y}{1 + y^2}\, dx \right] dy$ 5. $\displaystyle\int_0^1 \left[\int_{-y}^{3y} e^{2y^2}\, dx \right] dy$

6. $\displaystyle\int_{1/4}^1 \left[\int_0^{1/x^2} e^{1/x}\, dy \right] dx$

7. $\displaystyle\int_0^4 \left[\int_0^x \frac{2xy}{(1 + y^2)^2}\, dy \right] dx$

8. $\displaystyle\int_0^2 \left[\int_0^{3y^2} \sqrt{1 + y^3}\, dx \right] dy$

9. Sketch the tetrahedron bounded by the coordinate planes and the plane $4x + 2y + z = 12$. Then find the volume of this tetrahedron by integration.

10. Use integration to find the volume of the tetrahedron bounded by the coordinate planes and the plane $2x - 2y + z = 4$. Check by using a volume formula from geometry.

11. Consider the double integral $\iint_D 2x\, dy\, dx$, where $D = \{(x, y): 0 \le y \le x^3, 0 \le x \le 2\}$.
    (a) Write the corresponding iterated integral in the order $dy\, dx$ and evaluate.
    (b) Write the corresponding iterated integral in the order $dx\, dy$ and evaluate.

12. Evaluate the double integral $\iint_D 4xy^3\, dy\, dx$ where D is the region bounded by $y = \sqrt{x}$, $y = 0$, and $x = 4$ in two ways, that is, by using both orders of integration.

13. Find the volume of the solid under the plane $z = x + 2y$ and above the region
    $$D = \{(x, y): 0 \le y \le \sqrt{4 - x^2}, 0 \le x \le 2\}.$$

14. Find the volume of the solid under the plane $z = 4x$ and above the region
    $$D = \{(x, y): 0 \le y \le 4 - 2x, 0 \le x \le 2\}$$

15. Find the mass of the lamina
    $$L = \{(x, y): 0 \le y \le x, 0 \le x \le 1\}$$
    if its density at (x, y) is $k(x, y) = 2e^{x^3}y$.

16. Find the mass of the lamina bounded by $y = x^2$, $y = 4$, and $x = 0$ if its density at (x, y) is $k(x, y) = 6x\sqrt{y}$.

17. The elevation of a surface at (x, y) is $E(x, y) = 2x^2$.
    (a) Sketch this surface over the region
    $$D = \{(x, y): 0 \le y \le 6, -2 \le x \le 0\}$$
    (b) Find the average elevation of the surface over D.

18. A surface has equation $z = 4 - x^2 - y^2$. Find its average height (elevation) on the square bounded by $y = 0$, $y = 1$, $x = 0$, and $x = 1$.

19. Find the average value of $f(x, y) = y$ over the region
    $$D = \{(x, y): 0 \le y \le 3x - x^2, 0 \le x \le 3\}$$

20. Find the average value of $f(x, y) = x^2$ on the region
    $$D = \{(x, y): 0 \le y \le (x - 3)^2, 0 \le x \le 3\}$$

21. A triangular plate has its vertices at $(0, 0)$, $(4, 0)$, and $(4, 2)$. The temperature of the plate at the point (x, y) is $T(x, y) = 30e^{x-y-2}$ degrees Celsius. Find the average temperature of the plate.

22. A square plate has its vertices at $(0, 0)$, $(4, 0)$, $(4, 4)$, and $(0, 4)$. The temperature of the plate at the point (x, y) is $T(x, y) = 30\sqrt{x}/(y + 1)$. Find the average temperature of the plate.

23. Find the mass of the lamina bounded by $y = \sqrt{x}$, $y = 1$, and $x = 0$ if its density at (x, y) is
    $$k(x, y) = 2y/(x + 1).$$

24. Find the mass of the lamina bounded by $y = 1/x$, $y = 0$, $x = 1$, and $x = e$ if its density at (x, y) is $k(x, y) = 4xy$.

25. Evaluate $\int_2^\infty [\int_0^\infty (e^{-y}/x^2)\, dy]\, dx$.

26. Evaluate $\displaystyle\int_3^\infty \left[\int_2^\infty \frac{4}{x^2 y^2}\, dy \right] dx$.

27. Find the volume of the solid above the region
    $$D = \{(x, y): \sqrt{x} \le y \le e^x, 0 \le x \le 1\}$$
    and trapped between the surfaces $z = [\ln(1 + xy)]^4$ and $z = [\ln(1 + xy)]^4 + 2y$.

28. Find the volume of the solid above the region
    $$D = \{(x, y): (4 - y)/2 \le x \le 4 - y, 0 \le y \le 4\}$$
    and trapped between the surfaces $z = 2e^{\sqrt{x}} + 3x^2$ and $z = 2e^{\sqrt{x}}$.

The Mandelbrot set is a strangely beautiful but amazingly complex object. It appears to be a wildly irregular set; yet this irregularity repeats itself endlessly in a kind of infinite regress as one zooms in on any piece of the set. Will it or its relatives provide models for turbulence?

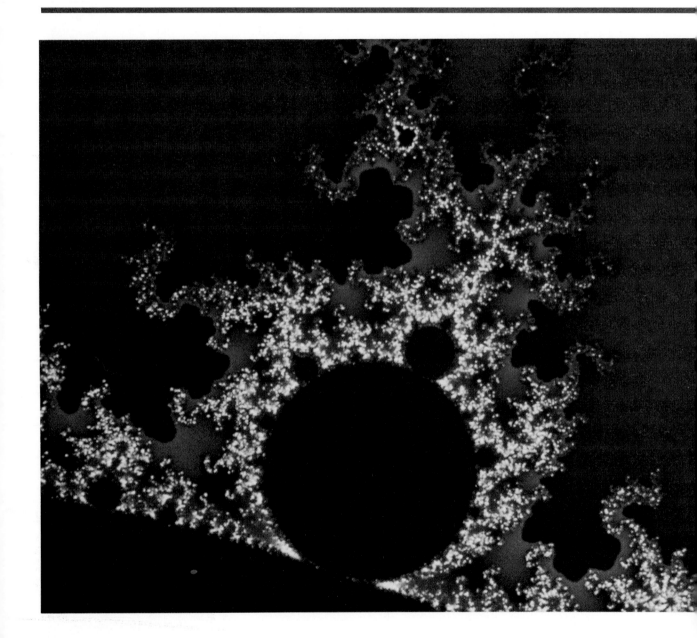

12

The Trigonometric Functions

Benoit Mandelbrot, mathematical maverick and scientific nomad, has had a career almost as chaotic as the discipline he invented—*fractal geometry*. He was born in Warsaw to Lithuanian Jewish parents, spent his formative years in France, earned a Ph.D. at the University of Paris with a thesis on linguistics and thermodynamics, taught at a number of universities, and was eventually appointed an IBM fellow at the Thomas J. Watson Research Center. At times, he has been Visiting Professor of Economics, Applied Mathematics, and Mathematics at Harvard, of Engineering at Yale, of Physiology at Albert Einstein College of Medicine, and of Mathematics at the University of Paris-Sud.

But Mandelbrot does not fit any of the standard categories. He is not a typical mathematician; he finds modern mathematics too constricting and too cluttered with algorithms and formulas. He thinks in pictures rather than formulas, but is not a typical geometer either. While his early work involved forays in every direction with little apparent unity, it now seems clear that all of his studies, travels, and papers were preparing him for the formulation of a new branch of mathematics, which he has described in two influential books—*Fractals: Form, Chance, and Dimension* and *The Fractal Geometry of Nature*. In these books and his other writings, Mandelbrot finds a strange regularity both in the pathological examples of earlier mathematicians and also in such diverse places as the chaotic outlines of a coastline, the wild fluctuations of the stock market, and the turbulent eddies in a rushing river. His marvelous computer pictures representing these phenomena have appeared in both popular books and serious scientific articles.

Is all this just a fad or does it have real significance? Time will tell, but one can hardly argue with Mandelbrot's own words. "Clouds are not spheres, mountains are not cones, coastlines are not circles and bark is not smooth, nor does lightning travel in a straight line." Perhaps Mandelbrot's fractals will give us the ultimate insight into the world of irregularity.

Benoit B. Mandelbrot
(1924–   )

ANGLES, ARCS, AND TRIANGLES

Many phenomena that occur in the real world follow a cyclic pattern; that is, as time goes by they repeat themselves over and over again. Examples are the number of daylight hours in a day, the voltage in an alternating current, the blood pressure in the aorta, the temperature of the human body, and the amount of business activity. When measured and displayed pictorially, each of these phenomena has a graph with features something like what is shown in Figure 1.

Functions with graphs like the one in Figure 1 are said to be *periodic*. The premier examples of periodic functions are the trigonometric functions, the subject of this chapter. Before we can describe them, we need to develop some preliminary material on angles and how they are measured.

Angles, Degrees, and Radians

Rotating a ray (half-line) about its endpoint (Figure 2) generates an **angle.** The original position of the ray is the **initial side** of the angle, the final position is its **terminal side,** and the point of intersection of the two sides is the **vertex** of the angle. An angle is in **standard position** in the coordinate system if its initial side coincides with the positive x-axis. It is a **positive** angle if it is generated by a counterclockwise rotation, and it is a **negative** angle if it is generated by a clockwise rotation (Figure 3).

The ancient Babylonians divided a circle into 360 equal parts, thereby introducing the **degree** measure of an angle. Note (Figure 4) that an angle with the same initial and terminal sides can have many different degree measures depending on how it was generated. Degree measure is commonly used in everyday life and in engineering but is a clumsy measure for the purposes of calculus.

In calculus and in all of advanced mathematics, we measure angles in radians. Recall that the circumference of a circle of radius r measures $2\pi r$; thus there are $2\pi \approx 6.28$ arcs of length r around the circumference. Each of these arcs determines an angle of size one **radian** (Figure 5). It follows

FIGURE 1

FIGURE 2

FIGURE 3

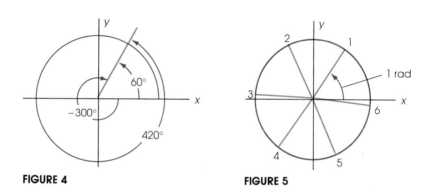

FIGURE 4

FIGURE 5

that 360 degrees (360°) and 2π radians measure the same size angle, namely, a complete counterclockwise revolution. Equivalently,

$$180° = \pi \text{ radians}$$

The boxed equality allows us to change from degree measure to radian measure, and vice versa. Dividing it by 2, 3, 4, and 6, respectively, gives us the conversions for the so-called special angles. They are shown in Figure 6.

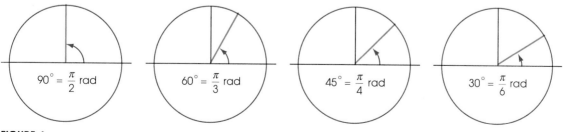

FIGURE 6

EXAMPLE A

Sketch angles in standard position of size $600°$ and $-315°$. Then change their measures to radians.

$$600° = \frac{10\pi}{3} \text{ rad} \qquad -315° = \frac{-7\pi}{4} \text{ rad}$$

FIGURE 7

Solution The two angles are shown in Figure 7. Here are their radian measures.

$$600° = 360° + 240° = 2\pi + \left(\frac{2}{3}\right)2\pi = \frac{10\pi}{3} \text{ radians}$$

$$-315° = -(180° + 135°) = -\left(\pi + \frac{3\pi}{4}\right) = -\frac{7\pi}{4} \text{ radians} \qquad ■$$

EXAMPLE B

Sketch an angle in standard position of size $11\pi/6$ radians and convert its measure to degrees.

$$\frac{11\pi}{6} \text{ rad}$$

FIGURE 8

Solution The angle is shown in Figure 8. Also,

$$\frac{11\pi}{6} \text{ radians} = \left(\pi + \frac{5\pi}{6}\right) \text{ radians} = 180° + 150° = 330° \qquad ■$$

Dividing the equality $180° = \pi$ radians by 180 gives

$$1° = \frac{\pi}{180} \text{ radians} \approx 0.01745329 \text{ radians}$$

and dividing by π gives

$$1 \text{ radian} = \frac{180°}{\pi} \approx 57.2957795°$$

Thus, to change degrees to radians, we multiply by $\pi/180$; to change radians to degrees, we multiply by $180/\pi$.

EXAMPLE C

Convert 24.56° to radians and 24.56 radians to degrees.

Solution

$$24.56° = 24.56\left(\frac{\pi}{180}\right) \approx 0.4286529 \text{ radians}$$

$$24.56 \text{ radians} = 24.56\left(\frac{180}{\pi}\right) \approx 1407.184345° \qquad ■$$

Measuring the Lengths of Arcs

Look again at our definition of radian measure. We have formulated it in such a way that if an angle of t radians has its vertex at the center of a circle of radius r, this angle will cut off an arc of length tr units on the circumference of the circle. In other words, if s denotes the length of arc cut off on a circle of radius r by a central angle of size t radians, then

$$s = tr$$

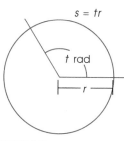

a formula worth remembering (Figure 9). We emphasize that to use this formula, you must measure the angle in radians.

FIGURE 9

EXAMPLE D

Find the length of the arc cut off on a circle of radius 10.8 centimeters by an angle of size 24.56°.

Solution From Example C, we learn that the given angle measures 0.4286529 radians. Thus, the indicated arc measures

$$s = tr \approx (0.4286529)(10.8) = 4.629451 \text{ centimeters} \qquad ■$$

EXAMPLE E

Find the distance traveled by a bicycle while its wheels of radius 15 inches complete 40 revolutions.

Solution The angle through which the wheels turn measures $40(2\pi)$ radians, which corresponds to a total arc length of

$$s = tr = 40(2\pi)(15) = 1200\pi \approx 3770 \text{ inches} \qquad ■$$

$$A = \frac{1}{2}tr^2$$

t rad

r

FIGURE 10

The Area of a Sector

Consider the sector of a circle of radius *r* cut off by a central angle of size *t* radians (Figure 10). Its area *A* is equal to the fraction $t/2\pi$ of the area of the whole circle; that is, $A = (t/2\pi)(\pi r^2)$. This simplifies to the following formula.

$$A = \frac{1}{2}tr^2$$

Again we emphasize that to use this formula, the angle of the sector must be measured in radians.

EXAMPLE F

Find the area of the sector of a circle of radius 25.7 centimeters determined by an angle of size 60°.

Solution An angle of size 60° measures $\pi/3$ radians. Thus,

$$A = \frac{1}{2}tr^2 = \frac{1}{2}\frac{\pi}{3}(25.7)^2 \approx 345.832 \text{ square centimeters} \qquad ■$$

Right Triangle Trigonometry

Trigonometry was invented to help with problems in surveying, navigation, and astronomy. In these subjects, it is necessary to be able to solve for unknown parts of triangles when certain information is given. While this is not the thrust of this course, we think it worthwhile to explain the fundamental idea.

Consider an arbitrary right triangle (one with a right angle, that is, a 90° angle). Let *t* denote one of the acute angles of the triangle and label the sides relative to *t* as adjacent side, opposite side, and hypotenuse (Figure 11). Then, using obvious abbreviations, we define the sine of *t*, cosine of *t*, and tangent of *t* by

$$\sin t = \frac{\text{opp}}{\text{hyp}} \qquad \cos t = \frac{\text{adj}}{\text{hyp}} \qquad \tan t = \frac{\text{opp}}{\text{adj}}$$

We will give more general definitions of these functions in later sections, but this is enough to be able to solve right triangles.

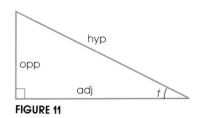

FIGURE 11

EXAMPLE G

Suppose we know the hypotenuse of a right triangle measures 12.56 meters and that one of the acute angles measures 24.56° ≈ 0.4286529 radians (Figure 12). Find the length of the side opposite the given angle.

FIGURE 12

Solution Let t denote the given angle and b the length of the opposite side. Then, $\sin t = b/12.56$; that is,

$$b = 12.56 \sin t$$

Now we could look for $\sin t$ in Table C or D at the back of the book. However, it is simpler and better to use a calculator. We may express the measure of t in either degrees or radians, provided we set our calculator to the corresponding mode. Suppose that we use the radian mode. Then we get

$$b = 12.56 \sin(0.4286529) \approx 12.56(0.415646) \approx 5.22 \text{ meters} \quad ■$$

Problem Set 12.1

Determine the radian measure of each of the angles in Problems 1–6.

1.

2.

3.

4.

5.

6.

Change the degree measures in Problems 7–14 to radian measures. Leave π in the first four answers.

7. 120°   8. 315°   9. −690°
10. 2640°   C11. 121.5°   C12. −39.4°
C13. −821.6° C14. $(\pi/180)°$

Each of the numbers in Problems 15–22 is a radian measure. Change it to a degree measure as in Examples B and C.

15. $7\pi/6$ 16. $-3\pi/2$ 17. $-13\pi/6$
18. $15\pi/4$ C19. −1.36   C20. 5.49
C21. 41.23   C22. $\pi\sqrt{3} - 2$

Find the length of arc and the area of the sector cut off by the central angle shown in Problems 23–26. See Examples D and F.

23.

24.

$\boxed{\text{c}}$25.

$\boxed{\text{c}}$26.

$\boxed{\text{c}}$27. Find the radius of each of the following circles.
(a)

(b)   s = 45 in.

$\boxed{\text{c}}$28. Find the radius of the circle if a central angle of 33°
cuts off a sector of area 25.4 square centimeters.

$\boxed{\text{c}}$29. Suppose that a tire on a truck has an outer diameter of
3.2 feet. If the tire is making 400 revolutions per
minute, how fast is the truck going in miles per hour?

30. Through how many radians does the minute hand of a
clock turn from 1.00 P.M. to 3.30 P.M.? The hour
hand during the same time period?

31. Find the value of x in each of the following.
(a)

$\boxed{\text{c}}$(b)   42.5 ft

32. For the triangle in Figure 13, find: (a) b, (b) $\sin t$,
(c) $\cos t$.

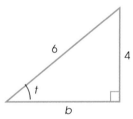

FIGURE 13

33. A guy wire is attached to the top of a 300-foot vertical pole and to a stake at ground level. This wire
makes an angle of 58° with the ground.
(a) How long is the guy wire?
(b) How far is the stake from the base of the pole?

34. A building casts a shadow on the ground 98 feet long
when the angle of elevation (angle above the horizontal) of the sun is 62°. How tall is the building?

35. Change 13° to radian measure and change 13 radians
to degree measure.

36. An arc of length 35 centimeters is cut off by a central
angle of t radians on a circle of radius 14 centimeters.
Determine t and then calculate the area of the corresponding sector of the circle.

37. The top section of a flagpole is painted red. From a
point on the ground 100 feet from the base of the
pole, I measure the angle of elevation to the tip of the
pole to be 44° and to the bottom of the painted section to be 41°. How long is the painted section?

38. A boy flying a kite has let out 400 feet of string. The taut string makes an angle of 32° with the horizontal. Assuming the boy grips the string at a point 5 feet above the ground, how high in the air is the kite?

39. There is a small grease spot on the belt that passes around the two pulleys of radii 5 inches and 9 inches shown in Figure 14. The larger pulley is turning at 40 revolutions per minute.

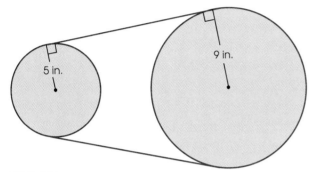

FIGURE 14

   (a) How fast (in inches per minute) is the grease spot moving?
   (b) How fast (in revolutions per minute) is the smaller pulley turning?

40. Find the length of the belt in Figure 14 assuming the centers of the pulleys are 15 inches apart.

41. Find the area of the shaded region in Figure 15.

FIGURE 15

42. Find x in Figure 16, given that $\overline{AB} = 2\overline{BC}$.

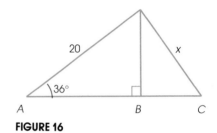

FIGURE 16

12.2
THE SINE AND COSINE FUNCTIONS

There are six trigonometric functions, but we will put most of our emphasis on the two called the sine and cosine. There are two reasons for this. First, the other four functions are simple algebraic combinations of sines and cosines. Second, these two functions have the major role in the applications we have in mind.

From now on (unless otherwise indicated), we will measure all angles in radians. Typically, we will use t to stand for both an angle and its radian measure. Thus, let t be an angle in standard position and let $P(x, y)$ be a point (other than the vertex) on the terminal side of t (Figure 17). Let r denote the distance from the origin to $P(x, y)$ and note that $r = \sqrt{x^2 + y^2}$, a result that follows from the Pythagorean Theorem. Then the **sine** of t and the **cosine** of t (abbreviated sin t and cos t) are defined as follows.

FIGURE 17

$$\sin t = \frac{y}{r} \quad \text{and} \quad \cos t = \frac{x}{r}$$

Because of similarity of triangles, it does not matter which point P we pick on the terminal side of angle t. The ratios y/r and y'/r' in Figure 18 are the same; so are x/r and x'/r'.

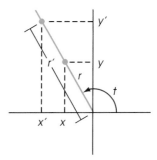

FIGURE 18

Evaluating Sines and Cosines

Our first two examples are based directly on the definitions given above.

EXAMPLE A

In Figure 19, we show angles t and u with corresponding points $P(3, 4)$ and $Q(-5, 12)$ on their terminal sides. Find the sines and cosines of both angles.

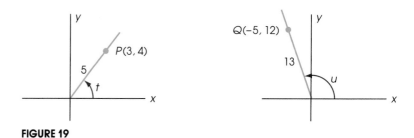

FIGURE 19

Solution First use the formula $r = \sqrt{x^2 + y^2}$ to find that P is 5 units and Q is 13 units from the origin. Thus

$$\sin t = \frac{4}{5} \qquad \cos t = \frac{3}{5}$$

$$\sin u = \frac{12}{13} \qquad \cos u = \frac{-5}{13}$$

∎

FIGURE 20

EXAMPLE B

Find $\sin(3\pi/4)$.

Solution The angle of radian measure $3\pi/4$ has its terminal side along the line $y = -x$ (Figure 20). Thus, $(-1, 1)$ is one point on its terminal side, and this point is a distance $\sqrt{2}$ from the origin. Our conclusion is that $\sin(3\pi/4) = 1/\sqrt{2} \approx 0.7071$. ∎

It should be clear that sine and cosine are genuine functions. In particular, sine associates with any real number t (the radian measure of an angle) a corresponding unique number $\sin t$. To say that $\sin t$ is well-defined is one thing; to find its value for a given t is quite another thing. Traditionally, this was done by making use of a table (like Table C or D at the back of the book), but now it is simpler to use a scientific calculator. About all that needs to be said is that for most calculations in this book, you will want to make sure that your calculator is in the radian mode.

EXAMPLE C

Evaluate $\sin(4.345)$ and $\cos(-5.123)$.

Solution We use a calculator to obtain, after rounding to four decimal places,

$$\sin(4.345) = -0.9333 \qquad \cos(-5.123) = 0.3992$$ ∎

While we encourage use of a calculator, we think it wise to list exact sine and cosine values for some special angles, and we advise you to memorize them.

t	0	$\pi/6$	$\pi/4$	$\pi/3$	$\pi/2$
$\sin t$	0	$1/2$	$\sqrt{2}/2$	$\sqrt{3}/2$	1
$\cos t$	1	$\sqrt{3}/2$	$\sqrt{2}/2$	$1/2$	0

The Graphs of Sine and Cosine

In order to graph the sine and cosine functions, we could now follow our standard procedure: Make a table of values, plot the corresponding points, and connect the points with a smooth curve. But before we do this, let us see what we can learn about these graphs by examining the definitions of sine and cosine. For each angle t, let us pick a point $P(x, y)$ on its terminal side exactly 1 unit from the origin. Then $\sin t = y$ and $\cos t = x$ (Figure 21).

As t increases from 0 to 2π, the point P moves around the unit circle. Correspondingly, $y = \sin t$ increases while P is in the first quadrant, then

THE TRIGONOMETRIC FUNCTIONS

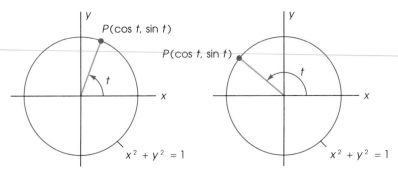

FIGURE 21

decreases as P moves through the second and third quadrants, and finally increases as P moves through the fourth quadrant. As t increases from 2π to 4π, P retraces its path and sin t exactly mimics its behavior between 0 and 2π. If you check, you will find that cos t also has this repetitive behavior. We say that sine and cosine are periodic functions with period 2π. In symbols, this means that

$$\sin(t + 2\pi) = \sin t \qquad \cos(t + 2\pi) = \cos t$$

In general, a function f is said to be **periodic** with **period** p if $f(t + p) = f(t)$ for all t, and if p is the smallest positive number for which this equality holds. The sine and cosine functions are periodic with period 2π.

In Figure 22, we display a table of values for $w = \sin t$ and $w = \cos t$ and then in Figure 23 we show their corresponding graphs. Note the periodic behavior of these graphs; note also that the graphs increase and decrease in the manner described above.

From the graphs, you can guess at a number of relationships. The graph of $w = \sin t$ appears to be symmetric with respect to the origin; that of $w = \cos t$ seems to be symmetric with respect to the y-axis. In algebraic symbols,

$$\sin(-t) = -\sin t \qquad \cos(-t) = \cos t$$

Also, the graph of $w = \cos t$ is exactly like that of $w = \sin t$, only shifted $\pi/2$ units to the left. Equivalently,

$$\cos t = \sin\left(t + \frac{\pi}{2}\right)$$

t	$\sin t$	$\cos t$
0	0	1
$\pi/4$	0.707	0.707
$\pi/2$	1	0
$3\pi/4$	0.707	−0.707
π	0	−1
$5\pi/4$	−0.707	−0.707
$3\pi/2$	−1	0
$7\pi/4$	−0.707	0.707
2π	0	1
$5\pi/2$	1	0
3π	0	−1
$7\pi/2$	−1	0
4π	0	1

FIGURE 22

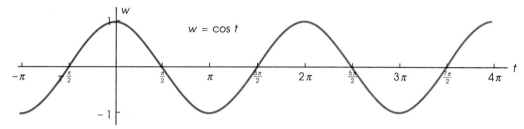

FIGURE 23

Finally, we mention three other important identities.

$$\sin\left(\frac{\pi}{2} - t\right) = \cos t \qquad \cos\left(\frac{\pi}{2} - t\right) = \sin t$$

$$\sin^2 t + \cos^2 t = 1$$

(Here, we follow standard practice in writing $\sin^2 t$ and $\cos^2 t$ in place of $(\sin t)^2$ and $(\cos t)^2$.) Each of these identities can be proved rigorously directly from the definitions of sine and cosine.

Some Modified Sine and Cosine Graphs

Let A and B be positive numbers. The graphs of $w = A \sin t$ and $w = \sin Bt$ are simple modifications of the graph of $w = \sin t$. The first is stretched or compressed in the vertical direction; the second is stretched or compressed in the horizontal direction.

EXAMPLE D

Sketch the graphs of $w = 3 \sin t$ and $w = \sin 2t$.

Solution We suggest that you construct tables of values to convince yourself that the graphs are as shown in Figure 24. ∎

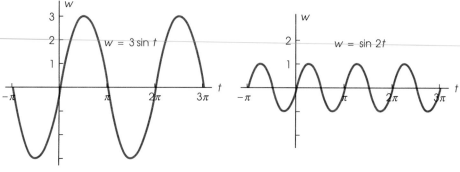

FIGURE 24

One-half the amount of vertical variation of a graph is called its **amplitude.** Thus (see Figure 24), the amplitude of the graph of $w = 3 \sin t$ is $\frac{6}{2} = 3$, while for $w = \sin 2t$, the amplitude is $\frac{2}{2} = 1$. The horizontal length of a complete cycle is the **period** of the graph. Thus, the graph of $w = 3 \sin t$ has period 2π, and the graph of $w = \sin 2t$ has period π. More generally, the graph of $w = A \sin Bt$, with both A and B positive, is a repetitive wave characterized by the two numbers:

$$\text{Amplitude} = A$$

$$\text{Period} = \frac{2\pi}{B}$$

This is also true of the graph of $w = A \cos Bt$.

EXAMPLE E

Sketch the graphs of $w = \frac{1}{2} \sin 3t$ and $w = \frac{3}{2} \cos \frac{1}{2}t$.

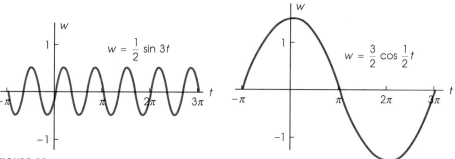

FIGURE 25

Solution We simply modify the graphs of $w = \sin t$ and $w = \cos t$ using the concepts of amplitude and period. They are shown in Figure 25. ∎

EXAMPLE F

Sketch the graph of $w = 2 \sin t + \cos 2t$.

Solution We first graph $w = 2 \sin t$ and then $w = \cos 2t$. Finally, we add the w-coordinates of the two graphs (the so-called method of *addition of ordinates*). The result is shown in Figure 26. ∎

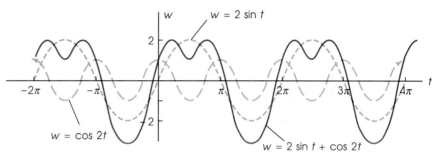

FIGURE 26

Problem Set 12.2

In Problems 1–6, find the values of r, sin t, and cos t.

1.

3.

2.

4.

5.

6.

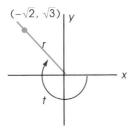

7. Let t be a second-quadrant angle (Figure 27) and let $P(x, 8)$ be a point on its terminal side r units from the origin. Find r, x, and $\cos t$, given that $\sin t = 0.8$.

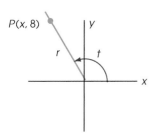

FIGURE 27

8. Let t be a fourth-quadrant angle (Figure 28) and let $P(21, y)$ be a point on its terminal side r units from the origin. Find r, y, and $\sin t$, given that $\cos t = \frac{7}{25}$.

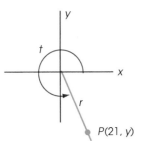

FIGURE 28

9. Find $\sin t$ and $\cos t$ if t has the given radian measure.
   (a) $5\pi/4$ (b) $-7\pi/4$ (c) $5\pi/2$ (d) $15\pi/4$

10. Find $\sin t$ and $\cos t$ if t has the given radian measure.
    (a) $-\pi/2$ (b) $11\pi/4$ (c) 7π (d) -12π

11. Use the fact that the right triangle with sides 1 and $\sqrt{3}$ and hypotenuse 2 is a 30°-60° right triangle to help you determine a point on the terminal side of each of the following angles. Then find the sine and cosine of each angle.
    (a) $150°$ (b) $-330°$ (c) $7\pi/6$ radians
    (d) $13\pi/6$ radians

12. Find $\sin t$ and $\cos t$ if t has the given radian measure. See Problem 11.
    (a) $2\pi/3$ (b) $5\pi/3$ (c) $-7\pi/3$ (d) $19\pi/3$

C 13. Use your calculator to find each of the following. When the degree symbol is missing, it is understood that the angle is measured in radians.
    (a) $\sin 322°$ (b) $\cos(-52.3°)$ (c) $\sin 4.32$
    (d) $\sin(-4.32)$ (e) $\sin^2(\pi - 1.43)$
    (f) $\cos(\sin 0.123)$

C 14. Use your calculator to find each of the following.
    (a) $\cos(-5.34)$ (b) $\cos 5.34$ (c) $\tan 1.57$
    (d) $\sin^2 3 + \cos^2 3$ (e) $\sin^3(\pi/11)$
    (f) $\cos(\sin 87°)$

15. Suppose $\sin t > 0$ and $\cos t < 0$. In what quadrant does the terminal side of the angle t lie?

16. If $\sin t$ and $\cos t$ are both negative, in what quadrant does the terminal side of the angle t lie?

17. If $\sin t = \frac{3}{5}$ and $\cos t < 0$, evaluate $\cos t$. *Hint:* $\sin^2 t + \cos^2 t = 1$.

18. If $\cos t = -\frac{5}{13}$ and $\sin t < 0$, evaluate $\sin t$.

19. Suppose that t is a second-quadrant angle with $\cos t = -\frac{3}{5}$. Use identities of this section to find:
    (a) $\sin t$ (b) $\cos(t + 2\pi)$ (c) $\cos(-t)$
    (d) $\sin(\pi/2 - t)$ (e) $\sin(-t)$ (f) $\sin(t + 4\pi)$

20. Find six values of t between 0 and 2π for which $\cos 3t = 0$.

21. Find the amplitude and period of the graph of the given equation. See Example E.
    (a) $w = 4 \cos 4t$ (b) $w = 2 \sin(\frac{1}{3}t)$
    (c) $w = -3 \sin(1.5t)$

22. Find the amplitude and period of the graph.
    (a) $w = 3 \sin(\frac{3}{2}t)$ (b) $w = \pi \cos \pi t$
    (c) $w = -\sqrt{2} \cos(\frac{1}{2}t)$

In Problems 23–28, sketch the graph of the given equation on the indicated interval. See Examples D and E.

23. $w = 4 \sin t$, $-\pi \leq t \leq 2\pi$
24. $w = -2 \sin t$, $-\pi \leq t \leq 2\pi$
25. $w = 2 \sin(\frac{1}{3}t)$, $-3\pi \leq t \leq 3\pi$
26. $w = 4 \sin 3t$, $-\pi \leq t \leq 2\pi$

27. $w = 2.5 \cos 2t, \; -\pi \le t \le 2\pi$

28. $w = 3 \cos(\frac{1}{2}t), \; -2\pi \le t \le 2\pi$

In Problems 29 and 30, sketch each graph using the method of addition of ordinates as in Example F.

29. $w = \sin 2t + 3 \cos t$ 30. $w = \sin(\frac{1}{2}t) + \frac{1}{2} \sin t$

31. Suppose that the strength I in amperes of an alternating current is given by $I = 40 \sin(120\pi t)$, where t is measured in seconds.
    (a) Determine the period of this alternating current.
    (b) Find the number of cycles (periods) per minute.
    (c) What is the maximum strength of this current?

32. An object moves back and forth along a straight line so that its x-coordinate in feet at time t seconds is given by $x = 4 \sin 30t$ (an example of what is called simple harmonic motion).
    (a) Determine the period of this motion.
    (b) When does x first attain its maximum value?
    (c) What is the total distance traveled by the object during the first π seconds, that is, during $0 \le t \le \pi$?

$\boxed{\text{C}}$ 33. Evaluate $[2 + \cos^2(2.85)]/\sin(\cos 4.32)$.

34. Find two possible values of $\sin t$ if $\cos t = \frac{2}{3}$.

35. Let $f(t) = A \cos Bt$. If $f(0) = 2.5$ and f has period $5\pi/4$, find A and B.

36. Sketch the graph of each of the following on the interval $-\pi \le t \le \pi$.
    (a) $w = 3 \cos t$ (b) $w = 2 \sin 3t$
    (c) $w = 3 \cos t + 2 \sin 3t$

37. Sketch the graph and determine the period of each of the following periodic functions.
    (a) $f(t) = |\sin t|$ (b) $g(t) = \sin t + |\sin t|$

38. Sketch the graph of $w = t + \sin 2t$ on $0 \le t \le 3\pi$.

39. The *greatest integer function* $f(x) = [x]$ assigns to x the greatest integer less than or equal to x. For example, $[2.4] = 2$ and $[-2.4] = -3$. Convince yourself that each of the following functions is periodic and determine its period.
    (a) $g(x) = [x] - x$ (b) $h(x) = [2x] - 2x$

40. The *distance to the nearest integer function* $f(x) = \langle x \rangle$ assigns to x the distance to the nearest integer. For example, $\langle 2.4 \rangle = 0.4$ and $\langle 2.7 \rangle = 0.3$. Convince yourself that each of the following is a periodic function and then determine its period.
    (a) $f(x) = \langle x \rangle$ (b) $g(x) = \langle 3x \rangle$
    (c) $h(x) = \langle \frac{1}{2}x \rangle$

12.3

DERIVATIVES OF THE TRIGONOMETRIC FUNCTIONS

Let us begin by considering the graph of $w = \sin t$. Suppose that at each point t, we estimate the slope $m(t)$ of the graph at that point. Then, suppose that we graph $w = m(t)$ superimposed on the graph of $w = \sin t$. Of course, this second graph will be the graph of the derivative of $\sin t$. The process just described has been carried out in Figure 29.

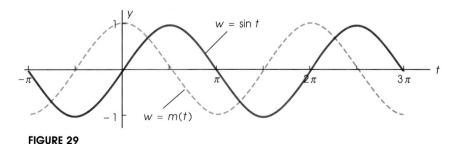

FIGURE 29

Figure 29 is very suggestive. It suggests the fundamental fact of this section, namely,

$$\frac{d}{dt}(\sin t) = \cos t$$

No one would call what we have just observed a proof. Actually, a rigorous proof is quite involved; such a proof can be found in most of the large engineering-type calculus books (for example, Purcell and Varberg, *Calculus with Analytic Geometry,* 5th edition, Prentice-Hall, pages 110–113). But to increase your confidence in the result, we give another demonstration that is much closer to being a proof.

A Geometric Argument for $\frac{d}{dt}(\sin t) = \cos t$

Let t be the radian measure of an angle in standard position which intersects the unit circle at a point $P(x, y)$ as in Figure 30. Then, as we learned in Section 12.2, $x = \cos t$ and $y = \sin t$. Give a small increment Δt to t, thus determining a second point Q on the unit circle whose coordinates are $\cos(t + \Delta t)$ and $\sin(t + \Delta t)$. Because our angles are measured in radians, the length of the arc from P to Q is just Δt. Draw in the vertical and horizontal segments QR and PR (Figure 30), thus forming region PQR which is very close to being a right triangle (its hypotenuse is slightly curved). Note that angle PQR is equal to angle QOT because their respective sides are perpendicular; thus angle PQR is equal to $t + \Delta t$. Finally let Δy be the length of QR. Then

$$\frac{\Delta y}{\Delta t} = \frac{\sin(t + \Delta t) - \sin t}{\Delta t} \approx \cos(t + \Delta t)$$

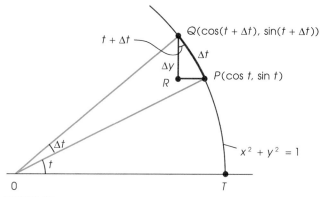

FIGURE 30

To tie down the argument, consider the right triangles formed by using the tangent lines at P and Q as in the diagrams below.

Note that $l < \Delta t < m$ and so

$$\frac{\Delta y}{m} < \frac{\Delta y}{\Delta t} < \frac{\Delta y}{l}$$

from which we conclude that

$$\cos(t + \Delta t) < \frac{\Delta y}{\Delta t} < \cos t$$

Now let $\Delta t \to 0$.

We conclude that

$$\frac{d}{dt}(\sin t) = \lim_{\Delta t \to 0} \frac{\sin(t + \Delta t) - \sin t}{\Delta t} = \lim_{\Delta t \to 0} \cos(t + \Delta t) = \cos t$$

More about Derivatives

Taking the formula for the derivative of $\sin t$ for granted, we can easily find the corresponding formula for the derivative of $\cos t$. Recall from Section 12.2 that

$$\cos t = \sin\left(\frac{\pi}{2} - t\right) \quad \text{and} \quad \cos\left(\frac{\pi}{2} - t\right) = \sin t$$

Applying the Chain Rule to the first of these equalities with $u = \pi/2 - t$ and $w = \cos t = \sin u$, we obtain

$$\frac{dw}{dt} = \frac{dw}{du}\frac{du}{dt} = \frac{d(\sin u)}{du}\frac{du}{dt} = (\cos u)(-1) = (-1)\cos\left(\frac{\pi}{2} - t\right) = -\sin t$$

Thus, we have our second important formula.

$$\frac{d}{dt}(\cos t) = -\sin t$$

Now we can find the derivatives of all kinds of functions which involve sines and cosines.

EXAMPLE A

Find $\dfrac{dw}{dt}$ for $w = \cos(t^2 + 3t)$.

Solution We apply the Chain Rule with $u = t^2 + 3t$ to obtain

$$\frac{dw}{dt} = \frac{dw}{du}\frac{du}{dt} = (-\sin u)(2t + 3) = -(2t + 3)\sin(t^2 + 3t) \qquad ∎$$

We have consistently used t as the independent variable in connection with sines and cosines. Any variable will do as well.

EXAMPLE B

Find $\dfrac{d}{dx}(\sin^3 x)$.

Solution We think of $\sin^3 x$ as u^3, where $u = \sin x$, and apply the Chain Rule. We obtain

$$\frac{d}{dx}(\sin^3 x) = \frac{d}{dx}(u^3) = 3u^2 \frac{du}{dx} = 3 \sin^2 x \cos x \qquad \blacksquare$$

EXAMPLE C

Find $f'(x)$ and $f''(x)$ for $f(x) = x \sin x$.

Solution We treat $x \sin x$ as a product.

$$f'(x) = x \cos x + \sin x$$

$$f''(x) = x(-\sin x) + \cos x + \cos x = -x \sin x + 2 \cos x \qquad \blacksquare$$

EXAMPLE D

Find an equation of the tangent line to the curve $y = 2 \cos^3 x$ at the point where $x = \pi/3$.

Solution The point in question has coordinates $(\pi/3, \frac{1}{4})$. Also

$$\frac{dy}{dx} = 6 (\cos^2 x)(-\sin x)$$

The slope m of the desired line is this derivative evaluated at $x = \pi/3$; that is, $m = 6(\frac{1}{2})^2(-\sqrt{3}/2) = -3\sqrt{3}/4$. The equation of this line in point-slope form is

$$y - \frac{1}{4} = -\frac{3\sqrt{3}}{4}\left(x - \frac{\pi}{3}\right) \qquad \blacksquare$$

Four Other Trigonometric Functions

Besides the sine and cosine, there are four other common trigonometric functions (tangent, cotangent, secant, and cosecant). They are defined as follows.

$$\tan t = \frac{\sin t}{\cos t} \qquad \cot t = \frac{\cos t}{\sin t} \qquad \sec t = \frac{1}{\cos t} \qquad \csc t = \frac{1}{\sin t}$$

Of these four functions, the tangent function is the most useful; we propose to begin our study of this function by drawing its graph. Since $\tan t$ is undefined whenever $\cos t = 0$, that is, when $t = \pm\pi/2, \pm 3\pi/2, \ldots$, the graph of $w = \tan t$ has vertical asymptotes at these points. With this information and a table of values (not shown), we are able to sketch the graph in Figure 31. Note that the tangent function is periodic with period π.

We notice from Figure 31 that the graph of $w = \tan t$ is increasing on each of its branches. We demonstrate that this must be so by showing that its derivative is always positive.

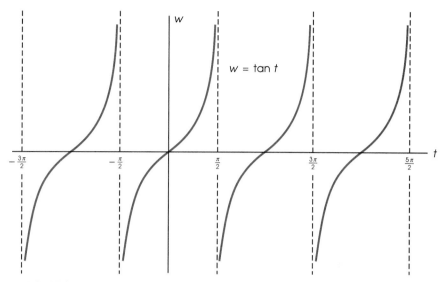

FIGURE 31

EXAMPLE E

Find a formula for $\dfrac{d}{dt} (\tan t)$.

Solution We use the quotient rule together with what we already know about the derivatives of sines and cosines.

$$\frac{d}{dt} (\tan t) = \frac{d}{dt} \left(\frac{\sin t}{\cos t} \right) = \frac{(\cos t)(\cos t) - (\sin t)(-\sin t)}{\cos^2 t}$$

$$= \frac{\sin^2 t + \cos^2 t}{\cos^2 t} = \frac{1}{\cos^2 t}$$

Finally, we obtain

$$\frac{d}{dt} (\tan t) = \sec^2 t$$

■

You will be asked to show in a similar way (Problems 29 and 30) that

$$\frac{d}{dt} (\cot t) = -\csc^2 t$$

$$\frac{d}{dt} (\sec t) = \sec t \tan t$$

$$\frac{d}{dt} (\csc t) = -\csc t \cot t$$

EXAMPLE F

Find $\dfrac{d}{dx}(\cot e^{2x})$.

Solution If we let $w = \cot u$, where $u = e^{2x}$, then by the Chain Rule

$$\frac{dw}{dx} = \frac{dw}{du}\frac{du}{dx} = (-\csc^2 u)(2e^{2x}) = -2e^{2x}\csc^2 e^{2x} \qquad ∎$$

Problem Set 12.3

*In Problems 1–20, find dw/dt as in Examples A and
B.*

1. $w = 3\sin 2t$
2. $w = 4\cos 5t$
3. $w = 2\cos(1 - 3t)$
4. $w = \frac{1}{2}\sin(-4t + 1)$
5. $w = \sin(3t^2)$
6. $w = \cos(t^3 + 2t)$
7. $w = \cos^3(t)$
8. $w = \sin^4(t)$
9. $w = \sin^2(2t^3 - 1)$
10. $w = \cos^3(1 - 3t^2)$
11. $w = \sin(\cos t)$
12. $w = \cos(\sin t)$
13. $w = t^2 \cos 5t$
14. $w = t^3 \sin 3t$
15. $w = (\sin 2t)/(\cos 4t)$
16. $w = e^t/(\sin 2t)$
17. $w = e^{\sin 2t}$
18. $w = e^{t+\cos t}$
19. $w = \ln(1 + \cos^2 t)$
20. $w = \ln(3 + 2\sin t)$

*In Problems 21–24, find f '(x) and f"(x). See Example
C.*

21. $f(x) = \cos^2 3x$
22. $f(x) = x^2 \sin x$
23. $f(x) = \ln(\sin x)$
24. $f(x) = \cos(x^3)$
25. Find $f''(\pi/2)$ if $f(x) = x\cos x + \sin x$.
26. Find $f''(0)$ if $f(x) = e^{2x}\cos x$.
27. Find an equation for the tangent line to the graph of
$y = 2\cos 2x - \sin^2 x$ at the point $(\pi/6, \frac{1}{4})$.
28. Find an equation for the tangent line to the graph of
$y = \cos^4 x$ at the point where $x = 3\pi/4$.
29. Use the identity $\cot x = (\cos x)/(\sin x)$ to show that
$d(\cot x)/dx = -\csc^2 x$.
30. Use the identities $\sec x = 1/\cos x$ and $\csc x = 1/\sin x$ to derive the formulas for the derivatives of
$\sec x$ and $\csc x$.
31. Find $f'(x)$ in each case.
(a) $f(x) = \tan^2(3x)$
(b) $f(x) = 4\tan(x^2)$
(c) $f(x) = \csc x \cot x$
(d) $f(x) = e^{\sec 2x}$
32. Find dy/dt in each case.
(a) $y = \sec t \tan t$
(b) $y = \sec^2 3t$
(c) $y = \cot(\tan t)$
(d) $y = \ln(\csc^2 t)$
33. Suppose that a particle moves along the x-axis in such
a way that its x-coordinate (in feet) after t seconds is
$x = 4\cos(\pi t/3)$.

(a) Determine the position and velocity of the particle after 1 second.
(b) Find the maximum velocity of the particle.
(c) At what $t > 0$ does the particle first reach maximum velocity?
34. Show that $y = A\cos(Bt + C)$ is a solution to the differential equation $d^2y/dt^2 = -B^2y$.

---

35. Find dy/dx.
(a) $y = \tan(1 + 2\sqrt{x})$ (b) $y = (x/\sin x)^2$
(c) $y = \cos^3(1/x)$ (d) $y = \sin(\sin(\sin x))$
36. If $f(x) = 3\sin(\pi x/3) - 2\cos(\pi x/6)$, find $f'(1)$.
37. A long sheet of tin 12 inches in width is to be bent to
form a rain gutter as shown in Figure 32. Determine
θ so that the gutter will carry the most water, that is,
so its cross-sectional area is a maximum.

FIGURE 32

38. Consider $f(x) = x + \cos x$.
(a) Find the critical points of f; that is, find the values of x where $f'(x) = 0$.
(b) Show that f has no local maxima or minima.
(c) Sketch the graph of f. *Hint:* The graph crosses the
line $y = x$ at $x = \pi/2, 3\pi/2, 5\pi/2$, and so on.
39. Evaluate $\lim_{h\to 0} \dfrac{\sin(2 + h) - \sin 2}{h}$ by recognizing
this limit as a certain derivative.
40. Find the maximum and minimum values of

$$f(x) = \frac{1 - \sin x}{1 + \cos x}$$

on the interval $[0, 3\pi/4]$

INTEGRATION OF THE TRIGONOMETRIC FUNCTIONS

Each of the differentiation formulas of the previous section yields a corresponding antidifferentiation (indefinite integration) formula. We state the three that we consider to be most important using u as the variable of integration.

$$\int \sin u \, du = -\cos u + C \qquad \int \cos u \, du = \sin u + C$$

$$\int \sec^2 u \, du = \tan u + C$$

To check that these formulas are correct, it is enough to differentiate each right side and observe that you get the corresponding integrand on the left side. We state these results using u as the variable because we want next to illustrate them in connection with changes of variables.

Change of Variables

You may want to review Section 7.2 before continuing.

EXAMPLE A

Find $\int 2x \sin(x^2) \, dx$.

Solution We make the substitution $u = x^2$; then $du = 2x \, dx$. Thus,

$$\int 2x \sin(x^2) \, dx = \int \sin u \, du = -\cos u + C = -\cos(x^2) + C \qquad \blacksquare$$

EXAMPLE B

Evaluate the definite integral $\int_0^{\pi/2} \sin^2 x \cos x \, dx$.

Solution Let $u = \sin x$ so that $du = \cos x \, dx$ and note that $u = 0$ when $x = 0$ and $u = 1$ when $x = \pi/2$. Thus,

$$\int_0^{\pi/2} \sin^2 x \cos x \, dx = \int_0^1 u^2 \, du = \left[\frac{u^3}{3} \right]_0^1 = \frac{1}{3}$$

Be sure to note that we changed the limits on the integral sign when we changed to the variable u. $\qquad \blacksquare$

EXAMPLE C

Find $\int (x^2 + 1) \cos (x^3 + 3x) \, dx$.

Solution Let $u = x^3 + 3x$ and $du = (3x^2 + 3) \, dx = 3(x^2 + 1) \, dx$. Then

$$\int (x^2 + 1) \cos(x^3 + 3x) \, dx = \frac{1}{3} \int (3x^2 + 3) \cos(x^3 + 3x) \, dx$$

$$= \frac{1}{3} \int \cos u \, du = \frac{1}{3} \sin u + C$$

$$= \frac{1}{3} \sin(x^3 + 3x) + C \qquad \blacksquare$$

EXAMPLE D

Evaluate $\int_{\pi/6}^{\pi/3} \tan x \, dx$.

Solution First, write $\tan x$ as $(\sin x)/(\cos x)$; then make the substitution $u = \cos x$, $du = -\sin x \, dx$.

$$\int_{\pi/6}^{\pi/3} \tan x \, dx = \int_{\pi/6}^{\pi/3} \frac{\sin x}{\cos x} \, dx = -\int_{\sqrt{3}/2}^{1/2} \frac{1}{u} \, du = \left[-\ln |u| \right]_{\sqrt{3}/2}^{1/2}$$

$$= -\ln \frac{1}{2} + \ln \frac{\sqrt{3}}{2} = \ln \sqrt{3} \approx 0.54931 \qquad \blacksquare$$

EXAMPLE E

FIGURE 33

Find the area A of the region under one hump of the sine curve (Figure 33).

Solution This area is given by the following definite integral.

$$A = \int_0^{\pi} \sin x \, dx = \left[-\cos x \right]_0^{\pi} = -\cos \pi + \cos 0 = 1 + 1 = 2 \qquad \blacksquare$$

Integration by Parts

The technique called *integration by parts* was introduced in Section 7.3, a section worth reviewing now. It involves a double substitution using the formula

$$\int u \, dv = uv - \int v \, du$$

Here is how we use this formula in some examples that involve the trigonometric functions.

EXAMPLE F

Find $\displaystyle\int x \cos x \, dx$.

Solution It should be clear that no simple substitution will work; maybe a double substitution will help. Let

$$u = x \qquad dv = \cos x \, dx$$

Then,

$$du = dx \qquad v = \sin x$$

and the boxed integration-by-parts formula gives

$$\int \overset{u}{x} \overset{dv}{\cos x \, dx} = \overset{u}{x} \overset{v}{\sin x} - \int \overset{v}{\sin x} \overset{du}{dx}$$

$$= x \sin x + \cos x + C$$

Note that integration by parts left us with an integration to do; it was a useful technique because the new integration was simpler than the original one. ■

EXAMPLE G

Find $\displaystyle\int x^2 \sin x \, dx$.

Solution The integral looks complicated, but perhaps integration by parts will help. Let

$$u = x^2 \qquad dv = \sin x \, dx$$

so

$$du = 2x \, dx \qquad v = -\cos x$$

Then the boxed formula gives

$$\int \overset{u}{x^2} \overset{dv}{\sin x \, dx} = \overset{u}{x^2} \overset{v}{(-\cos x)} - \int \overset{v}{(-\cos x)} \overset{du}{2x \, dx}$$

$$= -x^2 \cos x + 2 \int x \cos x \, dx$$

Note that, while we still have an integration to do, at least it appears to be simpler than the one we started with. In fact this second integration is exactly the one we did in Example F. Thus, we may finish by writing

$$\int x^2 \sin x \, dx = -x^2 \cos x + 2(x \sin x + \cos x + C)$$

$$= -x^2 \cos x + 2x \sin x + 2 \cos x + K \qquad ■$$

EXAMPLE H

Find $\int x \cos(x^2)\, dx$.

Solution Be sure to distinguish between this example and the previous two. The present example calls for the simple substitution $u = x^2$, $du = 2x\, dx$ (see Example A).

$$\int x \cos(x^2)\, dx = \frac{1}{2} \int 2x \cos(x^2)\, dx = \frac{1}{2} \int \cos u\, du = \frac{1}{2} \sin u + C$$

$$= \frac{1}{2} \sin(x^2) + C \qquad \blacksquare$$

Integration Using Tables

Simple substitution and integration by parts are powerful but limited techniques. Most functions that one could write down cannot be integrated using these two tools. Another tool that is often useful is a table of integrals; a sample from such a table is shown at the end of this book. We illustrated the use of this table in Sections 7.3 and 8.3. Here we illustrate its use in connection with the trigonometric functions.

EXAMPLE I

Evaluate $\int_0^\pi e^{3x} \sin 2x\, dx$.

Solution From Formula 67 in the table combined with the Fundamental Theorem of Calculus, we find that

$$\int_0^\pi e^{3x} \sin 2x\, dx = \left[\frac{e^{3x}}{9 + 4}(3 \sin 2x - 2 \cos 2x) \right]_0^\pi$$

$$= \frac{e^{3\pi}}{13}(3 \sin 2\pi - 2 \cos 2\pi) - \frac{e^0}{13}(3 \sin 0 - 2 \cos 0)$$

$$= \frac{-2e^{3\pi} + 2}{13} \qquad \blacksquare$$

EXAMPLE J

Evaluate $\int_0^{\pi/2} \sin 3x \cos x\, dx$.

Solution We appeal to Formula 32 at the back of the book with $a = 3$ and $b = 1$. We obtain

$$\int_0^{\pi/2} \sin 3x \cos x\, dx = \left[-\frac{\cos 2x}{2(2)} - \frac{\cos 4x}{2(4)} \right]_0^{\pi/2}$$

$$= -\frac{\cos \pi}{4} - \frac{\cos 2\pi}{8} + \frac{\cos 0}{4} + \frac{\cos 0}{8}$$

$$= \frac{1}{4} - \frac{1}{8} + \frac{1}{4} + \frac{1}{8} = \frac{1}{2} \qquad \blacksquare$$

Problem Set 12.4

Find the definite and indefinite integrals in Problems 1–24. See Examples A, B, C, and D.

1. $\displaystyle\int \cos 3x \, dx$

2. $\displaystyle\int \sin(x/4) \, dx$

3. $\displaystyle\int_0^{\pi/6} \cos 3x \, dx$

4. $\displaystyle\int_0^{4\pi} \sin(x/4) \, dx$

5. $\displaystyle\int_0^1 \sin(\pi x) \, dx$

6. $\displaystyle\int_0^{\pi/2} \sin x \cos x \, dx$

7. $\displaystyle\int x^2 \sin(x^3) \, dx$

8. $\displaystyle\int \sec^2(3x) \, dx$

9. $\displaystyle\int \cos^3 t \sin t \, dt$

10. $\displaystyle\int \tan^4 t \sec^2 t \, dt$

11. $\displaystyle\int \frac{1}{2\sqrt{x}} \cos \sqrt{x} \, dx$

12. $\displaystyle\int \frac{1}{x^2} \sin^2(1/x) \cos(1/x) \, dx$

13. $\displaystyle\int_0^{\sqrt{\pi/2}} t \tan^3(t^2) \sec^2(t^2) \, dt$

14. $\displaystyle\int_0^1 (x + 2) \cos^3(x^2 + 4x) \sin(x^2 + 4x) \, dx$

15. $\displaystyle\int_{\pi/8}^{\pi/6} \frac{\sec^2 2t}{\tan^2 2t} \, dt$

16. $\displaystyle\int_0^{\pi/2} \sin t \, e^{\cos t} \, dt$

17. $\displaystyle\int e^{2u} \cos(e^{2u}) \, du$

18. $\displaystyle\int \frac{\cos u}{\sin^3 u} \, du$

19. $\displaystyle\int \frac{\sin 3x}{\sqrt{1 + \cos 3x}} \, dx$

20. $\displaystyle\int \frac{\sec^2 2x}{(1 + \tan 2x)^2} \, dx$

21. $\displaystyle\int_{\pi/6}^{\pi/2} \cot x \, dx$

22. $\displaystyle\int_0^{\sqrt{\pi/2}} x \tan(x^2) \, dx$

23. $\displaystyle\int_{\pi/3}^{\pi/2} \frac{\sin t}{1 + \cos t} \, dt$

24. $\displaystyle\int_0^{\pi/4} \frac{\sec^2 t}{1 + \tan t} \, dt$

25. Find the area under the curve $y = 4 \sec^2 2x$ between $x = 0$ and $x = \pi/8$. See Example E.

26. Find the area under one arch of the curve $y = 3 \cos 2x$.

27. The region bounded by $y = (\cos x)^{1/2}$ and the x-axis between $x = -\pi/2$ and $x = \pi/2$ is revolved about the x-axis. Find the volume of the resulting solid.

28. Solve the differential equation $dy/dx = x \sin(2x^2)$ given that $y = 5$ when $x = \sqrt{\pi}/2$.

Find the indefinite integrals in Problems 29–34, some of which will require integration by parts. See Examples F, G, and H.

29. $\displaystyle\int x \sin 2x \, dx$

30. $\displaystyle\int x \sin(x^2) \, dx$

31. $\displaystyle\int x \cos(2x^2) \, dx$

32. $\displaystyle\int x^3 \cos(x^2) \, dx$

33. $\displaystyle\int x^2 \cos 2x \, dx$

34. $\displaystyle\int x^5 \sin(x^2) \, dx$

Use the tables at the end of the book to evaluate each of the definite integrals in Problems 35–38. See Examples I and J.

35. $\displaystyle\int_0^{\pi/4} \sec^3 t \, dt$

36. $\displaystyle\int_{\pi/3}^{\pi/2} \sin^3 t \, dt$

37. $\displaystyle\int_0^{\pi/2} \sin 4x \sin x \, dx$

38. $\displaystyle\int_0^{\pi/6} \sin^3 x \cos^2 x \, dx$

39. Find each integral.

(a) $\displaystyle\int \frac{1}{x^2} \cos \frac{1}{x} \, dx$

(b) $\displaystyle\int \frac{\cos(\ln x)}{x} \, dx$

(c) $\displaystyle\int_0^{\sqrt{\pi/3}} x \sin^3(x^2) \cos(x^2) \, dx$

(d) $\displaystyle\int_0^{\pi/4} x \sin 4x \, dx$

(e) $\displaystyle\int e^{\cos^2 x} \sin x \cos x \, dx$

(f) $\displaystyle\int_{\pi^2/9}^{\pi^2/4} \frac{\sin \sqrt{x}}{\sqrt{x}} \, dx$

40. Use what you know about the derivatives of sec x and csc x to find:

(a) $\displaystyle\int \sec x \tan x \, dx$ (b) $\displaystyle\int \csc x \cot x \, dx$

41. Find the area of each shaded region.

(a)

(b)

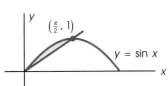

42. The wolf population of a certain region was 6400 on January 1, 1990, and t months later was expected to be changing at the rate of $450 \cos(\pi t/12)$ wolves per month. How many wolves does this model predict for the region 18 months later, that is, on July 1, 1991?

43. Evaluate the following iterated integrals.

(a) $\displaystyle\int_0^{1/2} \left[\int_0^x \sin(\pi x^2) \, dy \right] dx$

(b) $\displaystyle\int_0^1 \left[\int_0^{\pi/2} y \sin(xy) \, dy \right] dx$

44. Find the mass of the lamina L of density $k(x, y) = x$ if L is bounded by $y = \sin x$ and the x-axis between 0 and π.

12.5
APPLICATIONS OF THE TRIGONOMETRIC FUNCTIONS

Our task in this section is to model mathematically some of the common phenomena that are periodic, that is, phenomena that repeat themselves in well-defined cycles. We expect the trigonometric functions to play a principal role, since they are the premier examples of periodic functions. We turn to the health sciences for our first example.

Blood Pressure

As the heart beats, a person's blood pressure P ranges from a high (called the systolic pressure) to a low (called the diastolic pressure) and does so in a periodic manner that may be approximated by a sine wave.

EXAMPLE A

Suppose that Amy's blood pressure P (in appropriate units) at time t (in seconds) is given by

$$P = 110 + 30 \sin 6t$$

(a) Determine the amplitude and the period of this motion.
(b) Sketch the graph of P.
(c) Calculate the systolic pressure, the diastolic pressure, and the number of beats per minute.
(d) Find the *rate* at which the pressure P is changing at the end of 6 seconds, that is, at $t = 6$.

Solution

(a) The notions of amplitude and period for the motion $y = A \sin Bt$ were discussed at the end of Section 12.2. In the present case, the amplitude is 30 and the period is $2\pi/6 = \pi/3 \approx 1.047$.

(b) The graph of $P = 110 + 30 \sin 6t$ is just the graph of $P = 30 \sin 6t$ lifted up 110 units. Both graphs are shown in Figure 34.

(c) The systolic pressure is $110 + 30 = 140$, and the diastolic pressure is $110 - 30 = 80$. The period is $\pi/3$, which means that the frequency (number of cycles per second) is $3/\pi$ and so the number of cycles (beats) per minute is $60(3/\pi) \approx 57$.

(d) We are asked to calculate dP/dt at $t = 6$. Now, $dP/dt = (30 \cos 6t)(6)$. At $t = 6$, this has the value $180(\cos 36) \approx -23$. The pressure is falling at a rate of 23 units per second. ■

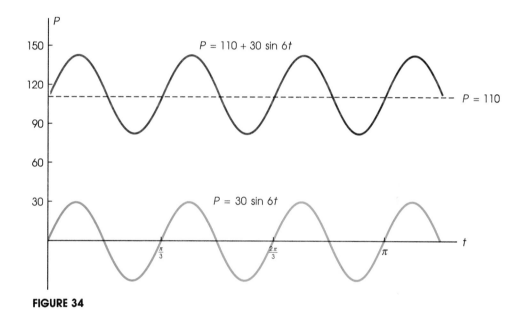

FIGURE 34

A Business Application

Consider a store that sells woolen sweaters. The *rate* at which the store sells these sweaters is likely to peak around the beginning of the year, fall to a low in midsummer, and then rise to the peak again at year end. A possible model for this behavior is the cosine function (Figure 35), but with an appropriately chosen amplitude and period. We will measure time t in months after January 1 and sales per month s in thousands of dollars. Then the equation describing the relation between s and t should have the form

$$s = d + A \cos Bt$$

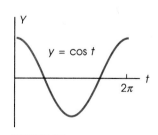

FIGURE 35

with A, B, and d to be chosen.

EXAMPLE B

Suppose the store described above sells sweaters at the rate of $10,000 per month during its peak selling time, but only $2000 per month during its slowest selling time.

(a) Determine A, B, and d in the model $s = d + A \cos Bt$.
(b) Sketch the graph of s against t for a 2-year interval.
(c) Determine the rate of sales on March 1 and August 15.
(d) Calculate total sales S over this 2-year interval.

Solution

(a) The amplitude (one-half the variation) in this business activity is $(10 - 2)/2 = 4$, and its period is 12. We conclude that $A = 4$ and that $2\pi/B = 12$, which makes $B = \pi/6$. The graph of $s = 4 \cos(\pi/6)t$ would have a high of 4 and a low of -4. If we lift this graph by an amount 6, that is, if we make $d = 6$, we should have what we want.

(b) The graph of $s = 6 + 4 \cos(\pi/6)t$ for a 2-year period (24 months) is shown in Figure 36. Note that it does have a high of 10 at the beginning and end of the year and a low of 2 at midyear.

(c) March 1 corresponds to $t = 2$; so on this date,

$$s = 6 + 4 \cos\left(\frac{\pi}{6}2\right) = 6 + 4\left(\frac{1}{2}\right) = 8$$

August 15 corresponds to $t = 8.5$, so on this date,

$$s = 6 + 4 \cos\left(\frac{\pi}{6}8.5\right) = 6 + 4 \cos 4.45059 = 4.965$$

Both of these answers are rates. On March 1, the store was selling sweaters at the rate of $8000 per month; on August 15, it was selling sweaters at the rate of $4965 per month.

(d) The total sales S is simply the area under the curve for s. (Think about why this is so; remember that s is the rate of selling). Let us find the sales for 1 year and double the result. Thus,

FIGURE 36

$$\frac{S}{2} = \int_0^{12} \left(6 + 4 \cos \frac{\pi}{6}t\right) dt = \int_0^{12} 6 \, dt + 4 \int_0^{12} \cos \frac{\pi}{6}t \, dt$$

The first integral has the value $6(12) = 72$. In the second integral, we make the substitution $u = (\pi/6)t$, so $du = (\pi/6) \, dt$. Then,

$$\int_0^{12} \cos \frac{\pi}{6}t \, dt = \frac{6}{\pi} \int_0^{12} \left(\cos \frac{\pi}{6}t\right)\frac{\pi}{6} \, dt = \frac{6}{\pi} \int_0^{2\pi} \cos u \, du$$

$$= \frac{6}{\pi}(\sin 2\pi - \sin 0) = 0$$

We conclude that $S/2 = 72$, and so $S = 144$. Total sweater sales during the 2-year period are \$144,000. ∎

Horizontal Shifts of Trigonometric Graphs

We have seen in our first two examples how to achieve a vertical shift (translation) of a graph. We simply added the number d to the y-value. But how can we achieve a horizontal shift? The answer is simple. To shift the graph of $y = f(t)$ by c units horizontally, simply replace t by $t - c$. This is illustrated with $c = \pi/6$ and $c = -\pi/3$ for the graph of $y = \sin t$ in Figure 37.

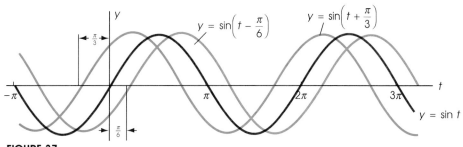

FIGURE 37

EXAMPLE C

Sketch the graph of $y = 3 \cos 2\left(t - \frac{\pi}{8}\right)$.

Solution This will have the same graph as $y = 3 \cos 2t$, but shifted $c = \pi/8$ units to the right. The amplitude is $A = 3$, and the period is $2\pi/B = 2\pi/2 = \pi$. With these parameters known, we can quickly sketch the graph shown in Figure 38 on the next page. ∎

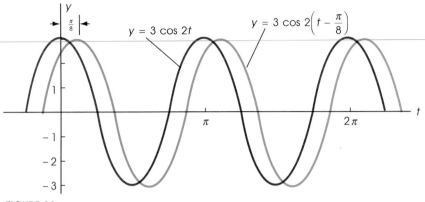

FIGURE 38

A Temperature Problem

If we were to keep track of noontime temperatures in St. Paul over many years, average the results, and smooth out the fluctuations, we would get a temperature graph something like the one shown in Figure 39.

Our goal is to give a model of the form $T = d + A \sin B(t - c)$ for this phenomenon (T being temperature in degrees Fahrenheit, and t being time in days after January 1).

FIGURE 39

EXAMPLE D

Determine A, B, c, and d so that

$$T = d + A \sin B(t - c)$$

is a model for the data shown in Figure 39. Then use this model to predict the noontime temperature on September 2, that is, on day 245.

Solution The temperature varies from a low of 5 to a high of 85, so the amplitude $A = (85 - 5)/2 = 40$ and the median temperature $d = (85 + 5)/2 = 45$. The period is 365, so $2\pi/B = 365$, which implies that $B = 2\pi/365$. The graph in Figure 38 appears to cross the 45°-axis about May 1, which means that we want the sine wave shifted about $c = 120$ units (days) to the right. We conclude that an appropriate model is

$$T = 45 + 40 \sin \frac{2\pi}{365}(t - 120)$$

The model predicts a noontime temperature on day 245 of

$$T = 45 + 40 \sin \frac{2\pi}{365}(245 - 120) \approx 45 + 40(0.836) \approx 78 \qquad \blacksquare$$

EXAMPLE E

For the model in Example D, determine (a) the rate of change in temperature (in degrees per day) on May 1, that is, on day 120, (b) the average noontime temperature T_{ave} during the first 3 months (90 days) of the year.

Solution

(a) $dT/dt = [40 \cos (2\pi/365)(t - 120)](2\pi/365)$. At $t = 120$, dT/dt has the value $(40)(2\pi/365) \approx 0.69°$/day.

(b) The average value of a function f over an interval is just its definite integral over that interval divided by the length of the interval (see Section 8.2). Thus,

$$T_{ave} = \frac{1}{90} \int_0^{90} \left[45 + 40 \sin \frac{2\pi}{365}(t - 120) \right] dt$$

$$= \frac{1}{90} \left[\int_0^{90} 45 \, dt + 40 \int_0^{90} \sin \frac{2\pi}{365}(t - 120) \, dt \right]$$

The first integral is easy to evaluate. In the second, let $u = (2\pi/365)(t - 120)$, whence $du = (2\pi/365) \, dt$. Note that $t = 0$ and $t = 90$ give $u = -240\pi/365$ and $u = -60\pi/365$. Thus,

$$T_{ave} = \frac{1}{90} \left[(45)(90) + (40) \left(\frac{365}{2\pi} \right) \int_{-240\pi/365}^{-60\pi/365} \sin u \, du \right.$$

$$= \frac{1}{90} \left\{ (45)(90) + (40) \left(\frac{365}{2\pi} \right) \left[-\cos u \right]_{-240\pi/365}^{-60\pi/365} \right\}$$

$$\approx 45 + \frac{4}{9} \left(\frac{365}{2\pi} \right) (-1.34454) \approx 10.3° \qquad \blacksquare$$

POINT OF INTEREST

Any physical phenomenon that can be modeled by the equation

$$y = d + A \sin B(t - c)$$

or the corresponding equation with sine replaced by cosine is said to be undergoing simple harmonic motion. This kind of behavior is particularly important in physics.

Problem Set 12.5

Find the indicated quantity in Problems 1–10.

1. $f'(0)$ given that $f(t) = 400 + 30 \sin(5\pi t/12)$

C 2. $f'(100)$ given that $f(t) = 120 \cos[(2\pi/365)(t - 72)]$

3. $\displaystyle\int_0^\pi 4 \cos(3t/2)\, dt$

C 4. $\displaystyle\int_2^4 2 \sin 3(t - 2)\, dt$

5. $\displaystyle\int_0^1 [400 + 30 \sin(3\pi t/4)]\, dt$

6. $\displaystyle\int_0^6 \{120 \cos[(\pi/4)(t - 2)]\}\, dt$

7. $f(\pi/2)$ given that $f'(x) = -10 + \cos 3x$ and $f(0) = 20$.

8. $f(\pi)$ given that $f'(x) = 5 + 6 \sin 2x$ and $f(0) = 2$.

9. The period and amplitude of
$$f(t) = 6 + 4 \cos[3(t - 2)].$$

10. The period and amplitude of
$$f(t) = -10 + 3 \sin[4\pi (t + 1)].$$

Sketch the graph of each pair of functions in Problems 11–14 using the same axes (see Examples A and C).

11. $f(t) = 3 \cos 2t$, $g(t) = 4 + 3 \cos 2t$

12. $f(t) = 2 \sin \frac{1}{2}t$, $g(t) = 5 + 2 \sin \frac{1}{2}t$

13. $f(t) = 4 \sin \dfrac{\pi}{2}t$, $g(t) = 4 \sin\left[\dfrac{\pi}{2}\left(t - \dfrac{1}{2}\right)\right]$

14. $f(t) = \dfrac{1}{2} \cos \dfrac{\pi}{3}t$, $g(t) = \dfrac{1}{2} \cos\left[\dfrac{\pi}{3}(t - 1)\right]$

15. Let $f(t) = d + A \sin Bt$. Find A, B, and d if f has a period of 6, a maximum value of 30, and a minimum value of 10.

16. Let $f(t) = d + A \cos Bt$. Find A, B, and d if f has a period of $\pi/3$, a maximum value of 10, and a minimum value of 4.

17. June Smith's blood pressure (in appropriate units) at time t (in seconds) is given by $P = 120 + 32 \sin 5t$. See Example A.
(a) Sketch the graph of this equation.
(b) Calculate the systolic pressure and diastolic pressure.
(c) How many times per minute does June's heart beat?

18. The population P of foxes in Jefferson County t months after January 1, 1980, followed a cyclic pattern which was modeled by the equation $P = 8000 + 2100 \sin(\pi t/12)$.
(a) Determine the length of a cycle (the period).

(b) When after January 1, 1980, did the population first reach its minimum value?
(c) How fast was the population growing on April 1, 1980?
(d) Determine the average population during 1980.

19. Based on experience in other stores, a new hardware store owner is told that he can expect lawn tool sales to follow a cyclic pattern given by

$$s(t) = 6 - 6 \cos\left(\frac{\pi}{26}t\right)$$

where $s(t)$ measures sales in hundreds of dollars per week and t is the number of weeks after the first of the year. See Example B.
(a) Sketch the graph of s for $0 \leq t \leq 104$.
(b) What does this model predict for total sales during 1 year?

20. Redo Problem 19 based on the model

$$s(t) = \frac{t}{26} + 6 - 6 \cos\left(\frac{\pi}{26}t\right)$$

21. In a certain city the length of daylight in minutes on a day t days after the beginning of the year is given approximately by (see Example B)

$$L(t) = 720 + 240 \sin\left[\frac{\pi}{182}(t - 81)\right]$$

(a) What day of the year has the most daylight?
(b) How fast is the length of daylight increasing on March 22, that is, on day 81?
C (c) Find the average length of daylight during the first 81 days of the year.

22. A weight attached to a vertical spring bobs up and down so that its y-coordinate satisfies $y = 4 \sin[3(t + \pi/6)]$. Here, y is measured in feet and time t in seconds.
(a) Determine the amplitude and period of the motion.
(b) When after $t = 0$ does the weight first reach its lowest point?
(c) What is the maximum velocity of the weight?
(d) Determine the total distance (up and down) traveled by the weight in 2π seconds.

23. Sketch the graph of each of the following using the same axes.
(a) $f(t) = 2 \sin \pi t$ (b) $g(t) = 2 \sin[\pi (t - \frac{1}{2})]$
(c) $h(t) = 3 + 2 \sin[\pi (t - \frac{1}{2})]$

24. Let $f(t) = A + B \cos[(\pi/12)(t - 6)]$. Find A and B if $f(0) = 20$ and $f(6) = 65$.

25. A small stream flows into a lake in such a way that its rate of flow (in cubic feet per day) is given by

$$f(t) = 7000 + 7000 \sin\left(\frac{\pi}{182}t\right)$$

where t measures the number of days after March 31.
(a) When is the rate of flow a maximum?
[C](b) Find the total amount of water that the stream pours into the lake during the month of April.

26. Suppose that for a person at rest the rate of airflow from the lungs in liters per second during a breathing cycle is given by $v = 0.815 \sin[(3\pi/8)t]$, where t is in seconds. Note that $v > 0$ corresponds to exhaling, $v < 0$ to inhaling.

(a) Determine the length of a complete breathing cycle and compare this with your own breathing cycle.
[C](b) How much air is inhaled (and exhaled) during one cycle?

27. Sketch the graph of $y = e^{-t/10} \sin \pi t$ for $t \geq 0$. Note that this gives an oscillating curve with the size of the oscillations decreasing as time increases—typical of many phenomena in real life.

28. Sketch as best you can the graph of $y = \sin(1/t)$ for $t > 0$. *Hint:* Begin by finding out where the graph crosses the t-axis.

Chapter 12 Review Problem Set

1. Change $2340°$ to radians.
2. Change $-1590°$ to radians.
3. Change $20\pi/3$ radians to degrees.
4. Change $-33\pi/4$ radians to degrees.
5. A central angle of a circle of radius 8 inches cuts off an arc of length 18 inches.
   (a) Find the radian measure of this angle.
   (b) Find the area of the sector of the circle interior to this angle.
[C]6. A central angle of $102.3°$ cuts off an arc on a circle of radius 15 centimeters. Determine the length of this arc and also the area of the corresponding sector of the circle.
7. If the point $(-5, -12)$ is on the terminal side of an angle t in standard position and if this point is r units from the origin, determine each of the following.
   (a) r (b) $\sin t$ (c) $\sec t$
   (d) $\sin(t + 4\pi)$ (e) $\sin(-t)$ (f) $1 - \cos^2 t$
8. Let $(-7, 24)$ be a point on the terminal side of an angle t in standard position and let r be its distance from the origin. Evaluate:
   (a) r (b) $\cos t$ (c) $\tan t$
   (d) $\cos(t - 2\pi)$ (e) $\cos(-t)$ (f) $\sin^2(t) - 1$
9. A 20-foot ladder leans against a vertical wall, making an angle of $76°$ with the level ground. How high up the wall is the upper end of the ladder?
10. Suppose that triangle ABC has $\angle C = 90°$, $\angle A = 32°$, and $\overline{AB} = 30$. Determine $\overline{AC}$ and $\overline{BC}$.
11. Calculate.
    (a) $\cos(10\pi/3)$ (b) $\tan(-495°)$
    [C](c) $[\sin^2(4.12)]/[1 + \tan(1.56)]$
    [C](d) $[\ln(\cos 1.25) + e^{-\sin 2.34}]^3$

12. Calculate:
    (a) $\sin(13\pi/6)$ (b) $\tan 855°$
    [C](c) $(\tan^3 6.45)/(2 + 3 \cos 3.46)$
    [C](d) $[(e^{-3.2\pi} \sin 13.2 + \ln 96.2]^{1/2}$
13. Find the largest and smallest values of
    $$f(t) = 5 - 3 \cos 2t.$$
14. Find the largest and smallest values of
    $$1/(6 - 4 \sin 3t)$$
15. Sketch carefully the graphs of $y = 2 \sin(\pi t)$ and $y = \sin[\pi(t - 1)]$ and $y = 2 \sin(\pi t) + \sin[\pi(t - 1)]$ using the same axes.
16. Sketch carefully the graphs of $y = 4 \cos 2t$, $y = \cos 2(t - \pi/6)$, and $y = 4 \cos 2t + \cos 2(t - \pi/6)$ using the same axes.
17. Find $f'(x)$.
    (a) $f(x) = 2 \cos 4x$ (b) $f(x) = 3 \sin \sqrt{x}$
    (c) $f(x) = 3\sqrt{\sin x}$
    (d) $f(x) = \cos[(x^2 + 1)^3]$
    (e) $f(x) = 3 \cos^3(x^2 + 1)$
    (f) $f(x) = 2x \tan(x^2)$
18. Find $g'(x)$.
    (a) $g(x) = 3 \sin 2x$ (b) $g(x) = 4 \cos(1/x)$
    (c) $g(x) = (3x + \sin x)^{3/2}$ (d) $g(x) = \sin\sqrt{x^3 + 4}$
    (e) $g(x) = \sqrt{\sin(x^3 + 4)}$ (f) $g(x) = x^2/\cos(x^3)$
19. Find an equation of the tangent line to the curve $y = \ln(e + \sin x)$ at the point $(0, 1)$.
20. Find an equation for the tangent line to the curve $y = \ln(3 \cos x + 1/e)$ at the point where $x = \pi/2$.
21. Find each indefinite integral.
    (a) $\int \cos(3x + 1)\, dx$ (b) $\int \sec^2(4x)\, dx$

(c) $\displaystyle\int \frac{\cos x}{1 + \sin x}\, dx$ (d) $\displaystyle\int \sin(5x)\,\cos^{3/2}(5x)\, dx$

22. Perform each integration.

(a) $\displaystyle\int \sin(3 - 2x)\, dx$ (b) $\displaystyle\int x\cos(x^2)\, dx$

(c) $\displaystyle\int \frac{\sec^2 x}{(5 + \tan x)^2}\, dx$ (d) $\displaystyle\int \cos 3x \sin^3 3x\, dx$

23. Evaluate each definite integral.

(a) $\displaystyle\int_0^{\pi/2} [2 + \sin(2t)]\, dt$ (b) $\displaystyle\int_0^{\pi/2} \cos^3 t \sin t\, dt$

24. Evaluate each definite integral.

(a) $\displaystyle\int_0^{\pi/6} \sin^4 t \cos t\, dt$

(b) $\displaystyle\int_0^{\pi/6} [(1 + \sin t)(1 - \sin t)]^{1/2}\, dt$

25. Find $f(x)$ if $f'(x) = x \sin(\pi x^2)$ and $f(0) = 4/\pi$.

26. Find y if $dy/dt = t \sec^2(3t^2)$ and $y = 7/6$ when $t = \sqrt{\pi/12}$.

27. The rate of change per month in a certain population P at time t in months is given by

$$P'(t) = 100\pi \cos \frac{\pi t}{18}$$

Also, $P(0) = 6400$.
(a) Find the formula for $P(t)$.
(b) What is the minimum population for $t \geq 0$?
(c) Determine the average population on the interval $0 \leq t \leq 6$.

28. Let $N(t)$ denote the number of wolves in a certain region t months after June 1, 1990. Suppose that $N(0) = 4000$ and that

$$N'(t) = (-1600\pi/18)\sin(\pi t/18)$$

(a) Find the formula for $N(t)$.
(b) Determine the minimum population of wolves during the interval $0 \leq t \leq 36$.
(c) Find the average population on the interval $0 \leq t \leq 36$.

29. Find the area under the curve $y = (\cos \sqrt{x})/\sqrt{x}$ between $x = \pi^2/36$ and $x = \pi^2/4$.

30. Find the area under the curve $y = 4x \sin(x^2)$ between $x = 0$ and $x = \sqrt{\pi}$.

31. The temperature in degrees Fahrenheit on a certain day was given by $T = 75 + 10 \sin[(\pi/12)(t - 9)]$, where t measures the number of hours after midnight.
(a) At what time of day did the temperature reach its maximum?
(b) During what hours was the temperature above 75°F?

(c) If it costs 15¢ to cool a large building 1° for an hour, what will it cost to keep the temperature in the building 75°F or below on the day described above?

32. The body temperature in degrees Fahrenheit of a hospital patient during a 24-hour period was given by $T(t) = 102.2 - 2.5 \sin(\pi t/12)$, where t denotes the number of hours after midnight.
(a) At what time of day did the temperature reach its maximum?
(b) During what time was the temperature above 102.2°?

33. Calculate $\int_0^1 [x \cos x + x \cos(x^2)]\, dx$

34. Calculate $\int_0^{\sqrt{\pi/3}} x \sin(x^2) \cos^2(x^2)\, dx$.

35. Find the slope of the tangent line to the curve

$$x^2 + y^2 + \sin(y^2) = 4 + 4y$$

at the point $(2, 0)$. Hint: Use implicit differentiation.

36. Find the slope of the tangent line to the curve

$$(x + 2y)^2 + 3 \sin 2y = 3y + x^2$$

at the point $(2, 0)$.

37. The graph of $y = x + (\pi/2) \sin^2 x$ has two points of inflection on the interval $[0, \pi]$. Find their coordinates.

38. Find the maximum value of $y = 3x + 6 \cos x$ on $[0, \pi]$.

39. The point $P(x, y)$ moves along the curve $y = 3 \cos 2x$ in such a way that dx/dt is always 4.
(a) Find dy/dt when $x = \pi/12$.
(b) As in Figure 40, let $A(x)$ be the area under this curve between 0 and x. Find dA/dt at $x = \pi/12$.

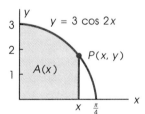

FIGURE 40

40. A point $P(x, y)$ moves along the line $y = 3$ in such a way that dx/dt is always 5 feet per second.
(a) How fast is the distance $\overline{OP}$ from the origin increasing when $x = 4$?
(b) If A is the point with coordinates $(0, 3)$, how fast is the area of the right triangle OAP increasing when $x = 4$?

41. A man plans to follow the path from A to C shown in Figure 41. If he runs from A to B at 6 miles per hour and then swims from B to C at 3 miles per hour, what angle θ will minimize the time it takes?

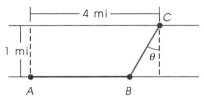

FIGURE 41

42. Let $y = \sqrt{3} \sin 2x + \cos 2x$ for $0 \leq x \leq \pi/4$.
   (a) Find the coordinates of the global maximum and minimum points on the graph of this equation.
   (b) Find the area under the graph between 0 and $\pi/4$.

Putting a space vehicle into the desired orbit around the earth or another planet is an intricate problem of optimal control theory involving the solving of difficult differential equations. Here Russian scientists excel and among the greatest was the blind Soviet mathematician Lev Semyonovich Pontryagin.

13

Differential Equations

Lev Semyonovich Pontryagin was born in 1908 to a rather poor Russian family in Moscow. As a boy, he showed little interest in mathematics, preferring to study gadgets like electric motors and electric arcs. But as a result of an accident (to be described later), and an apparent lack of ability in music, which he studied for some years, he turned to mathematics. At first, he found mathematics difficult, but gradually his innate talent for the subject surfaced and he made rapid progress. At the university, he took no notes but seemed to have complete recall of all that he heard in lectures. Later in life, he was to argue that taking "lecture notes does only harm, as it distracts the listener's attention from heeding the subject." University graduation was followed by graduate work (completed in 2 years) and employment at the Mathematics Institute of Moscow University.

Pontryagin's early work was in topology, an abstract generalization of geometry, which at that time seemed to have little relation to the problems of the real world. Here he made remarkable discoveries which he summarized in a book, *Topological Groups*, published in 1938. This book, which was quickly translated into English, propelled Pontryagin to the world stage and made his name a household word among mathematicians. It was the first of many well-written books and illustrates his attitude toward writing. "The more carefully a book is written, the thinner it becomes and the more work is spent in writing it."

Pontryagin admitted to suffering from many fears: fear that he would not be able to cope with a new situation, fear that he would not be able to solve a problem and thereby waste years of work, fear that his work would not be significant. This last fear caused him in 1952 to shift the direction of his work to differential equations and their applications to control devices. As a result, his name is now almost synonymous with the theory of optimal control. The practical significance of this theory led to his election, together with astronauts Gagarin and Tereshkova, to honorary membership in the International Astronautical Society. Pontryagin's achievements were remarkable, but what makes them astounding is that *he was totally blind from age 13*, the result of a tragic explosion. We must admire this great man while simultaneously deploring his apparent antisemitism.

L. S. Pontryagin
(1908–1988)

13.1
FIRST-ORDER EQUATIONS

A scientist trying to model a real-world problem is often led to an equation that relates a quantity to its rate of change. Recall, for example, the growth models in Sections 6.1 and 6.2 where the equations $dy/dt = ky$, $dy/dt = k(L - y)$, and $dy/dt = ky(L - y)$ played prominent roles. Any equation that relates an unknown function y to one or more of its derivatives is called a **differential equation.** Here are two more examples:

$$\frac{dy}{dx} = y \sin x$$

$$\frac{d^4y}{dx^4} + 4\frac{d^2y}{dx^2} + y = \cos x$$

The **order** of a differential equation is the order of the highest-order derivative that appears in the equation. Thus, our first equation is a first-order equation and the second is a fourth-order equation.

A **solution** to a differential equation in a domain D is a function $y = y(x)$ which when substituted in the equation reduces it to an identity in x over the set D. Note that $y = e^{-\cos x}$ provides a solution to our first equation, since

$$\frac{dy}{dx} = \frac{d(e^{-\cos x})}{dx} = e^{-\cos x} \sin x = y \sin x$$

and $y = -\frac{1}{2}\cos x$ is a solution to the second, since

$$\frac{d^4y}{dx^4} + 4\frac{d^2y}{dx^2} + y = \frac{d^4(-\frac{1}{2}\cos x)}{dx^4} + 4\frac{d^2(-\frac{1}{2}\cos x)}{dx^2} - \frac{1}{2}\cos x$$

$$= -\frac{1}{2}\cos x + 2\cos x - \frac{1}{2}\cos x = \cos x$$

Both solutions are solutions on the whole real line; that is, $D = (-\infty, \infty)$.

To **solve** a differential equation is to find all its solutions. To develop techniques for doing this is a major goal of this chapter. We begin with first-order equations.

Separable Equations

This is a familiar topic, since we discussed it under the rubric *separation of variables* in some detail in Section 7.4. Because it is such an important topic, we offer a review here. A differential equation is called **separable** if it has the form $dy/dx = f(x)/g(y)$ since, after multiplying by dx and by $g(y)$, it transforms to

$$g(y)\, dy = f(x)\, dx$$

in which the variables are separated. To solve such an equation, simply integrate both sides, recognizing that you can combine the two constants of integration on one side.

EXAMPLE A

Solve the differential equation $\dfrac{dy}{dx} = \dfrac{2x}{3y^2}$.

Solution We rewrite the equation in separated form and integrate.

$$3y^2\, dy = 2x\, dx$$

$$\int 3y^2\, dy = \int 2x\, dx$$

$$y^3 = x^2 + C$$

$$y = (x^2 + C)^{1/3} \qquad\blacksquare$$

We call the final result obtained in Example A the **general solution** to the equation. The fact that an arbitrary constant C appears in the general solution shows that there is a whole family of solutions to the given equation, one corresponding to each value of C. A sample of the graphs of this family is displayed in Figure 1.

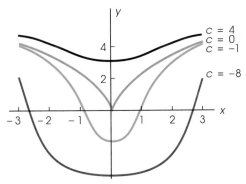

FIGURE 1

EXAMPLE B

Solve $\dfrac{dy}{dx} = \dfrac{4x^3 - 3x}{y + \sin y}$.

Solution

$$(y + \sin y)\, dy = (4x^3 - 3x)\, dx$$

$$\int (y + \sin y)\, dy = \int (4x^3 - 3x)\, dx$$

$$\frac{1}{2}y^2 - \cos y = x^4 - \frac{3}{2}x^2 + C$$

We would like to solve the last equation explicitly for y in terms of x, but this is impossible. We are content to say that the final result provides an **implicit solution** to the differential equation.   ■

Often a side condition is prescribed along with a differential equation. This in effect narrows down the general solution to a particular solution; that is, it specifies the arbitrary constant. We illustrate in our next example.

EXAMPLE C

Solve $\dfrac{dy}{dx} = \dfrac{y + 1}{x - 3}$ subject to the condition $y(1) = 0$.

Solution We use the following steps.

$$\frac{dy}{y + 1} = \frac{dx}{x - 3}$$

$$\int \frac{1}{y + 1}\, dy = \int \frac{1}{x - 3}\, dx$$

$$\ln|y + 1| = \ln|x - 3| + C$$

$$\ln\left|\frac{y + 1}{x - 3}\right| = C$$

$$\left|\frac{y + 1}{x - 3}\right| = e^C = K$$

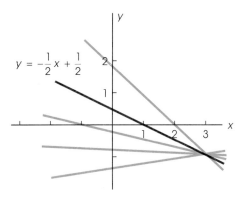

$$y = -\frac{1}{2}x + \frac{1}{2}$$

FIGURE 2

$$y + 1 = \pm K(x - 3) = L(x - 3)$$

$$y = L(x - 3) - 1$$

Here C, K, and L represent arbitrary constants. The last result gives a whole family of solutions (the family of lines shown in Figure 2) from which we shall pick the desired one by making use of the given condition $y(1) = 0$. This condition yields $0 = L(1 - 3) - 1$ or $L = -\frac{1}{2}$. We conclude that $y = -\frac{1}{2}x + \frac{1}{2}$. ■

Linear Equations

A differential equation of the form

$$\frac{dy}{dx} + p(x)y = q(x)$$

is called a **first-order linear equation.** In general, it will not be a separable equation; so if we are to solve it, we will need a new technique. The trick we use is to multiply both sides of the equation by the so-called **integrating factor** $e^{\int p(x)\,dx}$. This gives

$$e^{\int p(x)\,dx}\frac{dy}{dx} + e^{\int p(x)\,dx}p(x)y = q(x)e^{\int p(x)\,dx}$$

The reason for this trick is that the left side of the above equation is the derivative of the product $ye^{\int p(x)\,dx}$. Thus, we can rewrite our equation as

$$\frac{d}{dx}(ye^{\int p(x)\,dx}) = q(x)e^{\int p(x)\,dx}$$

Integration of both sides yields

$$ye^{\int p(x)\,dx} = \int q(x)e^{\int p(x)\,dx}\,dx$$

or

$$y = e^{-\int p(x)\,dx}\int q(x)e^{\int p(x)\,dx}\,dx$$

This is the general solution to the equation. One could memorize this formula and apply it whenever one is given a first-order linear equation. We prefer, however, to go through the process by which it was derived in each case, remembering only the form $e^{\int p(x)\,dx}$ of the integrating factor.

EXAMPLE D

Solve the differential equation $\dfrac{dy}{dx} + \dfrac{2}{x}y = \dfrac{\cos 5x}{x^2}$.

Solution Here $p(x) = 2/x$, and so the integrating factor is

$$e^{\int (2/x)\,dx} = e^{2\ln|x|} = e^{\ln|x|^2} = e^{\ln x^2} = x^2$$

Multiplying both sides of the differential equation by x^2 gives

$$x^2\frac{dy}{dx} + 2xy = \cos 5x$$

or

$$\frac{d}{dx}(x^2 y) = \cos 5x$$

Integration of both sides yields

$$x^2 y = \int \cos 5x\,dx = \frac{1}{5}\sin 5x + C$$

We conclude that the general solution is

$$y = x^{-2}\left(\frac{1}{5}\sin 5x + C\right) = \frac{\sin 5x}{5x^2} + \frac{C}{x^2} \qquad\blacksquare$$

EXAMPLE E

Find the solution to $\dfrac{dy}{dx} - 2y = xe^{3x}$ that satisfies the condition $y = 2$ when $x = 0$.

Solution The required integrating factor is $e^{\int -2\,dx} = e^{-2x}$. After multiplying both sides by e^{-2x}, our differential equation takes the form

$$e^{-2x}\frac{dy}{dx} - 2e^{-2x}y = xe^x$$

or

$$\frac{d}{dx}(e^{-2x}y) = xe^x$$

Next, we integrate both sides (using integration by parts on the right) to obtain

$$e^{-2x}y = \int xe^x\,dx = (x - 1)e^x + C$$

Thus, the general solution is

$$y = (x - 1)e^{3x} + Ce^{2x}$$

The condition $y = 2$ when $x = 0$ requires that $2 = -1 + C$, which implies that $C = 3$. We conclude that the particular solution we want is

$$y = (x - 1)e^{3x} + 3e^{2x} \qquad\blacksquare$$

EXAMPLE F

Find the general solution of

$$\frac{dy}{dx} + \frac{2x}{x^2 + 3}y = \frac{5x}{\sqrt{x^2 + 3}}$$

Solution After multiplying both sides by the integrating factor

$$e^{\int [2x/(x^2+3)]\,dx} = e^{\ln(x^2+3)} = x^2 + 3$$

the equation takes the form

$$\frac{d}{dx}[(x^2 + 3)y] = 5x(x^2 + 3)^{1/2}$$

Integration of both sides gives

$$(x^2 + 3)y = \frac{5}{2}\int 2x(x^2 + 3)^{1/2}\,dx = \frac{5}{3}(x^2 + 3)^{3/2} + C$$

Thus

$$y = \frac{5}{3}(x^2 + 3)^{1/2} + C(x^2 + 3)^{-1} \qquad \blacksquare$$

Other First-Order Equations

Separable equations and linear equations are not the only types of first-order differential equations. In fact, many equations that occur in applications are of neither type. Some of these can be solved by special tricks that we leave to more advanced books; others can be solved only by approximate numerical methods. We will discuss the latter methods in Section 14.5.

Problem Set 13.1

Find the general solution of each differential equation in Problems 1–14 (see Examples A and B).

1. $\dfrac{dy}{dx} = \dfrac{4x}{3y^2}$

2. $y^{-2/3}\dfrac{dy}{dx} = 4x^3$

3. $\dfrac{dy}{dt} = -4y$

4. $\dfrac{dx}{dt} = 3x$

5. $e^y\dfrac{dy}{dx} = 4x^3 + 2x$

6. $\dfrac{1}{x^2}\dfrac{dy}{dx} = y^3$

7. $\dfrac{dy}{dx} = \dfrac{3x^2 + 2\cos 2x}{2y + e^y}$

8. $\dfrac{dy}{dx} = e^{3x-2y}$

9. $\dfrac{dy}{dx} = e^{2x+y}$

10. $(1 + x)\dfrac{dy}{dx} = 2y$

11. $xy\dfrac{dy}{dx} = \ln x$

12. $\dfrac{dy}{dx} = \dfrac{1}{x\tan y}$

13. $x\dfrac{dy}{dx} = xy + y$

14. $(x^2 + 4x + 2)\dfrac{dy}{dx} = x + 2$

In Problems 15–20, solve the given differential equation subject to the stated condition. See Example C.

15. $\dfrac{dy}{dx} = \dfrac{y}{x^2};\ y\!\left(\dfrac{1}{2}\right) = e$

16. $\dfrac{dy}{dx} = \dfrac{5 - y}{\sqrt{x}};\ y(1) = 4$

17. $\dfrac{dy}{dt} = te^{-2y}$; $y(0) = 2$

18. $\dfrac{dy}{dt} = te^{-y} + e^{-y}$; $y(0) = 2$

19. $ye^{-3x^2} = \dfrac{y^2 + 4}{x}\dfrac{dy}{dx}$; $y(0) = 1$

20. $3xy^2\dfrac{dy}{dx} = \ln x$; $y(1) = 5$

Solve the linear differential equations in Problems 21–26. See Examples D and F.

21. $\dfrac{dy}{dx} = \dfrac{-3}{x}y + 4$ 22. $\dfrac{dy}{dx} + \dfrac{y}{x} = \dfrac{1}{x^2}$

23. $\dfrac{dy}{dx} - \dfrac{y}{x} = x^2e^x$ 24. $\dfrac{dy}{dx} + y = e^{3x}$

25. $\dfrac{dy}{dx} + y\tan x = \sec x$ 26. $\dfrac{dy}{dx} = -y + 3\cos x$

Solve each differential equation in Problems 27–30 subject to the stated condition. See Example E.

27. $\dfrac{dy}{dx} + \dfrac{y}{x} = x^2 + 1$; $y(2) = 5$

28. $\dfrac{dy}{dx} - \dfrac{y}{x+4} = \sqrt{x+4}$; $y(0) = 24$

29. $\dfrac{dy}{dx} + y\cos x = \cos x$; $y(0) = 5$

30. $\dfrac{dy}{dx} + \dfrac{5y}{x} = \dfrac{2}{x^6}$; $y(1) = 4$

Solve each differential equation in Problems 31–38.

31. $\dfrac{dy}{dx} = 2x - 3x^2 + e^{2x}$

32. $\dfrac{dy}{dx} = 3\sin 2x + \dfrac{4}{x} - 3\sqrt{x}$

33. $\dfrac{dy}{dt} = \dfrac{5t\sqrt{t}}{y}$ 34. $\dfrac{dy}{dt} = -\dfrac{y}{t} + e^{t^2}$

35. $\dfrac{dy}{dx} - \dfrac{2y}{x+1} = (x+1)^3$; $y(0) = 2$

36. $\dfrac{dy}{dx} + \dfrac{2xy}{1+x^2} = \dfrac{4x}{\sqrt{1+x^2}}$; $y(1) = 3$

37. $\dfrac{dy}{dx} = \dfrac{5(y+2)}{(x-2)(x+3)}$; $y(3) = -\dfrac{1}{2}$

38. $\dfrac{dy}{dx} = \dfrac{\ln x}{\sqrt{x}\sin y}$; $y(1) = \pi$

39. Find the xy-equation of the curve that passes through $(1,e)$ and whose slope at any point on the curve is equal to four times the y-coordinate of that point.

40. Suppose that the rate of change in y with respect to x varies directly as the product of x and the square of y. Express y explicitly in terms of x given that $y = 20$ when $x = 0$ and $y = 30$ when $x = 20$.

13.2
APPLICATIONS OF FIRST-ORDER EQUATIONS

There are two fundamental interpretations of the derivative dy/dx. Geometrically, dy/dx is the slope of the tangent line to a curve $y = f(x)$ at the point (x, y). Of more significance for applications is the interpretation of dy/dx as the rate of change in y with respect to x. Rates of change, especially rates of change with respect to time, pervade all of science. Moreover, a relationship involving rates of change, when viewed mathematically, is a differential equation. Thus, many of the problems of modern science automatically become the problem of solving differential equations. We plan to offer many illustrations in this section.

A Geometrical Application

We begin with an easy problem that illustrates the slope interpretation of the derivative.

RECALL

We studied the application of differential equations to problems of population growth and radioactive decay in Sections 6.1 and 6.2, sections worth reviewing now. Because these two applications were treated extensively there, we treat other applications in the present section.

EXAMPLE A

The slope of a curve at the point (x, y) in the plane is equal to xe^{-y}. Also, the curve goes through the point $(6, 0)$. Find the equation $y = f(x)$ of this curve.

Solution In mathematical terms, we are being asked to solve the differential equation $dy/dx = xe^{-y}$ subject to the condition that $y = 0$ when $x = 6$. Our procedure is to separate variables, integrate, evaluate the constant of integration, and finally solve explicitly for y.

$$e^y \, dy = x \, dx$$

$$\int e^y \, dy = \int x \, dx$$

$$e^y = \frac{1}{2}x^2 + C$$

$$1 = e^0 = \frac{1}{2}(6)^2 + C \qquad \text{so} \qquad C = -17$$

$$e^y = \frac{1}{2}x^2 - 17$$

$$y = \ln\left(\frac{1}{2}x^2 - 17\right)$$

Cooling and Heating

Cooling and heating problems are solved by appealing to Newton's Law of Cooling, which states: The time rate of change in the temperature T of an object placed in a medium of constant temperature M is proportional to $T - M$; that is,

$$\frac{dT}{dt} = k(T - M)$$

where k is a constant.

EXAMPLE B

A boiling solution (100°C) was set on a lab table where the room temperature was 20°C. It had cooled to 60°C after 5 minutes. Find a formula for the temperature T of the solution t minutes after being set on the table and determine how long it will take to cool to 22°C.

Solution The differential equation $\dfrac{dT}{dt} = k(T - 20)$ can be treated either as a separable or a linear equation. We choose to view it as a separable equation. Thus,

POINT OF INTEREST

Mathematical modeling is the name given to the process of describing a physical problem in mathematical symbols. Thus, the differential equation $dy/dt = k(T - M)$ together with side conditions may be called a *mathematical model* for the cooling or heating of a physical body.

$$\frac{dT}{T - 20} = k \, dt$$

$$\int \frac{1}{T - 20} \, dT = \int k \, dt$$

$$\ln(T - 20) = kt + C$$

$$T - 20 = e^{kt+C} = e^{C} e^{kt} = L e^{kt}$$

Now $T = 100$ at $t = 0$ implies that $L = 80$; so $T = 20 + 80e^{kt}$. The condition $T = 60$ at $t = 5$ will allow us to determine the constant k. Substitution gives

$$60 = 20 + 80e^{5k}$$

$$\frac{40}{80} = e^{5k}$$

$$\ln \frac{40}{80} = 5k$$

This implies $k = -0.13863$, and so

$$T = 20 + 80e^{-0.13863t}$$

answering our first question.

To find when $T = 22$, we substitute in this formula and solve for t.

$$22 = 20 + 80e^{-0.13863t}$$

$$e^{-0.13863t} = \frac{2}{80} = \frac{1}{40}$$

$$-0.13863t = \ln\left(\frac{1}{40}\right)$$

$$t = \frac{-\ln \frac{1}{40}}{0.13863} \approx 26.6 \text{ minutes} \qquad \blacksquare$$

Mixing Problems

Input rate

$y = y(t)$

Output rate

FIGURE 3

Many mixing problems when reduced to their essence can be viewed as consisting of a tank into which a substance is entering at a certain rate (the input rate) and leaving at another rate (the output rate). Let $y = y(t)$ denote the amount of the substance in the tank at time t. The situation can be pictured as shown in Figure 3 and modeled mathematically by the differential equation

$$\frac{dy}{dt} = \text{input rate} - \text{output rate}$$

If the input and output rates are given (or can be determined) and if we know y at some instant, we should be able to predict y at any time t.

EXAMPLE C

Initially a tank contains 1000 liters of pure water. A valve is opened, allowing a brine solution containing 0.1 kilogram of salt per liter to enter the tank at a rate of 8 liters per minute. The solution in the tank is stirred constantly, keeping its concentration uniform. If the brine also drains out of the tank at 8 liters per minute, determine (a) the amount of salt in the tank t minutes after opening the valve and (b) the instant when the concentration reaches 0.05 kilogram per liter.

Solution

(a) Let $y = y(t)$ denote the amount (number of kilograms) of salt in the tank at time t. Note that the input rate for salt is 8(0.1), whereas the output rate is eight times the concentration in the tank, that is, $8y/1000$. Thus, our mathematical model for this problem is the differential equation

$$\frac{dy}{dt} = 0.8 - 0.008y$$

subject to the condition that $y(0) = 0$. This equation is both separable and linear. For variety, we solve it as a linear equation with integrating factor $e^{\int 0.008\,dt} = e^{0.008t}$. The steps are as follows.

$$\frac{dy}{dt} + 0.008y = 0.8$$

$$e^{0.008t}\frac{dy}{dt} + 0.008ye^{0.008t} = 0.8e^{0.008t}$$

$$\frac{d}{dt}(ye^{0.008t}) = 0.8e^{0.008t}$$

$$ye^{0.008t} = \int 0.8e^{0.008t}\,dt = 100e^{0.008t} + C$$

$$y = 100 + Ce^{-0.008t}$$

The condition $y(0) = 0$ implies $C = -100$; so

$$y = 100 - 100e^{-0.008t} = 100(1 - e^{-0.008t})$$

answering the first question.

(b) To find out when the concentration reaches 0.05 kilogram per liter, we must solve the equation $y/1000 = 0.05$. We may rewrite this successively as

$$0.1(1 - e^{-0.008t}) = 0.05$$

$$1 - e^{-0.008t} = 0.5$$

$$e^{-0.008t} = 0.5$$

$$t = -\frac{\ln 0.5}{0.008} \approx 86.6 \text{ minutes}$$

Note that the limiting value for y as $t \to \infty$ is 100, which corresponds to a concentration of salt in the tank of 0.1 kilogram per liter. Just as we would expect, the concentration of salt in the tank ultimately reaches the concentration of the entering solution. ∎

EXAMPLE D

Same as Example C, but assume the solution drains out at a rate of 6 liters per minute.

Solution

(a) The amount of solution in the tank will now increase according to the formula $1000 + 2t$, which will eventually cause the tank to overflow. However, assuming the tank is very large, we can model our problem (at least initially) with the differential equation

$$\frac{dy}{dt} = 8(0.1) - \frac{6y}{1000 + 2t}, \qquad y(0) = 0$$

This is again a linear differential equation, now with integrating factor

$$e^{\int 6/(1000+2t)\,dt} = e^{3\ln(1000+2t)} = e^{\ln(1000+2t)^3} = (1000 + 2t)^3$$

Following our usual procedure for solving a linear equation, we find that

$$y(1000 + 2t)^3 = \int 0.8(1000 + 2t)^3 \, dt = 0.1(1000 + 2t)^4 + C$$

The condition $y(0) = 0$ makes $C = -10^{11}$ and yields the solution

$$y = 0.1(1000 + 2t) - 10^{11}(1000 + 2t)^{-3}$$

(b) To find out when the concentration reaches 0.05, we must solve the equation $y/(1000 + 2t) = 0.05$. This leads to

$$0.1 - 10^{11}(1000 + 2t)^{-4} = 0.05$$

$$(1000 + 2t)^{-4} = \frac{0.05}{10^{11}}$$

$$(1000 + 2t)^4 = \frac{10^{11}}{0.05} = 2 \times 10^{12}$$

$$1000 + 2t = \sqrt[4]{2 \times 10^{12}} = 1000\sqrt[4]{2}$$

$$t = \frac{1000(\sqrt[4]{2} - 1)}{2}$$

This gives $t \approx 94.6$ minutes. ∎

EXAMPLE E

Janet's liver has a capacity of 360 cubic centimeters. Blood enters her liver at a rate of 3 cubic centimeters per second and leaves at the same rate. To treat a medical problem, a drug is introduced into the blood entering her liver, quickly bringing the concentration of the drug in this blood to 0.2 gram per cubic centimeter. How long will it take for the concentration of the drug in the blood of her liver to reach 0.15 gram per cubic centimeter.

Solution We model this problem with the differential equation

$$\frac{dy}{dt} = (0.2)(3) - \frac{y}{360}(3), \qquad y(0) = 0$$

Here $y = y(t)$ is the amount of the drug in her liver t seconds after the drug treatment is begun. We may rewrite this equation in the form

$$\frac{dy}{dt} + \frac{y}{120} = 0.6$$

Using the integrating factor $e^{\int (1/120)\,dt} = e^{t/120}$, we find that

$$ye^{t/120} = \int 0.6e^{t/120}\,dt = (0.6)(120)e^{t/120} + C$$

or

$$y = 72 + Ce^{-t/120}$$

The condition $y(0) = 0$ implies that $C = -72$, so

$$y = 72(1 - e^{-t/120})$$

To find when the concentration is 0.15 is to solve the equation

$$0.15 = \frac{y}{360} = \frac{72}{360}(1 - e^{-t/120})$$

which transforms successively to

$$0.75 = 1 - e^{-t/120}$$

$$e^{-t/120} = 0.25$$

$$\frac{-t}{120} = \ln(0.25)$$

$$t = -120 \ln(0.25) \approx 166 \text{ seconds} \qquad ∎$$

Falling Bodies

According to Newton's Second Law, $F = ma$; that is, force is mass times acceleration. An object of mass m falling in a vacuum is subject only to a downward force due to gravity, which produces an acceleration g of approximately 9.81 meters per second per second. Thus, the mathematical model for such a falling body is $m(dv/dt) = -m(9.81)$; that is, $dv/dt = -9.81$. The minus sign is chosen to correspond to a positive upward direction; thus, the force of gravity causes the velocity v to decrease but to become larger and larger in magnitude (the speed increases).

EXAMPLE F

An object of mass 5 kilograms is dropped from a height of 1000 meters. Assuming air resistance can be ignored, when will it hit the ground and with what speed?

Solution Solving the simple separable differential equation $dv/dt = -9.81$ subject to $v = 0$ at $t = 0$ gives $v = -9.81t$. Now velocity is the rate of change in distance with respect to time. Thus, if $y = y(t)$ denotes the height of the object at time t, we have the second differential equation

$$\frac{dy}{dt} = -9.81t \qquad y(0) = 1000$$

Its solution is

$$y = -4.905t^2 + 1000$$

The object hits the ground when $-4.905t^2 + 1000 = 0$, that is, at $t \approx 14.28$ seconds. The corresponding velocity is $(-9.81)(14.28) \approx -140$ meters per second. Thus, the speed is 140 meters per second. ∎

Example F was really only a first approximation (and not a very good one unless the object is very dense), since air provides a resistance force proportional to the velocity. Taking this into account, Newton's Second Law takes the form

$$m\frac{dv}{dt} = -m(9.81) - kv$$

EXAMPLE G

Use the same data as in Example F but assume that the air provides a resisting force proportional to the velocity. Answer the same questions as in Example F with $k = 2.5$.

Solution After dividing by $m = 5$, the differential equation takes the form

$$\frac{dv}{dt} + 0.5v = -9.81 \qquad v(0) = 0$$

Using the integrating factor $e^{0.5t}$, we obtain the solution

$$v = -19.62 + Ce^{-0.5t}$$

Then, since $v(0) = 0$, it follows that $C = 19.62$ and

$$v = 19.62(-1 + e^{-0.5t})$$

An important consequence of this last formula is that v reaches a limiting velocity of -19.62 meters per second as $t \to \infty$.

Next, as in Example E, let $y = y(t)$ be the height of the object at time t. Then

$$\frac{dy}{dt} = -19.62 + 19.62e^{-0.5t} \qquad y(0) = 1000$$

This differential equation has the solution

$$y = -19.62t - 39.24e^{-0.5t} + 1039.24$$

Setting $y = 0$ gives, after a bit of algebra, $t + 2e^{-0.5t} = 52.97$. This equation could be solved numerically by methods to be described in Section 14.3. However, the second term on the left is clearly very small, so small that we can ignore it. Thus, the object will hit the ground at $t \approx 53$ seconds with a speed of about 19.6 meters per second. ∎

Problem Set 13.2

When we ask for a mathematical model of a given situation, we are simply asking for the differential equation (together with any side conditions) that describes the situation.

1. A quantity Q depends on time t in such a way that the time rate of change in Q is proportional to the product of t and the square root of Q. Moreover, $Q(1) = 4$. Write down a mathematical model for this situation.

2. The velocity of an object is proportional to the time elapsed and inversely proportional to the distance traveled. Write down a mathematical model for this situation using s for distance traveled and t for time elapsed.

3. The slope of a curve at the point (x, y) is equal to $6x^2y^2$, and the curve passes through the point $(0, 2)$. See Example A.
   (a) Construct a mathematical model for this situation.
   (b) Obtain an equation for this curve and express it in the form $y = f(x)$.

4. A curve passes through $(3\pi/2, 0)$ and has slope at (x, y) equal to $e^{-y} \cos x$.
   (a) Construct a mathematical model.
   (b) Express y in the form $y = f(x)$.

C5. A roast was taken out of an oven at 340°F and left to cool in a room with temperature 70°F. In 15 minutes. the temperature at the center of the roast was 300°F. See Example B.
   (a) Construct a mathematical model for this situation.
   (b) Find a formula for the temperature T at the center of the roast after t minutes.
   (c) How long will it take for this temperature to cool to 270°F?

C6. A frozen fish was taken from a freezer at a temperature of -18°C and brought into the kitchen where the temperature registered 22°C. It took 20 minutes for the temperature of the fish to rise to 0°C. When will its temperature reach 15°C?

7. A tank contains 30 gallons of a solution, with 12 pounds of a chemical in this solution. At a certain instant, Mary opens a valve, allowing a solution con-

taining the same chemical in a concentration of 1.5 pounds per gallon to enter the tank at a rate of 2 gallons per minute. While continuously stirring the solution in the tank, she drains off this solution also at 2 gallons per minute. See Example C.
   (a) Construct the mathematical model for this situation.
   (b) How many pounds of the chemical are in the tank after 25 minutes?

8. A tank initially contains 240 gallons of brine, with 60 pounds of salt in solution. Then brine containing 1 pound of salt per gallon begins to enter the tank at 3 gallons per minute, and the well-stirred solution flows out at the same rate.
   (a) Determine the amount of salt in the tank at the end of t hours.
   (b) If this process is continued indefinitely, how much salt will ultimately be in the tank?

$\boxed{C}$9. A tank contains 200 liters of pure water. Brine with a concentration of $\frac{1}{4}$ kilogram of salt per liter begins to flow into the tank at 4 liters per minute, and the well-stirred solution flows out of the tank at 5 liters per minute. See Example D.
   (a) Determine a mathematical model for this situation.
   (b) Write a formula for the amount of salt in the tank after t minutes, where $0 \leq t < 200$.
   (c) When will the concentration of salt in the tank be $\frac{1}{8}$ kilogram per liter?
   (d) When will the amount of salt in the tank be a maximum?

$\boxed{C}$10. A very large tank initially contains 100 liters of brine with 4 kilograms of salt in solution. A brine solution with a concentration of $\frac{1}{10}$ kilogram per liter begins to flow into the tank at the rate of 4 liters per minute. At the same time, the solution in the tank (kept uniform by stirring) is drained off at the rate of 2 liters per minute.
   (a) Determine a formula for tne amount of salt in the tank after t minutes.
   (b) What is the concentration of salt in the tank after 10 minutes?
   (c) How fast is the concentration of salt changing after 10 minutes?

$\boxed{C}$11. Refer to Example E and suppose the blood entering Janet's liver has a drug concentration of 0.3 gram per cubic centimeter and that the rate at which the blood enters and leaves the liver is 4 cubic centimeters per second.
   (a) Determine the concentration of the drug in the blood of her liver after t minutes.
   (b) How long will it take for this concentration to reach 0.2 gram per cubic centimeter?

12. The ozone in the air of a 2000-cubic-foot room is to be removed by passing air through a charcoal filter at

the rate of 300 cubic feet per minute. After passing through the filter, the purified air is recirculated in the room. Assuming the ozone in the room at any time is uniformly distributed, how long will it take to remove 90% of the ozone from the room?

13. A dense object is dropped from a helicopter flying at an altitude of 4800 feet. Assume that air resistance can be ignored and that the acceleration of gravity g is 32 feet per second per second. See Example F.
   (a) Construct the mathematical model for this situation.
   (b) Write a formula for the height s of the object after t seconds.
   (c) When will the object hit the ground?
   (d) With what speed will it hit the ground?

14. Redo Problem 13 assuming the object is thrown downward with a speed of 64 feet per second.

15. A steel ball of mass 40 pounds is dropped from a height of 12,000 feet. Suppose that the air offers a resistance proportional to the velocity leading to Newton's model

$$ m\frac{dv}{dt} = -m(32) - kv $$

with $k = 3.2$. See Example G.
   (a) Determine the height s of the ball t seconds later.
   (b) Estimate when the ball will hit the ground.

16. Suppose the ball in Problem 15 is shot from the ground straight upward at a speed of 288 feet per second and that $k = 3.2$ as in that problem. Determine the maximum height attained by the ball.

---

17. When an amount of sugar S_0 is placed in a large container of water, it will dissolve at a rate proportional to the amount of undissolved sugar. Construct a mathematical model for the amount of undissolved sugar.

18. When a cell is placed in a solution, the solute passes through the cell wall by diffusion. Fick's Law says that the concentration C of the solute inside the cell changes at a rate that is jointly proportional to the area A of the cell wall and the difference between the concentration inside the cell and outside the cell. Assuming that the concentration outside the cell is a constant p and that $C = 0$ at $t = 0$, construct a mathematical model that describes this situation.

19. The rate at which an epidemic spreads in a city is jointly proportional to the number of people who are infected and the number who are not infected. Construct a differential equation model for the number of people infected at time t days after the first reported infection for a city of M people.

20. A population P growing according to the Gompertz growth model is such that its rate of change with re-

spect to time is jointly proportional to the number present and the difference between the natural logarithm of this number and a constant M. Write down the differential equation satisfied by P.

21. An initial insect population of 6000 grew at the rate $40t + 30\,t^2$ after t days. Determine the population at the end of 10 days.

22. A radioactive substance disintegrates at a rate proportional to the amount present. If its half-life (time required to disintegrate to half its original size) is 120 years, how long will it take to disintegrate to 1% of its original size?

23. The residents of a small city have voted to discontinue the fluoridation of their water. Their water tower holds 500,000 gallons of water, which presently contain 20 pounds of fluoride. The city uses water at the rate of 25,000 gallons a day, and beginning today this water will be replaced by unfluoridated water at the same rate.

(a) Determine the differential equation that describes the amount A of fluoride in the tower t days from now.
(b) Solve this differential equation for A in terms of t.
(c) How long will it take to lower the amount of fluoride in the tower down to $\frac{1}{2}$ pound?

24. Suppose that tank A initially contains 100 gallons of brine with 50 pounds of dissolved salt and that tank B contains 200 gallons of brine also with 50 pounds of dissolved salt. Pure water flows into tank A at 2 gallons per minute, and then the well-stirred solution in tank A drains into tank B at 2 gallons per minute. Finally the well-stirred solution in tank B drains into a sewer at 2 gallons per minute. Find the formula for the amount of salt in tank B after t minutes.

13.3*
HOMOGENEOUS SECOND-ORDER LINEAR EQUATIONS

A second-order linear differential equation has the form

$$\frac{d^2y}{dx^2} + a_1(x)\frac{dy}{dx} + a_2(x)y = g(x)$$

Throughout this section, we make two simplifying assumptions. We assume that $a_1(x)$ and $a_2(x)$ are constants, and we assume that $g(x)$ is identically 0 (the **homogeneous** case). Thus, our task is to solve

$$\frac{d^2y}{dx^2} + a_1\frac{dy}{dx} + a_2y = 0$$

Recall that the general solution of a first-order equation is a family of functions involving one arbitrary constant. We guess that the general solution of a second-order equation will be a family of functions involving two arbitrary constants. In fact, we now state without proof a fundamental result on which all that we do in this section is based.

THEOREM

If $u_1(x)$ and $u_2(x)$ are independent solutions to the differential equation

$$\frac{d^2y}{dx^2} + a_1\frac{dy}{dx} + a_2y = 0$$

then $y = C_1u_1(x) + C_2u_2(x)$ is the general solution to this equation.

We need to explain what we mean by independent solutions. Two functions $u_1(x)$ and $u_2(x)$ are called **dependent** if there are constants a and b (not both 0) such that $au_1(x) + bu_2(x)$ is identically 0; if no such constants exist, $u_1(x)$ and $u_2(x)$ are said to be **independent.** Stated more simply, $u_1(x)$ and $u_2(x)$ are independent if neither function is a constant multiple of the other.

EXAMPLE A

Find the general solution of the differential equation

$$\frac{d^2y}{dx^2} - y = 0$$

Solution Note that $y = e^x$ and $y = e^{-x}$ are independent solutions (simply substitute in the equation to see this). It follows that $y = C_1e^x + C_2e^{-x}$ is the general solution to the equation.   ∎

Unfortunately, most equations are more complicated than this; we will not be able to find independent solutions by inspection. We need a method, a matter to which we now turn.

The Auxiliary Equation

Let's be brave and try $y = e^{rx}$ as a potential solution to

$$\frac{d^2y}{dx^2} + a_1\frac{dy}{dx} + a_2y = 0$$

This gives

$$r^2e^{rx} + a_1re^{rx} + a_2e^{rx} = 0$$

or equivalently

$$e^{rx}(r^2 + a_1r + a_2) = 0$$

This will be an identity (and consequently e^{rx} will be a solution to the differential equation) if and only if the quadratic expression in parentheses is 0. The algebraic equation

$$r^2 + a_1r + a_2 = 0$$

is called the **auxiliary equation** corresponding to the given differential equation. In general, its two roots r_1 and r_2 determine two independent functions e^{r_1x} and e^{r_2x} and provide the general solution

$$y = C_1e^{r_1x} + C_2e^{r_2x}$$

to the differential equation. This is all fine if the two roots of the auxiliary equation are real and distinct. The case of repeated roots and complex roots introduces complications that we will consider later.

The Case of Distinct Real Roots

We have said all that is necessary about the case of distinct real roots, but let us give some illustrations.

EXAMPLE B

Find the general solution of the differential equation

$$\frac{d^2y}{dx^2} - \frac{dy}{dx} - 6y = 0$$

Solution The auxiliary equation $r^2 - r - 6 = 0$ has two solutions, -2 and 3, determining the two independent solutions e^{-2x} and e^{3x} of the differential equation. The required general solution of the differential equation is

$$y = C_1e^{-2x} + C_2e^{3x} \qquad\blacksquare$$

As in the first-order case, a condition may be given that determines the arbitrary constants, but now it will take two of them.

EXAMPLE C

Solve the differential equation

$$\frac{d^2y}{dx^2} - 8\frac{dy}{dx} + 15y = 0$$

subject to the conditions $y(0) = 0$ and $y'(0) = 1$.

Solution The auxiliary equation $r^2 - 8y + 15 = 0$ with its two roots 3 and 5 determines the general solution

$$y = C_1e^{3x} + C_2e^{5x}$$

having derivative

$$y' = 3C_1e^{3x} + 5C_2e^{5x}$$

The two conditions result in the pair of equations

$$0 = C_1 + C_2$$

$$1 = 3C_1 + 5C_2$$

with the simultaneous solutions $C_1 = -\frac{1}{2}$ and $C_2 = \frac{1}{2}$. Thus, the required solution to the given differential equation with the given conditions is

$$y = -\frac{1}{2}e^{3x} + \frac{1}{2}e^{5x} \qquad\blacksquare$$

EXAMPLE D

Find the general solution of the differential equation

$$\frac{d^2y}{dx^2} - 2\frac{dy}{dx} - 2y = 0$$

Solution To solve the auxiliary equation $r^2 - 2r - 2 = 0$, we use the Quadratic Formula.

$$r = \frac{2 \pm \sqrt{4 + 8}}{2} = \frac{2 \pm 2\sqrt{3}}{2} = 1 \pm \sqrt{3}$$

We conclude that the general solution of the differential equation is

$$y = C_1 e^{(1+\sqrt{3})x} + C_2 e^{(1-\sqrt{3})x}$$ ∎

The Case of Repeated Real Roots

Suppose the auxiliary equation has r as a repeated root and thus determines only the single solution $y = e^{rx}$ to the differential equation. It turns out that $y = xe^{rx}$ is also a solution in this case.

EXAMPLE E

Find the general solution to $\dfrac{d^2y}{dx^2} + 6\dfrac{dy}{dx} + 9y = 0$.

Solution The auxiliary equation is

$$r^2 + 6r + 9 = (r + 3)(r + 3) = 0$$

with a repeated root of -3. Clearly $y = e^{-3x}$ is one solution to the differential equation, and we claim that $y = xe^{-3x}$ is another. To demonstrate the latter, note that

$$\frac{d^2}{dx^2}(xe^{-3x}) + 6\frac{d}{dx}(xe^{-3x}) + 9xe^{-3x}$$

$$= \frac{d}{dx}(-3xe^{-3x} + e^{-3x}) + 6(-3xe^{-3x} + e^{-3x}) + 9xe^{-3x}$$

$$= 9xe^{-3x} - 3e^{-3x} - 3e^{-3x} - 18xe^{-3x} + 6e^{-3x} + 9xe^{-3x} = 0$$

We conclude that $y = C_1 e^{-3x} + C_2 xe^{-3x}$ is the general solution we are seeking.   ∎

The Case of Complex Roots

We begin with a simple example and then generalize.

EXAMPLE F

Find the general solution to $\dfrac{d^2y}{dx^2} + \beta^2 y = 0$.

Solution The auxiliary equation is $r^2 + \beta^2 = 0$, which has the two imaginary roots $r = \pm\beta i$, so we face a new situation. However, the given differential equation is easily solved by trial and error. The facts that $d^2(\sin x)/dx^2 = -\sin x$ and $d^2(\cos x)/dx^2 = -\cos x$ suggest that $y = \sin \beta x$ and $y = \cos \beta x$ might be solutions, and indeed they are. Thus the desired general solution is $y = C_1 \sin \beta x + C_2 \cos \beta x$. ∎

But what if the auxiliary equation has the two roots $\alpha \pm \beta i$? Then we assert that $e^{\alpha x} \sin \beta x$ and $e^{\alpha x} \cos \beta x$ are two independent solutions of the corresponding differential equation.

EXAMPLE G

Solve $\dfrac{d^2y}{dx^2} + 4\dfrac{dy}{dx} + 13y = 0$.

Solution We solve the auxiliary equation $r^2 + 4r + 13 = 0$ by the Quadratic Formula, obtaining

$$r = \frac{-4 \pm \sqrt{16 - 52}}{2} = \frac{-4 \pm 6i}{2} = -2 \pm 3i$$

From what we said above, we infer that $e^{-2x} \sin 3x$ and $e^{-2x} \cos 3x$ are independent solutions of the differential equation, but we urge you to check this by substitution. We conclude that the general solution is

$$y = C_1 e^{-2x} \sin 3x + C_2 e^{-2x} \cos 3x$$ ∎

Higher-Order Linear Equations

The results we have obtained so far generalize in a rather obvious way to higher-order linear differential equations. We offer one example.

EXAMPLE H

Solve $\dfrac{d^3y}{dx^3} - 3\dfrac{d^2y}{dx^2} + 2\dfrac{dy}{dx} = 0$

Solution The auxiliary equation is

$$r^3 - 3r^2 + 2r = r(r-2)(r-1) = 0$$

with the three real roots 0, 1, and 2. They determine the three independent solutions e^{0x}, e^x, and e^{2x}. The general solution to the differential equation is

$$y = C_1 + C_2 e^x + C_3 e^{2x}$$

Problem Set 13.3

In Problems 1–6, find the general solution of the differential equation by antidifferentiating twice. This means that your answer will involve two arbitrary constants.

1. $\dfrac{d^2y}{dx^2} = 6x$

2. $\dfrac{d^2y}{dx^2} = 3x^2$

3. $\dfrac{d^2y}{dt^2} = \cos 3t$

4. $\dfrac{d^2y}{dt^2} = \sin 2t$

5. $\dfrac{d^2y}{dt^2} = e^{3t} + 2t$

6. $\dfrac{d^2y}{dt^2} = e^{-t} - 4$

In Problems 7–28, find the general solution to the given differential equation. See Examples B–G.

7. $\dfrac{d^2y}{dx^2} - 4y = 0$

8. $\dfrac{d^2y}{dx^2} - 5\dfrac{dy}{dx} - 6y = 0$

9. $\dfrac{d^2y}{dx^2} - 4\dfrac{dy}{dx} = 0$

10. $\dfrac{d^2y}{dx^2} + 5\dfrac{dy}{dx} = 0$

11. $\dfrac{d^2y}{dx^2} - 2\dfrac{dy}{dx} - 24y = 0$

12. $\dfrac{d^2y}{dx^2} + 8\dfrac{dy}{dx} + 15y = 0$

13. $2\dfrac{d^2y}{dx^2} + 5\dfrac{dy}{dx} - 3y = 0$

14. $6\dfrac{d^2y}{dx^2} - 5\dfrac{dy}{dx} - 6y = 0$

15. $\dfrac{d^2y}{dx^2} + \dfrac{dy}{dx} - y = 0$

16. $\dfrac{d^2y}{dx^2} - 4\dfrac{dy}{dx} - 2y = 0$

17. $\dfrac{d^2x}{dt^2} - 6\dfrac{dx}{dt} + 4x = 0$

18. $4\dfrac{d^2x}{dt^2} - 12\dfrac{dx}{dt} + 7x = 0$

19. $\dfrac{d^2x}{dt^2} - 8\dfrac{dx}{dt} + 16x = 0$

20. $\dfrac{d^2x}{dt^2} + 4\dfrac{dx}{dt} + 4x = 0$

21. $9\dfrac{d^2x}{dt^2} + 6\dfrac{dx}{dt} + x = 0$

22. $4\dfrac{d^2x}{dt^2} + 12\dfrac{dx}{dt} + 9x = 0$

23. $\dfrac{d^2y}{dx^2} + 25y = 0$

24. $\dfrac{d^2y}{dx^2} + 49y = 0$

25. $\dfrac{d^2y}{dx^2} + 2\dfrac{dy}{dx} + 5y = 0$

26. $\dfrac{d^2y}{dx^2} - 14\dfrac{dy}{dx} + 50y = 0$

27. $\dfrac{d^2y}{dx^2} - 10\dfrac{dy}{dx} + 34y = 0$

28. $\dfrac{d^2y}{dx^2} - 4\dfrac{dy}{dx} + 9y = 0$

In Problems 29–34, solve the differential equation subject to the stated conditions. See Example C.

29. $\dfrac{d^2y}{dx^2} - 4\dfrac{dy}{dx} - 5y = 0$; $y(0) = 3$ and $y'(0) = 9$

30. $\dfrac{d^2y}{dx^2} + 3\dfrac{dy}{dx} - 4y = 0$; $y(0) = 8$ and $y'(0) = -7$

31. $\dfrac{d^2y}{dx^2} - 4\dfrac{dy}{dx} = 0$; $y(0) = 5$ and $y'(0) = 12$

32. $\dfrac{d^2y}{dx^2} - 14\dfrac{dy}{dx} + 49y = 0$; $y(0) = 4$ and $y'(0) = 26$

33. $\dfrac{d^2x}{dt^2} + x = 0$; $x(0) = 0$ and $x(\pi/2) = 3$

34. $\dfrac{d^2x}{dt^2} + 4x = 0$; $x(0) = 2$ and $x'(0) = 4$

35. Find the general solution of

$$\dfrac{d^3y}{dx^3} - 6\dfrac{d^2y}{dx^2} + 11\dfrac{dy}{dx} - 6y = 0.$$

Hint: $r - 1$ is a factor of $r^3 - 6r^2 + 11r - 6$.

36. Find the general solution of
$$\frac{d^3y}{dx^3} + 5\frac{d^2y}{dx^2} + 6\frac{dy}{dx} = 0$$

37. Determine the general solution of each differential equation.

(a) $\dfrac{d^2x}{dt^2} - 9 = 0$ 　　(b) $\dfrac{d^2x}{dt^2} - 9x = 0$

(c) $\dfrac{d^2x}{dt^2} + 9x = 0$

(d) $\dfrac{d^2x}{dt^2} + 6\dfrac{dx}{dt} + 9x = 0$

(e) $\dfrac{d^2x}{dt^2} + 12\dfrac{dx}{dt} + 9x = 0$

(f) $\dfrac{d^2x}{dt^2} + 2\dfrac{dx}{dt} + 9x = 0$

38. Write down the differential equation whose auxiliary equation is $(r + 2)(r + 1)(r - 1)(r - 2) = 0$ and solve it.

39. Show that if $u(t)$ and $v(t)$ both satisfy the differential equation $\dfrac{d^2x}{dt^2} + a\dfrac{dx}{dt} + bx = 0$, then so does $C_1u(t) + C_2v(t)$. *Note:* This result also holds for higher-order linear differential equations, as you can show.

40. The differential equation
$$\frac{d^3x}{dt^3} - 3\frac{d^2x}{dt^2} + 3\frac{dx}{dt} - x = 0$$
has $(r - 1)^3 = 0$ as its auxiliary equation (check this).
(a) Show that e^t, te^t, and t^2e^t each satisfy this differential equation.
(b) Write the general solution of the differential equation.

41. Based on the text and Problems 39 and 40, guess at the general solution to the differential equation with the given auxiliary equation.
(a) $(r + 2)^3 = 0$
(b) $(r - 2)^2(r - 3) = 0$
(c) $(r - 1)^4 = 0$
(d) $(r - 1)^2(r - 2)^2 = 0$

42. Solve the differential equation $\dfrac{d^3x}{dt^3} - 4\dfrac{d^2x}{dt^2} = 0$ subject to the conditions $x(0) = 1$, $x'(0) = -1$, $x''(0) = -16$.

13.4*
NONHOMOGENEOUS SECOND-ORDER LINEAR EQUATIONS

We consider now the task of solving a differential equation of the form

$$\frac{d^2y}{dx^2} + a_1\frac{dy}{dx} + a_2y = F(x)$$

This is not an easy task, and we attempt it only because it is precisely equations of this type that play an important role in many areas of science, especially physics. It is fortunate that in the applications $F(x)$ tends to be of a particularly simple form. We shall concentrate on the case where $F(x)$ is a polynomial, an exponential, or a combination of sines and cosines. The method of solving the equation can be described as a sequence of three steps.

1. Find the general solution y_h of the homogeneous equation using what we learned in the last section.

2. Find a particular solution y_p of the nonhomogeneous equation by methods to be described in this section.

3. Add y_h and y_p to obtain the general solution of the nonhomogeneous equation.

EXAMPLE A

Find the general solution to

$$\frac{d^2y}{dx^2} + 2\frac{dy}{dx} - 8y = -5e^x$$

Solution The auxiliary equation is

$$r^2 + 2r - 8 = (r + 4)(r - 2) = 0$$

with roots -4 and 2. Thus the general solution to the homogeneous equation is

$$y_h = C_1 e^{-4x} + C_2 e^{2x}$$

The harder task is to find a particular solution to the nonhomogeneous equation. The fact that the right-hand side of the differential equation is $-5e^x$ suggests that we might try $y_p = Ae^x$, with A to be determined. Substituting Ae^x for y in the differential equation gives

$$Ae^x + 2Ae^x - 8Ae^x = -5e^x$$

which is an identity if $A = 1$. Thus $y_p = e^x$ is a particular solution, and we conclude that

$$y = C_1 e^{-4x} + C_2 e^{2x} + e^x$$

is the general solution we are seeking.   ∎

The example just finished can be turned into a general method.

The Method of Undetermined Coefficients

For the simple types of function $F(x)$ that we are considering, we offer the suggestions in the following table.

If $F(x) =$	Try $y_p =$
$b_m x^m + \cdots + b_1 x + b_0$	$B_m x^m + \cdots + B_1 x + B_0$
be^{ax}	Be^{ax}
$b \sin \beta x + c \cos \beta x$	$B \sin \beta x + C \cos \beta x$

Modification: If a term of $F(x)$ is already a solution to the homogeneous equation, multiply the trial solution by x (or perhaps a higher power of x).

To illustrate the use of the table, we suggest the appropriate trial solution in several cases.

1. $y'' - 2y' - 8y = x^2 + 5$, $y_p = B_2 x^2 + B_1 x + B_0$.
2. $y'' - 2y' - 8y = e^{3x}$, $y_p = Be^{3x}$.
3. $y'' + y = \sin 3x$, $y_p = B \sin 3x + C \cos 3x$.

4. $y'' - 2y' = x^2 + 5$, $y_p = B_2 x^3 + B_1 x^2 + B_0 x$ (5 is a solution to the homogeneous equation).

5. $y'' - 2y' - 8y = e^{-2x}$, $y_p = xe^{-2x}$ (e^{-2x} is a solution to the homogeneous equation).

6. $y'' + y = \sin x$, $y_p = Bx \sin x + Cx \cos x$ ($\sin x$ is a solution to the homogeneous equation).

EXAMPLE B

Find the general solution to the differential equation

$$\frac{d^2 y}{dx^2} - \frac{dy}{dx} - 6y = 18x^2 + 5$$

Solution The auxiliary equation $r^2 - r - 6 = 0$ has roots -2 and 3, and so

$$y_h = C_1 e^{-2x} + C_2 e^{3x}$$

For a particular solution, we try $y_p = Ax^2 + Bx + C$. Substitution of this expression in the differential equation gives

$$2A - (2Ax + B) - 6(Ax^2 + Bx + C) = 18x^2 + 5$$

Equating coefficients of x^2, x, and 1 on the two sides of this equation, we obtain

$$-6A = 18 \qquad -2A - 6B = 0 \qquad 2A - B - 6C = 5$$

so $A = -3$, $B = 1$, and $C = -2$. Hence, $y_p = -3x^2 + x - 2$, and the general solution to the differential equation is

$$y = C_1 e^{-2x} + C_2 e^{3x} - 3x^2 + x - 2 \qquad \blacksquare$$

EXAMPLE C

Solve the differential equation

$$y'' - y' - 6y = e^{-2x}$$

Solution This equation, like that in the previous example, has

$$y_h = C_1 e^{-2x} + C_2 e^{3x}$$

as the general solution to the homogeneous equation. Note that $F(x) = e^{-2x}$ already appears in y_h, so for our particular solution we try $y_p = Bxe^{-2x}$. Substitution in the differential equation gives

$$(4Bxe^{-2x} - 4Be^{-2x}) - (-2Bxe^{-2x} + Be^{-2x}) - 6Bxe^{-2x} = e^{-2x}$$

which implies that $B = -\frac{1}{5}$. Thus, the general solution to the given differential equation is

$$y = C_1 e^{-2x} + C_2 e^{3x} - \frac{1}{5}xe^{-2x} \qquad \blacksquare$$

EXAMPLE D

Solve the differential equation

$$\frac{d^2y}{dx^2} - \frac{dy}{dx} - 6y = 8e^{2x}$$

subject to the conditions $y(0) = 0$ and $y'(0) = 7$.

Solution The solution y_h to the homogeneous equation is as in Example B. As a particular solution, we try $y_p = Be^{2x}$. Substitution in the differential equation yields

$$4Be^{2x} - 2Be^{2x} - 6Be^{2x} = 8Be^{2x}$$

and this in turn implies that $B = -2$. We conclude that the general solution to our differential equation is

$$y = C_1e^{-2x} + C_2e^{3x} - 2e^{2x}$$

The condition $y(0) = 0$ gives

$$C_1 + C_2 = 2$$

Since $y' = -2C_1e^{-2x} + 3C_2e^{3x} - 4e^{2x}$, $y'(0) = 7$ gives

$$-2C_1 + 3C_2 = 11$$

This pair of equations has the common solution $C_1 = -1$ and $C_2 = 3$. Thus, the final solution to the given problem is

$$y = -e^{-2x} + 3e^{3x} - 2e^{2x} \qquad \blacksquare$$

EXAMPLE E

Solve the differential equation $y'' + y = \sin 2x$ subject to $y(0) = 1$ and $y(\pi/2) = 0$.

Solution The homogeneous equation has solution

$$y_h = C_1 \sin x + C_2 \cos x$$

We propose

$$y_p = B \sin 2x + C \cos 2x$$

as a particular solution. When we substitute this expression in the differential equation, we get

$$(-4B \sin 2x - 4C \cos 2x) + (B \sin 2x + C \cos 2x) = \sin 2x$$

This implies $-3B = 1$ and $-3C = 0$, so $B = -\frac{1}{3}$ and $C = 0$. The general solution to the differential equation is

$$y = C_1 \sin x + C_2 \cos x - \frac{1}{3} \sin 2x$$

POINT OF INTEREST

Examples D and E illustrate two types of problems, both extensively studied. The first is called an **initial value** problem because the two conditions are specified at the same point. The second is a **boundary value** problem; the two conditions are specified at different points.

To evaluate C_1 and C_2, we use the given conditions. First, $y(0) = 1$ means that $C_2 = 1$, whereas $y(\pi/2) = 0$ means that $C_1 = 0$. The required solution to the differential equation subject to the given conditions is

$$y = \cos x - \frac{1}{3} \sin 2x$$ ∎

Applications

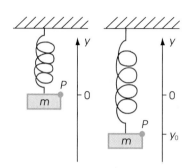

FIGURE 4

Typical applications for second-order linear equations involve the study of waves, vibrations, electric circuits, and other types of oscillatory phenomena. We use the motion of a simple coiled spring as an illustration. Consider a vertical spring to which a mass of size m is attached (Figure 4) so that the equilibrium position of the point P is at $y = 0$. If the point P is pulled down to a point $y = y_0$ and released, the mass m and point P will bob up and down in a predictable way that we wish to investigate.

We begin by appealing to Hooke's Law, which says that the force exerted by the spring on the mass m (tending to pull P toward its equilibrium position) is proportional to the amount the spring is stretched; that is, $F = -ky$, where y is the position of the point P. Here k is a positive constant called the **spring constant.** Combining this with Newton's Second Law ($F = ma$) and ignoring any forces due to friction, we obtain the simple differential equation

$$m \frac{d^2y}{dt^2} = -ky \qquad y(0) = y_0 \qquad y'(0) = 0$$

Rewriting this as $y'' + (k/m)y = 0$ and applying the method of Section 13.3, we get

$$y = C_1 \sin \sqrt{\frac{k}{m}} t + C_2 \cos \sqrt{\frac{k}{m}} t$$

from which

$$y' = \sqrt{\frac{k}{m}} \left(C_1 \cos \sqrt{\frac{k}{m}} t - C_2 \sin \sqrt{\frac{k}{m}} t \right)$$

The given conditions imply that $C_1 = 0$ and $C_2 = y_0$. Thus, the equation of motion for the point P is

$$y = y_0 \cos \sqrt{\frac{k}{m}} t$$

which goes by the name **simple harmonic motion.**

POINT OF INTEREST

Units in physical problems can be confusing. If force is in pounds, distance in feet, and time in seconds, then mass is in slugs. If force is in dynes, distance in centimeters, and time in seconds, then mass is in grams.

EXAMPLE F (NO FRICTION)

Consider a spring with an attached mass of 2 slugs hanging as in Figure 4. If the spring constant k is 8, and the mass is pulled down 1.5 feet and released, determine the equation of motion for the point P and draw its graph.

$y = -1.5 \cos 2t$

Simple harmonic motion

FIGURE 5

Solution From the discussion above, we conclude that

$$y = y_0 \cos \sqrt{\frac{k}{m}}\,t = -1.5 \cos \sqrt{\frac{8}{2}}\,t = -1.5 \cos 2t$$

The familiar cosine graph is shown in Figure 5. Note that under the assumption of no friction, the mass m will continue to bob up and down equally far forever.   ∎

To make a realistic model of a bobbing spring, we should introduce a frictional force representing the retarding force due to the air (or other medium) and the internal friction of the spring. Experiments suggest that this force is proportional to the velocity of the point P, that is, has the form $-q\,dy/dt$, where q is a positive constant called the **damping constant.** The differential equation to be studied now takes the form

$$\frac{d^2y}{dt^2} + \frac{q}{m}\frac{dy}{dt} + \frac{k}{m}y = 0 \qquad y(0) = y_0 \qquad y'(0) = 0$$

EXAMPLE G (FRICTION)

Assume the same data as in Example F but add in a frictional force with a damping constant $q = 0.2$. Find the equation of motion of the point P and draw its graph.

Solution The differential equation is $y'' + 0.1y' + 4y = 0$ with corresponding auxiliary equation $r^2 + 0.1r + 4 = 0$. Its roots are $r = (-0.1 \pm \sqrt{0.01 - 16})/2 \approx -0.05 \pm 2i$. The general solution to the differential equation is (see Example G in Section 13.3)

$$y = C_1 e^{-0.05t} \sin 2t + C_2 e^{-0.05t} \cos 2t$$

The conditions $y(0) = -1.5$ and $y'(0) = 0$ determine the constants $C_2 = -1.5$ and $C_1 = -0.0375$. Thus

$$y = e^{-0.05t}(-0.0375 \sin 2t - 1.5 \cos 2t) \approx e^{-0.05t}(-1.5 \cos 2t)$$

The graph is shown in Figure 6 and is a typical example of **damped harmonic motion.** Now, the amplitude of the motion decreases as time goes along.   ∎

$y = e^{-0.05t}$

$y = e^{-0.05t}(-1.5 \cos 2t)$

Damped harmonic motion

FIGURE 6

Last we add a forcing function $F(t)$, say moving the ceiling mount for the spring up and down, to see what this does to the equation of motion.

EXAMPLE H (FRICTION AND FORCING)

Assume the data in the previous example but add in a forcing function $F(t) = \sin t$ so that the differential equation takes the form

$$\frac{d^2y}{dt^2} + 0.1\frac{dy}{dt} + 4y = \sin t \qquad y(0) = -1.5 \qquad y'(0) = 0$$

Solve this differential equation.

Solution The solution y_h to the homogeneous equation is as in Example G. We propose trying $y_p = B \sin t + C \cos t$. Upon substituting in the differential equation and solving for B and C, we get $B \approx \frac{1}{3}$ and $C \approx -\frac{1}{90}$. Thus, the general solution is

$$y \approx e^{-0.05t}(C_1 \sin 2t + C_2 \cos 2t) + \frac{1}{3}\sin t - \frac{1}{90}\cos t$$

The two conditions imply that $C_1 \approx -0.20$ and $C_2 \approx -1.49$. We conclude that

$$y \approx e^{-0.05t}(-0.20 \sin 2t - 1.49 \cos 2t) + 0.33 \sin t - 0.01 \cos t \qquad ■$$

Problem Set 13.4

In Problems 1–18, find the general solution of the given differential equation. See Examples B and C.

1. $\dfrac{d^2y}{dx^2} - 4y = 9e^x$

2. $\dfrac{d^2y}{dx^2} - 9y = 8e^{-x}$

3. $\dfrac{d^2y}{dx^2} + 4y = 4x + 8$

4. $\dfrac{d^2y}{dx^2} + 9y = 6x - 18$

5. $\dfrac{d^2y}{dx^2} + 4y = 6 \cos x$

6. $\dfrac{d^2y}{dx^2} + 9y = 3 \sin x$

7. $\dfrac{d^2y}{dx^2} - 3\dfrac{dy}{dx} + 2y = 20 \sin 2x$

8. $\dfrac{d^2y}{dx^2} - \dfrac{dy}{dx} - 6y = e^{2x}$

9. $\dfrac{d^2y}{dx^2} - 3\dfrac{dy}{dx} + 2y = x^2 - 1$

10. $\dfrac{d^2y}{dx^2} - \dfrac{dy}{dx} - 6y = 3x^2$

11. $\dfrac{d^2y}{dx^2} - 6\dfrac{dy}{dx} + 8y = e^x$

12. $\dfrac{d^2y}{dx^2} - 7\dfrac{dy}{dx} + 10y = 130 \cos x$

13. $\dfrac{d^2y}{dx^2} + 2\dfrac{dy}{dx} + 2y = 10e^x$

14. $\dfrac{d^2y}{dx^2} - 4\dfrac{dy}{dx} + 8y = 10e^{3x}$

15. $\dfrac{d^2y}{dx^2} - 6\dfrac{dy}{dx} + 8y = 4e^{2x}$

16. $\dfrac{d^2y}{dx^2} - 7\dfrac{dy}{dx} + 10y = 6e^{5x}$

17. $\dfrac{d^2y}{dx^2} + 4y = \sin 2x$

18. $\dfrac{d^2y}{dx^2} + 9y = 12 \cos 3x$

In Problems 19–24, solve the differential equation subject to the given conditions (see Examples D and E).

19. $y'' - 4y = 9e^x$; $y(0) = 4$, $y'(0) = -1$ (see Problem 1)

20. $y'' + 4y = 4x + 8$; $y(\pi) = 2$, $y'(\pi) = 3$ (see Problem 3)

21. $y'' + 4y = 6 \cos x$; $y(\pi/2) = 4$, $y'(\pi/2) = 6$ (see Problem 5)

22. $y'' - 3y' + 2y = 20 \sin 2x$; $y(0) = 6$, $y'(0) = 2$ (see Problem 7)

23. $y'' + 4y = 6 \cos x$; $y(0) = 4$, $y(\pi/4) = 4$ (see Problem 5)

24. $y'' + 4y = 4x + 8$; $y(0) = 0$, $y(\pi/4) = 6$ (see Problem 3)

25. A mass of 2.5 slugs is attached to a spring with spring constant $k = 10$. The mass is pulled down 1 foot below the equilibrium position and released. See Example F.

(a) Write the differential equation and initial conditions that model this situation.

(b) Solve this differential equation.

(c) Sketch the graph of the solution.

26. Follow the instructions in Problem 25 using the same data with one change. Assume that the mass is given an initial downward velocity of 2 feet per second.

27. Follow the instructions in Problem 25 with the same data but add a frictional force with damping constant $q = 0.125$.

28. Suppose we have a spring-mass problem satisfying the data in Problem 25 with the addition of a frictional force as in Problem 27 and a forcing function $F(t) = \cos t$.

(a) Write the differential equation with initial conditions·that models this problem.

(b) Solve this differential equation.

Solve the differential equations in Problems 29–36.

29. $y' - 4y = 2 \sin x; \; y(0) = 1$

30. $y'' - 4y = 5 \sin x; \; y(0) = 1, \; y'(0) = 5$

31. $y'' + 4y = 6 \sin x; \; y(0) = 9, \; y'(0) = 3$

32. $y'' - 4y = e^{2x}; \; y(0) = 8, \; y'(0) = \frac{1}{4}$

33. $y'' + 4y = 8 \sin 2x; \; y(0) = 9, \; y'(0) = 3$

34. $y'' - 5y' + 4y = 2e^{2x}; \; y(0) = 9, \; y'(0) = 3$

35. $y'' - 4y' + 4y = 2 \sin x; \; y(0) = 9, \; y'(0) = 3$

36. $y''' - 5y'' + 4y' = 8e^{2x}; \; y(0) = 9, \; y'(0) = 3, \; y''(0) = -11$

37. A stockbroker studying the gold market over the first 20 weeks of a year constructed the model

$$\frac{d^2P}{dt^2} + 0.8 \frac{dP}{dt} - 0.2P = -80$$

$$P(0) = 401.50 \qquad P'(0) = -0.9$$

to represent the price P of an ounce of gold in dollars after t weeks.

(a) Solve this differential equation.

(b) Determine the price of gold after 2 weeks, 10 weeks, and 20 weeks.

(c) What price does this model predict for gold after 30 weeks? After 52 weeks?

(d) Do you have any faith in this model?

38. In an experiment, a biologist counted the number N of bacteria (in hundreds) at the end of each hour for 5 hours. Based on her results, she constructed the model

$$\frac{d^2N}{dt^2} + 1.5 \frac{dN}{dt} - N = 4t - 1$$

$$N(0) = 1.4, \; N'(0) = 4.2$$

(a) How many bacteria were there at the end of 5 hours?

(b) How many bacteria are predicted by this model at the end of 10 hours?

39. The differential equation $d^2y/dt^2 + 9y = 12 \cos 3t$ describes a spring-mass problem with the forcing function

$$F(t) = 12 \cos 3t$$

Solve this differential equation subject to the initial conditions $y(0) = 1$ and $y'(0) = 0$. Note that the amplitude of this motion increases with time, producing a phenomenon called resonance.

40. In Example H, estimate the ultimate amplitude of the motion.

Chapter 13 Review Problem Set

In Problems 1–24, determine the general solution of each differential equation.

1. $\dfrac{dy}{dx} = x^2 e^{-y}$

2. $\dfrac{dy}{dx} = y^3 e^{2x}$

3. $(x^2 + 1)\dfrac{dy}{dx} = xy$

4. $y^2 \dfrac{dy}{dx} = \dfrac{y^3 + 1}{3x}$

5. $\dfrac{dy}{dx} - \dfrac{y}{x} = x^2 + x$

6. $\dfrac{dy}{dx} + \dfrac{2y}{x} = \dfrac{4}{x} + 3x$

7. $\dfrac{dy}{dx} + y \tan x = \sec x \tan x$

8. $\dfrac{dy}{dx} + \dfrac{y \cos x}{1 + \sin x} = \cos x$

9. $\dfrac{d^2y}{dx^2} - 16y = 0$

10. $\dfrac{d^2y}{dx^2} - 4\dfrac{dy}{dx} + 3y = 0$

11. $\dfrac{d^2y}{dx^2} + 16y = 0$

12. $\dfrac{d^2y}{dt^2} + 25y = 0$

13. $\dfrac{d^2y}{dx^2} - \dfrac{dy}{dx} - 20y = 0$

14. $\dfrac{d^2y}{dt^2} - 2\dfrac{dy}{dt} - 3y = 0$

15. $\dfrac{d^2y}{dx^2} + 4\dfrac{dy}{dx} - y = 0$

16. $\dfrac{d^2y}{dx^2} + 4\dfrac{dy}{dx} - 3y = 0$

17. $\dfrac{d^2y}{dx^2} - 6\dfrac{dy}{dx} + 9y = 0$

18. $\dfrac{d^2x}{dt^2} - 8\dfrac{dx}{dt} + 16x = 0$ 19. $\dfrac{d^2y}{dx^2} + 16y = e^x$

20. $\dfrac{d^2y}{dx^2} + 4y = 2\sin x$ 21. $\dfrac{d^2y}{dx^2} - 16y = \cos x$

22. $\dfrac{d^2y}{dt^2} - 4y = 3e^{3t} - e^t$ 23. $\dfrac{d^2y}{dx^2} - \dfrac{dy}{dx} - 20y = x^2$

24. $\dfrac{d^2x}{dt^2} - 4\dfrac{dx}{dt} - 5x = 16e^t + 5t$

25. Solve the differential equation $dy/dx + 2y = 6xy$ subject to $y(1) = 4$.

26. Solve the differential equation $\dfrac{dy}{dx} + 3y^2 = y^2x$ subject to $y(2) = \dfrac{1}{2}$.

27. Find the xy-equation of the curve that passes through the point $(\tfrac{1}{2}, -\tfrac{1}{4})$ and whose slope at (x, y) is y^2/x^2.

28. Find the equation of the curve that passes through the point $(3, 1)$ and whose slope at (x, y) is $3x^2y - 4xy$.

29. Determine the solution to the differential equation in Problem 9 which satisfies $y(0) = 4$ and $y'(0) = 8$.

30. Determine the particular solution of the differential equation of Problem 10 for which $y(0) = 3$ and $y'(0) = 7$.

31. Determine the solution to the differential equation in Problem 17 that satisfies $y(0) = 3$ and $y'(0) = 8$.

32. Find the particular solution to the differential equation of Problem 18 that satisfies $x(0) = 3$ and $x'(0) = 2$.

C 33. A brick was heated in an oven to a temperature of 450°F and then placed on a table in a room where the temperature was 75°F. After 20 minutes, the brick had cooled to 200°F.
   (a) Determine the temperature of the brick t minutes after removal from the oven.
   (b) How long will it take the brick to cool to 120°F?

34. A beef roast at temperature 90°F was placed in a freezer that is set at 10°F. After 2 hours the roast has cooled to 40°F.
   (a) Find the temperature of the roast after it had been in the freezer 5 hours.
   (b) How long did it take for the roast to cool to 20°F?

C 35. A tank initially contains 1000 liters of brine with 25 kilograms of salt in solution. Brine containing $\tfrac{1}{4}$ kilogram of salt per liter flows into the tank at 6 liters per minute, and the well-stirred solution flows out at the same rate. Find the amount of salt in the tank after 45 minutes.

36. A vat contains 1500 liters of brine with 40 kilograms of salt in solution. Pure water flows into the vat at 5

liters per minute and the well-stirred solution flows out at the same rate.
   (a) Find the amount of salt in the vat after 30 minutes.
   (b) How long will it take for the concentration of salt in the vat to drop to 0.01?

C 37. An insect population whose present size is 360 will grow to the size $P(t)$ after t days. If $P'(t) = 2e^{0.4t}$, find the population after 12 days.

38. Let $P(t)$ denote the size of a growing bacteria population t hours from now. If $P(0) = 3000$ and $P'(t) = e^{0.213t}$, find the population at the end of 11 hours.

C 39. A ball is thrown downward from a tower 400 feet high with a speed of 32 feet per second. Assuming that air resistance can be ignored, when will the ball hit the ground and with what speed?

40. Suppose that the ball in Problem 39 is thrown upward from the top of the 400-foot tower with a speed of 32 feet per second. When will the ball hit the ground and with what speed?

41. What form would a particular solution of the following differential equation have?
$$y'' - 5y' + 6y = 3x^2 + 6e^{2x}$$

42. What form would a particular solution of the following differential equation have?
$$y'' - 5y' + 4y = 2x^3 + 5 + 4e^x - 3e^{2x}$$

43. Solve the differential equation
$$y'' - 4y = 16\cos t + 32e^{2t}$$

44. Solve the differential equation $y'' - 2y' - 3y = 8e^{3x}$.

45. Consider the differential equation
$$6x^2\dfrac{d^2y}{dx^2} + 7x\dfrac{dy}{dx} - 2y = 0$$
   (a) For what values of r is $y = x^r$ a solution?
   (b) Write the general solution to the differential equation.

46. Find the particular solution to the differential equation in Problem 44 for which $y(0) = -1$ and $y'(0) = -9$.

47. Solve the logistic differential equation $dy/dt = 4y(10 - y)$ subject to $y(0) = 1$.

48. Solve the differential equation
$$\dfrac{dy}{dt} = 6(y - 1)(y - 3)y$$
subject to $y(0) = 4$.

Just as the invention of machines fostered the industrial revolution of the 19th century, the computer has sparked the information revolution of our day. Government, business, manufacturing, and education are all inescapably beholden to these marvelous devices. Grace Hopper was one of the pioneers in their development.

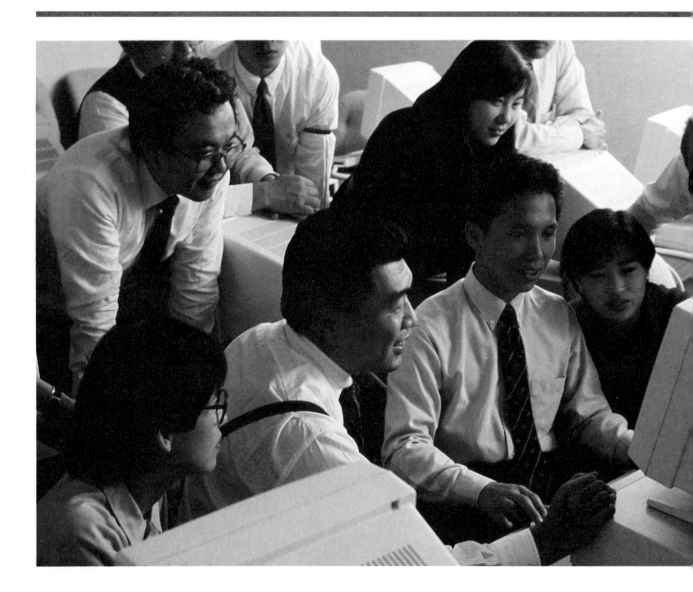

14

Numerical Calculus: Approximation

14.1 Numerical Integration

14.2 Approximating Functions by Polynomials

14.3 Solving Equations Numerically

14.4* The Method of Successive Substitutions

14.5* Numerical Solution of Differential Equations

Grace Murray Hopper
(1906–   )

No business, bank, or manufacturer today would think of operating without a computer. One can communicate with a computer in a variety of languages (BASIC, FORTRAN, Pascal, and so on). But in business, the language most often used is COBOL, invented by one of the most remarkable women of our time.

Even as a child, Grace Murray Hopper was interested in gadgets. At age 7 or 8, she took apart one of the clocks in the family home. Having trouble putting it back together, she took apart another, and before she was done she had all seven family clocks in pieces. It was an omen of the future, for Grace Hopper was to be a pioneer in working on the biggest gadgets there are, large-scale electronic computers. An intense curiosity has marked her long life; so has rebellion against the status quo, typified by her office clock which runs counterclockwise.

After graduating from Vassar College, she earned an M.A. and a Ph.D. from Yale University, both in mathematics. She taught mathematics at Vassar from 1931 until 1942, then briefly at Barnard College before joining the Navy during World War II. Her first Navy assignment was with Howard Aiken at Cambridge, Massachusetts where she became the third programmer on the first large-scale computer, the Mark I.

She was soon a computer expert. By 1959, she could see the potential of these giant number crunchers for business applications. Unfortunately, to speak to these early machines, one had to use a sophisticated and highly symbolic language. Most people communicate in words; Grace Hopper decided to develop a computer language that used ordinary English words and then let the computer translate into its own special symbolic language. Thus COBOL was born, the child of Grace Hopper.

When she retired from the Navy in 1986, Grace Hopper was the oldest officer on active duty. By this time, she had received most of the awards given in the computer world and had collected 49 honorary degrees. Retirement at age 79, Grace Hopper said, would not end her computer career. She hoped to find a job with a computer company.

14.1

NUMERICAL INTEGRATION

Our theme for this chapter is finding approximate solutions to a series of problems that arise in calculus. The methods we introduce depend for their implementation on our ability to perform large numbers of calculations quickly. In fact, it was the invention of the electronic calculator and computer that has made these methods practical. The teaching of calculus is likely to change dramatically in the years ahead as numerical methods, implemented on computers, become the preferred way of solving a host of calculus problems.

The first such problem that we consider is that of evaluating definite integrals by numerical (approximate) methods. Even if the exact method based on the Fundamental Theorem of Calculus is available, numerical methods are often preferable because they are so easy to implement on a computer. Moreover, many integrals that arise in practice (for example, $\int_a^b e^{-x^2/2}\, dx$) cannot be evaluated any other way, since their integrands do not have elementary antiderivatives.

Approximation by Riemann Sums

We begin with a definite integral for which the Fundamental Theorem of Calculus provides (in principle) the exact answer.

$$\int_1^2 \frac{1}{x}\, dx = \left[\ln x \right]_1^2 = \ln 2 - \ln 1 = \ln 2 \approx 0.69314718056$$

Even in this evaluation, we had to approximate at the last stage, for a calculator cannot give the exact value of ln 2 as a finite decimal. We have made a slight error, an error that we call *roundoff error* or *error of calculation*.

Suppose that we had forgotten the Fundamental Theorem of Calculus but desperately needed the value of $\int_1^2 (1/x)\, dx$. Recalling that the definite integral is a limit of Riemann sums (Section 8.4), we could divide the interval from 1 to 2 into five equal subintervals, calculate heights at the right endpoints, and approximate the integral by the sum of the areas of five corresponding rectangles as shown in Figure 1. This introduces another type of error, called the *discretization error* or *error of the method*, caused by the fact that we approximate the integral (area under the curve) by the area of five discrete rectangles.

Discretization error

$$\int_1^2 \frac{1}{x}\, dx \approx \left(\frac{1}{1.2} + \frac{1}{1.4} + \frac{1}{1.6} + \frac{1}{1.8} + \frac{1}{2} \right)(0.2)$$

Roundoff error

$$\approx (0.833333 + 0.714286 + 0.625 + 0.555556 + 0.5)(0.2)$$

$$= 0.645635$$

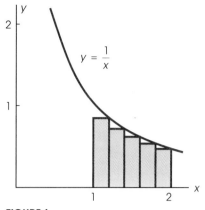

FIGURE 1

Note that this answer is much too small, a result we expected since the five rectangles certainly underapproximate the area under the curve (Figure 1). If we had divided the interval from 1 to 2 into 100 subintervals, the method of Riemann sums would produce a much better answer, namely, 0.690654. Dividing the interval into 1 million subintervals would reduce the discretization error, but because of the many calculations would at least potentially increase the roundoff error. We have not bothered to do this calculation since, even on a computer, it takes some time.

The Trapezoidal Rule

The approximation by rectangles illustrated above was rather crude; it would seem that we could do better if we approximated by trapezoids. From elementary school, we recall the formula for the area of a trapezoid shown in Figure 2. We apply this formula to each of the trapezoids in Figure 3 and add the results. This gives

$$\frac{h}{2}[f(x_0) + f(x_1)] + \frac{h}{2}[f(x_1) + f(x_2)] + \cdots + \frac{h}{2}[f(x_{n-1}) + f(x_n)]$$

and after simplification yields the **Trapezoidal Rule.**

THE TRAPEZOIDAL RULE

$$\int_a^b f(x)\,dx \approx [f(x_0) + 2f(x_1) + 2f(x_2) + \cdots + 2f(x_{n-1}) + f(x_n)]\frac{h}{2}$$

Here n is the number of subintervals, and $h = (b - a)/n$.

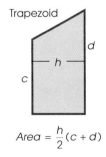

Trapezoid

$$Area = \frac{h}{2}(c + d)$$

FIGURE 2

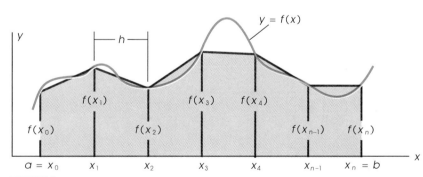

FIGURE 3

EXAMPLE A

Approximate $\int_1^2 \frac{1}{x}\,dx$ using the Trapezoidal Rule with $n = 5$.

Solution

$$\int_1^2 \frac{1}{x} \, dx \approx \left(\frac{1}{1} + \frac{2}{1.2} + \frac{2}{1.4} + \frac{2}{1.6} + \frac{2}{1.8} + \frac{1}{2} \right) \frac{1}{10} \approx 0.695635$$

Note that the trapezoidal approximation does give a better result than the right endpoint Riemann sum, as we expected. ■

The Parabolic Rule (Simpson's Rule)

Parabola

$$\text{Area} = \frac{h}{3}(c + 4d + e)$$

FIGURE 4

If the piecewise linear approximation of the Trapezoidal Rule gives a good value for a definite integral, it seems likely that a piecewise parabolic approximation should do even better. Consider first the problem of evaluating the area under a parabolic arch in terms of its heights at the two endpoints and the midpoint. You will be asked to establish the result shown in Figure 4 in Problem 22. Next divide the interval $[a, b]$ into n equal pieces, where n is an *even* number, and let $h = (b - a)/n$. Refer to Figure 5 and note that we can fit a parabolic arch through each triad of points on the curve, thus obtaining a piecewise parabolic approximation to the curve. When we apply the formula in Figure 4 to each of the $n/2$ parabolic pieces, we obtain an integral approximation called the **Parabolic Rule** [also called **Simpson's Rule** after its discoverer Thomas Simpson (1710–1761)].

THE PARABOLIC RULE

$$\int_a^b f(x) \, dx \approx [\, f(x_0) + 4f(x_1) + 2f(x_2) + 4f(x_3) + \cdots + 4f(x_{n-1}) + f(x_n)] \frac{h}{3}$$

The pattern of coefficients in this formula is 1, 4, 2, 4, 2, 4, 2, 4, . . . , 2, 4, 1.

FIGURE 5

EXAMPLE B

Use the Parabolic Rule with $n = 4$ to approximate $\displaystyle\int_1^2 \frac{1}{x} \, dx$.

Solution Note that $h = (2 - 1)/4 = \frac{1}{4}$. Thus,

$$\int_1^2 \frac{1}{x}\, dx \approx \left(\frac{1}{1} + \frac{4}{1.25} + \frac{2}{1.5} + \frac{4}{1.75} + \frac{1}{2}\right)\frac{1}{12} \approx 0.693254$$

This value, computed with $n = 4$, is considerably better than the corresponding value from Example A, which used the Trapezoidal Rule with $n = 5$.   ■

EXAMPLE C

Use the Parabolic Rule with $n = 10$ to calculate $\int_0^1 e^{-x^2}\, dx$.

Solution We approximate the integral by the following sum.

$$(e^{-0^2} + 4e^{-0.1^2} + 2e^{-0.2^2} + 4e^{-0.3^2} + 2e^{-0.4^2} + \cdots + 4e^{-0.9^2} + e^{-1^2})\frac{0.1}{3}$$

$$\approx 0.746825$$ ■

Comparison of the Trapezoidal and Parabolic Rules

n	Trapezoidal Rule	Parabolic Rule
4	0.697024	0.693254
10	0.693771	0.693150
100	0.693153	0.693147
1,000	0.693147	0.693147
10,000	0.693147	0.693147
100,000	0.693145	0.693145
200,000	0.693136	0.693145

FIGURE 6

Using a computer, we can approximate $\int_1^2 (1/x)\, dx$ using many different values of n; some results are shown in Figure 6. Keep in mind that the correct value to six decimal places is 0.693147.

Figure 6 confirms our feeling that the Parabolic Rule should give better results than the Trapezoidal Rule for the same n. It also shows the folly of taking too large an n. For while a larger and larger n reduces the discretization error, the huge number of calculations (each with a roundoff) tends to increase the total roundoff error. The last two lines of Figure 6 show roundoff error beginning to distort the results.

Why would one ever choose the Trapezoidal Rule in preference to the Parabolic Rule, since the latter tends to give better answers with essentially the same amount of work? One answer relates to our ability to give a bound on the errors we are making. No approximation result is very useful unless we have some idea of its accuracy. An important result proved in advanced calculus is the following.

DISCRETIZATION ERRORS

The discretization error E in approximating $\int_a^b f(x)\, dx$ is bounded as follows.

Trapezoidal Rule: $|E| \leq \dfrac{(b-a)^3}{12n^2}[\max |f''(x)|]$

Parabolic Rule: $|E| \leq \dfrac{(b-a)^5}{180n^4}[\max |f''''(x)|]$

Of course, in each case it is understood that the maximum is taken over the interval $a \leq x \leq b$.

The existence of n^4 rather than n^2 in the denominator of the error bound for the Parabolic Rule shows that this rule will in general be better than the Trapezoidal Rule. However, note that to find the actual error bound for the Parabolic Rule, we need to calculate and bound the fourth derivative, compared to the second derivative for the Trapezoidal Rule. This can sometimes be an overwhelming job.

EXAMPLE D

Use the Trapezoidal Rule and the Parabolic Rule to approximate $\int_1^3 \frac{1}{x}\, dx$ using $n = 10$. Then give bounds for their discretization errors.

Solution

$$\text{Trapezoidal Rule: } \int_1^3 \frac{1}{x}\, dx \approx 1.101562$$

$$\text{Parabolic Rule: } \int_1^3 \frac{1}{x}\, dx \approx 1.098661$$

Now $f(x) = 1/x = x^{-1}$, and so $f'(x) = -x^{-2}$, $f''(x) = 2x^{-3}$, $f'''(x) = -6x^{-4}$, and $f''''(x) = 24x^{-5}$. The absolute value of each of these derivatives achieves its maximum at the left end of the interval $[1, 3]$. Thus, we have the following error bounds.

$$\text{Trapezoidal Rule: } |E| \leq \frac{2^3}{1200}(2) = \frac{16}{1200} = 0.013333$$

$$\text{Parabolic Rule: } |E| \leq \frac{2^5}{1,800,000}(24) = 0.000427$$

Using the Trapezoidal Rule, we would report the value of the integral as 1.101562 ± 0.013333; using the Parabolic Rule, we would report 1.098661 ± 0.000427. The correct value obtained by use of the Fundamental Theorem of Calculus is 1.098612 (accurate to six decimal places). ∎

ⒸProblem Set 14.1

For each of the integrals in Problem 1–6, calculate each of the following indicated quantities.

(a) *The (exact) value of the integral using the Fundamental Theorem of Calculus.*

(b) *An approximate value using a right endpoint Riemann sum with $n = 6$.*

(c) *An approximate value using the Trapezoidal Rule with $n = 6$ (see Example A).*

(d) *An approximate value using the Parabolic Rule with $n = 6$ (see Example B).*

1. $\int_1^4 x^3\, dx$

2. $\int_0^3 x^4\, dx$

3. $\int_0^3 \sin x \, dx$

4. $\int_0^6 \frac{1}{x + 1} \, dx$

5. $\int_2^8 \frac{1}{(x + 1)^2} \, dx$

6. $\int_1^4 \ln x \, dx$

7. Approximate $\int_0^2 \sin(x^2) \, dx$ using the Trapezoidal Rule with (a) $n = 4$, (b) $n = 8$, and (c) $n = 16$. *Note:* $\sin(x^2)$ does not have an elementary antiderivative, so this integral cannot be evaluated by the Fundamental Theorem of Calculus.

8. Use the Trapezoidal Rule to approximate $\int_1^2 (\ln x)^2 \, dx$ with $n = 10$.

9. Redo Problem 7 using the Parabolic Rule in place of the Trapezoidal Rule.

10. Redo Problem 8 using the Parabolic Rule in place of the Trapezoidal Rule.

11. In Problem 5, you calculated $\int_2^8 \frac{1}{(x + 1)^2} \, dx$ using both the Trapezoidal Rule and the Parabolic Rule with $n = 6$. Give a good bound for the discretization error resulting from using (a) the Trapezoidal Rule and (b) the Parabolic Rule. See Example D.

12. In Problem 6, you calculated $\int_1^4 \ln x \, dx$ using both the Trapezoidal Rule and the Parabolic Rule using $n = 6$. Give a good bound for the discretization error resulting from using (a) the Trapezoidal Rule and (b) the Parabolic Rule.

Often in practical situations, we do not have a formula for $f(x)$ but may know its values at several equally spaced points along the interval $[a, b]$. If so, we can still approximate $\int_a^b f(x) \, dx$ using either the Trapezoidal Rule or the Parabolic Rule. This is the basis for Problems 13–16.

13. Approximate $\int_0^4 f(x) \, dx$ using the Trapezoidal Rule given the following set of data.

x	0	0.5	1	1.5	2	2.5	3	3.5	4
$f(x)$	0.4	0.9	1.2	1.6	1.6	1.3	0.8	0.7	0.1

14. Approximate $\int_0^8 f(x) \, dx$ using the Parabolic Rule given the following set of data.

x	0	1	2	3	4	5	6	7	8
$f(x)$	8	20	150	90	45	10	1	−2	−5

15. The Macintoshes have a lakefront lot with measurements, in feet, as shown in Figure 7. Use the Parabolic Rule to estimate the area of their lot.

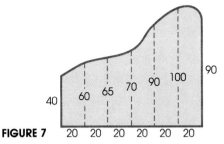

FIGURE 7

16. Suppose that after the takeoff of a rocket, the velocity readings during the first 20 seconds at 2-second intervals were 0, 22, 48, 100, 160, 232, 310, 400, 505, 610, and 730, all in feet per second. Use the Parabolic Rule to approximate the distance traveled. *Hint:* The distance traveled is $\int_0^{20} v(t) \, dt$, where $v(t)$ is the velocity at time t.

---

17. Approximate $\int_0^2 e^{\sqrt{x}} \, dx$ using the Parabolic Rule with $n = 10$.

18. Approximate $\int_1^2 \sqrt{\ln x} \, dx$ using the Trapezoidal Rule with $n = 10$.

19. The depths of a river 100 feet wide were taken at 10-foot intervals along a line perpendicular to the river banks and found to be 5, 6, 8, 8, 9, 10, 12, 8, 7, 7, and 4, all in feet. Use the Parabolic Rule to calculate the cross-sectional area of the river at this line. Then calculate the total volume of water that flows past this line in 1 hour if the river is flowing at 5 feet/second.

20. Temperature readings in degrees Celsius were taken at 3-hour intervals beginning at midnight, giving the following temperatures: 20°, 18°, 22°, 25°, 30°, 31°, 29°, 24°, 24°. Use the Parabolic Rule to estimate the average temperature during that day.

21. By considering the area of the unit circle, convince yourself that $\pi = 4 \int_0^1 \sqrt{1 - x^2} \, dx$. Now use the Parabolic Rule with $n = 10$ to approximate π.

22. Let $f(x) = Ax^2 + Bx + C$. Show that

$$\int_{m-h}^{m+h} f(x) \, dx = \frac{h}{3}[f(m - h) + 4f(m) + f(m + h)]$$

by showing that both sides have the value

$$\frac{h}{3}[A(6m^2 + 2h^2) + B(6m) + 6C]$$

This establishes the formula on which the Parabolic Rule is based.

23. Show that the Parabolic Rule is exact for any cubic function, that is, for any function of the form $f(x) = Ax^3 + Bx^2 + Cx + D$. *Hint:* This can be shown directly but is also a consequence of the error formula for the Parabolic Rule.

APPROXIMATING FUNCTIONS BY POLYNOMIALS

A polynomial in x is an expression of the form

$$P(x) = a_n x^n + a_{n-1} x^{n-1} + \cdots + a_2 x^2 + a_1 x + a_0$$

where the a's are real numbers and n is a nonnegative integer. Typical examples are $3x - 7$, $4x^2 + \pi x - 0.5$, and $x^3 - x + 8$. Actually, we will want to consider the slightly more general case of polynomials in $x - a$, where a is a constant. Examples are $3(x - a) + 4$ and $2(x - a)^3 + 5(x - a)^2 - 2(x - a) - 6$. Polynomial functions are considered to be the simplest of all functions because they involve just the three basic operations of addition, subtraction, and multiplication (raising to a positive integral power is repeated multiplication).

Long ago, mathematicians asked a natural question: Is it possible to approximate complicated functions like logarithmic functions and trigonometric functions by polynomial functions? The answer is yes. How to do this is the subject of this section.

Linear Approximation

Consider the graph of an arbitrary differentiable function (Figure 8). Near the point $(a, f(a))$, we may approximate the graph of $y = f(x)$ by a straight line passing through the given point and having the same slope as the curve, namely, $f'(a)$. The equation of this line is

$$y - f(a) = f'(a)(x - a)$$

and this leads to the *linear approximation*

$$\boxed{f(x) \approx f(a) + f'(a)(x - a)}$$

The linear polynomial $P_1(x) = f(a) + f'(a)(x - a)$ is called the **Taylor polynomial of order 1 based at** a for $f(x)$, being named after the English mathematician Brook Taylor (1685–1731).

FIGURE 8

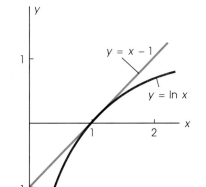

FIGURE 9

EXAMPLE A

Find the Taylor polynomial of order 1 based at $a = 1$ for the function $f(x) = \ln x$ and use it to approximate $\ln(1.04)$ and $\ln(0.76)$.

Solution Since $f(x) = \ln x$, $f'(x) = 1/x$, and so $f(1) = \ln 1 = 0$ and $f'(1) = 1$. Thus the desired linear polynomial is

$$P_1(x) = 0 + 1(x - 1) = x - 1$$

We conclude that near 1,

$$\ln x \approx x - 1$$

and

$$\ln(1.04) \approx 1.04 - 1 = 0.04$$

$$\ln(0.76) \approx 0.76 - 1 = -0.24$$

These values should be compared with the correct four-place values of $\ln(1.04) = 0.0392$ and $\ln(0.76) = -0.274$. As expected, the first approximation is much better, since 1.04 is much closer to 1 than is 0.76.

Figure 9 shows the graph of $y = \ln x$ and its linear approximation $y = P_1(x) = x - 1$ near $x = 1$.   ■

Quadratic Approximation

We consider next the problem of approximating $f(x)$ near $x = a$ by a quadratic polynomial $P_2(x)$ in $x - a$. Now we want three conditions to hold at a, namely,

$$P_2(a) = f(a) \qquad P_2'(a) = f'(a) \qquad P_3''(a) = f''(a)$$

The unique quadratic polynomial in $x - a$ that satisfies these three conditions (the **Taylor polynomial of order 2 based at a**) is

$$P_2(x) = f(a) + f'(a)(x - a) + \frac{f''(a)}{2}(x - a)^2$$

as you may check. The corresponding *quadratic approximation* for $f(x)$ is

$$f(x) \approx f(a) + f'(a)(x - a) + \frac{f''(a)}{2}(x - a)^2$$

EXAMPLE B

Find $P_2(x)$, the Taylor polynomial of order 2 based at 1, for $f(x) = \ln x$ and use it to approximate $\ln(1.04)$ and $\ln(0.76)$.

Solution In this case, $f(x) = \ln x$, $f'(x) = 1/x$, and $f''(x) = -1/x^2$, and so $f(1) = 0$, $f'(1) = 1$, and $f''(1) = -1$. Thus,

$$P_2(x) = 0 + 1(x - 1) - \frac{1}{2}(x - 1)^2$$

and

$$\ln x \approx (x - 1) - \frac{1}{2}(x - 1)^2$$

The required approximations are

$$\ln(1.04) \approx (1.04 - 1) - \frac{1}{2}(1.04 - 1)^2 = 0.0392$$

$$\ln(0.76) \approx (0.76 - 1) - \frac{1}{2}(0.76 - 1)^2 = -0.2688$$

As expected, these approximations are better than the linear ones in Example A.

Figure 10 shows $y = \ln x$ and its corresponding quadratic approximation near $x = 1$. ∎

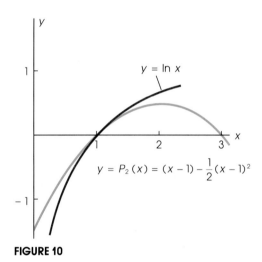

FIGURE 10

Taylor Polynomials of Order n

We could consider next cubic approximations to $f(x)$, but we choose rather to go directly to the general case. We seek a polynomial $P_n(x)$ of degree at most n which satisfies the $n + 1$ conditions $P_n(a) = f(a)$, $P_n'(a) = f'(a)$, $P_n''(a) = f''(a)$, . . . , $P_n^{(n)}(a) = f^{(n)}(a)$. The unique polynomial (the **Taylor polynomial of order n based at a**) satisfying these conditions is

$$P_n(x) = f(a) + f'(a)(x - a) + \frac{f''(a)}{2!}(x - a)^2 + \cdots + \frac{f^{(n)}(a)}{n!}(x - a)^n$$

Here $n!$, (read n **factorial**), is the standard symbol for the product of the first n positive integers, that is,

$$n! = n(n - 1)(n - 2)(n - 3) \cdots 3 \cdot 2 \cdot 1$$

The corresponding approximation is

$$f(x) \approx f(a) + f'(a)(x - a) + \frac{f''(a)}{2!}(x - a)^2 + \cdots + \frac{f^n(a)}{n!}(x - a)^n$$

EXAMPLE C

Find the Taylor polynomial of order 4 based at $x = 1$ for $f(x) = \ln x$ and use it to approximate $\ln(1.04)$ and $\ln(0.76)$.

Solution Note that $f(x) = \ln x$, $f'(x) = 1/x$, $f''(x) = -1/x^2$, $f^{(3)}(x) = 2/x^3$, and $f^{(4)}(x) = -6/x^4$. Thus $f(1) = 0$, $f'(1) = 1$, $f''(1) = -1$, $f^{(3)}(1) = 2$, and $f^{(4)}(1) = -6$, and

$$\ln x \approx 0 + (x - 1) - \frac{1}{2}(x - 1)^2 + \frac{1}{3}(x - 1)^3 - \frac{1}{4}(x - 1)^4$$

This gives the approximations

$$\ln(1.04) \approx (0.04) - \frac{1}{2}(0.04)^2 + \frac{1}{3}(0.04)^3 - \frac{1}{4}(0.04)^4 \approx 0.0392207$$

$$\ln(0.76) \approx (-0.24) - \frac{1}{2}(-0.24)^2 + \frac{1}{3}(-0.24)^3 - \frac{1}{4}(-0.24)^4 \approx -0.2742374$$

These are to be compared with the correct seven-place values of 0.0392207 and -0.2744368. We remark that these approximations involve the two kinds of errors we mentioned in Section 14.1. The first $\approx$ introduces a discretization error; the second $\approx$ introduces an error of calculation (accumulation of roundoff errors). ∎

Maclaurin Polynomials

In case $a = 0$, the Taylor polynomial of order n based at a simplifies to the **Maclaurin polynomial of order n,** named after the Scottish mathematician Colin Maclaurin (1693–1746). In this case, we have the following particularly useful approximation valid near $x = 0$.

$$f(x) \approx f(0) + f'(0)x + \frac{f''(0)}{2!}x^2 + \frac{f^{(3)}(0)}{3!}x^3 + \cdots + \frac{f^{(n)}(0)}{n!}x^n$$

EXAMPLE D

Find the Maclaurin polynomials of order n for e^x and for $\sin x$. Then approximate $e^{0.12}$ and $\sin(0.12)$ using $n = 5$.

Solution Figure 11 gives the successive derivatives of e^x and $\sin x$ and their corresponding values at $x = 0$. The resulting Maclaurin polynomials are

$$e^x \approx 1 + x + \frac{1}{2!}x^2 + \frac{1}{3!}x^3 + \cdots + \frac{1}{n!}x^n$$

and

$$\sin x \approx x - \frac{1}{3!}x^3 + \frac{1}{5!}x^5 + \cdots + \frac{(-1)^{(n-1)/2}}{n!}x^n \qquad (n \text{ odd})$$

The required approximations are

$$e^{0.12} \approx 1 + 0.12 + \frac{1}{2}(0.12)^2 + \frac{1}{6}(0.12)^3 + \frac{1}{24}(0.12)^4 + \frac{1}{120}(0.12)^5$$

$$\approx 1.12749684736$$

and

$$\sin(0.12) \approx 0.12 - \frac{1}{6}(0.12)^3 + \frac{1}{120}(0.12)^5 \approx 0.11971220736$$

We will have something to say about the accuracy of these results in Example F.

		At $x = 0$		At $x = 0$
$f(x)$	e^x	1	$\sin x$	0
$f'(x)$	e^x	1	$\cos x$	1
$f''(x)$	e^x	1	$-\sin x$	0
$f^{(3)}(x)$	e^x	1	$-\cos x$	-1
$f^{(4)}(x)$	e^x	1	$\sin x$	0
$f^{(5)}(x)$	e^x	1	$\cos x$	1
$\vdots$	$\vdots$	$\vdots$	$\vdots$	$\vdots$

FIGURE 11   ■

EXAMPLE E

For the function $f(x) = \cos x$, determine the Maclaurin polynomials $P_1(x)$, $P_2(x)$, . . . , $P_5(x)$, and $P_8(x)$. Then display their graphs in the same coordinate plane, thus demonstrating geometrically that these polynomials approximate $\cos x$ better and better as n gets larger.

Solution The results are shown in Figure 12.

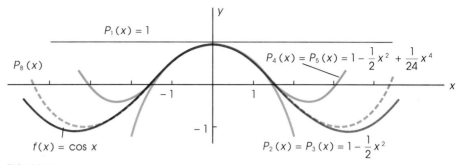

FIGURE 12

The Discretization Error in Taylor's Approximation

It is always important to give a good bound on the error that is made by an approximation. We do this for the discretization error in the following theorem. A proof of this theorem is quite technical, and we leave it for more advanced books.

DISCRETIZATION ERROR IN TAYLOR'S APPROXIMATION

The discretization error $E(x)$ made in approximating $f(x)$ by its Taylor polynomial of order n based at a satisfies

$$|E(x)| \le \frac{|x - a|^{n+1}}{(n + 1)!}[\max |f^{(n+1)}(x)|]$$

the maximum being taken over an interval containing x and a.

EXAMPLE F

Give a bound for the discretization errors made in Example D.

Solution The discretization error E in the approximation given for $e^{0.12}$ satisfies

$$|E| \le \frac{(0.12)^6}{6!}\left[\max_{0 \le x \le 0.12} e^x\right] = \frac{(0.12)^6}{720}e^{0.12} \le 0.000000005$$

Similarly the discretization error E in the approximation given for $\sin 0.12$ satisfies

$$|E| \le \frac{(0.12)^6}{6!}\left[\max_{0 \le x \le 0.12} |\sin x|\right] \le \frac{(0.12)^6}{720}\sin 0.12 \le 0.0000000005 \quad \blacksquare$$

CProblem Set 14.2

In each of Problems 1–6, a function is given. (a) Find the linear polynomial that best approximates the given function near x = 2 (see Example A). (b) Find the quadratic polynomial that best approximates the given function near x = 2 (see Example B). (c) Approximate f(2.01) using (a), using (b), and using your calculator directly.

1. $f(x) = \dfrac{1}{x + 3}$

2. $f(x) = \dfrac{1}{x^2 + 1}$

3. $f(x) = \sqrt{x + 2}$

4. $f(x) = \sqrt[3]{x + 6}$

5. $f(x) = x \ln(x + 1)$

6. $f(x) = \dfrac{x}{x^2 + 3}$

7. For the function $f(x) = \sqrt{x}$, find the Taylor polynomial of order 4 based at $a = 4$. Then use this polynomial to estimate $\sqrt{3.9}$ and $\sqrt{4.2}$. See Example C.

8. For $f(x) = x \ln(x + 1)$, find the Taylor polynomial of order 4 based at $a = 1$. Then use this polynomial to estimate $f(1.1)$.

Find the Maclaurin polynomial of order n for each function in Problems 9–12 (see Example D).

9. $f(x) = e^{-x}$

10. $f(x) = (e^x + e^{-x})/2$

11. $f(x) = 1/(1 - x)$

12. $f(x) = \ln(1 + x)$

13. Sketch carefully the graph of $f(x) = e^{-x}$ and its third-order Maclaurin approximation $P_3(x)$ on the interval $[-2, 2]$ using the same axes (see Problem 9).

14. Sketch carefully the graph of $f(x) = 1/(1 - x)$ and its third-order Maclaurin approximation $P_3(x)$ on the interval $(-1, 1)$ using the same axes (see Problem 11).

15. Give a good bound for the maximum discretization error that is made in approximating e^{-x} by its third-order Maclaurin polynomial $P_3(x)$ on the interval $[-0.5, 0.5]$. See Example F.

16. Give a good bound for the maximum discretization error that is made in approximating $1/(1 - x)$ by its

third-order Maclaurin polynomial $P_3(x)$ on the interval $[-0.5, 0.5]$.

17. In Problem 7, you estimated $\sqrt{4.2}$ by using a Taylor polynomial of order 4 based at $a = 4$. Give a good bound for the discretization error that you made.

18. Give a good bound for the discretization error that you made in Problem 8.

---

19. Find the Taylor polynomial of order 5 based at $a = \pi/2$ for $f(x) = \sin x$ and use it to estimate $\sin 1.68$.

20. Find the Taylor polynomial of order 4 based at $a = \pi$ for $f(x) = \cos x$ and use it to estimate $\cos 3.28$.

21. Give a good bound for the discretization error made in Problem 19.

22. Give a good bound for the discretization error made in Problem 20.

23. Find the Taylor polynomial of order 4 based at $a = 1$ for $f(x) = 2x^4 - 3x^2 + x - 1$ and show that it represents $f(x)$ exactly for all x.

24. Find a, b, c, and d so that

$$2(x - 2)^3 + (x - 2)^2 - 4(x - 2) + 5 =$$
$$a(x + 3)^3 + b(x + 3)^2 + c(x + 3) + d$$

for all x.

25. Find the Taylor polynomial of order 2 based at $a = e$ for $f(x) = \ln(\ln x)$ and use it to approximate $\ln(\ln 2.72)$. Recall that $e = 2.71828 \ldots$.

26. How large would n have to be in order for us to be sure that the error in calculating e using the Maclaurin polynomial $P_n(x)$ for e^x is less than 0.00001?

27. Use the Maclaurin polynomial of order 2 for $f(x) = \sqrt{\cos x}$ to get a good estimate for $\int_0^1 \sqrt{\cos x} \, dx$.

28. Use the Maclaurin polynomial for e^x to get a polynomial approximation of order 6 for e^{-x^2} (simply replace x by $-x^2$). Then approximate $\int_0^1 e^{-x^2} \, dx$ using your result.

14.3
SOLVING EQUATIONS NUMERICALLY

In mathematics and in science, we often need to find the roots (solutions) of an equation $f(x) = 0$. If the equation happens to be linear or quadratic, this is a simple matter. For example, the linear equation $3x + 8 = 0$ has the unique root $x = -\frac{8}{3}$. The quadratic equation $x^2 + x - 1 = 0$ has two roots, $x = (-1 \pm \sqrt{5})/2$, obtained by using the familiar Quadratic Formula. More complicated equations can rarely be solved explicitly; fortunately there are numerical methods for finding the root of an equation to any desired degree of accuracy.

All resourceful people are familiar with the method of solving problems that we call "trial and error." Given a cup of tea, we add sugar a bit at a time until it tastes just right. Given a stopper too large for a hole, we whittle it down until it fits. We change the solution step by step, improving the accuracy at each step, until we are satisfied. Mathematicians call this the *method of successive approximations* or the *method of iterations*. One such method is attributed to Isaac Newton (1642–1727).

Newton's Method

Let us look at the problem of solving the equation $f(x) = 0$ from a geometric point of view. A root r is the x-coordinate of the point where the graph of $y = f(x)$ crosses the x-axis (Figure 13). Given an arbitrary value x_1 somewhere near r, draw the tangent line at $(x_1, f(x_1))$ and find the point x_2 where this line crosses the x-axis. Note that x_2 is closer to r than was x_1. Next draw the tangent line at $(x_2, f(x_2))$ and find the point x_3 where this line crosses the x-axis. Continue this process, thereby obtaining a sequence x_1, x_2, x_3, . . . , which appears to be marching toward r (Figure 14). Stop when the desired accuracy is achieved.

This process can be mechanized, so it is easy to implement on a calculator or computer. An equation of the tangent at $(x_1, f(x_1))$ is

$$y - f(x_1) = f'(x_1)(x - x_1)$$

Its x-intercept x_2 is found by setting $y = 0$ and solving for x. The result is

$$x_2 = x_1 - \frac{f(x_1)}{f'(x_1)}$$

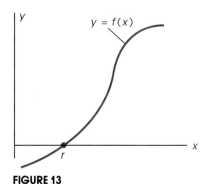

FIGURE 13

FIGURE 14

More generally, we have Newton's *algorithm*

$$x_{n+1} = x_n - \frac{f(x_n)}{f'(x_n)}$$

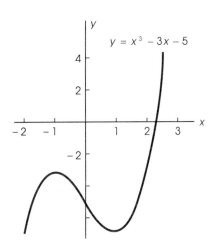

FIGURE 15

n	x_n
1	2.5
2	2.3015873
3	2.2792907
4	2.2790188
5	2.2790188

FIGURE 16

EXAMPLE A

Use Newton's Method to find the unique root of $f(x) = x^3 - 3x - 5 = 0$ accurate to six decimal places.

Solution $f'(x) = 3x^2 - 3$, and so the algorithm takes the form

$$x_{n+1} = x_n - \frac{x_n^3 - 3x_n - 5}{3x_n^2 - 3} = \frac{2x_n^3 + 5}{3x_n^2 - 3}$$

Figure 15 shows the graph of $y = f(x) = x^3 - 3x - 5$ and suggests $x_1 = 2.5$ as a seed value. Thus

$$x_2 = \frac{2(2.5)^3 + 5}{3(2.5)^2 - 3} \approx 2.3015873$$

$$x_3 = \frac{2(2.3015873)^3 + 5}{3(2.3015873)^2 - 3} \approx 2.2792907$$

and so on. Since we want six-decimal-place accuracy, we continue the process until we get a repetition of digits through seven decimal places. The results are summarized in Figure 16. Our six-decimal-place estimate for r is 2.279019. ∎

For most problems, the sequence of numbers obtained by Newton's Method converges very rapidly. There is even a theorem suggesting that the accuracy should improve about two decimal places at each step. However, there are pathological cases where the method does not converge at all, a matter we take up later.

EXAMPLE B

Find the root of $x = e^{-x}$ accurate to six decimal places.

Solution Rather than graphing $f(x) = x - e^{-x}$ to get an initial seed value x_1, we sketch the graphs of $y = x$ and $y = e^{-x}$ using the same axes (Figure 17). The x-coordinate of the intersection point is the root we are looking for; it suggests using $x_1 = 0.5$. Now $f'(x) = 1 + e^{-x}$, and so Newton's algorithm takes the form

$$x_{n+1} = x_n - \frac{x_n - e^{-x_n}}{1 + e^{-x_n}} = \frac{x_n + 1}{e^{x_n} + 1}$$

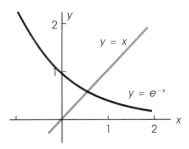

FIGURE 17

n	x_n
1	0.5
2	0.566311
3	0.5671432
4	0.5671433
5	0.5671433

FIGURE 18

Applying this algorithm over and over leads to the table shown in Figure 18. We conclude that $r \approx 0.567143$. ∎

EXAMPLE C

Solve $\sin x = \ln x$ accurate to six decimal places.

Solution Figure 19 shows the graphs of $y = \sin x$ and $y = \ln x$ and suggests $x_1 = 2$ as a seed value. Let $f(x) = \sin x - \ln x$ so that $f'(x) = \cos x - 1/x = (x \cos x - 1)/x$. Newton's algorithm takes the form

$$x_{n+1} = x_n - \frac{x_n(\sin x_n - \ln x_n)}{x_n \cos x_n - 1}$$

and leads to the table shown in Figure 20. We conclude that $r \approx 2.219107$. ∎

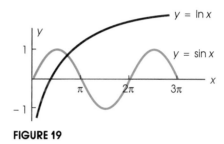

FIGURE 19

n	x_n
1	2
2	2.2359336
3	2.2191851
4	2.2191076
5	2.2191072
6	2.2191069
7	2.2191069

FIGURE 20

The Nonconvergent Possibility

We have hinted that Newton's Method can fail to converge to a root. One way for this to happen is illustrated in Figure 21. The problem here is that the seed value x_1 was not chosen close enough to the root r. A more serious problem is illustrated by the equation $f(x) = x^{1/3} = 0$. That Newton's Method fails to converge is shown geometrically in Figure 22, and numerically in Figure 23.

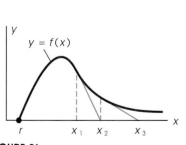

FIGURE 21

FIGURE 22

$$x_{n+1} = x_n - \frac{x_n^{1/3}}{\frac{1}{3} x_n^{-2/3}} = -2x_n$$

n	x_n
1	0.15
2	−0.30
3	0.60
4	−1.20

FIGURE 23

We could state conditions that would guarantee the convergence of Newton's Method. However, it seems best to say for any given problem: Try it. If the method fails to converge, it will be readily apparent from the calculated values of x_n.

An Application: Internal Rate of Return

Suppose that an initial investment of P dollars produces returns of various amounts at various times in the future. The *internal rate of return* or the *implied interest rate* is the interest rate for which the sum of the present values of the returns equals the initial investment.

EXAMPLE D

RECALL

Usually, we obtain the seed value for Newton's algorithm by drawing the graph of $y = f(x)$ and noting where it crosses the x-axis. This would have been the safe approach in Example D. However, 9% seems like a reasonable interest rate and suggests $x_1 = 1.09$.

What rate of interest compounded annually is implied in an offer to sell a lot for $50,000 cash or to sell it in three annual installments of $20,000 payable 1, 2, and 3 years from now?

Solution Assume the interest rate to be i (written as a decimal) and discount each of the future payments back to the present. Thus

$$50,000 = 20,000(1 + i)^{-1} + 20,000(1 + i)^{-2} + 20,000(1 + i)^{-3}$$

Divide both sides by 10,000, multiply both sides by $(1 + i)^3$, and substitute $x = 1 + i$ to obtain the equation

$$5x^3 = 2x^2 + 2x + 2$$

or equivalently

$$f(x) = 5x^3 - 2x^2 - 2x - 2 = 0$$

n	x_n
1	1.09
2	1.0970728
3	1.0970103
4	1.0970103

FIGURE 24

Now use $x_1 = 1.09$ as a seed value in the corresponding Newton's algorithm

$$x_{n+1} = x_n - \frac{5x_n^3 - 2x_n^2 - 2x_n - 2}{15x_n^2 - 4x_n - 2} = \frac{10x_n^3 - 2x_n^2 + 2}{15x_n^2 - 4x_n - 2}$$

This leads to the table shown in Figure 24. We conclude that $x \approx 1.09701$ and thus that $i = 0.09701 = 9.701\%$. ∎

[C]Problem Set 14.3

In Problems 1–4, use Newton's Method to find the indicated root of the given equation accurate to four decimal places.

1. Root near 1 of $x^3 + 3x - 5 = 0$
2. Root near -1 of $x^4 + 4x^3 + 2 = 0$

3. Root near -4 of $x^4 + 4x^3 + 2 = 0$
4. Root near 2 of $x^{10} - 2x - 1000 = 0$

Each equation in Problems 5–8 has exactly one real root. Use Newton's Method to find this root accurate to six decimal places. Begin by sketching a graph to obtain a seed value as in Example A.

5. $x^3 - 3x - 5 = 0$
6. $x^3 - 3x - 10 = 0$
7. $x^3 + x - 29 = 0$
8. $2x^3 - 6x - 13 = 0$

Follow the procedure in Examples B and C to find a seed value and then use Newton's Method to solve the equations in Problems 9–12. Give your answers correct to six decimal places.

9. $2x = \cos 3x$
10. $e^x = 2 - x^3$
11. $\ln x = 1/x$
12. $\ln x = 2 - x$

13. What rate of interest compounded annually is implied in an offer to sell a house for $100,000 cash or to sell it in four annual installments of $30,000 due at the end of each of the next 4 years? See Example D.

14. Jane Smith is offering to sell her lakefront lot for $80,000 cash. Alternatively she will accept $20,000 down and three equal payments of $24,000 at the end of each of the next 3 years. What rate of interest compounded annually is implied by this offer?

---

15. Find all three roots of $x^3 - 3x + 4 = x + 3$ accurate to five decimal places.

16. Find all real roots of $x^4 + x^3 - 3x^2 + 4x - 28 = 0$ accurate to five decimal places.

17. Find the two smallest positive roots of $\tan x = x + 1$ accurate to five decimal places.

18. Find the two largest negative roots of $x = 5 \cos x$ accurate to five decimal places.

19. Find all positive roots of $e^x = 1 + \cos x$ accurate to six decimal places.

20. Show that Newton's Method applied to the equation $x^2 - a = 0$ leads to the well-known square root algorithm $x_{n+1} = x_n/2 + a/2x_n$. Then use this algorithm to find $\sqrt{31}$ accurate to six decimal places.

21. Use Newton's Method to find $\sqrt[3]{7}$ accurate to six decimal places. *Hint:* Begin by asking what equation you want to solve.

22. Apply Newton's algorithm to the equation $x^{1/3} = 0$ starting with $x_1 = 1$. Then sketch the graph of $f(x) = x^{1/3}$ to see why the algorithm fails to produce a root.

23. A present debt of $3600 is to be paid off by payments of $1200 at the end of each of the next 4 years. What rate of interest compounded annually is implied by this scheme?

24. It is known that the concentration $f(t)$ of a certain drug in a patient's blood t hours after injection is given by $f(t) = te^{-t/3}$ milligrams per cubic centimeter. Another injection is to be given when the concentration falls to 0.01 milligrams per cubic centimeter. How many hours after the first injection will the second injection be given?

25. Find the point on the curve $y = x^2 - 4$ that is closest to the point $(4, 5)$. *Hint:* Minimize the square of the distance between the points $(x, x^2 - 4)$ and $(4, 5)$.

14.4*
THE METHOD OF SUCCESSIVE SUBSTITUTIONS

The last section offered one method for finding the roots of an equation. Here we offer another method, a method so simple that it has no right to work; yet it does work in a large variety of situations.

Suppose that the equation we want to solve can be put in the form $x = g(x)$. To solve this equation is to find a number r that is unchanged by the function g, that is, that satisfies $r = g(r)$. Such a number is called a *fixed point* of the function g. To solve the equation $x = g(x)$ is to find its fixed points.

Just as in Newton's Method in the last section, we need a seed value x_1 to get the process started. Based on a graph (or perhaps just on good intuition), we choose x_1 near the root r that we are aiming to approximate. Then we successively calculate $x_2 = g(x_1)$, $x_3 = g(x_2)$, and so on, using the general algorithm.

$$x_{n+1} = g(x_n)$$

If we are lucky, the sequence $x_1, x_2, x_3, \ldots,$ will converge to the root r; that is, it will get closer and closer to r as n gets larger and larger. We refer to this iterative process as the *method of successive substitutions* and call the boxed equation the *fixed-point algorithm*.

The Method Illustrated

We begin with the equation that was treated in Example B in the last section.

EXAMPLE A

Find the root of the equation $x = e^{-x}$ accurate to six decimal places.

Solution The equation has the required form $x = g(x) = e^{-x}$. To find a first guess for the root r, we graph $y = x$ and $y = e^{-x}$ (Figure 25) and note that the two graphs cross near $x = 0.5$, suggesting $x_1 = 0.5$ as the seed value. Applying the fixed-point algorithm, we obtain successively

$$x_2 = e^{-0.5} = 0.6065307$$

$$x_3 = e^{-0.6065307} = 0.5452392$$

and so on. A listing of the successive values is shown in Figure 26.

While it takes 27 steps to get a repetition of the first seven digits, the process does converge and to the right value, namely, 0.567143. Moreover, this process is particularly easy to implement on a calculator or a computer.

FIGURE 25

n	x_n	n	x_n	n	x_n
1	0.5	10	0.5675596	19	0.5671408
2	0.6065307	11	0.5669072	20	0.5671447
3	0.5452392	12	0.5672772	21	0.5671425
4	0.5797031	13	0.5670674	22	0.5671438
5	0.5600646	14	0.5671864	23	0.5671430
6	0.5711722	15	0.5671189	24	0.5671435
7	0.5648630	16	0.5671572	25	0.5671432
8	0.5684381	17	0.5671354	26	0.5671433
9	0.5664095	18	0.5671478	27	0.5671433

FIGURE 26   ∎

EXAMPLE B

Solve $x = 2 \cos x$.

Solution The graphs in Figure 27 suggest $x_1 = 1$ as a seed value. The fixed-point algorithm, $x_{n+1} = 2 \cos x_n$, leads to the results shown in Figure 28. Clearly the process is not converging even though our seed value is very near the root of the equation.

FIGURE 27

n	x_n	n	x_n
1	1	6	1.4394614
2	1.0806046	7	0.2619155
3	0.9415902	8	1.9317916
4	1.1770062	9	-0.7064109
5	0.7673820	10	1.5213931

FIGURE 28

Let's take a different tack. The given equation $x = 2 \cos x$ can be rewritten as $x = (x + 2 \cos x)/2$. The corresponding fixed-point algorithm is

$$x_{n+1} = \frac{x_n + 2 \cos x_n}{2}$$

Again we use $x_1 = 1$ as the seed value, and now the process converges, as we see in Figure 29. (The oscillation in the last digit is probably due to roundoff errors.) The desired root is $r \approx 1.029867$. ∎

n	x_n	n	x_n	n	x_n
1	1	7	1.0298054	13	1.0298665
2	1.0403023	8	1.0298883	14	1.0298666
3	1.0261107	9	1.0298588	15	1.0298665
4	1.0312046	10	1.0298693	16	1.0298666
5	1.0293881	11	1.0298655		
6	1.0300374	12	1.0298668		

FIGURE 29

But why did the method converge in one case but not in the other? The answer is simple, but it needs some discussion.

Convergence of the Method

To get a feeling for the convergence problem, let's look at it geometrically. Note that we can get x_2 from x_1 by locating $g(x_1)$, sending it horizontally to the line $y = x$, and projecting down to the x-axis (Figure 30). When we use this process repeatedly, we are faced with one of the situations illustrated in Figure 31.

What determines convergence or divergence of the sequence x_1, x_2, x_3, . . . ? A close look at Figure 31 reveals the answer. It depends on the slope of the curve $y = g(x)$ near the root r. If the magnitude of the slope $|g'(x)|$ near r is too large (larger than 1) the sequence diverges; if $|g'(x)|$ is small near r (strictly less than 1), the sequence converges. Here is the precise result, stated as a theorem.

FIGURE 30

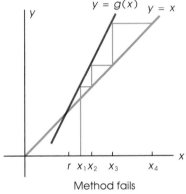

FIGURE 31

FIXED-POINT THEOREM

Suppose there is a constant $M < 1$ such that $|g'(x)| \leq M$ for all x in some interval $[a, b]$ containing a root r of the equation $x = g(x)$. If x_1 is chosen in $[a, b]$, then the algorithm

$$x_{n+1} = g(x_n)$$

generates a sequence that converges to r as n gets larger and larger.

Now we understand the two kinds of behavior in Example B. For the first case, $g(x) = 2 \cos x$, and so $g'(x) = -2 \sin x$. Thus, in a neighborhood of the root $r \approx 1.03$, $|g'(x)| \approx 2 \sin 1.03 \approx 1.7 > 1$. In the second case, $g(x) = (x + 2 \cos x)/2$, and so $g'(x) = \frac{1}{2} - \sin x$. Now in a neighborhood of 1.03, $|g'(x)| \approx |0.5 - \sin 1.03| \approx 0.4 < 1$. We expect divergence in the first case and are certain of convergence in the second case.

Referring back to Example A where $g(x) = e^{-x}$, we see that $|g'(x)| = e^{-x}$, which is certainly less than 1 in a neighborhood of the root

$r \approx 0.57$. Thus, we could be sure the method would converge provided we took our seed value x_1 somewhere near 0.57.

Finally, we remark that the smaller $|g'(x)|$ is, the more rapid the convergence will be. Thus, in using the method of successive substitutions, we should try to rearrange the equation to make $|g'(x)|$ as small as we can.

Another Example

To illustrate what we have just said, we consider a polynomial example.

EXAMPLE C

The equation $x^3 - 3x + 1 = 0$ has three real roots (Figure 32). Find the root that is near 1.5.

Solution The given equation can be written as

$$x = \frac{x^3 + 1}{3} = g(x)$$

but $g'(x) = x^2$, and so $g'(1.5) = 2.25 > 1$. Another way to write the equation is $x(x^2 - 3) = -1$, or equivalently

$$x = -\frac{1}{x^2 - 3} = g(x)$$

Now $g'(x) = 2x/(x^2 - 3)^2$ and $g'(1.5) \approx 5.33$, which is even worse. A third way to write the equation is

$$x - \frac{3}{x} + \frac{1}{x^2} = 0 \qquad \text{or} \qquad x = \frac{3}{x} - \frac{1}{x^2} = g(x)$$

This gives $g'(x) = (-3x + 2)/x^3$ and $g'(1.5) \approx -0.74$, which indicates that our method should work. The corresponding algorithm is

$$x_{n+1} = \frac{3}{x_n} - \frac{1}{x_n^2}$$

This together with the seed value $x_1 = 1.5$ gives the sequence shown in Figure 33. Note the slowness of convergence; it takes 45 steps to get a repetition of eight digits.

We can modify the algorithm just given to improve the speed of convergence. Rewrite the original equation as

$$2x = x + \frac{3}{x} - \frac{1}{x^2} \qquad \text{or} \qquad x = \frac{x}{2} + \frac{3}{2x} - \frac{1}{2x^2} = g(x)$$

and note that

$$g'(x) = \frac{1}{2} - \frac{3}{2x^2} + \frac{1}{x^3}$$

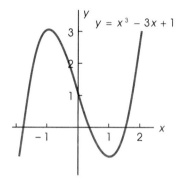

$y = x^3 - 3x + 1$

FIGURE 32

n	x_n
1	1.5
2	1.5555556
3	1.5153061
⋮	⋮
21	1.5320411
22	1.5321234
⋮	⋮
33	1.5320871
34	1.5320902
⋮	⋮
43	1.5320888
44	1.5320889
45	1.5320889

FIGURE 33

POINT OF INTEREST

The iterative processes described in this and the previous section are currently receiving intense study. Even when these processes diverge, they do so in a way that turns out to provide models for many physical phenomena. To get a glimpse of this exciting new area of research, we recommend the article: Douglas Hofstadter, "Strange Attractors: Mathematical Patterns Delicately Balanced Between Order and Chaos," *Scientific American,* Vol. 245 (November 1981), 22–43.

n	x_n
1	1.5
2	1.5277778
3	1.5314922
4	1.5320060
5	1.5320774
6	1.5320873
7	1.5320887
8	1.5320889
9	1.5320889

FIGURE 34

This means that $g'(1.5) \approx 0.13$, which is quite small. We expect that the algorithm

$$x_{n+1} = \frac{x_n}{2} + \frac{3}{2x_n} - \frac{1}{2x_n^2} = \frac{x_n^3 + 3x_n - 1}{2x_n^2}$$

should give rapid convergence. The results are shown in Figure 34.   ■

Problem Set 14.4

1. Find the positive root of the equation $2x = \cos x$ accurate to five decimal places. Use the method of successive substitutions starting with $x_1 = 0.4$. See Example A.

2. Find the positive root of the equation $x = \sin 2x$ accurate to five decimal places. Use the method of successive substitutions starting with $x_1 = 1$.

3. Find the smallest positive root of $x = 4 - \ln x$ accurate to four decimal places using the method of successive substitutions.

4. Follow the directions in Problem 3 for the equation $x = 2 + \ln(x^2 + 2)$.

5. The equations $x = 3 \sin x$ and $x = \frac{1}{2}(x + 3 \sin x)$ are equivalent and have a root near 2.3. See Example B.
   (a) Try the fixed-point algorithm based on the first equation to see that it diverges.
   (b) Use the fixed-point algorithm based on the second equation to obtain the root near 2.3 accurate to four decimal places.

6. The equations $x = \cot x$ and $x = \frac{1}{2}(x + \cot x)$ are equivalent and have a root near 0.85.
   (a) Try the fixed-point algorithm based on the first equation to see that it diverges.
   (b) Use the fixed-point algorithm based on the second equation to obtain the root near 0.85 accurate to four decimal places.

7. Let $f(x) = 3 \sin x$ and $g(x) = \frac{1}{2}(x + 3 \sin x)$. Calculate $f'(2.3)$ and $g'(2.3)$ and use the results to explain the behavior you found in Problem 5. See Example C.

8. Let $f(x) = \cot x$ and $g(x) = \frac{1}{2}(x + \cot x)$. Calculate $f'(0.85)$ and $g'(0.85)$ and use the results to explain the behavior you found in Problem 6.

9. The equation $x^3 + 3x - 5 = 0$ has its single real root near $x = 1$.
   (a) Show that $x = \frac{1}{2}\left(x + \dfrac{5}{x^2 + 3}\right)$ is a reformulation of this equation.
   (b) Use this formulation to find the root accurate to four decimal places.

10. The equation $x^4 + 4x^3 + 2 = 0$ has a root near $x = -0.9$.
    (a) Show that $x = -\sqrt[3]{2/(x + 4)}$ is a reformulation of this equation.
    (b) Use this formulation to find the root accurate to four decimal places.

11. Let $f(x) = x^5 + 3x^3 - 2$, $g(x) = -3x^{-1} + x^{-3} + 2x^{-4}$, and $h(x) = \sqrt[3]{(x + 2)/(x^2 + 3)}$.
    (a) Show that $x = f(x)$, $x = g(x)$, and $x = h(x)$ are equivalent equations.
    (b) Calculate $f'(1)$, $g'(1)$, and $h'(1)$.

(c) Use the information from parts (a) and (b) to find the root of $x = f(x)$ near $x = 1$ accurate to four decimal places.

12. Let $f(x) = x^{10} - 10x^4 + 9$, $g(x) = 10x^{-5} + x^{-8} - 9x^{-9}$, and $h(x) = \sqrt[4]{(x - 9)/(x^6 - 10)}$.
    (a) Show that $x = f(x)$, $x = g(x)$, and $x = h(x)$ are equivalent equations.
    (b) Calculate $f'(1)$, $g'(1)$, and $h'(1)$.
    (c) Use the information from parts (a) and (b) to find the root of $x = f(x)$ near $x = 1$ accurate to four decimal places.

Use the method of successive substitutions to find all solutions to each equation in Problems 13–16 accurate to four decimal places.

13. $x = \sqrt{3 + x}$

14. $x = 1 + \frac{1}{2} \sin x$

15. $2 \cos x - 5x = 0$

16. $\cos x = 3x^2$

17. The equation $x^3 + x = e^{-x}$ has a root near $x = 0.5$. This equation can be reformulated in the following four ways.

$$x = e^{-x} - x^3 = f(x) \qquad x = \sqrt[3]{e^{-x} - x} = g(x)$$

$$x = \frac{e^{-x}}{x^2 + 1} = h(x) \qquad x = \frac{1}{2}\left(x + \frac{e^{-x}}{x^2 + 1}\right) = k(x)$$

(a) For which of these formulations will the method of successive substitutions converge?
(b) Choose the best of these and find the root accurate to four decimal places.

18. It is found that t hours after a patient has received an injection of a certain drug, the concentration of that drug in the patient's blood is given by $C(t) = 2te^{-0.4t}$ milligrams per cubic centimeter. Another injection is to be given when the concentration falls to 0.05 milligram per cubic centimeter. Use the method of successive substitutions to calculate to three decimal places how many hours later the new injection should be given. *Hint:* The method will not work on $t = 0.025e^{0.4t}$. Try a logarithmic reformulation.

19. Suppose that an item costing P dollars is purchased on the installment plan with monthly payments of R dollars at the end of each of the next k months. If interest is at the rate of i per month, then $iP = R[1 - (1 + i)^{-k}]$. A new car costing \$10,000 is to be paid for by installments of \$280 at the end of each of the next 48 months. Determine the interest rate i accurate to five decimal places. *Hint:* Try $i_1 = 0.013$ as your seed value.

14.5*

NUMERICAL SOLUTION OF DIFFERENTIAL EQUATIONS

In Sections 13.1 and 13.2, we studied first-order differential equations with initial conditions. Here are three examples.

$$\frac{dy}{dx} = \frac{y + 1}{x - 3} \qquad y(1) = 0$$

$$\frac{dy}{dx} = 2y + xe^{3x} \qquad y(2) = 0$$

$$\frac{dy}{dx} = y^2 - yx^3 \qquad y(2) = 1$$

The first two were solved (Examples C and E in Section 13.1), but we would have a hard time solving the third one since it is nonseparable and nonlinear. Moreover, while a first-order differential equation that is either separable or linear is solvable in principle by methods we have learned, the practical difficulties involved in carrying out the necessary integrations may lead us to prefer the numerical methods of this section.

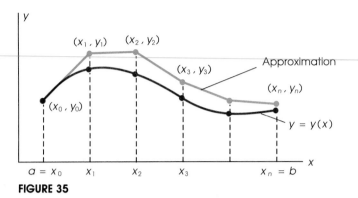

FIGURE 35

The three examples above have the form

$$y' = g(x, y) \qquad y(a) = y_0$$

We are going to solve this equation numerically on an interval $[a, b]$. By this statement, we mean to say that, for a given partition $x_0 = a$, $x_1 = a + h$, $x_2 = a + 2h$, . . . , $x_n = a + nh = b$ of the interval $[a, b]$, we will produce a set of values $y_0, y_1, y_2, . . . , y_n$ that will be good approximations to the true values $y(x_0)$, $y(x_1)$, $y(x_2)$, . . . , $y(x_n)$. Equivalently, we will produce a polygonal curve that will approximate the true solution curve (see Figure 35). To do this, we offer two methods: Euler's method and Euler's improved method.

Euler's Method

Euler's idea is very simple. Let us approximate the value of $y(x_1)$ by the value we would get if we pretended that the solution curve was a straight line of slope $y'(a) = y'(x_0) = g(x_0, y_0)$ on the interval $[x_0, x_0 + h]$. This leads to the approximation y_1 for $y(x_1)$ given by

$$y_1 = y_0 + g(x_0, y_0)h$$

as illustrated in Figure 36. Then using $g(x_1, y_1)$ as an approximation to the slope at x_1, we approximate $y(x_2)$ by y_2, where

$$y_2 = y_1 + g(x_1, y_1)h$$

In general, we move along a step at a time, approximating $y(x_k)$ by y_k using the algorithm

$$y_{k+1} = y_k + g(x_k, y_k)h$$

FIGURE 36

Let's see how this works in an example where we know the true solution to the differential equation.

EXAMPLE A

Use Euler's method to solve the differential equation

$$y' = -y \qquad y(0) = 3$$

on the interval [0, 2] using $h = 0.25$ (so $n = 8$) and compare the resulting values with the true values.

Solution Recall that the actual solution to this equation is $y = 3e^{-x}$. We illustrate the first few steps of the numerical solution. Note that $g(x, y) = -y$, so our algorithm is simply

$$y_{k+1} = y_k - y_k h = y_k(1 - h) = y_k(0.75)$$
$$y_0 = 3$$
$$y_1 = 3(0.75) = 2.25$$
$$y_2 = 2.25(0.75) = 3(0.75)^2 = 1.6875$$

and so on. We summarize in a table and graph the results (Figure 37).

k	x_k	y_k	$y(x_k) = 3e^{-x_k}$
0	0.00	3.0000	3.0000
1	0.25	2.2500	2.3364
2	0.50	1.6875	1.8196
3	0.75	1.2656	1.4171
4	1.00	0.9492	1.1036
5	1.25	0.7119	0.8595
6	1.50	0.5339	0.6694
7	1.75	0.4005	0.5213
8	2.00	0.3003	0.4060

FIGURE 37

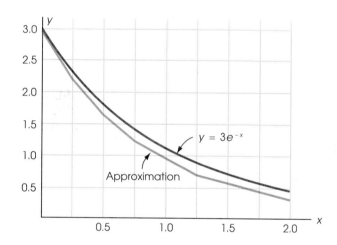

As is usual in numerical methods, there are two sources of error. First, there is the error of the method (the discretization error). In Euler's method, this results from our pretending that the slope $g(x, y)$ of the solution curve on each subinterval is constant and that it can be calculated at x_k using the approximate value y_k rather than $y(x_k)$. Second, there is the error of calculation caused by the fact that we round off our answers at each step of the process. These errors tend to compete with each other. We can reduce the error of the method by making the step size h smaller, but this increases the number of calculations and potentially the size of the roundoff error. In Example A, we can certainly improve the results by reducing the step size. If we use $h = \frac{1}{256}$, the value of y_{512} (the approximation at 2) is

0.4044, which compares very favorably with the true value 0.4060. It probably would not make sense to take a step size very much smaller than this.

EXAMPLE B

Use Euler's method to solve the differential equation

$$y' = 2y + xe^{3x} \qquad y(0) = 2$$

on the interval $[0, 1]$ using $h = 0.125$ and compare the resulting values with the true values.

Solution Here $g(x, y) = 2y + xe^{3x}$, so the Euler algorithm is

$$y_{k+1} = y_k + (2y_k + x_k e^{3x_k})h = (1.25)y_k + (0.125)x_k e^{3x_k}$$

This gives

$$y_0 = 2$$
$$y_1 = (1.25)2 + (0.125)(0)e^0 = 2.5$$
$$y_2 = (1.25)(2.5) + (0.125)(0.125)e^{0.375} = 3.1477342$$
$$y_3 = (1.25)(3.1477342) + (0.125)(0.25)e^{0.75} = 4.0008241$$

and so on. Continuing in this way, we obtain the data in Figure 38 in the column headed y_k. The data $y(x_k)$ in the last column come from the actual solution $y = (x - 1)e^{3x} + 3e^{2x}$ (see Example E in Section 13.1). ∎

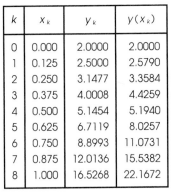

k	x_k	y_k	$y(x_k)$
0	0.000	2.0000	2.0000
1	0.125	2.5000	2.5790
2	0.250	3.1477	3.3584
3	0.375	4.0008	4.4259
4	0.500	5.1454	5.1940
5	0.625	6.7119	8.0257
6	0.750	8.8993	11.0731
7	0.875	12.0136	15.5382
8	1.000	16.5268	22.1672

FIGURE 38

You can see that the approximations in Example B are not very good. Of course, we could do much better by taking a smaller step size. But maybe we can find a better method, a matter we take up next.

Euler's Improved Method

Approximating the slope of the curve on an interval by the slope at the left endpoint clearly leads to errors. A better approximation for this slope would be the average of the slopes at the left and right endpoints of the interval. This suggests the new algorithm

$$\hat{y}_{k+1} = \hat{y}_k + \frac{h}{2}[g(x_k, \hat{y}_k) + g(x_{k+1}, y_{k+1})]$$

Here y_k denotes the value from Euler's (simple) method, and $\hat{y}_k$ is the corresponding value for Euler's improved method.

EXAMPLE C

For $k = 1$ to 10, find the values y_k, $\hat{y}_k$, and $y(x_k)$ for the differential equation $y' = 1 + y^2$, $y(0) = 0$, on the interval $[0, 1]$ using the step size $h = 0.1$.

Solution While this equation is separable, our knowledge of integration theory is insufficient to solve the equation by the methods of Section 13.1. However, note that $y = \tan x$ is the true solution, as you can check by substitution. Euler's algorithm for this differential equation is

$$y_{k+1} = y_k + h(1 + y_k^2) = y_k + (0.1)(1 + y_k^2)$$

and Euler's improved algorithm is

$$\hat{y}_{k+1} = \hat{y}_k + \frac{h}{2}(1 + \hat{y}_k^2 + 1 + y_{k+1}^2)$$

$$= \hat{y}_k + (0.05)(2 + \hat{y}_k^2 + y_{k+1}^2)$$

When we apply these algorithms, we get the table in Figure 39. You can see from this that Euler's improved method does a better job than Euler's simple method. ∎

Finally, we offer an example where we do not know the true solution.

EXAMPLE D

For $k = 1$ to 10, find the values y_k and $\hat{y}_k$ for the nonlinear nonseparable equation

$$y' = x^2 + y^2 \qquad y(0) = 0$$

on the interval $[0, 1]$ using $h = 0.1$.

Solution Euler's algorithm for this equation is

$$y_{k+1} = y_k + h(x_k^2 + y_k^2) = y_k + (0.1)(x_k^2 + y_k^2)$$

and Euler's improved algorithm is

$$\hat{y}_{k+1} = \hat{y}_k + \frac{h}{2}(x_k^2 + \hat{y}_k^2 + x_{k+1}^2 + y_{k+1}^2)$$

$$= \hat{y}_k + (0.05)(x_k^2 + x_{k+1}^2 + \hat{y}_k^2 + y_{k+1}^2)$$

Using these algorithms, we obtain the results in Figure 40. ∎

Computers

The calculations in this section were done on a hand-held calculator with an eight digit display. The approximations we got were not very good, mainly because of the large step size used. To decrease the step size significantly,

k	x_k	Euler (simple) y_k	Euler (improved) $\hat{y}_k$	True $y(x_k)$
0	0.0	0.0000	0.0000	0.0000
1	0.1	0.1000	0.1005	0.1003
2	0.2	0.2010	0.2030	0.2027
3	0.3	0.3050	0.3097	0.3093
4	0.4	0.4143	0.4231	0.4228
5	0.5	0.5315	0.5462	0.5463
6	0.6	0.6598	0.6829	0.6841
7	0.7	0.8033	0.8385	0.8423
8	0.8	0.9678	1.0204	1.0296
9	0.9	1.1615	1.2400	1.2602
10	1.0	1.3964	1.5143	1.5574

FIGURE 39

k	x_k	y_k	$\hat{y}_k$
0	0.0	0.0000	0.0000
1	0.1	0.0000	0.0005
2	0.2	0.0010	0.0030
3	0.3	0.0050	0.0095
4	0.4	0.0140	0.0220
5	0.5	0.0300	0.0426
6	0.6	0.0551	0.0733
7	0.7	0.0914	0.1165
8	0.8	0.1413	0.1747
9	0.9	0.2072	0.2509
10	1.0	0.2925	0.3488

FIGURE 40

we would need to use a computer. Fortunately, the algorithms in this section are easy to program on a computer, and therefore we have included some computer problems in the problem set.

[C] Problem Set 14.5

In Problems 1–4 find the exact solution to the given differential equation. Then use Euler's simple method to solve the same differential equation on the indicated interval with the given h. Show the values of y_k and $y(x_k)$ in a table as in Examples A and B.

1. $y' = y$, $y(0) = 4$; $[0, 1]$, $h = 0.1$
2. $y' = \sqrt{xy}$, $y(0) = 3$; $[0, 1]$, $h = 0.1$
3. $y' = y \tan x$, $y(0) = 2$; $[0, 1]$, $h = 0.1$
4. $y' = \dfrac{2xy}{1 + x^2}$, $y(0) = 3$; $[0, 2]$, $h = 0.2$

5. Solve $y' = 2y + 3e^x$, $y(0) = 0$, exactly and then numerically on $[0, 1]$ with $h = 0.2$ using both Euler's simple method and Euler's improved method. Show the values of y_k, $\hat{y}_k$, and $y(x_k)$ in a table as in Example C.

6. Solve $y' = x^2 - y/x$, $y(2) = 1$, exactly and then numerically on $[2, 4]$ with $h = 0.2$ using both Euler's simple method and Euler's improved method. Show the values of y_k, $\hat{y}_k$, and $y(x_k)$ in a table as in Example C.

7. Review Example D and then estimate $y(1)$ for the nonlinear differential equation $y' = 2x + y^2$, $y(0) =$ 0, using Euler's improved method with $h = 0.1$. *Hint:* This means you will have to find y_{10}.

8. Follow the directions in Problem 7 for the differential equation $y' = y^2 - 2xy$, $y(0) = 1$.

9. Solve the differential equation $y' = -y + e^{-x}$, $y(0) = 0$, exactly and then by Euler's simple method on $[0, 2]$ using $h = 0.2$. Show your results in a table.

10. Use Euler's simple method to solve the differential equation $y' = x^2 + y^2 + xy$, $y(0) = 2$ on $[0, 1]$ using $h = 0.1$.

[PC] 11. Solve the equation in Problem 9 on $[0, 2]$ by Euler's simple method using $h = 0.005$. For your answer, give the estimated value of $y(2)$.

[PC] 12. Use a computer to solve the equation in Problem 10 on $[0, 1]$ using Euler's simple method with $h = 0.01$. For your answer, give the estimated value of $y(1)$.

[PC] 13. Follow the directions in Problem 12 for the differential equation $y' = \sin(x + y)$, $y(0) = 0$.

[PC] 14. Follow the directions in Problem 12 for the differential equation $y' = x^2 + e^y$, $y(0) = 4$.

[C] Chapter 14 Review Problem Set

1. Consider the definite integral $\displaystyle\int_0^1 \dfrac{x}{1 + x^2}\, dx$.
   (a) Evaluate using the Fundamental Theorem of Calculus.
   (b) Approximate using a right endpoint Riemann sum with eight equal subintervals.
   (c) Approximate using the Trapezoidal Rule with $n = 8$.
   (d) Approximate using the Parabolic Rule with $n = 8$.

2. Consider the definite integral $\displaystyle\int_3^4 \dfrac{4}{x - 2}\, dx$.
   (a) Evaluate using the Fundamental Theorem of Calculus.
   (b) Approximate using a right endpoint Riemann sum with eight subintervals.

(c) Approximate using the Trapezoidal Rule with $n = 8$.
   (d) Approximate using the Parabolic Rule with $n = 8$.

3. Find the trapezoidal approximation to $\int_1^3 \sin^2 x\, dx$ with $n = 8$ and give a good bound for the discretization error you have made.

4. Find the trapezoidal approximation to $\int_0^3 \cos^2 x\, dx$ with $n = 6$ and give a good bound for the discretization error you have made.

5. Consider the definite integral $\displaystyle\int_1^2 \dfrac{e^{\sqrt{x}}}{\sqrt{x}}\, dx$.
   (a) Approximate this integral using the Parabolic Rule with $n = 6$.

(b) Evaluate this integral exactly.

6. Use the Fundamental Theorem of Calculus to find $\int_1^3 \frac{e^{1/x}}{x^2}\,dx$ and compare with the parabolic approximation using $n = 4$.

7. Consider $f(x) = \ln(x + 4)$.
    (a) Find the Maclaurin polynomial $P_4(x)$ for $f(x)$.
    (b) Use this polynomial to approximate $\ln(3.9)$.
    (c) Use this polynomial to approximate $\int_0^2 \ln(x + 4)\,dx$.

8. Let $f(x) = x/(x + 2)$.
    (a) Find the Maclaurin polynomial $P_4(x)$ for $f(x)$.
    (b) Evaluate $f(0.1) - P_4(0.1)$.
    (c) Use the polynomial from (a) to approximate $\int_0^1 f(x)\,dx$.

9. Consider $f(x) = \cos x$.
    (a) Find the Taylor polynomial of order 3 based at $a = \pi/3$ for this function.
    (b) Use this polynomial to approximate $\cos(\pi/3 + 0.02)$.
    (c) Give a good bound for the discretization error made in part (b).

10. Let $g(x) = \sin x$.
    (a) Find the Taylor polynomial of order 3 based at $a = \pi/4$ for this function.
    (b) Use this polynomial to approximate $\sin\!\left(\dfrac{\pi}{4} - 0.05\right)$.
    (c) Give a good bound for the discretization error made in part (b).

11. Use Newton's Method to find the x-coordinate, accurate to five decimal places, of the point where the line $y = 3 - x$ intersects the curve $y = e^x$.

12. Use Newton's Method to solve $\cos 2x = 2x - 1$, accurate to four decimal places.

13. Use Newton's Method to find all real solutions, accurate to five decimal places, of the equation $x^4 - x - 3 = 0$.

14. Use Newton's Method to find all real solutions of $x^3 - 10x + 4 = 0$, accurate to four decimal places.

15. Use the method of successive substitutions to find, ac-curate to four decimal places, the x-coordinate of the point where the line $y = x - 2$ intersects the curve $y = e^{-x}$.

16. Use the method of successive substitutions to solve the equation $2x = \cos x$, accurate to five decimal places.

17. Use the method of successive substitutions to find, accurate to four decimal places, the positive root of the equation $x^2 + \sin x = 4$.

18. Use the method of successive substitutions to solve the equation $x^3 + \cos x = 6$, accurate to four decimal places.

19. Use Euler's simple method with $h = 0.2$ to approximate $y(2)$ for the differential equation $y' = y\cos(x + y)$, $y(0) = 2$.

20. Use Euler's simple method with $h = 0.1$ to approximate $y(1)$ for the differential equation $y' = x\sin(x + y)$, $y(0) = 1$.

21. Determine the value of x between 0 and $\pi/2$, where $f(x) = x - \frac{1}{2}x^2 + \cos x$ attains a local maximum value, accurate to four decimal places. *Hint:* Use Newton's Method.

22. Find the maximum value, accurate to four decimal places, of $f(x) = \sin x - \frac{1}{2}x^2 + 3$ on $[0, \pi/2]$. Note that you will first need to find the x-value where this maximum occurs.

23. Find the Maclaurin polynomial $P_3(x)$ for $f(x) = \sqrt{1 + x^2}$ and use it to approximate $\sqrt{1.004}$.

24. Find the Maclaurin polynomial $P_4(x)$ for $f(x) = 1/\sqrt{1 + x^2}$ and use it to approximate $1/\sqrt{1.0009}$.

25. Evaluate $\int_0^1 \sqrt{1 + x^2}\,dx$ three ways.
    (a) By using the result of Problem 12.
    (b) By using the Parabolic Rule with $n = 8$.
    (c) By using a table of integrals.

26. Evaluate $\int_0^1 (1 + x^2)^{-1/2}\,dx$ three ways.
    (a) By using the result of Problem 24.
    (b) By using the Parabolic Rule with $n = 4$.
    (c) By using a table of integrals.

In August, 1989, π was calculated to over 1 billion places using formulas similar to those of Srinivasa Ramanujan. Who cares? In addition to satisfying people's innate curiosity, such calculations test the reliability of super computers and raise interesting questions about the randomness of digits in the expansion of π. For example, does 2 occur 10% of the time?

1415926535	8979323846	2643383279	5028841971	6939937510
5820974944	5923078164	0628620899	8628034825	3421170679
8214808651	3282306647	0938446095	5058223172	5359408128
4811174502	8410270193	8521105559	6446229489	5493038196
4428810975	6659334461	2847564823	3786783165	2712019091
4564856692	3460348610	4543266482	1339360726	0249141273
7245870066	0631558817	4881520920	9628292540	9171536436
7892590360	0113305305	4882046652	1384146951	9415116094
3305727036	5759591953	0921861173	8193261179	3105118548
0744623799	6274956735	1885752724	8912279381	8301194912
9833673362	4406566430	8602139494	6395224737	1907021798
6094370277	0539217176	2931767523	8467481846	7669405132
0005681271	4526356082	7785771342	7577896091	7363717872
1468440901	2249534301	4654958537	1050792279	6892589235
4201995611	2129021960	8640344181	5981362977	4771309960
5187072113	4999999837	2978049951	0597317328	1609631859
5024459455	3469083026	4252230825	3344685035	2619311881
7101000313	7838752886	5875332083	8142061717	7669147303
5982534904	2875546873	1159562863	8823537875	9375195778
1857780532	1712268066	1300192787	6611195909	2164201989

15

Infinite Series

It could be argued that Ramanujan does not belong on our list of prominent applied mathematicians. This brilliant Indian mathematician cared little for the applications of mathematics; rather he devoted his immense talents to developing some of the most abstruse mathematical formulas yet devised by humankind. We include him in our list because much of his mathematics had to do with infinite series, because some of his strange formulas are finding unexpected applications, and more especially because his story is one of the most unusual in all of mathematical history.

Born to an impoverished family, Ramanujan grew up in southern India. His formal education was severely limited, partly because he chose to devote all of his energies to mathematics. In this subject, he educated himself by studying two books which today are considered rather elementary. Working on his own, he began writing down formulas that he discovered in the now famous "Notebooks." His results were deep and almost eerie—largely unintelligible to the Indian mathematical community. Fortunately, he was encouraged to send some of them to the most prominent British mathematician of the day, G. H. Hardy. At first, Hardy thought he had received another piece of crank mail, but upon close study he saw that he was looking at the work of a genius. Without hesitation, Hardy invited the young Indian to come to Cambridge University. Thus began a 5-year collaboration that produced a series of remarkable papers and led Hardy to rate Ramanujan as the greatest mathematician of the day.

Srinivasa Ramanujan
(1887–1920)

But as recognition came, Ramanujan's physical and mental health deteriorated. He was in and out of sanatoriums and was said to be suffering from tuberculosis. Returning to India at the end of World War I, the still young man gave his decreasing strength to his beloved "Notebooks" with even more intensity. At age 32, he died, leaving behind a bewildering mass of handwritten formulas and theorems that mathematicians are only now unravelling 70 years later. A typical formula, and one that was used in the mid 1980's to calculate π to 17 million digits, is

$$\frac{1}{\pi} = \frac{\sqrt{8}}{9801} \sum_{n=0}^{\infty} \frac{(4n)! \, (1103 + 23{,}369n)}{(n!)^4 (396)^{4n}}$$

15.1
INFINITE SEQUENCES AND SERIES

Infinite sequences appeared in Sections 14.3 and 14.4 but were treated very informally. It is time to study this important notion more carefully. Our symbol for a sequence is

$$a_1, a_2, a_3, a_4, a_5, \ldots$$

or the more compact $\{a_n\}$. The a's are called the terms of the sequence; thus a_1 is the first term, a_2 is the second term, and so on. Of course, we may use any letter of the alphabet in place of a, and occasionally we will start our sequence with a_0 rather than a_1.

The terms of a sequence may be events, people, or telephone poles, but in this discussion they will be numbers. Usually, a sequence is given by an *explicit formula* as in $a_n = 2^n$, $b_n = 3n/(2n + 1)$, and $c_n = 1 + (-1)^n/n$. Note that these three formulas generate sequences which begin as follows.

Formula	First few terms
$a_n = 2^n$	$2, 4, 8, 16, 32, 64, \ldots$
$b_n = \dfrac{3n}{2n + 1}$	$1, \dfrac{6}{5}, \dfrac{9}{7}, \dfrac{12}{9}, \dfrac{15}{11}, \dfrac{18}{13}, \ldots$
$c_n = 1 + (-1)^n/n$	$0, \dfrac{3}{2}, \dfrac{2}{3}, \dfrac{5}{4}, \dfrac{4}{5}, \dfrac{7}{6}, \ldots$

In contrast to the sequences above, the sequences of Chapter 14 were determined by *recursion formulas* in which a first term x_1 was given and the general term x_{n+1} was related to the previous term x_n. As an example, consider the sequence $\{d_n\}$ determined by $d_1 = 1$ and $d_{n+1} = (d_n^2 + 5)/2d_n$. This gives us the sequence

$$1, 3, 2.3333333, 2.2380952, 2.2360689, 2.236080, \ldots$$

So what is an (infinite) sequence? For us, a sequence is an ordered collection of numbers in which there is a first term, a second term, a third term, and so on indefinitely. In technical language, a sequence is a function whose domain is the set of positive integers. Like other functions, a sequence can be graphed. Figures 1 and 2 show the graphs of sequences $\{b_n\}$ and $\{c_n\}$.

Convergence or Divergence

A sequence must either converge or diverge. The sequence $\{a_n\}$ is said to **converge** to the number L if $\lim_{n \to \infty} a_n = L$, that is, if a_n gets closer and closer to L as n gets larger and larger without bound. If the sequence does not converge to any number L, we say that it **diverges.** Refer to the examples above. Notice that the sequence $\{a_n\}$ is marching off to infinity; it is not getting closer and closer to any number (∞ is not a number). Therefore, the

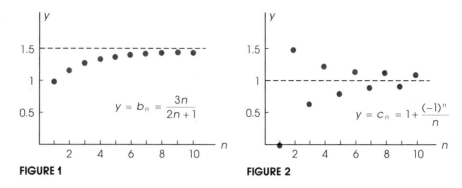

FIGURE 1

FIGURE 2

sequence $\{a_n\}$ diverges. Sequence $\{b_n\}$ converges to $\frac{3}{2}$, and sequence $\{c_n\}$ converges to 1. Sequence $\{d_n\}$ appears to be converging, but we would need to study it carefully to be sure (actually, the given recursion formula is Newton's algorithm for finding the square root of 5).

EXAMPLE A

Let p be any positive number. Show that the sequence $\{a_n\}$ with formula $a_n = 1/n^p$ converges to 0.

Solution To get a feeling for the problem, consider the cases $p = 1$ and $p = 2$ which correspond to the sequences

$$1, \frac{1}{2}, \frac{1}{3}, \frac{1}{4}, \frac{1}{5}, \cdots$$

and

$$1, \frac{1}{4}, \frac{1}{9}, \frac{1}{16}, \frac{1}{25}, \cdots$$

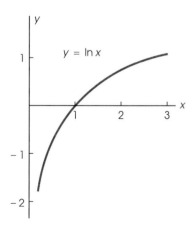

FIGURE 3

These sequences clearly converge to 0. But what if $p = 0.1$ or $p = 0.0001$? These cases are not obvious, even after playing with them on a calculator. We need a method.

Look at the natural logarithm of the sequence $\{1/n^p\}$; that is, look at the sequence $\{p \ln(1/n)\}$. Recall the graph of $y = \ln x$ (Figure 3). Since $1/n \to 0$ as $n \to \infty$, $p \ln(1/n) \to -\infty$ as $n \to \infty$. Again appealing to the graph in Figure 3, we see that, if the natural logarithm of something is moving toward $-\infty$, that something must be approaching 0. ∎

EXAMPLE B

Show that the sequence $\{2n^2/(3n^2 + n)\}$ converges to $\frac{2}{3}$. That is, show that

$$\lim_{n \to \infty} \frac{2n^2}{3n^2 + n} = \frac{2}{3}$$

Solution We first divide numerator and denominator by n^2 (the highest power of n in the fraction) and then apply the limit theorems we learned earlier (Section 2.1). We obtain

$$\lim_{n\to\infty}\frac{2n^2}{3n^2+n}=\lim_{n\to\infty}\frac{2}{3+1/n}=\frac{2}{3+0}=\frac{2}{3}$$ ∎

Infinite Series

We all know what is meant by the sum of a finite set of numbers. But what is meant by the sum of infinitely many numbers? For example, what is meant by the sum

$$S=\frac{1}{2}+\frac{1}{4}+\frac{1}{8}+\frac{1}{16}+\frac{1}{32}+\cdots$$

where the dots indicate that there are infinitely many terms to be added? Actually, it is not too hard to make sense out of this particular sum. Consider a segment of string of length 1 (Figure 4). Divide it into two pieces each of length $\frac{1}{2}$, then divide the right half into two pieces of length $\frac{1}{4}$, then divide the right quarter into two pieces of length $\frac{1}{8}$, and so on. If this process is continued indefinitely, we obtain an infinite sequence of pieces of lengths $\frac{1}{2}, \frac{1}{4}, \frac{1}{8}, \ldots$, which together form the string of length 1. It seems that we should say that the sum S is 1.

FIGURE 4

More generally, consider

$$a_1+a_2+a_3+a_4+\cdots$$

which we also denote with the symbol $\sum_{k=1}^{\infty} a_k$ and call an **infinite series.** We want, if possible, to attach a number S to this series and call it the **sum** of the series. Here is how we do it. Let

$$S_n=a_1+a_2+a_3+\cdots+a_n=\sum_{k=1}^{n} a_k$$

be called the **nth partial sum** of the series and consider the sequence $\{S_n\}$. Note that

$$S_1=a_1$$
$$S_2=a_1+a_2$$
$$S_3=a_1+a_2+a_3$$

and so on. If this sequence converges to S (that is, if $\lim_{n\to\infty} S_n=S$), then we say the series converges and has sum S. If $\{S_n\}$ diverges, we say that the series diverges.

EXAMPLE C

Show that the series $\sum_{k=1}^{\infty} \left(\dfrac{1}{k+1} - \dfrac{1}{k+2} \right)$ converges and find its sum.

Solution According to the definition given above, we must examine the nth partial sum S_n. To do this, we write out the terms as follows.

$$S_n = \left(\frac{1}{2} - \frac{1}{3} \right) + \left(\frac{1}{3} - \frac{1}{4} \right) + \left(\frac{1}{4} - \frac{1}{5} \right) + \cdots + \left(\frac{1}{n+1} - \frac{1}{n+2} \right)$$

After removing parentheses and canceling like terms, we obtain

$$S_n = \frac{1}{2} - \frac{1}{n+2}$$

and since $\lim_{n \to \infty} S_n = \frac{1}{2}$, we conclude that our series converges and has the sum $\frac{1}{2}$. ∎

EXAMPLE D

Show that the series

$$1 + 1 + 1 + 1 + \cdots$$

diverges.

Solution $S_n = 1 + 1 + 1 + \cdots + 1 = n$, so S_n marches off to infinity as n gets larger and larger. The series diverges. ∎

Example D suggests that if the nth term does not approach 0 as $n \to \infty$, there is no hope of convergence of a series. We state the result formally.

---

nth TERM TEST FOR DIVERGENCE

If $\sum_{k=1}^{\infty} a_k$ converges, then $\lim_{n \to \infty} a_n = 0$. Equivalently, if $\lim_{n \to \infty} a_n$ does not exist or is not 0, then $\sum_{k=1}^{\infty} a_k$ diverges.

Proof Suppose $\sum_{k=1}^{\infty} a_k$ converges. Then $\lim_{n \to \infty} S_n = S$ for some number S. Thus,

$$\lim_{n \to \infty} a_n = \lim_{n \to \infty} (S_n - S_{n-1}) = \lim_{n \to \infty} S_n - \lim_{n \to \infty} S_{n-1} = S - S = 0$$ ∎

EXAMPLE E

Show that $\sum_{k=1}^{\infty} \dfrac{k+1}{10k}$ diverges.

Solution

$$\lim_{n \to \infty} a_n = \lim_{n \to \infty} \frac{n+1}{10n} = \frac{1}{10} \neq 0$$ ∎

INFINITE SERIES

A common mistake is to try to make the nth Term Test say something it does not say. It *does not say* that if $\lim_{n\to\infty} a_n = 0$, then the series converges. In fact, many series which satisfy $\lim_{n\to\infty} a_n = 0$ diverge. We illustrate this with the famous **harmonic series**

$$1 + \frac{1}{2} + \frac{1}{3} + \frac{1}{4} + \frac{1}{5} + \cdots$$

EXAMPLE F

Show that the harmonic series diverges even though $\lim_{n\to\infty} a_n = \lim_{n\to\infty} (1/n) = 0$.

Solution We will need to play a little trick, grouping terms in a clever way (two terms, four terms, eight terms, and so on).

$$S_n = 1 + \frac{1}{2} + \frac{1}{3} + \frac{1}{4} + \frac{1}{5} + \frac{1}{6} + \cdots + \frac{1}{n}$$

$$= 1 + \frac{1}{2} + \left(\frac{1}{3} + \frac{1}{4}\right) + \left(\frac{1}{5} + \frac{1}{6} + \frac{1}{7} + \frac{1}{8}\right) + \cdots + \frac{1}{n}$$

$$> 1 + \frac{1}{2} + \frac{1}{2} + \frac{1}{2} + \cdots + \frac{1}{n}$$

It is clear that, by taking n large enough, we can introduce as many $\frac{1}{2}$s as we wish into this last sum and thereby make S_n arbitrarily large. The series diverges. ∎

POINT OF INTEREST
We have shown that the harmonic series diverges. However, its partial sums grow very, very slowly. In fact, it takes over 272 million terms for S_n to reach 20, and over 10^{43} terms for S_n to reach 100. Thus, a computer adding up terms would never discover the fact that the series diverges.

A major goal for this chapter is to develop tests that will help us decide whether a given series converges or diverges. The nth Term Test is one such test. Here is another test. Consider a series where all the a_k's are nonnegative and, as usual, let $S_n = a_1 + a_2 + \cdots + a_n$. Then $S_n = S_{n-1} + a_n$, which means that S_n increases as n gets larger and larger. It follows that S_n can do one of two things as $n \to \infty$. Either S_n marches off to infinity or it is bounded above by some constant M. In the first case, the corresponding series diverges; in the second case, the corresponding series converges with a sum less than or equal to M. We summarize what we have just said in the Bounded Sum Test.

BOUNDED SUM TEST

Let $\sum_{k=1}^{\infty} a_k = a_1 + a_2 + a_3 + \cdots$ be a series of nonnegative terms and let $S_n = a_1 + a_2 + \cdots + a_n$. If the sequence $\{S_n\}$ is unbounded, the series diverges. If there is a constant M such that $S_n \leq M$ for all n, then the series converges and has sum $S \leq M$.

EXAMPLE G

Show that the series

$$1 + \frac{1}{2!} + \frac{1}{3!} + \frac{1}{4!} + \frac{1}{5!} + \cdots$$

converges and has a sum of at most 2.

Solution

$$k! = 1 \cdot 2 \cdot 3 \cdot 4 \cdots k \geq 1 \cdot 2 \cdot 2 \cdot 2 \cdots 2 = 2^{k-1}$$

and so $1/k! \leq 1/2^{k-1}$. Thus

$$S_n = 1 + \frac{1}{2!} + \frac{1}{3!} + \frac{1}{4!} + \cdots + \frac{1}{n!}$$

$$\leq 1 + \frac{1}{2} + \frac{1}{4} + \frac{1}{8} + \cdots + \frac{1}{2^{n-1}}$$

$$\leq 1 + 1 \qquad \text{(see Figure 4)}$$

Since the partial sums are bounded above, we conclude that the series converges and has a sum of at most 2. ∎

Problem Set 15.1

1. Write the first five terms of each sequence.
   (a) $a_n = 2n + 3$
   (b) $a_n = 19 - n^2$
   (c) $a_n = (\frac{2}{3})^{n-1}$
   (d) $a_n = \frac{3n + 1}{2n - 3}$
   (e) $a_1 = 2, a_{n+1} = 3a_n - 2$
   (f) $a_1 = 2, a_{n+1} = (-1)^n a_n$

2. Write the first four terms of each sequence.
   (a) $a_n = 3^{n-1}$
   (b) $a_n = 1.7 + (0.3)n$
   (c) $a_n = (2n - 1)^2$
   (d) $a_n = \frac{4n}{5 - 3n}$
   (e) $a_1 = -2, a_{n+1} = -2a_n + 1$
   (f) $a_1 = 4, a_{n+1} = (\frac{1}{2}a_n)^2$

3. Determine whether the given sequence converges or diverges. If it converges, find its limit. See Example A.
   (a) $a_n = \frac{1}{n\sqrt{n}}$
   (b) $b_n = \frac{1}{n^{0.03}} - 4$
   (c) $c_n = \frac{n}{\sqrt[3]{n}}$
   (d) $d_n = \ln\left(2 + \frac{1}{\sqrt{n}}\right)$

4. Follow the directions in Problem 3.
   (a) $a_n = n^{-5/3}$
   (b) $b_n = \ln(\sqrt{n})$

   (c) $c_n = \frac{n^{3/4}}{n^{2/3}}$
   (d) $d_n = \cos\left(\frac{4}{n} - 2\right)$

5. Find the limit as $n \to \infty$ of each sequence. See Example B.
   (a) $a_n = \frac{3 - 2n}{3n + 1}$
   (b) $b_n = \frac{n^2}{2n^2 + 3}$
   (c) $c_n = \frac{\sqrt{n} - 4}{3 - 2\sqrt{n}}$
   (d) $d_n = \frac{2^n + 1}{2^n + 1000}$
   (e) $e_n = \sqrt{\frac{2n^2 - 4}{n^2 + 2n}}$
   (f) $f_n = \frac{\sqrt{2n^2 - 4}}{3n + 2}$

6. Find the limit as $n \to \infty$ of each sequence.
   (a) $a_n = \frac{4n - 1}{n + 60}$
   (b) $b_n = \frac{2 - n^3}{2n^3 + n - 4}$
   (c) $c_n = \frac{3n\sqrt{n}}{2 + n\sqrt{n}}$
   (d) $d_n = \frac{10n}{n^2 + 1}$
   (e) $e_n = \sqrt[3]{\frac{n^3 + 3n}{1 - 8n^3}}$
   (f) $f_n = \frac{\sqrt[3]{n^3 + 3n}}{1 + 2n}$

In Problems 7–14, determine whether the given series converges or diverges. If it converges, find its sum. See Examples C and D.

7. $\left(2 - \dfrac{3}{2}\right) + \left(\dfrac{3}{2} - \dfrac{4}{3}\right) + \left(\dfrac{4}{3} - \dfrac{5}{4}\right) + \cdots +$

$\left(\dfrac{n+1}{n} - \dfrac{n+2}{n+1}\right) + \cdots$

8. $\left(\dfrac{1}{4} - \dfrac{3}{7}\right) + \left(\dfrac{3}{7} - \dfrac{5}{10}\right) + \left(\dfrac{5}{10} - \dfrac{7}{13}\right) + \cdots +$

$\left(\dfrac{2n-1}{3n+1} - \dfrac{2n+1}{3n+4}\right) + \cdots$

9. $\displaystyle\sum_{k=2}^{\infty} \left[\dfrac{1}{k^2} - \dfrac{1}{(k+1)^2}\right]$

10. $\displaystyle\sum_{k=2}^{\infty} \left(\dfrac{2k+1}{k} - \dfrac{2k+3}{k+1}\right)$

11. $0.01 + 0.01 + 0.01 + \cdots$

12. $2 - 2 + 2 - 2 + 2 - \cdots$

13. $\displaystyle\sum_{k=1}^{\infty} (-1)^{k+1} k$

14. $\displaystyle\sum_{k=1}^{\infty} \cos(k\pi)$

In Problems 15–20, use the nth Term Test for Divergence to show that the series diverges. See Example E.

15. $\displaystyle\sum_{k=1}^{\infty} \ln\left(2 + \dfrac{1}{k}\right)$

16. $\displaystyle\sum_{k=1}^{\infty} \dfrac{3k-1}{4k}$

17. $\displaystyle\sum_{k=1}^{\infty} \sin k$

18. $\displaystyle\sum_{k=1}^{\infty} 2^{1/k}$

19. $\displaystyle\sum_{k=1}^{\infty} \dfrac{k}{100k + 1000}$

20. $\displaystyle\sum_{k=1}^{\infty} \left(0.01 + \dfrac{(-1)^k}{k}\right)$

In Problems 21–26, determine whether the given series converges or diverges. Give a reason for your answer. See Examples F and G.

21. $\dfrac{1}{2} + \dfrac{1}{4} + \dfrac{1}{6} + \dfrac{1}{8} + \cdots + \dfrac{1}{2n} + \cdots$

22. $\dfrac{1}{\ln 3} + \dfrac{1}{\ln 9} + \dfrac{1}{\ln 27} + \cdots + \dfrac{1}{\ln 3^n} + \cdots$

23. $\dfrac{1}{3} + \dfrac{1}{5} + \dfrac{1}{9} + \dfrac{1}{17} + \cdots + \dfrac{1}{2^n + 1} + \cdots$

24. $3 + \dfrac{3}{2!} + \dfrac{3}{3!} + \dfrac{3}{4!} + \cdots + \dfrac{3}{n!} + \cdots$

25. $\displaystyle\sum_{k=1}^{\infty} a_k$, where $a_k \geq 0$ and $k^2 < S_k < (k+1)^2$

26. $\displaystyle\sum_{k=1}^{\infty} a_k$, where $a_k \geq 0$ and $S_k > \ln k$

27. Determine whether the given sequence converges or diverges. If it converges, find its limit.

(a) $a_n = \dfrac{3n^2}{n^2 + 2}$

(b) $a_n = 2 + (-1)^n$

(c) $a_n = (\ln 2.7)^n$

(d) $a_n = 1 + \dfrac{1}{2} + \dfrac{1}{3} + \cdots + \dfrac{1}{n}$

(e) $a_n = 2^{1/n}$

(f) $a_n = 1 + \dfrac{1}{2} + \dfrac{1}{4} + \cdots + \dfrac{1}{2^{n-1}}$

28. Write an explicit formula for the nth term a_n of each of the following sequences based on the pattern that you see.

(a) $1, 4, 9, 16, \ldots$

(b) $\dfrac{2}{3}, \dfrac{3}{5}, \dfrac{4}{7}, \dfrac{5}{9}, \ldots$

(c) $3, 7, 11, 15, \ldots$

(d) $1, \dfrac{2}{3}, \dfrac{3}{9}, \dfrac{4}{27}, \dfrac{5}{81}, \ldots$

29. Let $a_n = (-1)^n/n$. Calculate S_5, the sum of the first five terms of this sequence.

30. Let $a_1 = 3$ and $a_{n+1} = 3a_n - 4$ for $n \geq 1$. Calculate S_6.

31. Determine whether the given series converges or diverges. Give a reason for your answer.

(a) $\displaystyle\sum_{k=1}^{\infty} 3(\tfrac{1}{2})^k$

(b) $\displaystyle\sum_{k=1}^{\infty} \dfrac{2k}{k + 100}$

(c) $\displaystyle\sum_{k=1}^{\infty} (\tfrac{1}{3})^k$

(d) $\displaystyle\sum_{k=1}^{\infty} \dfrac{3}{2k}$

32. Let $\sum_{k=1}^{\infty} a_k$ be the series for which $S_n = \dfrac{5n+1}{n+5}$.

(a) Calculate $a_1, a_2, a_3,$ and a_4.

(b) Calculate the sum of this series.

C 33. The famous Fibonacci sequence is defined by $f_1 = f_2 = 1$ and $f_{n+2} = f_{n+1} + f_n$ for $n \geq 1$.

(a) Calculate $f_3, f_4, f_5, \ldots, f_{11}$.

(b) It is known that the sequence $a_n = f_{n+1}/f_n$ converges. Determine $\lim_{n\to\infty} a_n$ accurate to three decimal places.

C 34. Consider the sequence determined by $a_1 = 2$ and $a_{n+1} = \sqrt{4a_n - 3}$ for $n \geq 1$. It is known that this sequence converges. Determine $\lim_{n\to\infty} a_n$ accurate to three decimal places.

35. Suppose that 300 milligrams of a chemical have been injected into a person's bloodstream and that the body eliminates 20% of the chemical present in the blood each hour. Find a formula for a_n, the amount of chemical remaining in the bloodstream after n hours.

36. Suppose that a cylinder is filled with air and that at each stroke a pump removes $\tfrac{2}{3}$ of the air. How many strokes will it take to reduce that amount of air in the cylinder to $\tfrac{1}{1000}$ of its initial value?

15.2
GEOMETRIC SERIES

Perhaps the most important of all series is the geometic series. A **geometric series** is a series of the form

$$a + ar + ar^2 + ar^3 + ar^4 + \cdots$$

where a is a constant $(a \neq 0)$ and r is another constant called the *ratio*. Here are three examples.

(a) $\quad \dfrac{1}{2} + \dfrac{1}{4} + \dfrac{1}{8} + \dfrac{1}{16} + \dfrac{1}{32} + \cdots \quad \left(a = \dfrac{1}{2}, r = \dfrac{1}{2}\right)$

(b) $\quad 0.3 + 0.03 + 0.003 + 0.0003 + \cdots \quad (a = 0.3, r = 0.1)$

(c) $\quad 1 + \dfrac{3}{2} + \dfrac{9}{4} + \dfrac{27}{8} + \dfrac{81}{16} + \cdots \quad \left(a = 1, r = \dfrac{3}{2}\right)$

You should recognize the first two of these examples. We studied series (a) in the previous section, deciding that it converged with sum 1. Series (b) is the equivalent of the repeating decimal $0.3333333\ldots$, which you should recognize as the decimal expansion of $\frac{1}{3}$. After we have introduced the necessary apparatus, we will see that series (b) does indeed converge with sum $\frac{1}{3}$.

The Partial Sums of a Geometric Series

Let S_n denote the sum of the first n terms of the geometric series above. Multiply this sum by r and subtract from S_n as shown below.

$$S_n = a + ar + ar^2 + ar^3 + \cdots + ar^{n-2} + ar^{n-1}$$
$$rS_n = \qquad ar + ar^2 + ar^3 + \cdots + ar^{n-2} + ar^{n-1} + ar^n$$
$$\overline{S_n - rS_n = a \hspace{6cm} - ar^n}$$

After factoring both the left and the right sides, we obtain

$$S_n(1 - r) = a(1 - r^n)$$

or after division by $1 - r$

$$\boxed{S_n = \frac{a(1 - r^n)}{1 - r} = \frac{a(r^n - 1)}{r - 1} \qquad r \neq 1}$$

> **CAUTION**
> If $r = 1$, the formula for S_n that we have given is clearly incorrect since it would involve division by 0. The correct formula in this case is very simple.
> $$S_n = a + a + a + \cdots + a$$
> $$= na$$

We call this the partial sum formula for a geometric series.

EXAMPLE A

Find the sum of the first 10 terms of the series

$$1 + \frac{3}{2} + \frac{9}{4} + \frac{27}{8} + \cdots$$

Solution Note that $a = 1$ and $r = \frac{3}{2} = 1.5$.

$$S_{10} = \frac{1(1.5^{10} - 1)}{1.5 - 1} = 2(1.5^{10} - 1) \approx 113.33008$$ ■

A series of equal payments made at regular intervals is called an **annuity.** The payments that a typical homeowner makes in paying off a mortgage provide one example of an annuity. The payments that a person makes into a retirement fund form another example.

EXAMPLE B

Suppose that Amy puts \$100 into a retirement fund at the end of each month for the next 10 years. How much will her retirement fund be worth at the end of that time if interest is figured at 9% compounded monthly?

Solution We begin by recalling that interest rates are stated as annual rates. Thus an annual rate of 9% should be interpreted as a rate of $\frac{9}{12}\% = 0.0075$ per month. The first \$100 Amy puts in the fund will have a term of 119 months and hence will be worth $\$100(1.0075)^{119}$ ten years from now. The second payment will have a term of 118 months and will be worth $\$100(1.0075)^{118}$ ten years from now. This pattern continues until the final payment, which is made right at the end of the 10-year period and hence will be worth only \$100 then. The total value 10 years from now of the 120 payments form a geometric sum with $r = 1.0075$, namely,

$$100 + 100(1.0075) + 100(1.0075)^2 + \cdots + 100(1.0075)^{119}$$

We evaluate this sum by use of the boxed formula.

$$S_{120} = \frac{100(1.0075^{120} - 1)}{1.0075 - 1} = \$19,351.43$$ ■

Convergence of Geometric Series

Consider the infinite geometric series

$$a + ar + ar^2 + ar^3 + \cdots$$

To test its convergence, examine the limit of its partial sums, that is, examine $\lim_{n \to \infty} S_n$. Now,

$$\lim_{n\to\infty} S_n = \lim_{n\to\infty} \frac{a(1 - r^n)}{1 - r} = \lim_{n\to\infty} \left(\frac{a}{1 - r} - \frac{ar^n}{1 - r} \right)$$

$$= \frac{a}{1 - r} + \frac{a}{1 - r} \lim_{n\to\infty} r^n$$

Thus, our problem reduces to studying the behavior of r^n as n gets larger and larger. To get a feeling for this, we suggest you try evaluating 0.99^{10}, 0.99^{100}, 0.99^{1000} on your calculator. Then try the same thing with 0.99 replaced by 1.01. Then try some other values for r. Perhaps you will come to the correct conclusion that r^n does not have a finite limit as $n \to \infty$ if $|r| \geq 1$, but that $\lim_{n\to\infty} r^n = 0$ if $|r| < 1$. Indeed, we have the following theorem.

THEOREM

The geometric series

$$a + ar + ar^2 + ar^3 + \cdots$$

converges and has sum $a/(1 - r)$ if $|r| < 1$. Otherwise, the series diverges.

Let us apply this theorem to the three series introduced at the beginning of this section.

EXAMPLE C

Determine the convergence or divergence of the three series below and find the sum in the case of convergence.

(a) $\dfrac{1}{2} + \dfrac{1}{4} + \dfrac{1}{8} + \dfrac{1}{16} + \dfrac{1}{32} + \cdots$

(b) $0.3 + 0.03 + 0.003 + 0.0003 + \cdots$

(c) $1 + \dfrac{3}{2} + \dfrac{9}{4} + \dfrac{27}{8} + \dfrac{81}{16} + \cdots$

Solution The first series has a ratio of $\frac{1}{2}$; the second has a ratio of 0.1. Thus, series (a) and (b) converge. The first series has sum $a/(1 - r) = \frac{1}{2}/(1 - \frac{1}{2}) = 1$, and the second has sum $a/(1 - r) = 0.3/(1 - 0.1) = 0.3/0.9 = \frac{1}{3}$. These are the results we expected. Series (c) diverges, since its ratio $r = \frac{3}{2} \geq 1$. ∎

EXAMPLE D

Write the repeating decimal $0.15151515 \ldots$ as a ratio of two integers.

Solution First we note that this repeating decimal corresponds to the infinite geometric series

$$0.15 + 0.0015 + 0.000015 + \cdots$$

with ratio 0.01. Its sum is

$$\frac{a}{1-r} = \frac{0.15}{1-0.01} = \frac{0.15}{0.99} = \frac{15}{99} = \frac{5}{33}$$ ∎

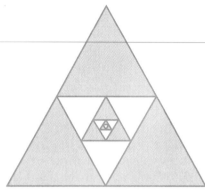

FIGURE 5

A Geometric Application

Figure 5 shows an equilateral triangle with an infinite descending pattern. Each triangle in the descending chain has its vertices at the midpoints of the sides of the previous bigger triangle.

EXAMPLE E

Determine the fraction of the total area in Figure 1 that is colored.

Solution Assume that the biggest triangle has an area of 1 unit and consider the sequence consisting of every other triangle in the descending chain; that is, consider those triangles in the chain which have their bases horizontal and point upward. The areas of these triangles form the sequence $1, (\frac{1}{4})^2, (\frac{1}{4})^4, (\frac{1}{4})^6, \ldots$, and each of them (ignoring the rest of the chain) is three-fourths colored. Thus, the colored region has an area of

$$\frac{3}{4}(1) + \frac{3}{4}\left(\frac{1}{4}\right)^2 + \frac{3}{4}\left(\frac{1}{4}\right)^4 + \frac{3}{4}\left(\frac{1}{4}\right)^6 + \cdots$$

This is a geometric series with $a = \frac{3}{4}$ and $r = \frac{1}{16}$ and so has sum $a/(1-r) = \frac{3}{4}/(1 - \frac{1}{16}) = \frac{4}{5}$. Thus, four-fifths of the area in Figure 5 is colored.   ∎

Other Applications

Suppose the government decides to give a rebate to its taxpayers. This will generate spending far in excess of the rebate. For the part of the rebate spent by one individual becomes income for another who then spends part of it, and so on. This phenomenon is called the **multiplier effect** and is important in economic studies.

EXAMPLE F

Suppose that, on the average, individuals spend 95% of their income and save the rest. If the government gives a \$10 billion tax rebate, how much spending will result?

Solution The original recipients of the rebate will spend \$10(0.95) billion. This amount then becomes income for new spenders, resulting in further spending of \$10(0.95)(0.95), and so on. The total amount of spending resulting from the rebate is

$$10(0.95) + 10(0.95)^2 + 10(0.95)^3 + \cdots = \frac{10(0.95)}{1 - 0.95} = 190$$

A rebate of $10 billion results in spending of $190 billion. ■

EXAMPLE G

The ABC Corporation has promised to pay Podunk University $1000 now and at the end of each year forever. What is the total present value of this sequence of payments (called a **perpetuity**) if interest is assumed to be 8% compounded annually?

Solution We have to discount each of these payments to the present. The present value of the first payment is 1000, the present value of the second payment is $1000(1.08)^{-1}$, the present value of the third payment is $1000(1.08)^{-2}$, and so on. The total present value of all the payments is

$$1000 + 1000(1.08)^{-1} + 1000(1.08)^{-2} + \cdots = \frac{1000}{1 - (1.08)^{-1}}$$

$$= \$13,500 \quad ■$$

EXAMPLE H

To counteract a deficiency, a patient is to be given an injection of 5 units of a certain drug each morning for the rest of her life. The drug is eliminated from the body exponentially, so that the amount remaining from an injection n days after it is administered is $5(\frac{2}{3})^n$. What will be the ultimate level of the drug in the patient's body right after an injection.

Solution Assume the treatment program has gone on for a long, long time. Then the number of units of the drug in the body right after an injection is

$$5 + 5\left(\frac{2}{3}\right) + 5\left(\frac{2}{3}\right)^2 + 5\left(\frac{2}{3}\right)^3 + \cdots = \frac{5}{1 - \frac{2}{3}} = 15 \quad ■$$

Series Involving Sums

Occasionally, we are faced with a series of the form $\Sigma_{k=1}^{\infty} (a_k + b_k)$ or something even more complicated. The following theorem will often be of great help. We state it without proof, though the proof is quite easy.

THEOREM

Suppose that $\Sigma_{k=1}^{\infty} a_k$ and $\Sigma_{k=1}^{\infty} b_k$ both converge. Then $\Sigma_{k=1}^{\infty} (a_k + b_k)$ also converges. Moreover,

$$\sum_{k=1}^{\infty} (ca_k + db_k) = c \sum_{k=1}^{\infty} a_k + d \sum_{k=1}^{\infty} b_k$$

EXAMPLE I

Evaluate $\sum_{k=1}^{\infty} [5(\frac{1}{4})^k - 2(\frac{2}{3})^k]$.

Solution Appealing to the theorem above, we have

$$\sum_{k=1}^{\infty} \left[5\left(\frac{1}{4}\right)^k - 2\left(\frac{2}{3}\right)^k \right] = 5\sum_{k=1}^{\infty} \left(\frac{1}{4}\right)^k - 2\sum_{k=1}^{\infty} \left(\frac{2}{3}\right)^k$$

$$= 5\left(\frac{1}{3}\right) - 2\left(2\right) = -\frac{7}{3} \qquad \blacksquare$$

Problem Set 15.2

In Problems 1–8, find the sum S_{12} of the first 12 terms of each geometric series. See Example A.

1. $1 + \frac{1}{3} + \frac{1}{9} + \frac{1}{27} + \cdots$

2. $4 + 3 + \frac{9}{4} + \frac{27}{16} + \cdots$

3. $3 - \frac{3}{2} + \frac{3}{4} - \frac{3}{8} + \cdots$

4. $3 - 2 + \frac{4}{3} - \frac{8}{9} + \cdots$

C 5. $200 + 200(1.08) + 200(1.08)^2 + 200(1.08)^3 + \cdots$

C 6. $100 + 100(1.025) + 100(1.025)^2 + \cdots$

C 7. $100(1.09)^{-1} + 100(1.09)^{-2} + 100(1.09)^{-3} + \cdots$

C 8. $200 + 200(1.04)^{-1} + 200(1.04)^{-2} + \cdots$

C 9. Jennifer plans to save for her daughter's college education by making 20 semiannual deposits of $500 in a savings account that accumulates at 8% compounded semiannually. If she makes the first deposit 6 months from now, how much will be in the account at the end of 10 years? See Example B.

C 10. Joseph McHenry will make five annual deposits of $20,000 in a savings account in order to be able to pay for a new tractor at the end of 5 years. The first deposit will be made 1 year from now. If the account accumulates at 7% compounded annually, how much will he be able to spend for the new tractor?

C 11. Find the present value of the set of 20 semiannual deposits in Problem 9. *Hint:* The present value ($500 discounted 6 months) of the first deposit is $500(1.04)^{-1}$, the present value of the second deposit is $500(1.04)^{-2}$, so on. Thus the present value of the 20 deposits is the sum of the first 20 terms of a geometric series with ratio $(1.04)^{-1}$.

C 12. Find the present value of the set of five annual deposits in Problem 10.

C 13. Mary Simpson is taking out a 30-year mortgage loan on a house with an interest charge of 9% compounded monthly. Her monthly payments are $365, starting 1 month from now. Find the (present) value of this loan.

C 14. The Heppners have purchased a summer home by putting down $4000 cash, with the balance due in monthly payments of $180 spread out over 12 years. The first monthly payment is due 1 month from the date of purchase. What is the actual purchase price of the house, assuming the interest rate is 12% compounded monthly?

In Problems 15–22, determine whether the given series converges or diverges, and if it converges find its sum. See Examples C and I.

15. $\sum_{k=0}^{\infty} \left(\frac{2}{3}\right)^k$

16. $\sum_{k=0}^{\infty} \left(\frac{3}{4}\right)^k$

17. $\sum_{k=0}^{\infty} \left(\frac{\pi}{3}\right)^k$

18. $\sum_{k=0}^{\infty} (\ln 2)^k$

19. $\sum_{k=0}^{\infty} \left[3\left(\frac{2}{3}\right)^k - 2\left(\frac{3}{4}\right)^k \right]$

20. $\sum_{k=0}^{\infty} \left[4\left(\frac{1}{5}\right)^k + 3\left(\frac{5}{6}\right)^k \right]$

21. $\sum_{k=0}^{\infty} [(0.54)^k + (-0.2)^k]$

22. $\sum_{k=1}^{\infty} (-\sin 1)^k$

Write each repeating decimal in Problems 23–26 as a ratio of two integers (see example D).

23. $0.242424\ldots$ 24. $0.717171\ldots$

25. $1.234234234\ldots$ 26. $0.271717171\ldots$

27. What fraction of the area of the triangle in Figure 6 is represented by the total area of the infinite sequence of squares? See Example E.

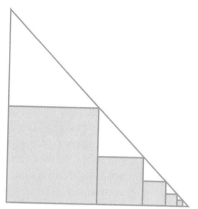

FIGURE 6

28. Betty will paint a pattern on a large square canvas by first painting the top right quarter of the canvas, then the top right quarter of each of the three unpainted quarters, and so on (Figure 7). What percentage of the area of the canvas will ultimately be painted?

FIGURE 7

29. Suppose that, on the average, individuals spend 88% of their income and save the rest. If the federal government gives a $12 billion rebate, how much spending will ultimately result? See Example F.

30. A rich alumnus is setting up a trust that will provide an income for his alma mater of $50,000 per year forever starting 1 year from now. How much money must he put in the trust right now if interest is 7.5% compounded annually? See Example G.

31. Marty has won a lottery prize and may choose between receiving $140,000 cash or a set of monthly payments of $675 per month forever, the first payment due 1 month from now. Which option should she choose if interest is figured at 6% compounded monthly?

32. Suppose that each day a patient is given 4 milligrams of a certain drug. It is estimated that each day the body eliminates 35% of the drug present at the beginning of that day. Calculate the limiting amount of the drug present in the patient's body right after treatment, assuming this regimen has continued for a long time? See Example H.

33. Calculate the sum of each of the following convergent geometric series.

(a) $\sum_{k=1}^{\infty} 36(0.01)^k$ (b) $\sum_{k=1}^{\infty} (0.36)^k$

(c) $\sum_{k=1}^{\infty} \left(\frac{\sqrt{5}}{3}\right)^k$ (d) $\sum_{k=1}^{\infty} (-1)^{k+1} \left(\frac{3}{4}\right)^k$

34. A ball is dropped from a height of 16 feet. Each time the ball bounces it rises to three-fifths of the distance it has just fallen. Find the total distance the ball travels (up and down) in coming to rest.

C 35. How much money do I need to set aside on my fifty-ninth birthday in a fund that earns 8% interest compounded annually in order to give me 26 annual payments of $30,000 each starting on my sixtieth birthday? How much more would I have to set aside so these payments could continue forever?

C 36. Maggie wants to accumulate $40,000 in a fund by the end of 15 years by making equal monthly deposits beginning 1 month from now. If the fund earns interest at 6% compounded monthly, find the size of the monthly deposit.

C 37. How much should I be willing to pay today for a bond redeemable in 10 years for $10,000, which pays semiannual dividends of $450 each, the first one payable 6 months from now. Assume that I can invest money at 7% compounded semiannually.

C 38. Starting now, a house insurance company expects to sell 4500 new policies each year indefinitely into the future. It estimates that 15% of the policies in force at the beginning of a year will be canceled by the end of that year.
(a) How many policies should it have in force by the end of 20 years?
(b) Determine the limiting number of policies it will have in force.

39. Each month 900 small-mouth bass are added to a small lake. However, at the end of each month, one-fourth of the bass in the lake at the beginning of the month have died or been caught by fisherpersons. In the long run, how many bass will be in the lake? Assume there were no bass in the lake when this process started.

40. Amy walked 1 mile north, $\frac{1}{2}$ mile east, $\frac{1}{4}$ mile south, $\frac{1}{8}$ mile west, $\frac{1}{16}$ mile north, and so on, thus describing a spiral path with infinitely many segments. How far was her final destination from her starting point?

15.3

POSITIVE SERIES

Geometric series are particularly nice. We can easily decide when they converge and when they diverge. Moreover, in the case of convergence, we even have a formula for the sum of the series. The general situation is much more complicated. Write down a series at random, and it will be a rare coincidence if you can both determine that it converges and actually calculate its sum. On the other hand, if we require that the terms of a series be positive (or at least nonnegative), we can often decide whether the series converges or diverges though we will usually not be able to calculate the sum exactly. For this purpose, we will develop a number of tests. Our first test should seem pretty reasonable.

The Comparison Test

A positive series with terms less than the corresponding terms of a known convergent series ought to converge. Similarly, a series with terms larger than the terms of a known divergent series ought to diverge. What ought to be true is true.

THEOREM (COMPARISON TEST)

Suppose $0 \le a_k \le b_k$ for $k = 1, 2, 3, \ldots$.

(i)   If $\sum_{k=1}^{\infty} b_k$ converges, then $\sum_{k=1}^{\infty} a_k$ converges.

(ii)   If $\sum_{k=1}^{\infty} a_k$ diverges, then $\sum_{k=1}^{\infty} b_k$ diverges.

Proof

(i) Since $\sum_{k=1}^{\infty} b_k$ converges, it has a sum; call the sum B. Then

$$S_n = a_1 + a_2 + \cdots + a_n \le b_1 + b_2 + \cdots + b_n \le B$$

and this is true for all n. By the Bounded Sum Test in Section 15.1, the series $\sum_{k=1}^{\infty} a_k$ converges.

(ii) Since $\sum_{k=1}^{\infty} a_k$ diverges, its sequence of partial sums $\{S_n\}$ is unbounded. But

$$T_n = b_1 + b_2 + \cdots + b_n \ge a_1 + a_2 + \cdots + a_n = S_n$$

and so the sequence $\{T_n\}$ is also unbounded. We conclude, again by the Bounded Sum Test, that $\sum_{k=1}^{\infty} b_k$ diverges.   ■

EXAMPLE A

Show that the series $\sum_{k=1}^{\infty} 1/k^k$ converges.

Solution To get a feeling for this series, let us write out the first few terms.

$$1 + \frac{1}{2^2} + \frac{1}{3^3} + \frac{1}{4^4} + \cdots$$

The terms of this series are clearly dominated by the terms of the convergent geometric series (ratio $r = \frac{1}{2}$)

$$1 + \frac{1}{2} + \frac{1}{2^2} + \frac{1}{2^3} + \cdots$$

that is, $1/k^k \leq 1/2^{k-1}$. Thus, the given series converges. We are unable to say anything about its sum except that it is less than the sum of the dominating geometric series which has a sum of 2. ∎

The Ratio Test

Suppose that a series $\Sigma_{k=1}^{\infty} a_k$ of positive terms is such that the ratio a_{k+1}/a_k approaches a limit r as $k \to \infty$. Then for large k, $a_{k+1} \approx ra_k$, which means that the series behaves much like a geometric series and should converge or diverge according as $r < 1$ or $r > 1$. This intuitive reasoning can be made completely rigorous (see, for example, Purcell and Varberg, *Calculus with Analytic Geometry*, 5th edition, Prentice-Hall, pp. 490–491) and leads to the following theorem.

RATIO TEST

Let $\Sigma_{k=1}^{\infty} a_k$ be a series of positive terms and suppose that $\lim_{k \to \infty} a_{k+1}/a_k$ exists and equals r.

  (i) If $r < 1$, the series converges.

 (ii) If $r > 1$, the series diverges.

(iii) If $r = 1$, the test is inconclusive.

It turns out that this test is often useful when the formula for the kth term involves $k!$ and/or b^k, where b is a constant.

EXAMPLE B

Determine the convergence or divergence of the series $\Sigma_{k=1}^{\infty} (3^k/k!)$.

Solution

$$r = \lim_{k \to \infty} \frac{a_{k+1}}{a_k} = \lim_{k \to \infty} \frac{3^{k+1}}{(k+1)!} \frac{k!}{3^k} = \lim_{k \to \infty} \frac{3}{k+1} = 0$$

We conclude by the Ratio Test that the series converges. ∎

EXAMPLE C

Test $\displaystyle\sum_{k=1}^{\infty} \frac{3^k}{k^4}$ for convergence or divergence.

INFINITE SERIES

Solution

$$r = \lim_{k \to \infty} \frac{a_{k+1}}{a_k} = \lim_{k \to \infty} \frac{3^{k+1}}{(k+1)^4} \frac{k^4}{3^k} = \lim_{k \to \infty} 3\left(\frac{k}{k+1}\right)^4 = 3$$

We conclude from the Ratio Test that the series diverges.   ∎

EXAMPLE D

Test $\sum_{k=1}^{\infty} \frac{1}{k^2}$ for convergence or divergence.

Solution

$$r = \lim_{k \to \infty} \frac{a_{k+1}}{a_k} = \lim_{k \to \infty} \frac{k^2}{(k+1)^2} = 1$$

The Ratio Test provides no information about convergence or divergence in this case.   ∎

> **CAUTION**
> We emphasize that obtaining $r = 1$ in applying the Ratio Test yields no information about convergence or divergence. $\Sigma\,(1/k)$ and $\Sigma\,(1/k^2)$ both give $r = 1$. The first diverges (harmonic series); the second actually converges (see Example E).

Example D illustrates the limitations of the Ratio Test for a rather simple series. In fact, the Ratio Test fails for all series in which the kth term is simply a rational expression in k. For such series, another test, called the Integral Test, is often very useful.

The Integral Test

We studied the notions of convergence and divergence for infinite integrals in Section 9.4, a section worth reviewing now. There is a very close relationship between infinite integrals and infinite series, as the next theorem makes clear.

> **INTEGRAL TEST**
> Let $f(x)$ be a continuous, positive, nonincreasing function on the interval $1 \le x < \infty$ and suppose $a_k = f(k)$ for all positive integers k. Then the series $\sum_{k=1}^{\infty} a_k$ and the integral $\int_1^{\infty} f(x)\,dx$ converge or diverge together.

Proof   Refer to the diagrams in Figure 8 and note that the area of each rectangle is its height since the width in each case is 1. It follows from these diagrams that

$$\sum_{k=2}^{n} a_k \le \int_1^n f(x)\,dx \le \sum_{k=1}^{n-1} a_k$$

Thus, if $\int_1^{\infty} f(x)\,dx$ converges with value B, then

$$S_n = a_1 + \sum_{k=2}^{n} a_k \le a_1 + \int_1^n f(x)\,dx \le a_1 + B$$

and the Bounded Sum Test says that $\sum_{k=1}^{\infty} a_k$ converges.

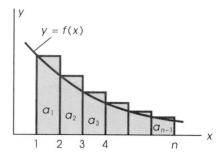

FIGURE 8

On the other hand, if $\int_1^\infty f(x)\,dx$ diverges, then $\int_1^n f(x)\,dx$ grows without bound. But then $S_{n-1} = \sum_{k=1}^{n-1} a_k$ also grows without bound and the series $\sum_{k=1}^\infty a_k$ diverges. ∎

EXAMPLE E

Show that $\sum\limits_{k=1}^\infty \dfrac{1}{k^p}$ converges or diverges according as $p > 1$ or $p \le 1$. Thus,

$\sum\limits_{k=1}^\infty \dfrac{1}{k^2}$ converges but $\sum\limits_{k=1}^\infty \dfrac{1}{k}$ diverges.

Solution Back in Section 9.4, we learned that the infinite integral $\int_1^\infty (1/x^p)\,dx$ converges or diverges according as $p > 1$ or $p \le 1$. The Integral Test now yields the desired conclusion. ∎

EXAMPLE F

Use the Integral Test to show that $\sum\limits_{k=1}^\infty \dfrac{2k}{4 + k^2}$ diverges.

Solution We consider the corresponding integral. Now

$$\int_1^\infty \frac{2x}{4 + x^2}\,dx = \lim_{b\to\infty} \int_1^b \frac{2x}{4 + x^2}\,dx = \lim_{b\to\infty} \left[\ln(4 + x^2)\right]_1^b$$

$$= \lim_{b\to\infty} \left[\ln(4 + b^2) - \ln 5\right] = \infty$$

Since the integral diverges, so does the series. ∎

EXAMPLE G

Determine the convergence or divergence of $\sum\limits_{k=1}^\infty \dfrac{k + 3}{k^2}$.

Solution Since $\dfrac{k + 3}{k^2} \ge \dfrac{k}{k^2} = \dfrac{1}{k}$ and $\sum\limits_{k=1}^\infty \dfrac{1}{k}$ diverges, the Comparison Test implies that the given series diverges. ∎

EXAMPLE H

Determine the convergence or divergence of $\sum_{k=2}^{\infty} \dfrac{\sqrt{k-2}}{k^2}$.

Solution Since $\dfrac{\sqrt{k-2}}{k^2} \le \dfrac{1}{k^{3/2}}$, the Comparison Test combined with Example E implies that the series converges. ∎

Problem Set 15.3

Recall that the harmonic series diverges and that a geometric series converges if the ratio $r < 1$ but diverges otherwise. Use these facts together with the Comparison Test to decide whether each of the following series converges or diverges (see Example A).

1. $\dfrac{1}{2} + \dfrac{1}{4} + \dfrac{1}{10} + \cdots + \dfrac{1}{3^{k-1} + 1} + \cdots$

2. $\dfrac{1}{\ln 3} + \dfrac{1}{2 \ln 3} + \dfrac{1}{4 \ln 3} + \cdots + \dfrac{1}{2^{k-1} \ln 3} + \cdots$

3. $1 + \dfrac{\sqrt{2}}{2} + \dfrac{\sqrt{3}}{3} + \cdots + \dfrac{\sqrt{k}}{k} + \cdots$

4. $\dfrac{\ln 2}{2} + \dfrac{\ln 3}{3} + \dfrac{\ln 4}{4} + \cdots + \dfrac{\ln k}{k} + \cdots$

5. $\sum_{k=1}^{\infty} \dfrac{3^k}{4^k + 1}$

6. $\sum_{k=1}^{\infty} \dfrac{k+1}{k^2}$

Use the Ratio Test to determine the convergence or divergence of the series in Problems 7–14 (see Examples B, C, and D).

7. $\sum_{k=1}^{\infty} \dfrac{2^k}{k!}$

8. $\sum_{k=1}^{\infty} \dfrac{5^{k+2}}{k!}$

9. $\sum_{k=1}^{\infty} \dfrac{2^k}{k^3}$

10. $\sum_{k=1}^{\infty} \dfrac{3^{(k+1)/2}}{k^6}$

11. $\sum_{k=1}^{\infty} \dfrac{k^2}{k^4 + 1}$

12. $\sum_{k=1}^{\infty} \dfrac{2^k + 3^k}{4^k}$

13. $\dfrac{1}{2} + \dfrac{1}{4\sqrt{2}} + \dfrac{1}{8\sqrt{3}} + \dfrac{1}{16\sqrt{4}} + \cdots$

14. $2! + \dfrac{4!}{2^3} + \dfrac{6!}{3^3} + \dfrac{8!}{4^3} + \cdots$

Use the result of Example E to determine the convergence of each series in Problems 15–18.

15. $\sum_{k=1}^{\infty} \dfrac{1}{\sqrt{k}}$

16. $\sum_{k=1}^{\infty} \dfrac{1}{k\sqrt[3]{k}}$

17. $\sum_{k=1}^{\infty} \dfrac{k}{\sqrt{k^5}}$

18. $\sum_{k=1}^{\infty} \dfrac{k}{k^{2.01}}$

Use the Integral Test as in Example F to determine the convergence or divergence of each series in Problems 19–22.

19. $\sum_{k=1}^{\infty} \dfrac{2}{2k - 1}$

20. $\sum_{k=1}^{\infty} \dfrac{1}{(3k + 1)^{3/2}}$

21. $\sum_{k=1}^{\infty} \dfrac{2}{(k + 1)^{1.2}}$

22. $\sum_{k=2}^{\infty} \dfrac{1}{k \ln k}$

As in Examples G and H, determine the convergence of the series in Problems 23–26.

23. $\sum_{k=1}^{\infty} \dfrac{k^2 + 4}{k^3}$

24. $\sum_{k=1}^{\infty} \dfrac{\sqrt{k + 1}}{k^{3/2}}$

25. $\sum_{k=1}^{\infty} \dfrac{k^2}{k^4 + 1}$

26. $\sum_{k=1}^{\infty} \dfrac{\sin^2 k}{k^2}$

Determine the convergence or divergence of the series in Problems 27–40 using any of the tests so far developed in this chapter. Give a reason for your answer.

27. $\sum_{k=1}^{\infty} \dfrac{k^2 - 2}{k^2 + 1}$

28. $\sum_{k=1}^{\infty} \sin \dfrac{k\pi}{2}$

29. $\sum_{k=1}^{\infty} \dfrac{1}{k^{1.3}}$

30. $\sum_{k=2}^{\infty} \dfrac{1}{\sqrt{k} - 1}$

31. $\sum_{k=1}^{\infty} \dfrac{k^2}{(k + 1)!}$

32. $\sum_{k=1}^{\infty} \dfrac{k!}{10^k}$

33. $\sum_{k=1}^{\infty} \dfrac{\cos k\pi}{2^k}$

34. $\sum_{k=1}^{\infty} \dfrac{|\sin(k\pi/2)|}{k^2}$

35. $\displaystyle\sum_{k=2}^{\infty} \frac{1}{\ln k}$

36. $\displaystyle\sum_{k=2}^{\infty} \frac{1}{k (\ln k)^2}$

43. $\displaystyle\sum_{k=1}^{\infty} \frac{\sqrt{k}}{k^2 + 1}$

44. $\displaystyle\sum_{k=1}^{\infty} \cos\left(1 + \frac{1}{k^2}\right)$

37. $\displaystyle\sum_{k=1}^{\infty} ke^{-k^2}$

38. $\displaystyle\sum_{k=1}^{\infty} e^{1/k}$

45. $\displaystyle\sum_{k=0}^{\infty} \frac{5}{1.1^k}$

46. $\displaystyle\sum_{k=1}^{\infty} \frac{(1.1)^k}{k^2}$

39. $\displaystyle\sum_{k=1}^{\infty} \left[4\left(\frac{1}{3}\right)^k - \frac{2}{k^2} \right]$

40. $\displaystyle\sum_{k=1}^{\infty} \left[\frac{10}{k!} + \frac{|\sin(k^2)|}{4k^2} \right]$

47. $\displaystyle\sum_{k=9}^{\infty} \frac{3}{\sqrt{k} - 2}$

48. $\displaystyle\sum_{k=1}^{\infty} \frac{2^k}{k!}$

49. $\displaystyle\sum_{k=1}^{\infty} \frac{k!}{k^{100}}$

50. $\displaystyle\sum_{k=2}^{\infty} \frac{5}{k (\ln k)^{3/2}}$

Use any of the results established so far in this chapter to determine the convergence or divergence of the series in Problems 41–58. Give a reason for your conclusion. For some problems, it may be useful to know that $\Sigma\, ca_k$ converges if and only if $\Sigma\, a_k$ converges ($c \neq 0$).

51. $\displaystyle\sum_{k=1}^{\infty} \frac{3^{k+1}(2k + 1)}{k 4^k}$

52. $\displaystyle\sum_{k=1}^{\infty} ke^{-k}$

53. $\displaystyle\sum_{k=1}^{\infty} \frac{k^2 + 1}{k 3^k}$

54. $\displaystyle\sum_{k=1}^{\infty} \frac{k!}{3^k}$

55. $\displaystyle\sum_{k=1}^{\infty} \left(\frac{1}{3^k + k} - \frac{1}{k^2} \right)$

56. $\displaystyle\sum_{k=1}^{\infty} \left(\frac{1}{2^k} + \frac{1}{k} \right)$

41. $\displaystyle\sum_{k=1}^{\infty} \frac{k + 1}{2k + 1}$

42. $\displaystyle\sum_{k=1}^{\infty} \frac{4}{k^2 + 1}$

57. $\displaystyle\sum_{k=1}^{\infty} \left(1 + \frac{1}{k} \right)^k$

58. $\displaystyle\sum_{k=1}^{\infty} \frac{k!}{k^k}$

15.4
POWER SERIES

All series in the previous section were assumed to have positive (or at least nonnegative) terms. That is too strong an assumption for the subject of this section. Thus, consider a series

$$a_1 + a_2 + a_3 + a_4 + \cdots$$

where the terms are of arbitrary sign. To analyze convergence of such a series is difficult, but there is one result that should seem reasonable. If, after changing the signs of all the negative terms, the series converges, then certainly the original series must have converged. Here is the precise statement.

ABSOLUTE CONVERGENCE TEST

If a series converges absolutely, it converges. That is, if $\Sigma_{k=1}^{\infty} |a_k|$ converges, then $\Sigma_{k=1}^{\infty} a_k$ converges.

Proof A formal proof depends on a trick. Let $b_k = a_k + |a_k|$, so that $0 \leq b_k \leq 2|a_k|$. Now $\Sigma_{k=1}^{\infty} 2|a_k|$ converges (why?), and consequently $\Sigma_{k=1}^{\infty} b_k$ converges (Comparison Test). But

$$S_n = \sum_{k=1}^{n} a_k = \sum_{k=1}^{n} (b_k - |a_k|) = \sum_{k=1}^{n} b_k - \sum_{k=1}^{n} |a_k|$$

Since both partial sums on the right have finite limits as $n \to \infty$, it follows that S_n has a finite limit as $n \to \infty$. Thus, the series $\Sigma_{k=1}^{\infty} a_k$ converges. ∎

EXAMPLE A

Show that $\sum_{k=1}^{\infty} [\sin(k\pi/3)]/k^2$ converges.

Solution

$$\left| \frac{\sin(k\pi/3)}{k^2} \right| \le \frac{1}{k^2}$$

Now $\displaystyle\sum_{k=1}^{\infty} \frac{1}{k^2}$ converges and, by the Comparison Test, so does

$\displaystyle\sum_{k=1}^{\infty} \left| \frac{\sin(k\pi/3)}{k^2} \right|$. The claimed assertion follows from the Absolute Convergence Test. ∎

There is a nice result for series of alternating signs.

ALTERNATING SERIES TEST

Let $b_1 - b_2 + b_3 - b_4 + b_5 - b_6 + b_7 - \cdots$ be an alternating series with $b_k > b_{k+1} > 0$. If $\lim_{k\to\infty} b_k = 0$, then the series converges.

Proof Let S_n have its usual meaning, so that

$$S_1 = b_1$$
$$S_2 = b_1 - b_2$$
$$S_3 = b_1 - b_2 + b_3$$
$$S_4 = b_1 - b_2 + b_3 - b_4$$

and so on. Refer to the geometric description in Figure 9 and note that the sequence of even-numbered terms S_2, S_4, S_6, . . . , is increasing and bounded above and so has a limit, say S_*. The sequence of odd-numbered terms is decreasing and bounded below and also has a limit, say S^*. Now (see Figure 10),

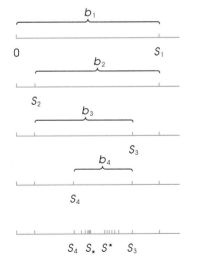

FIGURE 9

n even

$S_n \qquad S_* \; S^* \qquad S_{n+1}$

n odd

$S_{n+1} \qquad S_* \; S^* \qquad S_n$

FIGURE 10

$$\left|S^* - S_*\right| \le \left|S_{n+1} - S_n\right| = b_{n+1}$$

for all n and $\lim_{n\to\infty} b_{n+1} = 0$ by hypothesis. It follows that $S^* = S_*$, and therefore the given series converges to their common value. ∎

EXAMPLE B

Show that the alternating harmonic series

$$1 - \frac{1}{2} + \frac{1}{3} - \frac{1}{4} + \frac{1}{5} - \frac{1}{6} + \cdots$$

converges.

Solution The hypotheses of the Alternating Series Test are satisfied. ∎

Power Series

So far, we have been studying what may be called *series of constants*, that is, series of the form $\sum_{k=1}^{\infty} a_k$. Now we turn our attention to *power series*, that is, to series of the form

$$\sum_{k=0}^{\infty} c_k x^k = c_0 + c_1 x + c_2 x^2 + c_3 x^3 + \cdots$$

Of course, as soon as we substitute a value for x, we are on familiar territory; we have a series of constants.

Given a power series, the first and most obvious question to ask is: For what values of x does the series converge? There is one such series for which we already know the answer. Consider the power series

$$1 + x + x^2 + x^3 + x^4 + \cdots$$

This is a geometric series with ratio x, and so it converges precisely on the interval $|x| < 1$, that is, on the interval $(-1, 1)$. Here is another example.

EXAMPLE C

For what values of x does the series

$$1 + x + \frac{1}{2!} x^2 + \frac{1}{3!} x^3 + \frac{1}{4!} x^4 + \cdots$$

converge?

Solution Consider the corresponding series of absolute values and apply the Ratio Test.

$$r = \lim_{k\to\infty} \frac{a_{k+1}}{a_k} = \lim_{k\to\infty} \frac{\left|x^{k+1}\right|}{(k+1)!} \frac{k!}{\left|x^k\right|} = \lim_{k\to\infty} \frac{\left|x\right|}{k+1} = 0$$

We conclude that the series of absolute values converges for every real number x. The Absolute Convergence Test implies that the given series converges for all real numbers. ■

   In both of the examples we have considered, the power series converged on an interval centered at the origin. In the first example, the interval had radius 1; in the second, the interval had an infinite radius. Here is the general theorem we have illustrated.

THEOREM

The convergence set for a power series $\sum_{n=0}^{\infty} c_k x^k$ is always an interval of one of the following types.

   (i)   A degenerate interval—the single point $x = 0$.
   (ii) An interval of the form $(-R, R)$, plus possibly one or both endpoints.
   (iii) The whole real line.

We say that the intervals in the three cases have radius 0, R, and ∞, respectively.

Proof   Suppose that the series converges at $x = b \neq 0$. Then $\lim_{k \to \infty} c_k b^k = 0$, so that there is certainly a number K such that $|c_k b^k| < 1$ for $k \geq K$. Thus for any x for which $|x| < |b|$,

$$|c_k x^k| = |c_k b^k| \left|\frac{x}{b}\right|^k < \left|\frac{x}{b}\right|^k$$

this holding for $k \geq K$. But $\sum |x/b|^k$ is a geometric series with a ratio of less than 1 and so converges. We conclude that the series $\sum_{k=0}^{\infty} |c_k x^k|$ converges, and hence so does $\sum_{k=0}^{\infty} c_k x^k$.

   On the other hand, suppose a power series diverges at $x = d$. Then it must diverge for all x such that $|x| > |d|$. For suppose it converged at some b such that $|b| > |d|$. Then by what we just proved, it would converge at d contrary to our supposition.

   These two paragraphs together eliminate all but the three possibilities we have enumerated. Note that our proof shows that a *power series converges absolutely on the interior of the interval of convergence.* ■

EXAMPLE D

Find the interval of convergence of the power series $\displaystyle\sum_{k=1}^{\infty} \frac{k}{2^k} x^k$.

Solution   We apply the Ratio Test to the series of absolute values.

$$r = \lim_{k \to \infty} \frac{a_{k+1}}{a_k} = \lim_{k \to \infty} \frac{|(k+1)x^{k+1}|}{2^{k+1}} \frac{2^k}{|kx^k|} = \lim_{k \to \infty} \frac{|x|}{2} \frac{(k+1)}{k} = \frac{|x|}{2}$$

We conclude that the given power series converges absolutely (hence converges) for $|x|/2 < 1$, that is, for $|x| < 2$. The endpoints of this interval must be checked separately (recall that the Ratio Test gives no information when $r = 1$). Substitution of $x = 2$ and $x = -2$ leads to two series that diverge (check this). The interval of convergence is $(-2, 2)$. ■

Maclaurin Series

The next obvious problem would be to determine the function $S(x)$, which is the sum of a power series on its interval of convergence. Unfortunately, we are unable to make much progress on this problem except for the geometric series

$$1 + x + x^2 + x^3 + x^4 + \cdots$$

which has sum $S(x) = 1/(1 - x)$ for $|x| < 1$ as we learned in Section 15.2.

There is a related question for which we can give some answers. Given a ries? For example, can $f(x) = e^x$ be represented in the form

$$e^x = c_0 + c_1 x + c_2 x^2 + c_3 x^3 + \cdots$$

To answer this question, we appeal to Section 14.2 where we learned that e^x can be written as an nth-degree polynomial plus an error term $E(x)$. If this error term tends to 0 as n tends to ∞, then e^x can be represented by its Maclaurin series. It can be shown that this does happen for all x and that the following series representations are valid.

$$e^x = 1 + x + \frac{1}{2!}x^2 + \frac{1}{3!}x^3 + \frac{1}{4!}x^4 + \cdots \qquad (-\infty < x < \infty)$$

$$\sin x = x - \frac{1}{3!}x^3 + \frac{1}{5!}x^5 - \frac{1}{7!}x^7 + \cdots \qquad (-\infty < x < \infty)$$

$$\cos x = 1 - \frac{1}{2!}x^2 + \frac{1}{4!}x^4 - \frac{1}{6!}x^6 + \cdots \qquad (-\infty < x < \infty)$$

$$\ln(1 + x) = x - \frac{1}{2}x^2 + \frac{1}{3}x^3 - \frac{1}{4}x^4 + \cdots \qquad (-1 < x \le 1)$$

$$\frac{1}{1 - x} = 1 + x + x^2 + x^3 + x^4 + \cdots \qquad (-1 < x < 1)$$

EXAMPLE E

Express the number e as an infinite series.

Solution

$$e = e^1 = 1 + 1 + \frac{1}{2!} + \frac{1}{3!} + \frac{1}{4!} + \cdots$$

■

EXAMPLE F

Represent e^{x^2} in a power series.

Solution Replace x by x^2 in the power series for e^x. We obtain

$$e^{x^2} = 1 + x^2 + \frac{1}{2!}x^4 + \frac{1}{3!}x^6 + \cdots$$

■

Power series are sometimes referred to as infinite-degree polynomials. While this is inaccurate terminology, it is suggestive. In particular it suggests that the way to find the power series representation of $e^x \ln(1 + x)$ is to represent each function as a power series and then multiply them like polynomials.

EXAMPLE G

Find the terms through degree 4 in the power series representation of $e^x \ln(1 + x)$.

Solution As suggested above, we write down the two power series and multiply, collecting terms of like degree.

$$0 + x - \frac{x^2}{2} + \frac{x^3}{3} - \frac{x^4}{4} + \cdots$$

$$1 + x + \frac{x^2}{2} + \frac{x^3}{6} + \frac{x^4}{24} + \cdots$$

$$\overline{}$$

$$0 + (0 + 1)x + \left(0 + 1 - \frac{1}{2}\right)x^2 + \left(0 + \frac{1}{2} - \frac{1}{2} + \frac{1}{3}\right)x^3$$

$$+ \left(0 + \frac{1}{6} - \frac{1}{4} + \frac{1}{3} - \frac{1}{4}\right)x^4 + \cdots$$

$$= 0 + x + \frac{1}{2}x^2 + \frac{1}{3}x^3 + 0x^4 + \cdots$$

■

Problem Set 15.4

Use the Absolute Convergence Test to show that each of the following series converges. See Example A.

1. $1 - \dfrac{1}{2^2} + \dfrac{1}{3^2} + \dfrac{1}{4^2} - \dfrac{1}{5^2} - \dfrac{1}{6^2} - \dfrac{1}{7^2} + \dfrac{1}{8^2} + \dfrac{1}{9^2}$
   $+ \dfrac{1}{10^2} + \dfrac{1}{11^2} - \cdots$

2. $1 + \dfrac{1}{3} - \dfrac{1}{9} - \dfrac{1}{27} + \dfrac{1}{81} + \dfrac{1}{243} - \dfrac{1}{729}$
   $- \dfrac{1}{2187} + \cdots$

3. $\displaystyle\sum_{k=1}^{\infty} \frac{\sin(k\pi/3)}{k^2}$

4. $\displaystyle\sum_{k=1}^{\infty} \frac{2^k \cos k}{k!}$

Use the Alternating Series Test to show that each of the following series converges. See Example B.

5. $\sum_{k=1}^{\infty} (-1)^{k+1} \dfrac{1}{\sqrt{k}}$

6. $\sum_{k=2}^{\infty} (-1)^k \dfrac{1}{k \ln k}$

7. $\sum_{k=1}^{\infty} (-1)^{k+1} \dfrac{1}{\sqrt[3]{k} + 5}$

8. $\sum_{k=2}^{\infty} (-1)^k \dfrac{2k + 1}{k^{3/2}}$

9. $\sum_{k=1}^{\infty} (-1)^{k+1} \dfrac{k}{2^k}$

10. $\sum_{k=1}^{\infty} (-1)^{k+1} k e^{-k^2}$

An infinite series that converges but does not converge absolutely is said to **converge conditionally.** *For example, the alternating harmonic series* $1 - \frac{1}{2} + \frac{1}{3} - \frac{1}{4} + \frac{1}{5} - \frac{1}{6} + \frac{1}{7} - \cdots$ *converges conditionally. In Problems 11–20, determine whether the given series converges absolutely, converges conditionally, or diverges.*

11. $\sum_{k=1}^{\infty} (-1)^{k+1} \dfrac{k}{2k - 1}$

12. $\sum_{k=1}^{\infty} (-1)^{k+1} \dfrac{4}{k^{1.1}}$

13. $\sum_{k=1}^{\infty} (-1)^{k+1} \dfrac{1}{3k}$

14. $\sum_{k=1}^{\infty} (-1)^{k+1} \dfrac{\sin(k\pi/2)}{k}$

15. $\sum_{k=1}^{\infty} (-1)^{k+1} \dfrac{k^4}{2^k}$

16. $\sum_{k=2}^{\infty} (-1)^k \dfrac{1}{k \ln k}$

17. $\sum_{k=2}^{\infty} (-1)^k \dfrac{k}{k^2 - 1}$

18. $\sum_{k=1}^{\infty} (-1)^{k+1} \sin\left(1 + \dfrac{1}{k}\right)$

19. $\sum_{k=2}^{\infty} (-1)^k \dfrac{1}{k (\ln k)^2}$

20. $\sum_{k=1}^{\infty} \dfrac{\cos(k\pi)}{\sqrt{k}}$

In Problems 21–28, find the interval of convergence of the given power series. Be sure to check any endpoints. See Examples C and D.

21. $\sum_{k=1}^{\infty} \dfrac{2^k x^k}{k!}$

22. $\sum_{k=0}^{\infty} \dfrac{x^k}{(2k + 1)!}$

23. $\sum_{k=1}^{\infty} \dfrac{x^k}{k}$

24. $\sum_{k=1}^{\infty} k^2 x^k$

25. $\sum_{k=1}^{\infty} 2^k x^k$

26. $\sum_{k=1}^{\infty} \dfrac{x^k}{4^k}$

27. $1 + 3x + \dfrac{3^2 x^2}{2!} + \dfrac{3^3 x^3}{3!} + \dfrac{3^4 x^4}{4!} + \cdots$

28. $\dfrac{1}{2} + \dfrac{x}{2^2 + 1} + \dfrac{x^2}{3^2 + 1} + \dfrac{x^3}{4^2 + 1} + \cdots$

Use the procedures of Examples E and F to represent each of the following in a series.

29. $\sin 2$

30. e^2

31. $\cos 1$

32. $\ln 1.5$

33. $\dfrac{1}{1 - 3x}$

34. $\sin 2x$

35. $\ln\left(1 + \dfrac{x}{2}\right)$

36. e^{-x}

37. $\cos(3x^2)$

38. $\dfrac{1}{1 - 5x^2}$

39. e^{2x}

40. $\cos(x^3)$

Find the terms through degree 4 of the power series representation of each of the following. See Example G.

41. $e^x \cos x$

42. $\dfrac{\cos x}{1 - x}$

43. $\dfrac{\sin x}{1 - x^2}$

44. $\sin x \ln(1 + x^2)$

45. $\sin 2x + \cos(x^2)$

46. $\ln(1 + x) - \dfrac{2}{1 + x}$

C 47. Find $\int_0^{0.5} e^{-x^2} \, dx$ accurate to four decimal places, by using the series representation of e^{-x^2}.

C 48. Find $\int_0^{0.5} \sin(x^2) \, dx$ accurate to six decimal places, by using the series representation of $\sin(x^2)$.

49. Use the fact that $\ln[(1 + x)/(1 - x)] = \ln(1 + x) - \ln(1 - x)$ to find the power series representation of $\ln[(1 + x)/(1 - x)]$. Note that this representation is valid for $-1 < x < 1$.

50. Show that any positive number M can be written in the form $(1 + x)/(1 - x)$ for some x satisfying $-1 < x < 1$.

C 51. Problems 49 and 50 together allow us to find the logarithm of any positive number. Use these results to find $\ln 3$ accurate to four decimal places.

52. By relating it to something familiar, find the sum of the following series.

$$2x - \dfrac{4x^2}{2} + \dfrac{8x^3}{3} - \dfrac{16x^4}{4} + \cdots$$

53. Find the interval of convergence of each of the following series.

(a) $\sum_{k=1}^{\infty} \dfrac{k}{3^k} x^k$

(b) $\sum_{k=1}^{\infty} \dfrac{4^k}{k!} x^k$

(c) $\sum_{k=1}^{\infty} \dfrac{x^k}{k^2}$

54. Find the first six terms in the power series for $x^2 e^{-x} + (1 - x)^{-1}$.

55. It is known that a power series can be differentiated term by term on the interior of its interval of convergence. Use this fact to find the power series for each of the following.

(a) $f(x) = \dfrac{1}{(1 + x)^2}$ Hint: Differentiate the series for $\dfrac{1}{1 + x}$.

(b) $g(x) = \dfrac{1}{(1 + x)^3}$

56. Use the idea in Problem 55 to evaluate $\sum_{k=1}^{\infty} kx^{k-1}$.

Chapter 15 Review Problem Set

1. Determine whether the given sequence converges or diverges and in the case of convergence find the limit.
(a) $a_n = (1.01)^n$
(b) $a_n = (3n - 1)/(2n + 1)$
(c) $a_n = \sin\left(2 + \dfrac{1}{n}\right)$ (d) $a_n = (-1)^n + \dfrac{1}{n}$

2. Follow the instructions of Problem 1.
(a) $a_n = (0.98)^n$ (b) $a_n = \dfrac{1 - 4n - 3n^2}{n^2 + 5}$
(c) $a_n = \cos(2 + n)$ (d) $a_n = (-1)^n n^{-2} + \sqrt{n}$

3. Find the fifth term of the sequence for which $a_1 = 3$ and $a_{n+1} = 2a_n + 4$ for $n \geq 1$.

4. Find the fifth term in the sequence for which $a_1 = 5$ and $a_{n+1} = 2a_n - 6$ for $n \geq 1$.

5. Write the repeating decimal $0.427427427 \ldots$ as the ratio of two integers.

6. Write the repeating decimal $0.027027027 \ldots$ as the ratio of two integers in reduced form.

7. Suppose the partial sum $S_n = a_1 + a_2 + \cdots + a_n$ of a series is given by $S_n = 2n/(n + 4)$.
(a) Find a_1, a_2, and a_3.
(b) Find the sum of the infinite series $\sum_{k=1}^{\infty} a_k$.

8. Suppose that $S_n = n^2/(2n^2 + 4)$.
(a) Find a_1, a_2, and a_3.
(b) Find $\sum_{k=1}^{\infty} a_k$.

In Problems 9–24, determine whether the given series converges or diverges and give a reason for your answer.

9. $\sum_{k=0}^{\infty} (e - 2)^k$

10. $\sum_{k=0}^{\infty} (\pi - 2)^k$

11. $\sum_{k=1}^{\infty} \dfrac{2k + 1}{3k - 2}$

12. $\sum_{k=1}^{\infty} \cos\left(3 + \dfrac{1}{k}\right)$

13. $\sum_{k=1}^{\infty} \dfrac{k!}{10^k}$

14. $\sum_{k=1}^{\infty} \dfrac{8^k}{k!}$

15. $\sum_{k=1}^{\infty} \dfrac{k}{k^2 + 1}$

16. $\sum_{k=1}^{\infty} \dfrac{k^2}{k^3 + 1}$

17. $\sum_{k=1}^{\infty} \dfrac{\cos k\pi}{k^{1/4}}$

18. $\sum_{k=1}^{\infty} \dfrac{(-1)^k}{k^{1/3}}$

19. $\sum_{k=1}^{\infty} k^2\left(\dfrac{2}{3}\right)^k$

20. $\sum_{k=1}^{\infty} \dfrac{1}{k}\left(\dfrac{5}{4}\right)^k$

21. $\sum_{k=0}^{\infty} \dfrac{1}{2^k + k}$

22. $\sum_{k=1}^{\infty} \dfrac{1}{k^{7/6} + k}$

23. $\sum_{k=1}^{\infty} \dfrac{2k + 1}{(k^2 + k + 3)^2}$

24. $\sum_{k=1}^{\infty} \dfrac{e^k}{(e^k + 1)^2}$

25. Find the interval of convergence for each of the following power series.
(a) $\sum_{k=1}^{\infty} \dfrac{2k}{k + 1} x^k$ (b) $\sum_{k=1}^{\infty} \dfrac{3^k}{k^2} x^k$

26. Find the interval of convergence for each of the following series.
(a) $\sum_{k=1}^{\infty} \dfrac{3^k}{k^2} x^k$ (b) $\sum_{k=1}^{\infty} \dfrac{100^k}{k!} x^k$

27. Make use of known power series to write the first four nonzero terms of the (Maclaurin) power series for each function.
(a) $f(x) = e^{2x}$ (b) $(x + 2) \cos x$
(c) $e^x \sin x$

28. Make use of known power series to write the first four nonzero terms of the power series in x for each function.
(a) $f(x) = \ln(1 + 3x)$ (b) $f(x) = (x^2 - 1)e^{2x}$
(c) $f(x) = (\sin x)/(1 - x^2)$

29. Find the sum of the series $\sum_{k=1}^{\infty} [2(\tfrac{3}{4})^k - 3(\tfrac{1}{2})^k]$.

30. Evaluate $\sum_{k=0}^{\infty} \left[3\left(\dfrac{5}{6}\right)^k + 14\left(-\dfrac{3}{4}\right)^{k+1}\right]$.

C 31. Sue will deposit $1000 in a fund at the end of each year for the next 15 years. How much will she have in the fund at the end of that time if the fund earns interest at 8.5% compounded annually?

C 32. The Smiths expect to purchase a small motor home 5 years from now. To prepare for this, they will make five annual deposits of $6000 each beginning 1 year from now. How much money will be in the fund at the end of the 5-year period if interest is 8.25% compounded annually?

33. A patient is given 10 milligrams of a certain drug each day. It is estimated that each day the body eliminates 40% of the drug present in the body that day. Calculate the limiting amount of the drug in the body after a long period of time, assuming this amount is measured right after the drug is administered.

C 34. How much money must Kevin set aside on his 64th birthday in a fund that earns 9% interest compounded annually in order to give him 20 annual payments of $24,000 each starting on his 65th birthday? How much more would he have to set aside so these payments could continue forever?

C 35. Find $\int_0^{0.6} x^2 \cos x \, dx$ in three ways.
   (a) Using the first four nonzero terms in the power series representation of $x^2 \cos x$.
   (b) Using the Parabolic Rule with $h = 0.1$.
   (c) Using a table of integrals.

C 36. Calculate $\int_0^{0.5} (x + 1) \ln(x + 1) \, dx$ in three ways.
   (a) Using the first four nonzero terms in the power series representation of $(x + 1) \ln(x + 1)$.
   (b) Using the Trapezoidal Rule with $h = 0.1$.
   (c) Using a table of integrals.

37. Find the terms through degree 5 in the power series representation of $x/(1 + x^2)^2$ by differentiating the power series for $1/(1 + x^2)$.

38. Find the terms through degree 6 in the power series representation of $x^2 \cos x^3$ by differentiating another power series.

Repeatedly tossing two dice raises many interesting probability questions. Call getting a total of 7 a success and anything else a failure. What is the probability of success in one toss? How is S_n, the number of successes in n tosses, distributed when n is very large? This last question leads to the normal law, a law that Mark Kac said cast a spell over his whole life.

16

Calculus and Probability

Mark Kac grew up in Krzemieniec (Kremenetz), a city now well inside Russia but which historically, and again from 1919 to 1939, was part of Poland. In 1914 when Kac was born, the city was predominantly Jewish, as were both of his parents. Kac's father was the first Jew from Krzemieniec to receive an advanced degree, and he obtained two of them—one from the University of Leipzig and the other from the University of Moscow. Probably it was from his father that the young Mark inherited an intense love of learning, but it was his mother (also well educated) who made him curious about everything in the world around him.

Kac chose the University of Lwow for his undergraduate and graduate study, and there he met Hugo Steinhaus. It was Steinhaus who inspired the young Kac to work on hard mathematical problems, who introduced him to probability, and who eventually supervised his doctoral work. While probability is a very old subject, it was only in the 1930s that it began to achieve respectability as a legitimate mathematical discipline. Before that it was regarded as a somewhat trivial and even disreputable study of games of chance. But in 1933, a Russian mathematician, A. N. Kolmogorov, published a book which put the subject on a firm foundation and attracted a host of young mathematicians to its study. Mark Kac was to become one of the guiding lights in the development of this new discipline.

In 1938, a postdoctoral fellowship brought Kac to the United States and Johns Hopkins University. This event undoubtedly saved his life, for most of his relatives were killed during the Holocaust. From Johns Hopkins, he moved successively to Cornell University, Rockefeller University, and the University of Southern California, along the way writing a host of influential papers on probability theory and winning many honors. He was especially interested in concrete applied problems, and here the normal law (which crops up in astronomy, physics, biology, statistics, and most other areas of science) fascinated him. Kac believed that mathematics must never be separated from its roots in the real world.

Mark Kac
(1914–1984)

16.1

DISCRETE DISTRIBUTIONS

This chapter introduces a very large subject and one that has a huge array of applications in both the physical and the social sciences. We assume that the reader is familiar with the basic concept of probability, but we think it wise to review the terminology we will use. We imagine an **experiment** which can result in any one of a number of **outcomes.** The set consisting of all possible outcomes of the experiment is the **sample space.** Any collection of outcomes, that is, any subset of the sample space, is called an **event.** Associated with each event is a nonnegative number called its **probability,** which measures the likelihood of the event occurring when the experiment is performed.

EXAMPLE A

Consider the experiment of tossing a fair die (Figure 1). Illustrate the boldface terms in the paragraph above for this experiment.

Solution We may think of the possible outcomes as consisting of the numbers 1, 2, 3, 4, 5, and 6, corresponding to the number of dots that appear on the top of the die. Thus, the sample space is the set $S = \{1, 2, 3, 4, 5, 6\}$, and typical events are $A = \{1, 3, 5\}$ and $B = \{5, 6\}$. The event A can be described in words as "getting an odd number," the event B as "getting a number at least 5." Since the die is fair, we assign probability $\frac{1}{6}$ to each outcome and then compute the probability of each event by adding the probabilities of the outcomes that make up the event. Thus $\text{prob}(A) = \frac{3}{6} = \frac{1}{2}$, $\text{prob}(B) = \frac{2}{6} = \frac{1}{3}$, and $\text{prob}(S) = \frac{6}{6} = 1$. ■

FIGURE 1

If we use S to denote the sample space, $\emptyset$ to denote the empty event (also called the impossible event), and A^C to denote the event complementary to A, then probability satisfies the following nice properties.

1. $0 \le \text{prob}(A) \le 1$ for every event A.
2. $\text{prob}(\emptyset) = 0$ and $\text{prob}(S) = 1$.
3. $\text{prob}(A \cup B) = \text{prob}(A \text{ or } B) = \text{prob}(A) + \text{prob}(B)$ provided A and B are disjoint (cannot occur simultaneously).
4. $\text{prob}(A^C) = 1 - \text{prob}(A)$.

Random Variables

A **random variable** is simply a function defined on a sample space. We will use capital boldface letters like **X, Y,** and **Z** for random variables. The simplest random variable in Example A is the variable $\mathbf{X}(x) = x$. Thus, **X** represents the number of spots showing on the die. Another random variable **Y** could be defined as the square of the number of spots plus 1 so that $\mathbf{Y}(1) = 1^2 + 1 = 2$, $\mathbf{Y}(2) = 2^2 + 1 = 5$, and so on. All random variables discussed in this section will be **discrete random variables,** meaning that they will assume only a finite (or countable) number of values. Continuous random variables will be treated in the following sections.

EXAMPLE B

Let an experiment consist of tossing two fair dice. Describe an appropriate sample space. If Y is the random variable corresponding to the total number of spots showing, determine $\text{prob}\{Y = 7\}$ and $\text{prob}\{Y > 3\}$.

Solution We take the sample space to consist of the 36 outcomes $(1, 1)$, $(1, 2)$, $(1, 3)$, . . . , $(6, 5)$, $(6, 6)$, each being assigned probability $\frac{1}{36}$. The event $\{Y = 7\}$ consists of the six outcomes $(1, 6)$, $(2, 5)$, $(3, 4)$, $(4, 3)$, $(5, 2)$ and $(6, 1)$, so

$$\text{prob}\{Y = 7\} = \frac{6}{36} = \frac{1}{6}$$

Also,

$$\text{prob}\{Y > 3\} = 1 - \text{prob}\{Y \le 3\} = 1 - \frac{3}{36} = \frac{11}{12}$$

Here we have used the fact that the event $\{Y \le 3\}$ consists of the three outcomes $(1, 1)$, $(1, 2)$, and $(2, 1)$. ∎

Consider a discrete random variable X taking on values $x_1, x_2, x_3, \ldots,$ x_n. Associated with X is its **probability function** f_i defined by $f_i = \text{prob}\{X = x_i\}$. To picture a probability function, we draw its bar graph, which consists of a set of bars of height f_i at x_i.

EXAMPLE C

Draw the bar graph for the probability function corresponding to the random variable Y in Example B.

Solution The random variable Y takes on the values 2, 3, . . . , 12 with the probabilities f_i shown in the table on the left in Figure 2. The bar graph for f_i appears to the right of the table. ∎

The Mean of a Discrete Random Variable

Suppose that we have a random variable X defined on some sample space which takes on the values $x_1, x_2, \ldots, x_n$ with probabilities $f_1, f_2, \ldots, f_n$, respectively. Assume further that these are the only values X takes on with nonzero probabilities; in other words, assume that $f_1 + f_2 + \cdots + f_n = 1$. We define the **mean** (also called the **expected value**) μ of X by

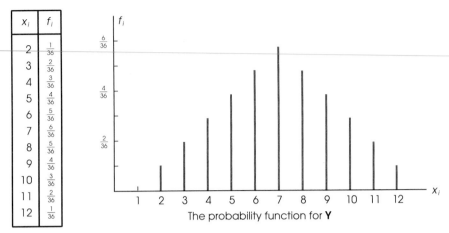

x_i	f_i
2	$\frac{1}{36}$
3	$\frac{2}{36}$
4	$\frac{3}{36}$
5	$\frac{4}{36}$
6	$\frac{5}{36}$
7	$\frac{6}{36}$
8	$\frac{5}{36}$
9	$\frac{4}{36}$
10	$\frac{3}{36}$
11	$\frac{2}{36}$
12	$\frac{1}{36}$

The probability function for **Y**

FIGURE 2

$$\mu = \sum_{i=1}^{n} x_i f_i$$

EXAMPLE D

Suppose that a die has been so loaded that the numbers 1, 2, 3, 4, 5, and 6 come up 10%, 10%, 20%, 30%, 20%, and 10% of the time, respectively. Let **X** correspond to the number of spots that show. Find the mean of **X**.

Solution The given percentages correspond to probabilities and add to 100% or 1. Therefore,

$$\mu = 1(0.1) + 2(0.1) + 3(0.2) + 4(0.3) + 5(0.2) + 6(0.1) = 3.7 \quad \blacksquare$$

$\mu = 3.7$

FIGURE 3

We give two interpretations of the mean in Example D. Suppose we place weights of size 0.1, 0.1, 0.2, 0.3, 0.2, and 0.1 along a teeterboard at points 1, 2, 3, 4, 5, and 6, respectively, as shown in Figure 3. Then $\mu = 3.7$ is the point at which this distribution of weights will balance. The mean of a distribution is its balance point.

As a second interpretation, suppose you would win $1.00 if the loaded die came up 1, $2.00 if the die came up 2, and so on. Suppose further that you tossed this die over and over. How much could you expect to win on the average? The answer is that you could expect to win $3.70 on an average toss (thus the name *expected value* as a synonym for *mean*). To see why this is so, reason as follows. In 100 tosses of this die, you could expect about 10 ones, 10 twos, 20 threes, 30 fours, 20 fives, and 10 sixes. Therefore, your expected winnings on an average toss would be your total expected winnings divided by 100, that is,

$$\frac{1(10) + 2(10) + 3(20) + 4(30) + 5(20) + 6(10)}{100}$$

$$= 1(0.1) + 2(0.1) + 3(0.2) + 4(0.3) + 5(0.2) + 6(0.1)$$

$$= 3.7$$

The Variance of a Discrete Random Variable

Again assume that **X** takes on the values $x_1, x_2, x_3, \ldots, x_n$ with probabilities $f_1, f_2, f_3, \ldots, f_n$, respectively, with the sum of the f_i's equal to 1. We define the **variance** σ^2 of **X** by

$$\sigma^2 = \sum_{i=1}^{n} (x_i - \mu)^2 f_i$$

The variance of a random variable measures how its values spread out about the mean. If σ^2 is large, then these values are widely dispersed; if σ^2 is small, these values cluster closely about the mean. The formula we have given is messy to use, especially if μ is a fraction. We derive an equivalent form which makes calculations easier. Note that

$$\sum_{i=1}^{n} (x_i - \mu)^2 f_i = \sum_{i=1}^{n} (x_i^2 f_i - 2\mu x_i f_i + \mu^2 f_i)$$

$$= \sum_{i=1}^{n} x_i^2 f_i - 2\mu \sum_{i=1}^{n} x_i f_i + \mu^2 \sum_{i=1}^{n} f_i = \sum_{i=1}^{n} x_i^2 f_i - 2\mu\mu + \mu^2$$

Thus

$$\sigma^2 = \left(\sum_{i=1}^{n} x_i^2 f_i \right) - \mu^2$$

the formula we will ordinarily use in all calculations.

EXAMPLE E

Find the variance of the random variable **X** in Example D.

Solution Using the boxed formula above, we obtain

$$\sigma^2 = 1(0.1) + 4(0.1) + 9(0.2) + 16(0.3) + 25(0.2) + 36(0.1) - 3.7^2$$

$$= 2.01$$

∎

EXAMPLE F

Find the mean and variance of the random variable **Y** in Examples B and C.

Solution From the bar graph in Figure 2, we can see that the balance point of the distribution is 7. Of course, we could also calculate μ using the formula for it. In any case, $\mu = 7$. For the variance, we obtain

$$\sigma^2 = 4\left(\frac{1}{36}\right) + 9\left(\frac{2}{36}\right) + 16\left(\frac{3}{36}\right) + 25\left(\frac{4}{36}\right) + 36\left(\frac{5}{36}\right) + 49\left(\frac{6}{36}\right)$$

$$+ 64\left(\frac{5}{36}\right) + 81\left(\frac{4}{36}\right) + 100\left(\frac{3}{36}\right) + 121\left(\frac{2}{36}\right) + 144\left(\frac{1}{36}\right) - 49$$

$$= \frac{1974}{36} - 49 = 5.83 \qquad \blacksquare$$

Discrete Random Variables with Infinitely Many Values

Our study of infinite series (Chapter 15) allows us to consider discrete random variables with infinitely many values, but as you might expect the calculations are more complicated.

EXAMPLE G

Consider the random variable **X** which takes on the infinite set of values 1, 2, 3, . . . , with probabilities $\frac{1}{2}, \frac{1}{4}, \frac{1}{8}, . . .$, respectively.

(a) Show that this is an honest probability distribution; that is, show that $\frac{1}{2} + \frac{1}{4} + \frac{1}{8} + \cdots = 1$.
(b) Find the mean of this random variable; that is, calculate $\mu = 1(\frac{1}{2}) + 2(\frac{1}{4}) + 3(\frac{1}{8}) + \cdots$.

Solution From our knowledge of power series (Section 15.4), we recall that

$$\frac{x}{1 - x} = x + x^2 + x^3 + x^4 + \cdots \qquad (-1 < x < 1)$$

and substituting $x = \frac{1}{2}$ gives the result in part (a). Next differentiate both sides of the above equality in x to obtain

$$\frac{1}{(1 - x)^2} = 1 + 2x + 3x^2 + 4x^3 + \cdots$$

Then multiply both sides of this result by x to get

$$\frac{x}{(1 - x)^2} = x + 2x^2 + 3x^3 + 4x^4 + \cdots$$

Finally substitute $x = \frac{1}{2}$, giving 2 as the required value of μ. $\blacksquare$

SHARP CURVE

Calculation of the mean of a discrete random variable with infinitely many values is a nontrivial task, since it involves summing an infinite series. We used a trick to do this in Example G. We don't mean to imply that it is easy to think of tricks like this or that there is a trick for every problem.

Problem Set 16.1

1. An experiment consists of tossing a fair die and a fair coin. Typical outcomes are (4, H) and (2, T).
   (a) Describe the sample space by listing the 12 possible outcomes.
   (b) Calculate prob(A), where A is the event of getting an even number and a head.
   (c) Calculate prob(B), where B is the event of getting a number less than 5 and a tail.

2. An experiment consists of tossing three fair coins. Typical outcomes are THT and HTT.
   (a) Describe the sample space by listing all possible outcomes.
   (b) Calculate prob(A), where A is the event of getting exactly two heads.
   (c) Calculate prob(B), where B is the event of getting more tails than heads.

3. An experiment consists of tossing four fair coins leading to a sample space with 16 possible outcomes of which HTHH and HHHT are examples. Let X be the random variable which counts the number of heads. Find prob$\{X = 3\}$ and prob$\{X \leq 1\}$.

4. An experiment consists of tossing two regular tetrahedra (with four identical triangular faces as in Figure 4). Suppose further that the faces have 1, 2, 3, and 4 dots, respectively. The sample space for this experiment has 16 outcomes of which (2, 4) and (3, 1) are typical examples, these corresponding to the numbers of dots on the faces that are down. Let X denote the sum of these two numbers. Find prob$\{X = 4\}$ and prob$\{X > 4\}$.

FIGURE 4

5. Draw the bar graph for the probability function corresponding to the random variable X in Problem 3.

6. Draw the bar graph for the probability function corresponding to the random variable X in Problem 4.

7. Find the mean of the random variable X in Problem 3. See Example D.

8. Find the mean of the random variable X in Problem 4.

9. Find the variance of the random variable X in Problem 3. See Example E.

10. Find the variance of the random variable X in Problem 4.

11. Aaron is to toss a fair die. He will receive $5 for a six, $2 for a four or five; otherwise, he must pay $1. What are his expected winnings? *Hint:* See the discussion following Example D.

12. You are to draw 1 ball from an urn containing 6 red, 2 blue, and 12 white balls and are to receive $3 if it is red and $10 if it is blue, but you must pay $2 if it is white. What are your expected winnings?

13. A die has been loaded in such a way that the numbers 1, 2, 3, 4, 5, and 6 come up 10%, 10%, 10%, 20%, 20%, and 30% of the time, respectively. Let X correspond to the number that shows when the die is tossed. See Example F.
    (a) Find prob$\{X > 4\}$ and prob$\{2 < X < 6\}$.
    (b) Find the mean and variance of X.

14. A regular octahedron with eight identical triangular faces (Figure 5) numbered 1, 2, . . . , 8 has been loaded so that these numbers come up 6%, 8%, 8%, 10%, 15%, 15%, 18%, and 20% of the time, respectively. Let X correspond to the number that shows on the top face when the octahedron is tossed.
    (a) Find prob$\{X > 4\}$ and prob$\{2 < X < 7\}$.
    (b) Find the mean and variance of X.

FIGURE 5

15. An experiment consists of tossing three fair dice, leading to a sample space with $6 \cdot 6 \cdot 6 = 216$ outcomes, a typical one being (1, 5, 2). Let X count the number of dots that show. Calculate prob$\{X = 3\}$, prob$\{X = 4\}$, and prob$\{X > 4\}$.

16. An experiment consists of choosing at random a shortest polygonal path from A to B in Figure 6. An outcome (shortest path) can be labeled by a sequence of three E's and three N's, corresponding to three

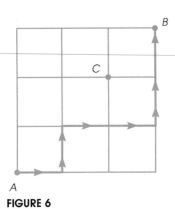

FIGURE 6

steps east and three steps north in some order. For example, the path with the arrows would be labeled ENEENN.
(a) List the 20 possible outcomes.
(b) Find the probability that such a path will go through C.

17. Consider the random variable $\mathbf{X}$ which takes on the values $1, 2, 3, \ldots,$ with probabilities $\frac{2}{3}, \frac{2}{9}, \frac{2}{27}, \ldots,$ respectively. See Example G.
(a) Show that this is a valid probability distribution; that is, show that $\frac{2}{3} + \frac{2}{9} + \frac{2}{27} + \cdots = 1$.
(b) Find the mean of $\mathbf{X}$.

18. A Poisson random variable $\mathbf{X}$ has the probability function $f_i = e^{-p}p^i/i!$ for $i = 0, 1, 2, 3, \ldots$. Here, p is a positive constant, $i!$ denotes i factorial, and $0! = 1$.

(a) Show that f_i is a probability function; that is, show

$$\sum_{i=0}^{\infty} \frac{e^{-p}p^i}{i!} = 1$$

(b) Find the mean of $\mathbf{X}$.

19. Find the probability that a number drawn at random from the integers $1, 2, 3, \ldots, 100$ will (a) be divisible by 3, (b) have a 9 in its decimal representation, (c) have the sum of its digits greater than 9.

20. An experiment consists of tossing a fair die and a fair tetrahedron (Figure 4). We keep track of outcomes by listing the number of dots on the faces that are down. For example, $(5, 2)$ labels the outcome corresponding to the die having face 5 down and tetrahedron face 2 down. Let $\mathbf{X}$ count the total number of dots down; thus, $\mathbf{X} = 7$ for the outcome $(5, 2)$.
(a) Calculate prob$\{\mathbf{X} > 8\}$.
(b) Calculate the mean of $\mathbf{X}$.
(c) Calculate the variance of $\mathbf{X}$.

21. An experiment consists of tossing 5 coins, leading to 32 possible outcomes of which HHTHT is a typical outcome. Let $\mathbf{X}$ count the number of heads that appear. Find the mean and variance of $\mathbf{X}$.

22. A fair coin is tossed over and over. Let $\mathbf{X} = i$ if the first head occurs on the ith toss, $i = 1, 2, 3, \ldots$. Determine $f_i = $ prob$\{\mathbf{X} = i\}$ and then find the mean of $\mathbf{X}$.

16.2
CONTINUOUS DISTRIBUTIONS

We turn now to the part of probability theory that is most intimately related to calculus, the study of continuous distributions. To introduce the idea, consider a simple spinner like the one shown in Figure 7 which has a real number scale from 0 to 10 wrapped around its circumference. If the pointer is spun, it can (theoretically) stop at any real number on the interval 0 to 10. The probability of its stopping on any two subintervals of equal length should be the same. A bit of thought suggests that the probability of its stopping on any interval of length L should be $L/10$. Consider the function $f(x) = \frac{1}{10}$ with its domain the interval $[0, 10]$; its graph is shown in Figure 8. The probability that the pointer will stop on the interval $[c, d]$ is the area under the graph between c and d. Thus, if $\mathbf{X}$ denotes the stopping point,

$$\text{prob}\{c \leq \mathbf{X} \leq d\} = \int_c^d f(x)\, dx = \int_c^d \frac{1}{10}\, dx = \frac{d - c}{10}$$

We make several observations about this experiment. The sample space is the interval $[0, 10]$. The random variable $\mathbf{X}$ can take on any value in this in-

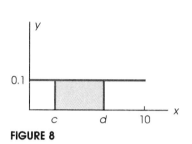

FIGURE 7

FIGURE 8

terval. The probability of a typical event $\{c \le \mathbf{X} \le d\}$ is given by an integral rather than by a sum as in the discrete case. Finally, note that the probability of any single outcome (stopping at a specific point) is 0, in marked contrast to the discrete case. This example leads us to formulate some new definitions.

Continuous Random Variables

A random variable whose values lie on an interval of the real line is called a **continuous random variable.** Associated with a continuous random variable $\mathbf{X}$ is a function called its **probability density function** with the property that

$$\text{prob}\{c \le \mathbf{X} \le d\} = \int_c^d f(x)\, dx$$

We shall assume that a probability density function f has the whole real line as its domain, $f(x)$ being 0 outside the interval on which $\mathbf{X}$ has its values. In addition it must satisfy two properties.

1. It must be nonnegative, that is, $f(x) \ge 0$.
2. The area under its graph must be 1, that is $\int_{-\infty}^{\infty} f(x)\, dx = 1$.

(The corresponding conditions in the discrete case were that $f_i > 0$ and $\Sigma f_i = 1$.) Any function that satisfies the above two properties can serve as a probability density function.

EXAMPLE A

Show that the function $f(x) = \frac{1}{2} \sin x$ on the interval $[0, \pi]$ and $f(x) = 0$ elsewhere can serve as a probability density function (Figure 9).

Solution Clearly $f(x)$ is nonnegative. Moreover,

$$\int_{-\infty}^{\infty} f(x)\, dx = \int_0^{\pi} \frac{1}{2} \sin x\, dx = \left[-\frac{1}{2} \cos x \right]_0^{\pi} = 1 \qquad \blacksquare$$

FIGURE 9

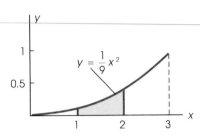

FIGURE 10

EXAMPLE B

Let $f(x) = kx^2$ on the interval $[0, 3]$ and $f(x) = 0$ elsewhere. Determine k so that f is a probability density function and display the corresponding graph. If X is a random variable with this density function, calculate prob$\{1 \leq X \leq 2\}$.

Solution We require that

$$1 = \int_0^3 kx^2 \, dx = \left[\frac{kx^3}{3}\right]_0^3 = 9k$$

so $k = \frac{1}{9}$. The resulting function is graphed in Figure 10. Also,

$$\text{prob}\{1 \leq X \leq 2\} = \int_1^2 \frac{x^2}{9} \, dx = \left[\frac{x^3}{27}\right]_1^2 = \frac{7}{27}$$ ∎

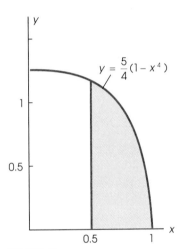

FIGURE 11

EXAMPLE C

The ABC Clinic claims that the waiting time X (in hours) between a scheduled appointment and actually seeing a doctor has the probability density function $f(x) = \frac{5}{4}(1 - x^4)$ on the interval $[0, 1]$ and $f(x) = 0$ elsewhere. Show that $f(x)$ is a probability density function, draw its graph, and calculate the probability of having to wait more than 30 minutes.

Solution Clearly $f(x)$ is nonnegative. Moreover,

$$\int_0^1 \frac{5}{4}(1 - x^4) \, dx = \frac{5}{4}\left[x - \frac{1}{5}x^5\right]_0^1 = \frac{5}{4}\left(1 - \frac{1}{5}\right) = 1$$

The graph is shown in Figure 11. Also,

$$\text{prob}\left\{X > \frac{1}{2}\right\} = \int_{1/2}^1 \frac{5}{4}(1 - x^4) \, dx = \frac{5}{4}\left[x - \frac{1}{5}x^5\right]_{1/2}^1 = 0.38$$ ∎

The Mean and Variance of a Continuous Random Variable

If you refer back to the definitions of the mean and the variance for a discrete random variable in Section 16.1, you should not be too surprised at the definitions we now give for the mean and variance of a continuous random variable. Let X have density function $f(x)$. Then its **mean** μ and **variance** σ^2 are given by

$$\mu = \int_{-\infty}^\infty xf(x) \, dx$$

and

$$\sigma^2 = \int_{-\infty}^{\infty} (x - \mu)^2 f(x) \, dx$$

Just as in the discrete case, there is an alternate and better formula for σ^2. Moreover, the justification for the new formula is the same as in the discrete case, with integrals replacing sums.

$$\sigma^2 = \int_{-\infty}^{\infty} x^2 f(x) \, dx - \mu^2$$

EXAMPLE D

Let **X** be *uniformly distributed* on the interval [0, 10]. This simply means that **X** has the constant density function $f(x) = \frac{1}{10}$ on [0, 10], as in the spinner problem with which we began this section (Figures 7 and 8). Evaluate the means and variance of **X**.

Solution If you think of the mean as the balance point of the distribution, you will not be surprised when we conjecture $\mu = 5$. But to make sure, we do the required integration.

$$\mu = \int_0^{10} x \frac{1}{10} \, dx = \left[\frac{x^2}{20} \right]_0^{10} = 5$$

Also

$$\sigma^2 = \int_0^{10} x^2 \frac{1}{10} \, dx - 25 = \left[\frac{x^3}{30} \right]_0^{10} - 25 = 8.33 \qquad \blacksquare$$

EXAMPLE E

Find the mean and variance for the random variable **X** with probability density function $f(x) = x^2/9$ on [0, 3] (see Example B and Figure 10).

Solution One look at the graph in Figure 10 should convince you that the mean (balance point) is greater than 1.5. In fact,

$$\mu = \int_0^3 x \frac{x^2}{9} \, dx = \left[\frac{x^4}{36} \right]_0^3 = \frac{9}{4} = 2.25$$

and

$$\sigma^2 = \int_0^3 x^2 \frac{x^2}{9} \, dx - (2.25)^2 = \left[\frac{x^5}{45} \right]_0^3 - (2.25)^2 = 0.3375 \qquad \blacksquare$$

EXAMPLE F

Find the mean and variance for the waiting time random variable in Example C.

Solution

$$\mu = \int_0^1 x \frac{5}{4}(1 - x^4) \, dx = \frac{5}{4}\left[\frac{x^2}{2} - \frac{x^2}{6}\right]_0^1 = \frac{5}{12}$$

Thus the mean waiting time is $\frac{5}{12}$ hour or 25 minutes. Finally

$$\sigma^2 = \int_0^1 x^2 \frac{5}{4}(1 - x^4) \, dx - \left(\frac{5}{12}\right)^2$$

$$= \frac{5}{4}\left[\frac{x^3}{3} - \frac{x^7}{7}\right]_0^1 - \frac{25}{144} \approx 0.0645 \qquad \blacksquare$$

The Standard Deviation

The **standard deviation** of a random variable X is defined to be the positive square root of its variance and is denoted by the symbol σ. Because it has the same units as X, it is more useful than the variance for some purposes. For example, in Example F the standard deviation of waiting times is $\sigma = \sqrt{0.0645} \approx 0.254$ hours or about 15.2 minutes. The standard deviation will play an important role in the next section where we will also be able to give it a more intuitive meaning.

Problem Set 16.2

1. Show that $f(x) = \frac{3}{14}(x^2 + 1)$ on the interval [0, 2] and $f(x) = 0$ elsewhere can serve as a probability density function for a random variable X. See Example A.

2. Show that $f(x) = 4(x - x^3)$ on [0, 1] and $f(x) = 0$ elsewhere can serve as a probability density function for a random variable X.

In Problems 3–6, determine the value of k so that f is a probability function. In each case, assume that $f(x) = 0$ outside the indicated interval. See Example B.

3. $f(x) = k(x^2 - 2x)$; [3, 4]
4. $f(x) = k(3x - x^2)$; [0, 3]
5. $f(x) = kx^3$; [1, 3]   6. $f(x) = k\sqrt{x}$; [0, 4]
7. Draw the graph of the probability function in Problem 1 and then calculate prob[$X \geq 1.5$].

8. Draw the graph of the probability function in Problem 2 and then calculate prob[$0.6 \leq X \leq 0.8$].

9. The West Side Dental Clinic estimates that the waiting time X (in hours) between the scheduled appointment and actually seeing a dentist has the probability density function $f(x) = \frac{4}{3}(1 - x^3)$ on the interval [0, 1] and $f(x) = 0$ elsewhere. Calculate the probability that a patient will have to wait more than 45 minutes. See Example C.

10. A psychologist has found that the time T (in minutes) that a 10-year old child needs to solve a certain puzzle has probability density function $f(t) = 32/[15(1 + t)^3]$ on [0, 3] and $f(t) = 0$ elsewhere. Find the probability that it will take a child no more than 2 minutes to solve the puzzle.

11. Let X be uniformly distributed on the interval [0, 6].
    (a) Write the probability density function for X.
    (b) Find the mean and variance for X.

12. Let **X** be uniformly distributed on the interval $[1, 4]$.
    (a) Write the probability density function for **X**.
    (b) Find the mean and variance for **X**.

13. Find the mean and the variance for **X** in Problem 1.

14. Find the mean and the variance for **X** in Problem 2.

15. Find the mean waiting time in Problem 9.

16. Find the mean puzzle-solving time in Problem 10.

---

17. Find the value of k so that $f(x) = k \ln x$ on the interval $[1, e^2]$ and $f(x) = 0$ elsewhere is a probability density function.

18. Consider the spinner in Figure 7 and let **Y** be the natural logarithm of the number where the pointer stops. Find prob$[0.6 < \mathbf{Y} < 0.7]$.

19. A certain fast food restaurant guarantees that no one will have to wait more than 6 minutes to be served from its luncheon menu and estimates the waiting time to have probability density function $f(t) = (36 - t^2)/144$ for the time interval $[0, 6]$.
    (a) Find the probability that a customer will have to wait more than 2 minutes.
    (b) Find the mean waiting time.

20. A fast food franchiser claims that the fraction **X** of new restaurants it sponsors that are still in business after 10 years has the probability density function

$f(x) = 20x^3(1 - x)$ for $0 \le x \le 1$. Find the probability that at least 70% of its restaurants will still be in business after 10 years.

21. John Nelson arrives home for dinner at a time **T** that has the density function graphed in Figure 12. What is his average (mean) arrival time?

FIGURE 12

22. A dart is thrown at random at a circular dart board of radius 10 inches and always hits the board. Let **X** denote the distance from the center where it lands.
    (a) Find the probability density function for **X** assuming that "at random" means regions of equal area have equal probability of being hit.
    (b) On the average, how far from the center will a dart land?

16.3

THE EXPONENTIAL AND NORMAL DISTRIBUTIONS

We consider now two distributions that have been used to model many real-world phenomena. Both are continuous distributions which range over infinite intervals. This means that the integrals that will arise in this section will be improper integrals, a subject discussed in Section 9.4.

The Exponential Distribution

Let **X** measure the time between random events (for example, the time between telephone calls arriving at the switchboard in an office). Note that **X** can have a very small value but can also be arbitrarily large. Experience suggests that **X** can be treated as a random variable which is **exponentially distributed.** This means that its probability density function has the form

$$f(x) = ke^{-kx}$$

on $[0, \infty)$. Here k is a positive constant which depends on the particular type of phenomenon being studied. The graph of a typical exponential density

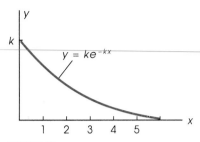

FIGURE 13

function is shown in Figure 13. To establish that it really is a density function requires that we show that the area under its graph is 1. This in turn involves showing that the improper integral $\int_0^\infty ke^{-kx}\,dx = 1$. Now,

$$\int_0^\infty ke^{-kx}\,dx = \lim_{b \to \infty} -\int_0^b -ke^{-kx}\,dx = \lim_{b \to \infty} \left[-e^{-kx}\right]_0^b$$

$$= \lim_{b \to \infty} (-e^{-kb} + e^0) = 0 + 1 = 1$$

We will need to know the mean and standard deviation of an exponentially distributed random variable.

THEOREM

A random variable $\mathbf{X}$ with probability density function $f(x) = ke^{-kx}$ on $[0, \infty)$ has mean $\mu = 1/k$ and standard deviation $\sigma = 1/k$.

Proof We use integration by parts.

$$\mu = \int_0^\infty x(ke^{-kx})\,dx = \lim_{b \to \infty} -\int_0^b x(-ke^{-kx})\,dx$$

Now let $u = x$ and $dv = -ke^{-kx}\,dx$, so that $du = dx$ and $v = e^{-kx}$. We obtain

$$\mu = \lim_{b \to \infty} \left\{ -\left[xe^{-kx}\right]_0^b + \int_0^b e^{-kx}\,dx \right\}$$

$$= \lim_{b \to \infty} -be^{-kb} + 0 + \lim_{b \to \infty} -\frac{1}{k}\int_0^b (-ke^{-kx})\,dx$$

$$= 0 + 0 + \lim_{b \to \infty} \left[-\frac{1}{k}e^{-kx}\right]_0^b = \frac{1}{k}$$

Finding σ involves evaluating $\int_0^\infty x^2(ke^{-kx})\,dx - 1/k^2$ and then taking its square root. This is a slightly more complicated integration-by-parts problem; we omit the details. ∎

EXAMPLE A

The mean time between the arrival of telephone calls at the UVW switchboard is 5 minutes. Find the probability that two calls will arrive within 3 minutes of each other.

Solution We assume that $\mathbf{X}$, the time between calls, is exponentially distributed with $\mu = 1/k = 5$. Thus, $f(x) = 0.2e^{-0.2x}$ on $[0, \infty)$, the function graphed in Figure 14. We must calculate the area under this graph between 0 and 3.

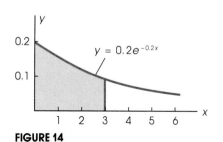

FIGURE 14

$$\text{prob}\{0 \leq \mathbf{X} \leq 3\} = \int_0^3 0.2e^{-0.2x}\, dx = \left[-e^{-0.2x} \right]_0^3$$

$$= 1 - e^{-0.6} \approx 0.4512 \qquad \blacksquare$$

In addition to modeling the time between random events, the exponential distribution is often used to model the time one must wait for service (say at a checkout counter) and the useful lifetime of parts of a machine (for example, a transistor in a radio). The problem set will explore these applications.

The Normal Distribution

Certainly, the most important of all continuous distributions is the normal distribution. Occasionally, you will hear people say that a random variable **X** is normally distributed if the graph of its probability density function is a bell-shaped curve. For calculus students, we can be much more precise. We say that **X** is **normally distributed** if its probability density function has the form

$$f(x) = \frac{1}{\sigma\sqrt{2\pi}} e^{-(1/2)[(x-\mu)/\sigma]^2}$$

This density function has two parameters (constants) μ and σ, which may be chosen arbitrarily though σ must be positive. As the notation suggests, these constants turn out to be the mean and standard deviation, respectively, a fact that we now state as a theorem without proof. (The proof is difficult and beyond the level of this course.)

THEOREM

A random variable **X** with the two-parameter normal probability density function defined above on $(-\infty, \infty)$ has mean μ and standard deviation σ.

The graph of such a density function is a bell-shaped curve centered at μ and with inflection points at $\mu \pm \sigma$. We have shown three such curves on the left in Figure 15 illustrating how changing σ affects the shape of the curve. On the right in Figure 15, we show the fraction of the area (probability) that lies in each of several intervals of length σ. For example, the probability that a normally distributed variable will have a value within one standard deviation of the mean is 0.6826, that it will have a value within two standard deviations of the mean is 0.9544. While this last sentence is accurate only for normally distributed variables, it is often used as a "rule of thumb" for other variables, thereby giving us an intuitive feeling for the significance of the standard deviation.

FIGURE 15

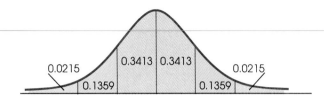

EXAMPLE B

Let $\mathbf{Z}$ have the **standard normal** distribution, meaning that $\mathbf{Z}$ is normally distributed with $\mu = 0$ and $\sigma = 1$. Calculate prob$\{0 \le \mathbf{Z} \le 1.5\}$.

Solution

$$\text{prob}\{0 \le \mathbf{Z} \le 1.5\} = \int_0^{1.5} \frac{1}{\sqrt{2\pi}} e^{-x^2/2} \, dx$$

FIGURE 16

Unfortunately, this integral cannot be evaluated using the Fundamental Theorem of Calculus because $e^{-x^2/2}$ does not have an elementary antiderivative. The integral could be evaluated by the numerical methods in Section 14.1. However, definite integrals of the standard normal density function occur so often that extensive tables have been prepared. One such table appears as Table E in the appendix. Note that this table gives the value of the shaded area shown in Figure 16 for various values of z. We find 0.4332 as the required value. ∎

EXAMPLE C

Suppose that $\mathbf{Z}$ is a standard normal variable ($\mu = 0$, $\sigma = 1$). Evaluate prob$\{0.5 \le \mathbf{Z} \le 1.2\}$ and prob$\{\mathbf{Z} \ge -1.3\}$.

FIGURE 17

Solution We note that

$$\text{prob}\{0.5 \le \mathbf{Z} \le 1.2\} = \text{prob}\{0 \le \mathbf{Z} \le 1.2\} - \text{prob}\{0 \le \mathbf{Z} \le 0.5\}$$

Both of the latter probabilities are in Table E. We get $0.3849 - 0.1915 = 0.1934$ as our answer.

For the second problem, refer to Figure 17 and use the symmetry of the normal curve to conclude that

$$\text{prob}\{\mathbf{Z} \ge -1.3\} = \text{prob}\{0 \le \mathbf{Z} \le 1.3\} + \text{prob}\{\mathbf{Z} \ge 0\}$$
$$= 0.4032 + 0.5 = 0.9032 \qquad ∎$$

But what if we have a nonstandard normal variable? Can we still use Table E to calculate probabilities in this case? The answer is yes, provided we first transform our variable to a standard variable (sometimes called a **Z**-variable). To explain, suppose **X** is normal with mean μ and standard deviation σ. Then it can be shown that $\mathbf{Z} = (\mathbf{X} - \mu)/\sigma$ is normal with mean 0 and standard deviation 1. Here is how we use this fact.

EXAMPLE D

Let **X** be normally distributed with mean $\mu = 2$ and standard deviation $\sigma = 3$. Calculate prob$\{3.5 \le \mathbf{X} \le 5.6\}$.

Solution We rewrite the double inequality by subtracting 2 and dividing by 3 as follows.

$$\text{prob}\{3.5 \le \mathbf{X} \le 5.6\} = \text{prob}\left\{\frac{3.5 - 2}{3} \le \frac{\mathbf{X} - 2}{3} \le \frac{5.6 - 2}{3}\right\}$$

$$= \text{prob}\{0.5 \le \mathbf{Z} \le 1.2\} = 0.1934$$

In the last step, we used a calculation already made in Example C. ∎

Applications of the Normal Distribution

Many random variables that occur in the real world are at least approximately normally distributed. It is claimed by some that grades should be normally distributed. Heights and weights of adult males (or females) are thought to be normally distributed. If a quantity (say the diameter of a shaft or the wavelength of a signal) is repeatedly measured, the numbers obtained will be approximately normally distributed. A machine that is supposed to package sugar in 1-pound boxes will actually produce boxes of varying weights but normally distributed about 1.

EXAMPLE E

Assume that IQ scores for adults are normally distributed with mean 100 and standard deviation 10. If genius level is defined to be 135 and above, what proportion of the adult population is of genius level?

Solution Let **X** denote IQ score. Then

$$\text{prob}\{\mathbf{X} \ge 135\} = \text{prob}\left\{\frac{\mathbf{X} - 100}{10} \ge \frac{135 - 100}{10}\right\}$$

$$= \text{prob}\{\mathbf{Z} \ge 3.5\} = 0.5 - \text{prob}\{0 \le \mathbf{Z} \le 3.5\}$$

$$= 0.5 - 0.4998 = 0.0002$$ ∎

EXAMPLE F

A bakery that makes 1-pound loaves of bread actually produces loaves of varying weights but with mean 1 and standard deviation 0.015. Assuming that loaf weight X is normally distributed, find the probability that a customer will get a loaf weighing less than 0.99 pound.

Solution

$$\text{prob}\{X < 0.99\} = \text{prob}\left\{\frac{X - 1}{0.015} < \frac{0.99 - 1}{0.015}\right\} = \text{prob}\{Z < -0.67\}$$

$$= \text{prob}\{Z > 0.67\} = 0.5 - \text{prob}\{0 \le Z \le 0.67\}$$

$$= 0.5 - 0.2486 = 0.2514$$

So a customer will get a loaf weighing less than 0.99 pound about one-fourth of the time. Of course, it is also true that this customer would get a loaf weighing more than 1.01 pounds about one-fourth of the time.   ∎

Problem Set 16.3

1. Let X have the exponential density function $f(x) = 0.3e^{-0.3x}$ on $[0, \infty)$. Calculate (a) prob$\{\frac{10}{3} \le X \le \frac{20}{3}\}$ and (b) prob$\{X \ge 5\}$.

2. Let Y have the exponential density function $f(y) = 0.02e^{-0.02y}$ on $[0, \infty)$. Find (a) prob$\{50 \le Y \le 100\}$ and (b) prob$\{Y \ge 150\}$.

3. The time T in minutes between two consecutive cars pulling up to a certain toll booth is exponentially distributed with the mean time being $\frac{1}{2}$ minute. What fraction of successive cars will be separated by more than 1 minute? See Example A.

4. The time T in minutes that a customer stands in line at the Honey Spot Cafeteria before reaching the food counter is exponentially distributed. If the mean waiting time is 5 minutes, what is the probability that a person will have to stand in line more than 15 minutes?

5. Let T correspond to the time in minutes it takes a rat to find its way through a maze and suppose that T is exponentially distributed with mean 4 minutes. Calculate the probability that a rat will require less than 6 minutes to traverse the maze.

6. An insurance company estimates that the lifetime T in years of a certain 1990 model automobile has the probability density function $f(t) = \frac{1}{6}e^{-t/6}$ on $[0, \infty)$. What fraction of these cars will still be in use 10 years after purchase?

7. Let Z be a standard normal variable ($\mu = 0$, $\sigma = 1$). Use Table E to calculate prob$\{0 \le Z \le 1.7\}$. See Example B.

8. If Z is a standard normal variable, calculate prob$\{0 \le Z \le 2.3\}$.

9. If Z is a standard normal variable, calculate prob$\{-0.5 \le Z \le 1.5\}$ and prob$\{Z \ge 1.5\}$. See Example C.

10. If Z is a standard normal variable, calculate prob$\{Z > -0.3\}$ and prob$\{-0.3 < Z \le 1.2\}$.

11. Let X be normally distributed with $\mu = 1$ and $\sigma = 2$. Calculate prob$\{2.6 \le X \le 3.8\}$ and prob$\{X \ge 0.2\}$. See Example D.

12. Let Y be normally distributed with $\mu = 0.08$ and $\sigma = 0.02$. Calculate prob$\{0.05 \le Y \le 0.12\}$ and prob$\{Y \le 0.07\}$.

13. A league bowler has found that his bowling scores over a 5-year period were approximately normally distributed with mean 180 and standard deviation 20. What fraction of his scores were between 170 and 210 inclusive? See Example E.

14. The scores on a certain psychological test are said to be normally distributed with mean 65 and standard deviation 10. What fraction of the scores are between 51 and 88?

15. The amount in ounces of soda in a machine-filled 12-ounce bottle is claimed to be normally distributed with mean $\mu = 12.1$ and standard deviation $\sigma = 0.2$. If so, what fraction of the bottles contain at least 12 ounces?

16. The actual number of ounces of pancake mix in a certain brand of machine-packed 12-ounce boxes is said to be normally distributed with mean $\mu = 12.1$ and standard deviation $\sigma = 0.3$. What fraction of the boxes contain at least 11.9 ounces?

17. Let $\mathbf{X}$ be normally distributed with mean $\mu = 195$ and standard deviation $\sigma = 20$. Evaluate (a) prob$\{\mathbf{X} = 195\}$, (b) prob$\{\mathbf{X} \le 225\}$, and (c) prob$\{145 \le \mathbf{X} \le 245\}$.

18. Based on your knowledge of the normal distribution, evaluate each of the following integrals.

(a) $\dfrac{1}{\sqrt{2\pi}} \displaystyle\int_{-1}^{2} e^{-x^2/2} \, dx$ (b) $\dfrac{1}{\sqrt{2\pi}} \displaystyle\int_{-\infty}^{\infty} xe^{-x^2/2} \, dx$

(c) $\dfrac{1}{\sqrt{2\pi}} \displaystyle\int_{-\infty}^{\infty} x^2 e^{-x/2} \, dx$

(d) $\dfrac{1}{2\sqrt{2\pi}} \displaystyle\int_{-\infty}^{\infty} (x-1)^2 e^{-(x-1)^2/8} \, dx$

19. At a busy intersection in a large city, the number of days between two successive accidents is exponentially distributed with a mean of 9.5 days. Find the probability that two successive accidents will occur within 4 days.

20. To meet government specifications, a gizmo must measure 10 ± 0.05 centimeters long. A certain company has a machine that will produce a normal distribution of gizmos with mean 10 and standard deviation 0.030. If it bids on a contract to make gizmos, what fraction of its output should it expect to discard?

21. The daily number of pieces of mail handled by a certain post office is normally distributed with $\mu = 12,500$ and $\sigma = 500$. Find the probability of needing to handle more than 13,200 pieces on a randomly selected day.

22. A random variable $\mathbf{X}$ has density function $f(x) = cxe^{-2x}$ on $[0, \infty)$ and $f(x) = 0$ elsewhere. Determine (a) the constant c and (b) the mean of $\mathbf{X}$.

23. The *median* of a continuous random variable with probability density function $f(x)$ is the number m such that $\int_{-\infty}^{m} f(x) \, dx = 0.5$. Find the median of an exponentially distributed random variable with mean 4.

16.4*
CONTINUOUS BIVARIATE DISTRIBUTIONS

Often a scientist wants to study two random variables $\mathbf{X}$ and $\mathbf{Y}$ at the same time. For example, one might want to study height and weight for people or SAT scores and high school rank for high school seniors. If both variables are continuous, then there is a nonnegative function $f(x, y)$ called the **joint probability density function** satisfying

$$\int_{-\infty}^{\infty}\int_{-\infty}^{\infty} f(x, y) \, dy \, dx = 1$$

Moreover, if E is a set in the xy-plane, then

$$\text{prob}\{(\mathbf{X}, \mathbf{Y}) \text{ is in } E\} = \iint_{E} f(x, y) \, dy \, dx$$

Examples of Joint Density Functions

By analogy with the single variable case, we say that $(\mathbf{X}, \mathbf{Y})$ is **uniformly distributed** on D if its joint probability density function is constant on D.

EXAMPLE A

Let $D = \{(x, y): -5 \le x \le 5, -3 \le y \le 3\}$ and suppose $(\mathbf{X}, \mathbf{Y})$ is uniformly distributed on D. Determine the joint density function $f(x, y)$ and calculate prob$\{\mathbf{X}^2 + \mathbf{Y}^2 \le 6\}$.

Solution Before making any calculations, let us note an obvious fact that has not been mentioned explicitly before, namely, that the double integral of the constant 1 over a set D is just the area $A(D)$ of D. Now since $f(x, y)$ is a constant (say k) on D, and since we require that $\int_D\!\int k\,dy\,dx = 1$, we must have

$$1 = \iint_D k\,dy\,dx = k\iint_D 1\,dy\,dx = kA(D) = k(10)(6) = 60k$$

This implies $k = \frac{1}{60}$, and so $f(x, y) = \frac{1}{60}$ on D and $f(x, y) = 0$ elsewhere.

To answer the second question, we note that $x^2 + y^2 \le 6$ corresponds to the interior and boundary of a circle of radius $\sqrt{6}$ centered at the origin (Figure 18). Calling this set C, we have

$$\text{prob}\{\mathbf{X}^2 + \mathbf{Y}^2 \le 6\} = \iint_C \frac{1}{60}\,dy\,dx = \left(\frac{1}{60}\right)A(C) = \left(\frac{1}{60}\right)(6\pi) = \frac{\pi}{10} \quad\blacksquare$$

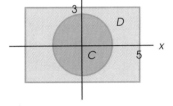

FIGURE 18

EXAMPLE B

Suppose that $f(x, y) = \frac{1}{2}$ on the triangle D with vertices at $(0, 0)$, $(2, 0)$, and $(0, 2)$ and $f(x, y) = 0$ elsewhere. [Note that $A(D) = 2$, so $f(x, y)$ is a joint density function.] Calculate (a) $\text{prob}\{\mathbf{Y} \ge 1\}$ and (b) $\text{prob}\{\mathbf{X} \ge \mathbf{Y}\}$.

Solution The sets E and F over which we must integrate for parts (a) and (b), respectively, are shown in Figure 19. We obtain

$$\text{(a)} \quad \text{prob}\{\mathbf{Y} \ge 1\} = \iint_E \frac{1}{2}\,dy\,dx = \frac{1}{2}A(E) = \frac{1}{2}\left(\frac{1}{2}\right) = \frac{1}{4}$$

$$\text{(b)} \quad \text{prob}\{\mathbf{X} \ge \mathbf{Y}\} = \iint_F \frac{1}{2}\,dy\,dx = \frac{1}{2}A(F) = \frac{1}{2}(1) = \frac{1}{2} \qquad\blacksquare$$

EXAMPLE C

Suppose that $f(x, y) = kx^2y$ on the triangle D in Example B and is 0 elsewhere. Determine k and calculate $\text{prob}\{\mathbf{X} \ge \mathbf{Y}\}$.

Solution D is the large triangle in Figure 19. To determine k, we write

$$1 = \iint_D kx^2y\,dy\,dx = k\int_0^2 \left[\int_0^{2-x} x^2y\,dy\right]dx = k\int_0^2 \left[\frac{1}{2}x^2y^2\right]_0^{2-x}dx$$

$$= k\int_0^2 \frac{1}{2}x^2(2-x)^2\,dx = k\int_0^2 \left(\frac{1}{2}x^4 - 2x^3 + 2x^2\right)dx$$

$$= k\left[\frac{1}{10}x^5 - \frac{1}{2}x^4 + \frac{2}{3}x^3\right]_0^2 = \frac{8}{15}k$$

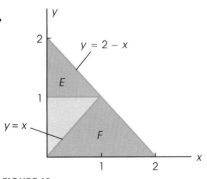

FIGURE 19

Thus, $k = \frac{15}{8}$.

To calculate prob$\{X \geq Y\}$, we must integrate $f(x, y) = \frac{15}{8}x^2y$ over the set F in Figure 19. It should be clear that it is preferable to do the x-integration first (with y fixed, x goes from $x = y$ to $x = 2 - y$)

$$\text{prob}\{X \geq Y\} = \iint_F \frac{15}{8}x^2y \, dy \, dx = \frac{15}{8} \int_0^1 \left[\int_y^{2-y} x^2y \, dx \right] dy$$

$$= \frac{15}{24} \int_0^1 \left[x^3y \right]_y^{2-y} dy = \frac{5}{8} \int_0^1 \left[(2 - y)^3y - y^4 \right] dy$$

$$= \frac{5}{8} \int_0^1 (8y - 12y^2 + 6y^3 - 2y^4) \, dy$$

$$= \frac{5}{8} \left[4y^2 - 4y^3 + \frac{3}{2}y^4 - \frac{2}{5}y^5 \right]_0^1 = \frac{11}{16} \qquad \blacksquare$$

EXAMPLE D

A key electronic part in an airplane is backed up by a similar second part made by another manufacturer. The two parts operate simultaneously, but one will keep the system working. Studies have shown that the life spans X and Y (in years) of the two parts are exponentially distributed, with $k = 0.3$ and $k = 0.2$, respectively. This means that the joint probability density of (X, Y) is $f(x, y) = (0.3e^{-0.3x})(0.2e^{-0.2y})$, $x \geq 0$, $y \geq 0$. Find the probability that an airplane will have a complete failure in this system during the first 2 years of service.

Solution We must calculate prob$\{0 \leq X \leq 2, 0 \leq Y \leq 2\}$. This is equal to

$$\int_0^2 \left[\int_0^2 (0.3e^{-0.3x})(0.2e^{-0.2y}) \, dy \right] dx$$

$$= \int_0^2 (0.3e^{-0.3x}) \left[-e^{-0.2y} \right]_0^2 dx$$

$$= (-e^{-0.4} + 1) \int_0^2 (0.3e^{-0.3x}) \, dx = (-e^{-0.4} + 1) \left[-e^{-0.3x} \right]_0^2$$

$$= (-e^{-0.4} + 1)(-e^{-0.6} + 1) \approx 0.15 \qquad \blacksquare$$

POINT OF INTEREST

In writing the joint density $f(x, y)$ as the product of the two individual density functions, we have assumed that the random variables X and Y are independent. Roughly speaking, this means that the values of one do not affect the values of the other, an assumption that seems reasonable in Example D.

EXAMPLE E

Amy needs to make transactions at two different windows at her bank. She estimates that each transaction will take 3 minutes but knows that she will have to stand in line at both windows. Suppose that the waiting times X and Y at the two windows are exponentially distributed, with $k = 0.3$ and $k = 0.2$, respectively. What are her chances of getting in and out of the bank in less than 14 minutes?

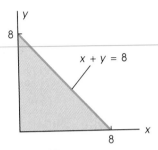

FIGURE 20

Solution Subtracting 6 minutes for the two transactions, Amy is allowing 8 minutes to stand in line. She wants to know prob$\{X + Y < 8\}$. Note that the joint probability density function for **X** and **Y** is the same as in Example D, but now we must integrate over the triangular set where $x + y < 8$ (see Figure 20). The required probability can be calculated as follows.

$$\int_0^8 \left[\int_0^{8-x} (0.3e^{-0.3x})(0.2e^{-0.2y}) \, dy \right] dx$$

$$= \int_0^8 (0.3e^{-0.3x}) \left[-e^{-0.2y} \right]_0^{8-x} dx$$

$$= \int_0^8 (0.3e^{-0.3x})[-e^{-1.6+0.2x} + 1] \, dx$$

$$= \int_0^8 [(3e^{-1.6})(-0.1e^{-0.1x}) + (0.3e^{-0.3x})] \, dx$$

$$= 3e^{-1.6} \left[e^{-0.1x} \right]_0^8 + \left[-e^{-0.3x} \right]_0^8$$

$$= 3e^{-1.6}(e^{-0.8} - 1) + (-e^{-2.4} + 1) \approx 0.58 \quad \blacksquare$$

Means, Variances, and Covariances

Suppose we have a bivariate random variable (X, Y) with joint probability density function $f(x, y)$. Let μ_x, σ_x^2, and μ_y, σ_y^2 be the corresponding means and variances of **X** and **Y**, respectively. Then

$$\mu_x = \int_{-\infty}^{\infty} \int_{-\infty}^{\infty} x f(x, y) \, dy \, dx \qquad \sigma_x^2 = \int_{-\infty}^{\infty} \int_{-\infty}^{\infty} x^2 f(x, y) \, dy \, dx - \mu_x^2$$

with corresponding expressions for μ_y and σ_y^2.

Next, we introduce two new concepts, important in statistics. The **covariance** of **X** and **Y**, denoted by σ_{xy}, is defined as

$$\sigma_{xy} = \int_{-\infty}^{\infty} \int_{-\infty}^{\infty} xy \, f(x, y) \, dy \, dx - \mu_x \mu_y$$

Finally, the **correlation coefficient** of **X** and **Y**, denoted by ρ_{xy}, is given by

$$\rho_{xy} = \frac{\sigma_{xy}}{\sigma_x \sigma_y}$$

It can be shown that the correlation coefficient satisfies $|\rho_{xy}| \le 1$. It is a measure of the relationship between **X** and **Y**. A value of ρ_{xy} near 0 suggests little (linear) relationship; a value near 1 or −1 suggests a strong relationship.

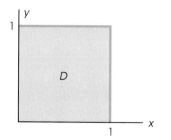

FIGURE 21

EXAMPLE F

Suppose that the bivariate random variable $(\mathbf{X}, \mathbf{Y})$ has the joint probability density function $f(x, y) = \frac{3}{2}(x^2 + y^2)$ on the square D in Figure 21 and $f(x, y) = 0$ elsewhere (you can check that $\int_D\int (x^2 + y^2)\, dy\, dx = \frac{2}{3}$ so that $f(x, y)$ is a density function). Find μ_x, μ_y, σ_x^2, σ_y^2, σ_{xy}, and ρ_{xy}.

Solution Because of the symmetry in x and y, $\mu_x = \mu_y$ and $\sigma_x = \sigma_y$. Now,

$$\mu_x = \frac{3}{2}\int_0^1\left[\int_0^1 x(x^2 + y^2)\, dy\right] dx = \frac{3}{2}\int_0^1\left[x^3y + \frac{1}{3}xy^3\right]_0^1 dx$$

$$= \frac{3}{2}\int_0^1\left(x^3 + \frac{1}{3}x\right) dx = \frac{3}{2}\left[\frac{1}{4}x^4 + \frac{1}{6}x^2\right]_0^1 = \frac{5}{8}$$

Also,

$$\frac{3}{2}\int_0^1\left[\int_0^1 x^2(x^2 + y^2)\, dy\right] dx$$

$$= \frac{3}{2}\int_0^1\left[x^4y + \frac{1}{3}x^2y^3\right]_0^1 dx$$

$$= \frac{3}{2}\int_0^1\left[x^4 + \frac{1}{3}x^2\right] dx = \frac{3}{2}\left[\frac{1}{5}x^5 + \frac{1}{9}x^3\right]_0^1 = \frac{7}{15}$$

and so

$$\sigma_x^2 = \frac{7}{15} - \left(\frac{5}{8}\right)^2 = \frac{73}{960}$$

Next,

$$\frac{3}{2}\int_0^1\left[\int_0^1 xy(x^2 + y^2)\, dy\right] dx$$

$$= \frac{3}{2}\int_0^1\left[\frac{1}{2}x^3y^2 + \frac{1}{4}xy^4\right]_0^1 dx$$

$$= \frac{3}{2}\int_0^1\left(\frac{1}{2}x^3 + \frac{1}{4}x\right) dx = \frac{3}{2}\left[\frac{1}{8}x^4 + \frac{1}{8}x^2\right]_0^1 = \frac{3}{8}$$

which implies that

$$\sigma_{xy} = \frac{3}{8} - \left(\frac{5}{8}\right)\left(\frac{5}{8}\right) = -\frac{1}{64}$$

Finally,

$$\rho_{xy} = \frac{-\frac{1}{64}}{\frac{73}{960}} = -\frac{15}{73}$$ ∎

Problem Set 16.4

1. Suppose $(\mathbf{X}, \mathbf{Y})$ is uniformly distributed on the rectangle $D = \{(x, y): 0 \le x \le 4, 0 \le y \le 2\}$.
   (a) Determine the joint density function $f(x, y)$.
   (b) Calculate prob$\{\mathbf{X} \ge 2, \mathbf{Y} \le 1\}$.
   (c) Calculate prob$\{\mathbf{X} + \mathbf{Y} \le 2\}$.

2. Suppose $(\mathbf{X}, \mathbf{Y})$ is uniformly distributed on the rectangle $D = \{(x, y): -3 \le x \le 3, 0 \le y \le 1\}$.
   (a) Determine the joint density function $f(x, y)$.
   (b) Calculate prob$\{\mathbf{X} \le -2, \mathbf{Y} \ge 0.5\}$.
   (c) Calculate prob$\{\mathbf{X} + \mathbf{Y} \ge 4\}$.

3. Suppose that $f(x, y) = \frac{1}{4}$ on the triangle D in Figure 22 and $f(x, y) = 0$ elsewhere. Note that $f(x, y)$ is a joint density function since $\int_D\int \frac{1}{4} dy \, dx = 1$.
   (a) Calculate prob$\{\mathbf{Y} \ge 1\}$.
   (b) Calculate prob$\{\mathbf{Y} \ge \mathbf{X}/4\}$.

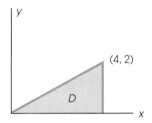

FIGURE 22

4. Suppose that $f(x, y) = \frac{1}{8}$ on the triangle D in Figure 23 and $f(x, y) = 0$ elsewhere. Note that $f(x, y)$ is a joint density function.
   (a) Calculate prob$\{\mathbf{X} \ge 3\}$.
   (b) Calculate prob$\{\mathbf{X} \le 4 - \mathbf{Y}\}$.

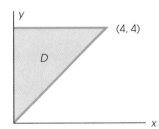

FIGURE 23

5. Let $f(x, y) = kxy^2$ on the triangle in Figure 22 and $f(x, y) = 0$ elsewhere. See Example C.
   (a) Determine k so that $f(x, y)$ is a joint density function.
   (b) Calculate prob$\{\mathbf{X} \le 3\}$.

6. Suppose $f(x, y) = kxy$ on the triangle D in Figure 23 and $f(x, y) = 0$ elsewhere.

   (a) Determine k so that $f(x, y)$ is a joint density function.
   (b) Calculate prob$\{\mathbf{Y} \ge 2\}$.

7. Refer to Example D but suppose that $\mathbf{X}$ and $\mathbf{Y}$ are distributed with joint exponential density $f(x, y) = (0.4e^{-0.4x})(0.1e^{-0.1y})$. Find the probability that an airplane will have a complete failure in the system (both parts fail) during the first 3 years of operation.

8. For the data in Problem 7, find the probability that neither part will fail during the first 3 years of operation.

9. Mabel has 10 minutes left from her lunch break in which to purchase two Twins baseball tickets and two Vikings football tickets. The waiting times $\mathbf{X}$ and $\mathbf{Y}$ in the two ticket lines are exponentially distributed, with means 4 minutes and 8 minutes, respectively. Assuming the lines are at adjoining windows, find the probability that Mabel will be able to make both purchases in the allotted time. See Example E.

10. Suppose that Mabel in Problem 11 stands in line for the baseball tickets while her friend Tim stands in line for the football tickets, both having only 10 minutes available. Find the probability of each of the following.
    (a) Both succeed in getting tickets.
    (b) Mabel succeeds but Tim fails to get tickets.
    (c) They wind up with no tickets at all.

11. The Duncans will buy a sports car and a station wagon on the same day. The lives $\mathbf{X}$ of the sports car and $\mathbf{Y}$ of the station wagon are exponentially distributed with means 10 years and 6 years, respectively. Find the probability of each of the following.
    (a) Both cars will be in operation 10 years from now.
    (b) Six years from now, only the sports car will be in operation.

12. In Problem 11, find the probability that the sports car will last longer than the station wagon.

13. Suppose that the bivariate variable $(\mathbf{X}, \mathbf{Y})$ has the joint density function $f(x, y) = \frac{5}{128}(x^4 + y^4)$ on the rectangle $D = \{(x, y): 0 \le x \le 2, 0 \le y \le 2\}$ and $f(x, y) = 0$ elsewhere. Find $\mu_x, \mu_y, \sigma_x^2, \sigma_y^2, \sigma_{xy}$, and ρ_{xy}. *Hint:* Making use of the symmetry in this problem can save a lot of work. See Example F.

14. Redo Problem 13 with $f(x, y) = \frac{6}{7}(x + y)^2$ on the rectangle $D = \{(x, y): 0 \le x \le 1, 0 \le y \le 1\}$.

15. Suppose $(\mathbf{X}, \mathbf{Y})$ is uniformly distributed on $D = \{(x, y): 0 \le y \le 16 - x^2, -4 \le x \le 4\}$ and $f(x, y) = 0$ elsewhere.

(b) Calculate prob$\{X + Y \geq 4\}$. *Hint:* Make a sketch.

16. Let **P** denote the price (in dollars) of a certain commodity and **S** the monthly sales (in units of 10,000). Suppose that the joint density function for (**P**, **S**) is $f(p, s) = kpe^{-ps}$ on
$$D = \{(p, s): 0.10 \leq p \leq 0.50, s \geq 0\}$$
and $f(p, s) = 0$ elsewhere.
   (a) Determine k.
   (b) Calculate prob$\{0.10 \leq \mathbf{P} \leq 0.20, \mathbf{S} \geq 10\}$.

17. If (**X**, **Y**) has joint density $f(x, y) = \frac{3}{4}(\sqrt{x} + \sqrt{y})$ on $D = \{(x, y): 0 \leq x \leq 1, 0 \leq y \leq 1\}$

and $f(x, y) = 0$ elsewhere, calculate
   (a) prob$\{0 < \mathbf{Y} \leq \mathbf{X}^2\}$ and (b) μ_x.

18. Pick a point (**X**, **Y**) at random on the unit square $D = \{(x, y): 0 \leq x \leq 1, 0 \leq y \leq 1\}$. Calculate prob$\{\mathbf{Y} \leq 4\mathbf{X}\}$.

19. Cut a straight wire of length 1 at random in two places. What is the probability the three pieces will form a triangle? *Hint:* This amounts to picking a point on the unit square in Problem 18 at random and finding the probability it satisfies certain conditions.

Chapter 16 Review Problem Set

1. An experiment consists of tossing three fair dice leading to a sample space with 216 outcomes of which (1, 2, 4) and (3, 2, 2) are typical examples. Let **Y** denote the total number of spots showing. Calculate (a) prob$\{\mathbf{Y} \leq 5\}$ and (b) prob$\{\mathbf{Y} > 5\}$.

2. In the experiment of Problem 1, calculate:
   (a) prob$\{\mathbf{Y} = 15\}$ (b) prob$\{\mathbf{Y} \geq 15\}$
   (c) prob$\{\mathbf{Y} < 15\}$

3. An experiment consists of spinning the spinner shown in Figure 24. Let **X** denote the number of the region where the spinner stops. Calculate the mean and variance of **X**.

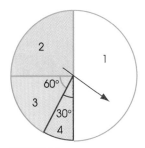

FIGURE 24

4. For the experiment of Problem 3, let $\mathbf{Y} = (\mathbf{X} - 2)^2$. Calculate the mean and variance of **Y**.

5. Let a fair die have two of its faces marked S and the other four faces marked F. An experiment consists of tossing this die four times, and **X** counts the number of S's that appear. Calculate
   (a) prob$\{\mathbf{X} = i\}$ for $i = 0, 1, 2, 3,$ and 4.
   (b) The mean of **X**.
   (c) The variance of **X**.

6. Consider the experiment of tossing a fair die. Let **X**

be the number of spots showing and $\mathbf{Y} = (\mathbf{X} - 2)^2$. Calculate the mean and the variance of **Y**.

7. Suppose that a random variable **X** has the probability density function $f(x) = k/x^2$ on [1, 3] and $f(x) = 0$ elsewhere. Show that $k = \frac{3}{2}$ and then calculate (a) prob$\{\mathbf{X} \leq 1.5\}$, (b) the mean of **X**, and (c) the variance of **X**.

8. Suppose that a random variable **X** has the probability density function $f(x) = k\sqrt{x}$ on [1, 4] and $f(x) = 0$ elsewhere. Calculate
   (a) k (b) prob$\{\mathbf{X} \geq 1.44\}$
   (c) the mean of **X**   (d) the variance of **X**

9. A popular movie is showing at the Royal Theater. The time **T**, in minutes, one must stand in line to buy a ticket has the probability density function $f(t) = 125t/[2(25 + t^2)^2]$ on [0, 10] and $f(t) = 0$ elsewhere. Find the probability of standing in line more than 5 minutes.

10. Suppose the time **T** to stand in line for play tickets is a random variable with probability density function $f(t) = 16/[3(4 + t)^2]$ on [0, 12] and $f(t) = 0$ elsewhere. Find the probability of standing in line no more than 6 minutes.

11. A random variable **X** has the probability density function $f(x) = k(4 + x)^{-3/2}$ on [0, ∞) and $f(x) = 0$ elsewhere. Determine k and then calculate prob$\{\mathbf{X} \geq 12\}$.

12. A random variable **X** has probability density function $f(x) = ke^x(1 + e^x)^{-2}$ on [0, ∞) and $f(x) = 0$ elsewhere. Determine k and then calculate prob$\{\mathbf{X} \leq 2\}$.

13. The time **T** it takes a 10-year-old child to solve a certain puzzle is exponentially distributed with mean 12 minutes. Calculate the probability it will take a randomly selected 10-year-old child more than 10 minutes to solve the puzzle.

14. Suppose that the lifetime in years of the Kensington garage door opener is a random variable with probability density function $f(t) = \frac{1}{10} e^{-t/10}$ on $[0, \infty)$. What fraction of these openers will still be in use after 8 years?

15. If **X** is normally distributed with mean 0 and standard deviation 2, find (a) prob$\{0.6 \leq \mathbf{X} \leq 1.4\}$ and (b) prob$\{-0.6 \leq \mathbf{X} \leq 2.6\}$.

16. Past experience suggests that scores on the final examination in Psychology 101 are normally distributed with mean 72 and standard deviation 8. Out of this year's class of 240 students, how many should we expect to score (a) between 54 and 86, (b) above 84?

17. The weights of a certain group of 1000 professional football players are approximately normally distributed, with mean 230 pounds and standard deviation 25 pounds. Estimate the number of players in the group that weigh more than 270 pounds.

18. The lifetime of a certain type of automobile battery is normally distributed with mean 62 months and standard deviation 4.5 months. Find the probability that such a battery will fail in less than 58 months.

19. Suppose that the bivariate variable $(\mathbf{X}, \mathbf{Y})$ has the joint probability density function $f(x, y) = \frac{3}{8} xy$ on $D = \{(x, y): 0 \leq y \leq 4 - 2x, 0 \leq x \leq 2\}$ and $f(x, y) = 0$ elsewhere. Calculate (a) prob$\{\mathbf{X} \geq 1\}$ and (b) prob$\{\mathbf{Y} \geq 2\mathbf{X}\}$.

20. Suppose that the bivariate random variable $(\mathbf{X}, \mathbf{Y})$ has the joint probability density function $f(x, y) = \frac{15}{64} x \sqrt{y}$ on

$$D = \{(x, y): 0 \leq y \leq x^2, 0 \leq x \leq 2\}$$

Calculate (a) prob$\{\mathbf{X} \geq 1\}$ and (b) prob$\{\mathbf{Y} \geq \mathbf{X}\}$.

21. A softball game is scheduled to start at 7:00 P.M. and is forfeited if a team is not ready to play by 7:10 P.M. The pitcher and catcher for the Logan Bears are consistently late. The number of minutes **P** that the pitcher is late and **C** that the catcher is late are exponentially distributed, with mean 8 minutes and 6 minutes, respectively. Assuming that all the other players consistently arrive on time and that they must have both their pitcher and catcher to play, calculate the probability that the Logan Bears will have to forfeit any given game.

22. A farmer bought a small garden tractor and a large all-purpose tractor on the same day. The life **X** of the small tractor and the life **Y** of the large tractor are each exponentially distributed with means 12 years and 7 years, respectively. Find the probability of each of the following.
(a) Both tractors will still be in operation 8 years later.
(b) The large tractor will last longer than the small one.

23. Suppose that $(\mathbf{X}, \mathbf{Y})$ is uniformly distributed on

$$D = \{(x, y): x^2 \leq y \leq 4, -2 \leq x \leq 2\}.$$

(a) Determine the joint density function $f(x, y)$.
(b) Calculate μ_x, μ_y, σ_x^2, σ_y^2, and σ_{xy}.

24. Suppose that $(\mathbf{X}, \mathbf{Y})$ is uniformly distributed on

$$D = \{(x, y): 3x \leq y \leq 4 - x^2, 0 \leq x \leq 1\}$$

(a) Determine the joint density function $f(x, y)$.
(b) Calculate μ_x, μ_y, σ_x^2, σ_y^2, and σ_{xy}.

25. Let $f(x) = k/(1 + x)$ on $[0, e^2 - 1]$ and $f(x) = 0$ elsewhere.
(a) Determine k so that $f(x)$ is a probability density function for a random variable **X**.
(b) Find the mean and variance of **X**.

26. Suppose that **X** has probability density function $f(x, y) = k/x^2$ on $[1, 3]$ and $f(x) = 0$ elsewhere.
(a) Determine k.
(b) Find the mean and variance of **X**.

27. On $D = \{(x, y): 0 \leq y \leq x, 0 \leq x \leq 1\}$ suppose that $f(x, y) = kye^{x^3}$ and that $f(x, y) = 0$ elsewhere.
(a) Determine k so that $f(x, y)$ is the joint density function for a bivariate random variable $(\mathbf{X}, \mathbf{Y})$.
(b) Calculate prob$\{\mathbf{Y} \leq \frac{1}{2}\mathbf{X}\}$.

28. Let $(\mathbf{X}, \mathbf{Y})$ be a bivariate random variable with probability density function $f(x, y) = kxy/(1 + x^4)$ on $D = \{(x, y): 0 \leq y \leq x, 0 \leq x \leq 1\}$ and $f(x, y) = 0$ elsewhere.
(a) Determine k.
(b) Calculate prob$\left\{\frac{1}{2}\mathbf{X} \leq \mathbf{Y} \leq \mathbf{X}\right\}$.

Appendix

Basic Algebra Review
A1 NUMBERS

The fundamental characters in this book are the **real numbers.** Among the real numbers are the *integers:* . . . , $-3, -2, -1, 0, 1, 2, 3,$ . . . ; the *rational numbers* which consist of ratios of integers such as $\frac{2}{3}$, $\frac{21}{13}$, $\frac{-7}{2}$, and $\frac{5}{1}$; and the *irrational numbers* such as $\sqrt{2}$, $\sqrt{3}$, π, and $\sqrt{5}/3$. The real numbers can be made to correspond to points on a line as indicated in Figure 1. Each point on the line has a real number label, and each real number determines a point on the line.

FIGURE 1

Another way to view the real numbers is as unending decimals. For example,

$$3 = 3.0000 \ldots \qquad = 3.\overline{0}$$

$$\tfrac{2}{3} = 0.6666 \ldots \qquad = 0.\overline{6}$$

$$\tfrac{122}{99} = 1.232323 \ldots \qquad = 1.\overline{23}$$

$$\sqrt{2} = 1.4142135623 \ldots$$

$$\pi = 3.1415926535 \ldots$$

The bar over a group of digits indicates that this group of digits repeats indefinitely. The rational numbers correspond to real numbers whose decimal expansions ultimately repeat; the irrational numbers have decimal expansions with no repeating pattern. The real numbers consist of the rational numbers and the irrational numbers.

While the real numbers consist of unending decimals, we normally work with rounded versions of them. For example, we often use 3.14159 as an approximation for π. Our rule for rounding is to round up if the first neglected digit is 5 or more; otherwise, we round down. Thus, π rounded to 5 digits is 3.1416, to 4 digits is 3.142, and to 3 digits is 3.14. Electronic calculators typically round to either 8 or 10 digits.

We operate on numbers using the five operations of addition, subtraction, multiplication, division, and exponentiation. It is important that we understand these operations and follow the established agreements as to the order in which they are performed. In particular, we have the following hierarchy of operations.

HIERARCHY OF OPERATIONS

In an expression without parentheses

  (i) exponentiations are performed first.
  (ii) multiplications and divisions are performed next going from left to right.
  (iii) additions and subtractions are performed last, again going from left to right.

An expression with parentheses is evaluated beginning inside the innermost parentheses and moving outward.

These rules are the same rules followed by most scientific calculators, though you must learn the special features of your own calculator to use it with confidence.

EXAMPLE A

Perform the following calculations both by hand and using your calculator.

(a) $3 \cdot 6 + 2 \cdot 5 - \frac{6}{3}$ (b) $3(6 + 2)(5 - 2^3)/3$
(c) $3 \cdot 6 + 2 \cdot 5^2 - 24/2^3$ (d) $3(6 - 2)^3 - (\frac{6}{3})^4$
(e) $3[2^3 - 2(4 - 3^2) + 2]$ (f) $-2[-3 - (-2)^3]^2$

Solution (a) 26, (b) -24, (c) 65, (d) 176, (e) 60, (f) -50 ■

The ability to handle numerical fractions is critical in understanding how to handle the more abstract algebraic fractions that will play an important role in calculus. Let us review the basic principles.

RULES FOR FRACTIONS

  (i) $\dfrac{a}{b} = c$ if and only if $a = bc$.

  (ii) $\dfrac{a}{1} = a, \dfrac{0}{a} = 0$ provided $a \neq 0$, and $\dfrac{a}{0}$ is undefined.

  (iii) $\dfrac{ac}{bc} = \dfrac{a}{b}$ provided $c \neq 0$.

  (iv) $\dfrac{-a}{b} = \dfrac{a}{-b} = -\dfrac{a}{b}$ and $\dfrac{-a}{-b} = \dfrac{a}{b}$.

  (v) $\dfrac{a}{c} + \dfrac{b}{c} = \dfrac{a + b}{c}$ and $\dfrac{a}{c} - \dfrac{b}{c} = \dfrac{a - b}{c}$

  (vi) $\dfrac{a}{c} \cdot \dfrac{b}{d} = \dfrac{ab}{cd}$ and $\dfrac{a}{c} \Big/ \dfrac{b}{d} = \dfrac{ad}{bc}$

We note that all three parts of Rule (ii) follow from Rule (i). Note in particular that $a/0$ cannot be defined, for if $a/0 = c$, Rule (i) would say that $a = 0 \cdot c = 0$, which is nonsense. To put it bluntly, *we must never divide by zero.*

EXAMPLE B

Simplify the following numerical expressions.

(a) $2(\frac{1}{6} - \frac{7}{6} + \frac{2}{6})$

(b) $\frac{2}{3} - \frac{3}{8} + \frac{5}{6}$

(c) $2 - \dfrac{\frac{2}{3}}{4}$

(d) $\dfrac{1 - \frac{1}{8}}{\frac{3}{4} - \frac{2}{3}}$

Solution

(a) $2(\frac{1}{6} - \frac{7}{6} + \frac{2}{6}) = 2(\frac{-4}{6}) = 2(\frac{-2}{3}) = -\frac{4}{3} = -1.333\ldots = -1.\overline{3}$

(b) $\frac{2}{3} - \frac{3}{8} + \frac{5}{6} = \frac{16}{24} - \frac{9}{24} + \frac{20}{24} = \frac{27}{24} = \frac{9}{8} = 1.125$

(c) $2 - \dfrac{\frac{2}{3}}{4} = 2 - \frac{2}{3}\frac{1}{4} = 2 - \frac{1}{6} = \frac{12}{6} - \frac{1}{6} = \frac{11}{6} = 1.8333\ldots = 1.8\overline{3}$

(d) $\dfrac{1 - \frac{1}{8}}{\frac{3}{4} - \frac{2}{3}} = \dfrac{\frac{8}{8} - \frac{1}{8}}{\frac{9}{12} - \frac{8}{12}} = \dfrac{\frac{7}{8}}{\frac{1}{12}} = \frac{7}{8}\frac{12}{1} = \frac{21}{2} = 10.5$ ∎

We use the symbols $<$ and $\leq$ to abbreviate "is less than" and "is less than or equal to." Geometrically, $a < b$ means that a is to the left of b on the real line (Figure 2). It is correct to say that $3 < 5$, $3 \leq 5$, and $3 \leq 3$. Equivalently, we may write these three facts as $5 > 3$, $5 \geq 3$, and $3 \geq 3$. The rules for $\leq$ are summarized below. There are corresponding rules for $<$, $>$, and $\geq$.

FIGURE 2

RULES FOR $\leq$

(i)   If $a \leq b$ and $b \leq c$, then $a \leq c$.

(ii) If $a \leq b$, then $a + c \leq b + c$.

(iii) If $a \leq b$ and $c \geq 0$, then $a \cdot c \leq b \cdot c$.

(iv) If $a \leq b$ and $c < 0$, then $a \cdot c \geq b \cdot c$.

(v)   If $0 < a \leq b$, then $\dfrac{1}{a} \geq \dfrac{1}{b}$.

(vi) $\dfrac{a}{b} \leq \dfrac{c}{d}$ provided $bd > 0$ and $ad \leq bc$.

EXAMPLE C

Arrange the numbers $3.1415\overline{8}$, $\frac{22}{7}$, $\frac{60}{19}$, and π in increasing order.

Solution Let us first note from Rule (vi) that $\frac{22}{7} < \frac{60}{19}$, since $(22)(19) <$ $(7)(60)$. Next use long division (or a calculator) to write $\frac{22}{7}$ as a decimal $(\frac{22}{7} = 3.\overline{142857})$ and compare with $3.1415\overline{8}$ and with $\pi = 3.141592635$ . . . . We conclude that

$$3.1415\overline{8} < \pi < \tfrac{22}{7} < \tfrac{60}{19} \qquad\blacksquare$$

Occasionally, we will use the symbol $|a|$, read the *absolute value of a*. The absolute value of a is the distance from a to 0 on the real line (Figure 3). Algebraically, $|a| = a$ if $a \geq 0$, and $|a| = -a$ if $a < 0$. Here are the main facts about absolute value.

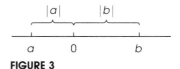

FIGURE 3

PROPERTIES OF ABSOLUTE VALUE

(i) $|a \cdot b| = |a| \cdot |b|$

(ii) $\left|\dfrac{a}{b}\right| = \dfrac{|a|}{|b|}$

(iii) $|a + b| \leq |a| + |b|$

A2 EXPONENTS

We used positive integral exponents in Section A1 when we wrote 2^3 in place of $2 \cdot 2 \cdot 2$ and 3^2 in place of $3 \cdot 3$. In general, if n is a positive integer,

$$a^n = \underbrace{a \cdot a \cdot a \ldots a}_{n \text{ factors}}$$

This definition leads immediately to five rules for exponents.

RULES FOR EXPONENTS

(i) $a^m a^n = a^{m+n}$ (ii) $a^m/a^n = a^{m-n}$

(iii) $(a^m)^n = a^{mn}$ (iv) $(ab)^n = a^n b^n$

(v) $(a/b)^n = a^n/b^n$

In stating these rules, we temporarily assume that m and n are positive integers with $m > n$. But it is precisely because we want these rules to be universally true that we are led to extend the notion of exponent. First, if Rule (ii) is to be true in general, $1 = a^2/a^2 = a^{2-2} = a^0$, which forces $a^0 = 1$. Also by Rule (ii), $1/a^n = a^0/a^n = a^{0-n} = a^{-n}$, which forces $a^{-n} = 1/a^n$. Then by Rule (iii), $(a^{1/n})^n = a^1$, which means that $a^{1/n}$ should mean the same thing as $\sqrt[n]{a}$. Finally, $a^{m/n} = (a^{1/n})^m = (\sqrt[n]{a})^m$. We summarize in the following definitions in which we suppose m and n to be positive integers.

ZERO, NEGATIVE, AND RATIONAL EXPONENTS

(vi) $a^0 = 1$ (vii) $a^{-n} = 1/a^n$

(viii) $a^{1/n} = \sqrt[n]{a}$ (ix) $a^{m/n} = (\sqrt[n]{a})^m$

With these understandings, Rules (i) to (v) are valid in general. There is one small caveat. In this course, we do not attach meaning to $a^{m/n}$ if $a < 0$ and n is an even integer, since this would lead us into a discussion of imaginary numbers, numbers not needed in this course.

EXAMPLE D

Simplify each of the following, writing your answer without any negative exponents.

(a) 2^{-3} (b) $8^{2/3}$ (c) $(2^{1/2}3^{3/2})^2$

(d) $a^{-2}b^3a^4b^{-5}$ (e) $\dfrac{a^2b^{-3}c^3}{a^4b^5c}$ (f) $\left(\dfrac{a^{1/3}b^2a^{4/3}}{b^4}\right)^3$

Solution

(a) $2^{-3} = 1/2^3 = \frac{1}{8}$

(b) $8^{2/3} = (\sqrt[3]{8})^2 = 2^2 = 4$

(c) $(2^{1/2}3^{3/2})^2 = (2^{1/2})^2(3^{3/2})^2 = 2 \cdot 3^3 = 2 \cdot 27 = 54$

(d) $a^{-2}b^3a^4b^{-5} = a^{-2}a^4b^3b^{-5} = a^2b^{-2} = a^2/b^2$

(e) $\dfrac{a^2b^{-3}c^3}{a^4b^5c} = \dfrac{a^2}{a^4} \cdot \dfrac{b^{-3}}{b^5} \cdot \dfrac{c^3}{c} = a^{2-4}b^{-3-5}c^{3-1} = a^{-2}b^{-8}c^2 = \dfrac{c^2}{a^2b^8}$

(f) $\left(\dfrac{a^{1/3}b^2a^{4/3}}{b^4}\right)^3 = \left(\dfrac{a^{5/3}}{b^2}\right)^3 = \dfrac{a^5}{b^6}$ ■

A3 ALGEBRAIC EXPRESSIONS

We illustrate the problem of multiplying two binomials (two-term expressions) with a simple example.

$$(2x + y)(3x + 4y) = (2x + y)(3x) + (2x + y)(4y)$$
$$= (2x)(3x) + (y)(3x) + (2x)(4y) + (y)(4y)$$
$$= 6x^2 + 3xy + 8xy + 4y^2$$
$$= 6x^2 + 11xy + 4y^2$$

There is a short cut called the FOIL method. Multiply the Firsts, the Outers, the Inners, and the Lasts; add the results. Let us apply this in another example.

$$(3a - 2b)(4a + b) = (3a)(4a) + (3a)(b) + (-2b)(4a) + (-2b)(b)$$

$$ \text{F} \qquad\quad \text{O} \qquad\quad \text{I} \qquad\quad \text{L}$$

$$= 12a^2 - 5ab - 2b^2$$

EXAMPLE E

Expand each of the following.

(a) $(2x - 3y)^2$
(b) $(4m - 3n)(4m + 3n)$
(c) $(2xy - 3z^2)(5xy - 2z^2)$

Solution

(a) $(2x - 3y)^2 = (2x - 3y)(2x - 3y) = 4x^2 - 12xy + 9y^2$
(b) $(4m - 3n)(4m + 3n) = 16m^2 + 12mn - 12mn - 9n^2$
$$= 16m^2 - 9n^2$$
(c) $(2xy - 3z^2)(5xy - 2z^2) = 10x^2y^2 - 19xyz^2 + 6z^4$ ■

Parts (a) and (b) in Example E are important enough to be emphasized.

$$(a + b)^2 = a^2 + 2ab + b^2$$
$$(a - b)(a + b) = a^2 - b^2$$

Factoring is the reverse of expanding (multiplying) and is equally important. Becoming proficient at it is mainly a matter of practice.

EXAMPLE F

Factor each of the following.

(a) $9x^2 - 16$
(b) $4x^2 + 4xy + y^2$
(c) $x^2 - x - 6$
(d) $3x^2 + 4xy - 4y^2$

Solution

(a) $9x^2 - 16 = (3x - 4)(3x + 4)$
(b) $4x^2 + 4xy + y^2 = (2x + y)^2$
(c) $x^2 - x - 6 = (x - 3)(x + 2)$
(d) $3x^2 + 4xy - 4y^2 = (3x - 2y)(x + 2y)$ ∎

We handle algebraic fractions just like numerical fractions.

EXAMPLE G

Perform the indicated operations and simplify.

(a) $\dfrac{3}{x + y} + \dfrac{a}{x + y} - \dfrac{5}{x + y}$ (b) $\dfrac{3}{x^2y} + \dfrac{2}{xy^2}$

(c) $\dfrac{2}{x + y} - \dfrac{4y}{x^2 - y^2} + \dfrac{2}{x - y}$ (d) $\dfrac{x^2 - x - 6}{6x} \cdot \dfrac{2x^2}{x^2 + 5x + 6}$

(e) $\dfrac{x^2 + 6x + 8}{x - 3} \Big/ \dfrac{x + 2}{x^2 - 9}$ (f) $\dfrac{1 - \dfrac{x}{x + y}}{y - \dfrac{y^2}{x + y}}$

Solution

(a) $\dfrac{3}{x + y} + \dfrac{a}{x + y} - \dfrac{5}{x + y} = \dfrac{3 + a - 5}{x + y} = \dfrac{a - 2}{x + y}$

(b) $\dfrac{3}{x^2y} + \dfrac{2}{xy^2} = \dfrac{3y}{x^2y^2} + \dfrac{2x}{x^2y^2} = \dfrac{3y + 2x}{x^2y^2}$

(c) $\dfrac{2}{x + y} - \dfrac{4y}{x^2 - y^2} + \dfrac{2}{x - y} = \dfrac{2(x - y)}{(x - y)(x + y)} - \dfrac{4y}{(x - y)(x + y)}$

$+ \dfrac{2(x + y)}{(x - y)(x + y)}$

$= \dfrac{2x - 2y - 4y + 2x + 2y}{(x - y)(x + y)}$

$= \dfrac{4x - 4y}{(x - y)(x + y)} = \dfrac{4(x - y)}{(x - y)(x + y)}$

$= \dfrac{4}{x + y}$

(d) $\dfrac{x^2 - x - 6}{6x} \cdot \dfrac{2x^2}{x^2 + 5x + 6} = \dfrac{(x - 3)(x + 2)}{6x} \cdot \dfrac{2x^2}{(x + 2)(x + 3)}$

$= \dfrac{x(x - 3)}{3(x + 3)}$

(e) $\dfrac{x^2 + 6x + 8}{x - 3} \Big/ \dfrac{x + 2}{x^2 - 9} = \dfrac{(x + 4)(x + 2)}{x - 3} \cdot \dfrac{(x - 3)(x + 3)}{x + 2}$

$= (x + 4)(x + 3) = x^2 + 7x + 12$

(f) $\dfrac{1 - \dfrac{x}{x+y}}{y - \dfrac{y^2}{x+y}} = \dfrac{\dfrac{x+y}{x+y} - \dfrac{x}{x+y}}{\dfrac{xy+y^2}{x+y} - \dfrac{y^2}{x+y}} = \dfrac{x+y-x}{x+y} \cdot \dfrac{x+y}{xy+y^2-y^2}$

$$= \dfrac{y}{xy} = \dfrac{1}{x}$$

Occasionally, we will need to work with radical expressions.

EXAMPLE H

Simplify the following.

(a) $(\sqrt{x} - \sqrt{2y})(\sqrt{x} + \sqrt{2y})$ (b) $\sqrt{x+y} - \dfrac{x}{\sqrt{x+y}}$

Solution

(a) $(\sqrt{x} - \sqrt{2y})(\sqrt{x} + \sqrt{2y})$

(b) $(\sqrt{x} - \sqrt{2y})(\sqrt{x} + \sqrt{2y}) = (\sqrt{x})^2 - (\sqrt{2y})^2 = x - 2y$

$$= \dfrac{y}{\sqrt{x+y}}$$

A4 EQUATIONS

Solving equations is one of the principal tasks of mathematics. For a linear (first-degree) equation, this is a simple task. We illustrate by solving $3x - 4 = 7$.

$$3x - 4 = 7$$

$3x = 11$ (adding 4 to each side)

$x = \frac{11}{3}$ (dividing both sides by 3)

Our procedure was to perform the same operations on both sides of the equation a step at a time until we isolated x on one side and a number (the solution) on the other. But what operations are legal? That is, what operations on both sides of an equation result in an equation with the same solutions as the original equation?

OPERATIONS THAT PRESERVE SOLUTIONS

  (i) Adding the same quantity to (or subtracting the same quantity from) both sides of an equation.
  (ii) Multiplying or dividing both sides of an equation by the same nonzero quantity.

Solving quadratic (second-degree) equations is harder but always possible in principle. The simplest method, the method of factoring, depends on the following property of numbers.

$$ab = 0 \text{ if and only if } a = 0 \text{ or } b = 0 \text{ (or both)}$$

Thus, for example, to solve the quadratic equation $x^2 - x - 6 = 0$, write it in factored form and then set each factor equal to 0.

$$x^2 - x - 6 = 0$$
$$(x + 2)(x - 3) = 0$$
$$x + 2 = 0 \quad \text{or} \quad x - 3 = 0$$
$$x = -2 \qquad x = 3$$

We conclude that the equation has two solutions, namely, -2 and 3.

EXAMPLE I

Solve each of the following equations.

(a) $x^2 - 2x = 15$ (b) $6x^2 + 13x - 5 = 0$

(c) $4y^2 - 4y + 1 = 0$ (d) $3x^2 - 4 = 0$

Solution

(a)
$$x^2 - 2x = 15$$
$$x^2 - 2x - 15 = 0$$
$$(x + 3)(x - 5) = 0$$
$$x = -3, 5$$

(b)
$$6x^2 + 13x - 5 = 0$$
$$(2x + 5)(3x - 1) = 0$$
$$x = -\tfrac{5}{2}, \tfrac{1}{3}$$

(c)
$$4y^2 - 4y + 1 = 0$$
$$(2y - 1)(2y - 1) = 0$$
$$y = \tfrac{1}{2}$$

Note that while a quadratic equation normally has two solutions, part (c) illustrates a case where there is only one solution.

(d)
$$3x^2 - 4 = 0$$
$$(\sqrt{3x} + 2)(\sqrt{3x} - 2) = 0$$
$$x = -2/\sqrt{3}, 2/\sqrt{3}$$

An alternative way to solve this last equation would be to first write

$$3x^2 = 4$$
$$x^2 = \tfrac{4}{3}$$

This means that x is a square root of $\tfrac{4}{3}$. There are two such square roots, namely, $-2/\sqrt{3}$ and $2/\sqrt{3}$. ∎

Many quadratic equations are too complicated to factor by inspection. For them we can always resort to the *Quadratic Formula*.

> **THE QUADRATIC FORMULA**
> The solutions of $ax^2 + bx + c = 0$ are given by the formula
> $$x = \frac{-b \pm \sqrt{b^2 - 4ac}}{2a}$$

EXAMPLE J

Solve by the Quadratic Formula.

(a) $2x^2 - 3x - 1 = 0$ (b) $6x^2 + 13x - 5 = 0$

Solution

(a) $a = 2$, $b = -3$, and $c = -1$. Thus,
$$x = \frac{3 \pm \sqrt{9 + 8}}{4} = \frac{3 \pm \sqrt{17}}{4}$$

(b) $a = 6$, $b = 13$, and $c = -5$. Thus
$$x = \frac{-13 \pm \sqrt{169 + 120}}{12} = \frac{-13 \pm \sqrt{289}}{12} = \frac{-13 \pm 17}{12}$$

Using the plus sign gives $\frac{4}{12} = \frac{1}{3}$, whereas using the minus sign gives $\frac{-30}{12} = -\frac{5}{2}$. Note that these are the same answers we got by factoring in part (b) of Example I. ∎

A5 INEQUALITIES

Solving a linear inequality is very similar to solving a linear equation. About all we need to remember is that multiplying (or dividing) both sides of an inequality by a negative number reverses the direction of the inequality.

EXAMPLE K

Solve $-2x + 3 \le 14$.

Solution

$$-2x + 3 \le 14$$
$$-2x \le 11$$
$$x \ge -\tfrac{11}{2}$$

Note that the solution is a whole set of numbers: all numbers greater than or equal to $-\frac{11}{2}$. The solution set is shown in Figure 4. ∎

$-\frac{11}{2}$

FIGURE 4

As in solving quadratic equations, so in solving quadratic inequalities, it is best to start by factoring.

EXAMPLE L

Solve the following inequalities.

(a) $x^2 - x - 6 \leq 0$ (b) $x^2 - 6x + 8 > 0$

Solution

(a)
$$x^2 - x - 6 \leq 0$$
$$(x + 2)(x - 3) \leq 0$$

FIGURE 5

The expression on the left can change sign only at -2 and 3. These points, the so-called *split points* for the given inequality, divide the real line into three intervals as shown in Figure 5. By checking with one value from each of these intervals (say -3, 0, and 4), we discover that only the points on the middle interval satisfy the inequality. The solution set consists of the interval $-2 \leq x \leq 3$.

(b)
$$x^2 - 6x + 8 > 0$$
$$(x - 2)(x - 4) > 0$$

FIGURE 6

The split points are 2 and 4, dividing the real line into the three intervals shown in Figure 6. A check at 0, 3, and 5 shows that the first and last of the three intervals satisfy the inequality. The solution set consists of the two intervals $x < 2$ and $x > 4$. ∎

Basic Algebra Review Problem Set

Evaluate the following expressions without using a calculator.

1. $2 \cdot 5 - 36/3^2 + 4^2$
2. $4(-2)^2 - (2 - 4)^2 - 2^2$
3. $5^2/(3^2 + 4^2) + 2^3$
4. $3[22 - (5 - 2)^3]^2 + \sqrt{3^2 + 4^2}$
5. $[(21^2 - 8^2)/(21 + 8)]^2$
6. $\{2[31 - (15 - 9)^2]\}^2$

Write each rational number as a repeating decimal (using the bar notation). Then round this number correctly to three decimal places.

7. $\frac{5}{12}$ 8. $\frac{11}{24}$

9. $\frac{5}{7}$ 10. $\frac{5}{13}$

Simplify each of the following, first writing the exact fractional answer and then its correct four-decimal approximation.

11. $\frac{2}{5} - \frac{5}{12} + \frac{11}{20}$ 12. $\frac{8}{3}(\frac{5}{12} - \frac{3}{4} + \frac{7}{24})$

13. $(\frac{2}{3} - \frac{4}{5})/\frac{7}{15}$ 14. $(\frac{3}{4} - \frac{5}{6})/(\frac{3}{4} + \frac{5}{6})$

15. $\frac{3}{2}[\frac{3}{4} - (\frac{2}{3} - \frac{1}{2})]$ 16. $\frac{5}{8} + \frac{2}{3}/(\frac{4}{9} - \frac{7}{18})$

17. $-\frac{3}{4} - \dfrac{\frac{5}{6}}{2}$ 18. $9 + 9/(1 - \frac{5}{2})^2 + \frac{2}{3}$

In each case, arrange the numbers in increasing order.

19. $\frac{13}{40}, \frac{19}{59}, 0.334, 0.3\overline{3}$ 20. $0.5\overline{3}, 0.5\overline{2}, \frac{13}{25}, \frac{41}{79}$

Use your calculator to evaluate each expression, rounding your answer to four decimal places.

21. $\dfrac{(3.134)^2 + \sqrt{11.321}}{5.432 + 21.351 - 4.652}$

22. $\dfrac{2.5(3.123 - 41.534)^2}{5.123 - (6.541)^3}$

23. $\dfrac{52.12 - (5.432/0.521)^2}{6.312}$

24. $\dfrac{5 \times 10^{-4} + 6 \times 10^{-3}}{4.1 \times 10^{-3} - 2.2 \times 10^{-4}}$

25. $\dfrac{|41.24 - (7.325)^2|}{4.412 \times 10^2}$

26. $2.32 + \sqrt{54.12}/|7.31 - 10.75|$

50. $\dfrac{3}{a^2 b^2} - \dfrac{2}{ab^3} + \dfrac{1}{a^2 b}$

51. $\dfrac{2}{a - 3} + \dfrac{2a}{a^2 - 9} - \dfrac{4}{a + 3}$

52. $\dfrac{4}{x + 3} - \dfrac{x}{x^2 + x - 6} + \dfrac{2}{x - 2}$

53. $\dfrac{3x}{x^2 + 7x + 12} \cdot \dfrac{x^2 + x - 6}{9x^2}$

54. $\dfrac{2xy}{x^2 - 2xy - 8y^2} \Big/ \dfrac{x^2}{x^2 - 16y^2}$

55. $\dfrac{1 - x^2/y^2}{x - y}$

56. $\dfrac{(x/y)^2 - (y/x)^2}{\dfrac{x^2 - y^2}{x^2 y^2}}$

Simplify, writing your answer with no negative exponents.

27. $\sqrt{3}(3^{3/2} - 3^{1/2})$

28. $(27)^{2/3}(25)^{3/2}$

29. $\dfrac{x^{-2} y^3 x^{2/3} y^{3/4}}{x^{-3} y^4}$

30. $\left(\dfrac{a^{2/3} b^{5/6}}{c^{-1}}\right)^6$

31. $\dfrac{(a^{-5/3} b^{-3/4})^2}{(a^{-2} b^{-2/3})^3}$

32. $\dfrac{(\sqrt{a}\,\sqrt[3]{a}\,\sqrt[4]{a})^{12}}{a^{12}}$

Simplify.

57. $(\sqrt{a} + \sqrt{b^2})(\sqrt{a} - \sqrt{b^2})$

58. $(\sqrt{x} - 2\sqrt{y})^2$

59. $2\sqrt{x + h} - \dfrac{4x}{2\sqrt{x + h}}$

60. $\left(\sqrt{x + 2 + h} - \dfrac{x}{\sqrt{x + 2 + h}}\right)\Big/ \sqrt{x + 2 + h}$

61. $\sqrt{4x^4 + 9}$

62. $\sqrt{4x^4 + 12x^2 + 9}$

Expand each of the following.

33. $(3xy - 2)^2$

34. $(1 - 4xy)^2$

35. $(3xy - 2)(3xy + 2)$

36. $(1 - 4xy)(1 + 4xy)$

37. $(2x - 3y)(x + 4y)$

38. $(5x - 1)(4x - 2)$

39. $(a^{1/2} - b^{1/2})^2$

40. $(a^{1/3} + a^{2/3})^2$

41. $(a^{1/2} - a^{-1/2})^2$

42. $(x - y)(x + y)(x^2 + y^2)$

Solve the following equations.

63. $3x - 2 = 8x + 4$

64. $5 = 11x - 3$

65. $x^2 + 5x = 0$

66. $x^2 + 5x - 36 = 0$

67. $a^2 + 2a = 8$

68. $4x^2 - 12x = -9$

69. $y^4 - 1 = 0$

70. $x^3 - x^2 - 6x = 0$

71. $6y^2 - y - 2 = 0$

72. $4y^2 - 19y + 12 = 0$

73. $5x^2 = 15$

74. $(x + 1)^2 = 4$

75. $x^2 - 2x - 2 = 0$

76. $x^2 + 4x - 4 = 0$

77. $2a^2 - 5a + 1 = 0$

78. $3a^2 + 7a + 3 = 0$

Factor each expression.

43. $25a^2 - 9$

44. $25a^2 - 10a + 1$

45. $5a^2 - 2a - 3$

46. $a^4 - 1$

47. $3x^2 y^2 + 5xyz - 2z^2$

48. $x^2 + 4x + 4 - y^2$

Solve the following inequalities. Then show the solution set on the real line.

79. $3x - 4 \geq x + 3$

80. $5x - 7 < 8x + 1$

81. $x^2 + x - 6 < 0$

82. $x^2 + 3x - 18 \geq 0$

83. $2x^2 - x - 3 > 0$

84. $3x^2 - 4x - 4 \leq 0$

85. $(x + 2)(x + 1)(x - 2) \leq 0$

86. $x^3 - 2x^2 - 8x > 0$

Perform the indicated operations and simplify.

49. $\dfrac{2}{x - y} + \dfrac{3}{y - x} - \dfrac{4}{x - y}$

Table A
THE NATURAL EXPONENTIAL FUNCTION

x	e^x	e^{-x}
0.00	1.0000	1.0000
0.01	1.0101	0.9900
0.02	1.0202	0.9802
0.03	1.0305	0.9704
0.04	1.0408	0.9608
0.05	1.0513	0.9512
0.06	1.0618	0.9418
0.07	1.0725	0.9324
0.08	1.0833	0.9231
0.09	1.0942	0.9139
0.10	1.1052	0.9048
0.11	1.1163	0.8958
0.12	1.1275	0.8869
0.13	1.1388	0.8781
0.14	1.1503	0.9694
0.15	1.1618	0.8607
0.16	1.1735	0.8521
0.17	1.1853	0.8437
0.18	1.1972	0.8353
0.19	1.2092	0.8270
0.20	1.2214	0.8187
0.21	1.2337	0.8106
0.22	1.2461	0.8025
0.23	1.2586	0.7945
0.24	1.2712	0.7866
0.25	1.2840	0.7788
0.26	1.2969	0.7711
0.27	1.3100	0.7634
0.28	1.3231	0.7558
0.29	1.3364	0.7483
0.30	1.3499	0.7408
0.31	1.3634	0.7334
0.32	1.3771	0.7261
0.33	1.3910	0.7189
0.34	1.4049	0.7118
0.35	1.4191	0.7047
0.36	1.4333	0.6977
0.37	1.4477	0.6907
0.38	1.4623	0.6839
0.39	1.4770	0.6771
0.40	1.4918	0.6703
0.41	1.5068	0.6637
0.42	1.5220	0.6570
0.43	1.5373	0.6505
0.44	1.5527	0.6440

x	e^x	e^{-x}
0.45	1.5683	0.6376
0.46	1.5841	0.6313
0.47	1.6000	0.6250
0.48	1.6161	0.6188
0.49	1.6323	0.6126
0.50	1.6487	0.6065
0.51	1.6653	0.6005
0.52	1.6820	0.5945
0.53	1.6989	0.5886
0.54	1.7160	0.5827
0.55	1.7333	0.5769
0.56	1.7507	0.5712
0.57	1.7683	0.5655
0.58	1.7860	0.5599
0.59	1.8040	0.5543
0.60	1.8221	0.5488
0.61	1.8044	0.5434
0.62	1.8589	0.5379
0.63	1.8776	0.5326
0.64	1.8965	0.5273
0.65	1.9155	0.5220
0.66	1.9348	0.5169
0.67	1.9542	0.5117
0.68	1.9739	0.5066
0.69	1.9937	0.5016
0.70	2.0138	0.4966
0.71	2.0340	0.4916
0.72	2.0544	0.4868
0.73	2.0751	0.4819
0.74	2.0959	0.4771
0.75	2.1170	0.4724
0.76	2.1383	0.4677
0.77	2.1598	0.4630
0.78	2.1815	0.4584
0.79	2.2034	0.4538
0.80	2.2255	0.4493
0.81	2.2479	0.4449
0.82	2.2705	0.4404
0.83	2.2933	0.4360
0.84	2.3164	0.4317
0.85	2.3396	0.4274
0.86	2.3632	0.4232
0.87	2.3869	0.4190
0.88	2.4109	0.4148
0.89	2.4351	0.4107

x	e^x	e^{-x}
0.90	2.4596	0.4066
0.91	2.4843	0.4025
0.92	2.5093	0.3985
0.93	2.5345	0.3946
0.94	2.5600	0.3906
0.95	2.5857	0.3867
0.96	2.6117	0.3829
0.97	2.6379	0.3791
0.98	2.6645	0.3753
0.99	2.6912	0.3716
1.00	2.7183	0.3679
1.05	2.8577	0.3499
1.10	3.0042	0.3329
1.15	3.1582	0.3166
1.20	3.3201	0.3012
1.25	3.4903	0.2865
1.30	3.6693	0.2725
1.35	3.8574	0.2592
1.40	4.0552	0.2466
1.45	4.2631	0.2346
1.50	4.4817	0.2231
1.55	4.7115	0.2122
1.60	4.9530	0.2019
1.65	5.2070	0.1920
1.70	5.4739	0.1827
1.75	5.7546	0.1738
1.80	6.0496	0.1653
1.85	6.3598	0.1572
1.90	6.6859	0.1496
1.95	7.0287	0.1423
2.00	7.3891	0.1353
2.05	7.7679	0.1287
2.10	8.1662	0.1225
2.15	8.5849	0.1165
2.20	9.0250	0.1108
2.25	9.4877	0.1054
2.30	9.9742	0.1003
2.35	10.486	0.0954
2.40	11.023	0.0907
2.45	11.588	0.0863
2.50	12.182	0.0821
2.55	12.807	0.0781
2.60	13.464	0.0743
2.65	14.154	0.0707
2.70	14.880	0.0672

x	e^x	e^{-x}
2.75	15.643	0.0639
2.80	16.445	0.0608
2.85	17.288	0.0578
2.90	18.174	0.0550
2.95	19.106	0.0523
3.00	20.086	0.0498
3.05	21.115	0.0474
3.10	22.198	0.0450
3.15	23.336	0.0429
3.20	24.533	0.0408
3.25	25.790	0.0388
3.30	27.113	0.0369
3.35	28.503	0.0351
3.40	29.964	0.0334
3.45	31.500	0.0317
3.50	33.115	0.0302
3.55	34.813	0.0287
3.60	36.598	0.0273
3.65	38.475	0.0260
3.70	40.447	0.0247
3.75	42.521	0.0235
3.80	44.701	0.0224
3.85	46.993	0.0213
3.90	49.402	0.0202
3.95	51.935	0.0193
4.00	54.598	0.0183
4.10	60.340	0.0166
4.20	66.686	0.0150
4.30	73.700	0.0136
4.40	81.451	0.0123
4.50	90.017	0.0111
4.60	99.484	0.0101
4.70	109.95	0.0091
4.80	121.51	0.0082
4.90	134.29	0.0074
5.00	148.41	0.0067
5.20	181.27	0.0055
5.40	221.41	0.0045
5.60	270.43	0.0037
5.80	330.30	0.0030
6.00	403.43	0.0025
7.00	1096.6	0.0009
8.00	2981.0	0.0003
9.00	8103.1	0.0001
10.00	22026.	0.00005

THE NATURAL LOGARITHM FUNCTION

	0.00	0.01	0.02	0.03	0.04	0.05	0.06	0.07	0.08	0.09
1.0	0.0000	0.0100	0.0198	0.0296	0.0392	0.0488	0.0583	0.0677	0.0770	0.0862
1.1	0.0953	0.1044	0.1133	0.1222	0.1310	0.1398	0.1484	0.1570	0.1655	0.1740
1.2	0.1823	0.1906	0.1989	0.2070	0.2151	0.2231	0.2311	0.2390	0.2469	0.2546
1.3	0.2624	0.2700	0.2776	0.2852	0.2927	0.3001	0.3075	0.3148	0.3221	0.3293
1.4	0.3365	0.3436	0.3507	0.3577	0.3646	0.3716	0.3784	0.3853	0.3920	0.3988
1.5	0.4055	0.4121	0.4187	0.4253	0.4318	0.4383	0.4447	0.4511	0.4574	0.4637
1.6	0.4700	0.4762	0.4824	0.4886	0.4947	0.5008	0.5068	0.5128	0.5188	0.5247
1.7	0.5306	0.5365	0.5423	0.5481	0.5539	0.5596	0.5653	0.5710	0.5766	0.5822
1.8	0.5878	0.5933	0.5988	0.6043	0.6098	0.6152	0.6206	0.6259	0.6313	0.6366
1.9	0.6419	0.6471	0.6523	0.6575	0.6627	0.6678	0.6729	0.6780	0.6831	0.6881
2.0	0.6931	0.6981	0.7031	0.7080	0.7130	0.7178	0.7227	0.7275	0.7324	0.7372
2.1	0.7419	0.7467	0.7514	0.7561	0.7608	0.7655	0.7701	0.7747	0.7793	0.7839
2.2	0.7885	0.7930	0.7975	0.8020	0.8065	0.8109	0.8154	0.8198	0.8242	0.8286
2.3	0.8329	0.8372	0.8416	0.8459	0.8502	0.8544	0.8587	0.8629	0.8671	0.8713
2.4	0.8755	0.8796	0.8838	0.8879	0.8920	0.8961	0.9002	0.9042	0.9083	0.9123
2.5	0.9163	0.9203	0.9243	0.9282	0.9322	0.9361	0.9400	0.9439	0.9478	0.9517
2.6	0.9555	0.9594	0.9632	0.9670	0.9708	0.9746	0.9783	0.9821	0.9858	0.9895
2.7	0.9933	0.9969	1.0006	1.0043	1.0080	1.0116	1.0152	1.0188	1.0225	1.0260
2.8	1.0296	1.0332	1.0367	1.0403	1.0438	1.0473	1.0508	1.0543	1.0578	1.0613
2.9	1.0647	1.0682	1.0716	1.0750	1.0784	1.0818	1.0852	1.0886	1.0919	1.0953
3.0	1.0986	1.1019	1.1053	1.1086	1.1119	1.1151	1.1184	1.1217	1.1249	1.1282
3.1	1.1314	1.1346	1.1378	1.1410	1.1442	1.1474	1.1506	1.1537	1.1569	1.1600
3.2	1.1632	1.1663	1.1694	1.1725	1.1756	1.1787	1.1817	1.1848	1.1878	1.1909
3.3	1.1939	1.1970	1.2000	1.2030	1.2060	1.2090	1.2119	1.2149	1.2179	1.2208
3.4	1.2238	1.2267	1.2296	1.2326	1.2355	1.2384	1.2413	1.2442	1.2470	1.2499
3.5	1.2528	1.2556	1.2585	1.2613	1.2641	1.2669	1.2698	1.2726	1.2754	1.2782
3.6	1.2809	1.2837	1.2865	1.2892	1.2920	1.2947	1.2975	1.3002	1.3029	1.3056
3.7	1.3083	1.3110	1.3137	1.3164	1.3191	1.3218	1.3244	1.3271	1.3297	1.3324
3.8	1.3350	1.3376	1.3403	1.3429	1.3455	1.3481	1.3507	1.3533	1.3558	1.3584
3.9	1.3610	1.3635	1.3661	1.3686	1.3712	1.3737	1.3762	1.3788	1.3813	1.3838
4.0	1.3863	1.3888	1.3913	1.3938	1.3962	1.3987	1.4012	1.4036	1.4061	1.4085
4.1	1.4110	1.4134	1.4159	1.4183	1.4207	1.4231	1.4255	1.4279	1.4303	1.4327
4.2	1.4351	1.4375	1.4398	1.4422	1.4446	1.4469	1.4493	1.4516	1.4540	1.4563
4.3	1.4586	1.4609	1.4633	1.4656	1.4679	1.4702	1.4725	1.4748	1.4770	1.4793
4.4	1.4816	1.4839	1.4861	1.4884	1.4907	1.4929	1.4952	1.4974	1.4996	1.5019
4.5	1.5041	1.5063	1.5085	1.5107	1.5129	1.5151	1.5173	1.5195	1.5217	1.5239
4.6	1.5261	1.5282	1.5304	1.5326	1.5347	1.5369	1.5390	1.5412	1.5433	1.5454
4.7	1.5476	1.5497	1.5518	1.5539	1.5560	1.5581	1.5602	1.5623	1.5644	1.5665
4.8	1.5686	1.5707	1.5728	1.5748	1.5769	1.5790	1.5810	1.5831	1.5851	1.5872
4.9	1.5892	1.5913	1.5933	1.5953	1.5974	1.5994	1.6014	1.6034	1.6054	1.6074
5.0	1.6094	1.6114	1.6134	1.6154	1.6174	1.6194	1.6214	1.6233	1.6253	1.6273
5.1	1.6292	1.6312	1.6332	1.6351	1.6371	1.6390	1.6409	1.6429	1.6448	1.6467
5.2	1.6487	1.6506	1.6525	1.6544	1.6563	1.6582	1.6601	1.6620	1.6639	1.6658
5.3	1.6677	1.6696	1.6715	1.6734	1.6752	1.6771	1.6790	1.6808	1.6827	1.6845
5.4	1.6864	1.6882	1.6901	1.6919	1.6938	1.6956	1.6974	1.6993	1.7011	1.7029

$$\ln (N \cdot 10^m) = \ln N + m \ln 10, \quad \ln 10 = 2.3026$$

	0.00	0.01	0.02	0.03	0.04	0.05	0.06	0.07	0.08	0.09
5.5	1.7047	1.7066	1.7084	1.7102	1.7120	1.7138	1.7156	1.7174	1.7192	1.7210
5.6	1.7228	1.7246	1.7263	1.7281	1.7299	1.7317	1.7334	1.7352	1.7370	1.7387
5.7	1.7405	1.7422	1.7440	1.7457	1.7475	1.7492	1.7509	1.7527	1.7544	1.7561
5.8	1.7579	1.7596	1.7613	1.7630	1.7647	1.7664	1.7682	1.7699	1.7716	1.7733
5.9	1.7750	1.7766	1.7783	1.7800	1.7817	1.7834	1.7851	1.7867	1.7884	1.7901
6.0	1.7918	1.7934	1.7951	1.7967	1.7984	1.8001	1.8017	1.8034	1.8050	1.8066
6.1	1.8083	1.8099	1.8116	1.8132	1.8148	1.8165	1.8181	1.8197	1.8213	1.8229
6.2	1.8245	1.8262	1.8278	1.8294	1.8310	1.8326	1.8342	1.8358	1.8374	1.8390
6.3	1.8406	1.8421	1.8437	1.8453	1.8469	1.8485	1.8500	1.8516	1.8532	1.8547
6.4	1.8563	1.8579	1.8594	1.8610	1.8625	1.8641	1.8656	1.8672	1.8687	1.8703
6.5	1.8718	1.8733	1.8749	1.8764	1.8779	1.8795	1.8810	1.8825	1.8840	1.8856
6.6	1.8871	1.8886	1.8901	1.8916	1.8931	1.8946	1.8961	1.8976	1.8991	1.9006
6.7	1.9021	1.9036	1.9051	1.9066	1.9081	1.9095	1.9110	1.9125	1.9140	1.9155
6.8	1.9169	1.9184	1.9199	1.9213	1.9228	1.9242	1.9257	1.9272	1.9286	1.9301
6.9	1.9315	1.9330	1.9344	1.9359	1.9373	1.9387	1.9402	1.9416	1.9430	1.9445
7.0	1.9459	1.9473	1.9488	1.9502	1.9516	1.9530	1.9544	1.9559	1.9573	1.9587
7.1	1.9601	1.9615	1.9629	1.9643	1.9657	1.9671	1.9685	1.9699	1.9713	1.9727
7.2	1.9741	1.9755	1.9769	1.9782	1.9796	1.9810	1.9824	1.9838	1.9851	1.9865
7.3	1.9879	1.9892	1.9906	1.9920	1.9933	1.9947	1.9961	1.9974	1.9988	2.0001
7.4	2.0015	2.0028	2.0042	2.0055	2.0069	2.0082	2.0096	2.0109	2.0122	2.0136
7.5	2.0149	2.0162	2.0176	2.0189	2.0202	2.0215	2.0229	2.0242	2.0255	2.0268
7.6	2.0282	2.0295	2.0308	2.0321	2.0334	2.0347	2.0360	2.0373	2.0386	2.0399
7.7	2.0412	2.0425	2.0438	2.0451	2.0464	2.0477	2.0490	2.0503	2.0516	2.0528
7.8	2.0541	2.0554	2.0567	2.0580	2.0592	2.0605	2.0618	2.0631	2.0643	2.0656
7.9	2.0669	2.0681	2.0694	2.0707	2.0719	2.0732	2.0744	2.0757	2.0769	2.0782
8.0	2.0794	2.0807	2.0819	2.0832	2.0844	2.0857	2.0869	2.0882	2.0894	2.0906
8.1	2.0919	2.0931	2.0943	2.0956	2.0968	2.0980	2.0992	2.1005	2.1017	2.1029
8.2	2.1041	2.1054	2.1066	2.1078	2.1090	2.1102	2.1114	2.1126	2.1138	2.1150
8.3	2.1163	2.1175	2.1187	2.1190	2.1211	2.1223	2.1235	2.1247	2.1258	2.1270
8.4	2.1282	2.1294	2.1306	2.1318	2.1330	2.1342	2.1353	2.1365	2.1377	2.1389
8.5	2.1401	2.1412	2.1424	2.1436	2.1448	2.1459	2.1471	2.1483	2.1494	2.1506
8.6	2.1518	2.1529	2.1541	2.1552	2.1564	2.1576	2.1587	2.1599	2.1610	2.1622
8.7	2.1633	2.1645	2.1656	2.1668	2.1679	2.1691	2.1702	2.1713	2.1725	2.1736
8.8	2.1748	2.1759	2.1770	2.1782	2.1793	2.1804	2.1815	2.1827	2.1838	2.1849
8.9	2.1861	2.1872	2.1883	2.1894	2.1905	2.1917	2.1928	2.1939	2.1950	2.1961
9.0	2.1972	2.1983	2.1994	2.2006	2.2017	2.2028	2.2039	2.2050	2.2061	2.2072
9.1	2.2083	2.2094	2.2105	2.2116	2.2127	2.2138	2.2148	2.2159	2.2170	2.2181
9.2	2.2192	2.2203	2.2214	2.2225	2.2235	2.2246	2.2257	2.2268	2.2279	2.2289
9.3	2.2300	2.2311	2.2322	2.2332	2.2343	2.2354	2.2364	2.2375	2.2386	2.2396
9.4	2.2407	2.2418	2.2428	2.2439	2.2450	2.2460	2.2471	2.2481	2.2492	2.2502
9.5	2.2513	2.2523	2.2534	2.2544	2.2555	2.2565	2.2576	2.2586	2.2597	2.2607
9.6	2.2618	2.2628	2.2638	2.2649	2.2659	2.2670	2.2680	2.2690	2.2701	2.2711
9.7	2.2721	2.2732	2.2742	2.2752	2.2762	2.2773	2.2783	2.2793	2.2803	2.2814
9.8	2.2824	2.2834	2.2844	2.2854	2.2865	2.2875	2.2885	2.2895	2.2905	2.2915
9.9	2.2925	2.2935	2.2946	2.2956	2.2966	2.2976	2.2986	2.2996	2.3006	2.3016

Table C

TRIGONOMETRIC FUNCTIONS, RADIAN MEASURE

Radians	sin	cos	tan	Radians	sin	cos	tan
0.00	0.0000	1.0000	0.0000	0.40	0.3894	0.9211	0.4228
0.01	0.0100	1.0000	0.0100	0.41	0.3986	0.9171	0.4346
0.02	0.0200	0.9998	0.0200	0.42	0.4078	0.9131	0.4466
0.03	0.0300	0.9996	0.0300	0.43	0.4169	0.9090	0.4586
0.04	0.0400	0.9992	0.0400	0.44	0.4259	0.9048	0.4708
0.05	0.0500	0.9988	0.0500	0.45	0.4350	0.9004	0.4831
0.06	0.0600	0.9982	0.0601	0.46	0.4439	0.8961	0.4954
0.07	0.0699	0.9976	0.0701	0.47	0.4529	0.8916	0.5080
0.08	0.0799	0.9968	0.0802	0.48	0.4618	0.8870	0.5206
0.09	0.0899	0.9960	0.0902	0.49	0.4706	0.8823	0.5334
0.10	0.0998	0.9950	0.1003	0.50	0.4794	0.8776	0.5463
0.11	0.1098	0.9940	0.1104	0.51	0.4882	0.8727	0.5594
0.12	0.1197	0.9928	0.1206	0.52	0.4969	0.8678	0.5726
0.13	0.1296	0.9916	0.1307	0.53	0.5055	0.8628	0.5859
0.14	0.1395	0.9902	0.1409	0.54	0.5141	0.8577	0.5994
0.15	0.1494	0.9888	0.1511	0.55	0.5227	0.8525	0.6131
0.16	0.1593	0.9872	0.1614	0.56	0.5312	0.8473	0.6269
0.17	0.1692	0.9856	0.1717	0.57	0.5396	0.8419	0.6410
0.18	0.1790	0.9838	0.1820	0.58	0.5480	0.8365	0.6552
0.19	0.1889	0.9820	0.1923	0.59	0.5564	0.8309	0.6696
0.20	0.1987	0.9801	0.2027	0.60	0.5646	0.8253	0.6841
0.21	0.2085	0.9780	0.2131	0.61	0.5729	0.8196	0.6989
0.22	0.2182	0.9759	0.2236	0.62	0.5810	0.8139	0.7139
0.23	0.2280	0.9737	0.2341	0.63	0.5891	0.8080	0.7291
0.24	0.2377	0.9713	0.2447	0.64	0.5972	0.8021	0.7445
0.25	0.2474	0.9689	0.2553	0.65	0.6052	0.7961	0.7602
0.26	0.2571	0.9664	0.2660	0.66	0.6131	0.7900	0.7761
0.27	0.2667	0.9638	0.2768	0.67	0.6210	0.7838	0.7923
0.28	0.2764	0.9611	0.2876	0.68	0.6288	0.7776	0.8087
0.29	0.2860	0.9582	0.2984	0.69	0.6365	0.7712	0.8253
0.30	0.2955	0.9553	0.3093	0.70	0.6442	0.7648	0.8423
0.31	0.3051	0.9523	0.3203	0.71	0.6518	0.7584	0.8595
0.32	0.3146	0.9492	0.3314	0.72	0.6594	0.7518	0.8771
0.33	0.3240	0.9460	0.3425	0.73	0.6669	0.7452	0.8949
0.34	0.3335	0.9428	0.3537	0.74	0.6743	0.7385	0.9131
0.35	0.3429	0.9394	0.3650	0.75	0.6816	0.7317	0.9316
0.36	0.3523	0.9359	0.3764	0.76	0.6889	0.7248	0.9505
0.37	0.3616	0.9323	0.3879	0.77	0.6961	0.7179	0.9697
0.38	0.3709	0.9287	0.3994	0.78	0.7033	0.7109	0.9893
0.39	0.3802	0.9249	0.4111	0.79	0.7104	0.7038	1.009

Table C   Trigonometric Functions, Radian Measure   545

Radians	sin	cos	tan	Radians	sin	cos	tan
0.80	0.7174	0.6967	1.030	1.20	0.9320	0.3624	2.572
0.81	0.7243	0.6895	1.050	1.21	0.9356	0.3530	2.650
0.82	0.7311	0.6822	1.072	1.22	0.9391	0.3436	2.733
0.83	0.7379	0.6749	1.093	1.23	0.9425	0.3342	2.820
0.84	0.7446	0.6675	1.116	1.24	0.9458	0.3248	2.912
0.85	0.7513	0.6600	1.138	1.25	0.9490	0.3153	3.010
0.86	0.7578	0.6524	1.162	1.26	0.9521	0.3058	3.113
0.87	0.7643	0.6448	1.185	1.27	0.9551	0.2963	3.224
0.88	0.7707	0.6372	1.210	1.28	0.9580	0.2867	3.341
0.89	0.7771	0.6294	1.235	1.29	0.9608	0.2771	3.467
0.90	0.7833	0.6216	1.260	1.30	0.9636	0.2675	3.602
0.91	0.7895	0.6137	1.286	1.31	0.9662	0.2579	3.747
0.92	0.7956	0.6058	1.313	1.32	0.9687	0.2482	3.903
0.93	0.8016	0.5978	1.341	1.33	0.9711	0.2385	4.072
0.94	0.8076	0.5898	1.369	1.34	0.9735	0.2288	4.256
0.95	0.8134	0.5817	1.398	1.35	0.9757	0.2190	4.455
0.96	0.8192	0.5735	1.428	1.36	0.9779	0.2092	4.673
0.97	0.8249	0.5653	1.459	1.37	0.9799	0.1994	4.913
0.98	0.8305	0.5570	1.491	1.38	0.9819	0.1896	5.177
0.99	0.8360	0.5487	1.524	1.39	0.9837	0.1798	5.471
1.00	0.8415	0.5403	1.557	1.40	0.9854	0.1700	5.798
1.01	0.8468	0.5319	1.592	1.41	0.9871	0.1601	6.165
1.02	0.8521	0.5234	1.628	1.42	0.9887	0.1502	6.581
1.03	0.8573	0.5148	1.665	1.43	0.9901	0.1403	7.055
1.04	0.8624	0.5062	1.704	1.44	0.9915	0.1304	7.602
1.05	0.8674	0.4976	1.743	1.45	0.9927	0.1205	8.238
1.06	0.8724	0.4889	1.784	1.46	0.9939	0.1106	8.989
1.07	0.8772	0.4801	1.827	1.47	0.9949	0.1006	9.887
1.08	0.8820	0.4713	1.871	1.48	0.9959	0.0907	10.98
1.09	0.8866	0.4625	1.917	1.49	0.9967	0.0807	12.35
1.10	0.8912	0.4536	1.965	1.50	0.9975	0.0707	14.10
1.11	0.8957	0.4447	2.014	1.51	0.9982	0.0608	16.43
1.12	0.9001	0.4357	2.066	1.52	0.9987	0.0508	19.67
1.13	0.9044	0.4267	2.120	1.53	0.9992	0.0408	24.50
1.14	0.9086	0.4176	2.176	1.54	0.9995	0.0308	32.46
1.15	0.9128	0.4085	2.234	1.55	0.9998	0.0208	48.08
1.16	0.9168	0.3993	2.296	1.56	0.9999	0.0108	92.62
1.17	0.9208	0.3902	2.360	1.57	1.0000	0.0008	1256.
1.18	0.9246	0.3809	2.427				
1.19	0.9284	0.3717	2.498				

Table D
TRIGONOMETRIC FUNCTIONS, DEGREE MEASURE

Angle	sin	tan	cot	cos	—	Angle	sin	tan	cot	cos	—
0.0°	0.0000	0.0000	—	1.0000	90.0°	22.5°	0.3827	0.4142	2.4142	0.9239	67.5°
0.5°	0.0087	0.0087	114.59	1.0000	89.5°	23.0°	0.3907	0.4245	2.3559	0.9205	67.0°
1.0°	0.0175	0.0175	57.290	0.9998	89.0°	23.5°	0.3987	0.4348	2.2998	0.9171	66.5°
1.5°	0.0262	0.0262	38.188	0.9997	88.5°	24.0°	0.4067	0.4452	2.2460	0.9135	66.0°
2.0°	0.0349	0.0349	28.636	0.9994	88.0°	24.5°	0.4147	0.4557	2.1943	0.9100	65.5°
2.5°	0.0436	0.0437	22.904	0.9990	87.5°	25.0°	0.4226	0.4663	2.1445	0.9063	65.0°
3.0°	0.0523	0.0524	19.081	0.9986	87.0°	25.5°	0.4305	0.4770	2.0965	0.9026	64.5°
3.5°	0.0610	0.0612	16.350	0.9981	86.5°	26.0°	0.4384	0.4877	2.0503	0.8988	64.0°
4.0°	0.0698	0.0699	14.301	0.9976	86.0°	26.5°	0.4462	0.4986	2.0057	0.8949	63.5°
4.5°	0.0785	0.0787	12.706	0.9969	85.5°	27.0°	0.4540	0.5095	1.9626	0.8910	63.0°
5.0°	0.0872	0.0875	11.430	0.9962	85.0°	27.5°	0.4617	0.5206	1.9210	0.8870	62.5°
5.5°	0.0958	0.0963	10.385	0.9954	84.5°	28.0°	0.4695	0.5317	1.8807	0.8829	62.0°
6.0°	0.1045	0.1051	9.5144	0.9945	84.0°	28.5°	0.4772	0.5430	1.8418	0.8788	61.5°
6.5°	0.1132	0.1139	8.7769	0.9936	83.5°	29.0°	0.4848	0.5543	1.8040	0.8746	61.0°
7.0°	0.1219	0.1228	8.1443	0.9925	83.0°	29.5°	0.4924	0.5658	1.7675	0.8704	60.5°
7.5°	0.1305	0.1317	7.5958	0.9914	82.5°	30.0°	0.5000	0.5774	1.7321	0.8660	60.0°
8.0°	0.1392	0.1405	7.1154	0.9903	82.0°	30.5°	0.5075	0.5890	1.6977	0.8616	59.5°
8.5°	0.1478	0.1495	6.6912	0.9890	81.5°	31.0°	0.5150	0.6009	1.6643	0.8572	59.0°
9.0°	0.1564	0.1584	6.3138	0.9877	81.0°	31.5°	0.5225	0.6128	1.6319	0.8526	58.5°
9.5°	0.1650	0.1673	5.9758	0.9863	80.5°	32.0°	0.5299	0.6249	1.6003	0.8480	58.0°
10.0°	0.1736	0.1763	5.6713	0.9848	80.0°	32.5°	0.5373	0.6371	1.5697	0.8434	57.5°
10.5°	0.1822	0.1853	5.3955	0.9833	79.5°	33.0°	0.5446	0.6494	1.5399	0.8387	57.0°
11.0°	0.1908	0.1944	5.1446	0.9816	79.0°	33.5°	0.5519	0.6619	1.5108	0.8339	56.5°
11.5°	0.1994	0.2035	4.9152	0.9799	78.5°	34.0°	0.5592	0.6745	1.4826	0.8290	56.0°
12.0°	0.2079	0.2126	4.7046	0.9781	78.0°	34.5°	0.5664	0.6873	1.4550	0.8241	55.5°
12.5°	0.2164	0.2217	4.5107	0.9763	77.5°	35.0°	0.5736	0.7002	1.4281	0.8192	55.0°
13.0°	0.2250	0.2309	4.3315	0.9744	77.0°	35.5°	0.5807	0.7133	1.4019	0.8141	54.5°
13.5°	0.2334	0.2401	4.1653	0.9724	76.5°	36.0°	0.5878	0.7265	1.3764	0.8090	54.0°
14.0°	0.2419	0.2493	4.0108	0.9703	76.0°	36.5°	0.5948	0.7400	1.3514	0.8039	53.5°
14.5°	0.2504	0.2586	3.8667	0.9681	75.5°	37.0°	0.6018	0.7536	1.3270	0.7986	53.0°
15.0°	0.2588	0.2679	3.7321	0.9659	75.0°	37.5°	0.6088	0.7673	1.3032	0.7934	52.5°
15.5°	0.2672	0.2773	3.6059	0.9636	74.5°	38.0°	0.6157	0.7813	1.2799	0.7880	52.0°
16.0°	0.2756	0.2867	3.4874	0.9613	74.0°	38.5°	0.6225	0.7954	1.2572	0.7826	51.5°
16.5°	0.2840	0.2962	3.3759	0.9588	73.5°	39.0°	0.6293	0.8098	1.2349	0.7771	51.0°
17.0°	0.2924	0.3057	3.2709	0.9563	73.0°	39.5°	0.6361	0.8243	1.2131	0.7716	50.5°
17.5°	0.3007	0.3153	3.1716	0.9537	72.5°	40.0°	0.6428	0.8391	1.1918	0.7660	50.0°
18.0°	0.3090	0.3249	3.0777	0.9511	72.0°	40.5°	0.6494	0.8541	1.1708	0.7604	49.5°
18.5°	0.3173	0.3346	2.9887	0.9483	71.5°	41.0°	0.6561	0.8693	1.1504	0.7547	49.0°
19.0°	0.3256	0.3443	2.9042	0.9455	71.0°	41.5°	0.6626	0.8847	1.1303	0.7490	48.5°
19.5°	0.3338	0.3541	2.8239	0.9426	70.5°	42.0°	0.6691	0.9004	1.1106	0.7431	48.0°
20.0°	0.3420	0.3640	2.7475	0.9397	70.0°	42.5°	0.6756	0.9163	1.0913	0.7373	47.5°
20.5°	0.3502	0.3739	2.6746	0.9367	69.5°	43.0°	0.6820	0.9325	1.0724	0.7314	47.0°
21.0°	0.3584	0.3839	2.6051	0.9336	69.0°	43.5°	0.6884	0.9490	1.0538	0.7254	46.5°
21.5°	0.3665	0.3939	2.5386	0.9304	68.5°	44.0°	0.6947	0.9657	1.0355	0.7193	46.0°
22.0°	0.3746	0.4040	2.4751	0.9272	68.0°	44.5°	0.7009	0.9827	1.0176	0.7133	45.5°
22.5°	0.3827	0.4142	2.4142	0.9239	67.5°	45.0°	0.7071	1.0000	1.0000	0.7071	45.0°
—	cos	cot	tan	sin	Angle	—	cos	cot	tan	sin	Angle

THE STANDARD NORMAL DISTRIBUTION

z	.00	.01	.02	.03	.04	.05	.06	.07	.08	.09
0.0	.0000	.0040	.0080	.0120	.0160	.0199	.0239	.0279	.0319	.0359
0.1	.0398	.0438	.0478	.0517	.0557	.0596	.0636	.0675	.0714	.0753
0.2	.0793	.0832	.0871	.0910	.0948	.0987	.1026	.1064	.1103	.1141
0.3	.1179	.1217	.1255	.1293	.1331	.1368	.1406	.1443	.1480	.1517
0.4	.1554	.1591	.1628	.1664	.1700	.1736	.1772	.1808	.1844	.1879
0.5	.1915	.1950	.1985	.2019	.2054	.2088	.2123	.2157	.2190	.2224
0.6	.2257	.2291	.2324	.2357	.2389	.2422	.2454	.2486	.2517	.2549
0.7	.2580	.2611	.2642	.2673	.2704	.2734	.2764	.2794	.2823	.2852
0.8	.2881	.2910	.2939	.2967	.2995	.3023	.3051	.3078	.3106	.3133
0.9	.3159	.3186	.3212	.3238	.3264	.3289	.3315	.3340	.3365	.3389
1.0	.3413	.3438	.3461	.3485	.3508	.3531	.3554	.3577	.3599	.3621
1.1	.3643	.3665	.3686	.3708	.3729	.3749	.3770	.3790	.3810	.3830
1.2	.3849	.3869	.3888	.3907	.3925	.3944	.3962	.3980	.3997	.4015
1.3	.4032	.4049	.4066	.4082	.4099	.4115	.4131	.4147	.4162	.4177
1.4	.4192	.4207	.4222	.4236	.4251	.4265	.4279	.4292	.4306	.4319
1.5	.4332	.4345	.4357	.4370	.4382	.4394	.4406	.4418	.4429	.4441
1.6	.4452	.4463	.4474	.4484	.4495	.4505	.4515	.4525	.4535	.4545
1.7	.4554	.4564	.4573	.4582	.4591	.4599	.4608	.4616	.4625	.4633
1.8	.4641	.4649	.4656	.4664	.4671	.4678	.4686	.4693	.4699	.4706
1.9	.4713	.4719	.4726	.4732	.4738	.4744	.4750	.4756	.4761	.4767
2.0	.4772	.4778	.4783	.4788	.4793	.4798	.4803	.4808	.4812	.4817
2.1	.4821	.4826	.4830	.4834	.4838	.4842	.4846	.4850	.4854	.4857
2.2	.4861	.4864	.4868	.4871	.4875	.4878	.4881	.4884	.4887	.4890
2.3	.4893	.4896	.4898	.4901	.4904	.4906	.4909	.4911	.4913	.4916
2.4	.4918	.4920	.4922	.4925	.4927	.4929	.4931	.4932	.4934	.4936
2.5	.4938	.4940	.4941	.4943	.4945	.4946	.4948	.4949	.4951	.4952
2.6	.4953	.4955	.4956	.4957	.4959	.4960	.4961	.4962	.4963	.4964
2.7	.4965	.4966	.4967	.4968	.4969	.4970	.4971	.4972	.4973	.4974
2.8	.4974	.4975	.4976	.4977	.4977	.4978	.4979	.4979	.4980	.4981
2.9	.4981	.4982	.4982	.4983	.4984	.4984	.4985	.4985	.4986	.4986
3.0	.4987	.4987	.4987	.4988	.4988	.4989	.4989	.4989	.4990	.4990

Also, for z = 4.0, 5.0, and 6.0, the probabilities are 0.49997, 0.4999997, and 0.499999999.

Answers to Problems

(The student edition contains answers to odd-numbered problems only.)

Problem Set 1.1   (p. 6)

1. $y = 2x^2$
2. $s = 50t$
3. $V = 2x^3$
4. $S = 10x^2$
5. (a) 11;   (b) -17;   (c) -3;   (d) 5;   (e) $-\frac{3}{2}$
6. (a) 7;   (b) -3;   (c) 52;   (d) 0;   (e) 3.5
7. (a) 1;   (b) -5;   (c) 233;   (d) $\frac{5}{4}$;   (e) $\frac{9}{4}$
8. (a) -7;   (b) 2;   (c) 57;   (d) $-\frac{103}{16}$;   (e) -3
9. (a) -2;   (b) $-2h^2 - 5h - 2$;   (c) $-2h - 5$;
   (d) $3 - 4x - 2d$
10. (a) 3;   (b) $3 - 2h - h^2$;   (c) $-2 - h$;
    (d) $4 - 2x - p$
11. $-4/[a(a + h)]$
12. $(-4a - 2h)/[a^2(a + h)^2]$
13. $-2 + 6x + 3h$ 14. $3c^2 + 3ch + h^2$
15. (a) 0.5;   (b) 2;   (c) undefined;
    (d) $2(2x - 4)^{-2}$;   (e) $\frac{1}{2}x^4$
16. (a) $-\frac{1}{4}$;   (b) $-\frac{1}{4}$;   (c) undefined;
    (d) $2/(x^2 - 2x - 8)$;   (e) $2(x - 5)/(27 - 5x)$
17. (a) $\sqrt{14}$;   (b) 0;   (c) $\frac{1}{2}$;   (d) $\sqrt{x^2 - 2}$
18. (a) 4;   (b) 6.25;   (c) 4.5;   (d) $x + 2 + x^{-1}$
19. $f(g(x)) = 2x^2 - 4x + 3; g(f(x)) = -2x^2$
20. $g(f(x)) = 9x^2 + 3x; f(g(x)) = 3x^2 - 3x + 1$
21. $g(f(x)) = x^6 + 2x^3 + 1; f(g(x)) = x^6 + 1$
22. $g(f(x)) = x \ (x \neq 0); f(g(x)) = x \ (x \neq 2)$
23. $g(h(f(x))) = 4/x^2; f(h(g(x))) = (7 - x^2)/(x^2 + 1)$
24. $h(g(f(x))) = 2/x^2; f(g(h(x))) = 1/16x^2$
25. (a) -0.4313416;   (b) 0.6296419;   (c) -0.5075616

26. (a) 0.9346489;   (b) 1.7471371;   (c) 0.7631220
27. $f(x) = \ln x, g(x) = \cos x, h(x) = \sqrt{x}$
28. $f(x) = \sqrt{x}, g(x) = \ln x, h(x) = 1/x$
29. (a) All real numbers;   (b) all real numbers except 2;
    (c) all real numbers except ± 3;
    (d) all real numbers except ± 1;
    (e) all real numbers greater than 1.
30. (a) 4;   (b) $\frac{5}{2}$;   (c) $\frac{9}{34}$;   (d) $\frac{1}{2}$;   (e) $\frac{25}{9}$;
    (f) $1/(x^2 + 2x + 2)$;   (g) $x^4 + 4x^3 + 8x^2 + 8x + 4$
31. $4x^3 + 6x^2h + 4xh^2 + h^3$
32. $f(x) = 180 + 0.2x$ 33. $f(x) = 700 + 60x$
34. $878 + 14.8x$ 35. $f(x) = 0.015x - 180$
36. $196x - 56x^2 + 4x^3$ 37. $\sqrt{4x^2 + 12x + 5}$
38. (a) 169.956; 399.707;   (b) $f(x) = (\sqrt{x} + 3)^3 - 10$
39. (a) 0; 54.48%; 96.30%;   (b) 37
40. $-x$

Problem Set 1.2   (p. 13)

1. f, h, G, H 2. f, h, H
3. (a) Odd;   (b) even;   (c) neither;   (d) neither;
   (e) even;   (f) odd.
4. (a) Neither;   (b) odd;   (c) even;   (d) even;
   (e) odd;   (f) odd.

5.

6.

15.

16.

7.

8.

17.

18.

9.

10.

19.

20.

11.

12.

21.

22.

13.

14.

23.

24.

25.

26.

34.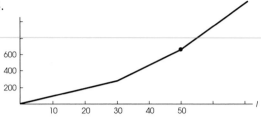

Thousands

35. (a) Odd; (b) odd; (c) even; (d) even; (e) even;
    (f) even; (g) neither; (h) even.

36. (a)   (b)

27.

28.

(c)   (d)

29.

30.

(e)

31.

32.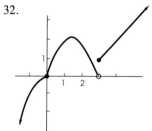

37. The graph of $y = f(x) + k$ is the graph of $y = f(x)$
    translated $|k|$ units vertically. The translation is down-
    ward if $k < 0$, and upward if $k > 0$.

33.

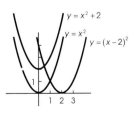

The graph of $y = f(x - k)$ is the graph of $y = f(x)$ translated $|k|$ units horizontally. The translation is to the left if $k < 0$, and to the right if $k > 0$.

38. $R(35) = \$4725$; $R(70) = \$9450$

39. $R(x) = \begin{cases} 160\,x & \text{if } x \le 30 \\ 220\,x - 2x^2 & \text{if } x > 30 \end{cases}$;
$R(35) = \$5250$; $R(70) = \$5600$

40.

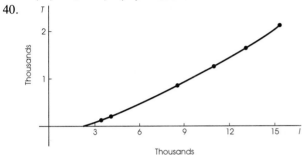

41. (a) \$908.67;   (b) \$1559.10;   (c) \$8305.40

43. (a) \$1312.65;   (b) \$1858.50;   (c) \$7474.18

Problem Set 1.3   (p. 21)

1. 4
2. $\frac{3}{2}$
3. $-\frac{2}{7}$
4. 1
5. -22.56
6. 1.7031
7. $y - 1 = 3(x - 2)$
8. $y - 5 = \frac{2}{3}(x - 3)$
9. $y - 1 = 4(x - 3)$
10. $y - 3 = \frac{3}{2}(x - 2)$
11. $y + 2 = \frac{9}{2}(x - 1)$
12. $y - 3 = -1(x - 2)$ or $x + y = 5$
13. $4x - y = 3$
14. $2x + 3y = 12$
15. $5x - y = 11$
16. $2x + 3y = 10$
17. $2x - y = 5$
18. $x + 3y = 7$
19. $4x + y = 12$
20. $2x - 5y = 29$
21. $3x - 4y = 17$
22. $2x + 5y = -6$
23. $3x - 2y = -9$
24. $2x - y = 10$
25. $f(x) = 3x$; $f(5) = 15$
26. $f(x) = 2x + 9$; $f(5) = 19$
27. $f(x) = -\frac{2}{5}x + \frac{16}{5}$, $f(5) = \frac{6}{5}$
28. $f(x) = \frac{9}{5}x - \frac{19}{5}$; $f(5) = \frac{26}{5}$
29. (a) $-\frac{2}{3}, 2$;   (b) $-\frac{4}{3}, -12$;   (c) $\frac{3}{4}, -\frac{11}{4}$;   (d) $\frac{3}{2}, -\frac{8}{3}$
30. $2x + 3y = 6$
31. (a) $3x + 2y = -4$;   (b) $2x + y = -1$;
   (c) $0 \cdot x + y = -5$;   (d) $5x + 3y = -5$

32. Points of the line are $(a, 0)$ and $(0, b)$, so the slope is $-b/a$. And $y = -(b/a)x + b$ is the slope-intercept form. Thus, $(x/a) + (y/b) = 1$.

33. $f(x) = -1800x + 20,000$

34. (a) -1;   (b) $k = -2$;   (c) $k = 4$;   (d) $k = -\frac{1}{2}$

35. (a) $32 = b$; $a = 1.8$; $F = 1.8C + 32$;
   (b) 89.6;   (c) 27.22

36. (a) $y = \frac{1}{2}x + \frac{3}{2}$;   (b) $x = 1, y = 2$;   (c) $3\sqrt{5}$

37. 8393 ft   38. 1000 ft²

Problem Set 1.4   (p. 26)

1. 196.4 cm   2. 4.72 kcal/liter oxygen
3. (a) $S(x) = 0.04x + 1200$;   (b) \$2640
4. (a) $F(x) = 30x + 1100$;   (b) \$2240;
   (c) $x = 50$ dresses
5. (a) $F(x) = 0.12x + 150$;   (b) \$1110
6. $A(x) = 10,000 + 0.09x$
7. (a) $V(n) = -6700n + 75,000$;
   (b)

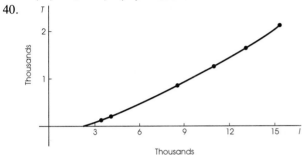

   (c) $-\$6700$; annual decrease in value.
8. (a) $C(n) = -22,000n + 120,000$;   (b) \$43,000
   (c) $\frac{20}{11} \approx 1.82$ yr (1 yr, 299 days)
9. $C(x) = 42,000 + 250x$ 10. $C(x) = 28,000 + 300x$
11. $R(x) = 375x$; 336 sets
12. $R(x) = 475x$; 160 freezers
13. (a) $C(x) = 40x + 20,000$;   (b) $R(x) = 100x$;
   (c) $P(x) = 60x - 20,000$;   (d) $P(1000) = \$40,000$
14. (a) 500 pairs;   (b) \$10,000;
   (c) $-15p^2 + 2600p - 100,000$
15. $s(t) = 480 - 60t$ 16. $s(t) = 54t + 300$
17. (a) Sarah: $S(t) = 55t$ for $t \ge 0$; John: $J(t) = 0$ for $0 \le t < 2$; $70t - 140$ for $t \ge 2$;
   (b) $t = \frac{28}{3}$ (at 9:20 P.M.)

18. (a) $R(t) = 4 + 3.2t$;　(b) 11.25 in.
19. (a) $L(T) = 0.00228T + 64$;　(b) 64.228 ft;
　　(c) 0.00228 ft/°F
20. (a) $P = 10T - 80$;　(b) 90%;
　　(c) 10% germination/°C
21. (a) $\frac{9}{5}$;　(b) $C = \frac{5}{9}(F - 32)$; $\frac{5}{9}$ °C/°F
22. -88 ft/s^2
23. (a) $f(n) = 48{,}000n + 1{,}200{,}000$;
　　(b) 48,000 cases/yr
24. (a) $P(t) = 0.0135t + 0.12$;　(b) 0.0135/yr
25. (a) $C(x) = 35x + 12{,}500$;　(b) \$12,500;
　　(c) \$35/suit;　(d) \$35
26. (a) $S(t) = 600 + 105t$;　(b) 105 mi/h

Problem Set 1.5　(p. 32)

1. -2
2. $\frac{1}{2}$
3. $-\frac{1}{3}$
4. $\frac{1}{7}$
5. 2
6. -2.5
7. 0
8. $-\frac{2}{5}$
9. (a) A, E;　(b) C;　(c) B, D
10. (a) Nowhere;　(b) A, C, E;　(c) B, D
11.

12.

13. $y - 9 = 6(x - 3)$
14. $y - 25 = -10(x + 5)$
15. 48
16. -32
17. 5
18. 1024
19. $30x^5 + 12x^2 - 12$
20. $12x^3 - 6x^2 + 5$
21. $8x^3 - 9x^2 + 10x$
22. $12x^5 - 22x$
23. $-15x^2 + x$
24. $9x^{11} - 2x^5$
25. $y + 64 = 160(x + 2)$
26. $y + 15 = -23(x - 3)$
27. $y + 30 = 53(x - 2)$
28. $y - 17 = 116(x + 2)$
29. $(2, -8)$ and $(-2, 20)$
30. $(3, 2)$ and $(-3, -46)$
31. (a) $6x^2 + 6x(\Delta x) + 2(\Delta x)^2$;　(b) $6x^2$
32. (a) $6x + 3(\Delta x) - 1$;　(b) $6x - 1$
33. $y = 3(x - 2)$; $y - 20 = 3(x + 2)$
34. $y = \frac{1}{7}x + 2$

35. All values of x.
36. $(-1, \infty)$
37. $x = 3$
38. $(\frac{1}{2}, \frac{1}{4})$
39. 19.384

Problem Set 1.6　(p. 39)

1. (a) 16π in.^{3}/in.;　(b) 1.6π in.^{3}
2. (a) 72π in.^{2}/in.;　(b) 45.24 in.^{2}
3. 1100 gal/min
4. 5°F/h
5. (a) -1;　(b) 4
6. (a) $\frac{1}{2}$;　(b) $-\frac{1}{2}$
7. (a)

(b)

t	0	1	2	4
v	0	-3	0	24

;
(c) at $t = 0$ and $t = 2$

8. (a)

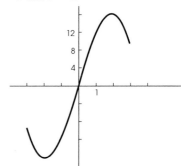

(b)

t	-3	-1	0	1	3
v	-15	9	12	9	-15

;
(c) $t = \pm 2$
9. (a) 6.1 ft/s;　(b) 6.01 ft/s;　(c) 6.001 ft/s;
　　(d) 6 ft/s
10. (a) -0.1;　(b) -0.01;　(c) -0.001;　(d) 0
11. (a) 64 ft/s; -64 ft/s;　(b) 4 s;　(c) 256 ft
12. (a) -160; -224 ft/s;　(b) 4 s;　(c) 440 ft; -416 ft/s
13. -80 ft/s
14. -112 ft/s
15. (a) 1500 people/yr;　(b) 1965;　(c) 3000 people/yr
16. (a) -260 farms/yr;　(b) about May 1978;
　　(c) -667 farms/yr
17. (a) \$136;　(b) \$0.90;　(c) \$0.90;　(d) 125 bats

18. (a) \$200/chair;   (b) \$200
19. (a) $P'(x) = 3.5 - \frac{1}{150}x$; $7 - \frac{1}{100}x$; $3.5 - \frac{1}{300}x$;
    (b) \$0.17/wagon;   (c) \$2.00/wagon;
    (d) \$1.83/wagon; \$0.17/wagon
20. (a) $20 - 0.04x$; $50 - 0.2x$; $30 - 0.16x$;   (b) \$12;
    (c) \$26/pair;   (d) $-$\$10/pair;   (e) at 187 pairs/wk
21. (a) About \$8000;   (b) about 38%
22. (a) \$14,000;   (b) about 45%
23. (a) -3 cm/s; 12 cm/s;   (b) 10 s;   (c) 2 s;
    (d) 17 cm/s
24. (a) $t = 1.5$ s;   (b) 196 ft;   (c) $t = 5$ s;
    (d) at 112 ft/s
25. (a) $t = 0$, $t = 1.5$;   (b) $t = \frac{3}{4}$ s;
    (c) at $t = \frac{1}{2}$ and $t = \frac{3}{4}$
26. (a) $32x - 0.008x^2$;   (b) \$25/pair;   (c) $7 - 0.016x$;
    (d) \$3.80;   (e) 437 pairs
27. (a) 25,000 bacteria;   (b) 900 bacteria/h;   (c) 1.4 h
28. (a) $4 + 0.1t$;   (b) $V(t) = \frac{4}{3}\pi(3 + 0.1t)^3$;
    (c) 15.4 cc/wk
29. $f'(t) > 0$ and $f''(t) < 0$; customer thought $f'(t) < 0$.

Chapter 1   Review Problem Set (p. 42)

1. (a) $[\frac{2}{3}, \infty)$;   (b) 3; $\sqrt{7}$;   (c) $2x + h - 3$ $(h \neq 0)$
2. (a) $[2, \infty)$;   (b) 18; 8;   (c) $2x + h$ $(h \neq 0)$
3. (a) Neither;   (b) odd;   (c) even.
4. (a) Even;   (b) neither;   (c) neither.
5. (a)

   (b)

   (c)

6. (a)

   (b)

   (c)

7. $h(x) = \sqrt{x}$, $g(x) = x^2 + 3x$, $f(x) = 2x + 1$
8. $f(x) = x^2 + 3x + 5$; $g(x) = \sqrt{x}$; $h(x) = 4/x$
9. (a) $\frac{3}{2}$; -6;   (b) 3; 5;   (c) $-\frac{4}{3}$; 4
10. (a) $\frac{1}{4}$; $\frac{3}{2}$;   (b) $\frac{6}{5}$; -3;   (c) $\frac{1}{2}$; -2
11. $y = \frac{2}{3}x + \frac{9}{5}$ 12. $y - 4 = 3(x + 2)$
13. (a) 11;   (b) 0;   (c) $-\frac{3}{2}$
14. (a) -2;   (b) -3;   (c) -1
15. $y + 7 = -14(x - 1)$ 16. $y + 6 = -16(x - 2)$
17. (a) $F(x) = 180 + 0.15x$;   (b) \$1080
18. $F(x) = 320,000 + 2000x$
19. (a) $S(x) = 1800 + 0.05x$;   (b) \$10,000
20. (a) $F(x) = 1275 + 250x$;
    (b) \$2275 will be paid by the insurance company, so Howard will have to pay the remaining \$925.
21. $f(t) = -11,000t + 90,000$
22. (a) $F(x) = 0.16x + 8000$;   (b) \$8000
23. (a) $1 < x < 3$;   (b) $x = 0.5$ and $x = 3.5$;
    (c) $2 < x < 4$;   (d) $x = 2$ and $4 < x < 5$;
    (e) $x = 1$
24.

25. (a) 96 ft/s; -128 ft/s;   (b) after 6 s;   (c) 576 ft;
    (d) after 12 s;   (e) 192 ft/s

26. (a) 32 ft/s; −96 ft/s;   (b) after 2 s;   (c) after 8 s;
    (d) 192 ft/s

27. (a) 6%/yr;   (b) about 10%/yr

28. (a) 50 mi/h;   (b) about 45 mi/h;
    (c) after about 40 min;   (d) about 100 mi/h;
    (e) 30 mi/h

29. (a) $27; $31; $4;   (b) about $27;   (c) 1500 tricycles

30. (a) $22.00/pair; $26.80/pair; $4.80/pair;
    (b) makes 1500 pairs/wk

31. (a) $A(x) = \frac{1}{8}(8 + \pi)x^2$; $P(x) = \frac{1}{2}(6 + \pi)x$;

  (b) $dA/dx = \dfrac{5}{2}(8 + \pi) \approx 27.85$;

     $dP/dx = \dfrac{1}{2}(6 + \pi) \approx 4.57$

32. (a) $A(x) = \frac{1}{4}(8x^2 + \sqrt{3}x^2)$; $P(x) = 7x$;
  (b) $A'(12) = 48 + 6\sqrt{3} \approx 58.39$; $P'(12) = 7$;
  (c) about 58.39 ft²

Problem Set 2.1   (p. 52)

1. (a) 11;   (b) 22;   (c) 9;   (d) 17
2. (a) 11;   (b) 13;   (c) 13;   (d) 25
3. 2   4. 2
5. 2   6. Does not exist.
7. Does not exist.   8. 5
9. (a)

  (b) 9

10. (a)

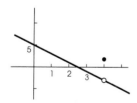

  (b) −3

11. (a) 3;   (b) does not exist;   (c) 4;   (d) 2;   (e) 3;
    (f) 1

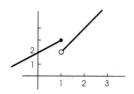

12. (a) 2;   (b) 2;   (c) 4;   (d) does not exist;   (e) 5;
    (f) 2

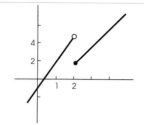

13. 4   14. 9
15. −5   16. 8
17. 11   18. 13
19. 2   20. $\frac{3}{2}$
21. $\frac{5}{8}$ 22. $\frac{3}{2}$
23. Does not exist.   24. 0
25. $-\frac{1}{7}$ 26. 0
27. $\frac{3}{2}$ 28. 1

29. $\displaystyle\lim_{x \to 2} (5x^3 - 4x^2 + 3x + 5)$

    $\overset{(4)}{=} \displaystyle\lim_{x \to 2} (5x^3 - 4x^2) + \lim_{x \to 2} (3x + 5)$

    $\overset{(4)}{=} \displaystyle\lim_{x \to 2} 5x^3 - \lim_{x \to 2} 4x^2 + \lim_{x \to 2} 3x + \lim_{x \to 2} 5$

    $\overset{(3,\,1)}{=} 5 \displaystyle\lim_{x \to 2} x^3 - 4 \lim_{x \to 2} x^2 + 3 \lim_{x \to 2} x + 5$

    $\overset{(7)}{=} 5(\displaystyle\lim_{x \to 2} x)^3 - 4(\lim_{x \to 2} x)^2 + 3 \lim_{x \to 2} x + 5$

    $\overset{(2)}{=} 5(2)^3 - 4(2)^2 + 3(2) + 5 = 35$

30. $\displaystyle\lim_{x \to 3} (2x^4 - 7x^2 + 9x - 1)$

    $\overset{(4)}{=} \displaystyle\lim_{x \to 3} (2x^4) - \lim_{x \to 3} (7x^2) + \lim_{x \to 3} (9x) - \lim_{x \to 3} (1)$

    $\overset{3,\,1}{=} 2 \displaystyle\lim_{x \to 3} (x^4) - 7 \lim_{x \to 3} (x^2) + 9 \lim_{x \to 3} (x) - 1$

    $\overset{(7,\,2)}{=} 2(\displaystyle\lim_{x \to 3} x)^4 - 7(\lim_{x \to 3} x)^2 + 9(3) - 1$

    $\overset{(2)}{=} 2(3)^4 - 7(3)^2 + 27 - 1 = 125$

31. −1   32. 40
33. 5   34. −1
35. 64   36. 81
37. −2   38. −3
39. 9   40. 1
41. $-\frac{4}{3}$ 42. 21

43. $\lim\limits_{x \to 2} \dfrac{3f(x) + 2x - 1}{2g(x)} \overset{6}{\equiv} \dfrac{\lim\limits_{x \to 2} [3f(x) + 2x - 1]}{\lim\limits_{x \to 2} 2g(x)}$

$\overset{4}{\equiv} \dfrac{\lim\limits_{x \to 2} 3f(x) + \lim\limits_{x \to 2} 2x - \lim\limits_{x \to 2} 1}{\lim\limits_{x \to 2} 2g(x)}$

$\overset{3,1}{\equiv} \dfrac{3 \lim\limits_{x \to 2} f(x) + 2 \lim\limits_{x \to 2} x - 1}{2 \lim\limits_{x \to 2} g(x)}$

$\overset{2}{\equiv} \dfrac{3(-3) + 2(2) - 1}{2(4)} = -\dfrac{3}{4}$

44. $\lim\limits_{x \to 5} \dfrac{f(x) \cdot g(x) + 30}{f(x) - 2x + 3}$

$\overset{6,4,5,3}{\equiv} \dfrac{\lim\limits_{x \to 5} f(x) \cdot \lim\limits_{x \to 5} g(x) + \lim\limits_{x \to 5} (30)}{\lim\limits_{x \to 5} f(x) - 2 \lim\limits_{x \to 5} x + \lim\limits_{x \to 5} 3}$

$\overset{2,1}{\equiv} \dfrac{(6)(-2) + 30}{(6) - 2(5) + 3} = \dfrac{18}{-1} = -18$

45. (a) 16; (b) $\frac{1}{16}$; (c) 3; (d) 125
46. (a) -32; (b) $\frac{1}{8}$; (c) 6; (d) 64
47. 4   48. 10
49. $-\frac{1}{49}$ 50. 0
51. 2   52. -2
53. Does not exist.   54. Does not exist.
55. Does not exist.   56. 1
57. 3   58. Does not exist.
59. (a) 1; (b) $\frac{1}{2}$ 60. (a) $\frac{1}{2}$; (b) 0.3535
61. (a) 0; (b) does not exist; (c) 36; (d) $\frac{1}{8}$

Problem Set 2.2   (p. 60)

1. $2x + 2$ 2. $6x$
3. $3x^2$ 4. $2x - 1$
5. $-4/x^2$ 6. $-1/(x + 1)^2$
7. $1/(x + 1)^2$ 8. $-1/x^2$
9. $1/[2\sqrt{x + 2}]$ 10. $1/[2\sqrt{x - 1}]$
11. $f(x) = x^{4/3}, f'(x) = \frac{4}{3}x^{1/3}$
12. $3\sqrt{x}/2$
13. $f(x) = x^{1.9}, f'(x) = 1.9x^{0.9}$
14. $2.5x^{1.5}$ 15. $2/\sqrt[3]{x^2}$
16. $3/\sqrt{x}$ 17. $-\frac{5}{2}x^{1/4}$
18. $2x^{\sqrt{2}-1}$ 19. $3.27x^{0.09}$
20. $4x^{2\pi-1}$ 21. $-11\sqrt{3}\, x^{\sqrt{3}-1}$
22. $8.44x^{1.11}$

23. (a) $2x^2$; (b) $x^{7/2}$; (c) $x^{1/2}$; (d) $x^{7/4}$; (e) $x^3/8$; (f) $x^3 - 2x^{7/2} + x^4$
24. (a) $x^6/27$; (b) $x^{9/4}$; (c) $x^{5/2}$; (d) $x^{11/6}$; (e) $x^2/27$; (f) $x^{2/3} + 2x^2 + x^{10/3}$
25. (a) $12x^3$; (b) $9x^{7/2}$; (c) $3x^{1/2}$; (d) $4x^3$
26. (a) $\frac{9}{64}x^8$; (b) $\frac{21}{2}x^{17/4}$; (c) $\frac{5}{4}x^{3/2}$; (d) 3
27. (a) 0; (b) $28x^6$; (c) $31x^{2.1}$; (d) $14x^{1/6}$; (e) $-(6\pi + 3)x^{2\pi}$; (f) 0
28. (a) $\frac{16}{3}$; (b) $-\frac{1}{16}$; (c) 0; (d) $-\frac{1}{48}$
29. $-2/x^3$ 30. $\frac{1}{3}x^{-2/3}$
31. 50π cm^3/cm
32. 7.5 mg/min; 202.5 mg/min
33. -0.8 dyn/cm   34. $\frac{15}{2}$ lines/h
35. (27, 243)   36. $y - \frac{1}{2} = 3(x - \frac{1}{4})$
37. (a) 0.24998; $\frac{1}{4}$; (b) -0.1249; $-\frac{1}{8}$
38. (a) $k \approx 1.047566 \times 10^{-4}$; (b) about \$7.67 million; (c) about \$0.33/person

Problem Set 2.3   (p. 67)

1. $6x^2 - 2x^{-1/3} - 10x^{-3}$
2. $-10x^{2/3} - 6x^{-4} + 3x^{1/2}$ 3. $6x^2 + \frac{5}{2}x^{-1/2} + 20x^{-6}$
4. $12x^3 - \frac{3}{4}x^{-3/4} + 15x^{-4}$ 5. $x^{-2/3} + 2x^{-3/2}$
6. $\frac{4}{3}x^{1/3} + \frac{2}{3}x^{-4/3}$ 7. $\frac{7}{2}x^{5/2} - 2 + \frac{5}{2}x^{-1/2}$
8. $\frac{8}{3}x^{1/3} + \frac{4}{3}x^{-2/3}$ 9. $y + 3 = 17(x + 1)$
10. $y - 36 = \frac{41}{2}(x - 4)$ 11. $y - 2 = \frac{3}{16}(x - 16)$
12. $y - 8 = \frac{7}{6}(x - 8)$
13. (a) $4x^3 - 10x$; (b) $4x^3 - 10x$
14. (a) $3t^2 - 8$; (b) $3t^2 - 8$
15. $10x^4 + 21x^2 + 18x - 15$
16. $24x^3 + 30x^2 - 14x - 15$
17. $2x + 6x^{1/2} + 5x^{-1/2} + 2$
18. $24x + 20 - 35x^{-2}$ 19. $\frac{10}{3}$
20. 19   21. -4
22. 24
23. (a) $2 + 3x^{-2}$; (b) $2 + 3x^{-2}$
24. (a) $(3x + 2)/[2x\sqrt{x}]$; (b) $(3x + 2)/[2x\sqrt{x}]$
25. $(4x^2 - 4x - 10)/(2x - 1)^2$
26. $(-4x^4 - 8x^2 + 8x)/(x^3 + 2x + 1)^2$
27. $6/[\sqrt{x}(3 + \sqrt{x})^2]$
28. $(6x^2 - 6x - 6)/(x^2 + 2x)^2$
29. $2 - 2x^{-1/2} - 15x^{-4}$
30. $-60x^4 - 16x^3 + 84x^2 + 30x + 5$
31. $-12x^{-5} - 6x^{-4} + 8$ 32. $9x^2 - 16x + 21$
33. $4x + 3 + 4/(x - 4)^2$ 34. $4x^3 - 6x^2 - 12x + 16$
35. $y = \frac{13}{3}(x - 1)$
36. $H'(x) = G(x)G'(x) + G(x)G'(x) = 2G(x)G'(x)$

37. (a) $6(3x + 1)$;   (b) $2(2x - 1)(x^2 - x + 4)$;
     (c) $16 + 20x^{-1/2}$;   (d) $2(4x + 3x^{-2})(4 - 6x^{-3})$
38. $\frac{10}{507}$ 39. 2.01 cm
40. (a) -8;   (b) -0.08
41. (a) $P(x) = -0.02x^2 + 220x - 3000$;
     (b) $-0.04x + 220$; $180/mower;
     (c) 5500; this production level produces maximum profit.
42. (a) $8380.13;   (b) $16.76;   (c) $13.50;
     (d) $-$0.0065/jacket

Problem Set 2.4   (p. 73)

1. $36(3x - 4)^{11}$ 2. $63(7x - 8)^8$
3. $8(3x^2 + x + 1)^7(6x + 1)$
4. $14(2x^2 + 5x - 6)^{13}(4x + 5)$
5. $5(x + 2x^{-2})^4(1 - 4x^{-3})$
6. $(x^{2/3} - 2x)^5(4x^{-1/3} - 12)$
7. $10(x^2 - 2\sqrt{x})^9(2x - x^{-1/2})$
8. $4(2x^{1/2} - 5x^{-1})^3(x^{-1/2} + 5x^{-2})$
9. $\frac{1}{2}(2 + 3x - x^2)^{-1/2}(3 - 2x)$
10. $3(9x - 1)^{-2/3}$
11. $\frac{2}{3}(2x^3 - 7x + 6)^{-1/3}(6x^2 - 7)$
12. $\frac{3}{4}(x^4 - 2x^3 + 2)^{-1/4}(4x^3 - 6x^2)$
13. $\frac{3}{4}(6x^2 + 1)(2x^3 + x)^{-1/4}$
14. $\frac{5}{2}(3 - 5x^4)\sqrt{(3x - x^5)^3}$
15. $-60(3x - 1)^{-6}$ 16. $-2(2x + 5)^{-3/2}$
17. $\frac{8}{3}(2x - 1)^{-5/3}$ 18. $-12(2x + 3)^{-5/2}$
19. $\frac{-35}{4374} \approx -0.008$ 20. $3x + 4y = 4$
21. $21(2x - 1)^2(3x + 2)^{-4}$
22. $-44(4x + 1)^3(3x - 2)^{-5}$
23. $\dfrac{5(x^2 + 4)^4 (2x^2 - 10x - 8)}{(2x - 5)^6}$
24. $\dfrac{3(3x - 1)^2(-3x^2 + 2x - 11)}{(x^2 + x - 4)^4}$
25. $2(2x - 1)^4(48x + 21)$ 26. $4(3x - 1)^7(27x + 17)$
27. $2x(16x^2 + 9)(2x^2 + 3)^{2/3}$
28. $25x^2(x^2 + 4)^{3/2}(2x^2 + 4)$
29. $2x(5x^2 + 14)(x^2 + 1)^2(x^2 + 4)$
30. $3(2x^3 + 1)^3(x^2 + x)^2(14x^5 + 10x^3 + 3x^2 + 1)$
31. $-2x(3x + 1)(3x - 1)^{-5}$
32. $[4(4x + 1)^3(-4x + 17)](5 + 2x)^{-7}$
33. 114   34. 56
35. $y - \frac{1}{6} = -\frac{7}{18}(x - 1)$ 36. $y - 12 = -\frac{162}{5}(x - 3)$
37. (a) $40, 80;
     (b) increasing production from 100 to 101 increases daily profit by about $41.00;   (c) $148.33

38. (a) 1600.004;   (b) 327.86 ft/s;   (c) 319.86 ft/s
39. 13.5   40. $\frac{3}{4}$
41. 322.5   42. $\frac{89}{27} \approx 3.296$
43. 2
44. $y - 32 = 1040(x - 2)$ or $1040x - y = 2048$
45. (a) $20,000(1 + \frac{1}{2}r)^7$;   (b) $7969.24; $30,072.61
46. (a) 23.96;   (b) -0.032; the population is declining.
47. (a) $Y = 1,000,000r/(8.5r + 40,000)$;
     (b) $dy/dr = -8,500/(8.5r + 40,000)^2 < 0$ for all r;
        $dy/dr = 4 \times 10^{10}/(8.5r + 40,000)^2 > 0$ for all r.
     (c) $dy/dr < 0$: the yield per plant decreases as the density increases; $dY/dr > 0$: the yield per hectare increases as the density increases. Within limits this does make sense.
48. (a) 50; 57, 78;   (b) $I'(10) = -6$; $I'(15) \approx 13.63$;
     (c) increases.

Problem Set 2.5   (p. 81)

1. $60x^2 + 18$ 2. 12
3. $-24/x^4$ 4. $\frac{9}{4}x^{-5/2}$
5. $-\frac{3}{4}x^{-3/2} - 24x$
6. $d^3y/dx^3 = 9x^2 - 72x^{-5}$ 7. $-\frac{9}{256}$
8. $-\frac{1}{36}$ 9. $-\frac{45}{1024}$
10. $\frac{1}{81}$
11. $6(x^2 - 3x + 4)(5x^2 - 15x +$
12. $2(15x^4 - 12x^2 + 12x + 1)$
13. $-\frac{1}{64}$ 14. $\frac{1}{2}$
15. (a) $s = -16t^2 + 96t + 256$;   (b) 384 ft;   (c) 32 ft/s;
     (d) after 3 s;   (e) after 8 s;   (f) -160 ft/s;
     (g) -32 ft/s² both times.
16. (a) $s = -16t^2 - 64t + 256$;   (b) 64 ft;
     (c) -128 ft/s;   (d) 256 ft;   (e) $-2 + 2\sqrt{5}$; 2.472 s;
     (f) 143.11 ft/s;   (g) -32 ft/s²
17. (a) -1 cm; 24 cm/s; 30 cm/s²;   (b) -13.5 cm/s;
     (c) 18 cm/s²;   (d) $-1 < t < 2$
18. (a) -7; -3; 3;   (b) -10 ft;   (c) 3 m/min;   (d) 0
19. (a) From 1/1/72 to 12/31/77;   (b) after 6/30/74;
     (c) 7/1/77;
     (d) 7/1/74;   (e)

20. (a) $0 < t < 10$; (b) $t = 4$; (c) $2 < t < 4$;
(d)

21. (a) $dI/dt = 0$ last year; $dI/dt > 0$; $d^2I/dt^2 > 0$ in the future;
(b) dP/dt is quite negative now for $0 < t < 2$; $d^2P/dt^2 > 0$ in the future; $dP/dt > 0$ after 2 yr;
(c) $dT/dt > 0$, $d^2T/dt^2 < 0$;
(d) $ds/dt > 0$, $d^2s/dt^2 > 0$ for $0 < t < 2$; $d^2s/dt^2 = 0$ for $t > 2$

22. (a) For all n; (b) $0 < t < 200$; (c) $n = 200, 400$

23. -90

24. $f'''(x) = \frac{-320}{9}x^{-11/3} + \frac{3}{8}x^{-5/2}; f'''(1) \approx 35.180\overline{5}$

25. 192

26. $-\frac{21}{256}$

27. 30

28. 9.36

29. (a) 64 ft/s; (b) 160 ft/s

30. (a) $0 < t < 7$; (b) $4 < t < 7$; (c) $t = 3$

Chapter 2 Review Problem Set (p. 82)

1. (a) 0; (b) 2; (c) 12; (d) 4; (e) 0;
(f) does not exist.

2. (a) 2; (b) 1; (c) -1; (d) 0; (e) 2;
(f) does not exist.

3. 2
4. 8
5. 10
6. $25 + \sqrt{3}$
7. 3
8. 12
9. $\frac{2}{5}$
10. -1
11. $\frac{2}{3}$
12. $\frac{7}{10}$
13. 0
14. 0
15. $-\frac{5}{3}$
16. $\frac{7}{13}$
17. $\frac{3}{2}$
18. $\frac{75}{8}$
19. $\frac{19}{9}$
20. $-\frac{25}{2}$
21. $\frac{8}{3}$
22. 3
23. 3
24. $15\sqrt{3}$
25. $-\frac{5}{3}$
26. 1
27. $\frac{1}{6}$
28. $\frac{1}{2}$
29. -1
30. 8
31. $2x - 3$
32. $4x + 3$
33. $-x^{-3/2}$
34. $x(x^2 + 1)^{-1/2}$
35. $x^{-2/3}$
36. $2x^{-3/4}$
37. $14x^{4/3}$
38. $10x^{2/3}$
39. $3x^{1/2}$
40. $20x^3$ $(x \ge 0)$
41. $\frac{3}{4}x^{-1/4}$
42. $\frac{3}{2}x^{1/2}$
43. $12(2x - 1)^5$
44. $9(3x + 1)^2$
45. $(x^2 - 2x + 5)^{-1/2}(x - 1)$
46. $(x + 4)(x^2 + 8x - 10)^{-1/2}$
47. $3(2x - 4x^{-1/2})^2(2 + 2x^{-3/2})$
48. $64x^{-5}(x^2 + 2)^3(x^2 - 2)$ 49. $4(x^2 - 1)(x^3 - 3x)^{1/3}$
50. $7(x^5 - 5x^2)^{2/5}(x^4 - 2x)$
51. $\dfrac{3(2x^2 + 1)^2(-2x^4 - 7x^2 + 16x + 2)}{(x^3 - 2x + 4)^4}$
52. $-15(3x - 2)^4(2x^3 - 2x - 3)(x^3 + 3)^{-6}$
53. $5(2x^2 + 4x + 1)(2x + 5)^{-1/2}$
54. $(4x + 1)^{-1/2}(20x^2 + 10x + 3)$
55. $6x(2x^2 + 5)^{1/2}$
56. $\frac{3}{2}(2x^2 + x + 1)^{1/2}(4x + 1)$
57. $40(x + 1)(x^2 + 2x)^3[(x^2 + 2x)^4 + 3]^4$
58. $40(2x + 1)^4[(2x + 1)^5 - 2]^3$
59. $\frac{3}{4}$
60. 160
61. $y + 1 = 13(x - 1)$
62. $y - 13 = -12(x - 1)$
63. $(16, 16)$
64. $(4, 16)$
65. (a) $1000(1 + \frac{1}{1200}x)^{119}$; (b) \$24,513.57; \$2433.11
66. About 15°F/hr.
67. (a) $s(1) = 14$ ft, $v(1) = 6$ ft/s, $a(1) = -3$ ft/s²;
(b) $a(4) = -\frac{3}{2}$ ft/s²; (c) $0 \le t < 4$
68. (a) $s(4) = 8$ ft, $v(4) = -16$ ft/s, $a(4) = -3.5$ ft/s²;
(b) $a(1) = -10$ ft/s²
69. (a) $0 \le t < 17$; (b) $0 \le t < 10$; (c) $t = 10$
70.

Problem Set 3.1　(p. 91)

1. Continuous.
2. Not continuous.
3. Not continuous.
4. Continuous.
5. Continuous.
6. Not continuous.
7. Continuous.
8. Not continuous.
9. $\lim\limits_{x\to 1} f(x) = 1 = f(1)$; f is not defined at $x = 4$.
10. $\lim\limits_{x\to 1} f(x) = 2 = f(1)$; f is not defined at $x = 2$.
11. $\lim\limits_{x\to 4} f(x) = \frac{2}{3} = f(4)$; f is not defined at $x = -2$.
12. $\lim\limits_{x\to 10} f(x) = \frac{20}{3} = f(10)$; f is not defined at $x = 5$.
13. $f(4) = 2$
14. $f(-2) = -\frac{11}{2}$
15. $f(1) = -1$
16. $f(3) = -\frac{5}{2}$
17. $x = -1, 6; f(6) = \frac{8}{7}$
18. Discontinuous at $x = 3, -1; \frac{5}{4}$.
19. Nonremovable discontinuities at $x = 1, 4$.
20. Nonremovable discontinuities at $x = 5, 6$.
21. -2 and $2; f(2) = f(-2) = 3$
22. -1 and $1; f(1) = 1; f(-1) = -1$
23. None.
24. None.
25. None.
26. Continuous everywhere in its domain of $[-1, 1]$.
27. Continuous everywhere in its domain, $x > 0$.
28. Nowhere on the domain.
29. None.
30. Nowhere.
31. $x = 1$
32. Discontinuous at $x = 2$.
33. None.
34. Everywhere.
35. $x = -1$
36. Everywhere
37. $x = 0$
38. $x = 0$
39. $x = -\frac{1}{2}$
40. $x = \frac{1}{2}$
41. $x = 2$
42. $x = -3$
43. $x = 2$
44. $x = 1$
45. (a) p, r, t;　(b) p, q, t;　(c) p, t;　(d) t
46. (a) p, q, s, t;　(b) p, r, s, t;　(c) p, s, t;　(d) p
47. Continuous, not differentiable.
48. Continuous; not differentiable.
49. Not continuous.
50. Not continuous.
51. Not continuous.
52. Continuous; not differentiable.
53. Continuous; not differentiable.
54. Continuous; not differentiable.
55. Discontinuous at every integer.

56.

57.

Problem Set 3.2　(p. 98)

1.

2.

3.

4.

5.

6.

7. $(-6, \infty)$
8. $[-6, \infty)$
9. $(-\infty, -\frac{7}{3}]$
10. $[-\frac{11}{8}, \infty]$
11. $(-\infty, 6)$
12. $(20, \infty)$

13. $[-7, 3]$

14. $(-4, 6)$

15. $(-8, 2)$

16. $(-4, 2]$

17. $(-\infty, -2] \cup [3, \infty)$

18. $(-\infty, -1) \cup (6, \infty)$

19. $(-4, \frac{1}{2})$

20. $[-5, \frac{2}{3}]$

21. Inc $(1, \infty)$, dec $(-\infty, 1)$
22. Inc $(-2, \infty)$, dec $(-\infty, -2)$
23. Inc $(-\infty, 2)$, dec $(2, \infty)$
24. Inc $(-\infty, 4)$, dec $(4, \infty)$
25. Inc $(-\infty, -2) \cup (2, \infty)$, dec $(-2, 2)$
26. Inc $(-\infty, -3) \cup (3, \infty)$, dec $(-3, 3)$
27. Inc $(-\infty, 1 - \sqrt{5}) \cup (1 + \sqrt{5}, \infty)$,
    dec $(1 - \sqrt{5}, 1 + \sqrt{5})$
28. Inc $(-\infty, 3) \cup (2, \infty)$, dec $(-3, 2)$
29. Up $(0, \infty)$, down $(-\infty, 0)$
30. Up $(0, \infty)$, down $(-\infty, 0)$
31. Up $(1, \infty)$, down $(-\infty, 1)$
32. Up $(\frac{1}{2}, \infty)$, down $(-\infty, \frac{1}{2})$
33. Up $(-\infty, -2) \cup (2, \infty)$, down $(-2, 2)$
34. Up $(-\infty, -2) \cup (2, \infty)$, down $(-2, 2)$

35.

36.

37.

38.

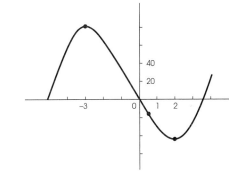

39. (a) $-2, 1, 3\frac{1}{2}$; (b) $(1, 3\frac{1}{2})$; (c) $(-2, 0) \cup (2, 4)$
40. (a) $-2, -\frac{1}{4}, 2\frac{1}{2}, 3$; (b) $(-3, -2) \cup (0, 2\frac{1}{2}) \cup (3, 4)$; (c) $-3, -1) \cup (1, 2\frac{1}{2})$
41. (a) $(0, 2)$; (b) $(3, 4)$; (c) $(2, 3)$
42. (a) $(3, 4)$; (b) $(-3, 0)$; (c) nowhere.
43.

44.

45.

46.

47.

48.

49.

50.

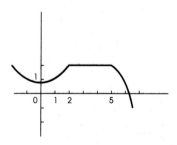

51. $(-\frac{29}{12}, \infty)$ 52. $(-\infty, 2.5)$
53. $(-\infty, 2] \cup [8, \infty)$ 54. $[-2, 9]$
55. $(-3, 2) \cup (2, 4)$ 56. $(-\infty, -1] \cup (2, \infty)$
57. $(-\infty, -3) \cup (2, \infty)$ 58. $(-\infty, 3)$

59. $f'(t) > 0$, but $f''(t) < 0$ from 1970 to 1990.

60. $g'(t) > 0$ and $g''(t) > 0$ from 1980 to 1990.

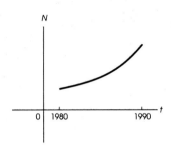

61. $f'(t) < 0$ and $f''(t) > 0$ for $0 < t < 10$; $f'(t) = 0$ for $t > 0$.

62. $g'(t) > 0$ and $g''(t) < 0$ for $0 < t < 15$; $g'(t) = 0$ for $t > 15$.

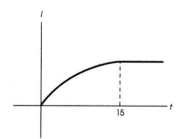

63. dy/dt increases as t increases.

64.

65.

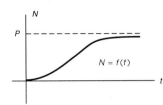

The maximum occurs when $x = \sqrt{8} \approx 2.83$.

Problem Set 3.3 (p. 107)

1. Global max at t, global min at u, local max at r, local min at p and s.
2. Global max at s; global min at r; local max at q; local min at p and u.
3. f has a global max value of $f(2) = 2$; no global min.
4. No global max. Global min value is $f(0) = 1$.
5. No global max. Global min value is $f(0) = 0$.
6. f has a global max value of $f(0) = 2$; no global min.
7. Global max value of $f(10) = 2$; global min value of $f(0) = 0$.
8. No global max. Global min value is $f(10) \approx 0.3$.
9. Max $f(0) = 0$; min $f(2) = -8$
10. Max $f(2) = 24$; min $f(-1) = -3$
11. Max $f(0) = 3$; min $f(3) = -6$
12. Max $f(-1) = 13$; min $f(2) = -5$
13. Max $f(3) = 19$; min $f(-2) = f(1) = -1$
14. Max $f(4) = f(-2) = 19$; min $f(2) = -13$
15. Max $f(-1) = 9$; min $f(-3) = -43$
16. Max $f(-4) = 90$; min $f(2) = -18$
17. Max $f(1) = 2$; min $f(-8) = -4$
18. Max $f(32) = 4$; min $f(0) = 0$
19. Max $f(3) = 3$; min $f(-6) = -6$
20. Max $f(5) = 8$; min $f(-3) = 0$
21. Max $f(-6) = f(3) = 0$; min $f(1) = -4$
22. Max $f(-4) = \frac{22}{3}$; min $f(5) = \frac{4}{3}$
23. Max $f(1) = \frac{3}{2}$; min $f(-1) = -\frac{3}{2}$
24. Max $f(2) = \frac{1}{2}$; min $f(-2) = -\frac{1}{2}$
25. Max $f(3) = \frac{9}{10}$; min $f(0) = 0$

26. Max $f(-4) = \frac{8}{5}$; min $f(0) = 0$
27. 0, 1; local max at 0, local min at 1
28. -1, 1; local max at -1, local min at 1
29. -2, 2; local max at 2, local min at -2
30. -3, 3; local max at -3, local min at 3
31. 1, 4; local max at 1, local min at 4
32. 2, 3; local max at 2; local min at 3
33. -2, 2; local max at -2, local min at 2
34. -1; local min at -1
35. 1, 3; local max at 1, local min at 3
36. -1; no local extrema
37. -2, 0, 2; local max at 0, local min at ± 2
38. -5, 0, 3; local max at 0, local min at -5 and 3
39. -2; local max at -2 40. 2; local min at 2
41. $(0, 0)$ 42. $(0, 0)$
43. $(2, -1)$ 44. $(-1, -4)$
45. $(-1, -3)$ and $(3, -179)$
46. $(0, -20)$ and $(-4, -372)$
47. $(1, 2)$ and $(-1, 2)$ 48. $(1, 0)$
49.

50.

51.

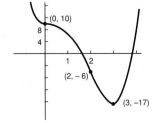

52. Min at $(1.5, -6.69)$; inflection points $(0, -5)$, $(1, -6)$

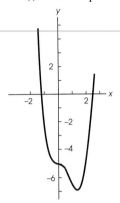

53. Max $f(1) = 10$; min $f(2) = 8$
54. Max $f(4) = 18$; min $f(1) = 5$
55. Max $f(0) = 6$; min $f(2) = 2$
56. Max $f(8) = 10$; min $f(1) = -1$
57. Local max $(0, 8)$; local min $(2, 4)$; inflection point $(1, 6)$

58. Local max $\left(-2, 11\frac{1}{3}\right)$; local min $\left(2, -31\frac{1}{3}\right)$; inflection point $(0, -10)$

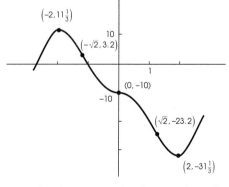

59. Local max $(0, 4)$; local min $(2, 2)$; inflection point $(1, 3)$

60. Local min $(3, 6)$; local max $(-3, -6)$; no inflection points

61.

62.

63.

64.

65.

$y = f(x)$

66.

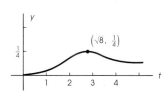

$\left(\sqrt{8}, \frac{1}{4}\right)$

$\frac{1}{4}$

67. Max concentration is $\frac{1}{4}$ at $t = 2\sqrt{2}$.
68. Min $x = \sqrt{2000} \approx 48$

Problem Set 3.4   (p. 116)

1. $f(x)$ gets closer and closer to 10 as x gets more and more negative.
2. $f(x)$ gets more and more negative as x gets closer and closer to -3.
3. $f(x)$ gets more and more positive as x gets more and more positive.
4. $f(x)$ gets more and more positive as x gets closer and closer to -4.
5. $f(x)$ gets more and more negative as x gets closer and closer to 2 from the right side.
6. $f(x)$ gets more and more positive as x gets closer and closer to 2 from the left side.

7. 0
8. 0
9. ∞
10. Does not exist.
11. ∞
12. $-\infty$
13. $-\infty$
14. ∞

15. ∞
16. ∞
17. 4
18. 2
19. $\frac{5}{4}$
20. 0
21. 0
22. 0
23. 0
24. 0
25. $-\frac{4}{3}$
26. -3
27. 2
28. -3
29. $\frac{1}{2}$
30. $\frac{1}{5}$
31. 0
32. 0
33. 5
34. 2
35. -1
36. 2
37. $-\frac{1}{2}$
38. $\frac{3}{4}$
39. $y = 0; x = 0$
40. $y = 3$; none.
41. $y = -1; x = -2$
42. $y = \frac{1}{2}; x = \frac{1}{4}$
43. $y = -2; x = 0, x = 5$
44. $y = \frac{1}{2}; x = -3, x = 3$
45. $y = 1; x = 3, x = -1$
46. $y = -\frac{1}{2}; x = 0$
47. None; none.
48. $y = \frac{3}{2}$; none.
49. $y = 1; x = -1$
50. $y = 1; x = -1$

51.

52.

53.

54.

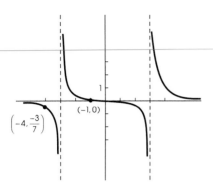

$\left(-4, \frac{-3}{7}\right)$

$(-1, 0)$

55.

56.

57.

58.

59.

60.

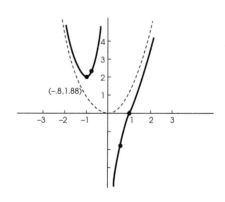

$(-.8, 1.88)$

61. $y = 0$; $x = 0$, $x = 3$, $x = -3$

62. $y = 0$; $x = 4$

63. $f(x) = \dfrac{4x^2}{x^2 - x - 6}$

64. $f(x) = \dfrac{(x + 1)(3 - 2x^2)}{x(x - 1)(x - 2)}$

65. (a) 0; (b) 2

66.

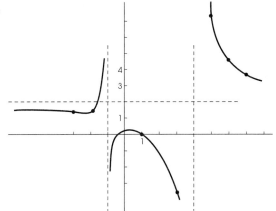

67. $250

68. $175/freezer

69. 12 min

70. (a) 1; (b) 2

71. ∞

72. 0

Chapter 3   Review Problem Set (p. 118)

1. (a) $\lim_{x \to 1} f(x) = \dfrac{7}{3} = f(1)$; (b) no; (c) 3

2. (a) $\lim_{x \to 2} f(x) = -1 = f(2)$; (b) no; (c) $\frac{5}{3}$

3. (a) $\left[-\frac{7}{5}, \infty\right)$; (b) $\left(-3, \frac{1}{2}\right)$

4. (a) $[20, \infty)$; (b) $\left(-\frac{2}{3}, 4\right)$

5. Continuous, differentiable.

6. Continuous, differentiable.

7. Neither continuous nor differentiable.

8. Neither continuous nor differentiable.

9. Continuous, not differentiable.

10. Continuous, differentiable.

11. Continuous, differentiable.

12. Continuous, differentiable.

13. Global max $f(1) = \frac{32}{3}$; global min $f(27) = 4$

14. Global max $f(1) = 8.8$; global min $f(32) = 3.6$

15. Local max at $x = -4$; local min at $x = 2$

16. Local max at $x = -3$; local min at $x = \frac{1}{2}$

17. $(-2, -162)$ and $(2, -18)$

18. $(2, -80)$ and $(4, 72)$

19. (a) Inc $(3, \infty)$; dec $(-\infty, 3)$;
(b) up $(-\infty, 0) \cup (2, \infty)$; down $(0, 2)$;
(c)

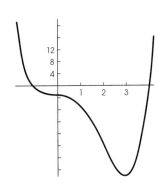

20. (a) Inc $(1, \infty)$, dec $(-\infty, 1)$;
(b) up $(-\infty, 0) \cup (\frac{1}{3}, \infty)$, down $(0, \frac{1}{3})$

21. ∞   22. ∞

23. $-\infty$ 24. $-\infty$

25. 0   26. ∞

27. -1 28. -2

29. (a) Inc $(5, \infty)$; dec $(0, 5)$; (b) up $(0, \infty)$;
(c) local min $f(5) = 10$; (d)

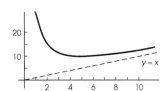

30. (a) Inc $(-\infty, -2) \cup (0, \infty)$; dec $(-2, 0)$;
(b) down $(-\infty, 0) \cup (0, \infty)$;
(c) local max $f(-2) = -3$; (d)

31.

32.

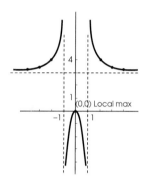

33. $f(x) = \dfrac{2x^3}{x(x+1)(x-2)}$

34. $f(x) = \dfrac{-2x^3}{x(x+2)(x-3)}$

35. (a) $(-\infty, -2) \cup (-2, \infty)$; (b) at $x = -2$; (c) 0;
    (d) at 0 and -2; (e) $\frac{-7}{600}$

36. (a) $(-\infty, 4) \cup (4, \infty)$; (b) $x = 4$; (c) 0;
    (d) not differentiable at $x = 2$ and $x = 4$; (e) $-\frac{4}{3}$

37. $f'(8) = \frac{1}{12}$ (this is $f'(8)$ where $f(x) = \sqrt[3]{x}$ so it has a value of $\frac{1}{12}$)

38. $f'(16) = -63.625$

Problem Set 4.1   (p. 129)

1. 5000
2. 8437.5
3. $\frac{9}{8}$
4. -48
5. 32
6. 187.5
7. 250π
8. 64π
9. $x = 4$, $y = 2$; max value of 8
10. Thus the rectangle has a max value of $6\sqrt{3}$ when $x = \sqrt{3}$, $y = 6$.
11. The area has a value of 3750 m² when $x = 75$ m and $y = 50$ m.
12. The area will have a max value of 7500 m² when $x = 150$ m and $y = 50$ m.

13. The rectangle has a max area of 2812.5 m² when $x = 75$ m and $y = 37.5$ m.
14. Max area of $(1500)^2/2\pi$ ft² when $x = 1500/\pi$ and $y = 750$ ft
15. Min cost of $\approx \$1357.65$ when $x = 20\sqrt{2}$ ft and $y = 30\sqrt{2}$ ft
16. $x = 5\sqrt{30}$ ft and $y = \frac{20}{3}\sqrt{30}$ ft
17. 20 in. (long) by 5 in. (wide) by 2 in. (deep).
18. The chart is 4 ft long, 2 ft wide, and $\frac{8}{3}$ ft deep; min surface area $= 48$ ft².
19. 20 ft by 20 ft for the base and 40 ft for the depth.
20. 2 ft by 2 ft for the base and 6 ft for the height.
21. $x = \frac{2000}{3}\sqrt{3} \approx 1155$ ft   22. 707 ft
23. 14 in. wide by 21 in. high
24. $\frac{100}{3}$ in. long; $\frac{50}{3}$ in. by $\frac{50}{3}$ in. at the ends.
25. $2\sqrt{3}/3$ ft by $2\sqrt{6}/3$ ft   26. $\frac{6}{15} \approx 2.68$ mi
27. (a) 17.596 cm (use it to make a circle);
    (b) 0 (use all 40 cm to make a circle).
28. 5.623 ft (wide) by 3.565 ft (high).
29. $r = 4\sqrt{6}$ in., $h = 8\sqrt{3}$ in.
30. $r = 4\sqrt{2}$ in., $h = 8$ in.
31. $(1 + \sqrt{2}, 2)$; $(1 - \sqrt{2}, 2)$
32. x-intercept is 8 and y-intercept is 4.
33. $4 + \sqrt[3]{16} \approx 6.52$; $y \approx 5.17$
34. Square base of 3 ft by 3 ft; 1 ft tall.

Problem Set 4.2   (p. 136)

1. (a) $-0.008x^2 + 8x - 1000$; (b) $-0.016x + 8$;
   (c) 500
2. (a) $-0.1x^2 + 84x - 6000$; (b) $-0.2x + 84$;
   (c) $x = 420$
3. (a) $-0.05x^2 + 30x - 1000$; (b) $-0.1x + 30$;
   (c) 300
4. (a) $-0.00\overline{16}x^2 + 3.5x - 400$; (b) $-0.00\overline{3}x + 3.5$;
   (c) $x = 1050$
5. $28
6. $98
7. 16
8. $x = 15$
9. (a) 31; (b) $36.96
10. $80,000
11. (a) 1000, $2; (b) 625   12. 2250 cabinets
13. (a) $C(x) = 10,000 + 120x$;
    (b) $R(x) = \frac{2}{3}(500x - x^2)$; (c) 200
14. (a) $R(x) = 15x - 0.001x^2$; (b) 7500 calculators
15. (a) $R(x) = 1200x - x^2$; (b) 550 stereos;
    (c) $272,500; (d) $650; (e) $800/unit; (f) $800
16. 6500 calculators

17. 145 mowers

18. $55,746; 1549 balls

19. 63,246; this might suggest two runs per year of 50,000 each.

20. 4000/run

21. 8.4%

22. 8.57%

23. $9\frac{2}{3}\%$

24. 8%

25. $600

26. $x \approx 7$

27. 5 wk

28. 3 wk   29. Sell immediately.   30. 8 wk and 2 days

31. (a) $P(x) = -Kx^3 + 0.04Kx^2 - x + 0.04$ (millions of $);
   (b) $P'(x) = -1 + 0.08\,kx - 3\,kx^2$;
   $x = (0.08/6) \pm \frac{1}{6}[0.0064 - (12/k)]^{-1/2}$
   (c) 9.53518%;   (d) $108,794.20

Problem Set 4.3   (p. 142)

1. 35

2. 50

3. (a) 2652 lb;   (b) 55 trees/acre

4. (a) $1620;   (b) $21

5. $3.70 and $3.80; $421.80

6. 525

7. (a) 63.25 mi/h;   (b) $286.49

8. 57.74 mi/h

9. (a) 44.72 mi/h;   (b) $69.77

10. 90.30 mi/h

11. (a) $\frac{8}{9}c$;   (b) $\frac{4}{9}c$

12. $x = \frac{2}{3}c$ for $R(x)$ max, $x = \frac{1}{3}c$ for max sensitivity.

13. 37,500; 28,125

14. 600,000 elk; 180,000 max annual kill

15. (a) $6\frac{2}{3}$ h;   (b) $3\frac{1}{3}$ h

16. (a) 5832 people;   (b) after 9 wk

17. (a) 10 additional trees per acre;   (b) 1156 lb

18. $2.75; $151,250

19. $90.40 resulting in 107 rented rooms

20. 6 days

21. (a) 12 days;   (b) 20 days

22. $1\frac{1}{4}$ bag/acre

23. 5.18 mi

24. $6.37

Problem Set 4.4   (p. 149)

1. $y = (3 - x^2)/2$

2. $(5x + 6)/2x$

3. $y = (2 - x)/(4 + 3x)$

4. $(2x + 1)/(x^2 - 3)$

5. $y = (x^2 - x - 3)/(1 - x)$

6. $(5x - 9)/(x^2 - 2)$

7. $y = (4x^2 + 7x + 2)/(x^2 + x)$

8. $(x^2 + 4x + 1)/(x + 2)$

9. $y = \pm\sqrt{x^2 - 2x + 4}$

10. $y = \pm\sqrt{(4 - x^2)/x}$

11. $y = 4 \pm \sqrt{x^2 + 5}$

12. $y = -1 + \sqrt{2x - 3}$; $y = -1 - \sqrt{2x - 3}$

13. $(-x^2 + 2x - 4)/(1 - x)^2$; $(1 - y - 2x)/(x - 1)$

14. $(-5x^2 + 18x - 10)/(x^2 - 2)^2$; $(5 - 2xy)/(x^2 - 2)$

15. $-2y/x$

16. $(4 - 9x^2y)/3x^3$

17. $2x(1 - y)/(x^2 + 4)$

18. $(2x - 3y + 2)/(3x - 2)$

19. $-y/(y^2 + x)$

20. $dy/dx = 6xy/(4y - 3x^2)$

21. $-(x + y)/(3y + x)$

22. $dy/dx = (y^2 - 2xy)/(x^2 - 2xy + 4)$

23. $-2x/y$

24. $20x/[y(3 + y^2)^4]$

25. $y - 5 = \frac{12}{5}(x + 12)$

26. $y - 3 = \frac{5}{3}(x - 5)$

27. $y - 1 = -\frac{1}{2}(x - 2)$

28. $y + 3 = \frac{1}{7}(x - 2)$

29. $y - 1 = -3(x - 4)$

30. $y - 1 = \frac{2}{7}(x + 2)$

31. 12 units/s

32. $\frac{1}{3}$

33. 1.5 ft/s

34. $\frac{25}{3}$ ft/s

35. (a) $8\sqrt{5}/5 \approx 3.58$ units/s;   (b) 6 units^2/s

36. 24.222 ft/s

37. Decreasing at $30,000/wk

38. $0.1345

39. (a) $\frac{4}{3}$ lb/wk;   (b) increase at $0.96/wk

40. (a) -300 hamburgers/mo;   (b) $-$250/mo;
   (c) increasing at $106/mo

41. $y - 1 = -\frac{5}{16}(x - 2)$

42. 0.8 ft/s

43. (a) 6.31 cm/min;   (b) 14.31 cm/min;
   (c) 27 cm^2/min

44. -18π cm^3/min

45. 51.2π cm^3/day

46. 0.199 ft/min

47. $-$5\frac{1}{3}$ hundred thousand/yr

48. $27,000/yr

Problem Set 4.5   (p. 156)

1. $dy = 3x^{1/2}\,dx$

2. $dy = -6x^{-3}\,dx$

3. $dy = (2x^{-1/2} - 5x^{-2})\,dx$

4. $dy = (-x^{-3/2} + 1)\,dx$

5. $dy = -15(4 - 3x)^4\,dx$

6. $dy = (x + \frac{3}{2})[(x^2 + 3x + 2)^{-1/2}]\,dx$

7. $\Delta y = 1.13$; $dy = 1.1$

8. $\Delta y = 0.04$; $dy = 0$

9. $\Delta y = 0.0049938$; $dy = 0.005$

10. $\Delta y = -0.0293383$; $dy = -0.03$

11. $\Delta y = -0.1184275$; $dy = -0.12$

12. $\Delta y = -1.501523$; $dy = -1.5$

13. $N \approx 3.00\overline{6}$ 14. $N \approx 5.006$

15. $N \approx 2.991$ 16. $N \approx 3.996$

17. $N \approx 29.525$ 18. $N \approx 130.25$

19. -937.5 in.^{3}; -150 in.^{2} 20. 1.778 beats/min

21. 108.72 ± 0.47

22. Volume $= 14{,}887 \pm 726$ unit^3; surface area $= 3631 \pm 118$ units^2

23. (a) 10.71%;   (b) 1.12%   24. 4.88%; 3.25%

25. 3%   26. <3%

27. (a) Inelastic;   (b) increased;   (c) $46.42

28. (a) Elastic;   (b) drop by $\approx$$125

29. (a) $15{,}000 - 1000p$;   (b) $p/(15 - p)$; (c) $7.5 < p < 15$

30. (a) $2p^2/(1875 - p^2)$; (b) elastic when $25 < p < \sqrt{1875}$ or $25 < p < \approx 43.3$; (c) -1.13%

31. (a) $(-3x^{-2} + 2x)\,dx$ (b) $[x(x^2 + 9)^{-1/2}]\,dx$; (c) $12(3t + 1)^3\,dt$;   (d) $-[8/(s - 2)^2]\,ds$

32. (a) 405.6;   (b) 436.789341

33. 42.8008   34. 291.6 in.^{3}

35. (a) $\sqrt{p}/[2(30 - \sqrt{p})]$;   (b) $400 < p < 900$; (c) $-$$15{,}000

36. (a) 0;   (b) -0.1875; (c) elastic when $600 < p < 1200$

Chapter 4 Review Problem Set (p. 158)

1. 5120   2. 225

3. $(9.5, \pm\sqrt{9.5})$ 4. $x = 6$

5. $x = 40/(4 + \pi) \approx 5.60$ ft; $y = 20/(4 + \pi) \approx 2.80$

6. $x = 80/(8 + 3\pi) \approx 4.59$ ft; $y \approx 4.10$ ft

7. Length: $20\sqrt[3]{6}$ ft; width: $10\sqrt[3]{6}$ ft; depth: $20\sqrt[3]{6}$ ft

8. 10 in. wide and 15 in. high.

9. (a) 119.52;   (b) $66.73

10. (a) 122 or 123 birdhouses;   (b) $79.60

11. (a) 200 sets;   (b) $6200

12. (a) 281.25 yields a maximum;   (b) 281 houses/wk

13. $600/mo   14. 33 trees/acre

15. 775/lot   16. 1154/lot

17. Either 17 or 18 days from now.

18. Increased by $12

19. About 61 mi/h   20. About 64.8 mi/h

21. $y - 2 = -\frac{9}{13}(x - 1)$ 22. $y - 2 = \frac{19}{20}(x - 3)$

23. (a) $160/3\pi \approx 16.98$ ft/min; (b) $160/27\pi \approx 1.89$ ft/min

24. 674.52 mi/h   25. 452.39 ± 4.52 ft^3

26. 15

27. (a) Elastic;   (b) decrease the price; (c) decrease by $500/wk

28. Allow grouse population to grow to 30,000. The size of the kill will then be 27,000.

29. (a) $-2, 3$;   (b) $f''(x) = 2 + 8x^{-3} > 0$ on $(0, \infty)$; (c) $3\sqrt[3]{4} \approx 4.762$;   (d) $f(\tfrac{1}{2}) = 8.25$; (e)

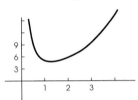

30. (a) $f'(\tfrac{1}{2}) = -15$; (b) $f''(x) = 8x^{-3} > 0$ on its domain; (c) $\lim_{x \to \infty} [f(x) - (x + 4)] = \lim_{x \to \infty} 4x^{-1} = 0$,   so $y = x + 4$ is an asymptote; (d) $f(2) = 8$; (e)

31.

32.

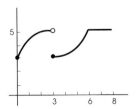

33. $(2\sqrt{2} - 3)/2\sqrt{2} \approx -0.0607$

34. $\frac{1}{48}$ 35. $47{,}000$

36. $\frac{800}{3}$

Problem Set 5.1   (p. 167)

1. 64
2. 9
3. 5
4. 1
5. $\frac{1}{2}$
6. 4
7. 3.388695
8. 4.106605
9. 0.0353581
10. 0.823192

11.

12.

13.

14.

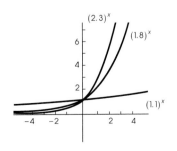

15. b
16. b^4
17. $1/c^2$
18. c
19. b^3
20. b^8
21. $1/\pi^8$
22. $1/\pi^3$
23. $(a + 1)^{1/6}$
24. $(b - 1)^{11/12}$
25. 3
26. 2
27. 5
28. -4
29. $\frac{3}{2}$
30. $\frac{3}{2}$
31. $-\frac{3}{2}$
32. -2
33. -0.3644157
34. 1.635584
35. -0.089754
36. 0.887908
37. 4.32
38. 1.129995
39. 2.485
40. 0.183
41. 1.911
42. 1.9555
43. 2.688
44. 5.401
45. 0.821
46. -0.787
47. $\log_b(x/64)$
48. $\log_b(xy/z)$
49. $\log_b(x^5/y^4)$
50. $\log_b(x + 3)^2(x - 3)^4$
51. $\log_b(x^3 z^6/y^3)$
52. $\log_b(z/\sqrt{x})$
53.

54.

55.

56.

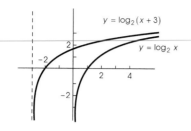

57. 4

58. $\frac{1}{9}$

59. 2

60. π

61. 9

62. 8

63. 23

64. $x = 9$

65. 2

66. $x = \frac{4}{3}$

67. 10^{1000}

68. $\log_5 117$

69. 13^{9000}

70. 17,353 mi

71. $2^{28} - 1$ cent = \$2,684,354.55

72. 3.689×10^{14} kg rice

Problem Set 5.2   (p. 173)

1. 2.6810215

2. 8.284252

3. −4.6459922

4. −2.216407

5. 6.8532634

6. 6.8532633

7. 1.3751734

8. 0.383139

9. 24.046754

10. 1.023267

11. 0.0432139

12. 22.459158

13. −4.1640482

14. 2107.021448

15. 2.4649735

16. −0.553981

17. 1.7858507

18. 4.647862

19. 0.2905096

20. 3.773707

21. 2.7588016

22. 1.600782

23. 2.1037232; −1.1037232

24. 0.921705; −1.921705   25. 2, 1, $\frac{1}{3}$, 0.01

26. 1.648721; 2.718282; 20.085537; 2.688117 × 10^{43}

27. $y - 3 = 3(x - e)/e$ 28. $y - 1 = 3(x - 1)$

29. $y - 4 = 4e(x - 1/e); y = 4ex$

30. $y - 1 = 9(x - 1)$ 31. (15, 12.124151)

32. $x < 2.857143$ 33. $x(1 + 2 \ln x)$

34. $(1 - 3 \ln x)/x^4$ 35. $(6 \ln x)/x$

36. $1/[2x(2 + \ln x)^{1/2}]$ 37. $-12/[x(\ln x)^4]$

38. $(x^2 + 2 - 2 \ln x)/x^2$

39. $6(x^2 + 2 \ln x)^2(x + x^{-1})$

40. $(1.2)(3x + \ln x)^2(3 + x^{-1})$

41. $(2x + 3)e^x$ 42. $(3x^2 + 6x)e^x$

43. $(6x + 2xe^x - x^2e^x)/(e^x + 3)^2$

44. $2e^x(3x - 4)/(3x - 1)^2$

45. $(e^x + x^4)/(5e^x + x^5)^{4/5}$

46. $3(e^x + 2x)^2(e^x + 2)$

47. (a) Inc $(-1, \infty)$, dec $(-\infty, -1)$;   (b) $f(-1) = -1/e$;
(c) up $(-2, \infty)$, down $(-\infty, -2)$;
(d)

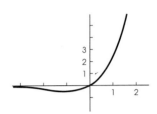

48. (a) Inc $(0, 2)$, dec $(-\infty, 0) \cup (2, \infty)$;
(b) no maximum value;
(c) up $(-\infty, 2 - \sqrt{2}) \cup (2 + \sqrt{2}, \infty)$; down
$(2 - \sqrt{2}, 2 + \sqrt{2})$;
(d)

49. (a) 2.9611202;   (b) 0.6714003

50. (a) 7.710692;   (b) $x = -2, 3$;   (c) 6.138036;
(d) 1.222870, −2.22287

51. $\ln[9x(x + 1)]$

52. (a) x^2;   (b) 8;   (c) $|x|$;   (d) 2

53. (a) 48;   (b) $-\frac{5}{54}$

54. (a) $(x^3 + 6x^2 + 6x)e^x$;
(b) $(x + \ln x)[12x + 15 + 6 \ln x + x^{-1}(3 \ln x) + 6x^{-1}]$

55. (a) Inc $(1, \infty)$, dec $(0, 1)$;   (b) $f(1) = e$;
(c) up $(0, \infty)$;
(d)

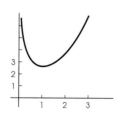

Problem Set 5.3   (p. 179)

1. $12e^{3x}$

2. $-6e^{-3x}$

3. $2xe^{x^2}$

4. $9x^2e^{x^3}$

5. $8/(2x + 5)$

6. $-6x/(1 - x^2)$

7. $(5x^2 - 1)/(5x^3 + x)$

8. $(x^2\sqrt{x} - 4)/[(2x)(x^2\sqrt{x} + 1)]$

9. $6e^{4\sqrt{x}}/\sqrt{x}$

10. $2e^{1/x}/x^2$

11. $\dfrac{3}{3x + 1} - \dfrac{2x}{x^2 + 2}; \dfrac{-3x^2 - 2x + 6}{(3x + 1)(x^2 + 2)}$

12. $(2x - 2)/[x(2x - 1)]$

13. $\dfrac{6e^{2x/(x+3)}}{(x + 3)^2}$

14. $\dfrac{2 - x}{x^3}e^{(1-x)/x^2}$

15. $\dfrac{4x}{2x^2 + 5} + \dfrac{3}{3x - 2}; \dfrac{18x^2 - 8x + 15}{(2x^2 + 5)(3x - 2)}$

16. $\dfrac{5}{5x - 3} + \dfrac{3x^2 + 2}{x^3 + 2x}; \dfrac{20x^3 - 9x^2 + 20x - 6}{(5x - 3)(x^3 + 2x)}$

17. $(3x^2 + 2x)e^{3x}$

18. $2 \ln 5x + 2 + x^{-1}$

19. $-4xe^{-2x}$

20. $\dfrac{4}{3x(4x + 1)} - \dfrac{\ln(4x + 1)}{3x^2}$

21. $\dfrac{2(x + 3)}{x^2 + 4} - \dfrac{3 \ln(x^2 + 4)}{x^2}$

22. $e^{2/x}(\frac{5}{2}x^{-1/2} - 10x^{-3/2})$

23. $2e^{x^2}(x^{-1} + x \ln x^2)$

24. $\left[\dfrac{2}{2x + 1} - 2 \ln(2x + 1)\right]\Big/ e^{2x+1}$

25. $(2 + 5e^{5t})/(2t + e^{5t})$

26. $(6t - e^{2t})/(t^2 - e^{2t})$

27. $e^t(t - 1)^2/(t^2 + 1)^2$

28. $3(t^3 + t^2 - 2)e^{3t}$

29. $1/(t \ln t)$

30. e^{t+e^t}

31. $1/[t \ln t \ln(\ln t)]$

32. $(2t - 1)/[(t^2 - t) \ln(t^2 - t)]$

33. $18e^{2x}(3e^{2x} - 1)^2$

34. $(1 + e^{2x})(2x + e^{2x})^{-1/2}$

35. $8x[\ln(x^2 + 4)]^3/(x^2 + 4)$

36. $5(2x + 3 \ln x)^4(2 + 3x^{-1})$

37. $-8(e^x + \ln x)^{-3}(e^x + x^{-1})$

38. $3(x + 1)e^x(2xe^x + 4)^{1/2}$

39. $x(e^x + 2)^2(3xe^x + 2e^x + 4)$

40. $\dfrac{(x + \ln x)\left[2 + \dfrac{2}{x} - x - \ln x\right]}{e^x}$

41. $4(2x + 1)e^{(2x+1)^2}$

42. $4e^{4x+2}$

43. $\dfrac{3(x + 4)[\ln(x + 4 \ln x)]^2}{x(x + 4 \ln x)}$

44. $\dfrac{3(1 + \ln x)[\ln(x \ln x)]^2}{x \ln x}$

45. (a) $6e^{3t} + 4e^{-t}$; (b) $6t/(t^2 - 4)$;
(c) $2t(e^t - \ln t) + t^2(e^t - t^{-1})$;
(d) $(2t^2 - 2t - 10)e^{2t}/(t^2 - 5)^2$;
(e) $4(2t + 1)e^{(2t+1)^2}$;
(f) $(8\sqrt{t} + 2)/(2t\sqrt{t} + t)$;
(g) $4[\ln(2t + \sqrt{t})]^3(4\sqrt{t} + 1)/(4t\sqrt{t} + 2t)$;
(h) $6(4t - e^{2/t})^5(4 + 2t^{-2}e^{2/t})$

46. (a) $2s - 10e^{2s}$; (b) $2(1 + 2s + 3s^2)/(s + s^2 + s^3)$;
(c) $(2s - 1 - e^s)/e^{2s}$;
(d) $[6s^2/(s^2 + 1)] + 3 \ln(s^2 + 1)$;
(e) $s/[3 + s^2 + 2(3 + s^2)^{1/2}]$;
(f) $-12(2 - s)^3e^{(2-s)^4}$;
(g) $(-20s^{-2} + 12e^{3s})(5s^{-1} + e^{3s})^3$;
(h) $3s^5\left[\dfrac{5s}{1 + 5s} + 2 \ln(1 + 5s)\right](\ln(1 + 5s))^2$

47. $2e^{-x^2}(2x^4 - 5x^2 + 1)$ 48. $\frac{3}{2}[(2 + \ln 2)^2](x - 1)$

49. $y = 3x$

50. (a) 51;   (b) 11 wk;   (c) 9.45

51. (a) \$16,404.98;   (b) \$2213.64/yr;   (c) about 7.7 yr

52. (a) 33.53;   (b) -3;   (c) 2.32 mo

53. (a) About \$813,011;   (b) \$2.04/dollar;
(c) An additional dollar of advertising will produce an additional profit of \$2.04 (\$1.04 after deducting the additional advertising cost).

54.

Problem Set 5.4   (p. 184)

1. $\ln 29/\ln 3 \approx 3.065044$

2. 1.639739

3. $e^{2.63 \ln 4} \approx 38.319319$

4. 8.804406

5. $\ln 0.432/\ln 6 \approx -0.4684388$

6. -1.443229

7. $e^{2.4 \ln 3.2} \approx 16.30647$

8. 1.653053

9. $e^{t \ln 4.71} \approx e^{1.5496879t}$

10. $e^{t\ln(0.0987)}$

11. $e^{-2t\ln 8.12} \approx e^{-4.1886603t}$

12. $(1/\ln 3)\ln t$

13. $3\ln t/\ln 6 \approx 1.6743319 \ln t$

14. $\sqrt{2}\ln t/\ln \pi$

15. $1/(x\ln 5)$

16. $1/(x\ln 9)$

17. $4^x \ln 4$

18. $10^x \ln 10$

19. $2/[(2x+1)\ln 3]$

20. $-3/[(1-3x)\ln 5]$

21. $-6^{4-5x^2}(10x\ln 6)$

22. $2e^{x^2+2x}(x+1)\ln 3$

23. $18x/[(3x^2+1)\ln 4]$

24. $10(3x-4)/[(x^2-2x)/\ln 2]$

25. $5^{4e^x+3}(4e^x\ln 5)$

26. $2^{x+e^{1/x}}(\ln 2)(1-e^{1/x}x^{-2})$

27. $2^x 3^{-2x}\ln\frac{2}{9}$

28. $3^{2x}[1/(x\ln 2)+(\ln 9)\log_2 3x]$

29. $(2x+1)^x[2x/(2x+1)+\ln(2x+1)]$

30. $(3x+2)^x[3x/(3x+2)+\ln(3x+2)]$

31. $2(x^2+5x)^{2x}[(2x+5)/(x+5)+\ln(x^2+5x)]$

32. $(3x-2x^2)^{3x}[(9-12x)/(3-2x)+3\ln(3x-2x^2)]$

33. $(5x)^{x^2-x-1}[x-x^{-2}+(2x+x^{-2})\ln(5x)]$

34. $4e^{-x}(2x+4)^{4e^{-x}}[1/(x+2)-\ln(2x+4)]$

35. $(e^x+4)^{\ln x}\left[\dfrac{e^x\ln x}{e^x+4}+\dfrac{\ln(e^x+4)}{x}\right]$

36. $e^x(3\ln x+2)^{e^x}\left[\dfrac{3}{x(3\ln x+2)}+\ln(3\ln x+2)\right]$

37. $x^{10}(2x-1)^6\left(\dfrac{10}{x}+\dfrac{12}{2x-1}\right)$

38. $-\dfrac{(3x+1)^3}{x^9}(12x+8)$

39. $\dfrac{(x+1)^3(x+2)^4}{(x+3)^5}\left(\dfrac{3}{x+1}+\dfrac{4}{x+2}-\dfrac{5}{x+3}\right)$

40. $-(2x-1)^4(x+3)^5(x-2)^2(28x^2-5x-57)$

41. $(x^2+1)^8(2x+5)^5\left(\dfrac{16x}{x^2+1}+\dfrac{10}{2x+5}\right)$

42. $-3(2x^2+11x-41)(3x+5)^3(x^2-x+9)^{-4}$

43. $\dfrac{\sqrt{x^2+2}\,(x+2)^4}{\sqrt[3]{3x-5}}\left(\dfrac{x}{x^2+2}+\dfrac{4}{x+2}-\dfrac{1}{3x-5}\right)$

44. $\dfrac{(3x+2)^{5/3}(x-1)^9}{e^{2x}}\left(\dfrac{5}{3x+2}+\dfrac{9}{x-1}-2\right)$

45. $6.6438562;\ 4.6051702;\ 2;\ 0.8691760$

46. $e^{-t\ln 10};\ e^{-t\ln 3};\ e^{t\ln 1.5};\ e^{t\ln 9}$

47. (a) 0;   (b) $(2e^{2x}+1)/[(e^{2x}+x)\ln 3]$;
   (c) $5^{3e^{2x}}(6e^{2x})(\ln 5)$;
   (d) $2x(4+3x^2)^{x^2}[3x^2/(4+3x^2)+\ln(4+3x^2)]$

48. $y-4=(32\ln 2)(x-2)$

49. -304

50. $0.227727282;\ 0.228576349$

51. $X+2Y=-5;\ m=-\frac{1}{2};\ b=-\frac{5}{2}$

53. (a) 3316;   (b) 4.899 hr

54. $10{,}000$ times as intense

55. Let L_w denote the loudness of a whisper in decibels. Then the ratio I_c/I_w of the intensity of the sound to that of a whisper is $10^{0.4L_w}$.

Chapter 5   Review Problem Set (p. 185)

1. (a) $\frac{1}{27}$;   (b) $\frac{1}{5}$;   (c) 3;   (d) $\frac{1}{3}$;   (e) $\frac{4}{3}$;   (f) $\pi+1$

2. (a) $\frac{1}{16}$;   (b) $\frac{1}{27}$;   (c) 4;   (d) $\frac{1}{2}$;   (e) $\frac{3}{2}$;   (f) $\frac{5}{2}$

3. (a) 118.9043;   (b) 8.611396;   (c) 7.550287;
   (d) 139.0456;   (e) 1.169610;   (f) 2.200973

4. (a) 11047.66;   (b) 41.76467;   (c) 7.5490949;
   (d) 32.34114;   (e) 59.81125;   (f) 2.577074

5.

6.

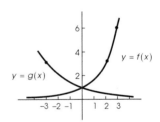

7. (a) $3\log_a 2+\log_a 5 \approx 1.895$;
   (b) $2\log_a 2+\log_a 3-2\log_a 5 \approx -0.377$;
   (c) $\frac{1}{2}\log_a 3+\frac{1}{2}\log_a 5-1 \approx -0.304$

8. (a) $\log_a 2+3\log_a 3 \approx 2.48$;
   (b) $\frac{1}{2}\log_a 7-\log_a 2-\log_a 3 \approx -0.510$;
   (c) $\log_a 3+\frac{3}{2}-\log_a 2 \approx 1.752$

9. $\ln(125x^2/y^5)$ 10. $\ln(\sqrt{15}x^3/y^2)$

11. (a) $x=(e^4-1)/3 \approx 17.86605$;
   (b) $x=2/(\ln\pi-3) \approx -1.07801$;
   (c) $x=-2.39911,\ 3.39911$;
   (d) $x\approx -2.14422,\ 2.14422$

12. (a) $x = (2 + e^3)/5$; (b) $x = (2 \ln 3 - 2 \ln \pi)/\ln 3$
    (c) $x = (1 - 3e^2)/(4e^2 - 2) \approx -0.77$, which is not greater than $-\frac{1}{2}$, so there are no solutions;
    (d) $\approx \pm 0.7170$

13. (a) $4xe^{x^2-3}$; (b) $-15/[x(4x + 3)]$;
    (c) $e^{x/(2x+1)}(2x + 1)^{-2}$;
    (d) $4x(x^2 + 3)/(x^2 + 6x + 10)$;
    (e) $2(x^2 + 4)/(2x + 1) + 2x \ln(2x + 1)$;
    (f) $e^{2x}(2x + 3)/(x + 2)^2$;
    (g) $8(3 + 2x \ln x)^3(1 + \ln x)$;
    (h) $\frac{3}{2}e^{-2x}(xe^{-2x} - 1)^{1/2}(1 - 2x)$

14. (a) $(x - 1)e^{x^2-2x+10}$; (b) $12(x^2 + 4)/(x^3 - 4x^2)$;
    (c) $18xe^{9x^2-4}$; (d) $x/[x^2 + 5 - 3(x^2 + 5)^{1/2}]$
    (e) $(2x - 1)^4[10 \ln(4 - 2x) + (2x - 1)/(x - 2)]$;
    (f) $0.5x^{-1/2}(4 - 3x)^{-2e^{2+\sqrt{x}}}(4 - 3x + 6\sqrt{x})$;
    (g) $5(x^2 + 2 \ln x)^4(2x + 2x^{-1})$;
    (h) $5xe^{4x}(x^2e^{4x} + 4)^{3/2}(2x + 1)$

15. (a) Inc $(1, \infty)$, dec $(-\infty, 1)$;
    (b) up $(0, \infty)$, down $(-\infty, 0)$;
    (c) $f(1) = -e \approx -2.72$;
    (d)

16. (a) Inc $(-\infty, 4)$, dec $(4, \infty)$;
    (b) up $(5, \infty)$, down $(-\infty, 5)$;
    (c) $g(4) = e^{-4} \approx 0.018$;
    (d)

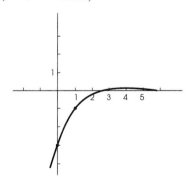

17. $x = -1$ and $x = 1$ 18. $x = -2$ and $x = 2$
19. (a) About 50; (b) decreasing by 3.2 points/mo;
    (c) 9 mo
20. (a) 1000; (b) 1778 bacteria/h; (c) 1.40 hr from now
21. $e^{t \ln(2\pi)} \approx e^{1.837877066t}$
22. $\exp([\ln(\sqrt{3} + 2)]t) \approx e^{1.316957887t}$

23. $(\ln 5)^{-1} \ln t \approx 0.6213349 \ln t$
24. $\dfrac{3}{\ln 8} \ln t \approx 1.442695041 \ln t$
25. (a) $2x^{-1/2}10^{4\sqrt{x}} \ln 10$; (b) $36x/[(3x^2 + 5) \ln 4]$
26. (a) $(4 + t^2)^{-1/2}7^{\sqrt{4+t^2}}t \ln 7$;
    (b) $(2t + 2)/[(t^2 + 2t - 4) \ln 5]$
27. (a) $4(3 \ln 2 + 2) \approx 16.3178$;
    (b) $155/(3\sqrt{3}) \approx 29.82976$
28. (a) $g'(2) = 2$; (b) $\frac{1007}{54}$
29. Let $Y = \ln y$, $X = x$; then $3x + Y = (4 + \ln\ 4)$. Slope is -3, Y-intercept is $4 + \ln 4 \approx 5.39$.
30. Slope is π and Y-intercept is $(2 + \ln 3)$.
31. $y - 1 = \frac{5}{4}(x - 2)$ 32. $y - 3 = 15(x - 1)$
33. $18e \approx 48.9291$
34. (a) 24; (b) $76(10)^{-1/2} \approx 24.0333$
35. $f'(5) = 0.4 \ln 5 \approx 0.6438$
36. $f'(2) = 12e^{12} \approx 1953057.497$
37. (a) $f(e) = e^\pi - \pi^e \approx 0.68 > 0$;
    (b) $f'(e) \approx 1.03 > 0$
38. (a) (e, ∞); (b) 1;
    (c) $f'(e^4) = 1/(4e^4 \ln 4) \approx 0.0033$
39. (a) $(0, 1) \cup (e, \infty)$; (b) $(\sqrt{e}, \infty)$; (c) $(0, e^{3/2})$
40. (a) (e, ∞); (b) $x > e^{1/4} \approx 1.2840$;
    (c) $f''(e) = -\frac{1}{4}e^{-2} \approx 0.0338$

Problem Set 6.1   (p. 194)

1. (a) $dy/dt = 5e^{0.03t}(0.03) = 0.03y$;
   (b) $dy/dt = 20e^{-0.6t}(-0.6) = -0.6y$
2. $dy/dt = y_0e^{kt}(k) = ky$ 3. $y = 16e^{0.8t}$; 9629.52
4. 24.6532   5. $y = 20e^{-0.05t}$; 13.4064
6. 38.1254
7. $y = 3.21e^{0.193t}$; 15.0333   8. 0.000030946
9. (a) 800; (b) $y = 100e^{0.009902t}$; (c) 328
10. (a) 9600; (b) $y = 600e^{0.213276056t}$; (c) 4749
11. (a) $y = 16,000e^{0.110146t}$; (b) ≈ 14.6118 h
12. (a) ≈ 9362; (b) 10.2678 h; 4.2678 h from now
13. About 35 yr   14. 37.88 yr
15. (a) $f(t) = 10.5e^{0.032965t}$; (b) about 21.7 million
16. 0.021770102
17. $k = -0.153691$; 17.1507 g
18. 3.21356 g   19. About 0.00015 s
20. 6.782 days   21. About 3841 yr
22. 5411 yr   23. About 4.74%
24. 440 yr   25. 160,000

26. 1 g
28. 5 yr
30. 103.44 thousand
32. (a) 348.2 mbar;   (b) 3973 ft
33. 2.8782 hr

27. 11,140 yr
29. $y = 0.723379e^{0.875469t}$
31. 0.0669 g

34. 6242

Problem Set 6.2   (p. 200)

1. $y = 8 - 6e^{-0.09t}$; 4.50351
2. 9.4557
3. $y = 20 - 8e^{-0.45t}$; 18.6776
4. 16.2858
5. (a) 84%;   (b) about 12.81 hr
6. (a) $f(t) = 24,000(1 - e^{-0.051082562t})$;   (b) 31.5 days
7. (a) $dt/at = K(T - T_m)$;   (b) 114.5°F
8. 71.7 min
9. (a) $W(t) = 72 - 72e^{-0.6t}$;   (b) 72 words/min;
   (c) 4.8 wk
10. 160.94 lb/acre   11. 18.2 cm
12. (a) 14.4 cm;
    (b) 7.37 mo; $[36/(1 + 17)]e^{-180k}$; $36/(1 + 17e)$
13. 34,866   14. 7.5×10^7
15. 21.4 days   16. 2.2953 yr
17. 3.477 yr   18. 32 sets
19. 63.69 h   20. 563 members
21. 23,030 children   22. 1377.9 persons/yr
23. 690 lb/acre   24. 663 lb/acre

Problem Set 6.3   (p. 206)

1. $1320   2. $1016
3. $1713   4. $2230
5. $3741.99   6. $4779.56
7. $1360.49   8. $1036.02
9. $3392.07   10. $3270.47
11. (a) $2367.36;   (b) $2411.71;   (c) $2435.19;
    (d) $2451.36;   (e) $2459.33
12. (a) $2158.92;   (b) $2191.12;   (c) $2208.04;
    (d) $2219.64;   (e) $2225.35
13. $7389.59   14. $6465.71
15. $2459.60   16. $2225.54
17. $6946.51   18. $7175.64
19. 10.3813%   20. 9.3807%
21. 9.3069%   22. 12.7475%
23. 8.2% compounded semiannually
24. 10% compounded quarterly

25. 9.6% compounded monthly
26. 10% compounded annually
27. About 76 mo   28. 2530 days
29. 7.2963 yr   30. 6305.5 days
31. (a) $4243.20;   (b) $5157.30;   (c) $5172.56;
    (d) $5173.08
32. $3768.18   33. $1770.13
34. $2390.44   35. 5727 days
36. 15.68616918   37. 8.3619%
38. 7.3465%   39. 7.3241%
40. 16.9421%   41. $4591.59
42. $7036.63

Problem Set 6.4   (p. 212)

1. $4033.76   2. $8560.03
3. $1126.31   4. $890.37
5. $1878.51   6. $1593.15
7. $2849.91   8. $2571.26
9. $3770.30   10. $6127.88
11. $6479.44   12. $5151.29
13. $3715.40   14. $5084.09
15. $3754.95   16. $9755.00
17. $5176.84   18. $3247.05
19. $11,481.70; 2.80193%   20. $47,943.31
21. The true growth rates are 3.84615% and 3.63636%, so
    the first is better.
22. $j \leq i/(i + 2)$
23. (a) $25,600;   (b) $13,107.20;   (c) $5368.71
24. (a) $38,580.25;   (b) $18,605.44;   (c) $8972.53
25. 5.885 yr   26. 3.8 yr
27. $3135.19   28. $518.17
29. $4651.24   30. $34,651.06
31. (a) $6304.83;   (b) $422.03
32. $1195.58   33. 9.5958%
34. 89%   35. 7.6 yr

Chapter 6   Review Problem Set (p. 213)

1. $y = 80e^{-0.029t}$; 43.51
2. $y = 1200e^{0.013t}$; 1458.37
3. $y = 12 - 9e^{-0.07t}$; 5.66
4. $y = 20 - 12e^{-0.045t}$; 12.685
5. $y = 20/(1 + 1.5e^{-t})$; 19.13
6. $30/(1 + 2e^{-1.2t})$; 29.96

7. (a) $y(t) = 2400e^{0.13412t}$; (b) 4693 bacteria;
(c) 368 bacteria/h; (d) 14.14 h

8. (a) $y(t) = 26e^{0.035357848t}$; (b) about 45.8 million

9. About 22,600 yr old   10. $t \approx 0.00015$ s

11. 44,197 people   12. 80.612 people

13. 7438 deer

14. $t \approx 9.47$ (about the middle of the 10th day)

15. (a) \$5400.00; (b) \$6476.77; (c) \$6658.92;
(d) \$6676.62

16. (a) \$7740.00; (b) \$8966.53; (c) \$9171.46;
(d) \$9244.95

17. (a) $t \approx 8.33$ yr (17 semiannual periods);
(b) $t \approx 8.154672712$ yr (8 yr, 56 days, 10 h, 55 min, 59 s)

18. (a) 57 mo;
(b) $t \approx 4.697759$ yr (4 yr, 8 mo, 11 days, 4 h, 38 min, 15 s)

19. 9.8157%

20. 9.2% compounded monthly

21. \$1775.12   22. \$4553.59

23. 7.7593%   24. 8.147% is realized.

25. \$3564.22 and \$7128.44

26. \$2719.80 and \$8159.40

27. (a) \$39,321.60;
(b) $n \approx 3.106$; about 3 yr, 39 days

28. \$2252.63   29. 9481 wolves

30. Virtually none left.   31. \$52,641.04

32. 9.5%

33. $dy/dt = k(L - y_0)e^{-kt} = k(L - y)$

34. $dy/dt = 10,000ke^{-100kt}(1 + e^{-100kt})^{-2}$
$= ky(100 - y)$

35. $y' = (2A + B + 2Bt)e^{2t}$; $y'' = (4A + 4B + 4Bt)e^{2t}$;
$\therefore y'' - 4y' + 4y = 0$

36. $y' = (3B + C)e^{3t} + 3Cte^{3t}$;
$y'' = (9B + 6C)e^{3t} + 9Cte^{3t}$;
$y'' = (27B + 27C)e^{3t} + 27Cte^{3t}$;
$\therefore y'' - 6y'' + 9y' = 0$

37. $dy/dt = kyt$; it will be faster.

38. (a) $dv/dt = kv^{-1/2}$
(b) $v = 4$ and $t = 0 \to B = 8$, so $v = (Kt + 8)^{2/3}$
$v = 9$ and $t = 2 \to K = 9.5$, so $v = (9.5t + 8)^{2/3}$

39. $d^2y/dt^2 = k(dy/dt)^2$

40. $t = 6 \ln(1.5)/\ln 2 \approx 3.5098$ h

Problem Set 7.1 (p. 222)

1. Antiderivative.   2. Constant.

3. $f(x)$ 4. $f(x) + C$

5. $4x^3$

6. $\frac{1}{5}x^5 + C$

7. $\frac{1}{4}x^4 + C$

8. $\frac{1}{6}x^6 + C$

9. $-\frac{1}{2}x^{-2} + C$

10. $-\frac{1}{6}x^{-6} + C$

11. $\frac{2}{7}x^{7/2} + C$

12. $\frac{4}{7}x^{7/4} + C$

13. $x^{\pi+1}/(\pi + 1) + C$

14. $x^{\sqrt{2}+1}/(\sqrt{2} + 1) + C$

15. $-\frac{1}{3}x^{-3} + C$

16. $\ln |x| + C$

17. $\ln |x| + C$

18. $\frac{2}{9}x^{9/2} + C$

19. $\frac{3}{2}x^2 + 2x + C$

20. $\frac{1}{3}x^3 + 2x^2 + C$

21. $x^5 - 3x^2 + C$

22. $\frac{3}{4}x^4 + x^2 - 3x + C$

23. $2x^{5/2} - 2x^{3/2} + C$

24. $3x^{4/3} + 3x + C$

25. $2x^{3/2} + 5 \ln |x| + C$

26. $-2x^{-2} + \frac{3}{2}x^{4/3} + C$

27. $-5x^{-1} + x^{-2} + C$

28. $-\frac{1}{2}x^{-4} - 3x + C$

29. $5e^x - 4 \ln |x| + C$

30. $-x^{-1} + 4e^x + C$

31. $\frac{9}{4}t^{4/3} + t^2 + C$

32. $3s^{4/3} - \ln |s| + C$

33. $\frac{1}{2}u^4 + 4u^{-1} + C$

34. $-v^{-5} + \frac{1}{6}v^6 + C$

35. $\frac{2}{3}x^3 + \frac{5}{2}x^2 - 3x + C$

36. $x^3 + \frac{5}{2}x^2 - 2x + C$

37. $\frac{4}{3}t^3 + 6t^2 + 9t + C$

38. $\frac{9}{5}u^5 - 2u^3 + u + C$

39. $x + 4 \ln |x| - 2x^{-1} + C$

40. $2x^{3/2} + 10x^{-1/2} + C$

41. $2x^2 - 2x + 3$

42. $3x - x^2 + 5$

43. $\frac{1}{3}x^3 - x^2 + 5x - 1$

44. $x^3 + \frac{1}{2}x^2 - 4x + 3$

45. $4e^x - 2$

46. $e^x - x^2 + 4$

47. $3 \ln |x| + 2x^3 + 3$

48. $\frac{1}{3}x^3 - 2 \ln x + 3$

49. (a) $x^2 - 4 \ln |x| + C$; (b) $\frac{1}{3}x^3 + \frac{4}{5}x^{5/2} + \frac{1}{2}x^2 + C$;
(c) $3x + 6\sqrt{x} + C$; (d) $3 \ln |x| + 2x^{-2} + C$;
(e) $6s + 4e^s + C$; (f) $e^t + t^{e+1}/(e + 1) + C$

50. $f(x) = x^3 - e^x + 5x + 5$

51. \$3,487,500

52. $g(t) = 8t^{3/2} - 24t + 26$

53. $s(t) = 2t^3 - 2t^2 + 2t + 12$

54. 40 ft/s; 100 ft   55. 26,800

56. $M(10) = 175$ g

57. (a) \$278,000; (b) \$10.86/book

58. 26

Problem Set 7.2 (p. 228)

1. $\frac{32}{5}x^{15} + C$ both ways.   2. $4x^{20} + C$; $4x^{20} + C$

3. $\frac{1}{3}(x^2 + 3)^3 + C$; $\frac{1}{3}x^6 + 3x^4 + 9x^2 + C$

4. $\frac{1}{3}(2x^2 - 1)^3 + C$; $\frac{8}{3}x^6 - 4x^4 + 2x^2 + C$

5. $\frac{1}{6}(x^3 - 4)^6 + C$ 6. $\frac{1}{4}(2x^3 + 1)^4 + C$

7. $\frac{1}{28}(x^4 + 5)^7 + C$ 8. $\frac{1}{30}(3x^2 - 4)^5 + C$

9. $\frac{1}{5}(x^2 + 6x + 4)^5 + C$ 10. $\frac{1}{4}(2x^2 + 8x - 3)^4 + C$

11. $\frac{1}{33}(x^3 + 3x - 2)^{11} + C$

12. $\frac{1}{24}(x^3 + 3x^2 - 11)^8 + C$

13. $\frac{2}{5}(\sqrt{x}+1)^5 + C$

14. $-\frac{1}{12}(3x^{-1}+1)^4 + C$

15. $\frac{1}{3}e^{3x} + C$

16. $-\frac{1}{2}e^{-2x} + C$

17. $\frac{3}{2}e^{2t+1} + C$

18. $\frac{4}{3}e^{3u-4} + C$

19. $\frac{1}{4}e^{2x^2+1} + C$

20. $-\frac{1}{6}e^{4-3x^2} + C$

21. $2e^{\sqrt{x}} + C$

22. $-\frac{1}{4}e^{4/x} + C$

23. $\ln|x^2+5x+4| + C$

24. $\ln|x^3-x+7| + C$

25. $\frac{1}{4}\ln|x^4+2x^2-2| + C$

26. $\frac{1}{8}\ln|2x^4-8x+5| + C$

27. $\frac{1}{3}\ln(3e^x+1) + C$

28. $\frac{1}{4}\ln|4e^x+8x| + C$

29. $\frac{2}{5}(4+\ln x)^{5/2} + C$

30. $\frac{2}{9}(2+3\ln x)^{3/2} + C$

31. $\frac{1}{10}[\ln(2x+1)]^5 + C$

32. $\frac{1}{4}(x\ln x)^4 + C$

33. $\frac{1}{4}[\ln(x^2+1)]^2 + C$

34. $\frac{1}{5}[\ln(\ln x)]^5 + C$

35. $\frac{1}{16}(x-4)^{16} + \frac{4}{15}(x-4)^{15} + C$

36. $\frac{1}{12}(x-2)^{12} + \frac{2}{11}(x-2)^{11} + C$

37. $\frac{1}{10}(2x+1)^{10} - \frac{1}{9}(2x+1)^9 + C$

38. $-\frac{1}{72}(1-3x)^8 + \frac{1}{81}(1-3x)^9 + C$

39. $\frac{1}{5}x^5 + \frac{10}{3}x^3 + 25x + C$

40. $\frac{4}{5}x^5 - 4x^3 + 9x + C$

41. $\frac{1}{2}x^2 + 4x^{3/2} + 9x + C$

42. $-9x^{-1} + 12\ln|x| + 4x + C$

43. $\frac{3}{2}x^2 + \ln|x| + 7x^{-1} + C$

44. $\frac{5}{2}x^2 - 2\ln|x| - \frac{7}{2}x^{-2} + C$

45. $2x + 10\sqrt{x} + C$

46. $\frac{1}{2}x^2 + 2x - 2x^{-1/2} + C$

47. $\frac{1}{2}x^2 + 2x + 3\ln|x-1| + C$

48. $\frac{1}{2}x^2 + 3x - 10\ln|x+2| + C$

49. $\frac{1}{2}x^2 + \frac{3}{2}\ln|x^2-1| + C$

50. $\frac{1}{2}x^2 + 2\ln(x^2+2) + C$

51. $\frac{8}{7}u^7 + \frac{12}{5}u^5 + 2u^3 + u + C$

52. $\frac{1}{48}(2t^2+1)^{12} + C$

53. $\frac{1}{3}e^{x^3} + C$

54. $\frac{1}{4}e^{4x} - 3e^{2x} + 9x + C$

55. $\frac{1}{12}(x+1)^{12} - \frac{1}{11}(x+1)^{11} + C$

56. $\frac{1}{4}(\ln x)^4 + C$

57. $\frac{1}{3}\ln|t^3+3t-4| + C$

58. $\frac{1}{5}t^5 + 4t^2 - 16t^{-1} + C$

59. $2\sqrt{t} + 2t + 3\ln|t| + C$

60. $\frac{1}{2}t^2 + 6t - 18\ln|t+3| + C$

61. $\frac{1}{2}[\ln(e^t+2)]^2 + C$

62. $u = t + 2;$
$-\dfrac{1}{t+2} + \dfrac{3}{2(t+2)^2} - \dfrac{1}{(t+2)^3} + \dfrac{1}{4(t+2)^4} + C$

63. $y = \ln|x^2+3x+1| + 3$

64. \$4377.33

65. (a) \$8495.26; (b) \$106.19

66. 2880

67. (a) $s(t) = 10(3-0.2t)^{5/2} - 50(3-0.2t)^{3/2} + 50$;
(b) $s(15) - s(0) = 103.92$ ft

Problem Set 7.3 (p. 235)

1. $(\frac{1}{2}x - \frac{1}{4})e^{2x} + C$

2. $(-\frac{1}{3}x - \frac{1}{9})e^{-3x} + C$

3. $x^2(\frac{1}{2}\ln x - \frac{1}{4}) + C$

4. $x^{3/2}(\frac{2}{3}\ln x - \frac{4}{9}) + C$

5. $-(1/x)(\ln x + 1) + C$

6. $\frac{2}{9}x^{9/2}(\ln x - \frac{2}{9}) + C$

7. $\frac{1}{7}x(1+x)^7 - \frac{1}{56}(1+x)^8 + C$

8. $[(2x+1)^7(14x-1)]/[(14)(16)] + C$

9. $-\frac{1}{2}x^2e^{-2x} - \frac{1}{2}xe^{-2x} - \frac{1}{4}e^{-2x} + C$

10. $\frac{1}{32}(8x^2-4x+1)e^{4x} + C$

11. $\frac{1}{2}x^2\ln(x+2) - \frac{1}{4}x^2 + x - 2\ln(x+2) + C$

12. $(x-2)\ln(x-2) - x + C$

13. $\dfrac{3}{x+1} - \dfrac{3}{x+2}$

14. $\dfrac{4}{x-3} - \dfrac{4}{x+3}$

15. $\dfrac{3}{x} + \dfrac{2}{x-3}$

16. $\dfrac{4}{x-1} - \dfrac{3}{x+2}$

17. $\dfrac{9}{x+1} - \dfrac{12}{x+2} + \dfrac{3}{x+3}$

18. $\dfrac{1}{x-2} + \dfrac{2}{x-1} + \dfrac{3}{x+2}$

19. $\dfrac{5}{x} + \dfrac{2}{x+3} - \dfrac{4}{x-2}$

20. $\dfrac{4}{x+3} - \dfrac{5}{x-3} + \dfrac{2}{x-1}$

21. $3\ln|x+1| - 3\ln|x+2| + C$

22. $4\ln|x-3| - 4\ln|x+3| + C$

23. $3\ln|x| + 2\ln|x-3| + C$

24. $4\ln|x-1| - 3\ln|x+2| + C$

25. $9\ln|x+1| - 12\ln|x+2| + 3\ln|x+3| + C$

26. $\ln|x-2| + 2\ln|x-1| + 3\ln|x+2| + C$

27. $5\ln|x| + 2\ln|x+3| - 4\ln|x-2| + C$

28. $4\ln|x+3| - 5\ln|x-3| + 2\ln|x-1| + C$

29. $2e^{2x} + C$

30. $-2e^{-3x} + C$

31. $-\frac{1}{5}(x+2)e^{-5x} - \frac{1}{25}e^{-5x} + C$

32. $e^{4x}(\frac{1}{4}x - \frac{13}{16}) + C$

33. $\frac{1}{8}e^{4x^2} + C$

34. $-\frac{1}{2}e^{x^2}(x^2+1) + C$

35. $2\sqrt{x}(\ln x - 2) + C$

36. $\frac{1}{5}(\ln x)^5 + C$

37. $\ln|x^2-x-2| + C$

38. $\ln|x^2+3x-10| + C$

39. $\frac{3}{7}\ln|x+5| + \frac{11}{7}\ln|x-2| + C$

40. $-\ln |x - 2| + 4 \ln |x + 1| + C$
41. $4 \ln (x + 1) + (x + 1)^{-1} + 3 \ln |x - 4| + C$
42. $-3 \ln |x - 2| - 5(x - 2)^{-1} - 7 \ln |x + 1| + C$
43. $8e^{16} - \frac{1}{2}e^{16} + \frac{1}{2} \approx 66{,}645{,}829.4$ ft
44. $f(x) = \frac{1}{3}x^3 \ln x - \frac{1}{9}x^3 + \frac{19}{9}$
45. 18.425 mg   46. 15,924

Problem Set 7.4   (p. 240)

1. $\frac{1}{3}y^3 = \frac{1}{2}y^2 + C$ or $2y^3 = 3y^2 + K$
2. $y^2 = 2 \ln |x| + C$
3. $y - y^3 = \frac{1}{3}x^3 + C$ or $3y - 3y^3 = x^3 + K$
4. $\frac{2}{5}y^{5/2} = x + \frac{1}{3}x^3 + C$
5. $-1/y = \frac{1}{2} \ln (1 + x^2) + C$ or
$$-1/y = \ln (K\sqrt{1 + x^2})$$
6. $e^y = (x + x^2) + C$
7. $\frac{1}{2} \ln (1 + y^2) = \frac{1}{4} \ln (1 + x^4) + \ln C$ or
$$1 + y^2 = K\sqrt{1 + x^4}$$
8. $-1/(y + 4) = 1/(2 - x) + C$
9. $y = K(x - 1)^2/x^2$
10. $3 \ln(x - 1) + 4 \ln(x + 1) = \frac{1}{2}y^2 + C$
11. $ye^y - e^y = x^3 + C$
12. $y^{3/2} = x^{3/2}(\ln x - \frac{2}{3}) + C'$
13. $y^2 = \frac{1}{2}x^2 + 14$ 14. $y^{4/3} = x^{4/3} + 15$
15. $y = 10/[(t^2 + 2)^5 - 22]$
16. $y = \frac{3}{8}(2t + 1)^{4/3} - \frac{3}{8}$
17. $y = 4e^{0.02t}$ 18. $y = 10e^{-0.18t}$
19. $y = 5e^{0.36(1-t)}$ 20. $y = 20e^{0.0125t - 0.05}$
21. $y = 12 - 4e^{-4t}$ 22. $y = 20 - 7e^{-0.9t}$
23. $y = 60/(5 + e^{-0.36t})$ 24. $y = 30/(5 + e^{-0.24t})$
25. (a) $x = 80p^{-3/4}$; (b) $x = \frac{80}{27} \approx 2.96$
26. 5.4549   27. $y = y_0 e^{-4t}$
28. $-\frac{1}{2}y^{-2} = \frac{1}{3}x^3 + C$ 29. $y^2 = 2e^x + C'$
30. $\ln y = \frac{1}{2}(\ln x)^2 + C$ 31. $-(1/y) = \frac{1}{2}e^{x^2} + C$
32. $\ln y = -xe^{-x} - e^{-x} + C$
33. $y = K(x - 2)^4/x^4$
34. $\ln[(y - 5)/(10 - y)] = 5x + C'$
36. $x \cdot p(x) = C$ 37. 333 rabbits
38. (a) $dp/dt = -k(1 - p)$; (b) 48.8%
39. (a) $dy/dt = ky(5000 - y)$; (b) 2848
40. (a) $dp/dt = 0.02P + 20{,}000$; (b) 18,393,721
41. $875.13   42. 2485

Chapter 7 Review Problem Set (p. 242)

1. (a) $x^8 - x^4 + \pi x + C$; (b) $2x^\pi - 4x^{5/2} + C$;
  (c) $-4 \ln |x| + \frac{1}{2}e^{4x} + C$; (d) $e^{x^3} - 2x + C$
2. (a) $3x^5 - 2x^4 + x^2 - \sqrt{3}x + C$;
  (b) $3x^{\sqrt{2}} - 4x^{7/2} + C$;
  (c) $2e^{2x} - \frac{2}{3}x^{3/2} - 5 \ln x + C$;
  (d) $\ln |x^3 + x - 4| - x^{-1} + C$
3. (a) $x^3 - 6x^2 + 5x + C$; (b) $2x^{3/2} - 5x^{-1} + C$;
  (c) $2x^2 + 4x + \ln |x| + C$; (d) $\frac{4}{3}e^{3x} - 9x^{2/3} + C$
4. (a) $x^4 - 2x^3 - 3x + C$;
  (b) $\frac{4}{3}x^3 + 4 \ln |x| - \frac{1}{3}x^{-3} + C$;
  (c) $2t^{5/2} + 8t^{1/2} + C$;
  (d) $\frac{1}{2}e^{2x} + 3x^{e+1}(e + 1)^{-1} + C$
5. $F(x) = -4x^{-1} + 4x + \frac{1}{3}x^3 - \frac{17}{3}$
6. $F(x) = \frac{1}{4}x^4 + 3x^2 + 12 \ln |x| - 4x^{-2} + 4x + \frac{3}{4}$
7. $f(t) = t^{-3} + 4t - 12$ 8. $f(t) = 4t^{5/2} + 2t - 3$
9. $C(x) = 280x - 6x^2 + 0.2x^3 + 700$
10. $C(x) = 2000 + 5x - 0.04x^2 + 0.04x^3$
11. 38,500 deer for January 1, 2005.
12. On January 1, 2000, the population will be about 26,400.
13. (a) $\frac{1}{5}(x^3 + 2x)^5 + C$; (b) $\frac{1}{8}(e^{2x} + 4)^4 + C$;
  (c) $\ln |x^2 + x - 7| + C$;
  (d) $-\frac{1}{8}(x^4 + 2x^2 - 7)^{-2} + C$; (e) $-(\ln x)^{-1} + C$;
  (f) $\frac{1}{11}(x + 3)^{11} - \frac{3}{10}(x + 3)^{10} + C$;
  (g) $\frac{1}{2}x^2 + 4x - 3 \ln |x - 3| + C$;
  (h) $\frac{1}{8}x^8 + \frac{4}{5}x^5 + 2x^2 + C$
14. (a) $\frac{1}{4}(2x^3 - 5x + 4)^4 + C$; (b) $\frac{1}{5}(5 + \ln t)^5 + C$;
  (c) $\ln(2x^2 - x + 8) + C$;
  (d) $-\frac{1}{3}(2t^3 + 6t - 5)^{-3} + C$; (e) $2(\ln x)^{3/2} + C$;
  (f) $\frac{1}{20}(x - 4)^{20} + \frac{4}{19}(x - 4)^{19} + C$;
  (g) $\frac{2}{3}x^3 + \frac{1}{2}x^2 - 2x + \frac{5}{2} \ln |2x - 1| + C$;
  (h) $\frac{1}{11}y^{11} + \frac{6}{7}y^7 + 3y^3 + C$
15. (a) $\frac{1}{4}xe^{4x} - \frac{1}{16}e^{4x} + C$;
  (b) $-2x^{-1/2} \ln x - 4x^{-1/2} + C$
16. (a) $-\frac{1}{3}xe^{-3x} - \frac{1}{9}e^{-3x} + C$;
  (b) $14x^{7/2} \ln x - 4x^{7/2} + C$
17. (a) $2 \ln |x - 4| + \ln |x + 3| + C$;
  (b) $\ln |x| + 2 \ln |x + 4| - \ln |x - 3| + C$
18. (a) $3 \ln |x + 5| - 2 \ln |x - 2| + C$;
  (b) $4 \ln |x + 3| - (2 \ln |x| + \ln |x - 3|) + C$
19. (a) $-y^{-1} = 2x^4 + C$; (b) $2y^{1/2} = \frac{1}{2}(\ln t)^2 + C$
20. (a) $2y^{1/2} = 2x^3 + C$;
  (b) $\ln(e^s + 2) = t - \ln |t + 1| + C$

21. (a) $-y^{-2} = e^x - 10$;
    (b) $y^{-1} = \frac{4}{3}(5 + x^2)^{3/2} - 34$
22. (a) $y^{-1} = \frac{1}{12}(x^3 + 2)^4 + \frac{5}{4}$;
    (b) $e^y = \ln(e^x + 2) + 1 - \ln 3$
23. $y^2 = 4(x^3 + 3)^{3/2} - \frac{7}{2}$.
24. $y = -\frac{1}{4}[(1 - 2\sqrt{x})^4 + 9]$.
25. (a) $dy/dt = ky(25{,}000 - y)$, $y_0 = 50$;
    (b) about 16,300 people.
26. (a) $dy/dt = ky(50{,}000 - y)$, $y_0 = 10$;
    (b) about 49,000 people.
27. (a) Concave up nowhere;   (b) $(-3, \infty)$;   (c) $(2, \infty)$;
    (d) $y = 1$ is a horizontal asymptote.
    (e)

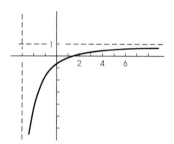

28. (a) Concave up nowhere;   (b) $(-1, \infty)$;
    (c) $(2, \ln 3 + 1)$;
    (d) $x > e^{-1} - 1 \approx -0.6321$;
    (e)

11. 3

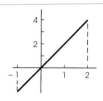

$\int_{-1}^{2} f(x)\,dx = 4 - 1 = 3$

12. -11.25

$\int_{-4}^{1} f(x)\,dx = .75 - 12 = -11.25$

13. 3

$\int_{0}^{6} f(x)\,dx = 4 - 1 = 3$

14. 4

$\int_{0}^{4} f(x)\,dx = \frac{1}{2} \cdot \frac{5}{2} \cdot 5 - \frac{1}{2} \cdot \frac{3}{2} \cdot 3 = \frac{25}{4} - \frac{9}{4} = 4$

15. -6

$\int_{0}^{2} f(x)\,dx = -6$

16. -24

$\int_{2}^{6} f(x)\,dx = -\frac{1}{2}(4)(12) = -24$

17. 4

$\int_{-1}^{3} f(x)\,dx = 5 - 1 = 4$

18. 2

$\int_{-2}^{4} f(x)\,dx = \int_{-2}^{0} f(x)\,dx = 2$

Problem Set 8.1   (p. 250)

1. 20   2. 24
3. $6 + \frac{25}{8}\pi \approx 15.82$ 4. 36.2916
5. $9\sqrt{3} \approx 15.59$ 6. 84.82
7. $\frac{3}{2}(\sqrt{7} + \sqrt{27}) \approx 11.763$ 8. $54\sqrt{3}$
9. $\frac{1}{2}\pi(5)^2 - \frac{1}{2}(6)(8) \approx 15.27$
10. 14

19. 5.5

$$\int_{-2}^{5} f(x)\,dx = 6 - \frac{1}{2} = \frac{11}{2}$$

20. −5

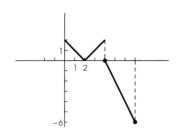

$$\int_{0}^{7} f(x)\,dx = 2\left(\frac{1}{2}\right)(2)(2) - \frac{1}{2}(3)(6) = 4 - 9 = -5$$

21. −1

$$\int_{0}^{5} f(x)\,dx = \frac{3}{2} - \frac{5}{2} = -1$$

22. $6\frac{1}{2}$

$$\int_{-2}^{5} f(x)\,dx = \int_{2}^{4} f(x)\,dx + \int_{4}^{5} f(x)\,dx = \frac{2+4}{2}\cdot 2 + \frac{1}{2} = 6\frac{1}{2}$$

23. $2\pi - 10$

24. $-(9 - 9\pi)/4$

25. 4.5

26. 2

27. 9

28. 4

29. $\frac{1}{5}\pi^5$

30. 1

31. $\frac{16}{3}$

32. 19.2

33. (a) Odd; (b) even; (c) neither; (d) odd.

34. (a) Even; (b) odd; (c) even; (d) even.

35. 0

36. $\frac{16}{3}$

37. 0.4

38. 0

39. 4

40. $\frac{32}{3}$

41. 0

42. 0

43. 4

44. 4

45. $\frac{5}{2}$

46. 4.5

47. $\frac{9}{2}\pi$

48. 4π

49. 6

50. 2

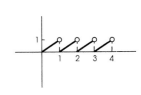

51. 6.5

52. 8

53. 6.5

54. 8.5

Problem Set 8.2   (p. 258)

1. 8

2. 4

3. $-\frac{3}{2}$

4. 3

5. $-\frac{9}{8}$

6. $-\frac{5}{3}$

7. 70

8. 15

9. 18

10. $\frac{14}{3}$

11. 0

12. 28.5

13. $\frac{64}{3}$

14. 4

15. 45

16. 1270

17. 1

18. $-\frac{4}{3}$

19. $5e - 6$

20. $5 - 4e^{-1}$

21. $\frac{76}{3}$

22. 9

23. $\frac{96}{5}$

24. $\frac{64}{5}$

25. 8

26. -2

27. 16

28. 2

29. 14

30. 0

31. -12

32. -14

33. 10

34. 6

35. 7

36. -4

37. $\frac{31}{6}$

38. $10\frac{1}{6}$

39. $\frac{7}{3}$

40. $\frac{7}{6}$

41. 3

42. 3

43. $\frac{9}{4}$

44. 0

45. 3

46. 2.340392219

47. 48

48. $\frac{3}{4} + 2 \ln 4$

49. $e^2/(e + 1)$

50. $8\frac{2}{7}$

51. 5

52. 26

53. -3

54. $e^2 - 1$

55. 1

56. 0.4

57. $e - 1$

58. $\frac{31}{5}$

59. (a) $3(e^{10} - 1) \approx 66{,}076.4$ in.³;
    (b) $10(e^3 - 1) \approx 190.9$ in.³

60. $e^2 + 1$

61. (a) 73.92°F;   (b) 72.44°F

62. (a) 6.928 s;   (b) $256\sqrt{3}$ ft ≈ 443.405 ft;
    (c) 64 ft/s

63. 460 lb

Problem Set 8.3   (p. 263)

1. $\frac{2}{3}$; $\frac{2}{3}$

2. 171; 171

3. $\frac{14}{3}$; $\frac{14}{3}$

4. $10\frac{1}{6}$; $\frac{61}{6}$

5. $\frac{1}{5}$

6. $-\frac{5}{4}$

7. $\frac{98}{3}$

8. 1

9. $5 \ln 4$

10. $3 \ln 5$

11. $\frac{1}{4} \ln 9$

12. $\frac{1}{2} \ln 17$

13. $\frac{3}{2}(e^2 - 1)$

14. $2(1 - e^{-4})$

15. $\frac{1}{2}(1 - e^{-4})$

16. $2(e^2 - e)$

17. $\frac{81}{8}$

18. $-\frac{1}{15}(4^5)$

19. 21

20. $\frac{3}{5}$

21. $\frac{1}{9}(\ln 10)^3$

22. 0.6221

23. $\frac{1}{2}(e^9 - 1)$

24. $4[(1 + e)^4 - 1]$

25. $\frac{3}{4}e^4 + \frac{1}{4}$

26. $-4e^{-3} + 1$

27. $\frac{6}{25}e^{5/2} + \frac{4}{25}$

28. $\frac{1}{9}(8e^3 - 2e^{3/2})$

29. $\frac{1}{30}$

30. $\frac{32}{7} \approx 4.571$

31. $\frac{15}{2} - 8 \ln 2$

32. 1.5258

33. $\frac{15}{2} - 8 \ln 2$

34. $\frac{7}{300} = 0.02\overline{3}$

35. $\frac{3}{2} - \frac{5}{4} \ln 7$

36. $\frac{116}{135}$

37. $\ln[(\sqrt{34} + 5)/9] \approx 0.1852$

38. 1

39. $\frac{5}{2}(\sqrt{34} - 4) + \frac{9}{2} \ln[(\sqrt{34} + 5)/9] \approx 5.4107$

40. $\frac{1}{2}(e^4 - 1)$

41. $e^2 + 1 \approx 8.3891$

42. $\frac{26}{3}$

43. $\frac{5}{4}e^6 + \frac{1}{4} \approx 504.5360$

44. $\frac{3}{4} - \frac{1}{4} \ln 4$

45. $1/[15(\sqrt[3]{2} - 1)] \approx 0.2565$

46. $\frac{1}{4}$ cm

Problem Set 8.4   (p. 270)

1. $\frac{21}{5}$

2. 30

3. 38

4. 38

5. $\frac{3}{4}$

6. $\frac{14}{5}$

7. 31

8. 53

9. $\sum\limits_{i=1}^{10} i^3$

10. $\sum\limits_{i=1}^{50} (2i)$

11. $\sum\limits_{i=1}^{100} (3/2^i)$

12. $\sum\limits_{i=1}^{70} (5/i)$

13. $\sum\limits_{i=4}^{21} x_i$

14. $\sum\limits_{i=1}^{n} t_1$

15. $\sum\limits_{i=1}^{n} g(t_i)\, \Delta t$

16. $\sum\limits_{i=1}^{m} f(x_i)\, \Delta x$

17. $\sum\limits_{i=1}^{n} 3x_i\, \Delta x$

18. $\sum\limits_{i=1}^{n} x_i^2\, \Delta x$

19. $\sum\limits_{i=1}^{n} (x_i^3 + 4)\, \Delta x$

20. $\sum\limits_{i=1}^{n} (x_i^2 + 5x_i)\, \Delta x$

21. 120

22. -2

23. 10

24. 104

25. 50

26. 34

27. 18

28. 28

29. $\frac{15}{2}$; $\frac{15}{2}$

30. 16; 16

31. 6; 6

32. $\frac{57}{2}$; $\frac{57}{2}$

33. 12; 12

34. $\frac{40}{3}$; $\frac{40}{3}$

35. $-\frac{2}{3}$; $-\frac{2}{3}$

36. $\frac{7}{6}$; $\frac{7}{6}$

37. 12; 12

38. 24; 24

39. 124,750

40. 2870

41. 4905

42. 316,210

43. 323,200

44. 348,450

45. 323,400

46. $\frac{100}{101}$

47. $\int_0^4 (x^2 + 3x)\, dx$

48. $\int_0^2 (3x^2 + x - 2)\, dx$

49. $-\ln 101 \approx -4.615$

50. $-\ln 20$

51. 6; 3.6

Chapter 8 Review Problem Set
(p. 271)

1. (a) $8 + \frac{9}{2}\pi \approx 22.1372$;   (b) 10.5

2. (a) -4 (b) 2

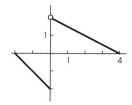

3. (a) 2.5   (b) $2\pi \approx 6.2832$

(c) -2

4. (a) 10

(b) $2\pi - 3.75$

5. (a) $\frac{10}{7}$;   (b) 0;   (c) -0.8;   (d) 0
6. (a) 0;   (b) 6;   (c) 6;   (d) 0
7. 14.5 ft   8. 27.5 ft
9. 52   10. 22.5
11. 12   12. 15
13. $\frac{1}{2}\ln 5 \approx 0.8047$ 14. $\frac{61}{3}$
15. $e^4 - e^2 \approx 47.2091$ 16. $\frac{1}{6}\ln 25 \approx 0.5365$
17. 4   18. $\frac{5}{7}$
19. $\frac{1}{8}(\ln 5)^4 \approx 0.8387$ 20. $2(e^2 - e) \approx 9.3415$
21. 26   22. $\ln 3 \approx 1.0986$
23. 1   24. 1
25. $e^3 + 12 \approx 32.0855$ 26. $e^2 + \frac{23}{3} \approx 15.0557$
27. $\frac{96}{7}$ 28. $-\frac{512}{21} \approx -24.3810$
29. $\sqrt{5}/18 \approx 0.1242$ 30. ≈ 225.8703
31. 29.5   32. $\frac{80}{3}$

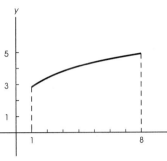

33. 94 in.   34. 250/yr
35. 100   36. 69

37. (a) $\sum_{i=1}^{97} i(i + 1)$;   (b) $\sum_{i=1}^{n} (2x_i + 3)\,\Delta x$

38. (a) $\sum_{i=1}^{100} i^2$;   (b) $\sum_{i=1}^{n} t_i^3\,\Delta t$

39. 54   40. 32
41. 76,527,700   42. 2090
43. 1   44. $1 + \ln x$; e
45. 14   46. 12

47. $\int_0^4 \frac{x}{1 + x^2}\,dx = [\frac{1}{2}\ln(1 + x^2)]_0^4 =$
    $\frac{1}{2}\ln 17 \approx 1.4166$

48. 10

49. (a)   (b) $(2, \frac{1}{4})$;
                                  (c) -0.005;
                                  (d) 0.999
                                  (e) 1

50. (a) R, the set of all real numbers;   (b) 1;
    (c) g is increasing on R;   (d) $(0, 0)$;   (e) 1

Problem Set 9.1   (p. 281)

1. 21

2. 9

3. $\frac{56}{3}$

4. $11\frac{2}{3}$

5. $\frac{9}{4}$

6. 16

7. $4(e^2 - 1)$

8. $6(1 - e^{-1/2})$

9. $\ln 3$

10. 3

11. $\frac{1}{2}\ln 5$

12. $\frac{8}{3}$

13. $18\frac{2}{3}$

14. $7\frac{1}{2}$

15.

$$A = \int_0^2 \left[x^2 + 2 - (-2x^2) \right] dx$$

$$= 12$$

16.

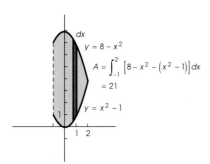

$$A = \int_{-1}^2 \left[8 - x^2 - (x^2 - 1) \right] dx$$

$$= 21$$

17.

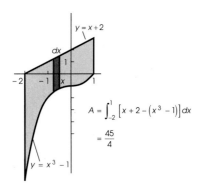

$$A = \int_{-2}^1 \left[x + 2 - (x^3 - 1) \right] dx$$

$$= \frac{45}{4}$$

18.

$$A = \int_{-2}^1 \left(4 - x^2 - 2x^3 \right) dx$$

$$= 16.5$$

19.

$$A = \int_{-1}^1 (2x + 3 - e^x)\, dx$$

$$= 6 - e + e^{-1}$$

20.

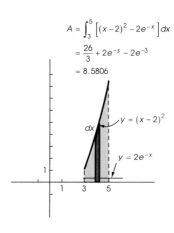

$$A = \int_3^5 \left[(x - 2)^2 - 2e^{-x} \right] dx$$

$$= \frac{26}{3} + 2e^{-s} - 2e^{-3}$$

$$\approx 8.5806$$

21.

$$A = \int_{-2}^2 \left(\frac{6}{x + 3} - 1 + x \right) dx$$

$$= 6\ln 5 - 4$$

22.

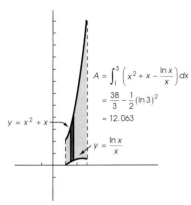

$$A = \int_1^3 \left(x^2 + x - \frac{\ln x}{x} \right) dx$$

$$= \frac{38}{3} - \frac{1}{2}(\ln 3)^2$$

$$\approx 12.063$$

$y = x^2 + x$

$y = \dfrac{\ln x}{x}$

23. $\displaystyle\int_{-1}^{3} [(2x + 5) - (x^2 + 2)]\,dx = \frac{32}{3}$

24. $\displaystyle\int_{-1}^{3} [(2x + 2) - (x^2 - 1)]\,dx = \frac{32}{3}$

25. $\displaystyle\int_{-2}^{3} [(5 - x^2) - (-x - 1)]\,dx = \frac{125}{6}$

26. $\displaystyle\int_{-3}^{4} [(9 - x^2) - (-x - 3)]\,dx = \frac{343}{6}$

27. $\displaystyle\int_{1}^{3} [(x - 1) - (x^2 - 3x + 2)]\,dx = \frac{4}{3}$

28. $\displaystyle\int_{-2}^{1} [(5x + 2) - (x^2 + 6x)]\,dx = 4.5$

29. $\displaystyle\int_{1}^{4} \left(2\sqrt{x} - \frac{2}{3}x - \frac{4}{3} \right) dx = \frac{1}{3}$

30. $\displaystyle\int_{1}^{9} \left[x^{1/2} - \frac{1}{4}(x + 3) \right] dx = \frac{4}{3}$

31. $\displaystyle\int_{-3}^{3} (6 - x^2 - x^2 + 12)\,dx = 72$

32. $\displaystyle\int_{-2}^{2} [(11 - x^2) - (x^2 + 3)]\,dx = \frac{64}{3}$

33. $2\displaystyle\int_{0}^{2} (4x - x^3)\,dx = 8$

34. $2\displaystyle\int_{0}^{1} [(2x) - (2x^3)]\,dx = 1$

35. $\displaystyle\int_{-2/3}^{0} (2x^3 + x^3 - x^2 - 2x)\,dx$

$+ \displaystyle\int_{0}^{1} (-x^3 + x^2 + 2x - 2x^3)\,dx = \frac{253}{324} \approx 0.7809$

36. $\displaystyle\int_{-1}^{0} [(x^3 - 2x) - x^2]\,dx + \int_{0}^{2} [x^2 - (x^3 - 2x)]\,dx$

$$= \frac{37}{12} \approx 3.083$$

37. $2\displaystyle\int_{0}^{2} [(4) - (y^2)]\,dy = \frac{32}{3}$

38. $2\displaystyle\int_{0}^{\sqrt{3}} [(-y^2) - (-3)]\,dy = 4\sqrt{3} \approx 6.9282$

39. $\displaystyle\int_{0}^{3} (4y - y^2 - y)\,dy = \frac{9}{2}$

40. $\displaystyle\int_{0}^{5} [(-y) - (y^2 - 6y)]\,dy = \frac{125}{6}$

41. $\displaystyle\int_{1}^{4} (y - 1 - y^2 + 4y - 3)\,dy = \frac{9}{2}$

42. $\displaystyle\int_{0}^{4} [(2y - 3) - (y^2 - 2y - 3)]\,dy = \frac{32}{3}$

43. 36

44. $\frac{32}{3}$

45. $\frac{253}{12} \approx 21.08\overline{3}$

46. $e^2 + 1 \approx 8.4$

47. $\ln 3$

48. 3

49. 22

50. 8

51. $\ln 2 - \frac{1}{2}$

52. $\frac{3}{2} \ln 2 - 1$

53. 20

54. 8

Problem Set 9.2 (p. 288)

1. $\frac{243}{5}\pi \approx 152.68$

2. $\frac{376}{15}\pi \approx 78.749$

3. $\frac{648}{5}\pi \approx 407.15$

4. $64\pi \approx 201.06$

5. $36\pi \approx 113.10$

6. $4\pi \approx 12.566$

7.

$$\pi\int_0^2 (4 - x^2)^2\,dx$$

$$= \frac{256\pi}{15}$$

$y = 4 - x^2$

8.

$$\pi\int_0^4 (4x - x^2)^2\,dx = \frac{512}{15}\pi$$

$$\approx 107.233$$

$y = 4x - x^2$

9.

$$\pi\int_1^4 \left(\frac{4}{x}\right)^2\,dx = 12\pi$$

$y = \dfrac{4}{x}$

10.

$$\pi \int_1^2 \frac{16}{x^4}\,dx = \frac{14\pi}{3}$$
$$\approx 14.661$$

11.

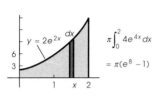

$$\pi \int_0^2 4e^{4x}\,dx$$
$$= \pi(e^8 - 1)$$

12.

$$\pi \int_{-2}^0 e^{-2x}\,dx = \frac{\pi}{2}\left(e^4 - 1\right)$$
$$\approx 84.192$$

13.

$$2\pi \int_0^1 \left[25 - \left(5x^2\right)^2\right]dx$$
$$= 40\pi$$

14.

$$\pi \int_0^4 (4 - x)\,dx = 8\pi$$
$$\approx 25.13$$

15.

$$\pi \int_0^1 \left(x^2 - x^4\right)dx$$
$$= \frac{2\pi}{15}$$

16.

$$\pi \int_{-(1/2)}^1 \left((x + 1)^2 - 4x^4\right)dx \approx 1.8\pi$$
$$\approx 5.65$$

17.

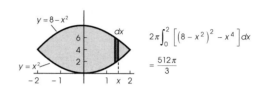

$$2\pi \int_0^2 \left[\left(8 - x^2\right)^2 - x^4\right]dx$$
$$= \frac{512\pi}{3}$$

18.

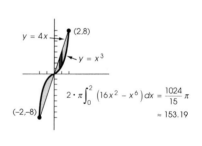

$$2 \cdot \pi \int_0^2 \left(16x^2 - x^6\right)dx = \frac{1024}{15}\pi$$
$$\approx 153.19$$

19.

$$\pi \int_0^4 \left(4y - y^2\right)^2 dy$$
$$= \frac{512\pi}{15}$$

20.

$$\pi \int_0^8 y^{2/3}\,dy \approx 19.2\pi = 60.319$$

21.

$$\pi \int_0^6 \left[\frac{1}{2}(6-y)\right]^2 dy = 18\pi$$

27. $\frac{423}{5}\pi \approx 265.87$ 28. $\frac{192}{5}\pi \approx 120.64$

29. $\frac{972}{5}\pi \approx 610.73$ 30. $\frac{704}{5}\pi \approx 442.336$

31. $72\pi \approx 226.19$ 32. $18\pi \approx 56.55$

33. $\frac{80}{3}\pi \approx 83.776$ 34. $\frac{8}{3}\pi \approx 8.38$

35. $\frac{7}{3}\pi \approx 7.33$ 36. $\frac{16}{3}\pi \approx 16.775$

37. $\frac{256}{3}\pi \approx 268.08$ 38. $\frac{1}{2}(e^2 - 1) \approx 3.19$

39. $3\pi e^4 \approx 514.575$

22.

$$\pi \int_0^2 x^2 dy = 6\pi$$
$$\approx 18.85$$

Problem Set 9.3   (p. 294)

1. 10,434   2. $13,600

3. 55,520   4. 320 million shares

5. (a) $6365;   (b) $42.43

6. (a) $31,345.95;   (b) $174.14

7. (a) $8175;   (b) $1810

8. (a) $35,784.03;   (b) $4438.08

9. 165 units   10. 368 units

11. $p = \$4$ 12. $p = \$6$

13. $p = \$100$ 14. $p = \$40$

15. $p = \$100$ 16. $p = \$320$

17. $CS = \$24$; $PS = \$24$

23.

$$\pi \int_0^8 \left(4 - y^{2/3}\right) dy$$
$$= \frac{64\pi}{5}$$

24.

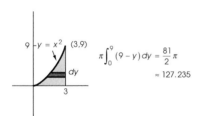

$$\pi \int_0^9 (9 - y) dy = \frac{81}{2}\pi$$
$$\approx 127.235$$

$CS = \$24, PS = \$24$

25.

$$\pi \int_0^2 \left(\frac{1}{2}y - \frac{1}{4}y^2\right) dy = \frac{\pi}{3}$$

18. $CS = \$18$; $PS = \$36$

26.

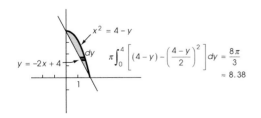

$$\pi \int_0^4 \left[(4-y) - \left(\frac{4-y}{2}\right)^2\right] dy = \frac{8\pi}{3}$$
$$\approx 8.38$$

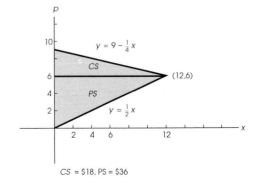

$CS = \$18, PS = \$36$

19. CS = \$6000; PS = \$2000

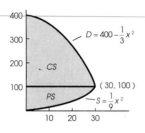

CS = \$6000, PS = \$2000

20. CS = \$1066.67; PS = \$533.33

CS = \$1066.67, PS = \$533.33

21. CS = \$166.67; PS = \$126.67

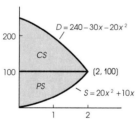

CS = \$166.67, PS = \$126.67

22. CS = \$413.33; PS = \$346.67

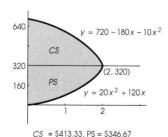

CS = \$413.33, PS = \$346.67

23. 144 units are demanded; \$3456
24. \$405.62   25. \$1272.59
26. \$243   27. $\frac{1}{2}$
28. 0.6   29. $\frac{1}{6}$
30. $\frac{2}{9}$
31. $(3 - e)/(e - 1) \approx 0.1640$
32. $2/(e^2 - 1) \approx 0.3130$ 33. \$34,995
34. 0.5815 billion barrels   35. On the 14th day
36. 2442 tons
37. (a) \$18,000;   (b) \$20,000
38. (a) 19 yr;   (b) \$384,123

Problem Set 9.4   (p. 300)

1. \$1481.64   2. \$5913.42
3. \$5934.22   4. \$6764.50
5. \$4393.24
6. Karen owes \$128.38 more than Robert.
7. \$161,383.94   8. \$144,523.83
9. \$170,918.48   10. \$141,224.64
11. $\frac{1}{2}$ 12. 1
13. $11.\overline{11}$ 14. $13.\overline{33}$
15. $10,000/81 \approx 123.4568$ 16. $177.\overline{77}$
17. \$88,888.89   18. \$16,000
19. \$101,234.57   20. \$23,111.11
21. \$9055.39   22. \$2460.66
23. \$3441.69   24. \$4039.21
25. \$6250   26. \$9375
27. Diverges.   28. Diverges.
29. Converges, $\frac{1}{2}$ 30. Converges, 1
31. Diverges.   32. Converges, $\frac{1}{2}$
33. $1/(p - 1)$ 34. $1/[2(p - 1)]$
35. \$10,135,135.14

Problem Set 9.5   (p. 306)

1. (a) $f(x) = \frac{1}{4}$ on $[1, 5]$; $f(x) = 0$ elsewhere;
   (b) 0.3
2. (a) 0.1;   (b) 0.4
3. $\frac{1}{3}$ 4. 0.75
5. (a) $k = \frac{1}{12}$;   (b) $\frac{3}{8}$
6. (a) $k = \frac{3}{14}$;   (b) 0.33825

7. (a) $1 = \int_{1}^{e^2} kx^{-1}\, dx = 2k$, so $k = \dfrac{1}{2}$;

   (b) $\frac{1}{2} \ln 2 \approx 0.347$;   (c) $a = e$

8. (a) $1 = \int_0^2 k(x+1)^{-2}\,dx = \frac{2}{3}k$, so $k = 1.5$;

(b) 0.25; (c) $b = \frac{1}{11}$

9. (a) $\frac{1}{144}\int_0^6 (36 - x^2)\,dx = 1$;

(b)

$f(x) = \frac{1}{144}(36 - x^2)$

(c) $\frac{325}{432} \approx 0.752$; (d) $\frac{324}{144} = 2.250$

10. (a) 2 min; (b) $\frac{5}{16} \approx 0.3125$; (c) $\frac{3}{4} = 0.75$

11. ≈ 0.087 (8.7%)

12. (a) $K = (\ln 2)/6$; (b) 0.2599

13. (a) 0.449 (44.9%); (b) 50 min

14. (a) 0.923 (92.3%); (b) 12.5 min

15. (a) $k = \frac{3}{4}$; (b) 0.061; (c) 1

16. (a) $k = 2$; (b) 2

17. (a) $k = 3$; (b) $\frac{3}{5}$

18. $b = 5$ 19. $1/\sqrt[3]{4} \approx 0.6306$

20. (a) Divergent; (b) convergent; 1;
    (c) convergent; $\frac{1}{3}$; (d) divergent.

21. $\mu = \dfrac{1}{k}$

22. (a) x^2;
    (b) $f(x) = zx$ for $0 \le x \le 1$, 0 elsewhere.

Chapter 9   Review Problem Set (p. 308)

1. $8 + 2\ln 3 \approx 10.1972$ 2. $18 + 2e \approx 23.4366$
3. 8   4. 5.75
5. 4.5

$y = 4 - x$
dx
$y = 3 - x^2$

$A = \int_{-1}^2 [4 - x - (3 - x^2)]\,dx = 4.5$

6. $e - 8 + 4\ln 4 \approx 0.2635$

$(\ln 4, 4)$
$y = e^x$

7. $\frac{4}{3}$

$(2, 2)$
$x = y^2 - y$
$x = y$
dy

$A = \int_0^2 [y - (y^2 - y)]\,dy = \frac{4}{3}$

8. $\frac{32}{3}$

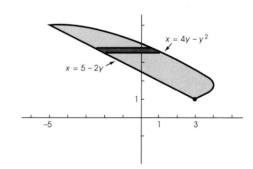

$x = 4y - y^2$
$x = 5 - 2y$

9. (a) $51.2\pi \approx 160.8495$;

dx
$y = x^2$

$V = 2\int_0^2 (\pi 4^2 - \pi x^4)\,dx = \frac{256\pi}{5}$

(b) $8\pi \approx 25.1327$

$$V = \int_0^4 \pi y \, dy = 8\pi$$

10. (a) $8\pi \approx 25.1327$; (b) $25.6\pi \approx 80.4248$

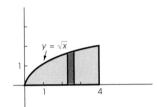

11. $\frac{32}{3}\pi \approx 33.5103$ 12. $14.4\pi \approx 45.2389$

13. (a) $30,640; (b) $14,360

14. (a) $17,090; (b) $14,814

15. $16,000   16. $5500

17. $CS = 37,944$; $PS = 54,000$

18. $CS = 4365$; $PS = 4365$

19. $\frac{1}{24}$ 20. 0.4

21. (a) 10,990 persons; (b) 110 persons/yr

22. (a) $P(t) = 4600t^2 + 15,000t + 12,000$;
    (b) 202,000 bacteria; (c) 87,733

23. $4416.10   24. $9462.90

25. $657,894.74   26. $903,614.46

27. (a) $k = \frac{1}{12}$; (b) $\frac{1}{16} = 0.0625$; (c) $\frac{32}{7}$

28. (a) $k = \frac{1}{80}$; (b) 0.95; (c) $\frac{484}{175} \approx 2.7657$

29. (a) $e^{-1} \approx 0.368$ (36.8%); (b) $\mu = 8$

30. $-e^{-0.6} + 1 \approx 0.0582$

31. (a) Area $= \int_a^b |f(x)| \, dx$;

    (b) volume $= \int_a^b \pi [f(x)]^2 \, dx$

32. (a) 25.5; (b) $79.2\pi \approx 248.8141$

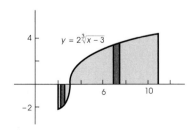

33. $a = 2$ 34. $k = 2$; $m = 1$

35. Work $= \int_a^b F(x) \, dx$ 36. 93 ft-lb

37. $\int_0^3 [1 + f(x)]^2 \, dx$

38. $(9 - 5^{3/2})/3 \approx 5.2732$

Problem Set 10.1 (p. 317)

1. (a) 13; (b) 4; (c) $3x^2$; (d) $h(2x - y + h)$

2. (a) 9; (b) $4 + 4\sqrt{2}$; (c) $x^2 + 4\sqrt{2}\sqrt{x}$;
   (d) $h(2x + h)$

3. (a) $\frac{1}{256}$; (b) $\frac{1}{8}$; (c) $1/16x^3$; (d) h/y^4

4. (a) $-\frac{9}{11}$; (b) $-\frac{3}{2}$; (c) $-1/x$;
   (d) $-5yh/[(x + h - 3y)(x - 3y)]$

5. (a) 4; (b) $7e^{-2}$; (c) $(3x + 1)e^{2x-4}$; (d) $3he^{y-4}$

6. (a) 11; (b) -1; (c) $x^2 \ln(2x^2 - 3) + 4x^2 - 5$;
   (d) $(x + h)^2 \ln[(x + h)y - 3] - x^2 \ln(xy - 3)$

7. (a) -3; (b) -30; (c) $3z^2$; (d) $2yz + xy - 3xz$

8. (a) 1; (b) 0; (c) $12z^3$; (d) $z^2(x + y) - 2xyz$

9. (a) 1; (b) -1; (c) $-2(z + 1)/(z + 4)$;
   (d) $(zy - 2x)/(yx - 2z)$

10. (a) $\sqrt{2}$; (b) 7; (c) $\sqrt{6}\,|z|$; (d) $(x^2 + y^2 + z^2)^{1/2}$

11. (a) 6; (b) 1176; (c) 0;
    (d) $(z + 2y)(y + 2x)(x + 2z)$

12. (a) e^5; (b) $10e^{23}$; (c) $z^2 + z$; (d) $(y^2 + x)e^{z-2y+4x}$

13. (a) $f(x, y) = 2x + 2y$, $g(x, y) = xy$; (b) 16, 12;
    (c) $y = 2x/(x - 2)$

14. (a) $f(x, y, z) = (x + 2z)(y + 2z)$;
    $g(x, y, z) = z(2x + 2y + 4z)$;
    (b) 192; 96; (c) $z = 3$

15. (a) $V(x, y) = x^2y$; $S(x, y) = 2x^2 + 4xy$;
    (b) it is doubled; (c) $C(x, y) = 0.72xy + 0.6x^2$

16. (a) $V(x, y, z) = xyz$ cm^3;
    $S(x, y, z) = 2(xy + xz + yz)$ cm^2;
    (b) $0.8xy + 0.5xz + 0.5yz$

17. (a) 22,680; 10,080;
    (b) multiply by $\sqrt[4]{2}$; multiply by $\sqrt[3]{8}$; double.

18. (a) 57,000; (b) 36,292; (c) 67,143

19. (a) $(4, 1, -8)$, $(2, -6, 12)$;
    (b) $3x + 2y = 6$, $3x + z = 6$, $2y + z = 6$;
    (c)

20. (a) $3x + y = 9$, $6x + y = 18$, $2y + z = 18$;
(b)

27.

21.

22.

28.

23.

24.

25.

26.

29.

30.

31.

32.

33.

34.

35.

36.

37. $P(x, y) = 10(3x + 3y - 200)$
38. (a) $R(p, q) = 300p + 400q + 11pq - 2p^2 - 2q^2$
    (b) $R(6, 7.5) = \$4110.50$; $R(5, 8) = \$4962.00$
39. $S(w, r) = 0.00003wr^2$; $S(2400, 60) = 259.2$ ft
40. $V(r, h) = \pi r^2 h + \frac{2}{3}\pi r^3 = \frac{1}{3}\pi r^2(3h + 2r)$

41. (a)   (b)

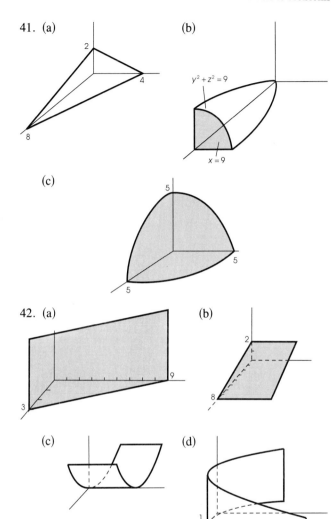

(c)

42. (a)   (b)

(c)   (d)

Problem Set 10.2   (p. 324)

1. $2xy^3$, $3x^2y^2$
2. $6(3x + 2y)$, $4(3x + 2y)$
3. $y^2 - 8x$, $2xy + 3$ 4. $6xy + 4$, $3x^2 - 12y$
5. $2\sqrt{y}$, $xy^{-1/2} - 8y^{-3}$
6. $2xy^{-1/2} - 2x^{-1/3}$, $-\frac{1}{2}x^2y^{-3/2}$
7. $-4x^{-2}y^{-2}$, $-8x^{-1}y^{-3}$
8. $-6y^{-1}x^{-3/2}$, $-12y^{-2}x^{-1/2}$
9. $2yxe^{2x}(x + 1)$, x^2e^{2x}

10. $2y^2(2x + 3y)^{-1}$, $2y \ln(2x + 3y) + 3y^2(2x + 3y)^{-1}$
11. $6x(x^2 + 2y)^2 \ln y$, $y^{-1}(x^2 + 2y)^3 + 6(x^2 + 2y)^2 \ln y$
12. $2xye^{x^2y-1}$, $e^{x^2y-1}(2y - x^2)$
13. $ye^{2y}(2xy + y^2)^{-1/2}$,
    $2(2xy + y^2)^{1/2}e^{2y} + (x + y)(2xy + 2)^{-1/2}e^{2y}$
14. $[y^2/x^2(x + 3)][x - (x + 3) \ln(x + 3)]$,
    $[2y \ln(x + 3)]/5$
15. $3y/[x(2x + y)]$, $(4x - y)/[y(2x + y)]$
16. $y + 1/[x \ln(xy^2)]$, $x + 2/[y \ln(xy^2)]$
17. $2xy + 6yz$
18. $4(3xyz^2 - 1)(xyz^3 - z)^3$
19. $xye^{z+2} + xy^{-1}$
20. $z^2(x^{-2}y^{-1} + z^{-1}) + 2z \ln(x^2y + z)$
21. $6, -9$ 22. $-13, 11$
23. $\frac{2}{7}, 3$ 24. $8; 0$
25. $1, 2$ 26. $4; 0$
27. -4 28. 6
29. $-\frac{3}{4}$ 30. $2e^3$
31. About 0.001356%   32. -0.000904%
33. About 208.14%   34. -60.36%
35. $32\pi; 12\pi$
36. (a) k/r^4;   (b) $-4kL/r^5$
37. (a) $294W^{-0.58}H^{0.73}$, $511W^{0.42}H^{-0.27}$;   (b) 1108;
    (c) 748.9
38. (a) 117.6 ft; 0.0588; 3.36;   (b) 3.36 ft
39. (a) $-4y$;   (b) $6y^2 - 4x$;   (c) $6y^2 - 4x$
40. (a) 10;   (b) $-4y$;   (c) $-4y$
41. (a) $y^4e^{xy^2}$;   (b) $2ye^{xy^2}(xy^2 + 1)$;
    (c) $2ye^{xy^2}(xy^2 + 1)$
42. (a) $(4y^2 - 4x^2y)/(x^2 + y)^2$;   (b) $4x^3/(x^2 + y)^2$;
    (c) $4x^3/(x^2 + y)^2$
43. (a) $24x^2(x^2 + 2y) + 6(x^2 + 2y)^2$;
    (b) $24x(x^2 + 2y)$;   (c) $24x(x^2 + 2y)$
44. $4(y^2 + e^x)^3e^x$; $8y(y^2 + e^x)^3$;
    (a) $4(y^2 + e^x)^2e^x(y^2 + 4e^x)$;   (b) $24ye^x(y^2 + e^x)^2$
    (c) $24ye^x(y^2 + e^x)^2$
45. Complementary.   46. Complementary.
47. Competitive.   48. Neither.
49. $R_x(20, 30) = 180$. Increasing research from \$20 to \$21 million while keeping advertising at \$30 million will increase revenue by about \$180 million. $R_y(20, 30) = 40$. Increasing advertising from \$30 to \$31 million while keeping research at \$20 million will increase revenue by about \$40 million.
50. $F_L(10, 20) \approx 204$. Increasing labor from 10 to 11 units while keeping capital at 20 units will increase production by about 204 skis/wk. $F_K(10, 20) \approx 60$. Increasing capital from 20 to 21 units while keeping labor at 10 units will increase production by about 60 skis/wk.

51. $T_x(4, 2) = -4$. As you move from $(4, 2)$ to $(4 + \Delta x, 2)$, the temperature drops about $4 \Delta x$ degrees. $T_y(4, 2) = -6$. As you move from $(4, 2)$ to $(4, 2 + \Delta y)$, the temperature drops by about $6 \Delta y$ degrees.

52. (a) $W(30, -20) \approx -114°C$. $W_v(30, -20) \approx -1.8°C$;
   (b) it indicates that when the wind velocity changes from 30 mi/h to 31 mi/h while the temperature remains at $-20°C$, the wind chill drops by about $1.8°C$.

53. (a) $A(x, y, z) = (6x + 9y + 12z)/(x + y + z)$;
   (b) $A(10, 20, 10) = -\frac{3}{40} = -7.5¢$;
   (c) increasing the first group of workers from 10 to 11 while keeping the second and third groups at 20 and 10 decreases the average hourly wage by about $7.5¢$.

54. $A_r(1000, 0.07, 10) = 10000e^{0.7} = \$20,137.53$. If the rate changes by Δr while keeping the principal at \$1000 and the time at 10 yr, the amount will change by about $\$20,137.53 \, \Delta r$. For example, if r changes to 8% so that Δr is 1%, then the accumulated amount will increase by about 1% of \$20,137.53 which is about \$201.37.

55. 34.5
56. -0.036
57. $220\pi \approx 691 \text{ cm}^3$
58. \$59

Problem Set 10.3 (p. 330)

1. Local min value: $f(2, -3) = -17$
2. Local min value: $f(1, -4) = -8$
3. Local min value: $f(-3, 5) = -17$
4. Local min value: $f(6, -2) = -41$
5. Local max value: $f(2, 2) = 13$
6. Local min value: $f(-1, -1) = -13$
7. Local max value: $f(2, 2) = 8$
8. Saddle point at $(18, 14)$
9. Saddle points at $(2, -2)$ and $(-2, 2)$
10. Saddle point at $(0, 0)$; local max value: $f(18, 6) = 112$
11. Saddle point at $(0, 0)$; local min at $(4, 4)$
12. $f(1, 2) = 6$ is a relative min value; $f(-1, -2) = 42$ is a relative max value; saddle points at $(1, -2)$ and $(-1, 2)$
13. Saddle point at $(0, 2)$
14. Saddle points at $(\sqrt{2}, -\sqrt{2})$ and $(-\sqrt{2}, \sqrt{2})$
15. 8 ft long, 8 ft wide, 4 ft high.
16. ≈ 24.33 ft long, 24.33 ft wide, 40.55 ft deep.
17. 28 in. long, 14 in. wide, 14 in. deep.
18. 4 ft long, 2 ft wide, 1 ft deep.
19. $p = \$53.65$, $q = \$67.67$
20. $p = \$450$, $q = \$575$

21. $x = 30$, $y = 25$
22. Sell candy A for \$1.72 and candy B for \$2.38.
23. Saddle point at $(0, 0)$; local min at $(8, 2)$ and $(-8, -2)$
24. $x = y = \sqrt[3]{12}$, $z = 2\sqrt[3]{12}$
25. $x = 6\frac{2}{3}$, $y = 6$; spend \$6,666,667 on research and \$6,000,000 on advertising.
26. $x = 12$, $y = 72$; spend \$12,000 on TV advertising and \$72,000 on radio advertising.
27. $R(5, \frac{1}{2}) = 50/e \approx 18.39$
28. 6.67 units of drug A and 3.33 units of drug B; $x = \frac{20}{3}$, $y = \frac{10}{3}$

Problem Set 10.4 (p. 337)

1. $f(8, 2) = 76$ 2. $f(2, 4) = 68$
3. $f(10.5, 11) = 689$ 4. $f(6, -2) = -19$
5. Max value: $f(4, 2) = f(-4, -2) = 18$; min value: $f(-4, 2) = f(4, -2) = 2$
6. Max value: $f(\sqrt{\frac{40}{3}}, \sqrt{\frac{20}{3}}) = f(-\sqrt{\frac{40}{3}}, -\sqrt{\frac{20}{3}}) = \frac{40}{3}\sqrt{\frac{20}{3}} \approx 34.427$
   Min value: $f(\sqrt{\frac{40}{3}}, -\sqrt{\frac{20}{3}}) = f(-\sqrt{\frac{40}{3}}, -\sqrt{\frac{20}{3}}) = -\frac{40}{3}\sqrt{\frac{20}{3}} \approx -34.427$
7. 75 m long, 37.5 m wide.   8. 150 m by 37.5 m
9. $f(2, 1, 3) = 14$ 10. $f(24, 12, 8) = 1056$
11. $f(-4, -2, -8) = -42$
12. $f(-1, -1, -1) = f(-1, 1, 1) = f(1, -1, 1) = f(1, 1, -1) = -1$
13. Square bottom 2 ft by 2 ft; 3 ft high.
14. Square base 6 ft by 6 ft; 7 ft high.
15. $x = 300$; $y = 120$; $\lambda \approx 0.368$
16. $x = 300$; $y = 150$; $\lambda \approx -0.6307$
17. $9\frac{3}{4}$ in. by $9\frac{3}{4}$ in.
18. Base radius at a height of 5.42 cm; 10.84 cm
19. $\frac{1}{2}\sqrt{14} \approx 1.87$ 20. $abc/27$
21. $x = 12$, $y = 14$, $z = 20$; \$740
22. $x = 92$, $y = 0$, $z = 0$; 8,300,000
23. $(\frac{5}{2}\sqrt{2}, \frac{5}{2}, \frac{5}{2})$ and $(-\frac{5}{2}\sqrt{2}, -\frac{5}{2}, -\frac{5}{2})$
24. $x = aM/C$, $y = bM/C$

Problem Set 10.5 (p. 342)

1. $y = \frac{7}{26}x + \frac{42}{13} = -0.269x + 3.231$
2. $y = \frac{37}{38}x + \frac{67}{38} = 0.974x + 1.76$
3. $y = 0.32x + 1.44$ 4. $y = -0.36x + 1.88$
5. $y = \frac{39}{35}x - \frac{13}{35} = 1.114x - 0.371$
6. $y = -\frac{3}{13}x + \frac{73}{26} = -0.2308x + 2.808$

7. $y = 0.8x + 1$ 8. $y = -1.045x + 8.748$

9. (a) $y = 0.1469x + 1.381$;   (b) \$4025 per capita;
   (c) \$7257 per capita

10. $y = 0.51x + 1.64$; \$4.7 billion; \$4.904 billion

11. $y = 15.5x + 180.5$; \$258,000

12. $y = 4.179x + 5.570$; 43 cases.

13. $x = -135.3p + 2514$; 890

14. $y = -770x + 7050$; 6th week.

15. $y = 1.58e^{0.68x}$

16. $y = 39.88e^{0.101x}$; 73.1 million

(c)

(d)

Chapter 10   Review Problem Set
(p. 351)

1. (a) -144;   (b) $\frac{1}{2}x^9$;   (c) 2

2. (a) 16;   (b) y^9;   (c) $16\sqrt[3]{2}$

3. $f(r, h) = 4\pi r^2 + 4\pi (r + h)^2 = 4\pi (2r^2 + 2rh + h^2)$;
   $f(3, 0.5) = 85\pi \approx 267.034$

4. $g(r, h) = \frac{4}{3}\pi [(r + h)^3 - r^3]$; $g(4, 0.3) = 20.676\pi \approx 64.956$

5. $g(a, b) = \frac{1}{4}ab^2\sqrt{3}$ 6. $V = \frac{1}{2}(a + b) dh$

7. (a)

(b)

8. (a)

(b)

(c)

(d)

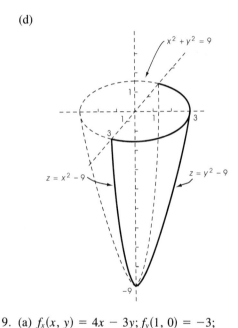

9. (a) $f_x(x, y) = 4x - 3y; f_y(1, 0) = -3;$
   (b) $f_x(x, y) = 2x^2(2x + 3y)^{-1} + 2x \ln(2x + 3y);$
      $f_y(1, 0) = \frac{3}{2};$
   (c) $f_x(x, y) = e^{x^2y}(2xy); f_y(1, 0) = e^0 = 1;$
   (d) $f_x(x, y) = 4(3x - 2x^{-1}y)^3(3 + 2x^{-2}y);$
      $f_y(1, 0) = -216$

10. (a) $f_y(x, y) = -6(x - 2y)^2; f_x(0, 2) = 48;$
    (b) $f_y(x, y) = 2ye^{2xy}(1 + xy); f_x(0, 2) = 16;$
    (c) $f_y(x, y) = 1 + 6y(x^2 + y^2 - x - 1)^{-1};$
       $f_x(0, 2) = -1;$
    (d) $f_y(x, y) = 4(xy + 8y^{-1})(x - 8y^{-2});$
       $f_x(0, 2) = 32$

11. -6

12. 3

13. $-\frac{4}{9}$

14. $-96e \approx -261$

15. $f_p(2000, 0.08) = (1.08)^{20} = \$4.66.$ If r remains at 8% and P changes by ΔP, then the accumulated amount changes by about \$4.66 ΔP. For example, if \$2001 is deposited, so that $\Delta P = 1$, then the accumulated amount will increase by about \$4.66. And $f_r(2000, 0.08) = 40,000(1.08)^{19} = \$172,628.04.$ If P remains at \$2000 and r changes by Δr, then the accumulated amount changes by about \$172,628.04 Δr. For example, if r increases to 9%, so that $\Delta r = 0.01$, then the accumulated amount will increase by about \$1726.28.

16. $S_x(2, 4, 10) = 1920; S_y(2, 4, 10) = 1920.$ The rate of change of the maximum safe load of a 2 in. by 4 in. by 10 ft horizontal beam is 1920 lb/in. with respect to its width and to its depth.

17. Saddle point at $(0, 0)$; local min at $(18, 6)$

18. Saddle point at $(0, 0)$; local min at $(3, 3)$

19. $x = \frac{1}{3}, y = \frac{2}{3}, z = \frac{2}{3}$

20. $w = h = 4, l = 8$ 21. $p = \$800, q = \$1000$

22. $(0.1096, 0.2193, 4.0601)$

23. $f(\frac{3}{2}\sqrt{2}, \sqrt{2}) = 3$

24. $f(3\sqrt{2}, 3) = f(-3\sqrt{2}, 3) = 54$ is the max value of $f(x, y)$.

25. (a) $y = 3.69x + 27.6;$ (b) $\approx \$6450$

26. (a) $y = 3.743x + 39.07;$ (b) $\approx \$8400$

27. (a) 18;   (b) $\frac{19}{60} \approx 0.3167$

28. (a) 36;   (b) $\frac{-473}{45} \approx -10.51$

Problem Set 11.1   (p. 353)

1. $3x^2 + 27$ 2. $14x$

3. $y^2 \ln \frac{5}{2}$ 4. $y(e^2 - 1)/2$

5. 28   6. 28

7. $9 \ln \frac{5}{2}$ 8. $12(e^2 - 1)$

9. 28   10. 27

11. $9 \ln \frac{5}{2}$

12. $12(e^2 - 1)$

13. $\frac{26}{3}$

14. 18

15. 4

16. 9

17. $\frac{1}{2}e^2 - \frac{3}{2}$

18. $(2e^2 - 3)/4$

19. $2 \ln 2 - \ln 3 = \ln \frac{4}{3}$

20. $\frac{16}{5}(4 - \sqrt{2})$

21. $\frac{13}{5}$

22. 21

23. $(e^{4x^2} - 1)/x^2$

24. $(e^{18y} - 1)/4$

25. $\frac{45}{2}$

26. 27

27. $\frac{1}{4} \ln \frac{5}{2}$

28. 81

29. 13.6071

30. $\frac{1}{2}(e^5 - e^4 - e + 1)$

31. $2e^2 + 2$

32. $e^3 - e^2 + e - 1$

33. $\frac{152}{9}$

34. $e^2 - e - 1$

15. $\frac{248}{15} \approx 16.5333$

16. $\frac{13}{4}$

Problem Set 11.2   (p. 359)

1. $\frac{15}{2}$

2. 36

3. 44

4. $\frac{1}{6}$

5. $3e - 3$

6. $\frac{1}{2}(e - 1)$

7. $\frac{8}{3}$

8. $\frac{2}{21}$

9. $\frac{8}{5}$

10. $\frac{512}{21}$

11. 64

12. 4

17. 37.2

18. 30

19. 204.8

20. $\frac{432}{5}$

13. $\frac{48}{7}$

14. $\frac{576}{5}$

21. $16 \ln 2 - 6$

22. $\frac{8}{3}$

23. $\int_0^8 \left[\int_{x/4}^{\sqrt[3]{x}} f(x, y)\, dy \right] dx$

24. $\int_0^1 \left[\int_y^{\sqrt[3]{y}} f(x, y)\, dx \right] dy$

25. $\int_0^1 \left[\int_{e^y}^e f(x, y)\, dx \right] dy$

26. $\int_0^e \int_1^{e^x} f(x, y)\, dy\, dx$

27. 12.4

28. $\frac{272}{3}$

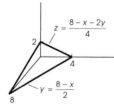

29. $\int_0^2 \left[\int_{x^2}^{8-x^2} 1\, dy \right] dx = \frac{32}{3}$ 30. 2

31. $\int_1^4 \left[\int_{4-y}^{4y-y^2} 1\, dx \right] dy = \frac{9}{2}$ 32. $\frac{7}{6}$

33. $2e - 1$ 34. $\frac{1}{4}$

35. $\frac{16}{3}$ 36. 7.56887841

Problem Set 11.3   (p. 364)

1. 9

2. $\frac{32}{3}$

3. 9

4. 6; 6

5. 4   6. 4

7. $\frac{256}{21} \approx 12.1905$ 8. $\frac{2349}{20}$

9. 80   10. $\frac{72}{5}$

11. $\frac{1}{2}(e^4 - 5)$ 12. 64

13. $\frac{128}{15} \approx 8.5333$ 14. 21.6

15. $128{,}000(1 - e^{-2})(1 - e^{-1}) \approx 69{,}960$

16. 293,333

17. $\frac{3}{2} \ln 3$ 18. $\frac{248}{45}$

19. 0   20. $\frac{4}{3}[1/(e^2 + 1)]$

21. $\frac{13}{3}$ 22. 20

23. $37°$ 24. $\frac{119}{3}$

25. 18   26. $\frac{1}{2}(e^3 - 1)(e^4 - 1)$

27. $18e^{10} - 16e^8 - 2 \approx 348{,}780$

28. $\frac{4}{91}$ 29. $\bar{x} = \frac{16}{3}, \bar{y} = \frac{16}{15}$

30. $\frac{10}{13}; \frac{117}{26}$

Chapter 11   Review Problem Set (p. 367)

1. $\frac{152}{3}$ 2. 23

3. 4   4. $\frac{9}{2} \ln 5$

5. $e^2 - 1$ 6. $e^4 - e$

7. $8 - \frac{1}{2} \ln 17$ 8. $\frac{52}{3}$

9. $V = 36$ 10. $\frac{8}{3}$

11. (a) $\int_0^2 \left[\int_0^{x^3} 2x\, dy \right] dx = \frac{64}{5}$;

(b) $\int_0^8 \left[\int_{\sqrt[3]{y}}^2 2x\, dx \right] dy = \frac{64}{5}$

12. 64   13. 8

14. $\frac{32}{3}$ 15. $\frac{1}{3}(e - 1)$

16. $\frac{192}{5}$

17. (a)   (b) $\frac{8}{3}$

18. $\frac{10}{3}$

19. $\frac{9}{10}$

20. $\frac{9}{10}$

21. $\frac{15}{2}(e^2 + e^{-2} - 2) \approx 41.4°$

22. $10 \ln 5°$

23. $-1 + 2 \ln 2$

24. 2

25. $\frac{1}{2}$

26. $\frac{2}{3}$

27. $\frac{1}{2}e^2 - 1$

28. 56

Problem Set 12.1 (p. 375)

1. $3\pi/2$
2. $-\pi$
3. 3π
4. $-5\pi/2$
5. $7\pi/4$
6. $-13\pi/4$
7. $2\pi/3$
8. $7\pi/4$
9. $-23\pi/6$
10. $44\pi/3$
11. 2.1206
12. -0.687659725
13. -14.3396
14. 0.0003046174
15. 210°
16. $-270°$
17. $-390°$
18. 675°
19. $-77.92°$
20. 314.5538295
21. 2362.30°
22. 197.1775863°
23. 18 cm, 108 cm²
24. 9.6 cm, 19.2 cm²
25. 83.78 in., 837.76 in.²
26. $5\pi/2$ in., $45\pi/4$ in.²
27. (a) 3.88; (b) 7.41 in.
28. 9.3915 cm
29. 45.70 mi/h
30. 5π radius, $5\pi/12$ radius
31. (a) 6.3 m; (b) 41.05 ft
32. (a) $2\sqrt{5}$; (b) $\frac{2}{3}$; (c) $\sqrt{5}/3$
33. (a) 353.75 ft; (b) 187.46 ft
34. 86.53 ft
35. 0.2269, 744.85°
36. 2.5 radians; 245 cm²
37. 9.64 ft
38. 217 ft
39. (a) 720π in./min; (b) 72 rpm
40. 70.737 in.
41. 2.572 in.²
42. 14.27

Problem Set 12.2 (p. 383)

1. $5, \frac{3}{5}, -\frac{4}{5}$
2. $r = 6, -\sqrt{2}/2, \sqrt{2}/2$
3. $2, -1, 0$
4. $r = 8, -1/2, -\sqrt{3}/2$
5. $\sqrt{74}, 7/\sqrt{74}, 5/\sqrt{74}$
6. $\sqrt{5}, \sqrt{3}/\sqrt{5}, -\sqrt{2}/\sqrt{5}$
7. $10, -6, -0.6$
8. $r = 75, -72, -\frac{24}{25}$

9. (a) $-\sqrt{2}/2, -\sqrt{2}/2$; (b) $\sqrt{2}/2, \sqrt{2}/2$
10. (a) $-1, 0$; (b) $\sqrt{2}/2, -\sqrt{2}/2$; (c) $0, -1$; (d) $0, 1$
11. (a) $1/2, -\sqrt{3}/2$; (b) $1/2, \sqrt{3}/2$; (c) $-1/2, -\sqrt{3}/2$; (d) $1/2, \sqrt{3}/2$
12. (a) $\sqrt{3}/2, -1/2$; (b) $-\sqrt{3}/2, 1/2$; (c) $-\sqrt{3}/2, 1/2$; (d) $\sqrt{3}/2, 1/2$
13. (a) -0.61566; (b) 0.61153; (c) -0.92400; (d) 0.92400; (e) 0.98031; (f) 0.99248
14. (a) 0.587212717; (b) 0.587212717; (c) 1255.765592; (d) 1; (e) 0.022362024; (f) 0.541455005
15. II
16. In quadrant III
17. $-\frac{4}{5}$
18. $-\frac{12}{13}$
19. (a) $\frac{4}{5}$; (b) $-\frac{3}{5}$; (c) $-\frac{3}{5}$; (d) $-\frac{3}{5}$; (e) $-\frac{4}{5}$; (f) $\frac{4}{5}$
20. $\pi/6, \pi/2, 5\pi/6, 7\pi/6, 3\pi/2, 11\pi/6$
21. (a) $4, \pi/2$; (b) $2, 6\pi$; (c) $3, 4\pi/3$
22. (a) $3, 4\pi/3$; (b) $\pi, 2$; (c) $\sqrt{2}, 4\pi$
23.

24.

25.

26.

27.

28.

29.

30.

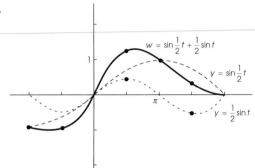

31. (a) $\frac{1}{60}$ s; (b) 36,000; (c) 40 amps
32. (a) $\pi/15$ s; (b) after $\pi/60$ (or 0.0524) s; (c) 960 ft
33. -7.81828 34. $\pm\sqrt{5}/3$
35. $A = 2.5, B = 1.6$
36.

37. (a)

(b)

38.

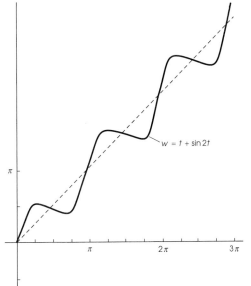

$w = t + \sin 2t$

39. (a) 1;   (b) $\frac{1}{2}$

40. (a) 1;   (b) $\frac{1}{3}$;   (c) 2

29. Use Quotient Rule to get
$[\sin x (-\sin x) - \cos x \cos x]/\sin^2 x$. Simplify.

30. $(d/dx)(\sec x) = (d/dx)(\cos x)^{-1} = (-1)(\cos x)^{-2}(-\sin x) = \tan x \sec x;$
$(d/dx)(\csc x) = (d/dx)(\sin x)^{-1} = (-1)(\sin x)^{-2} \cos x = -\cot x \csc x$

31. (a) $6 \tan 3x \sec^2 3x$;   (b) $8x \sec^2(x^2)$;
(c) $-\csc^3 x - \cot^2 x \csc x$;
(d) $(2 \sec 2x \tan 2x) e^{\sec 2x}$

32. (a) $\sec^3 t + \sec t \tan^2 t$;   (b) $6 \sec^2 3t \tan 3t$;
(c) $-\csc^2(\tan t) \sec^2 t$;   (d) $-2 \cot t$

33. (a) 2 ft, $-2\pi\sqrt{3}/3$ ft/s;
(b) $4\pi/3$ ft/s;   (c) $t = 4.5$

34. $d^2y/dt^2 = (-B^2)A \cos (Bt + c) = -B^2 y$

35. (a) $(1/\sqrt{x}) \sec^2 (1 + 2\sqrt{x})$;
(b) $(2x \sin x - 2x^2 \cos x)/\sin^3 x$;
(c) $3x^{-2} \cos^2(1/x) \sin(1/x)$;
(d) $\cos[\sin(\sin x)] \cos(\sin x) \cos x$

36. $2\pi/3$ 37. $\theta = \pi/3$

38. (a) $x = \pi/2, 5\pi/2, 9\pi/2, \ldots$;
(b) $f'(x)$ does not change sign at any critical point since $1 - \sin x \geq 0$ for all t;
(c)

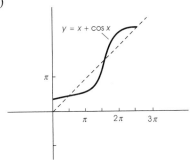

$y = x + \cos x$

39. $\cos 2 \approx -0.41615$ 40. $1; 0$

Problem Set 12.3   (p. 390)

1. $6 \cos 2t$
2. $-20 \sin 5t$
3. $6 \sin(1 - 3t)$
4. $-2 \cos(-4t + 1)$
5. $6t \cos(3t^2)$
6. $-(3t^2 + 2) \sin(t^3 + 2t)$
7. $-3 \cos^2 t \sin t$
8. $4 \sin^3 t \cos t$
9. $12t^2 \sin(2t^3 - 1) \cos(2t^3 - 1)$
10. $18t \cos^2(1 - 3t^2) \sin(1 - 3t^2)$
11. $-\sin t \cos(\cos t)$
12. $-\sin(\sin t) \cos t$
13. $-5t^2 \sin 5t + 2t \cos 5t$
14. $3t^2(t \cos 3t + \sin 3t)$
15. $(2 \cos 4t \cos 2t + 4 \sin 4t)/\cos^2 4t$
16. $[e^t(\sin 2t - 2 \cos 2t)]/\sin^2 2t$
17. $(2 \cos 2t)e^{\sin 2t}$
18. $(1 - \sin t)e^{t + \cos t}$
19. $(-2 \cos t \sin t)/(1 + \cos^2 t)$
20. $(2 \cos t)/(3 + 2 \sin t)$
21. $-6 \cos 3x \sin 3x$, $18 \sin^2 3x - 18 \cos^2 3x$
22. $f'(x) = x^2 \cos x + 2x \sin x$
23. $\cot x$, $-\csc^2 x$
24. $f'(x) = -3x^2 \sin(x^3)$, $-9x^4 \cos (x^3) - 6x \sin (x^3)$
25. -3 26. 3
27. $y - \frac{1}{4} = (-5\sqrt{3}/2)(x - \pi/6)$
28. $y - \frac{1}{4} = 1(x - 3\pi/4)$

Problem Set 12.4   (p. 395)

1. $\frac{1}{3} \sin 3x + C$
2. $-4 \cos \frac{1}{4}x + C$
3. $\frac{1}{3}$
4. 8
5. $2/\pi$
6. $\frac{1}{2}$
7. $-\frac{1}{3} \cos(x^3) + C$
8. $\frac{1}{3} \tan(3x) + C$
9. $-\frac{1}{4} \cos^4 t + C$
10. $\frac{1}{5} \tan^5 t + C$
11. $\sin \sqrt{x} + C$
12. $-\frac{1}{3} \sin^3(1/x) + C$
13. $\frac{1}{8}$
14. $-\frac{1}{8}(\cos^4 5 - 1)$
15. $(3 - \sqrt{3})/6 \approx 0.21132$
16. $e - 1$
17. $\frac{1}{2} \sin(e^{2u}) + C$
18. $-[1/(2 \sin^2 u)] + C$

19. $-\frac{2}{3}\sqrt{1 + \cos 3x} + C$

20. $-\frac{1}{2}[1/(1 + \tan 2x)] + C$

21. $\ln 2$ 22. $\frac{1}{4}\ln 2$

23. $\ln \frac{3}{2}$ 24. $\ln 2$

25. 2 26. 3

27. 2π 28. $y = -\frac{1}{4}\cos(2x^2) + 5$

29. $-\frac{1}{2}x\cos 2x + \frac{1}{4}\sin 2x + C$

30. $-\frac{1}{2}\cos(x^2) + C$ 31. $\frac{1}{4}\sin(2x^2) + C$

32. $\frac{1}{2}x^2\sin(x^2) + \frac{1}{2}\cos(x^2) + C$

33. $\frac{1}{2}x^2\sin 2x + \frac{1}{2}x\cos 2x - \frac{1}{4}\sin 2x + C$

34. $-\frac{1}{2}x^4\cos(x^2) + x^2\sin(x^2) + \cos(x^2) + C$

35. $\frac{1}{2}[\sqrt{2} + \ln(\sqrt{2} + 1)] \approx 1.14779$

36. $\frac{11}{24}$ 37. $-\frac{4}{15}$

38. $\frac{1}{15}(2 - 33\sqrt{3}/32)$

39. (a) $-\sin(1/x) + C$; (b) $\sin(\ln x) + C$; (c) $\frac{9}{128}$;
  (d) $\pi/16$; (e) $-\frac{1}{2}e^{\cos^2 x} + C$; (f) 1

40. (a) $\sec x + C$; (b) $-\csc x + C$

41. (a) $(\sqrt{3} - \pi)/3 \approx 0.68485$;
  (b) $(1 - \pi)/4 \approx 0.21460$

42. 4681

43. (a) $(2 - \sqrt{2})/4\pi \approx 0.04662$;
  (b) $(\pi - 2)/2 \approx 0.57080$

44. π

Problem Set 12.5 (p. 402)

1. $25\pi/2$ 2. -0.957558

3. $-\frac{8}{3}$ 4. 0.026553142

5. $400 + (40/\pi)[(1 + \sqrt{2})/2] \approx 421.73556$

6. $480/\pi$

7. $-5\pi - \frac{1}{3} + 20 \approx 3.95870$

8. $5\pi + 2$ 9. $2\pi/3, 4$

10. $\frac{1}{2}, 3$

11.

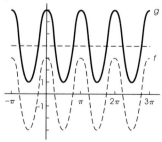

12. $f(t) = 2\sin\frac{1}{2}t$, $g(t) = 5 + 2\sin\frac{1}{2}t$

13.

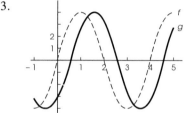

14. $f(t) = \frac{1}{2}\cos\frac{\pi}{3}t$, $g(t) = \frac{1}{2}\cos\frac{\pi}{3}(t - 1)$

15. $10, \pi/3, 20$ 16. $3; 6; 7$

17. (a)

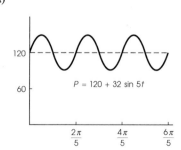

(b) $152, 88$;   (c) $150/\pi \approx 48$

18. (a) 24 mo;   (b) July 1, 1981;   (c) 389 foxes/mo

19. (a)   (b) $31,200

$$S = 6 - 6\cos\left(\frac{\pi}{26}t\right)$$

20. (a)   (b) 364

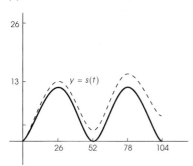

21. (a) Day 172;   (b) $240\pi/182 \approx 4.14$ min/day;
    (c) about 578 min

22. (a) 4; $2\pi/3$ s;   (b) after $\pi/3$ s;   (c) 12;   (d) 48 ft

23.   24. 20; 45

25. (a) June 30;   (b) 263,000 ft^3

26. (a) $\frac{16}{3}$;   (b) 1.383586972 liter

27.

28.

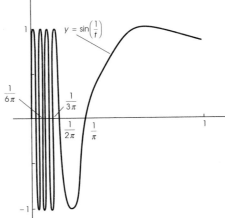

Chapter 12   Review Problem Set (p. 403)

1. 13π 2. -27.75073511

3. $1200°$ 4. $-1485°$

5. (a) 2.25;   (b) 72 in.^{2}

6. 26.782 cm^2; 200.866 cm^2

7. (a) 13;   (b) $-\frac{12}{13}$;   (c) $-\frac{13}{5}$;   (d) $-\frac{12}{13}$;   (e) $\frac{12}{13}$;
   (f) $\frac{144}{169}$

8. (a) 25;   (b) $-\frac{7}{25}$;   (c) $-\frac{24}{7}$;   (d) $-\frac{7}{25}$;   (e) $-\frac{7}{25}$;
   (f) $-\frac{49}{629}$

9. About 19.4 ft

10. 25.44144288; 15.89757793

11. (a) $-\frac{1}{2}$;   (b) 1;   (c) 0.00735;   (d) -0.29629

12. (a) $\frac{1}{2}$;   (b) -1;   (c) -0.005621527;
    (d) 2.136926496

13. 8, 2

14. Max $f(t) = \frac{1}{2}$; min $f(t) = \frac{1}{10}$

15.

16.

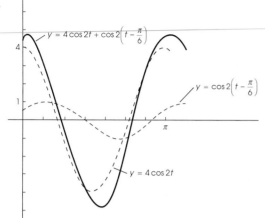

$y = 4\cos 2t + \cos 2\left(t - \frac{\pi}{6}\right)$

$y = \cos 2\left(t - \frac{\pi}{6}\right)$

$y = 4\cos 2t$

17. (a) $-8\sin 4x$;   (b) $(3/2\sqrt{x})\cos\sqrt{x}$;
    (c) $(3\cos x)/(2\sqrt{\sin x})$;
    (d) $-6x(x^2 + 1)^2 \sin[(x^2 + 1)^3]$;
    (e) $-18x\cos^2(x^2 + 1)\sin(x^2 + 1)$;
    (f) $4x^2\sec^2(x^2) + 2\tan(x^2)$

18. (a) $6\cos 2x$;   (b) $(4/x^2)\sin(1/x)$;
    (c) $\frac{3}{2}(3x + \sin x)^{1/2}(3 + \cos x)$;
    (d) $\frac{3}{2}x^2\cos\sqrt{x^3 + 4}/\sqrt{x^3 + 4}$;
    (e) $3x^2\cos(x^3 + 4)/2\sqrt{\sin(x^3 + 4)}$;
    (f) $[2x\cos(x^3) + 3x^4\sin(x^3)]/\cos^2(x^3)$

19. $y = (1/e)x + 1$

20. $y + 1 = -3e(x - \pi/2)$

21. (a) $\frac{1}{3}\sin(3x + 1) + C$;   (b) $\frac{1}{4}\tan 4x + C$;
    (c) $\ln(1 + \sin x) + C$;   (d) $-\frac{2}{25}\cos^{5/2}(5x) + C$

22. (a) $\frac{1}{2}\cos(3 - 2x) + C$;   (b) $\frac{1}{2}\sin(x^2) + C$;
    (c) $-1/(5 + \tan x) + C$;   (d) $\frac{1}{12}\sin^4 3x + C$

23. (a) $\pi + 1$;   (b) $\frac{1}{4}$ 24. (a) $\frac{1}{160}$;   (b) $\frac{1}{2}$

25. $[9 - \cos(\pi x^2)]/2\pi$ 26. $y = \frac{1}{6}\tan(3t^2) + 1$

27. (a) $P(t) = 6400 + 1800\sin(\pi t/18)$;
    (b) 4600;   (c) about 7259

28. (a) $N(t) = 1600\cos(\pi t/18) + 2400$;
    (b) 800;   (c) 2400

29. 1   30. 4

31. (a) 3:00 P.M.;   (b) 9:00 A.M. to 9:00 P.M.;   (c) $11.46

32. (a) 6:00 P.M.;
    (b) from 12:00 noon to 12:00 midnight.

33. $1.5\sin 1 + \cos 1 - 1 \approx 0.80251$

34. $\frac{7}{48}$ 35. $m = 1$

36. Zero   37. $(\pi/4, \pi/2), (3\pi/4, \pi)$

38. $(\pi + 6\sqrt{3})/2$ 39. (a) -12;   (b) $6\sqrt{3}$

40. (a) 4 ft/s;   (b) $\frac{15}{2}$ ft^2/s

41. $\theta = \pi/6$

42. (a) $(\pi/6, 2)$; $(0, 1)$;   (b) $(1 + \sqrt{3})/2$

Problem Set 13.1   (p. 413)

1. $y = (2x^2 + C)^{1/3}$ 2. $3y^{1/3} = x^4 + C$

3. $y = ke^{-4t}$ 4. $\ln x = 3t + C$

5. $y = \ln(x^4 + x^2 + C)$ 6. $-\frac{1}{2}y^{-2} = \frac{1}{3}x^3 + C$

7. $y^2 + e^y = x^3 + \sin 2x + C$

8. $\frac{1}{2}e^{2y} = \frac{1}{3}e^{3x} + C$ 9. $y = -\ln\frac{1}{2}(-e^{-2x} + K)$

10. $\ln y = 2\ln(1 + x) + \ln C$; $y = C(1 + x)^2$

11. $y^2 = (\ln x)^2 + K$ 12. $\sec y = Cx$

13. $y = Kxe^x$

14. $y = \frac{1}{2}\ln(x^2 + 4x + 2) + C$

15. $y = e^{3 - 1/x}$

16. $-\ln|5 - y| = 2\sqrt{x} - 2$ or $y = 5 - e^{2 - 2\sqrt{x}}$

17. $y = \frac{1}{2}\ln(t^2 + e^4)$

18. $e^y = \frac{1}{2}t^2 + t = e^2$ or $y = \ln(\frac{1}{2}t^2 + t + e^2)$

19. $e^{-3x^2} = -3y^2 - 24\ln|y| + 4$

20. Thus $y^3 = \frac{1}{2}(\ln x)^2 + 125$

21. $y = x + Cx^{-3}$ 22. $y = \frac{1}{x}(\ln x + C)$

23. $y = x^2e^x - xe^x + Cx$ 24. $y = \frac{1}{4}e^{3x} + Ce^{-x}$

25. $y = \sin x + C\cos x$

26. $y = \frac{3}{2}(\cos x + \sin x) + Ce^{-x}$

27. $y = \frac{1}{4}x^3 + \frac{1}{2}x + 4x^{-1}$

28. $y = 2(x + 4)^{3/2} + 2(x + 4)$

29. $y = 1 + 4e^{-\sin t}$

30. $yx^5 = 2\ln x + 4$; $y = (2\ln x + 4)/x^5$

31. $y = x^2 - x^3 + \frac{1}{2}e^{2x} + C$

32. $y = -\frac{3}{2}\cos 2x + 4\ln x - 2x^{3/2} + C$

33. $y^2 = 4t^{5/2} + K$

34. $ty = \frac{1}{2}e^{t^2} + C$

35. $y = \frac{1}{2}(x + 1)^4 + \frac{3}{2}(x + 1)^2$

36. $y = \frac{4}{3}\sqrt{1 + x^2} + [(18 + 8\sqrt{2})/3(1 + x^2)]$

37. $y = (7x - 24)/(x + 3)$

38. $-\cos y = 2\sqrt{x}\ln x - 4\sqrt{x} + 5$

39. $y = e^{4x - 3}$ 40. $y = 24{,}000/(1200 - x^2)$

Problem Set 13.2   (p. 421)

1. $dQ/dt = kt\sqrt{Q}$; $Q(1) = 4$

2. $ds/dt = k\frac{t}{s}$

3. (a) $dy/dx = 6x^2y^2$; $y(0) = 2$;
   (b) $y = 2/(1 - 4x^3)$

4. (a) $dy/dx = e^{-y} \cos x$, $y(3\pi/2) = 0$;
   (b) $y = \ln(2 + \sin x)$

5. (a) $dT/dt = k(T - 70)$; $T(0) = 340$, $T(15) = 300$;
   (b) $T = 70 + 270e^{-0.01069t}$;   (c) 28.1 min

6. 58 min

7. (a) $dy/dt = 3 - \frac{1}{15}y$, $y(0) = 12$;
   (b) about 38.77 lb

8. (a) $y = 240 - 180e^{-t/80}$;   (b) 240 lb of salt

9. (a) $dy/dt = 1 - [5y/(200 - t)]$; $y(0) = 0$;
   (b) $y = \frac{1}{4}(200 - t) - (50/200^5)(200 - t)^5$;
   (c) about 31.8 min;   (d) about 66.3 min

10. (a) $y = (0.4)[(100t + t^2)/(100 + 2t)] + [400/(100 + 2t)]$ kg;
    (b) $0.058\overline{3}$ kg/liter;   (c) $0.0013\overline{8}$ kg/liter/min

11. (a) $y/360 = \frac{108}{360}(1 - e^{-t/90})$;   (b) about 99 min

12. 46.05 min

13. (a) $dv/dt = -32$; $v(0) = 0$, $s(0) = 4800$;
    (b) $s = -16t^2 + 4800$;   (c) $10\sqrt{3} \approx 17.32$ s;
    (d) $320\sqrt{3} \approx 554$ ft/s

14. (a) $d^2s/dt^2 = -32$; $s(0) = 4800$, $v(0) = -64$;
    (b) $s = -16t^2 - 64t + 4800$;
    (c) after approximately 15.4356 s;   (d) 558 ft/s

15. (a) $s = 17,000 - 400t - 5000e^{-0.08t}$;   (b) about 42 s

16. 888.4 ft

17. If S denotes the amount of undissolved sugar at time t, then $dS/dt = kS$ and $S(0) = s_0$.

18. $dC/dt = kA(p - C)$, $C(0) = 0$

19. If P denotes the number infected at time t, then $dP/dt = kP(M - P)$ and $P(0) = 1$.

20. $dP/dt = kP \ln|P - M|$ 21. 18,000

22. 1731 yr

23. (a) $dA/dt = -0.05A$; $A(0) = 20$;   (b) $A = 20e^{-0.05t}$;
    (c) about 74 days

24. $y = -100e^{-0.02t} + 150e^{-0.01t}$

16. $y = C_1e^{(2+\sqrt{6})x} + C_2e^{(2-\sqrt{6})x}$

17. $x = C_1e^{(3+\sqrt{5})t} + C_2e^{(3-\sqrt{5})t}$

18. $x = C_1e^{(3+\sqrt{2})t/2} + C_2e^{(3-\sqrt{2})t/2}$

19. $x = C_1e^{4t} + C_2te^{4t}$ 20. $x = (C_1 + C_2t)e^{-2t}$

21. $x = C_1e^{t/3} + C_2te^{t/3}$ 22. $x = (C_1 + C_2t)e^{(-3/2)t}$

23. $y = C_1 \sin 5x + C_2 \cos 5x$

24. $y = C_1 \cos 7x + C_2 \sin 7x$

25. $y = C_1e^x \sin 2x + C_2e^x \cos 2x$

26. $y = e^{7x}(C_1 \cos x + C_2 \sin x)$

27. $y = C_1e^{5x} \sin 3x + C_2e^{5x} \cos 3x$

28. $y = e^{2x}(C_1 \cos \sqrt{5}x + C_2 \sin \sqrt{5}x)$

29. $y = e^x + 2e^{5x}$ 30. $y = 3e^{-4x} + 5e^x$

31. $y = 2 + 3e^{4x}$ 32. $y = (4 - 2x)e^{7x}$

33. $x = 3 \sin t$ 34. $x = 2 \cos 2t + 2 \sin 2t$

35. $y = C_1e^x + C_2e^{2x} + C_3e^{3x}$

36. $y = C_1 + C_2e^{-2x} + C_3e^{-3x}$

37. (a) $x = \frac{9}{2}t^2 + C_1t + C_2$;
    (b) $x = C_1e^{-3t} + C_2e^{3t}$;
    (c) $x = C_1 \sin 3t + C_2 \cos 3t$;
    (d) $x = C_1e^{-3t} + C_2te^{-3t}$;
    (e) $x = C_1e^{(-t+3\sqrt{3})t} + C_2e^{(-6-3\sqrt{3})t}$;
    (f) $x = C_1e^{-t} \sin 2\sqrt{2}\,t + C_2e^{-t} \cos 2\sqrt{2}\,t$

38. $(d^4y/dx^4) - 5(d^2y/dx^2) + 4y = 0$

39. Substitute $x = C_1u(t) + C_2v(t)$ in the differential equation. Then use the fact that u and v are solutions.

40. (a) e^t; te^t; t^2;   (b) $x = (C_1 + C_2t + C_3t^2)e^t$

41. (a) $x = C_1e^{-2t} + C_2te^{2t} + C_3t^2e^{-2t}$;
    (b) $x = C_1e^{2t} + C_2te^{2t} + C_3e^{3t}$;
    (c) $x = C_1e^t + C_2te^t + C_3t^2e^t + C_4t^3e^t$;
    (d) $x = C_1e^t + C_2te^t + C_3e^{2t} + C_4te^{2t}$

42. $x = 2 + 3t - e^{4t}$

Problem Set 13.4   (p. 435)

1. $y = C_1e^{-2x} + C_2e^{2x} - 3e^x$

2. $y_{\text{general}} = C_1e^{3x} + C_2e^{-3x} - e^{-x}$

3. $y = C_1 \sin 2x + C_2 \cos 2x + x + 2$

4. $y = C_1 \cos 3x + C_2 \sin 3x + \frac{2}{3}x - 2$

5. $y = C_1 \sin 2x + C_2 \cos 2x + 2 \cos x$

6. $y = C_1 \cos 3x + C_2 \sin 3x + \frac{3}{8} \sin x$

7. $y = C_1e^x + C_2e^{2x} - \sin 2x + 3 \cos 2x$

8. $y = C_1e^{3x} + C_2e^{-2x} - \frac{1}{4}e^{2x}$

9. $y = C_1e^x + C_2e^{2x} + \frac{1}{2}x^2 + \frac{3}{2}x + \frac{5}{4}$

10. $y = C_1e^{3x} + C_2e^{-2x} - \frac{1}{2}x^2 + \frac{1}{6}x - \frac{7}{36}$

11. $y = C_1e^{2x} + C_2e^{4x} + \frac{1}{3}e^x$

12. $y = C_1e^{2x} + C_2e^{5x} - 7 \sin x + 9 \cos x$

13. $y = C_1e^{-x} \sin x + C_2e^{-x} \cos x + 2e^x$

Problem Set 13.3   (p. 428)

1. $y = x^3 + C_1x + C_2$ 2. $y = \frac{1}{4}x^4 + C_1x + C_2$

3. $y = -\frac{1}{9} \cos 3t + C_1t + C_2$

4. $y = -\frac{1}{4} \sin 2t + C_1t + C_2$

5. $y = \frac{1}{9}e^{3t} + \frac{1}{3}t^3 + C_1t + C_2$

6. $y = e^{-t} - 2t^2 + C_1t + C_2$

7. $y = C_1e^{-2x} + C_2e^{2x}$ 8. $y = C_1e^{6x} + C_2e^{-x}$

9. $y = C_1 + C_2e^{4x}$ 10. $y = C_1 + C_2e^{-5x}$

11. $y = C_1e^{-4x} + C_2e^{6x}$ 12. $y = C_1e^{-5x} + C_2e^{-3x}$

13. $y = C_1e^{-3x} + C_2e^{x/2}$

14. $y = C_1e^{(2/3)x} + C_2e^{(3/2)x}$

15. $y = C_1e^{(-1+\sqrt{5})x/2} + C_2e^{(-1-\sqrt{5})x/2}$

14. $y = e^x(C_1 \cos 2x + C_2 \sin x) + 2e^{3x}$

15. $y = C_1 e^{2x} + C_2 e^{4x} - 2xe^{2x}$

16. $y = C_1 e^{2x} + C_2 e^{5x} + 2xe^{5x}$

17. $y = C_1 \sin 2x + C_2 \cos 2x - \frac{1}{4}x \cos 2x$

18. $y = C_1 \cos 3x + C_2 \sin 3x + 2x \sin 3x$

19. $y = 3e^{-2x} + 4e^{2x} - 3e^x$

20. $y = -\pi \cos 2x + \sin 2x + x + 2$

21. $y = -4 \sin 2x - 4 \cos 2x + 2 \cos x$

22. $y = e^{2x} + 2e^x - \sin 2x + 3 \cos 2x$

23. $y = (4 - \sqrt{2}) \sin 2x + 2 \cos 2x + 2 \cos x$

24. $y = -2 \cos 2x + [(16 - \pi)/4] \sin 2x + x + 2$

25. (a) $y'' + 4y = 0$; $y(0) = -1$, $y'(0) = 0$;

    (b) $y = -\cos 2t$

    (c)

26. (a) $2.5(d^2y/dt^2) = -10y$, $y(0) = -1$, $y'(0) = -2$;

    (b) $y = -\cos 2t - \sin 2t$;

    (c)

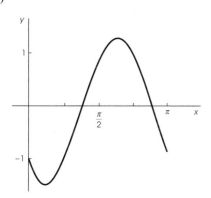

27. (a) $y'' + 0.05y' + 4y = 0$;

    $y(0) = -1$, $y'(0) = 0$;

    (b) $y = e^{-0.025t}(-0.0125 \sin 2t - \cos 2t)$;

    (c)

$y = e^{-0.025t}(-0.0125 \sin 2t - \cos 2t)$

28. (a) $y'' + 4y = \cos t$

    (b) $y = -\cos 2t$; $y_p = \frac{1}{3} \cos t$

29. $y = \frac{19}{17}e^{4x} - \frac{8}{17} \sin x - \frac{2}{17} \cos x$

30. $y = 2e^{2x} - e^{-2x} - \sin x$

31. $y = \frac{1}{2} \sin 2x + 9 \cos 2x + 2 \sin x$

32. $y = 4e^{2x} + 4e^{-2x} + \frac{1}{4}xe^{2x}$

33. $y = \frac{5}{2} \sin 2x + 9 \cos 2x - 2x \cos 2x$

34. $y = \frac{34}{3}e^x - \frac{11}{6}e^{4x} - \frac{1}{2}e^{2x}$

35. $y = 8.68e^{2x} - 14.6xe^{2x} + 0.24 \sin x + 0.32 \cos x$

36. $y = \frac{3}{2} + \frac{31}{3}e^x - \frac{5}{6}e^{4x} - 2e^{2x}$

37. (a) $P(t) = 400 + e^{-t} + 0.5e^{0.2t}$;

    (b) \$400.88, \$403.69, \$427.30;

    (c) \$601.71, \$16,829.81;   (d) no, at least for large t.

38. 2860; 60802   39. $y = \cos 3t + 2t \sin 3t$

40. $\frac{1}{3}$

Chapter 13 Review Problem Set (p. 436)

1. $y = \ln(\frac{1}{3}x^3 + C)$ 2. $y^2 = 1/(C_1 - e^{2x})$

3. $y = K\sqrt{x^2 + 1}$

4. $|y^3 + 1| = C|x|$; $y^3 + 1 = C_1 x$

5. $y = \frac{1}{2}x^3 + x^2 + Cx$

6. $y = 2 + (3x^2/x) + (C/x^2)$

7. $y = \frac{1}{2} \sin x \tan x + C \cos x$

8. $y = (\sin x + \frac{1}{2} \sin^2 x + C)/(1 + \sin x)$

9. $y = C_1 e^{-4t} + C_2 e^{4t}$ 10. $y = C_1 e^x + C_2 e^{3x}$

11. $y = C_1 \sin 4x + C_2 \cos 4x$

12. $y = C_1 \cos 5t + C_2 \sin 5t$

13. $y = C_1 e^{-4x} + C_2 e^{5x}$ 14. $y = C_1 e^{-x} + C_2 e^{3x}$

15. $y = C_1 e^{(-2+\sqrt{5})x} + C_2 e^{(-2-\sqrt{5})x}$

16. $y = C_1 e^{(-2+\sqrt{7})x} + C_2 e^{(-2-\sqrt{7})x}$

17. $y = C_1 e^{3x} + C_2 xe^{3x}$ 18. $(C_1 + C_2 t)e^{4t}$

19. $y = C_1 \sin 4x + C_2 \cos 4x + \frac{1}{17}e^x$

20. $y = C_1 \cos 2x + C_2 \sin 2x + \frac{2}{3} \sin x$

21. $y = C_1 e^{-4x} + C_2 e^{4x} - \frac{1}{17} \cos x$

22. $y = C_1 e^{2t} + C_2 e^{-2t} + \frac{3}{5}e^{3t} + \frac{1}{3}e^t$

23. $y = C_1 e^{-4x} + C_2 e^{5x} - 0.05x^2 + 0.005x - 0.00525$

24. $y = C_1 e^{-t} + C_2 e^{5t} - 2e^t - t + \frac{4}{5}$

25. $y = 4e^{3x^2 - 2x - 1}$

26. $y = -2/(x^2 - 6x + 4)$

27. $y = x/(1 - 6x)$ 28. $\ln y = x^3 - 2x^2 - 9$

29. $y = e^{-4x} + 3e^{4x}$ 30. $y = e^x + 2e^{3x}$

31. $y = 3e^{3x} - xe^{3x}$ 32. $x(t) = (3 - 10t)e^{4t}$

33. (a) $T = 75 + 375e^{-0.05493t}$; (b) $t \approx 38.6$ min

34. (a) $16.89°F$; (b) 4.24 hr

35. About 78.2 kg

36. (a) 36.19 kg; (b) after approximately $294\frac{1}{4}$ min
37. About 963   38. 3047
39. 4.099 s, 163.17 ft/s
40. Approximately 6.1 s, 163.2 ft/s
41. $y_p = Ax^2 + Bx + C + Dxe^{2x}$
42. $y_p = Ax^3 + Bx^2 + Cx + D + (E + Fx)e^x + Ge^{2x}$
43. $y = C_1e^{-2t} + C_2e^{2t} - \frac{16}{3}\cos t + 8te^{2t}$
44. $y = C_1e^{3x} + C_2e^{-x} + 2xe^{3x}$
45. (a) $r = -\frac{2}{3}, \frac{1}{2}$; (b) $y = C_1x^{-2/3} + C_2x^{1/2}$
46. $y = -3e^{3x} + 2e^{-x} + 2xe^{3x}$
47. $y = 10e^{40t}/(9 + e^{40t}) = 10/(1 + 9e^{-40t})$
48. $\ln\left[(\sqrt[3]{y}\sqrt[6]{y - 3}/\sqrt{y - 1})(\sqrt{3}/\sqrt[3]{4})\right] = 6t$

Problem Set 14.1   (p. 444)

1. (a) 63.75; (b) 80.4375; (c) 64.6875; (d) 63.75
2. (a) 48.6; (b) 71.09; (c) 50.84375; (d) 48.625
3. (a) 1.98999; (b) 1.98364; (c) 1.94836;
   (d) 1.99070
4. (a) 1.945910149; (b) 1.592857143;
   (c) 2.021428571; (d) 1.958730159
5. (a) 0.22222; (b) 0.17866; (c) 0.22804;
   (d) 0.22263
6. (a) 2.545177444; (b) 2.876286319;
   (c) 2.529712729; (d) 2.544648089
7. (a) 0.74427; (b) 0.79082; (c) 0.80135
8. 0.188894154
9. (a) 0.83801; (b) 0.80634; (c) 0.80486
10. 0.188320352
11. (a) $|E| \le [6^3/12(6^2)](6/3^4) = \frac{1}{27} \approx 0.03704$;
    (b) $|E| \le [6^5/180(6^4)](120/3^6) = \frac{4}{729} \approx 0.00549$
12. (a) 0.0625; (b) 0.00625
13. 4.175   14. 298
15. About 9067 ft^2 16. 5488 ft
17. 5.40031   18. 0.586287156
19. 790 ft^2, 14,220,000 ft^3 20. 24.917°C
21. 3.12701
23. $f^{(4)}(x) = 0$ for all x, which implies $E = 0$.

Problem Set 14.2   (p. 452)

1. (a) $\frac{1}{5} - \frac{1}{25}(x - 2)$; (b) $\frac{1}{5} - \frac{1}{25}(x - 2) + \frac{1}{125}(x - 2)^2$;
   (c) 0.1996, 0.1996008, 0.1996008
2. (a) $f(x) = \frac{1}{5} - 0.16(x - 2)$;
   (b) $f(x) = \frac{1}{5} - 0.16(x - 2) + 0.088(x - 2)^2$;
   (c) 0.1984; 0.1984088

3. (a) $2 + \frac{1}{4}(x - 2)$; (b) $2 + \frac{1}{4}(x - 2) - \frac{1}{64}(x - 2)^2$;
   (c) 2.0025, 2.0024984, 2.0024984
4. (a) $f(x) = 2 + \frac{1}{12}(x - 2)$;
   (b) $f(x) = 2 + \frac{1}{12}(x - 2) - (1/3^7)(x - 2)^2$;
   (c) 2.0008$\overline{3}$; 2.000833288
5. (a) $2.1972246 + 1.7652790(x - 2)$;
   (b) $2.1972246 + 1.7652790(x - 2) + $
   $$0.2222222(x - 2)^2;$$
   (c) 2.2148774, 2.2148996, 2.2148996
6. (a) $f(x) = \frac{2}{7} - \frac{1}{49}(x - 2)$;
   (b) $f(x) = \frac{2}{7} - \frac{1}{49}(x - 2) - 0.029154519(x - 2)^2$;
   (c) 0.285510204; 0.285507289
7. $2 + \frac{1}{4}(x - 4) - \frac{1}{64}(x - 4)^2 + $
   $\frac{1}{512}(x - 4)^3 - \frac{5}{16,384}(x - 4)^4$; 1.9748418, 2.0493901
8. $f(x) \approx \ln 2 + (\frac{1}{2} + \ln 2)(x - 1) + \frac{3}{8}(x - 1)^2 - $
   $$\frac{1}{12}(x - 1)^3 + \frac{5}{216}(x - 1)^4; 0.81613088$$
9. $1 - x + \frac{x^2}{2!} - \frac{x^3}{3!} + \cdots + (-1)^n \frac{x^n}{n!}$
10. $P_n(x) = 1 + \frac{x^2}{2!} + \frac{x^4}{4!} + \cdots + \frac{x^n}{n!}$, n even;
    $$P_n(x) = 1 + \frac{x^2}{2!} + \frac{x^4}{4!} + \cdots + \frac{x^{n-1}}{(n - 1)!} = P_{n-1}(x),$$
    n odd
11. $1 + x + x^2 + x^3 + \cdots + x^n$
12. $x - \frac{1}{2}x^2 + \frac{1}{3}x^3 - \frac{1}{4}x^4 + \cdots + [(-1)^{n-1}/n]x^n$
13.

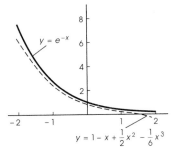

$y = e^{-x}$

$y = 1 - x + \frac{1}{2}x^2 - \frac{1}{6}x^3$

14. $P_3(x) = 1 + x + x^2 + x^3$

$y = \frac{1}{1 - x}$

$y = P_3(x)$

15. $|E| \le [(0.5)^4/24]e^{0.5} \approx 0.0042935$

16. 1

17. $\left|E(4.2)\right| \le [(0.2^5)/120][(\frac{105}{32})4^{-9/2}] \approx 1.709 \times 10^{-8}$

18. 9.375×10^{-8}

19. $1 - \frac{1}{2}(x - \pi/2)^2 + \frac{1}{24}(x - \pi/2)^4$; 0.9940432

20. $-1 + \frac{1}{2}(x - \pi)^2 - \frac{1}{24}(x - \pi)^4$; -0.990436994

21. $\left|E(1.68)\right| \le 2.3555 \times 10^{-9}$

22. 5.839641×10^{-8}

23. $-1 + 3(x - 1) + 9(x - 1)^2 + 8(x - 1)^3 + 2(x - 1)^4$.
One might expand this to see that it agrees with $f(x)$.
Alternatively, note that $f^{(5)}(x) = 0$ for all x, which implies that $E = 0$.

24. $2; -29; 136; d = -200$

25. $(1/e)(x - e) + (1/e^2)(x - 3)^2$; 0.0006317

26. 8 or larger

27. $\int_0^1 \left(1 - \frac{1}{4}x^2\right) dx = \frac{11}{12}$

28. 0.742857

Problem Set 14.3   (p. 456)

1. 1.1542

2. -0.8604

3. -3.9680

4. 1.9961

5. 2.279019

6. 2.612888

Seed value: 2.3

7. 2.963868

8. 2.391278

Seed value: 3

9. 0.304952

10. 0.586701

Seed value: 0.4

11. 1.763223

12. $x_1 = 0.5$; 1.557146

Seed value: 1.5

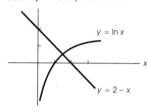

13. 7.71385%   14. 9.701% compounded annually

15. $-2.11491, 0.25410, 1.86081$ 16. $2.19258, -3.19258$

17. 1.13227, 4.53360

18. $-1.977383, -3.837467$

19. 0.601347

20. 5.567764

21. 1.912931

22. $x_{n+1} = -2x_n$

23. 12.5898%

24. Approximately $23\frac{1}{4}$ hr

25. (3.02668, 5.16079)

Problem Set 14.4   (p. 462)

1. 0.45018

2. 0.94775

3. 2.9263

4. 5.45973

5. (a) Using $x_1 = 2.3$, find x_2, x_3, x_4, x_5;   (b) 2.2789

6. (a) Using $x_1 = 0.85$, find x_2, x_3, x_4, x_5, x_6, and x_7.
   (b) 0.8603

7. $\left|f'(2.3)\right| \approx 1.999 > 1$, whereas
   $\left|g'(2.3)\right| \approx 0.499 < 1$

8. $1.772; -0.386$

9. (a) Algebra; (b) 1.1542
10. (a) $-\sqrt[3]{2}/(x+4)$; (b) -0.8604
11. (a) Algebra; (b) $f'(1) = 11, g'(1) = -8$, $h'(1) = -0.0505$; (c) 0.9125
12. (b) $f'(a) = -30, g'(1) = 23$, $h'(1) \approx 0.13$; (c) 0.9671
13. 2.3028   14. 1.4987
15. 0.3726   16. ± 0.5354
17. (a) $x = h(x)$ and $x = k(x)$; (b) 0.4921
18. 16.182 hr   19. $i = 0.01278$

Problem Set 14.5 (p. 468)

1. $y = 4e^x$

$y = 4e^x$

k	x_k	y_k	$y(x_k)$
0	0.0	4.0000	4.0000
1	0.1	4.4000	4.4207
2	0.2	4.8400	4.8856
3	0.3	5.3240	5.3994
4	0.4	5.8564	5.9673
5	0.5	6.4420	6.5949
6	0.6	7.0862	7.2885
7	0.7	7.7949	8.0550
8	0.8	8.5743	8.9022
9	0.9	9.4318	9.8384
10	1.0	10.3750	10.8731

2. $y = 3e^{2/3x^{3/2}}$

k	x_k	y_k	$y(x_k)$
0	0.0	3.0000	3.0000
1	0.1	3.0000	3.0639
2	0.2	3.0949	3.1843
3	0.3	3.2333	3.3473
4	0.4	3.4104	3.5511
5	0.5	3.6261	3.7974
6	0.6	3.8825	4.0896
7	0.7	4.1832	4.4329
8	0.8	4.5332	4.8339
9	0.9	4.9386	5.3006
10	1.0	5.4072	5.8432

3. $y = 2/\cos x$

$y = \dfrac{2}{\cos x}$

k	x_k	y_k	$y(x_k)$
0	0.0	2.0000	2.0000
1	0.1	2.0000	2.0100
2	0.2	2.0201	2.0407
3	0.3	2.0610	2.0935
4	0.4	2.1248	2.1714
5	0.5	2.2146	2.2790
6	0.6	2.3356	2.4233
7	0.7	2.4954	2.6149
8	0.8	2.7056	2.8706
9	0.9	2.9841	3.2175
10	1.0	3.3602	3.7016

4. $y = 3(1 + x^2)$

k	x_k	y_k	$y(x_k)$
0	0.0	3.0000	3.00
1	0.2	3.0000	3.12
2	0.4	3.2308	3.48
3	0.6	3.6764	4.08
4	0.8	4.3252	4.92
5	1.0	5.1691	6.00
6	1.2	6.2029	7.32
7	1.4	7.4232	8.88
8	1.6	8.8276	10.68
9	1.8	10.4146	12.72
10	2.0	12.1831	15.00

5. $y = 3e^x(e^x - 1)$

$y = 3\left(e^{2x} - e^x\right)$

k	x_k	y_k	$\hat{y}_k$	$y(x_k)$
0	0.0	0.0000	0.0000	0.0000
1	0.2	0.6000	0.7864	0.8113
2	0.4	1.5728	2.0722	2.2011
3	0.6	3.0970	4.1002	4.4940
4	0.8	5.4291	7.2204	8.1825
5	1.0	8.9361	11.9348	14.0123

6. Exact solution: $y = \frac{1}{4}x^3 - 2x^{-1}$

k	x_k	y_k	y_k	$y(x_k)$
0	2.0	1.0000	1.0000	1.0000
1	2.2	1.7000	1.7567	1.7529
2	2.4	2.5135	2.6321	2.6227
3	2.6	3.4560	3.6415	3.6248
4	2.8	4.5422	4.7992	4.7737
5	3.0	5.7858	6.1189	6.0833
6	3.2	7.2000	7.6140	7.5670
7	3.4	8.7980	9.2973	9.2378
8	3.6	10.5925	11.1816	11.1084
9	3.8	12.5960	13.2796	13.1917
10	4.0	14.8211	15.6036	15.5000

7. $y(1) \approx 1.2120$

k	x_k	y_k	$\hat{y}_k$
0	0.0	0.0000	0.0000
1	0.1	0.0000	0.0100
2	0.2	0.0200	0.0400
3	0.3	0.0600	0.0903
4	0.4	0.1204	0.1614
5	0.5	0.2019	0.2548
6	0.6	0.3059	0.3727
7	0.7	0.4353	0.5191
8	0.8	0.5942	0.7002
9	0.9	0.7895	0.9259
10	1.0	1.0319	1.2120

8. $y(1) \approx 1.4942$

k	x_k	y_k	y_k
0	0.0	1.0000	1.0000
1	0.1	1.1000	1.0995
2	0.2	1.1990	1.1969
3	0.3	1.2948	1.2895
4	0.4	1.3848	1.3744
5	0.5	1.4658	1.4481
6	0.6	1.5340	1.5061
7	0.7	1.5852	1.5439
8	0.8	1.6146	1.5562
9	0.9	1.6170	1.5379
10	1.0	1.5874	1.4850

9. $e^x y = x + C$

$y = xe^x$

k	x_k	y_k	$y(x_k)$
0	0.0	0.0000	0.0000
1	0.2	0.2000	0.1637
2	0.4	0.3237	0.2681
3	0.6	0.3931	0.3293
4	0.8	0.4242	0.3595
5	1.0	0.4292	0.3679
6	1.2	0.4170	0.3614
7	1.4	0.3938	0.3452
8	1.6	0.3644	0.3230
9	1.8	0.3319	0.2975
10	2.0	0.2986	0.2707

10. $y(1) \approx 10{,}769{,}344.65$

k	x_k	y_k
0	0.0	2.0
1	0.1	2.4
2	0.2	3.001
3	0.3	3.9656
4	0.4	5.6662
5	0.5	9.1194
6	0.6	17.9168
7	0.7	51.1291
8	0.8	316.7017
9	0.9	10,372.0967
10	1.0	10,769,344.65

Chapter 14   Review Problem Set (p. 468)

1. (a) 0.3465736; (b) 0.3765189; (c) 0.3452689;
   (d) 0.3465841
2. (a) 4 ln 2; (b) 2.651487; (c) 2.77649;
   (d) 2.77262
3. (a) 1.2909611; (b) $|E| \leq 0.0208333$
4. $1.43606667; |E| \leq (3^3)(2)/(12)(36) = 0.125$
5. (a) 2.7899674; (b) 2.7899371; (c) 1.322669403
6. $2e^{\sqrt{2}} - 2e \approx 2.7899371$
7. (a) $\ln 4 + \frac{1}{4}x - \frac{1}{32}x^2 + \frac{1}{192}x^3 - \frac{1}{1024}x^4$;
   (b) 1.3609766; (c) 3.2038387
8. (a) $P_4(x) = \frac{1}{2}x - \frac{1}{4}x^2 + \frac{1}{8}x^3 - \frac{1}{16}x^4$;
   (b) 2.97619×10^{-7}; (c) 0.1854
9. (a) $\frac{1}{2} - (\sqrt{3}/2)(x - \pi/3) -$
      $\frac{1}{4}(x - \pi/3)^2 + (\sqrt{3}/12)(x - \pi/3)^3$;
   (b) 0.48258065; (c) $|E| \leq 0.00000001$
10. (a) $(\sqrt{2}/2) + (\sqrt{2}/2)(x - \pi/4) -$
       $(\sqrt{2}/4)(x - \pi/4)^2 - (\sqrt{2}/12)(x - \pi/4)^3$;
    (b) 0.67088229; (c) 2.6×10^{-7}
11. 0.79206   12. 0.6417
13. −1.16404, 1.45263
14. $r_1 = -3.3460; r_2 = 2.9392; r_3 = 0.4067$
15. 2.1200   16. 0.45018
17. 1.7360   18. 1.8440
19. 0.5741   20. $y(1) \approx 1.4123$
21. 0.5110   22. 3.4005
23. $1 + \frac{1}{2}x^2 - \frac{1}{8}x^4$, 1.00019998
24. $P_4(x) = 1 - \frac{1}{2}x^2 + \frac{3}{8}x^4$
25. (a) 1.1416667; (b) 1.1477929; (c) 1.1477936
26. (a) $0.908\overline{3}$; (b) 0.88138; (c) 0.881373587

Problem Set 15.1   (p. 477)

1. (a) 5, 7, 9, 11, 13; (b) 18, 15, 10, 3, −6;
   (c) $1, \frac{2}{3}, \frac{4}{9}, \frac{8}{27}, \frac{16}{81}$; (d) $-4, 7, \frac{10}{3}, \frac{13}{5}, \frac{16}{7}$;
   (e) 2, 4, 10, 28, 82; (f) 2, −2, 2, −2, 2
2. (a) 1, 3, 9, 27; (b) 2, 2.3, 2.6, 2.9;
   (c) 1, 9, 25, 49; (d) $2, -8, -3, -\frac{16}{7}$;
   (e) 5, −9, 19, −37; (f) 4, 4, 4, 4
3. (a) Converges; 0; (b) converges; −4;
   (c) diverges; (d) converges; ln 2
4. (a) Converges; 0; (b) diverges;
   (c) diverges; (d) converges; cos 2
5. (a) $-\frac{2}{3}$; (b) $\frac{1}{2}$; (c) $-\frac{1}{2}$; (d) 1; (e) $\sqrt{2}$;
   (f) $\sqrt{2}/3$

6. (a) 4; (b) $-\frac{1}{2}$; (c) 3; (d) 0; (e) $-\frac{1}{2}$; (f) $\frac{1}{2}$
7. Converges; 1
8. Converges; the sum is $-\frac{5}{12}$.
9. Converges; $\frac{1}{4}$
10. Converges; the sum is $\frac{1}{2}$.
11. Diverges.   12. Series diverges.
13. Diverges.   14. Series diverges.
15. $\lim_{n \to \infty} \ln (2 + 1/n) = \ln 2 \neq 0$
16. Series diverges; $\frac{3}{4} \neq 0$
17. $\lim_{n \to \infty} \sin n$ does not exist.
18. Series diverges; $2^0 = 1 \neq 0$
19. $\lim_{n \to \infty} [0.01 + (-1)^n/n] = 0.01 \neq 0$
20. Series diverges; $0.01 \neq 0$
21. Diverges: $S_n = \frac{1}{2} + \frac{1}{4} + \frac{1}{6} + \cdots + 1/2n$
    $= \frac{1}{2}(1 + \frac{1}{2} + \frac{1}{3} + \cdots + 1/n) \to \infty$ as $n \to \infty$.
22. Diverges because harmonic series diverges.
23. Converges: $S_n = \frac{1}{3} + \frac{1}{5} + \frac{1}{9} + \cdots + 1/(2^n + 1)$
    $\leq \frac{1}{2} + \frac{1}{4} + \frac{1}{8} + \cdots + 1/2^n \leq 1$.
    Thus, by the Bounded Sum Test, the series converges.
24. Converges because
    $$1 + \frac{1}{2!} + \frac{1}{3!} + \frac{1}{4!} + \cdots \text{ converges.}$$
25. Diverges: $S_n > n^2$ and so $S_n \to \infty$ as $n \to \infty$.
26. Diverges because $\ln n \to \infty$ as $n \to \infty$.
27. (a) Converges: 3; (b) diverges; (c) converges; 0;
    (d) diverges; (e) converges; 1; (f) converges; 2
28. (a) n^2; (b) $(n + 1)/(2n + 1)$; (c) $4n - 1$;
    (d) $n/3^{n-1}$
29. $-\frac{47}{60}$ 30. 376
31. (a) Converges: $S_n = 3(\frac{1}{2} + \frac{1}{4} + \cdots + 1/2^n) \leq 3$
        (Bounded Sum Test);
    (b) diverges: $\lim_{n \to \infty} 2n/(n + 100) = 2 \neq 0$ (nth Term Test);
    (c) converges: $S_n = \frac{1}{3} + \frac{1}{9} + \cdots + 1/3^n \leq \frac{1}{2} + \frac{1}{4} + \cdots + 1/2^n \leq 1$ (Bounded Sum Test);
    (d) diverges: $S_n = \frac{3}{2}(1 + \frac{1}{2} + \frac{1}{3} + \cdots + 1/n)$
        (harmonic series)
32. (a) $1; \frac{4}{7}; \frac{3}{7}; \frac{1}{3}$; (b) 5
33. (a) 2, 3, 5, 8, 13, 21, 34, 55, 89; (b) 1.618
34. 3.000   35. $a_n = 300(0.8)^n$
36. 7 strokes

Problem Set 15.2   (p. 484)

1. $\frac{3}{2}[1 - (\frac{1}{3})^{12}] \approx 1.4999972$
2. $16[1 - (\frac{3}{4})^{12}]$
3. $2[1 - (\frac{1}{2})^{12}] \approx 1.9995117$

4. $\frac{9}{5}[1 - (\frac{2}{3})^{12}]$; 1.786126776

5. 3795.4253

6. $100[(1.025)^{12} - 1]/0.025$

7. 716.0725

8. $5200[(1.04)^{12} - 1]/(1.04)^{12}$; 1952.095342

9. $14,889.04   10. $115,014.78

11. $6795.16   12. $82,003.95

13. $45,362.88   14. $17,704.69

15. Converges; 3   16. Converges; 4

17. Diverges.

18. Converges; $1/(1 - \ln 2)$; 3.258891353

19. Converges; 1   20. Converges; 23

21. Converges; 3.0072464

22. Converges; $-\sin 1/(1 + \sin 1)$; -0.456955875

23. $\frac{8}{33}$ 24. $\frac{71}{99}$

25. $\frac{137}{111}$ 26. $\frac{269}{990}$

27. $\frac{2}{3}$ 28. 100%

29. $88 billion   30. $666,666.67

31. The first option is better by $5000.

32. 11.42857 mg

33. (a) $\frac{4}{11}$; (b) $\frac{9}{16}$; (c) $\sqrt{5}/(3 - \sqrt{5})$; (d) $\frac{3}{7}$

34. 64 ft

35. $324,299.34, $50,700.66

36. $137.54   37. $11,421.24

38. (a) 24,512 policies; (b) 25,500 policies

39. 3600   40. $\frac{2}{5}\sqrt{5}$ mi

Problem Set 15.3   (p. 490)

1. Converges, since $1/(3^{k-1} + 1) \le 1/3^{k-1}$ and $\Sigma (1/3^{k-1})$ converges.

2. Convergent.

3. Diverges, since $\sqrt{k}/k \ge 1/k$ and $\Sigma (1/k)$ diverges.

4. Divergent.

5. Converges, since $3^k/(4^k + 1) \le 3^k/4^k = (\frac{3}{4})^k$ and $\Sigma (\frac{3}{4})^k$ converges.

6. Divergent.

7. Converges, since $\lim_{k\to\infty} \dfrac{2^{k+1}}{(k + 1)!} \dfrac{k!}{2^k} = \lim_{k\to\infty} \dfrac{2}{k + 1} = 0$.

8. Convergent.

9. Diverges, since $\lim_{k\to 0} \dfrac{2^{k+1}}{(k + 1)^3} \dfrac{k^3}{2^k} =$

$$\lim_{k\to\infty} 2\frac{k^3}{(k + 1)^3} = 2.$$

10. Divergent.

11. Ratio Test fails, since $\lim_{k\to\infty} \dfrac{(k + 1)^2}{(k + 1)^4 + 1} \dfrac{k^4 + 1}{k^2} = 1$.
However, series converges by comparison with $\Sigma (1/k^2)$.

12. Convergent.

13. Converges, since $\lim_{k\to\infty} 2^k\sqrt{k}/2^{k+1}\sqrt{k + 1} = \frac{1}{2}$.

14. Divergent.   15. Diverges.

16. Convergent.   17. Converges.

18. Convergent.   19. Diverges.

20. Convergent.   21. Converges.

22. Divergent.   23. Diverges.

24. Divergent.   25. Converges.

26. Convergent.   27. Diverges, nth Term Test.

28. Divergent because $a_n \to 0$ as $n \to \infty$.

29. Converges. Example E of this section.

30. Divergent because $\dfrac{1}{\sqrt{k} - 1} > \dfrac{1}{\sqrt{k}}$ and $\displaystyle\sum_{k=2}^{\infty} \dfrac{1}{\sqrt{k}}$ diverges.

31. Converges. Ratio Test.

32. Diverges because $\lim_{n\to\infty} \dfrac{(x + 1)!/10^{n+1}}{n!/10^n} = \lim_{n\to\infty} \dfrac{n + 1}{10} = \infty$.

33. Converges. Geometric series with $r = -\frac{1}{2}$.

34. Convergent because $\dfrac{|\sin(k\pi/2)|}{k^2} \le 1/k^2$ (Example E).

35. Diverges. Compare with harmonic series.

36. Convergent because of the Integral Test.

37. Converges; Integral Test.

38. Divergent because $\lim_{n\to\infty} e^{1/n} = 1 \ne 0$.

39. Converges. Write as $4 \Sigma (\frac{1}{3})^k - 2 \Sigma (1/k^2)$. First series is geometric with $r = \frac{1}{3}$. Second series converges by Example E.

40. Convergent because of the theorem in Section 15.2 having to do with $\displaystyle\sum_{k=1}^{\infty} (ca_k + db_k)$.

41. Diverges; nth Term Test.

42. Convergent because $4/(k^2 + 1) < 4/k^2$ and $4\Sigma(1/k^2)$ converges.

43. Converges. Compare with $\Sigma (1/k^{3/2})$.

44. Divergent because $\cos(1 + 1/k^2) \to \cos 1 \ne 0$. $\cos n \to \infty$

45. Converges. Geometric series with $r = 1/1.1$.

46. Divergent because of the Ratio Test.

47. Diverges. Compare with $\Sigma (3/\sqrt{k})$.

48. Convergent because of the Ratio Test.

49. Diverges. Ratio Test.

50. Convergent because of the Integral Test.

51. Converges. Ratio Test.

52. Convergent because of the Integral Test.

53. Converges. Ratio Test.

54. Divergent because $\lim\limits_{n\to\infty}\left(\dfrac{(n+1)!\;3^n}{3^{n+1}\;n!}\right)=\infty$.

55. Converges. Write as difference of two series. First series converges by Ratio Test, second series by Example E.

56. Divergent because $(1/2^k)+(1/k)>1/k$.

57. Diverges. nth Term Test $[\lim\limits_{k\to\infty}(1+1/k)^k=e]$.

58. Convergent because of the Ratio Test; $\lim\limits_{n\to\infty}a_{n+1}/a_n=1/e<1$

Problem Set 15.4   (p. 496)

1. $\left|\pm1/k^2\right|=1/k^2$ and $\Sigma\,(1/k^2)$ converges.

2. $|a_n|=\dfrac{1}{3^{n-1}}\displaystyle\sum_{k=1}^{\infty}\dfrac{1}{3^{k-1}}$ converges (geometric series, $k<1$). $\therefore$ given series converges absolutely; hence, it converges.

3. $\left|[\sin(k\pi/3)/k^2]\right|\le1/k^2$ and $\Sigma\,(1/k^2)$ converges.

4. $\displaystyle\sum_{k=1}^{\infty}\dfrac{2^k\cos k}{k!}\;\left|\dfrac{2^k\cos k}{k!}\right|\le\dfrac{2^k}{k!};\;\displaystyle\sum_{k=1}^{\infty}\dfrac{2^k}{k!}$ converges, by the Ratio Test. $\therefore$ the given series converges (absolutely).

5. $1/\sqrt{k+1}\le1/\sqrt{k}$ and $\lim\limits_{k\to\infty}1/\sqrt{k}=0$

6. Alternating series, with $a_n\to0$, and $a_{n+1}<a_n$. $\therefore$ the series converges.

7. $1/\sqrt[3]{k+6}\le1/\sqrt[3]{k+5}$ and $\lim\limits_{k\to\infty}1/\sqrt[3]{k+5}=0$.

8. Alternating series. $\lim\limits_{n\to\infty}\dfrac{2n+1}{n^{3/2}}=0$;

$\dfrac{2n+3}{(n+1)^{3/2}}\dfrac{n^{3/2}}{2n+1}=\left(1+\dfrac{2}{2n+1}\right)\dfrac{1}{(1+1/n)^{3/2}}<$

$(2)\left(\dfrac{1}{e}\right)<1$.

$\therefore a_{n+1}<a_n$. So the series converges (absolutely).

9. $(k+1)/2^{k+1}<k/2^k$ if and only if $k+1<2k$. The latter is true for $k>1$. Also $\lim\limits_{k\to\infty}k/2^k=0$.

10. Alternating series $\lim\limits_{n\to\infty}n/e^{+n^2}=0$. If $f(x)=xe^{-x^2}$, $f'(x)=e^{-x^2}(1-2x)<0,\;x\ge1$. $\therefore f(x)$ is decreasing and so $a_{n+1}<a_n$. Hence the series converges (absolutely).

11. Diverges.   12. Converges absolutely.

13. Converges conditionally.

14. Converges conditionally.

15. Converges absolutely.

16. Converges conditionally.

17. Converges conditionally.

18. Diverges.

19. Converges absolutely.

20. Converges conditionally.

21. $-\infty<x<\infty$ 22. $-1\le x\le1$

23. $-1\le x<1$ 24. $-1<x<1$

25. $-\tfrac{1}{2}<x<\tfrac{1}{2}$ 26. $-4<x<4$

27. $-\infty<x<\infty$ 28. $-1\le x\le1$

29. $2-\dfrac{2^3}{3!}+\dfrac{2^5}{5!}-\dfrac{2^7}{7!}+\cdots$

30. $1+2+\dfrac{2^2}{2!}+\dfrac{2^3}{3!}+\cdots+\dfrac{2^n}{n!}+\cdots$

31. $1-\dfrac{1}{2!}+\dfrac{1}{4!}-\dfrac{1}{6!}+\cdots$

32. $(0.5)-\tfrac{1}{2}(0.5)^2+\tfrac{1}{3}(0.5)^3-\tfrac{1}{4}(0.5)^4+\cdots$

33. $1+3x+9x^2+27x^3+\cdots$

34. $2x-\tfrac{4}{6}x^3+\tfrac{32}{120}x^5-\tfrac{128}{5040}x^7+\cdots$

35. $\tfrac{1}{2}x-\tfrac{1}{8}x^2+\tfrac{1}{24}x^3-\tfrac{1}{64}x^4+\cdots$

36. $1-x+\dfrac{x^2}{2!}-\dfrac{x^3}{3!}+\dfrac{x^4}{4!}-\dfrac{x^5}{5!}+\cdots$

37. $1-\tfrac{9}{2}x^4+\tfrac{27}{8}x^8-\tfrac{81}{80}x^{12}+\cdots$

38. $1+5x^2+25x^4+125x^6+625x^8+\cdots$

39. $1+2x+2x^2+\tfrac{4}{3}x^3+\cdots$

40. $1-\dfrac{x^6}{2!}+\dfrac{x^{12}}{4!}-\dfrac{x^{18}}{6!}+\cdots$

41. $1+x-\tfrac{1}{3}x^3-\tfrac{1}{6}x^4+\cdots$

42. $1+x+\tfrac{1}{2}x^2+\tfrac{1}{3}x^3+\tfrac{13}{24}x^4+\cdots$

43. $x+\tfrac{5}{6}x^3+\cdots$ 44. x^3

45. $1+2x-\tfrac{4}{3}x^3-\tfrac{1}{3}x^4+\cdots$

46. $-2+3x-\tfrac{5}{2}x^2+\tfrac{7}{3}x^3-\tfrac{9}{4}x^4+\cdots$

47. 0.4613   48. 0.041481

49. $2x+\tfrac{2}{3}x^3+\tfrac{2}{5}x^5+\tfrac{2}{7}x^7+\cdots$

51. $\ln3=\ln[(1+\tfrac{1}{2})/(1-\tfrac{1}{2})]=$
$2[\tfrac{1}{2}+\tfrac{1}{3}(\tfrac{1}{2})^3+\tfrac{1}{5}(\tfrac{1}{2})^5+\tfrac{1}{7}(\tfrac{1}{2})^7+\cdots]\approx1.0986$

52. $\ln(1+2x)$

53. (a) $-3<x<3$; (b) $-\infty<x<\infty$;
(c) $-1\le x\le1$

54. $1+x+2x^2+0\cdot x^3+\tfrac{3}{2}x^4+\tfrac{5}{6}x^5+\tfrac{25}{24}x^6+\cdots$

55. $1/(1+x)^2=1-2x+3x^2-4x^3+5x^4+\cdots$
$1/(1+x)^3=1-3x+6x^2-10x^3+15x^4+\cdots$

56. $1/(1-x)^2$

Chapter 15   Review Problem Set (p. 498)

1. (a) Diverges; (b) converges; $\tfrac{3}{2}$;
(c) converges; $\sin2$; (d) diverges.

2. (a) Converges; 0; (b) converges; -3; (c) diverges; (d) diverges.

3. 108

4. $a_5 = -10$

5. $\frac{427}{999}$

6. $\frac{3}{111}$

7. (a) $\frac{2}{5}, \frac{4}{15}, \frac{4}{21}$; (b) 2

8. (a) $\frac{1}{6}; \frac{1}{6}; \frac{5}{66}$; (b) $\frac{1}{2}$

9. Converges. Geometric series with $r = e - 2 < 1$.

10. Diverges. It is a geometric series with $r > 1$.

11. Diverges. nth Term Test.

12. Diverges. $A_n \to \cos 3 \neq 0$

13. Diverges. Ratio Test.

14. Converges. $\lim_{n \to \infty} |a_{n+1}/a_n| = \lim_{n \to \infty} 8/(n + 1) = 0 < 1$

15. Diverges. Integral Test.

16. Diverges. $\dfrac{n^2}{n^3 + 1} > \dfrac{n^2}{2n^3} = \dfrac{1}{2n}$ and $\sum_{k=1}^{\infty} \dfrac{1}{2k} = \dfrac{1}{2} \sum_{1}^{\infty} \dfrac{1}{k}$ diverges.

17. Converges. Alternating Series Test.

18. Converges. It is a converging alternating series.

19. Converges. Ratio Test.

20. Diverges. $\dfrac{1}{n}\left(\dfrac{5}{4}\right)^n > \dfrac{1}{n}$ and $\sum^{\infty} \dfrac{1}{k}$ diverges.

21. Converges. Compare with $\Sigma (1/2^k)$.

22. Converges. $\dfrac{1}{n^{7/6} + n} < \dfrac{1}{n^{7/6}}$ and $\sum_{1}^{\infty} \dfrac{1}{k^{7/6}}$ converges.

23. Converges. Integral Test.

24. Converges. $f(x) = e^x/(e^x + 1)^2$ is decreasing as x increases. $\displaystyle\int_1^b (e^x + 1)^{-2} e^x \, dx = \dfrac{-1}{e^x + 1}\bigg]_1^b =$

$\dfrac{1}{e + 1} - \dfrac{1}{e^b + 1} \therefore \displaystyle\int_1^{\infty} (e^x + 1)^{-2} e^x \, dx = \dfrac{1}{e + 1}$

25. (a) $-1 < x < 1$; (b) $-\frac{1}{3} \leq x \leq \frac{1}{3}$

26. (a) $-\frac{1}{3} \leq x \leq \frac{1}{3}$; (b) converges for all values of x.

27. (a) $1 + 2x + 2x^2 + \frac{4}{3}x^3 + \cdots$;
    (b) $2 + x - x^2 - \frac{1}{2}x^3 + \cdots$;
    (c) $x + x^2 + \frac{1}{3}x^3 - \frac{1}{30}x^5 + \cdots$

28. (a) $3x - \frac{9}{2}x^2 + 9x^3 - \frac{81}{4}x^4 + \cdots$;
    (b) $-1 - 2x - x^2 + \frac{2}{3}x^3$;
    (c) $x + \frac{5}{6}x^3 + \frac{101}{120}x^5 + \frac{4241}{5040}x^7$

29. 3

30. 12

31. $28,232.27

32. $35,375.50

33. 25 mg

34. $219,085.10

35. (a) 0.06438907; (b) 0.06438565; (c) 0.06438908

36. (a) 0.14375; (b) 0.138743997; (c) 0.143648247

37. $x - 2x^3 + 3x^5 - 4x^7 + 5x^9 + \cdots$

38. x^2

Problem Set 16.1 (p. 507)

1. (a) $\{(1, H), (2, H), (3, H), (4, H), (5, H), (6, H),$ $(1, T), (2, T), (3, T), (4, T), (5, T), (6, T)\}$;
   (b) $\frac{3}{12} = \frac{1}{4}$; (c) $\frac{4}{12} = \frac{1}{3}$

2. (a) $\{HHH, HHT, HTH, THH, HTT, THT, TTH, TTT\}$;
   (b) prob $(B) = \frac{3}{8}$; (c) $\frac{1}{2}$

3. $\frac{1}{4}, \frac{5}{16}$

4. $\frac{3}{16}, \frac{5}{8}$

5.

6.

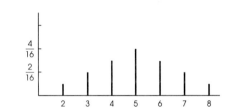

7. 2

8. 5

9. 1

10. 2.5

11. $1

12. $0.70

13. (a) $\frac{1}{2}, \frac{1}{2}$; (b) 4.2, 2.76

14. (a) 0.68, 0.48; (b) 5.37, 4.6331

15. $\frac{1}{216}, \frac{3}{216}, \frac{212}{216}$

16. (b) $\frac{3}{5}$

17. (a) $\frac{2}{3} + \frac{2}{9} + \frac{2}{27} + \cdots = \frac{2}{3}/(1 - \frac{1}{3}) = 1$; (b) $\frac{3}{2}$

18. (b) $\mu = p$

19. (a) $\frac{33}{100}$; (b) $\frac{19}{100}$; (c) $\frac{45}{100}$

20. (a) $\frac{1}{8}$; (b) 6; (c) $4.1\overline{6}$

21. $\frac{5}{2}, \frac{5}{4}$

22. $f_i = (\frac{1}{2})^i$; 2

Problem Set 16.2 (p. 512)

1. $\dfrac{3}{14} \displaystyle\int_0^2 (x^2 + 1) \, dx = \dfrac{3}{14}\left[\dfrac{1}{3}x^3 + x\right]_0^2 = 1$

2. 1

3. $k = \frac{3}{16}$

4. $k = \frac{6}{7}$

5. $k = \frac{1}{20}$

6. $k = \frac{3}{16}$

7. $\frac{7}{16}$ 8. 0.28

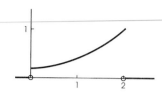

9. $\frac{27}{256} \approx 0.10547$ 10. 0.94815

11. (a) $f(x) = \frac{1}{6}$ on $[0, 6]$ and $f(x) = 0$ elsewhere;
    (b) $\mu = 3$, $\sigma^2 = 3$

12. (a) $f(x) = \frac{1}{3}$ for $1 \le x \le 4$, $f(x) = 0$ elsewhere;
    (b) 2.5, 0.75

13. $\mu = \frac{9}{7}$, $\sigma^2 = \frac{71}{245} \approx 0.28980$

14. $\frac{8}{15}$, $\frac{11}{225}$ 15. $\frac{2}{5}$ h $= 24$ min

16. $\frac{3}{5}$ min

17. $k = 1/(e^2 + 1) \approx 0.11920$

18. 0.1916332

19. (a) $\frac{14}{27} \approx 0.51852$;   (b) $\frac{9}{4} = 2.25$ min

20. 0.47178   21. $\frac{5}{2}$, $\frac{5}{4}$

22. (a) $f(x) = x/50$; $0 \le x \le 10$;   (b) $6\frac{2}{3}$ in.

Problem Set 16.3   (p. 518)

1. (a) 0.23254;   (b) 0.22313
2. (a) 0.232544158;   (b) 0.049787068
3. 0.13534   4. 0.049787068
5. 0.77687   6. 0.188876
7. 0.4554   8. 0.4893
9. 0.6247, 0.0668   10. 0.6179, 0.5028
11. 0.1311, 0.6554   12. 0.9104, 0.3085
13. 0.6247   14. 0.9085
15. 0.6915   16. 0.7475
17. (a) 0; (b) 0.9332; (c) 0.9876
18. (a) 0.8185; (b) 0; (c) 1; (d) 4
19. 0.34364   20. 0.0956
21. 0.0808   22. (a) $c = 4$; (b) 1
23. 2.77259

Problem Set 16.4   (p. 524)

1. (a) $f(x, y) = \frac{1}{8}$ on D and $f(x, y) = 0$ elsewhere;
   (b) $\frac{1}{4}$; (c) $\frac{1}{4}$

2. (a) $f(x, y) = \frac{1}{6}\{(x, y) \in D\}$, $f(x, y) = 0$ elsewhere;
   (b) $\frac{1}{12}$; (c) 0
3. (a) $\frac{1}{4}$; (b) $\frac{1}{2}$ 4. (a) $\frac{1}{16}$; (b) $\frac{1}{2}$
5. (a) $\frac{15}{128}$; (b) $\frac{243}{1024} \approx 0.23730$
6. (a) $k = \frac{1}{32}$; (b) 0.9375
7. 0.18112   8. 0.22313016
9. 0.50908
10. (a) 0.6549; (b) 0.2630; (c) 0.0235
11. (a) 0.06948; (b) 0.34692
12. $\frac{5}{8}$
13. $\mu_x = \mu_y = \frac{4}{3}$, $\sigma_x^2 = \sigma_y^2 = \frac{20}{63}$, $\sigma_{xy} = -\frac{1}{9}$, $\rho_{xy} = -\frac{7}{20}$
14. $\frac{9}{14}$, 0.067687075, 0.058956916, 0.871021772
15. (a) $f(x, y) = \frac{3}{256}$ on D and $f(x, y) = 0$ elsewhere;
    (b) $\frac{343}{512} \approx 0.66992$
16. (a) 2.5; (b) 0.058136039
17. (a) $\frac{19}{56}$; (b) $\frac{11}{20}$ 18. $\frac{7}{8}$
19. $\frac{1}{4}$

Chapter 16   Review Problem Set (p. 525)

1. (a) $\frac{5}{108}$; (b) $\frac{103}{108}$
2. (a) $\frac{5}{108}$; (b) $\frac{5}{54}$; (c) $\frac{49}{54}$
3. $\frac{11}{6}$, $\frac{35}{36}$ 4. $\mu = \frac{5}{3}$, $\sigma^2 = 1.\overline{8}$
5. (a) $\frac{16}{81}$, $\frac{32}{81}$, $\frac{24}{81}$, $\frac{8}{81}$, $\frac{1}{81}$; (b) $\frac{4}{3}$; (c) $\frac{8}{9}$
6. $\mu_y = \frac{31}{6}$, $\sigma_y^2 = 32.47\overline{2}$
7. (a) $\frac{1}{2}$; (b) $\frac{3}{2} \ln 3 \approx 1.64792$;
   (c) $3 - \frac{9}{4}(\ln 3)^2 \approx 0.28436$
8. (a) $k = \frac{3}{14}$; (b) 0.896; (c) $\frac{186}{70} \approx 2.657$;
   (d) ≈ 0.7151
9. $\frac{3}{8}$ 10. 0.8
11. 1, $\frac{1}{2}$ 12. $k = 2$
13. $e^{-5/6} \approx 0.43460$ 14. $\approx 44.93\%$
15. (a) 0.1401; (b) 0.5211
16. (a) ≈ 227 students; (b) ≈ 16 students
17. 55   18. 0.1870
19. (a) $\frac{5}{16}$; (b) $\frac{1}{2}$ 20. (a) $\frac{31}{32}$;
                                         (b) ≈ 0.50832
21. 0.42127   22. (a) 0.1637; (b) $\frac{7}{19}$
23. (a) $f(x, y) = \frac{3}{32}$ on D and $f(x, y) = 0$ elsewhere;
    (b) $\mu_x = 0$, $\mu_y = \frac{12}{5}$, $\sigma_x^2 = \frac{4}{5}$, $\sigma_y^2 = \frac{192}{175} \approx 1.09714$,
    $\sigma_{xy} = 0$
24. (a) $f(x, y) = \frac{6}{13}$ for $(x, y) \in D$ and $f(x, y) = 0$ elsewhere;
    (b) $\mu_x = \frac{9}{26} \approx 0.3462$, $\sigma_x^2 = 0.0571$, $\mu_y = \frac{158}{65} \approx 2.4308$, $\sigma_y^2 = 1.5232$, $\sigma_{xy} = 0.06243$

25. (a) $k = \frac{1}{2}$;
   (b) $\mu = \frac{1}{2}(e^2 - 3) \approx 2.19453$, $\sigma^2 = \frac{1}{2}(e^2 - 1) \approx$
   3.19453
26. (a) $k = \frac{3}{2}$;   (b) $\mu = \frac{3}{2}\ln 3$, $\sigma^2 = 0.2844$
27. (a) $6/(e - 1)$;   (b) $\frac{1}{4}$
28. (a) $k = 8/\ln 2 \approx 11.5416$;   (b) $\frac{3}{4}$

Basic Algebra Review Problem Set (p. 538)

1. 22
2. 8
3. 9
4. 80
5. 169
6. 100
7. 0.416, 0.417
8. 0.458$\overline{3}$, 0.458
9. 0.$\overline{714285}$, 0.714
10. 0.$\overline{384615}$, 0.385
11. $-\frac{8}{15}$, 0.5333
12. $-\frac{1}{9}$, -0.1111
13. $-\frac{2}{7}$, -0.2857
14. $-\frac{1}{19}$, -0.0526
15. $\frac{7}{8}$, 0.8750
16. $\frac{101}{8}$, 12.6250
17. $-\frac{7}{6}$, -1.1667
18. $\frac{41}{3}$, 13.6667
19. $\frac{19}{59}$, $\frac{13}{40}$, 0.3$\overline{3}$, 0.334
20. $\frac{41}{79}$, $\frac{13}{25}$, 0.5$\overline{2}$, 0.5$\overline{3}$
21. 0.5958
22. -13.4259
23. -8.9645
24. 1.6753
25. 0.0281
26. 4.4586
27. 6
28. 1125
29. $x^{5/3}/y^{1/4}$
30. $a^4 b^5 c^6$
31. $a^{8/3} b^{1/2}$
32. a
33. $9x^2 y^2 - 12xy + 4$
34. $1 - 8xy + 16x^2 y^2$
35. $9x^2 y^2 - 4$
36. $1 - 16x^2 y^2$
37. $2x^2 + 5xy - 12y^2$
38. $20x^2 - 14x + 2$
39. $a - 2a^{1/2}b^{1/2} + b$
40. $a^{2/3} + 2a + a^{4/3}$
41. $a - 2 + 1/a$
42. $x^4 - y^4$
43. $(5a - 3)(5a + 3)$
44. $(5a - 1)^2$
45. $(5a + 3)(a - 1)$
46. $(a - 1)(a + 1)(a^2 + 1)$
47. $(3xy - z)(xy + 2z)$
48. $(x + 2 - y)(x + 2 + y)$
49. $-5/(x - y)$
50. $(3b - 2a + b^2)/(a^2 b^3)$
51. $18/(a^2 - 9)$
52. $(5x - 2)/(x^2 + x - 6)$
53. $(x - 2)/[3x(x + 4)]$
54. $2y(x + 4y)/[x(x + 2y)]$
55. $-(x + y)/y^2$
56. $x^2 + y^2$
57. $a - b^2$
58. $x - 4\sqrt{x}\,\sqrt{y} + 4y$

59. $2h/\sqrt{x + h}$
60. $(2 + h)/(x + 2 + h)$
61. $\sqrt{4x^4 + 9}$
62. $2x^2 + 3$
63. $-\frac{6}{5}$
64. $\frac{8}{11}$
65. $-5, 0$
66. $-9, 4$
67. $-4, 2$
68. $\frac{3}{2}$
69. $-1, 1$
70. $-2, 0, 3$
71. $-\frac{1}{2}, \frac{2}{3}$
72. $\frac{3}{4}, 4$
73. $\pm\sqrt{3}$
74. $-3, 1$
75. $1 \pm \sqrt{3}$
76. $-2 \pm 2\sqrt{2}$
77. $(5 \pm \sqrt{17})/4$
78. $(-7 \pm \sqrt{13})/6$
79. $x \geq \frac{7}{2}$

80. $x > -\frac{8}{3}$

81. $-3 < x < 2$

82. $x \leq -6$ or $x \geq 3$

83. $x < -1$ or $x > \frac{3}{2}$

84. $-\frac{2}{3} \leq x \leq 2$

85. $x \leq -2$ or $-1 \leq x \leq 2$

86. $-2 < x < 0$ or $x > 4$

Index

Basic Identities

$$\tan t = \frac{\sin t}{\cos t} \qquad \cot t = \frac{\cos t}{\sin t} \qquad \cot t = \frac{1}{\tan t}$$

$$\sec t = \frac{1}{\cos t} \qquad \csc t = \frac{1}{\sin t}$$

$$\sin^2 t + \cos^2 t = 1 \qquad 1 + \cot^2 t = \csc^2 t$$

$$1 + \tan^2 t = \sec^2 t$$

Cofunction Identities

$$\sin\left(\frac{\pi}{2} - t\right) = \cos t \qquad \cos\left(\frac{\pi}{2} - t\right) = \sin t \qquad \tan\left(\frac{\pi}{2} - t\right) = \cot t$$

Odd-even Identities

$$\sin(-t) = -\sin t \qquad \cos(-t) = \cos t \qquad \tan(-t) = -\tan t$$

Addition Formulas

$$\sin(s + t) = \sin s \cos t + \cos s \sin t \qquad \sin(s - t) = \sin s \cos t - \cos s \sin t$$

$$\cos(s + t) = \cos s \cos t - \sin s \sin t \qquad \cos(s - t) = \cos s \cos t + \sin s \sin t$$

$$\tan(s + t) = \frac{\tan s + \tan t}{1 - \tan s \tan t} \qquad \tan(s - t) = \frac{\tan s - \tan t}{1 + \tan s \tan t}$$

Double Angle Formulas

$$\sin 2t = 2 \sin t \cos t$$

$$\cos 2t = \cos^2 t - \sin^2 t = 1 - 2\sin^2 t = 2\cos^2 t - 1$$

$$\tan 2t = \frac{2 \tan t}{1 - \tan^2 t}$$

Half Angle Formulas

$$\sin \frac{t}{2} = \pm\sqrt{\frac{1 - \cos t}{2}} \qquad \cos \frac{t}{2} = \pm\sqrt{\frac{1 + \cos t}{2}}$$

$$\tan \frac{t}{2} = \frac{1 - \cos t}{\sin t}$$

Product Formulas

$$2 \sin s \cos t = \sin(s + t) + \sin(s - t)$$

$$2 \cos s \sin t = \sin(s + t) - \sin(s - t)$$

$$2 \cos s \cos t = \cos(s + t) + \cos(s - t)$$

$$2 \sin s \sin t = \cos(s - t) - \cos(s + t)$$

Factoring Formulas

$$\sin s + \sin t = 2 \sin \frac{s+t}{2} \cos \frac{s-t}{2}$$

$$\sin s - \sin t = 2 \cos \frac{s+t}{2} \sin \frac{s-t}{2}$$

$$\cos s + \cos t = 2 \cos \frac{s+t}{2} \cos \frac{s-t}{2}$$

$$\cos s - \cos t = -2 \sin \frac{s+t}{2} \sin \frac{s-t}{2}$$

Laws of Sines and Cosines

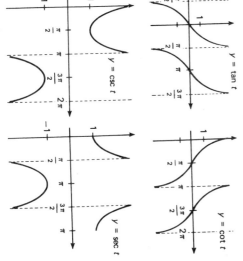

$$\frac{\sin \alpha}{a} = \frac{\sin \beta}{b} = \frac{\sin \gamma}{c}$$

$$a^2 = b^2 + c^2 - 2bc \cos \alpha$$

Graphs

$$\sin t = \sin \theta = y = \frac{b}{r} \qquad \cos t = \cos \theta = x = \frac{a}{r}$$

$$\tan t = \tan \theta = \frac{y}{x} = \frac{b}{a} \qquad \cot t = \cot \theta = \frac{x}{y} = \frac{a}{b}$$

$y = \sin t$

$y = \cos t$

$y = \tan t$

$y = \cot t$

$y = \csc t$

$y = \sec t$

Inverse Trigonometric Functions

$$y = \sin^{-1} x \Leftrightarrow x = \sin y, \; -\pi/2 \leq y \leq \pi/2$$

$$y = \cos^{-1} x \Leftrightarrow x = \cos y, \; 0 \leq y \leq \pi$$

$$y = \tan^{-1} x \Leftrightarrow x = \tan y, \; -\pi/2 < y < \pi/2$$

$$y = \sec^{-1} x \Leftrightarrow x = \sec y, \; 0 \leq y \leq \pi, \; y \neq \pi/2$$

$$\sec^{-1} x = \cos^{-1}(1/x)$$

Hyperbolic Functions

$$\sinh x = \frac{1}{2}(e^x - e^{-x}) \qquad \cosh x = \frac{1}{2}(e^x + e^{-x})$$

$$\tanh x = \frac{\sinh x}{\cosh x} \qquad \coth x = \frac{\cosh x}{\sinh x}$$

$$\text{sech } x = \frac{1}{\cosh x} \qquad \text{csch } x = \frac{1}{\sinh x}$$

Series

$$\frac{1}{1 - x} = 1 + x + x^2 + x^3 + \cdots \qquad -1 < x < 1$$

$$\ln(1 + x) = x - \frac{x^2}{2} + \frac{x^3}{3} - \frac{x^4}{4} + \cdots \qquad -1 < x \leq 1$$

$$\tan^{-1} x = x - \frac{x^3}{3} + \frac{x^5}{5} - \frac{x^7}{7} + \cdots \qquad -1 \leq x \leq 1$$

$$e^x = 1 + x + \frac{x^2}{2!} + \frac{x^3}{3!} + \cdots$$

$$\sin x = x - \frac{x^3}{3!} + \frac{x^5}{5!} - \frac{x^7}{7!} + \cdots$$

$$\cos x = 1 - \frac{x^2}{2!} + \frac{x^4}{4!} - \frac{x^6}{6!} + \cdots$$

$$(1 + x)^p = 1 + \binom{p}{1}x + \binom{p}{2}x^2 + \binom{p}{3}x^3 + \cdots \qquad -1 < x < 1$$

$$\binom{p}{k} = \frac{p(p - 1)(p - 2)\cdots(p - k + 1)}{k!}$$

FORMULA CARD for CALCULUS

3.14159

FORMULA CARD
to accompany
CALCULUS WITH APPLICATIONS
Dale Varberg/Walter Fleming
Prentice Hall, Englewood Cliffs, NJ 07632

DERIVATIVES

$$D_x x^r = rx^{r-1}$$

$$D_x \sin x = \cos x$$

$$D_x \tan x = \sec^2 x$$

$$D_x \sec x = \sec x \tan x$$

$$D_x \sinh x = \cosh x$$

$$D_x \cosh x = \sinh x$$

$$D_x \tanh x = \text{sech}^2 x$$

$$D_x \ln x = \frac{1}{x}$$

$$D_x e^x = e^x$$

$$D_x \sin^{-1}x = \frac{1}{\sqrt{1 - x^2}}$$

$$D_x \tan^{-1}x = \frac{1}{1 + x^2}$$

$$D_x |x| = \frac{|x|}{x}$$

$$D_x \cos x = -\sin x$$

$$D_x \cot x = -\csc^2 x$$

$$D_x \csc x = -\csc x \cot x$$

$$D_x \coth x = -\text{csch}^2 x$$

$$D_x \text{sech } x = -\text{sech } x \tanh x$$

$$D_x \text{csch } x = -\text{csch } x \coth x$$

$$D_x \log_a x = \frac{1}{x \ln a}$$

$$D_x a^x = a^x \ln a$$

$$D_x \cos^{-1}x = \frac{-1}{\sqrt{1 - x^2}}$$

$$D_x \sec^{-1}x = \frac{1}{|x| \sqrt{x^2 - 1}}$$

INTEGRALS

1. $\int u \, dv = uv - \int v \, du$

2. $\int u^n \, du = \frac{1}{n+1} u^{n+1} + C, \ n \neq -1$

3. $\int \frac{1}{u} \, du = \ln |u| + C$

4. $\int e^u \, du = e^u + C$

5. $\int a^u \, du = \frac{a^u}{\ln a} + C$

6. $\int \sin u \, du = -\cos u + C$

7. $\int \cos u \, du = \sin u + C$

8. $\int \sec^2 u \, du = \tan u + C$

9. $\int \csc^2 u \, du = -\cot u + C$

10. $\int \sec u \tan u \, du = \sec u + C$

11. $\int \csc u \cot u \, du = -\csc u + C$

12. $\int \tan u \, du = \ln |\sec u| + C$

13. $\int \cot u \, du = \ln |\sin u| + C$

14. $\int \sec u \, du = \ln |\sec u + \tan u| + C$

15. $\int \csc u \, du = \ln |\csc u - \cot u| + C$

16. $\int \frac{1}{\sqrt{a^2 - u^2}} \, du = \sin^{-1} \frac{u}{a} + C$

17. $\int \frac{1}{a^2 + u^2} \, du = \frac{1}{a} \tan^{-1} \frac{u}{a} + C$

18. $\int \frac{1}{a^2 - u^2} \, du = \frac{1}{2a} \ln \left| \frac{u + a}{u - a} \right| + C$

19. $\int \frac{1}{u\sqrt{u^2 - a^2}} \, du = \frac{1}{a} \sec^{-1} \left| \frac{u}{a} \right| + C$

GEOMETRY

Triangles

Pythagorean Theorem

$$a^2 + b^2 = c^2$$

Right triangle

Angles $\alpha + \beta + \gamma = 180°$
Area $A = \frac{1}{2}bh$

Any triangle

Circles

Circumference $C = 2\pi r$
Area $A = \pi r^2$

Cylinders

Surface area $S = 2\pi r^2 + 2\pi rh$
Volume $V = \pi r^2 h$

Cones

Surface area $S = \pi r^2 + \pi r\sqrt{r^2 + h^2}$
Volume $V = \frac{1}{3}\pi r^2 h$

Spheres

Surface area $S = 4\pi r^2$
Volume $V = \frac{4}{3}\pi r^3$

Conversions

1 inch = 2.54 centimeters
1 liter = 1000 cubic centimeters
1 kilogram = 2.20 pounds
1 kilometer = .62 miles
1 liter = 1.057 quarts
1 pound = 453.6 grams
π radians = 180 degrees

Table of Integrals

1. $\displaystyle\int u\,dv = uv - \int v\,du$

2. $\displaystyle\int u^n\,du = \frac{1}{n+1}u^{n+1} + C \quad \text{if } n \neq -1$

3. $\displaystyle\int \frac{du}{u} = \ln|u| + C$

4. $\displaystyle\int e^u\,du = e^u + C$

5. $\displaystyle\int a^u\,du = \frac{a^u}{\ln a} + C$

6. $\displaystyle\int \sin u\,du = -\cos u + C$

7. $\displaystyle\int \cos u\,du = \sin u + C$

8. $\displaystyle\int \sec^2 u\,du = \tan u + C$

9. $\displaystyle\int \csc^2 u\,du = -\cot u + C$

10. $\displaystyle\int \sec u \tan u\,du = \sec u + C$

11. $\displaystyle\int \csc u \cot u\,du = -\csc u + C$

12. $\displaystyle\int \tan u\,du = \ln|\sec u| + C$

13. $\displaystyle\int \cot u\,du = \ln|\sin u| + C$

14. $\displaystyle\int \sec u\,du = \ln|\sec u + \tan u| + C$

15. $\displaystyle\int \csc u\,du = \ln|\csc u - \cot u| + C$

16. $\displaystyle\int \frac{du}{\sqrt{a^2 - u^2}} = \sin^{-1}\frac{u}{a} + C$

17. $\displaystyle\int \frac{du}{a^2 + u^2} = \frac{1}{a}\tan^{-1}\frac{u}{a} + C$

18. $\displaystyle\int \frac{du}{a^2 - u^2} = \frac{1}{2a}\ln\left|\frac{u+a}{u-a}\right| + C$

19. $\displaystyle\int \frac{du}{u\sqrt{u^2 - a^2}} = \frac{1}{a}\sec^{-1}\left|\frac{u}{a}\right| + C$

TRIGONOMETRIC FORMS

20. $\displaystyle\int \sin^2 u\,du = \frac{1}{2}u - \frac{1}{4}\sin 2u + C$

21. $\displaystyle\int \cos^2 u\,du = \frac{1}{2}u + \frac{1}{4}\sin 2u + C$

22. $\displaystyle\int \tan^2 u\,du = \tan u - u + C$

23. $\displaystyle\int \cot^2 u\,du = -\cot u - u + C$

24. $\displaystyle\int \sin^3 u\,du = -\frac{1}{3}(2 + \sin^2 u)\cos u + C$

25. $\displaystyle\int \cos^3 u\,du = \frac{1}{3}(2 + \cos^2 u)\sin u + C$

26. $\displaystyle\int \tan^3 u\,du = \frac{1}{2}\tan^2 u + \ln|\cos u| + C$

27. $\displaystyle\int \cot^3 u\,du = -\frac{1}{2}\cot^2 u - \ln|\sin u| + C$

28. $\displaystyle\int \sec^3 u\,du = \frac{1}{2}\sec u \tan u + \frac{1}{2}\ln|\sec u + \tan u| + C$

29. $\displaystyle\int \csc^3 u\,du = -\frac{1}{2}\csc u \cot u + \frac{1}{2}\ln|\csc u - \cot u| + C$

30. $\displaystyle\int \sin au \sin bu\,du = \frac{\sin(a-b)u}{2(a-b)} - \frac{\sin(a+b)u}{2(a+b)} + C \quad \text{if } a^2 \neq b^2$

31. $\displaystyle\int \cos au \cos bu\,du = \frac{\sin(a-b)u}{2(a-b)} + \frac{\sin(a+b)u}{2(a+b)} + C \quad \text{if } a^2 \neq b^2$

32. $\displaystyle\int \sin au \cos bu\,du = -\frac{\cos(a-b)u}{2(a-b)} - \frac{\cos(a+b)u}{2(a+b)} + C \quad \text{if } a^2 \neq b^2$

33. $\displaystyle\int \sin^n u\,du = -\frac{1}{n}\sin^{n-1} u \cos u + \frac{n-1}{n}\int \sin^{n-2} u\,du$

34. $\displaystyle\int \cos^n u\,du = \frac{1}{n}\cos^{n-1} u \sin u + \frac{n-1}{n}\int \cos^{n-2} u\,du$

35. $\displaystyle\int \tan^n u\,du = \frac{1}{n-1}\tan^{n-1} u - \int \tan^{n-2} u\,du \quad \text{if } n \neq 1$

36. $\displaystyle\int \cot^n u\,du = \frac{-1}{n-1}\cot^{n-1} u - \int \cot^{n-2} u\,du \quad \text{if } n \neq 1$

37. $\displaystyle\int \sec^n u\,du = \frac{1}{n-1}\sec^{n-2} u \tan u + \frac{n-2}{n-1}\int \sec^{n-2} u\,du \quad \text{if } n \neq 1$

38. $\displaystyle\int \csc^n u\,du = \frac{-1}{n-1}\csc^{n-2} u \cot u + \frac{n-2}{n-1}\int \csc^{n-2} u\,du \quad \text{if } n \neq 1$

39a. $\displaystyle\int \sin^n u \cos^m u\,du = -\frac{\sin^{n-1} u \cos^{m+1} u}{n+m} + \frac{n-1}{n+m}\int \sin^{n-2} u \cos^m u\,du \quad \text{if } n \neq -m$

39b. $\displaystyle\int \sin^n u \cos^m u\,du = \frac{\sin^{n+1} u \cos^{m-1} u}{n+m} + \frac{m-1}{n+m}\int \sin^n u \cos^{m-2} u\,du \quad \text{if } m \neq -n$

40. $\displaystyle\int u \sin u\,du = \sin u - u \cos u + C$

41. $\displaystyle\int u \cos u\,du = \cos u + u \sin u + C$

42. $\displaystyle\int u^n \sin u\,du = -u^n \cos u + n\int u^{n-1} \cos u\,du$

43. $\displaystyle\int u^n \cos u\,du = u^n \sin u - n\int u^{n-1} \sin u\,du$

FORMS INVOLVING $\sqrt{u^2 \pm a^2}$

44 $\displaystyle\int \sqrt{u^2 \pm a^2}\, du = \frac{u}{2}\sqrt{u^2 \pm a^2} \pm \frac{a^2}{2}\ln|u + \sqrt{u^2 \pm a^2}| + C$

45 $\displaystyle\int \frac{du}{\sqrt{u^2 \pm a^2}} = \ln|u + \sqrt{u^2 \pm a^2}| + C$

46 $\displaystyle\int \frac{\sqrt{u^2 + a^2}}{u}\, du = \sqrt{u^2 + a^2} - a\ln\left(\frac{a + \sqrt{u^2 + a^2}}{u}\right) + C$

47 $\displaystyle\int \frac{\sqrt{u^2 - a^2}}{u}\, du = \sqrt{u^2 - a^2} - a\sec^{-1}\frac{u}{a} + C$

48 $\displaystyle\int u^2 \sqrt{u^2 \pm a^2}\, du = \frac{u}{8}(2u^2 \pm a^2)\sqrt{u^2 \pm a^2} - \frac{a^4}{8}\ln|u + \sqrt{u^2 \pm a^2}| + C$

49 $\displaystyle\int \frac{u^2\, du}{\sqrt{u^2 \pm a^2}} = \frac{u}{2}\sqrt{u^2 \pm a^2} \mp \frac{a^2}{2}\ln|u + \sqrt{u^2 \pm a^2}| + C$

50 $\displaystyle\int \frac{du}{u^2 \sqrt{u^2 \pm a^2}} = \mp \frac{\sqrt{u^2 \pm a^2}}{a^2 u} + C$

51 $\displaystyle\int \frac{\sqrt{u^2 \pm a^2}}{u^2}\, du = -\frac{\sqrt{u^2 \pm a^2}}{u} + \ln|u + \sqrt{u^2 \pm a^2}| + C$

52 $\displaystyle\int \frac{du}{(u^2 \pm a^2)^{3/2}} = \frac{\pm u}{a^2 \sqrt{u^2 \pm a^2}} + C$

53 $\displaystyle\int (u^2 \pm a^2)^{3/2}\, du = \frac{u}{8}(2u^2 \pm 5a^2)\sqrt{u^2 \pm a^2} + \frac{3a^4}{8}\ln|u + \sqrt{u^2 \pm a^2}| + C$

FORMS INVOLVING $\sqrt{a^2 - u^2}$

54 $\displaystyle\int \sqrt{a^2 - u^2}\, du = \frac{u}{2}\sqrt{a^2 - u^2} + \frac{a^2}{2}\sin^{-1}\frac{u}{a} + C$

55 $\displaystyle\int \frac{\sqrt{a^2 - u^2}}{u}\, du = \sqrt{a^2 - u^2} - a\ln\left|\frac{a + \sqrt{a^2 - u^2}}{u}\right| + C$

56 $\displaystyle\int \frac{u^2\, du}{\sqrt{a^2 - u^2}} = -\frac{u}{2}\sqrt{a^2 - u^2} + \frac{a^2}{2}\sin^{-1}\frac{u}{a} + C$

57 $\displaystyle\int u^2 \sqrt{a^2 - u^2}\, du = \frac{u}{8}(2u^2 - a^2)\sqrt{a^2 - u^2} + \frac{a^4}{8}\sin^{-1}\frac{u}{a} + C$

58 $\displaystyle\int \frac{du}{u^2 \sqrt{a^2 - u^2}} = -\frac{\sqrt{a^2 - u^2}}{a^2 u} + C$

59 $\displaystyle\int \frac{\sqrt{a^2 - u^2}}{u^2}\, du = -\frac{\sqrt{a^2 - u^2}}{u} - \sin^{-1}\frac{u}{a} + C$

60 $\displaystyle\int \frac{du}{u\sqrt{a^2 - u^2}} = -\frac{1}{a}\ln\left|\frac{a + \sqrt{a^2 - u^2}}{u}\right| + C$

61 $\displaystyle\int \frac{du}{(a^2 - u^2)^{3/2}} = \frac{u}{a^2 \sqrt{a^2 - u^2}} + C$

62 $\displaystyle\int (a^2 - u^2)^{3/2}\, du = \frac{u}{8}(5a^2 - 2u^2)\sqrt{a^2 - u^2} + \frac{3a^4}{8}\sin^{-1}\frac{u}{a} + C$

EXPONENTIAL AND LOGARITHMIC FORMS

63 $\displaystyle\int u e^u\, du = (u - 1)e^u + C$

64 $\displaystyle\int u^n e^u\, du = u^n e^u - n\int u^{n-1} e^u\, du$

65 $\displaystyle\int \ln u\, du = u\ln u - u + C$

66 $\displaystyle\int u^n \ln u\, du = \frac{u^{n+1}}{n+1}\ln u - \frac{u^{n+1}}{(n+1)^2} + C$

67 $\displaystyle\int e^{au}\sin bu\, du = \frac{e^{au}}{a^2 + b^2}(a\sin bu - b\cos bu) + C$

68 $\displaystyle\int e^{au}\cos bu\, du = \frac{e^{au}}{a^2 + b^2}(a\cos bu + b\sin bu) + C$

INVERSE TRIGONOMETRIC FORMS

69 $\displaystyle\int \sin^{-1} u\, du = u\sin^{-1} u + \sqrt{1 - u^2} + C$

70 $\displaystyle\int \tan^{-1} u\, du = u\tan^{-1} u - \frac{1}{2}\ln(1 + u^2) + C$

71 $\displaystyle\int \sec^{-1} u\, du = u\sec^{-1} u - \ln|u + \sqrt{u^2 - 1}| + C$

72 $\displaystyle\int u\sin^{-1} u\, du = \frac{1}{4}(2u^2 - 1)\sin^{-1} u + \frac{u}{4}\sqrt{1 - u^2} + C$

73 $\displaystyle\int u\tan^{-1} u\, du = \frac{1}{2}(u^2 + 1)\tan^{-1} u - \frac{u}{2} + C$

74 $\displaystyle\int u\sec^{-1} u\, du = \frac{u^2}{2}\sec^{-1} u - \frac{1}{2}\sqrt{u^2 - 1} + C$

75 $\displaystyle\int u^n \sin^{-1} u\, du = \frac{u^{n+1}}{n+1}\sin^{-1} u - \frac{1}{n+1}\int \frac{u^{n+1}}{\sqrt{1 - u^2}}\, du + C \quad \text{if } n \neq -1$

76 $\displaystyle\int u^n \tan^{-1} u\, du = \frac{u^{n+1}}{n+1}\tan^{-1} u - \frac{1}{n+1}\int \frac{u^{n+1}}{1 + u^2}\, du + C \quad \text{if } n \neq -1$

77 $\displaystyle\int u^n \sec^{-1} u\, du = \frac{u^{n+1}}{n+1}\sec^{-1} u - \frac{1}{n+1}\int \frac{u^n}{\sqrt{u^2 - 1}}\, du + C \quad \text{if } n \neq -1$